TETRAHEDRON ORGANIC CHEMISTRY SERIES
Series Editors: J E Baldwin, FRS & R M Williams

VOLUME 19

High-Resolution NMR Techniques in Organic Chemistry

Related Pergamon Titles of Interest

BOOKS

Tetrahedron Organic Chemistry Series:
CARRUTHERS: Cycloaddition Reactions in Organic Synthesis
DEROME: Modern NMR Techniques for Chemistry Research
FINET: Ligand Coupling Reactions with Heteroatomic Compounds
GAWLEY & AUBÉ: Principles of Asymmetric Synthesis
HASSNER & STUMER: Organic Syntheses Based on Name Reactions and Un-named Reactions
McKILLOP: Advanced Problems in Organic Reaction Mechanisms
OBRECHT & VILLALGORDO: Solid-Supported Combinatorial and Parallel Synthesis of Small-Molecular-Weight Compound Libraries
PERLMUTTER: Conjugate Addition Reactions in Organic Synthesis
SESSLER & WEGHORN: Expanded, Contracted & Isomeric Porphyrins
TANG & LEVY: Chemistry of C-Glycosides
WONG & WHITESIDES: Enzymes in Synthetic Organic Chemistry

JOURNALS

BIOORGANIC & MEDICINAL CHEMISTRY
BIOORGANIC & MEDICINAL CHEMISTRY LETTERS
TETRAHEDRON
TETRAHEDRON: ASYMMETRY
TETRAHEDRON LETTERS

Full details of all Elsevier Science publications/free specimen copy of any Elsevier Science journal are available on request from your nearest Elsevier Science office

High-Resolution NMR Techniques in Organic Chemistry

TIMOTHY D W CLARIDGE

Dyson Perrins Laboratory, Oxford

1999

PERGAMON

An imprint of Elsevier Science

Amsterdam - Lausanne - New York - Oxford - Shannon - Singapore - Tokyo

ELSEVIER SCIENCE Ltd
The Boulevard, Langford Lane
Kidlington, Oxford OX5 1GB, UK

First edition 1999

Library of Congress Cataloging in Publication Data
A catalog record from the Library of Congress has been applied for.

British Library Cataloguing in Publication Data
A catalogue record from the British Library has been applied for.

ISBN: 0 08 042799 5 (hardbound)
ISBN: 0 08 042798 7 (paperback)

♾ The paper used in this publication meets the requirements of ANSI/NISO Z39.48-1992 (Permanence of Paper).
Printed in The Netherlands

Foreword

It is now more than ten years since the late Andy Derome's book "*Modern NMR Techniques for Chemistry Research*" was published. This book had a tremendous impact on organic chemists' use of the developing field of pulsed NMR spectroscopy, bringing to the chemist the power of these methods in a practical form for the elucidation of molecular structure. It is probably true that NMR spectroscopy has had the greatest impact on chemical research since the development of the accurate balance. Interestingly both of these techniques, i.e. weighing and NMR, are based on nuclear properties.

In this new book, Tim Claridge carries on from Andy Derome and brings together the more recent methods in a practical context. This book will be essential for those in academe or industry who have a direct intent in applying NMR to chemical problems.

J.E. Baldwin
Dyson Perrins Laboratory
Oxford 1999

Preface

From the initial observation of proton magnetic resonance in water and in paraffin, the discipline of nuclear magnetic resonance (NMR) has seen unparalleled growth as an analytical method and now, in numerous different guises, finds application in chemistry, biology, medicine, materials science and geology. Despite its inception in the laboratories of physicists, it is in the chemical laboratory that NMR spectroscopy has found greatest use and, it may be argued, has provided the foundations on which modern organic chemistry has developed. Modern NMR is now a highly developed, yet still evolving, subject that all organic chemists need to understand, and appreciate the potential of, if they are be effective in and able to progress their current research. An ability to keep abreast of developments in NMR techniques is, however, a daunting task, made difficult not only by the sheer number of available techniques but also by the way in which these new methods first appear. These are spread across the chemical literature in both specialised magnetic resonance journals or those dedicated to specific areas of chemistry, as well as the more general entities. They are often referred to through esoteric acronyms and described in a seemingly complex mathematical language that does little to endear them to the research chemist. The myriad of sequences can be wholly bewildering for the uninitiated and can leave one wondering where to start and which technique to select for the problem at hand. In this book I have attempted to gather together the most valuable techniques for the research chemist and to describe the operation of these using pictorial models. Even this level of understanding is perhaps more than some chemists may consider necessary, but only from this can one fully appreciate the capabilities and (of equal if not greater importance) the limitations of these techniques. Throughout, the emphasis is on the more recently developed methods that have, or undoubtedly will, establish themselves as the principle techniques for the elucidation and investigation of chemical structures in solution.

NMR spectroscopy is, above all, a practical subject that is most rewarding when one has an interesting sample to investigate, a spectrometer at one's disposal and the knowledge to make the most of this (sometimes alarmingly!) expensive instrumentation. As such, this book contains a considerable amount of information and guidance on how one implements and executes the techniques that are described and thus should be equally at home in the NMR laboratory or on the chemist's or spectroscopist's desk.

This book is written from the perspective of an NMR facility manager in an academic research laboratory and as such the topics included are naturally influenced by the areas of chemistry I encounter. The methods are chosen, however, for their wide applicability and robustness, and because, in many cases, they have already become established techniques in NMR laboratories in both academic and industrial establishments. This is not intended as a review of all recent developments in NMR techniques. Not only would this be too immense to fit within a single volume, but the majority of the methods would have little significance for most research chemists. Instead, this is a distillation of the very many methods developed over the years, with only the most appropriate fractions retained. It should find use in academic and industrial research laboratories alike, and could provide the foundation for graduate level courses on NMR techniques in chemical research.

Acknowledgements

The preparation of this book has benefited from the co-operation, assistance, patience, understanding and knowledge of many people, for which I am deeply grateful. I must thank my colleagues, both past and present, in the NMR group of the Dyson Perrins Laboratory, in particular Elizabeth McGuinness and Tina Jackson for their first-class support and assistance, and Norman Gregory and Dr. Guo-Liang Ping for the various repairs, modifications and improvements they have made to the instruments used to prepare many of the figures in this book. Most of these figures have been recorded specifically for the book and have been made possible by the generosity of various research groups and individuals through making their data and samples available to me. For this, I would like to express my gratitude to Dr. Harry Anderson, Prof. Jack Baldwin, Dr. Paul Burn, Dr. John Brown, Dr. Duncan Carmichael, Dr. Antony Fairbanks, Prof. George Fleet, Dr. David Hodgson, Dr. Mark Moloney, Dr. Jo Peach and Prof. Chris Schofield, and to the members of their groups, too numerous to mention, who kindly prepared the samples; they will know who they are and I am indebted to each of them. I am similarly grateful to Prof. Jack Baldwin for allowing me to use the department's instrumentation for the collection of these illustrative spectra.

I would like to thank Drs. Carolyn Carr and Nick Rees for their assistance in proof-reading the manuscript and for being able to spot those annoying little mistakes that I read past time and time again but still did not register. Naturally I accept responsibility for those that remain and would be grateful to hear of these, whether factual or typographical. I also thank Eileen Morrell and Sharon Ward of Elsevier Science for their patience in waiting for this project to be completed and for their relaxed attitude as various deadlines failed to be met.

I imagine everyone entering into a career in science has at some time been influenced or even inspired by one or a few individual(s) who may have acted as teacher, mentor or perhaps role-model. Personally, I am indebted to Dr. Jeremy Everett and to John Tyler, both formally of (what was then) Beecham Pharmaceuticals, for accepting into their NMR laboratory for a year a 'sandwich' student who was initially supposed to gain industrial experience elsewhere as a chromatographer analysing horse urine! My fortuitous escape from this and subsequent time at Beechams proved to be a seminal year for me and I thank Jeremy and John for their early encouragement that ignited my interest in NMR. My understanding of what this could really do came from graduate studies with the late Andy Derome, and I, like many others, remain eternally grateful for the insight and inspiration he provided.

Finally, I thank my wife Rachael for her undying patience, understanding and support throughout this long and sometimes tortuous project, one that I'm sure she thought, on occasions, she would never see the end of. I can only apologise for the neglect she has endured but not deserved.

Tim Claridge
Oxford, May 1999

Contents

Foreword V
Preface VII
Acknowledgements IX

Chapter 1. Introduction

1.1. The development of high-resolution NMR 1
1.2. Modern high-resolution NMR and this book 4
1.2.1. What this book contains 5
1.2.2. Pulse sequence nomenclature 7
1.3. Applying modern NMR techniques 8
References 12

Chapter 2. Introducing high-resolution NMR

2.1. Nuclear spin and resonance 13
2.2. The vector model of NMR 16
2.2.1. The rotating frame of reference 16
2.2.2. Pulses 18
2.2.3. Chemical shifts and couplings 20
2.2.4. Spin-echoes 21
2.3. Time and frequency domains 24
2.4. Spin relaxation 25
2.4.1. Longitudinal relaxation: establishing equilibrium 26
2.4.2. Measuring T_1 with the inversion-recovery sequence 27
2.4.3. Transverse relaxation: loss of magnetisation in the x–y plane 30
2.4.4. Measuring T_2 with a spin-echo sequence 31
2.5. Mechanisms for relaxation 35
2.5.1. The path to relaxation 35
2.5.2. Dipole–dipole relaxation 37
2.5.3. Chemical shift anisotropy relaxation 38
2.5.4. Spin-rotation relaxation 39
2.5.5. Quadrupolar relaxation 40
References 43

Chapter 3. Practical aspects of high-resolution NMR

3.1. An overview of the NMR spectrometer 45
3.2. Data acquisition and processing 48
3.2.1. Pulse excitation 48
3.2.2. Signal detection 51
3.2.3. Sampling the FID 52
3.2.4. Quadrature detection 59
3.2.5. Phase cycling 63
3.2.6. Dynamic range and signal averaging 65
3.2.7. Window functions 70
3.2.8. Phase correction 73
3.3. Preparing the sample 75
3.3.1. Selecting the solvent 75
3.3.2. Reference compounds 77

3.3.3. Tubes and sample volumes 78
3.3.4. Filtering and degassing 80
3.4. Preparing the spectrometer 81
3.4.1. The probe 82
3.4.2. Tuning the probe 83
3.4.3. The field-frequency lock 85
3.4.4. Optimising the field homogeneity: shimming 87
3.5. Spectrometer calibrations 94
3.5.1. Radiofrequency pulses 94
3.5.2. Pulsed field gradients 99
3.5.3. Sample temperature 104
3.6. Spectrometer performance tests 105
3.6.1. Lineshape and resolution 106
3.6.2. Sensitivity 107
3.6.3. Solvent presaturation 109
References 110

Chapter 4. One-dimensional techniques

4.1. The single-pulse experiment 111
4.1.1. Optimising sensitivity 112
4.1.2. Quantitative measurements and integration 114
4.2. Spin decoupling methods 116
4.2.1. The basis of spin decoupling 117
4.2.2. Homonuclear decoupling 117
4.2.3. Heteronuclear decoupling 120
4.3. Spectrum editing with spin-echoes 125
4.3.1. The J-modulated spin-echo 125
4.3.2. APT 128
4.4. Sensitivity enhancement and spectrum editing 129
4.4.1. Polarisation transfer 130
4.4.2. INEPT 132
4.4.3. DEPT 139
4.4.4. PENDANT 142
4.5. Observing quadrupolar nuclei 143
References 145

Chapter 5. Correlations through the chemical bond I: Homonuclear shift correlation

5.1. Introducing two-dimensional methods 148
5.1.1. Generating a second dimension 149
5.2. Correlation spectroscopy (COSY) 153
5.2.1. Correlating coupled spins 155
5.2.2. Interpreting COSY 156
5.2.3. Peak fine structure 159
5.3. Practical aspects of 2D NMR 160
5.3.1. 2D lineshapes and quadrature detection 161
5.3.2. Axial peaks 167
5.3.3. Instrumental artefacts 168
5.3.4. 2D data acquisition 170
5.3.5. 2D data processing 172
5.4. Coherence and coherence transfer 174
5.4.1. Coherence-transfer pathways 177
5.5. Gradient-selected spectroscopy 178
5.5.1. Signal selection with pulsed field gradients 179
5.5.2. Phase-sensitive experiments 183
5.5.3. PFGs in high-resolution NMR 184
5.5.4. Practical implementation of PFGs 186
5.6. Alternative COSY sequences 187
5.6.1. Which COSY approach? 188
5.6.2. Double-quantum filtered COSY (DQF-COSY) 189
5.6.3. COSY-β 197

5.6.4. Delayed-COSY: detecting small couplings 199
5.6.5. Relayed-COSY 200
5.7. Total correlation spectroscopy (TOCSY) 201
5.7.1. The TOCSY sequence 202
5.7.2. Using TOCSY 205
5.7.3. Implementing TOCSY 208
5.8. Correlating dilute spins: INADEQUATE 211
5.8.1. 2D INADEQUATE 212
5.8.2. 1D INADEQUATE 213
5.8.3. Implementing INADEQUATE 215
5.8.4. Variations on INADEQUATE 216
References 218

Chapter 6. Correlations through the chemical bond II: Heteronuclear shift correlation

6.1. Introduction 221
6.2. Sensitivity 222
6.3. Heteronuclear single-bond correlation spectroscopy 224
6.3.1. Heteronuclear multiple-quantum correlation (HMQC) 224
6.3.2. Heteronuclear single-quantum correlation (HSQC) 229
6.3.3. Practical implementations 230
6.3.4. Hybrid experiments 238
6.4. Heteronuclear multiple-bond correlation spectroscopy 244
6.4.1. The HMBC sequence 245
6.4.2. Applying HMBC 248
6.5. Traditional X-detected correlation spectroscopy 251
6.5.1. Single-bond correlations 252
6.5.2. Multiple-bond correlations and small couplings 254
References 256

Chapter 7. Separating shifts and couplings: J-resolved spectroscopy

7.1. Introduction 259
7.2. Heteronuclear J-resolved spectroscopy 260
7.2.1. Measuring long-range proton–carbon coupling constants 263
7.2.2. Practical considerations 266
7.3. Homonuclear J-resolved spectroscopy 267
7.3.1. Tilting, projections and symmetrisation 268
7.3.2. Applications 270
7.3.3. Practical considerations 273
7.4. 'Indirect' homonuclear J-resolved spectroscopy 273
References 274

Chapter 8. Correlations through space: The nuclear Overhauser effect

8.1. Introduction 277
8.2. Definition of the NOE 279
8.3. Steady-state NOEs 279
8.3.1. NOEs in a two-spin system 279
8.3.2. NOEs in a multispin system 288
8.3.3. Summary 294
8.3.4. Applications 296
8.4. Transient NOEs 301
8.4.1. NOE kinetics 302
8.4.2. Measuring internuclear separations 303
8.5. Rotating-frame NOEs 304
8.6. Measuring steady-state NOEs: NOE difference 306
8.6.1. Optimising difference experiments 307
8.7. Measuring transient NOEs: NOESY 313
8.7.1. The 2D NOESY sequence 314
8.7.2. 1D NOESY sequences 320
8.7.3. Applications 323

8.7.4. Measuring chemical exchange: EXSY 326
8.8. Measuring rotating-frame NOEs: ROESY 328
8.8.1. The 2D ROESY sequence 329
8.8.2. 1D ROESY sequences 332
8.8.3. Applications 332
8.9. Measuring heteronuclear NOEs 335
8.10. Experimental considerations 336
References 337

Chapter 9. Experimental methods

9.1. Composite pulses 341
9.1.1. A myriad of pulses 344
9.1.2. Inversion vs. refocusing 344
9.2. Broadband decoupling and spin-locks 346
9.2.1. Spin-locks 347
9.2.2. Adiabatic pulses 348
9.3. Selective excitation and shaped pulses 348
9.3.1. Shaped soft pulses 350
9.3.2. DANTE sequences 354
9.3.3. Excitation sculpting 355
9.3.4. Practical considerations 357
9.4. Solvent suppression 359
9.4.1. Presaturation 361
9.4.2. Zero excitation 362
9.4.3. Pulsed field gradients 363
9.5. Recent methods 366
9.5.1. Heterogeneous samples and MAS 366
9.5.2. Diffusion-ordered spectroscopy 368
References 371

Appendix. Glossary of acronyms 373

Index 375

Chapter 1

Introduction

From the initial observation of proton magnetic resonance in water [1] and in paraffin [2], the discipline of nuclear magnetic resonance (NMR) has seen unparalleled growth as an analytical method and now, in numerous different guises, finds application in chemistry, biology, medicine, materials science and geology. Despite its inception in the laboratories of physicists, it is in the chemical laboratory that NMR spectroscopy has found greatest use. To put into context the range of techniques now available in the modern organic laboratory, including those described in this book, we begin with a short overview of the evolution of high-resolution (solution-state) NMR spectroscopy and some of the landmark developments that have shaped the subject.

1.1. THE DEVELOPMENT OF HIGH-RESOLUTION NMR

It is now a little over fifty years since the first observations of nuclear magnetic resonance were made in both solid and liquid samples, from which the subject has evolved to become the principal structural technique of the research chemist. During this time, there have been a number of key advances in high-resolution NMR that have guided the development of the subject [3,4] (Table 1.1) and consequently the work of organic chemists and their approaches to structure elucidation. The seminal step occurred during the early 1950s when it was realised that the resonant frequency of a nucleus is influenced by its chemical environment, and that one nucleus could further influence the resonance of another through intervening chemical bonds. Although these observations were seen as unwelcome chemical complications by the investigating physicists, a few pioneering chemists immediately realised the significance of these chemical shifts and spin–spin couplings within the context of structural chemistry. The first high-resolution proton NMR spectrum (Fig. 1.1) clearly

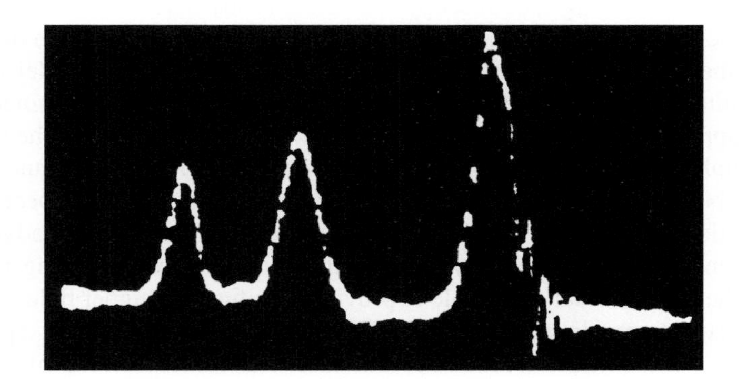

Figure 1.1. The first published 'high-resolution' proton NMR spectrum, recorded at 30 MHz, displaying the proton chemical shifts in ethanol (reproduced with permission from reference [5]).

Table 1.1. A summary of some key developments that have had a major influence on the practice and application of high-resolution NMR spectroscopy in chemical research

Decade	Notable advances
1940s	First observation of nuclear magnetic resonance in solids and liquids (1945)
1950s	Development of chemical shifts and spin–spin coupling constants as structural tools
1960s	Use of signal averaging for improving sensitivity Application of the pulse-Fourier transform (FT) approach The nuclear Overhauser effect employed in structural investigations
1970s	Use of superconducting magnets and their combination with the FT approach Computer controlled instrumentation
1980s	Development of multipulse and two-dimensional NMR techniques Automated spectroscopy
1990s	Routine application of pulsed field gradients for signal selection Development of coupled analytical methods e.g. LC–NMR
2000+	Use of high-sensitivity superconducting probes? Use of flow-injection NMR ('tubeless' NMR) in routine structural characterisation? . . . ?

demonstrated how the features of an NMR spectrum, in this case chemical shifts, could be directly related to chemical structure and it is from this that NMR has evolved to attain the significance it holds today.

The 1950s also saw a variety of instrumental developments that were to provide the chemist with even greater chemical insight. These included the use of sample spinning for averaging to zero field inhomogeneities which provided a substantial increase in resolution, so revealing fine splittings from spin–spin coupling. Later, spin-decoupling was able to provide more specific information by helping the chemists understand these interactions. With these improvements, sophisticated relationships could be developed between chemical structure and measurable parameters, leading to such realisations as the dependence of vicinal coupling constants on dihedral angles (the now well-known Karplus relationship). The application of computers during the 1960s was also to play a major role in enhancing the influence of NMR on the chemical community. The practice of collecting the same continuous-wave spectrum repeatedly and combining them with a CAT (computer of average transients) led to significant gains in sensitivity and made the observation of smaller sample quantities a practical realisation. When the idea of stimulating all spins simultaneously with a single pulse of radio frequency energy, collecting the time-domain response and converting this to the required frequency-domain spectrum by a process known as Fourier transformation (FT) was introduced, more rapid signal averaging became possible. This approach provided an enormous increase in signal-to-noise ratio, and was to change completely the development of NMR spectroscopy. The mid 1960s also saw the application of the nuclear Overhauser effect to conformational studies. Although described during the 1950s as a means of enhancing the sensitivity of nuclei through the simultaneous irradiation of electrons, the Overhauser effect has since found widest application in sensitivity enhancement between nuclei, or in the study of the spatial proximity of nuclei, and remains one of the most important tools of modern NMR. By the end of the 1960s, the first commercial FT spectrometer was available, operating at 90 MHz for protons. The next great advance in field strengths was provided by the introduction of superconducting magnets during the 1970s, which were able to provide significantly higher fields than the electromagnets previously employed. These, combined with the FT approach, made the observation of carbon-13 routine and provided the organic chemist with another probe of molecular structure. This also paved the way for the routine observation of a whole variety of previously inaccessible nuclei of low natural abundance and low magnetic moment. It was also in the early 1970s that the concept of spreading the information contained within the NMR spectrum into two separate frequency dimensions was presented in a lecture. However, because of instrumental limitations, the quality of the first two-dimensional spectra were considered too poor to be published, and not until the mid 1970s, when instrument stability had improved and developments in computers made the necessary complex calculations feasible, did the development of 2D methods begin in earnest. These methods, together with the various multipulse one-dimensional methods that also became possible with the FT approach, were not to have significant impact on the wider chemical community until the 1980s, from which point their development was nothing less than explosive. This period saw an enormous number of new pulse techniques presented which were capable of performing a variety of 'spin gymnastics' and so providing the chemist with ever more structural data, on smaller sample quantities and in less time. No longer was it necessary to rely on empirical correlations of chemical shifts and coupling constants with structural features, but instead a collection of spin interactions (through-bond, through-space, chemical exchange) could be mapped and used to determine structures more reliably and more rapidly. The evolution of new pulse methods has continued throughout the 1990s, alongside which has emerged a fundamentally different way of extracting the desired information from molecular systems. Pulsed field gradient selected experiments have now become routine structural tools, providing better quality spectra, often in shorter times, than was previously possible. These have come into widespread use not so much from a recent theoretical breakthrough (their use for signal selection was first demonstrated in 1980) but again as a result of progressive technological developments defeating practical difficulties. Similarly, the recent emergence of coupled analytical methods, such as liquid chromatography and NMR (LC–NMR), has come about after the experimental complexities of interfacing these very different techniques have been overcome, and these methods are proving particularly popular in the pharmaceutical industry, for example.

As the new century begins, it remains to be seen what new developments emerge and the impact these have on structural organic chemistry and the operation of the research laboratory. Already new breeds of 'shielded' superconducting magnets are available which have significantly smaller 'stray fields' and therefore demand less laboratory space. Advances in high-temperature superconducting materials may lead to still smaller and more economical liquid-nitrogen cooled magnets and possibly to yet higher field strengths. Could these advances, together with the ongoing miniaturisation of all electronic equipment, mean bench-top high-resolution, high-field instruments become standard in the future? The use of superconducting materials in probe detection coils is already providing significant improvements instrument sensitivity, and although such probes are now becoming commercially available, it remains to be seen whether they become an integral part of routine spectrometers. The advent of 'tubeless' NMR in high-throughput applications is also an emerging method utilising flow-injection technology, and may in due course force the demise of the NMR tube in routine structural characterisations after nearly half a century of service. Undoubtedly, all developments will be paralleled by advances in computational power and the added flexibility and opportunities for data handling this provides.

Modern NMR spectroscopy is now a highly developed and technologically advanced subject. With so many advances in NMR methodology in recent years it is understandably an overwhelming task for the research chemist, and even the dedicated spectroscopist, to appreciate what modern NMR has to offer.

This text aims to assist in this task by presenting the principal modern NMR techniques to the wider audience.

1.2. MODERN HIGH-RESOLUTION NMR AND THIS BOOK

There can be little doubt that NMR spectroscopy now represents the most versatile and informative spectroscopic technique employed in the modern chemical research laboratory, and that an NMR spectrometer represents one of the largest single investments in analytical instrumentation the laboratory is likely to make. For both these reasons it is important that the research chemist is able to make the best use of the available spectrometer(s) and to harness modern developments in NMR spectroscopy in order to promote their chemical or biochemical investigations. Even the most basic modern spectrometer is equipped to perform a myriad of pulse techniques capable of providing the chemist with a variety of data on molecular structure and dynamics. Not always do these methods find their way into the hands of the practising chemist, remaining instead in the realms of the specialist, obscured behind esoteric acronyms or otherwise unfamiliar NMR jargon. Clearly this should not be so and the aim of this book is to gather up the most useful of these modern NMR methods and present them to the wider audience who should, after all, find greatest benefit from their application.

The approach taken throughout is non-mathematical and is based firmly on using pictorial descriptions of NMR phenomena and methods wherever possible. In preparing this work, I have attempted to keep in mind what I perceive to be the requirements of three major classes of potential readers:

- those who use solution-state NMR as tool in their own research, but have little or no direct interaction with the spectrometer,
- those who have undertaken training in directly using a spectrometer to acquire their own data, but otherwise have little to do with the upkeep and maintenance of the instrument, and
- those who make use spectrometers and are nominally responsible for the day-to-day upkeep of the instrument (although they may not consider themselves dedicated NMR spectroscopists).

The first of these could well be research chemists and students in an academic or industrial environment who need to know what modern techniques are available to assist them in their efforts, but otherwise feel they have little concern for the operation of a spectrometer. Their data is likely to be collected under fully-automated conditions, or provided by a central analytical facility. The second may be a chemist in an academic environment who has hands-on access to a spectrometer and has his or her own samples which demand specific studies that are perhaps not available from fully automated instrumentation. The third class of reader may work in a small chemical company or academic chemistry department which has invested in NMR instrumentation but does not employ a dedicated NMR spectroscopist for its upkeep, depending instead on, say, an analytical or synthetic chemist for this. This, it appears (in the UK at least), is often the case for new start-up chemical companies. With these in mind, the book contains a fair amount of practical guidance on both the execution of NMR experiments and the operation and upkeep of a modern spectrometer. Even if you see yourself the first of the above categories, some rudimentary understanding of how a spectrometer collects the data of interest and how a sequence produces, say, the 2D correlation spectrum awaiting analysis on your desk, can be enormously helpful in correctly extracting the information it contains or in identifying and eliminating artefacts that may arise from instrumental imperfections or the use of less than optimal conditions for your sample. Although not specifically aimed at dedicated spectroscopists, the book may still contain new information or may serve as a reminder of what was once understood but has somehow faded away. The text should be suitable for (UK) graduate level courses on NMR spectroscopy, and sections of the book may also be appropriate for use in advanced undergraduate courses. The book does not, however, contain descriptions of the basic NMR phenomena such as chemical shifts and coupling constants, and neither does it contain extensive discussions on how these may be correlated with chemical structures. These topics are already well documented in various introductory texts [6–9] and it is assumed that the reader is already familiar with such matters.

The emphasis is on techniques in solution-state (high-resolution) spectroscopy, principally those used throughout organic chemistry as appropriate for this series, although extensions of the discussions to inorganic systems should be straightforward, relying on the same principles and similar logic. Likewise, greater emphasis is placed on small to mid-sized molecules (with masses up to a few thousand say) since it is on such systems that the majority of organic chemistry research is performed. That is not to say that the methods described are not applicable to macromolecular systems, and appropriate considerations are given when these deserve special comment. Biological macromolecules are not covered however, but are addressed in a number of specialised texts [10–12].

1.2.1. What this book contains

The aim of this text is to present the most important NMR methods used for organic structure elucidation, to explain the information they provide, how they operate and to provide some guidance on their practical implementation. The choice of experiments is naturally a subjective one, partially based on personal experience, but also taking into account those methods most commonly encountered in the chemical literature and those recognised within the NMR community as being most informative and of widest applicability. The operation of many of these is described using pictorial models (equations appear infrequently, and are only included when they serve a specific purpose) so that the chemist can gain some understanding of the methods they are using without recourse to uninviting mathematical descriptions. The sheer number of available NMR methods may make this seem an overwhelming task, but in reality most experiments are composed of a smaller number of comprehensible building blocks pieced together, and once these have been mastered an appreciation of more complex sequences becomes a far less daunting task. For those readers wishing to pursue a particular topic in greater detail, the original references are given but otherwise all descriptions are self-contained.

Following this introductory section, Chapter 2 introduces the basic model used throughout the book for the description of NMR methods and describes how this provides a simple picture of the behaviour of chemical shifts and spin-spin couplings during pulse experiments. Following this, the model is used to visualise nuclear spin relaxation, a factor of central importance for the optimum execution of all NMR experiments (indeed, it seems early attempts to observe NMR failed most probably because of a lack of understanding at the time of the relaxation behaviour of the chosen samples!). Methods for measuring relaxation rates also provide a simple introduction to multipulse NMR sequences. Chapter 3 describes the practical aspects of performing NMR spectroscopy. This is a chapter to dip into as and when necessary and is essentially broken down into self-contained sections relating to the operating principles of the spectrometer and the handling of NMR data, how to correctly prepare the sample and the spectrometer before attempting experiments, how to calibrate the instrument and how to monitor and measure its performance, should you have such responsibilities. It is clearly not possible to describe all aspects of experimental spectroscopy in a single chapter, but this (together with some of the descriptions in Chapter 9) should contain sufficient information to enable the execution of most modern experiments. These descriptions are kept general and in these I have deliberately attempted to avoid the use of a dialect specific to a particular instrument manufacturer. Chapter 4 contains the most widely used one-dimensional techniques, ranging from the optimisation of the single-pulse experiment, through to the multiplicity editing of heteronuclear spectra and the concept of polarisation transfer, another central feature of pulse NMR methods. This includes the universally employed methods for the editing of carbon spectra according to the number of attached protons. Specific requirements for the observation of certain quadrupolar nuclei that posses extremely broad resonances are also considered. The introduction of two-dimensional methods is presented in Chapter 5, specifically in the context of determining homonuclear correlations. This is so that the 2D concept is introduced with a realistically useful experiment in mind, although it also becomes apparent that the general principles involved are the same for any two-dimensional experiment. Following this, a variety of correlation techniques are presented for identifying scalar (J) couplings between homonuclear spins, which for the most part means protons. Included in these descriptions is an introduction to the operation and use of pulsed field gradients, the most recent revolution in NMR spectroscopy, again with a view to presenting them with a specific application in mind so that their use does not seem detached from reality. Again, these discussions aim to make it clear that the principles involved are quite general and can be applied to a variety of experiments, and those that benefit most from the gradient methodology are the heteronuclear correlation techniques of Chapter 6. These correlations are used to map coupling interactions between, typically, protons and a heteroatom either through a single bond or across multiple bonds. In this chapter, most attention is given to the modern correlation methods based on proton excitation and detection, so called 'inverse' spectroscopy. These provide significant gains in sensitivity over the traditional methods that use detection of the less sensitive heteroatom, which nevertheless warrant description because of specific advantages they provide for certain molecules. Chapter 7 considers methods for separating chemical shifts and coupling constants in spectra, which are again based on two-dimensional methods. Chapter 8 moves away from through-bond couplings and onto through-space interactions in the form of the nuclear Overhauser effect (NOE). The principles behind the NOE are presented initially for a simple two-spin system, and then for more realistic multi-spin systems. The practical implementation of both 1D and 2D NOE experiments is described, including the newly developed pulsed field gradient 1D NOE methods. Rotating-frame NOE methods are also described as these will inevitably take on ever greater importance in the world of organic NMR as interest in larger molecules continues to grow in areas such as host–guest complexes, oligomeric structures and so on. The final chapter considers additional experimental methods which do not, on their own, constitute complete NMR experiments but are the tools with which modern methods are constructed. These are typically used as elements within the sequences described in the preceding chapters and include such topics as broadband decoupling, selective excitation of specific regions of a spectrum and solvent suppression. Finally, the chapter concludes with a brief overview of some newly established methods, including the separation of the spectra of solutes based on their diffusion properties and the direct high-resolution analysis of materials attached to solid-state resins. At the end of the book is a glossary of some of the acronyms that permeate the language of modern NMR, and, it might be argued, have come to characterise the subject. Whether you love them or hate them they are clearly here to stay and although they provide a ready reference when speaking of pulse experiments to those in the know, they can also serve to confuse the uninitiated and leave them bewildered in the face of this NMR jargon. The glossary provides an immediate breakdown of the acronym together with a reference to its location in the book.

1.2.2. Pulse sequence nomenclature

Virtually all NMR experiments can described in terms of a pulse sequence which, as the name suggests, is a notation which describes the series of radiofrequency or field gradient pulses used to manipulate nuclear spins and so tailor the experiment to provide the desired information. Over the years a largely (although not completely) standard pictorial format has evolved for representing these sequences, not unlike the way a musical score is used to encode a symphony [1]. As these crop up repeatedly throughout the text, the format and conventions used in this book deserve explanation. Only the definitions of the various pictorial components of a sequence are given here, their physical significance in an NMR experiment will become apparent in later chapters.

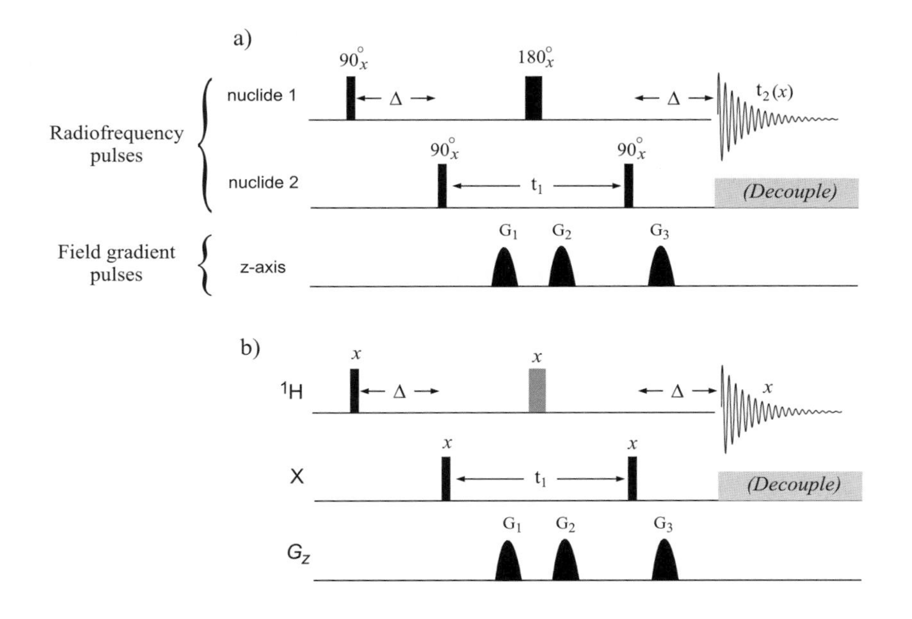

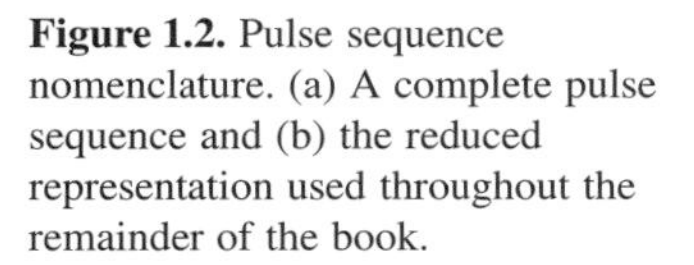
Figure 1.2. Pulse sequence nomenclature. (a) A complete pulse sequence and (b) the reduced representation used throughout the remainder of the book.

An example of a reasonably complex sequence is shown in Fig. 1.2 (a heteronuclear correlation experiment from Chapter 6) and illustrates most points of significance. Fig. 1.2a represents a more detailed account of the sequence, whilst Fig. 1.2b is the reduced equivalent used throughout the book for reasons of clarity. Radiofrequency (rf) pulses applied to each nuclide involved in the experiment are presented on separate rows running left to right, in the order in which they are applied. Most experiments nowadays involve protons, and often one (and sometimes two) additional nuclides. In organic chemistry this is naturally most often carbon but can be any other, so is termed the X-nucleus. These rf pulses are most frequently applied as so-called *90°* or *180°* pulses (the significance of which is detailed in the following chapter), which are illustrated by a thin black and thick grey bars respectively. Pulses of other angles are

[1] Indeed, just as a skilled musician can read the score and 'hear' the symphony in their head, an experienced spectroscopist can often read the pulse sequence and picture the general form of the resulting spectrum.

marked with an appropriate Greek symbol that is described in the accompanying text. All rf pulses also have associated with them a particular phase, indicated by x, y, $-x$ or $-y$ above each pulse. Pulses that act only on an small number of resonances, so-called *selective* or *shaped* pulses, are differentiated as rounded rather than rectangular bars as this reflects the manner in which these pulses are applied experimentally. Segments that make use of a long series of very many closely spaced pulses, such as the decoupling shown in Fig. 1.2, are shown as a solid grey box, with the bracket indicating the use of the decoupling sequence is optional. Below the row(s) of radiofrequency pulses are shown field gradient pulses (G_z), whenever these are used, again drawn as shaped pulses.

The operation of very many NMR experiments is crucially dependent on the experiment being tuned to the value of specific coupling constants. This is achieved by defining certain delays within the sequence according to these values, these delays being indicated by the general symbol Δ. Other time periods within a sequence which are not tuned to J-values but are chosen according to other criteria, such as spin recovery (relaxation) rates, are given the symbol τ. The symbols t_1 and t_2 are reserved for the time periods which ultimately correspond to the frequency axes f_1 and f_2 of two-dimensional spectra, one of which (t_2) will always correspond to the data acquisition period when the NMR response is actually detected. The acquisition period is illustrated by a simple decaying sine-wave in all experiments to represent the so-called Free Induction Decay or FID. Again it should be stressed that although these sequences can have rather foreboding appearances, they are generally built-up from much smaller and simpler segments that have well defined and easily understood actions. A little perseverance can clarify what might at first seem a total enigma.

1.3. APPLYING MODERN NMR TECHNIQUES

The tremendous growth in available NMR pulse methods over the last two decades can be bewildering and may leave one wondering just where to start or how best to make use of these new developments. The answer to this is not straightforward since it depends so much on the chemistry undertaken, on the nature of the molecule being handled and on the information required of it. It is also dependent on the amount of material and by the available instrumentation and its capabilities. The fact that NMR itself finds application in so many research areas means defined rules for experiment selection are largely intractable. A scheme that is suitable for tackling one type of problem may be wholly inappropriate for another. Nevertheless, it seems inappropriate that a book of this sought should contain no guidance on experiment selection other than the descriptions of the techniques in the following chapters. Here I attempt to broach this topic in rather general terms and present some loose guidelines to help weave a path through the maze of available techniques. Even so, only with a sound understanding of modern techniques can one truly be in a position to select the optimum experimental strategy for *your* molecule or system, and it is this understanding I hope to develop in the remaining chapters.

Most NMR investigations will begin with the analysis of the proton spectrum of the sample of interest, with the usual analysis of the chemical shifts, coupling constants and relative signal intensities, either manually or with the assistance of the various sophisticated computer-based structure-spectrum data bases now available. Beyond this, one encounters a plethora of available NMR methods to consider employing. The key to selecting appropriate experiments for the problem at hand is an appreciation of the type of information the principal NMR techniques can provide. Although there exist a huge number of pulse sequences, there are a relatively small number of what might be called core experiments, from which most others are derived by minor variation, of which only a rather small fraction ever find widespread use in the research laboratory. To begin, it is perhaps instructive to realise that virtually all NMR methods exploit only three basic phenomena:

- Through-bond interactions: scalar (J) coupling via bonding electrons.
- Through-space interactions: the nuclear Overhauser effect mediated through dipole coupling and spin relaxation.
- Chemical exchange: the physical exchange of one spin for another at a specific location.

When attempting to analyse the structure of a molecule and/or its behaviour in solution by NMR spectroscopy, one must therefore consider how to exploit these phenomena to gain the desired information, and from this select the appropriate technique(s). Thus, when building up the structure of a molecule one typically first searches for evidence of scalar coupling between nuclei as this can be used to indicate the location of chemical bonds. When the location of all bonding relationships within the molecule have been established, the gross structure of the molecule is defined. Spatial proximities between nuclei, and between protons in particular, can be used to define stereochemical relationships within a molecule and thus address questions of configuration and conformation. The unique feature of NMR spectroscopy, and the principal reason for its superiority over any other solution-state technique for structure elucidation, is its ability to define relationships between *specific nuclei* within a molecule or even between molecules. Such exquisite detail is generally obtained by correlating one nucleus with another by exploiting the above phenomena. Despite the enormous power of NMR, there are, in fact, rather few types of correlation available to the chemist to employ for structural and conformational analysis. The principal spin interactions and the main techniques used to map these, which are frequently two-dimensional methods, are summarised in Table 1.2, and further elaborated in the chapters that follow.

The homonuclear correlation experiment, COSY, identifies those nuclei that share a J-coupling, which, for protons, operate over two, three, and, less frequently, four bonds. This information can therefore be used to indicate the presence of a bonding pathway. The correlation of protons that exist within the same coupled network or chain of spins, but do not themselves share a J-coupling, can be made with the TOCSY experiment. This can be used to identify groups of nuclei that sit within the same isolated spin system, such as the amino acid residue of a peptide or the sugar ring of an oligosaccharide. One-bond heteronuclear correlation methods (HMQC or HSQC) identify the heteroatoms to which the protons are directly attached and can, for example, provide carbon assignments from previously established proton assignments. Proton chemical shifts can also be dispersed according to the shift of the attached heteroatom, so aiding the assignment of the proton spectrum itself. Long-range heteronuclear correlations over typically two- or three-bonds (HMBC) provide a wealth of information on the skeleton of the molecule and can be used to infer the location of carbon-carbon or carbon-heteroatom bonds. These correlations can be particularly valuable when proton–proton correlations are absent. The INADEQUATE experiment identifies connectivity between like nuclei of low-natural abundance, for which it is favoured over COSY. This can therefore correlate directly-connected carbon centres, but as this relies on the presence of neighbouring carbon-13 nuclei it suffers from appallingly low sensitivity and thus finds little use. Modern variants that use proton detection have greatly improved performance but are still likely to be less used than the heteronuclear correlation techniques. Measurements based on the nuclear Overhauser effect (NOE) are most often applied after the gross structure is defined and NMR assignments established.

Table 1.2. The principal correlations established through NMR techniques

Correlation	Principal technique(s)	Comments	Chapter
H J H X–X	^{1}H–^{1}H COSY	Proton J-coupling typically over 2 or 3 bonds.	5
H J H J H X–X–X	^{1}H–^{1}H TOCSY	Relayed proton J-couplings within a coupled spin system. Remote protons may be correlated provided there is a continuous coupling network in between them.	5
H J H X–X	^{1}H–X HMQC ^{1}H–X HSQC	One-bond heteronuclear couplings via proton observation.	6
H J X---X	^{1}H–X HMBC	Long-range heteronuclear couplings via proton observation. Typically over 2 or 3 bonds when X = ^{13}C.	6
J X–X	X–X COSY X–X INADEQUATE	COSY only used when X-spin natural abundance > 20%. Sensitivity problems when X has low natural abundance.	5
NOE H H X------X	^{1}H–^{1}H NOE difference 1/2D NOESY 1/2D ROESY	Through-space correlations. ROESY most applicable to 'mid-sized' molecules with masses ca. 1–2 kDa.	8
NOE H X------X	^{1}H–X NOE difference 2D HOESY	Sensitivity limited by X-spin observation. Care required to make NOEs specific in presence of proton decoupling.	8
exchange A B	1D saturation or inversion transfer 2D EXSY	Interchange of spins at chemically distinct locations. Exchange must be slow on NMR timescale for separate resonances to be observed. Intermediate to fast exchange requires lineshape analysis.	8

The correlated spins are shown in bold type for each correlation, with X indicating any spin-$^1/_2$ nucleus. The acronyms are explained in the glossary.

These define the 3D stereochemistry of a molecule since this effect maps through-space proximity between nuclei. The vast majority of such experiments investigate proton–proton NOEs, although in exceptional cases heteronuclear NOEs involving a proton and a heteroatom have been applied successfully. The final group of experiments correlate nuclei involved in chemical exchange processes that are slow on the NMR timescale and thus give rise to distinct resonances for each exchanging species or site.

The greatest use of NMR in the chemical research laboratory is in the routine characterisation of synthetic starting materials, intermediates and final products. In these circumstances it is often not so much full structure *elucidation* that is required, rather it is structure *confirmation* since the synthetic reagents are known, which naturally limit what the products may be, and because the desired synthetic target is usually defined. Routine analysis of this sought typically follows a general procedure similar to that summarised in Table 1.3, which is supplemented with data from the other analytical techniques, most notably mass spectrometry and infra-red spectroscopy. This general protocol has been greatly influenced by a number of developments over the last decade, which can perhaps be viewed as now characterising modern NMR methodology, thus:

- The *indirect* observation of heteronuclides via proton detection.
- The *routine* application of 2D methods on a daily basis.
- The application of pulsed field gradients for clean signal selection.
- The wider use of sophisticated data processing procedures which enhance the information content or quality of spectra.

Table 1.3. A typical protocol for the routine structure confirmation of synthetic organic materials

Procedure	Technique	Information
1D ^{1}H spectrum	1D	Information from chemical shifts, coupling constants, integrals
2D ^{1}H–^{1}H correlation	COSY	Identify J-coupling relationships between protons
1D ^{13}C (with spectrum editing)	1D, (DEPT or APT)	Carbon count and multiplicity determination (C, CH, CH_2, CH_3). Can often be avoided by using proton-detected heteronuclear 2D experiments.
1D heteronuclide spectra e.g. ^{31}P, ^{19}F etc.	1D	Chemical shifts and homonuclear/heteronuclear coupling constants
2D ^{1}H–^{13}C one-bond correlation (with spectrum editing)	HMQC or HSQC (with editing)	Carbon assignments transposed from proton assignments. Proton spectrum dispersed by ^{13}C shifts. Carbon multiplicities from edited HSQC (faster than above 1D approach).
2D ^{1}H–^{13}C long-range correlation	HMBC	Correlations identified over two- and three- bonds. Correlations established across heteroatoms e.g. N and O. Structural fragments pieced together.
Through-space NOE correlation	1D or 2D NOE	Stereochemical analysis: configuration and conformation

Not all these steps may be necessary, and the direct observation of a heteronuclide, such as carbon or nitrogen, can often be replaced through its indirect observation with the more sensitive proton-detected heteronuclear shift correlation techniques.

The first item in this list arises from the near universal adoption of techniques based on proton observation whenever possible. This is principally to help overcome limitations from the relatively low sensitivity associated with NMR observations and as a result provide data on smaller sample quantities and/or provide data in shorter times. This is a recurring theme in this book. The second significant feature is the increased use of a range of 2D methods for routine structural characterisation as computational power and data storage capacities increase, now making these as widely used as 1D observations. The third issue is the development of pulsed field gradients and the enormous impact these have had on the practical implementation of modern techniques, another recurring theme of this book. These points may be illustrated by considering the role of conventional 1D carbon-13 spectra in the modern NMR laboratory. Traditionally the carbon spectrum has been used as a routine tool in characterisation, supplemented by spectrum editing methods which indicate the multiplicities of each carbon centre, that is, identify it as belonging to a quaternary, methine, methylene or methyl group. The greatest problem with any carbon-13 detected experiment is its inherently low sensitivity. Nowadays, the collection of a 2D proton-detected ^{1}H–^{13}C shift correlation experiment requires significantly less time than the 1D carbon spectrum of the same sample, particularly with the gradient-selected technique, and provides both carbon shift data (of protonated centres) and correlation information. This is a far more powerful tool for routine structure confirmation than the 1D carbon experiment alone. In addition, editing can be introduced to the 2D experiment to differentiate methine from methylene correlations for example, providing yet more data in a single experiment and in less time. Even greater gains can be made in the indirect observation of heteronuclides of still lower intrinsic sensitivity, for example nitrogen-15, and when considering the observation of low-abundance nuclides it is sensible to first consider adopting a proton-detected method for this.

Even when dealing with unknown materials or with molecules of high structural complexity, the scheme of Table 1.3 still represents an appropriate general protocol to follow. In such cases, the basic experiments of this table are still likely to be employed, but may require data to be collected under a variety of experimental conditions (solvent, temperature, pH etc.) and/or may require additional support from other methods or extended versions of these techniques before a complete picture emerges. The following chapters aim to explain the primary NMR techniques and some of their more useful variants, and to describe their practical implementation, so that the research chemist may realise the full potential that modern NMR spectroscopy has to offer.

REFERENCES

[1] F. Bloch, W.W. Hansen and M.E. Packard, *Phys. Rev.*, 1946, **69**, 127.
[2] E.M. Purcell, H.C. Torrey and R.V. Pound, *Phys. Rev.*, 1946, **69**, 37–38.
[3] J.W. Emsley and J. Feeney, *Prog. Nucl. Magn. Reson. Spectrosc.*, 1995, **28**, 1–9.
[4] J.N. Shoolery, *Prog. Nucl. Magn. Reson. Spectrosc.*, 1995, **28**, 37–52.
[5] J.T. Arnold, S.S. Dharmatti and M.E. Packard, *J. Chem. Phys.*, 1951, **19**, 507.
[6] L.M. Harwood and T.D.W. Claridge, Introduction to Organic Spectroscopy, Oxford University Press, Oxford, 1997.
[7] H. Günther, NMR Spectroscopy, 2nd edn., Wiley, Chichester, 1995.
[8] H. Friebolin, Basic One- and Two-dimensional NMR Spectroscopy, 3rd edn., VCH Publishers, Weinheim, 1998.
[9] R. Abraham, J. Fisher and P. Loftus, Introduction to NMR Spectroscopy, Wiley, Chichester, 1988.
[10] J. Cavanagh, W.J. Fairbrother, A.G. Palmer and N.J. Skelton, Protein NMR Spectroscopy. Principles and Practice, Academic Press, San Diego, 1996.
[11] J.N.S. Evans, Biomolecular NMR Spectroscopy, Oxford University Press, Oxford, 1995.
[12] H.J. Dyson and P.E. Wright, in: Two-dimensional NMR Spectroscopy; Applications for Chemists and Biochemists, eds. W.R. Croasmun and R.M.K. Carlson, VCH Publishers, New York, 1994.

Chapter 2

Introducing high-resolution NMR

For anyone wishing to gain a greater understanding of modern NMR techniques together with an appreciation of the information they can provide and hence their potential applications, it is necessary to develop an understanding of some elementary principles. These, if you like, provide the foundations from which the descriptions in subsequent chapters are developed and many of the fundamental topics presented in this introductory chapter will be referred to throughout the remainder of the text. In keeping with the style of the book, all concepts are presented in a pictorial manner and avoid the more mathematical descriptions of NMR. Following a reminder of the phenomenon of nuclear spin, the Bloch vector model of NMR is introduced. This presents a convenient and comprehensible picture of how spins behave in an NMR experiment and provides the basic tool with which most experiments will be described.

2.1. NUCLEAR SPIN AND RESONANCE

The nuclei of all atoms may be characterised by a nuclear spin quantum number, I, which may have values greater than or equal to zero and which are multiples of $^1/_2$. Those with $I = 0$ possess no nuclear spin and therefore cannot exhibit nuclear magnetic resonance so are termed 'NMR silent'. Unfortunately, from the organic chemists' point of view, the nucleus likely to be of most interest, carbon-12, has zero spin, as do all nuclei with atomic mass and atomic number both even. However, the vast majority of chemical elements have at least one nuclide that does possess nuclear spin which is, in principle at least, observable by NMR (Table 2.1) and as a consolation the proton is a high-abundance NMR-active isotope. The property of nuclear spin is fundamental to the NMR phenomenon. The spinning nuclei possess angular momentum, P, and, of course, charge and the motion of this charge gives rise to an associated magnetic moment, μ, (Fig. 2.1) such that:

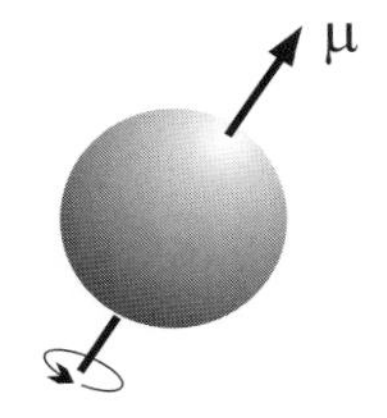

Figure 2.1. A nucleus carries charge and when spinning possesses a magnetic moment, μ.

$$\mu = \gamma P \tag{2.1}$$

where the term γ is the *magnetogyric ratio* [1] which is constant for any given nuclide and may be viewed as a measure of how 'strongly magnetic' this is. Both angular momentum and the magnetic moment are vector quantities, that is, they have both magnitude and direction. When placed in an external, static magnetic field (denoted B_0, strictly the magnetic flux density) the microscopic magnetic moments align themselves relative to the field in a

[1] The term gyromagnetic ratio is also in widespread use for γ, although this does not conform to IUPAC recommendations [1].

Table 2.1. Properties of selected spin-$^1/_2$ nuclei

Isotope	Natural abundance (%)	NMR frequency (MHz)	Relative sensitivity
^{1}H	99.98	400.0	1.0
^{3}H	0	426.7	1.2 [a]
^{13}C	1.11	100.6	1.76×10^{-4}
^{15}N	0.37	40.5	3.85×10^{-6}
^{19}F	100.00	376.3	0.83
^{29}Si	4.7	79.5	3.69×10^{-4}
^{31}P	100.00	161.9	6.63×10^{-2}
^{77}Se	7.58	76.3	5.25×10^{-4}
^{103}Rh	100.00	12.6	3.11×10^{-5}
^{113}Cd	12.16	88.7	1.33×10^{-3}
^{119}Sn	8.58	149.1	4.44×10^{-3}
^{183}W	14.40	16.6	1.03×10^{-5}
^{195}Pt	33.80	86.0	3.36×10^{-3}
^{207}Pb	22.60	83.7	2.07×10^{-3}

Frequencies are given for a 400 MHz spectrometer (9.4 T magnet) and sensitivities are given relative to proton observation and include terms for both intrinsic sensitivity of the nucleus and its natural abundance.
Properties of quadrupolar nuclei are given in Table 2.3 below.
[a] Assuming 100% ^{3}H labelling.

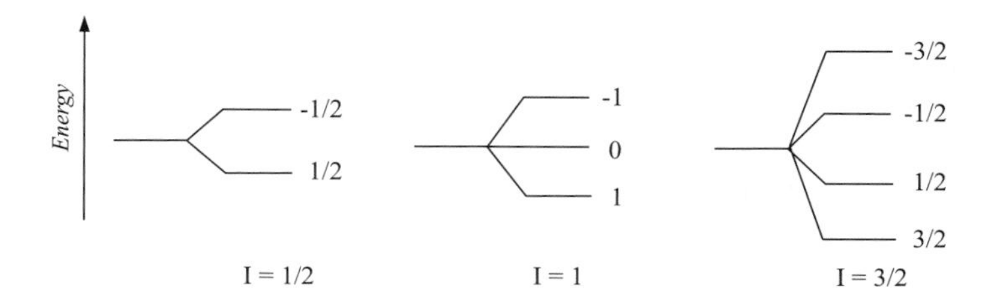

Figure 2.2. Nuclei with a magnetic quantum number I may take up 2I + 1 possible orientations relative to the applied static magnetic field B_0. For spin-$^1/_2$ nuclei, this gives the familiar picture of the nucleus behaving as a microscopic bar magnet having two possible orientations, α and β.

discrete number of orientations because the energy states involved are quantised. For a spin of magnetic quantum number I there exist 2I + 1 possible spin states, so for a spin-$^1/_2$ nucleus such as the proton, there are 2 possible states denoted $+^1/_2$ and $-^1/_2$, whilst for I = 1, for example deuterium, the states are +1, 0 and −1 (Fig. 2.2) and so on. For the spin-half nucleus, the two states correspond to the popular picture of a nucleus taking up two possible orientations with respect to the static field, either parallel (the α-state) or antiparallel (the β-state), the former being of lower energy. The effect of the static field on the magnetic moment can be described in terms of classical mechanics, with the field imposing a torque on the moment which therefore traces a circular path about the applied field (Fig. 2.3). This motion is referred to as precession, or more specifically *Larmor precession* in this context. It is

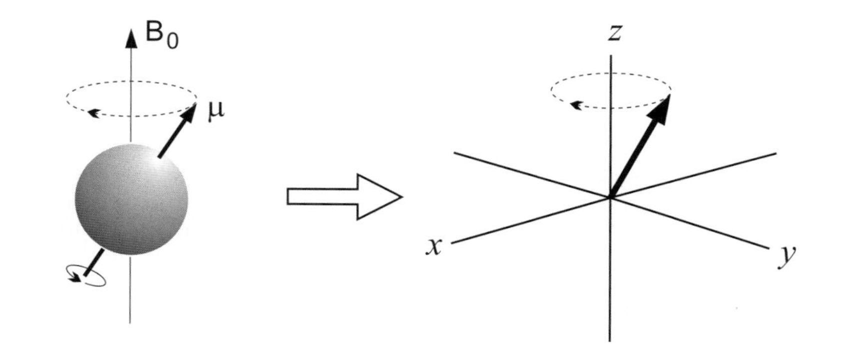

Figure 2.3. A static magnetic field applied to the nucleus causes it to precess at a rate dependent on the field strength and the magnetogyric ratio of the spin. The field is conventionally applied along the *z*-axis of a Cartesian co-ordinate frame and the motion of the nucleus represented as a vector moving on the surface of a cone.

analogous to the familiar motion of a gyroscope in the Earth's gravitational field, in which the gyroscope spins about its own axis, and this axis in turn precesses about the direction of the field. The rate of the precession as defined by the angular velocity (ω rad s^{-1} or ν Hz) is:

$$\omega = -\gamma B_0 \text{ rad s}^{-1}$$

or

$$\nu = \frac{-\gamma B_0}{2\pi} \equiv -\bar{\gamma} B_0 \text{ Hz} \quad (2.2)$$

and is known as the *Larmor frequency* of the nucleus. The direction of motion is determined by the sign of γ and may be clockwise or anticlockwise, but is always the same for any given nuclide. Nuclear magnetic resonance occurs when the nucleus changes its spin state, driven by the absorption of a quantum of energy. This energy is applied as electromagnetic radiation, whose frequency must match that of the Larmor precession for the resonance condition to be satisfied, the energy involved being given by:

$$\Delta E = h\nu = \frac{h\gamma B_0}{2\pi} \quad (2.3)$$

where h is Plank's constant. In other words, the resonant frequency of a spin is simply its Larmor frequency. Modern high-resolution NMR spectrometers currently employ field strengths up to 18.8 T (tesla) which, for protons, correspond to resonant frequencies up to 800 MHz, which fall within the radiofrequency region of the electromagnetic spectrum. For other nuclei at similar field strengths, resonant frequencies will differ from those of protons (due to the dependence of ν on γ) but it is common practice to refer to a spectrometer's operating frequency in terms of the resonant frequencies of protons. Thus, one may refer to using a '400 MHz spectrometer', although this would equally operate at 100 MHz for carbon-13 since $\gamma_H/\gamma_C \approx 4$. It is also universal practice to define the direction of the static magnetic field as being along the *z*-axis of a set of Cartesian co-ordinates, so that a *single* precessing spin-$^1/_2$ nucleus will have a component of its magnetic moment along the *z*-axis (the longitudinal component) and an orthogonal component in the *x–y* plane (the transverse component) (Fig. 2.3).

Now consider a collection of similar spin-$^1/_2$ nuclei in the applied static field. As stated, the orientation parallel to the applied field, α, has slightly lower energy than the anti-parallel orientation, β, so at equilibrium there will be an excess of nuclei in the α state as defined by the Boltzmann distribution:

$$\frac{N_\alpha}{N_\beta} = e^{\Delta E/RT} \quad (2.4)$$

where $N_{\alpha,\beta}$ represents the number of nuclei in the spin orientation, R the gas constant and T the temperature. The differences between spin energy levels are rather small so the corresponding population differences are similarly small and only about 1 part in 10^4 at the highest available field strengths. This is why NMR is so very insensitive relative to other techniques such a IR and UV, where the ground- and excited-state energy differences are substantially greater. The tiny population *excess* of nuclear spins can be represented as a collection of spins distributed randomly about the precessional cone and parallel to the *z*-axis. These give rise to a resultant *bulk magnetisation vector* $\mathbf{M}_0$ along this axis (Fig. 2.4). It is important to realise that this *z*-magnetisation arises because of *population differences* between the possible spin states, a point we return to in the following section. Since there is nothing to define a preferred orientation for the spins in the transverse direction, there exists a random distribution of individual magnetic moments about the cone and hence there is no *net magnetisation* in the transverse (*x–y*) plane. Thus we can reduce our picture

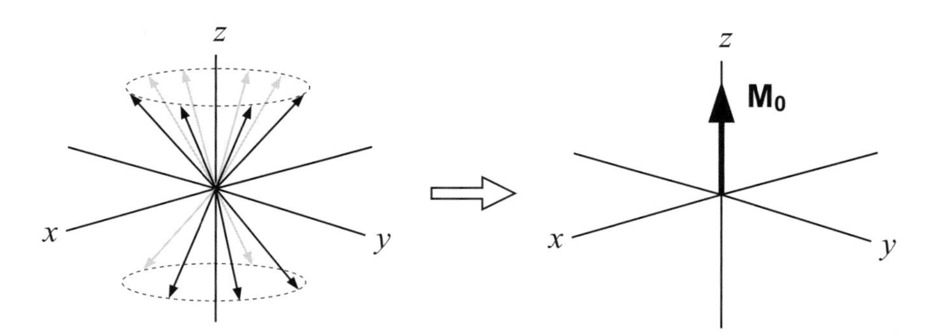

Figure 2.4. In the vector model of NMR many like spins are represented by a bulk magnetisation vector. At equilibrium the excess of spins in the α state places this parallel to the $+z$-axis.

of many similar magnetic moments to one of a single bulk magnetisation vector $\mathbf{M}_0$ that behaves according to the rules of classical mechanics. This simplified picture is referred to as the Bloch vector model (after the pioneering spectroscopist Felix Bloch), or more generally as the *vector model* of NMR.

2.2. THE VECTOR MODEL OF NMR

Having developed the basic model for a collection of nuclear spins, we can now describe the behaviour of these spins in pulsed NMR experiments. There are essentially two parts to be considered, firstly the application of the radiofrequency (rf) pulse(s), and secondly the events that occur following this. The essential requirement to induce transitions between energy levels, that is, to cause nuclear magnetic resonance to occur, is the application of a time-dependent magnetic field oscillating at the Larmor frequency of the spin. This field is provided by the magnetic component of the applied rf, which is designated the B_1 field to distinguish it from the static B_0 field. This rf is transmitted via a coil surrounding the sample, the geometry of which is such that the B_1 field exists in the *transverse* plane, perpendicular to the static field. In trying to consider how this oscillating field operates on the bulk magnetisation vector, one is faced with a mind-boggling task involving simultaneous rotating fields and precessing vectors. To help visualise these events it proves convenient to employ a simplified formalism, know as the *rotating frame* of reference, as opposed to the so-called *laboratory frame* of reference described thus far.

2.2.1. The rotating frame of reference

To aid the visualisation of processes occurring during an NMR experiment a number of simple conceptual changes are employed. Firstly, the oscillating B_1 field is considered to be composed of two counter-rotating magnetic vectors in the x–y plane, the resultant of which corresponds exactly to the applied oscillating field (Fig. 2.5). It is now possible to simplify things considerably by eliminating one of these and simultaneously freezing the motion of the other by picturing events in the *rotating frame* of reference (Fig. 2.6). In this, the set of x, y, z co-ordinates are viewed as rotating along with the nuclear precession, in the same sense and at the same rate. Since the frequency of oscillation of

Figure 2.5. The rf pulse provides an oscillating magnetic field along one axis (here the x-axis) which is equivalent to two counter-rotating vectors in the transverse plane.

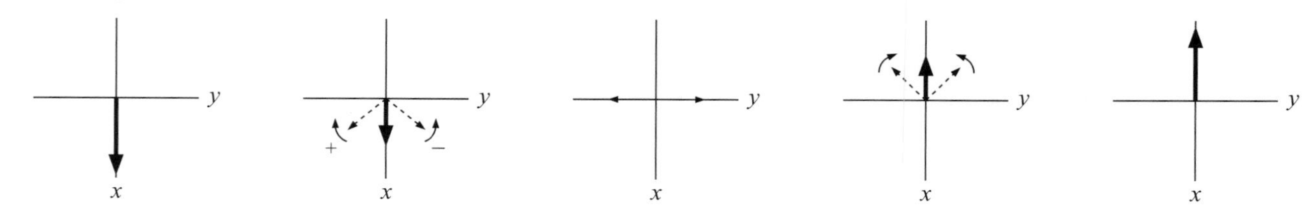

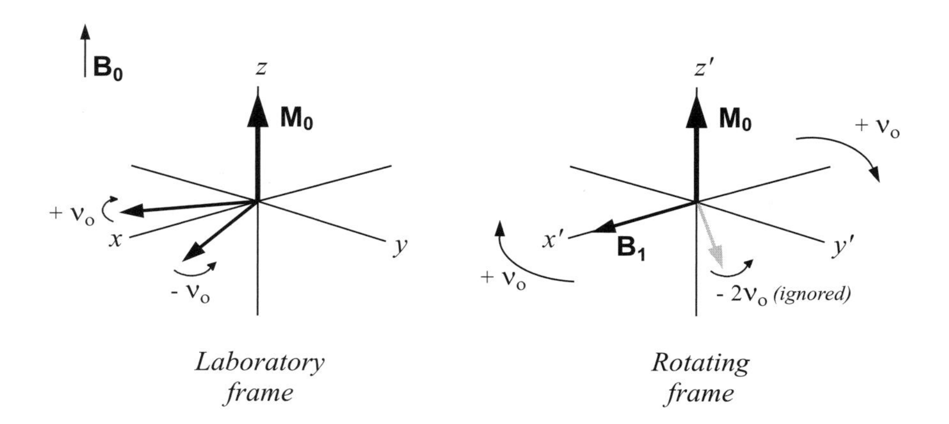

Figure 2.6. The laboratory and rotating frame representations. In the laboratory frame the co-ordinate system is viewed as being static, whereas in the rotating frame it rotates at a rate equal to the applied rf frequency, ν_0. In this representation the motion of one component of the applied rf is frozen whereas the other is far from the resonance condition and may be ignored. This provides a simplified model for the description of pulsed NMR experiments.

the rf field exactly matches that of nuclear precession (which it must for the magnetic resonance condition to be satisfied) then the rotation of one of the rf vectors is now static in the rotating frame whereas the other is moving at *twice* the frequency in the *opposite* direction. This latter vector is far from resonance and is simply ignored. Similarly, the precessional motion of the spins has been frozen as these are moving with the same angular velocity as the rf vector and hence the co-ordinate frame. Since this precessional motion was induced by the static magnetic field B_0, this is also no longer present in the rotating frame representation.

The concept of the rotating frame may be better pictured with the following analogy. Suppose you are at a fairground and are standing watching a child going round on the carousel. You see the child move towards you then away from you as the carousel turns, and are thus aware of the circular path he follows. This corresponds to observing events from the so-called *laboratory frame* of reference (Fig. 2.7a). Now imagine what you see if you step onto the carousel as it turns. You are now travelling with the same angular velocity and in the same sense as the child so his motion is no longer apparent. His precession has been frozen from your point of view and you are now observing

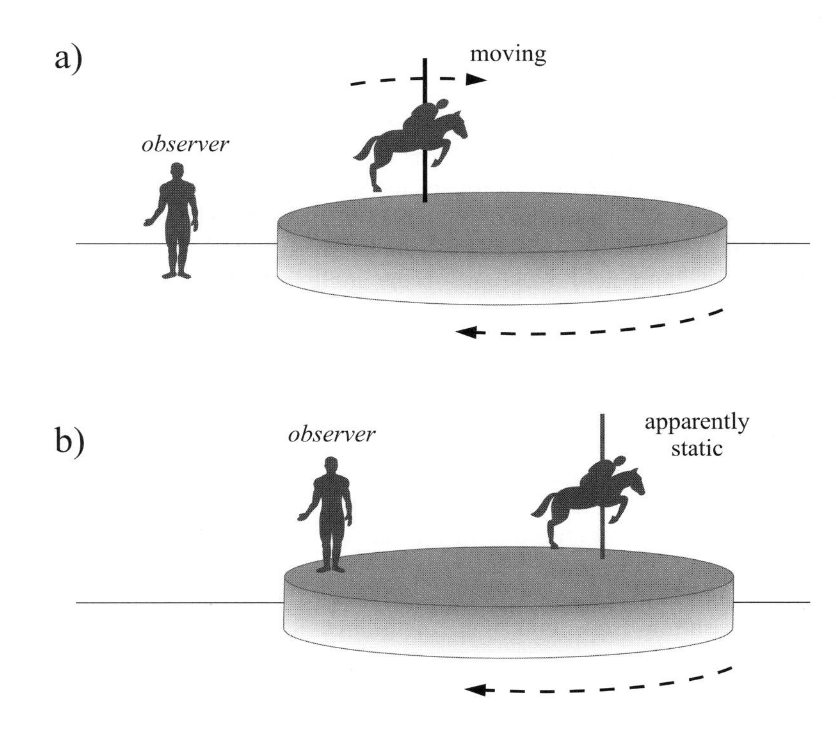

Figure 2.7. A fairground carousel can be viewed from (a) the laboratory or (b) the rotating frame of reference.

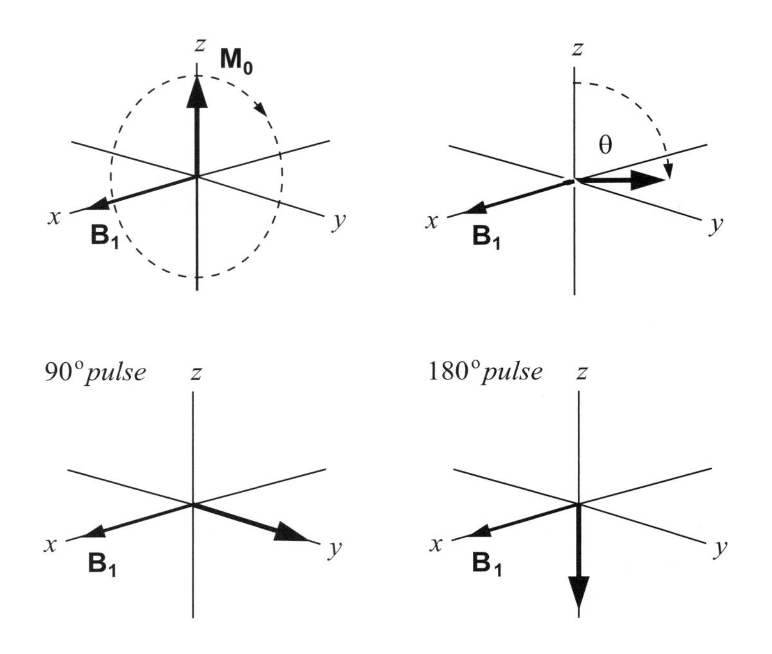

Figure 2.8. An rf pulse applies a torque to the bulk magnetisation vector and drives it toward the x–y plane from equilibrium. θ is the pulse tip or flip angle which is most frequently 90 or 180 degrees.

events in the *rotating frame* of reference (Fig. 2.7b). Obviously the child is still moving in the 'real' world but your perception of events has been greatly simplified. Likewise, this transposition simplifies our picture of events in an NMR experiment.

Strictly one should use the different co-ordinate labelling scheme for the laboratory and the rotating frames, such as x, y, z and x', y' and z' respectively, as in Fig. 2.6. However, since we shall be dealing almost exclusively with a rotating frame description of events, the simpler x, y, z notations will be used throughout the remainder of the book, and explicit indication provided where the laboratory frame of reference is used.

2.2.2. Pulses

We are now in a position to visualise the effect of applying an rf pulse to the sample. The 'pulse' simply consists of turning on rf irradiation of a defined amplitude for a time period t_p, and then switching it off. As in the case of the static magnetic field, the rf electromagnetic field imposes a torque on the bulk magnetisation vector in a direction that is perpendicular to the direction of the B_1 field (the 'motor rule') which rotates the vector from the z-axis toward the x–y plane (Fig. 2.8). Thus, applying the rf field along the x-axis will drive the vector toward the y-axis. The rate at which the vector moves is proportional to the strength of the rf field (γB_1) and so the angle θ through which the vector turns, colloquially known as the pulse flip or tip angle (but more formally as the nutation angle) will be dependent on the amplitude and duration of the pulse:

$$\theta = 360\gamma B_1 t_p \text{ degrees} \tag{2.5}$$

If the rf was turned off just as the vector reached the y-axis, this would represent a 90 degree pulse, if it reached the $-z$-axis, it would be a 180° pulse, and so on. Returning to consider the individual magnetic moments that make up the bulk magnetisation vector for a moment, we see that the 90° pulse corresponds to *equalising the populations* of the α and β states, as there is now no net z magnetisation. However, there *is* a net magnetisation in the x–y plane, resulting from 'bunching' of the individual magnetisation vectors caused by the application of the rf pulse. The spins are said to posses *phase coherence* at this point, forced upon them by the rf pulse (Fig. 2.9). Note that

Figure 2.9. Following a 90° pulse, the individual spin vectors bunch along the y-axis and are said to posses phase coherence.

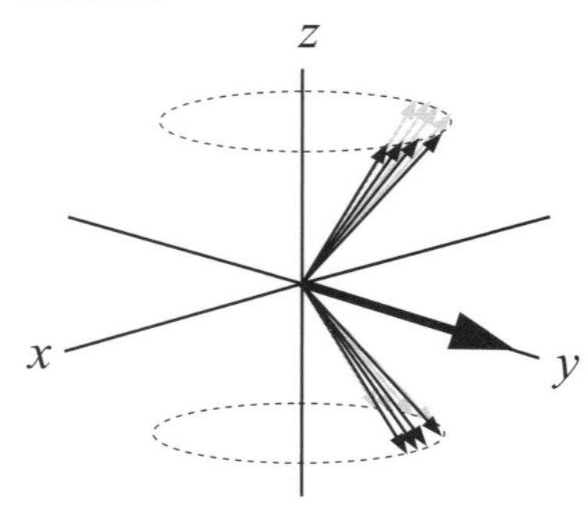

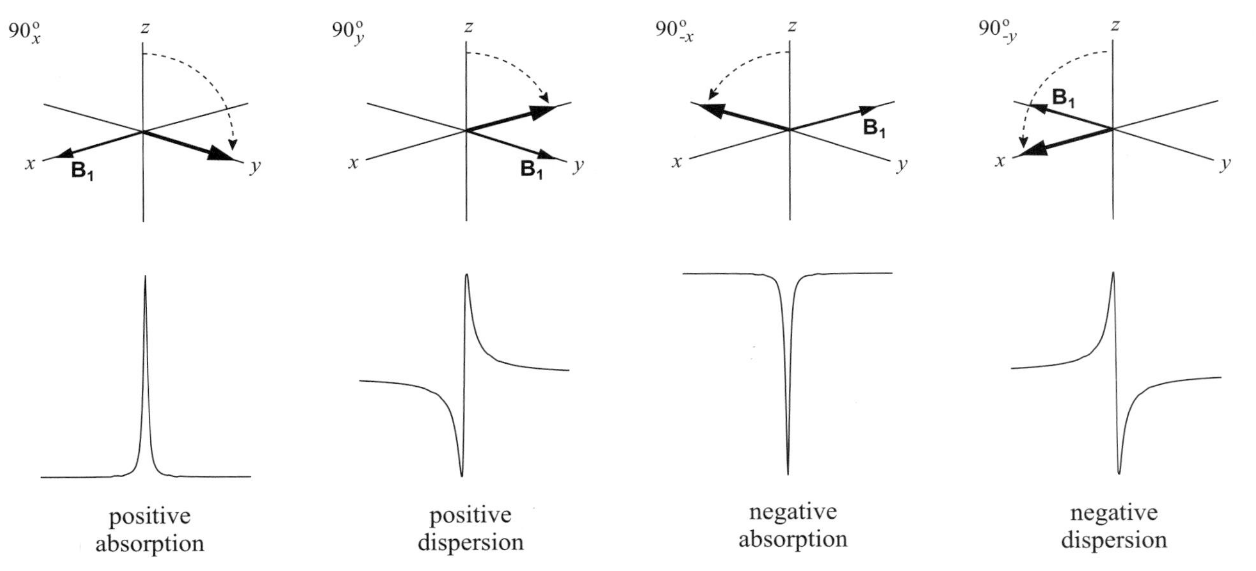

Figure 2.10. Excitation with pulses of varying rf phase. The differing initial positions of the excited vectors produce NMR resonances with similarly altered phases (here the $+y$-axis is arbitrarily defined as representing the positive absorption display).

this equalising of populations is *not* the same as the *saturation* of a resonance, a condition that will be encountered in a variety of circumstances in this book. Saturation corresponds again to equal spin populations but with the phases of the individual spins distributed randomly about the transverse plane such that there exists no net transverse magnetisation and thus no observable signal. In other words, under conditions of saturation the spins lack phase coherence. The 180° pulse *inverts the populations* of the spin states, since there must now exist more spins in the β than in the α orientation to place the bulk vector anti-parallel to the static field. Only magnetisation in the x–y plane is ultimately able to induce a signal in the detection coil (see below) so that the 90° and 270° pulse will produce the maximum signal intensity, but the 180° and 360° pulse will produce none (this provides a useful means of 'calibrating' the pulses, as in the following chapter). The vast majority of the multi-pulse experiments described in this book, and indeed throughout NMR, use only 90° and 180° pulses.

The example above made use of a $90°_x$ pulse, that is a 90° pulse in which the B_1 field was applied along the x-axis. It is, however, possible to apply the pulse with arbitrary phase, say along any of the axes x, y, $-x$, or $-y$ as required, which translates to a different starting phase of the excited magnetisation vector. The spectra provided by these pulses show resonances whose *phases* similarly differ by 90°. The detection system of the spectrometer designates one axis to represent the positive *absorption* signal (defined by a receiver reference phase, Section 3.2.2) meaning only magnetisation initially aligned with this axis will produce a pure absorption-mode resonance. Magnetisation that differs from this by +90° is said to represent the pure *dispersion* mode signal, that which differs by 180° is the negative absorption response and so on (Fig. 2.10). Magnetisation vectors initially between these positions result in resonances displaying a mixture of absorption and dispersion behaviour. For clarity and optimum resolution, all NMR resonances are displayed in the favoured absorption-mode whenever possible (which is achieved through the process known as phase correcton). Note that in all cases the *detected* signals are those *emitted* from the nuclei as described below, and a negative phase signal does not imply a change from emission to absorption of radiation (the absorption corresponds to the initial excitation of the spins).

The idea of applying a sequence of pulses of different phase angles is of central importance to all NMR experiments. The process of repeating a

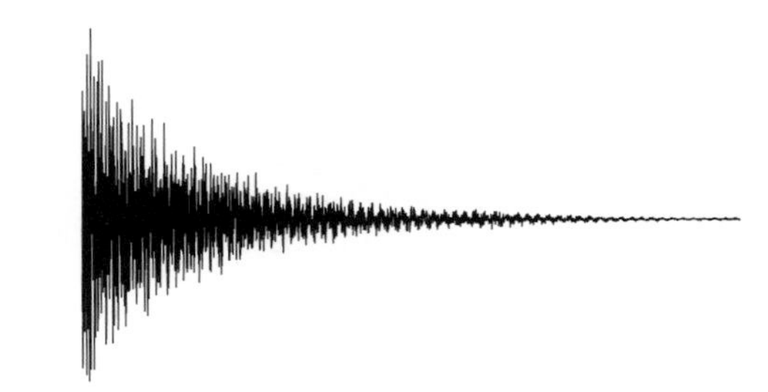

Figure 2.11. The detected NMR response, a Free Induction Decay (FID). The signal fades as the nuclear spins relax back towards thermal equilibrium.

multipulse experiment with different pulse phases and combining the collected data in an appropriate manner is termed *phase cycling*, and is one of the most widely used procedures for selecting the signals of interest in an NMR experiment and rejecting those that are not required. We shall encounter this concept further in Chapter 3 (and indeed throughout the rest of the book!).

Now consider what happens immediately after the application of, for example, a $90°_x$ pulse. We already know that in the rotating frame the precession of the spins is effectively frozen because the B_1 frequency ν_0 and hence the rotating frame frequency exactly match the spin Larmor frequency. Thus, the bulk magnetisation vector simply remains static along the $+y$-axis. However, if we step back from our convenient 'fiction' and return to consider events in the laboratory frame, we see that the vector starts to precess about the z-axis at its Larmor frequency. This rotating magnetisation vector will produce a weak oscillating voltage in the coil surrounding the sample, in much the same way that the rotating magnet in a bicycle dynamo induces a voltage in the coils that surround it. These are the electrical signals we wish to detect and it is these that ultimately produce the observed NMR signal. However, magnetisation in the x–y plane corresponds to deviation from the equilibrium spin populations and, just like any other chemical system that is perturbed from its equilibrium state, the system will adjust to re-establish this condition, and so the transverse vector will gradually disappear and simultaneously grow along the z-axis. This return to equilibrium is referred to as *relaxation*, and it causes the NMR signal to decay with time, producing the observed Free Induction Decay or FID (Fig. 2.11). The process of relaxation has wide-ranging implications for the practice of NMR and this important area is also addressed in this introductory chapter.

2.2.3. Chemical shifts and couplings

So far we have only considered the rotating frame representation of a collection of like spins, involving a single vector which is stationary in the rotating frame since the reference frequency ν_0 exactly matches the Larmor frequency of the spins (the rf is said to be *on-resonance* for these spins). Now consider a sample containing two groups of chemically distinct but uncoupled spins, A and X, with chemical shifts of ν_A and ν_X Hz respectively, which differ by ν Hz. Following excitation with a single $90°_x$ pulse, both vectors start in the x–y plane along the y-axis of the rotating frame. Choosing the reference frequency to be on-resonance for the A spins ($\nu_0 = \nu_A$) means these remain along the y-axis as before (ignoring the effects of relaxation for the present). If the X spins have a greater chemical shift than A ($\nu_X > \nu_A$) then the X vector will be moving faster than the rotating frame reference frequency by ν Hz so will move *ahead* of A (Fig. 2.12). Conversely, if $\nu_X < \nu_A$ it will be moving more slowly and will lag behind. Three sets of uncoupled spins can be represented by three rotating vectors and so on, such that differences in chemical shifts between spins are simply represented by vectors precessing at different rates in the rotating frame, according to their offsets from the reference

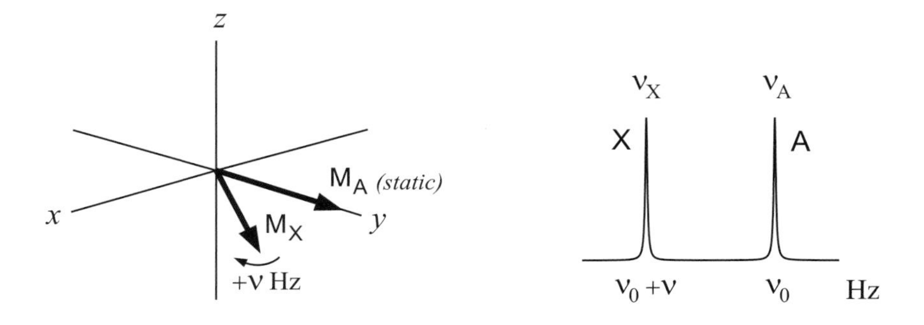

Figure 2.12. Chemical shifts in the rotating frame. Vectors evolve according to their offsets from the reference (transmitter) frequency ν_0. Here this is on-resonance for spins A ($\nu_0 = \nu_A$) whilst spins X move ahead at a rate of $+\nu$ Hz ($= \nu_X - \nu_0$).

frequency ν_0. By using the rotating frame to represent these events we need only consider the chemical shift *differences* between the spins of interest, which will be in the kilohertz range, rather than the absolute frequencies, which are of the order of many megahertz. As we shall discover in Section 2.2, this is exactly analogous to the operation of the detection system of an NMR spectrometer, in which a reference frequency is subtracted from the acquired data to produce signals in the kHz region suitable for digitisation. Thus, the 'trick' of using the rotating frame of reference in fact equates directly to a real physical process within the instrument.

When considering the effects of coupling on a resonance it is convenient to remove the effects of chemical shift altogether by choosing the reference frequency of the rotating frame to be the chemical shift of the multiplet of interest. This again helps clarify our perception of events by simplifying the rotation of the vectors in the picture. In the case of a doublet, the two lines are represented by two vectors precessing at $+J/2$ and $-J/2$ Hz, whilst for a triplet, the central line remains static and the outer two move at $+J$ and $-J$ Hz (Fig. 2.13). In many NMR experiments it is desirable to control the orientation of multiplet vectors with respect to one another, and, as we shall see, a particularly important relationship is when two vectors are *antiphase* to one another, that is, sitting in opposite directions. This can be achieved simply by choosing an appropriate delay period in which the vectors evolve, which is $1/2J$ for a doublet but $1/4J$ for the triplet.

2.2.4. Spin-echoes

Having seen how to represent chemical shifts and J-couplings with the vector model, we are now in a position to see how we can manipulate the effects of these properties in simple multi-pulse experiments. The idea here is to provide a simple introduction to using the vector model to understand what is happening during a pulse sequence. In many experiments, there exist time delays in which magnetisation is simply allowed to precess under the influence of chemical shifts and couplings, usually with the goal of producing a defined state of magnetisation before further pulses are applied or data is acquired. To illustrate these points, we consider one of the fundamental building blocks of numerous NMR experiments, the spin-echo.

Consider first two groups of chemically distinct protons, A and X, that share a mutual coupling J_{AX}, which will be subject to the simple two-pulse sequence in Fig. 2.14. For simplicity we shall consider the effect of chemical shifts and couplings separately, starting with the chemical shifts and again assuming the reference frequency to be that of the A spins (Fig. 2.15). The initial $90°_x$ creates transverse A and X magnetisation, after which the X vector precesses during the first time interval, Δ. The following $180°_y$ pulse (note this is now along the *y-axis*) rotates the X magnetisation through 180° about the y-axis, and so places it back in the x–y plane, but now lagging *behind* the A vector. A-spin magnetisation remains along the y-axis so is invariant to this pulse. During

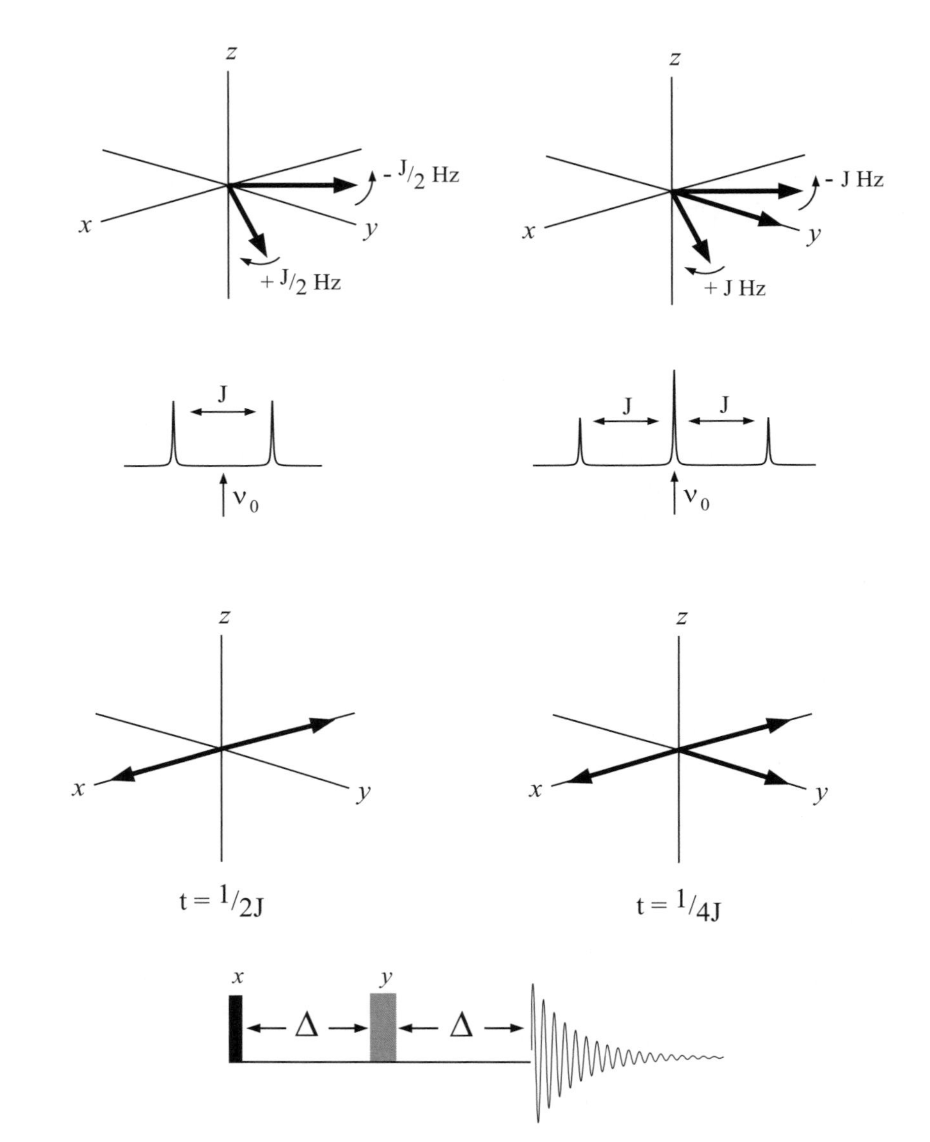

Figure 2.13. Scalar couplings in the rotating frame. Multiplet components evolve according to their coupling constants. The vectors have an antiphase disposition after an evolution period of 1/2J and 1/4J s for doublets and triplets respectively.

Figure 2.14. The basic spin-echo pulse sequence.

the second time period, Δ, the X magnetisation will precess through the same angle as in the first period and at the end of the sequence finishes where it began and *at the same position as the A vector*. Thus, after the time period 2Δ no phase difference has accrued between the A and X vectors despite their different shifts, and it were as if the A and X spins had the same chemical shift throughout the 2Δ period. We say *the spin-echo has refocused the chemical shifts*, the dephasing and rephasing stages gives rise to the echo terminology.

Consider now the effect on the coupling between the two spins with reference to the multiplet of spin A, safe in the knowledge that we can ignore the effects of chemical shifts. Again, during the first period Δ the doublet components will move in opposite directions, and then have their positions

Figure 2.15. Chemical shift evolution is refocused with the spin-echo sequence.

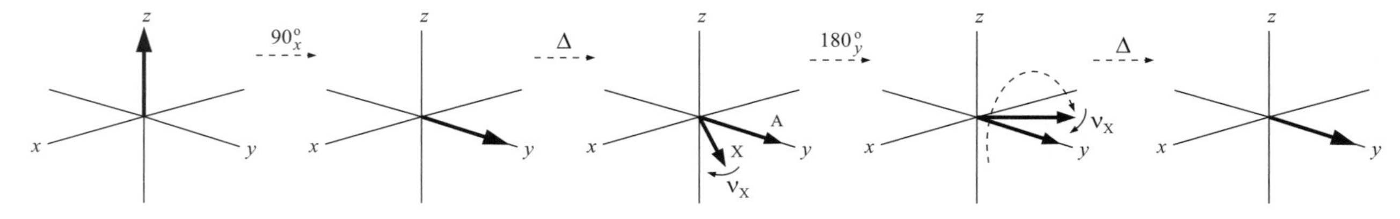

interchanged by the application of the $180°_y$ pulse. At this point it would seem obvious to assume that the two halves of the doublets would simply refocus as in the case of the chemical shift differences above, but we have to consider the effect of the 180° pulse on the J-coupled partner also, in other words, the effect on the X-spins. To appreciate what is happening, we need to remind ourselves of what it is that gives rise to two halves of a doublet. These result from spin A being coupled to its partner, X, which can have one of two orientations (α or β) with respect to the magnetic field. When spin X has one orientation, spin A will resonate as the high-frequency half of its doublet, whilst with X in the other, A will resonate as the low-frequency half. As there are approximately the same number of X spins in α and β orientations, the two halves of the A doublet will be of equal intensity (obviously there are not exactly equal numbers of α and β X spins, otherwise there would be no NMR signal to observe, but the difference is so small as to be negligible for these arguments). The effect of applying the 180° pulse on the X spins is to invert the relative orientations, so that any A spin that was coupled to Xα is now coupled to Xβ, and vice versa. This means the faster moving vector now becomes the slower and vice versa, the overall result being represented in Fig. 2.16a. The two halves of the doublet therefore *continue to dephase*, so that by the end of the 2Δ period, the J-coupling, in contrast to the chemical shifts, have continued to evolve so that *homonuclear couplings are not refocused by a spin-echo*. The reason for adding the term *homonuclear* to the previous statement is because it does not necessarily apply to the case of heteronuclear spin-echoes, that is, when we are dealing with two different nuclides, such as ^{1}H and ^{13}C for example. This is because in a heteronuclear system one may choose to apply the 180° pulse on only one channel, thus only one of the two nuclides will experience this pulse and refocusing of the heteronuclear coupling will occur in this case (Fig. 2.16b). If two simultaneous 180° pulses are applied to both nuclei via two different frequency sources, continued defocusing of the heteronuclear coupling occurs exactly as for the homonuclear spin-echo above.

The use of the $180°_y$ pulse instead of a $180°_x$ pulse in the above sequences was employed to provide a more convenient picture of events, yet it is important to realise that the refocusing effects on chemical shift and couplings described above would also have occurred with a $180°_x$ pulse except that the refocused vectors would now lie along the $-y$-axis instead of the $+y$-axis. One further feature of the spin-echo sequence is that it will also refocus the deleterious effects that arise from inhomogeneities in the static magnetic field, as these may be viewed as just another contribution to chemical shift differences throughout the sample. The importance of the spin-echo in modern NMR techniques can

Figure 2.16. The influence of spin-echoes on scalar coupling as illustrated for two coupled spins A and X. (a) A homonuclear spin-echo (in which both spins experience a 180° pulse) allows the coupling to evolve throughout the sequence. (b) A heteronuclear spin-echo (in which only one spin experiences a 180° pulse) causes the coupling to refocus. If both heteronuclear spins experience 180° pulses, the heteronuclear coupling evolves as in (a) (see text).

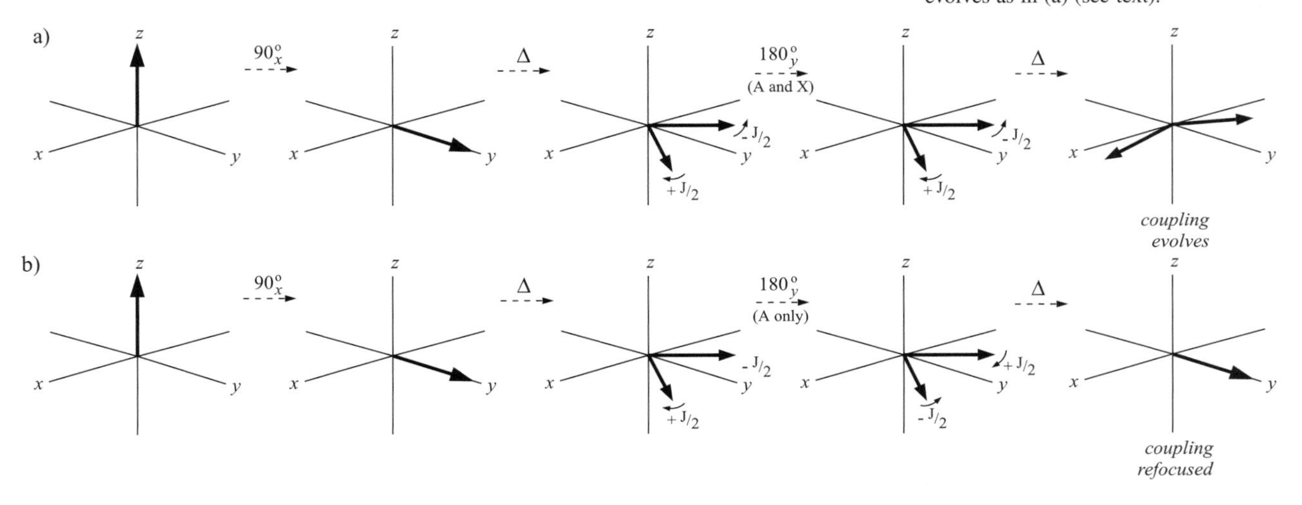

hardly be over emphasised. It allows experiments to be performed without having to worry about chemical shift differences within a spectrum and the complications these may introduce (phase differences for example). This then allows us to manipulate spins according to their couplings with neighbours and it is these interactions that are exploited to the full in many of the modern NMR techniques described later.

2.3. TIME AND FREQUENCY DOMAINS

It was shown in the previous section that the emitted rf signal from excited nuclear spins (the FID) is detected as a time dependent oscillating voltage which steadily decays as a result of spin relaxation. In this form the data is of little use to us because it is a time domain representation of the nuclear precession frequencies within the sample. What we actually want is a display of the frequency components that make up the FID as it is these we relate to transition energies and ultimately chemical environments. In other words, we need to transfer our *time domain* data into the *frequency domain.*

The time and frequency domains are related by a simple function, one being the inverse of the other (Fig. 2.17). The complicating factor is that a genuine FID is usually composed of potentially hundreds of components of differing frequencies and amplitude, in addition to noise and other possible artefacts, and in such cases the extraction of frequencies by direct inspection is impossible. By far the most widely used method to produce the frequency domain spectrum is the mathematical procedure of Fourier transformation, which has the general form:

$$f(\omega) = \int_{-\infty}^{+\infty} f(t)e^{i\omega t}\,dt \qquad (2.6)$$

where $f(\omega)$ and $f(t)$ represent the frequency and time domain data respectively. In the very early days of pulse-FT NMR the transform was often the rate limiting step in producing a spectrum, although with today's computers and the use of a fast Fourier transform procedure (the Cooley–Tukey algorithm) the time requirements are of little consequence. Fig. 2.18 demonstrates this procedure for very simple spectra. Clearly even for these rather simple spectra of only a few lines the corresponding FID rapidly becomes too complex for direct interpretation, whereas this is impossible for a genuine FID of any complexity (see Fig. 2.11 for example).

The details of the Fourier transform itself are usually of little consequence to anyone using NMR, although there is one notable feature to be aware of. The term $e^{i\omega t}$ can equally be written $\cos \omega t + i \sin \omega t$ and in this form it is apparent that the transformation actually results in *two* frequency domain spectra that differ only in their signal *phases*. The two are cosine and sine functions so are 90° out-of-phase relative to one another and are termed the 'real' and 'imaginary' parts of the spectrum (because the function contains complex

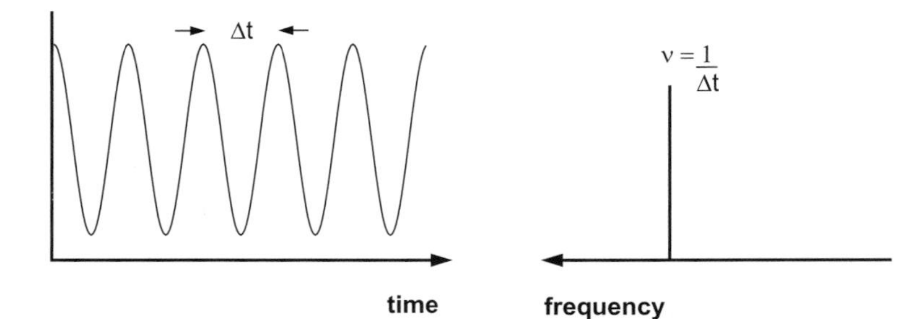

Figure 2.17. Time and frequency domains share a simple inverse relationship.

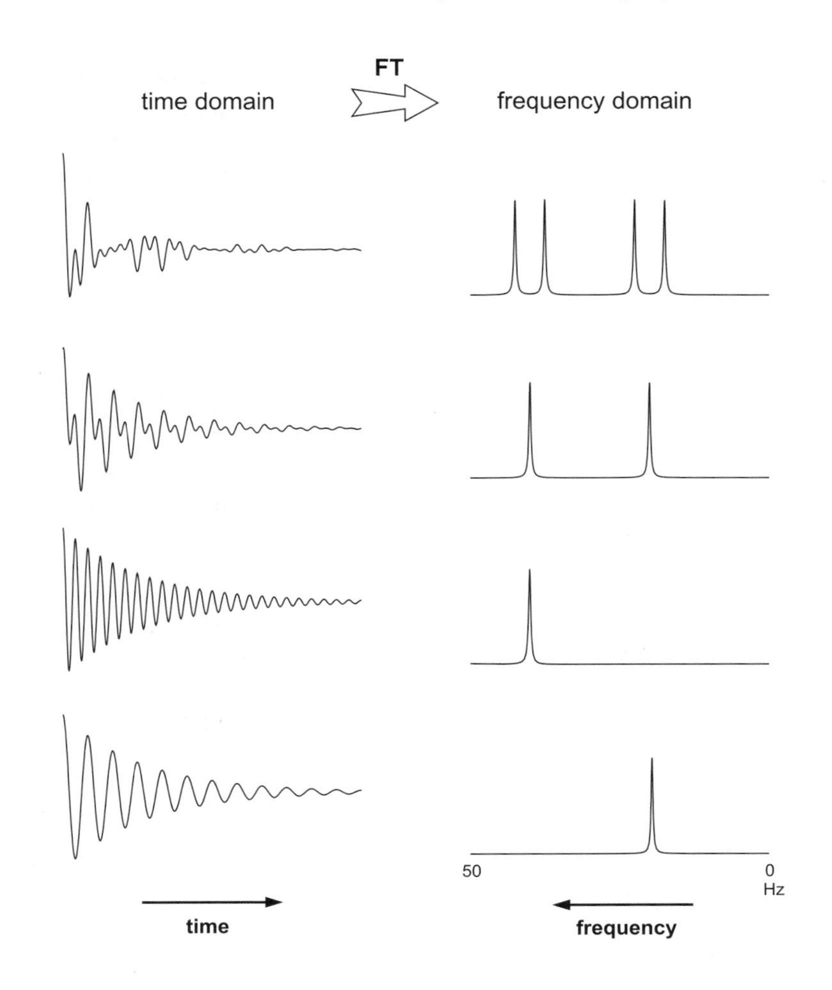

Figure 2.18. Fourier transformation of time domain FIDs produces the corresponding frequency domain spectra.

numbers). Generally we are presented with only the 'real' part of the data (although the 'imaginary' part can usually be displayed) and with appropriate phase adjustment we choose this to contain the preferred pure absorption mode data and the imaginary part to contain the dispersion mode representation. The significance of this phase relationship will be pursued in Chapters 3 and 5.

2.4. SPIN RELAXATION

The action of an rf pulse on a sample at thermal equilibrium perturbs the nuclear spins from this restful state. Following this perturbation, one would intuitively expect the system to re-establish the equilibrium condition, and lose the excess energy imparted to the system by the applied pulse. A number of questions then arise; where does this energy go, how does it get there (or in other words what mechanisms are in place to transfer the energy), and how long does all this take? While some appreciation of all these points is desirable, it is the last of the three that has greatest bearing on the day-to-day practice of NMR. The lifetimes of excited nuclear spins are extremely long when compared to, say, the excited electronic states of optical spectroscopy. These may be a few seconds or even minutes for nuclear spins as opposed to less than a picosecond for electrons, a consequence of the low transition energies associated with nuclear resonance. These extended lifetimes are crucial to the success of NMR spectroscopy as an analytical tool in chemistry. Not only do these mean that NMR resonances are rather narrow relative to those of rotational, vibrational or electronic transitions (as a consequence of the

Heisenberg Uncertainty principle), but it also provides time to manipulate the spin systems after their initial excitation, performing a variety of spin gymnastics and so tailoring the information available in the resulting spectra. This is the world of multi-pulse NMR experiments, into which we shall enter shortly, and knowledge of relaxation rates has considerable bearing on the design of these experiments, on how they should be implemented and on the choice of experimental parameters for optimum results. Even in the simplest possible case of a single pulse experiment, relaxation rates influence both achievable resolution and sensitivity (as mentioned in Chapter 1, the earliest attempts to observe the NMR phenomenon probably failed because of a lack of understanding of the spin relaxation properties of the samples used).

Relaxation rates of nuclear spins can also be related to aspects of molecular structure and behaviour in favourable circumstances, in particular internal molecular motions. It is true to say, however, that the relationship between relaxation rates and structural features are not as well defined as those of the chemical shift and spin–spin coupling constants, and are not used on a routine basis. The problem of reliable interpretation of relaxation data arises largely from the numerous extraneous effects that influence experimental results, meaning empirical correlations for using such data are not generally available and this aspect of NMR will not be pursued further in this book.

2.4.1. Longitudinal relaxation: establishing equilibrium

Immediately after pulse excitation of nuclear spins the bulk magnetisation vector is moved away from the thermal equilibrium $+z$-axis, which corresponds to a change in the spin populations. The recovery of the magnetisation along the z-axis, termed *longitudinal relaxation*, therefore corresponds to the equilibrium populations being re-established, and hence to an overall loss of energy of the spins (Fig. 2.19). The energy lost by the spins is transferred into the surroundings in the form of heat, although the energies involved are so small that temperatures changes in the bulk sample are undetectable. This gives rise to the original term for this process as *spin–lattice relaxation* which originated in the early days of solid-state NMR where the excess energy was described as dissipating into the surrounding rigid lattice.

The Bloch theory of NMR assumes that the recovery of the $+z$-magnetisation, M_z, follows exponential behaviour, described by:

$$\frac{dM_z}{dt} = \frac{(M_0 - M_z)}{T_1} \tag{2.7}$$

where M_0 is the magnetisation at thermal equilibrium, and T_1 is the (first-order) time constant for this process. Although exponential recovery was proposed as an hypothesis, it turns out to be an accurate model for the relaxation of spin-$^1/_2$ nuclei in most cases. Starting from the position of no net z-magnetisation (for example immediately after the sample has been placed in the magnet or after a 90° pulse) the longitudinal magnetisation at time t will be:

$$M_z = M_0(1 - e^{-t/T_1}) \tag{2.8}$$

as illustrated in Fig. 2.20. It should be stressed that T_1 is usually referred to as the longitudinal relaxation time throughout the NMR community (and,

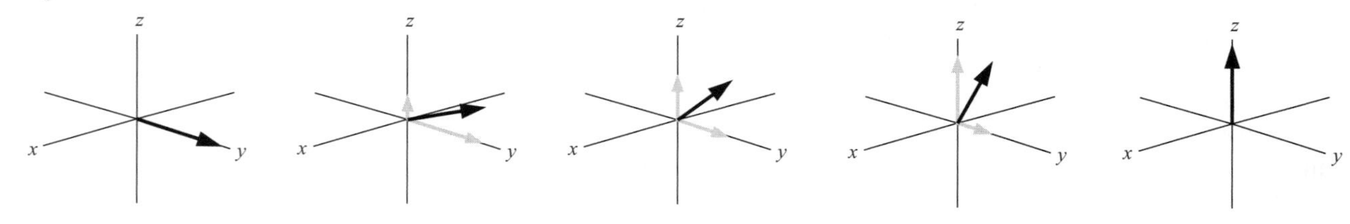

Figure 2.19. Longitudinal relaxation. The recovery of a magnetisation vector (shown on resonance in the rotating frame) diminishes the transverse (x–y) and re-establishes the longitudinal (z) components.

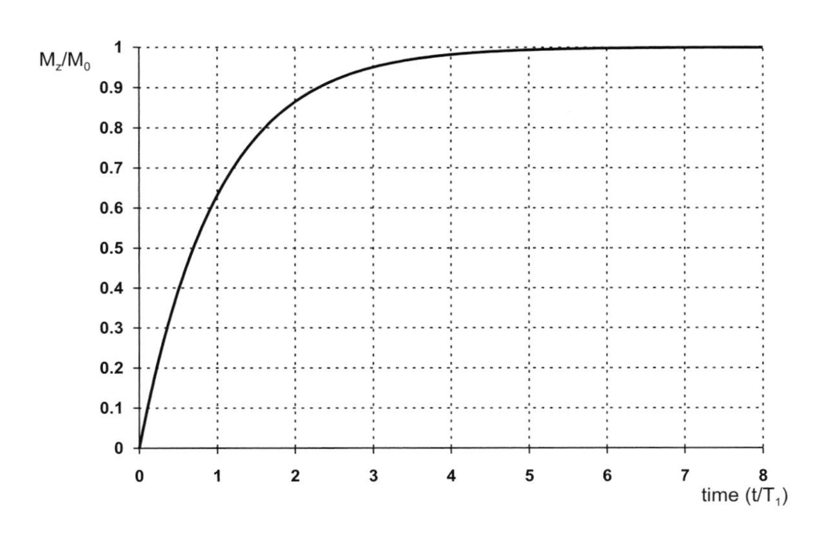

Figure 2.20. The exponential growth of longitudinal magnetisation is dictated by the time constant T_1 and is essentially complete after a period of $5T_1$.

following convention, throughout the remainder of this book), whereas, in fact, it is a *time constant* rather that a direct measure of the *time* required for recovery. Similarly, when referring to the rate at which magnetisation recovers, $1/T_1$ represents the relaxation *rate constant* (s^{-1}) for this process.

For medium sized organic molecules (those with a mass of a few hundred) proton T_1s tend to fall in the range 0.5–5 s, whereas carbon T_1s tend to range from a few seconds to many tens of seconds. For spins to relax fully after a 90° pulse, it is necessary to wait a period of at least $5T_1$ (at which point magnetisation has recovered by 99.33%) and thus it may be necessary to wait many minutes for full recovery. This is rarely the most time efficient way to collect NMR spectra and Section 4.1 describes the correct approach. The reason such long periods are required lies not in the fact that there is nowhere for the excess energy to go, since the energies involved are so small they can be readily taken up in the thermal energy of the sample, but rather that there is no efficient means for transferring this energy. The time required for *spontaneous* emission in NMR is so long (roughly equivalent to the age of the Universe!) that this has no effect on the spin populations, so *stimulated* emission must be operative for relaxation to occur. Recall that the fundamental requirement for inducing nuclear spin transitions, and hence restoring equilibrium populations in this case, is a magnetic field oscillating at the Larmor frequency of the spins and the long relaxation times suggests such suitable fields are not in great abundance. These fields can arise from a variety of sources with the oscillations required to induce relaxation coming from local molecular motions. Although the details of the various relaxation mechanisms can become rather complex, a qualitative appreciation of these, as in Section 2.5 below, is important for understanding many features of NMR spectra. At a practical level, some knowledge of T_1s in particular is crucial to the optimum execution of almost every NMR experiment, and the simple sequence below offers both a gentle introduction to multipulse NMR techniques as well as presenting a means of deducing this important parameter.

2.4.2. Measuring T_1 with the inversion-recovery sequence

There are a number of different experiments devised for the determination of the longitudinal relaxation times of nuclear spins [2] although only the most commonly applied method, inversion recovery, will be considered here. The full procedure is described first, followed by the 'quick-and-dirty' approach which is handy for experimental set-up.

Figure 2.21. The inversion recovery sequence.

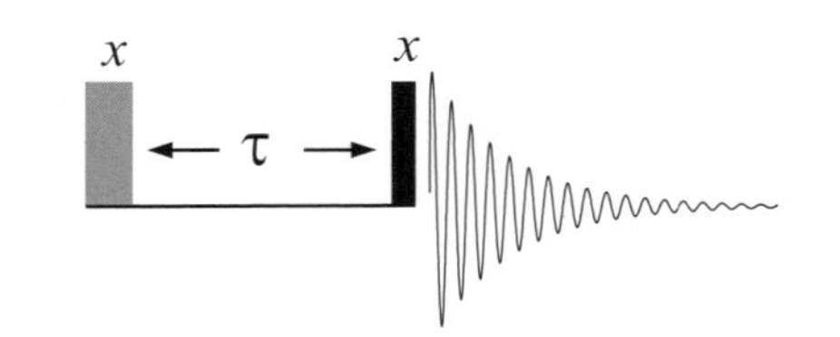

In essence, all one needs to do to determine T_1s is to perturb a spin system from thermal equilibrium and then devise some means of following its recovery as a function of time. The inversion-recovery experiment is a simple two-pulse sequence (Fig. 2.21) that, as the name implies, creates the initial population disturbance by *inverting* the spin populations through the application a 180° pulse. The magnetisation vector, initially aligned with the $-z$-axis, will gradually shrink back toward the x–y plane, pass through this plane and eventually make a full recovery along the $+z$-axis at a rate dictated by the quantity of interest, T_1. Since magnetisation along the z-axis is unobservable, the recovery is monitored by placing the vector back in the x–y plane with a 90° pulse after a suitable period, τ, following the initial inversion (Fig. 2.22).

If τ is zero, the magnetisation vector terminates with full intensity along the $-y$-axis producing an inverted spectrum using conventional spectrum phasing, that is, defining the $+y$-axis to represent positive absorption. Repeating the experiment with increasing values of τ allows one to follow the relaxation of the spins in question (Fig. 2.23). Finally, when τ is sufficiently long ($\tau_\infty > 5T_1$) complete relaxation will occur between the two pulses and the maximum positive signal is recorded. The intensity of the detected magnetisation, M_t, follows:

$$M_t = M_0(1 - 2e^{-\tau/T_1}) \tag{2.9}$$

where M_0 corresponds to equilibrium magnetisation, such as that recorded at τ_∞. Note here the additional factor of two relative to Eq. 2.8 as the recovery starts from inverted magnetisation. The relaxation time is determined by fitting the signal intensities to this equation, algorithms for which are found in many NMR software packages. The alternative traditional method of extracting T_1 from such an equation is to analyse a semi-logarithmic plot of $\ln(M_0 - M_t)$ vs. τ whose slope is $1/T_1$. The most likely causes of error in the use of the inversion recovery method are inaccurate recording of M_0 if full equilibration is not achieved, and inaccuracies in the 180° pulse causing imperfect initial inversion. The scaling factor (2 in Eq. 2.9) can be made variable in fitting routines to allow for incomplete inversion.

Figure 2.22. The inversion recovery process. With short recovery periods the vector finishes along the $-y$-axis, so the spectrum is inverted, whilst with longer periods a conventional spectrum of scaled intensity is obtained.

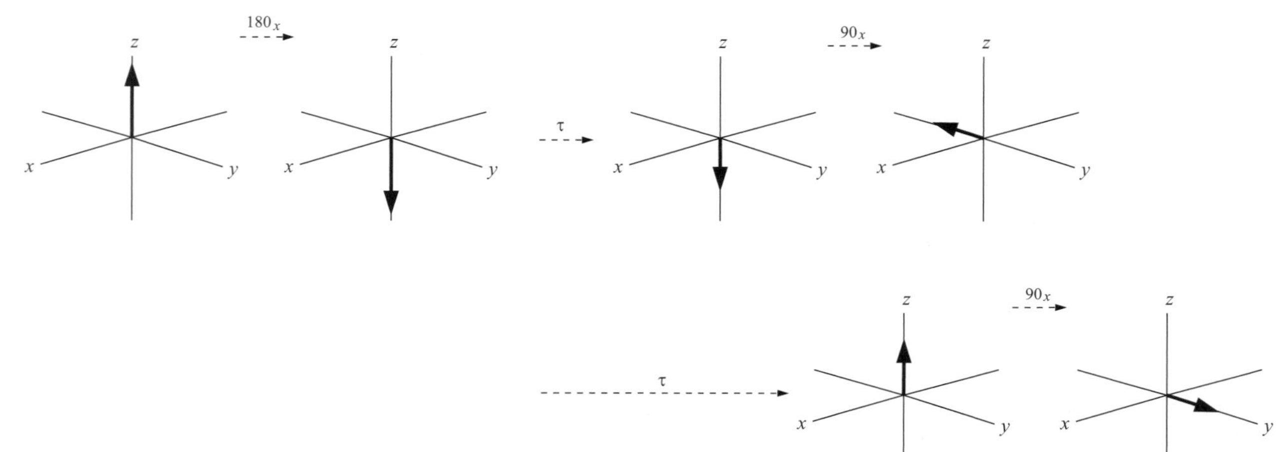

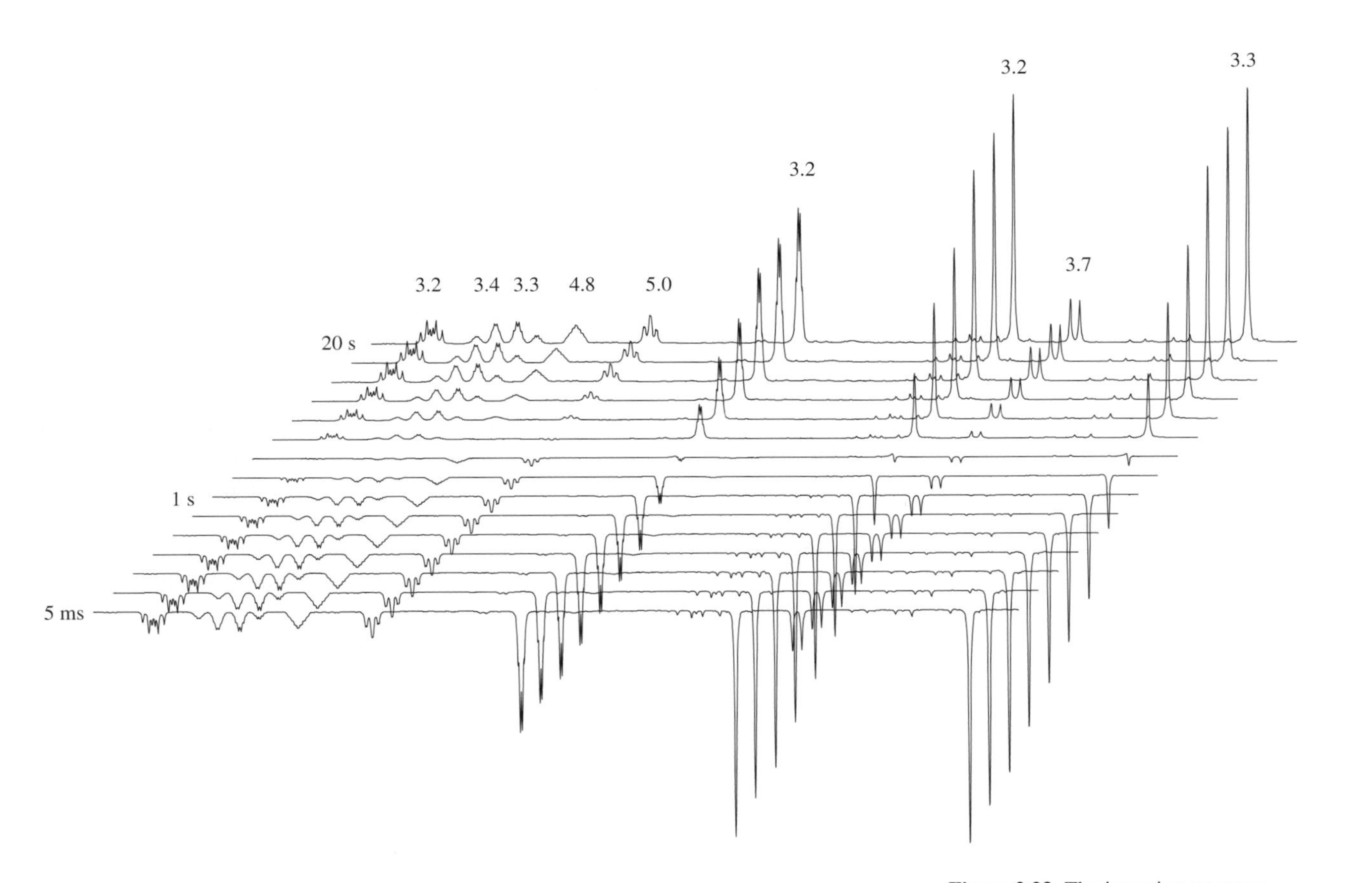

Figure 2.23. The inversion recovery experiment performed on α-pinene **2.1** in non-degassed $CDCl_3$ (^1H spectra, aliphatic region only shown). Recovery delays in the range 5 ms to 20 s were used and the T_1s (calculated from fitting peak intensities as described in the text) are shown for each resonance.

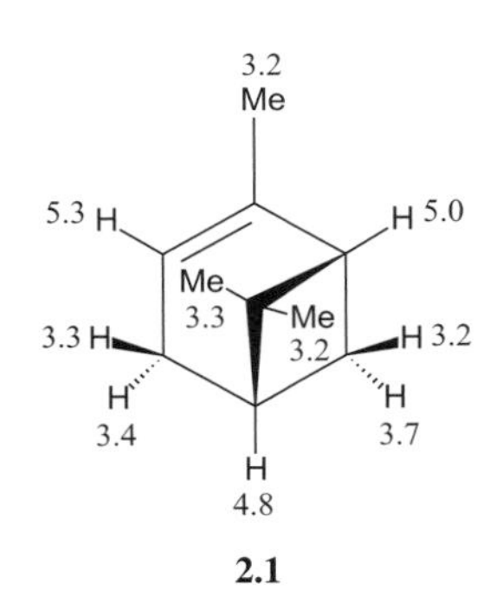

2.1

The quick T_1 estimation

In many practical cases it is sufficient to have just an estimation of relaxation times in order to calculate the optimum experimental timings for the sample at hand. In these instances the procedure described above is overly elaborate and since our molecules are likely to contain nuclei exhibiting a range of T_1s, accurate numbers will be of little use in experiment set-up. This 'quick and dirty' method is sufficient to provide estimates of T_1 and again makes use of the inversion recovery sequence. Ideally the sample in question will be sufficiently strong to allow rather few scans per τ value, making the whole procedure quick to perform. The basis of the method is the disappearance of signals when the longitudinal magnetisation passes through the *x–y* plane on its recovery (at time τ_{null}), because at this point the population difference is zero ($M_t = 0$). From the above equation, it can be shown that:

$$\tau_{null} = T_1 \ln 2 \tag{2.10}$$

hence

$$T_1 = \frac{\tau_{null}}{\ln 2} = 1.443\tau_{null} \tag{2.11}$$

Thus, the procedure is to run an experiment with $\tau = 0$ and adjust (phase) the spectrum to be negative absorption. After having waited $>5\ T_1$ repeat the experiment with an incremented τ using the same phase adjustments, until the signal passes through the null condition (see Fig. 2.23), thus defining τ_{null}, which may be a different value for each resonance in the spectrum. Errors may be introduced from inaccurate 180° pulses, from off-resonance effects (see Section 3.2) and from waiting for insufficient periods between acquisitions, so the fact that these values are estimates cannot be overemphasised.

One great problem with these methods is the need to know something about the T_1's in the sample even before these measurements. Between each new

τ value one must wait for the system to come to equilibrium, and if signal averaging were required one would also have to wait this long between each repetition! Unfortunately, it is the weak samples that require signal averaging that will benefit most from a properly executed experiment. To avoid this it is wise is to develop a feel for the relaxation properties of the types of nuclei and compounds you commonly study so that when you are faced with new material you will have some 'ball park' figures to provide guidance. Influences on the magnitude of T_1 are considered in Section 2.5.

2.4.3. Transverse relaxation: loss of magnetisation in the *x–y* plane

Referring back to the situation immediately following a 90° pulse in which the transverse magnetisation is on-resonance in the rotating frame, there exists another way in which observable magnetisation can be lost. Recall that the bulk magnetisation vector results from the addition of many microscopic vectors for the individual nuclei that are said to possess phase coherence following the pulse. In a sample of like spins one would anticipate that these would remain static in the rotating frame, perfectly aligned along the *y*-axis (ignoring the effects of longitudinal relaxation). However, this only holds if the magnetic field experienced by each spin in the sample is *exactly* the same. If this is not the case, some spins will experience a slightly greater local field than the mean causing them to have a higher frequency and to creep ahead, whereas others will experience a slightly smaller field and start to lag behind. This results in a fanning-out of the individual magnetisation vectors, which ultimately leads to no *net* magnetisation in the transverse plane (Fig. 2.24). This is another form of relaxation referred to as *transverse relaxation* which is again assumed to occur with an exponential decay now characterised by the time constant T_2.

Magnetic field differences in the sample can be considered to arise from two distinct sources. The first is simply from static magnetic field inhomogeneity throughout the sample volume which is really an instrumental imperfection and it is this one aims to minimise for each sample when optimising or 'shimming' the static magnetic field. The second is from the local magnetic fields arising from intramolecular and intermolecular interactions in the sample, which represent 'genuine' or 'natural' transverse relaxation processes. The relaxation time constant for the two sources combined is designated T_2^* such that:

$$\frac{1}{T_2^*} = \frac{1}{T_2} + \frac{1}{T_{2(\Delta B_0)}} \tag{2.12}$$

where T_2 refers to the contribution from genuine relaxation processes and $T_{2(\Delta B_0)}$ to that from field inhomogeneity. The decay of transverse magnetisation is manifested in the observed free induction decay. Moreover, the widths of NMR resonances are inversely proportional to T_2^* since a short T_2^* corresponds to a faster blurring of the transverse magnetisation which in turn corresponds

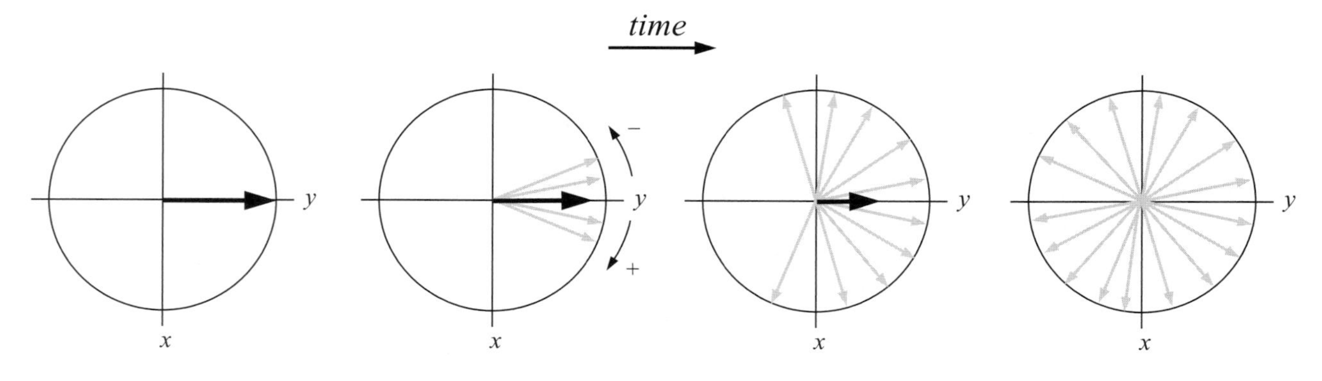

Figure 2.24. Transverse relaxation. Local field differences within the sample cause spins to precess with slightly differing frequencies, eventually leading to zero net transverse magnetisation.

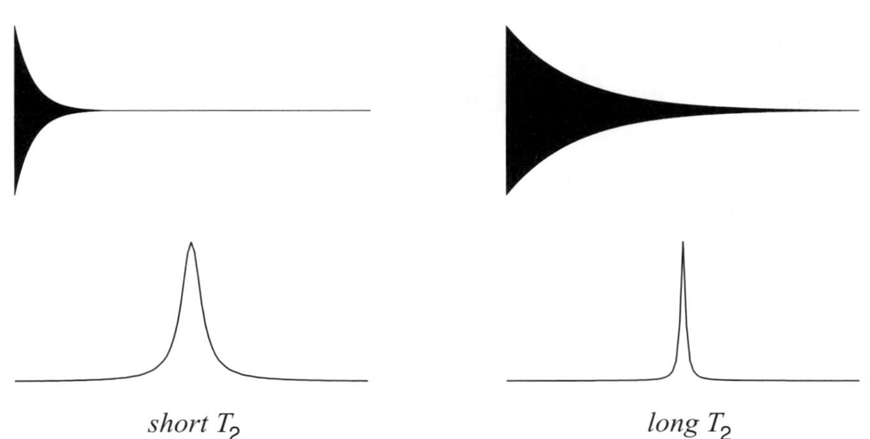

Figure 2.25. Rapidly relaxing spins produce fast decaying FIDs and broad resonances, whilst those which relax slowly produce longer FIDs and narrower resonances.

to a greater frequency difference between the vectors and thus a greater spread (broader line) in the frequency dimension (Fig. 2.25). For (single) exponential relaxation the lineshape is Lorentzian with a half-height linewidth, $\Delta\nu_{1/2}$ (Fig. 2.26) of

$$\Delta\nu_{1/2} = \frac{1}{\pi T_2^*} \tag{2.13}$$

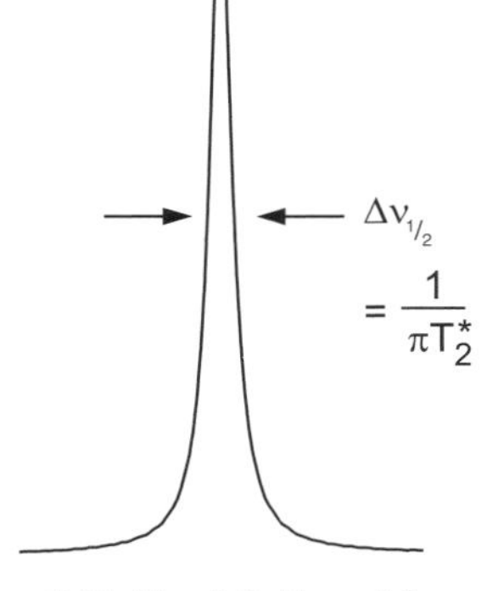

Figure 2.26. The definition of the half-height linewidth of a resonance.

For most spin-$^1/_2$ nuclei in small, rapidly tumbling molecules in low-viscosity solutions, it is field inhomogeneity that provides the dominant contribution to observed linewidths, and it is rarely possible to obtain genuine T_2 measurements directly from these. However, nuclei with spin > $^1/_2$ (quadrupolar nuclei) may be relaxed very efficiently by interactions with local electric field gradients and so have broad lines and short T_2s that can be determined directly from linewidths.

Generally speaking, relaxation mechanisms that operate to restore longitudinal magnetisation also act to destroy transverse magnetisation, and since there clearly can be no magnetisation remaining in the *x–y* plane when it has all returned to the +*z*-axis, T_2 can never be longer than T_1. However, additional mechanisms may also operate to reduce T_2, so that it may be shorter. Again, for most spin-$^1/_2$ nuclei in small, rapidly tumbling molecules, T_1 and T_2 have the same value, whilst for large molecules that tumble slowly in solution (or for solids) T_2 is often very much shorter than T_1 (see Section 2.5). Whereas longitudinal relaxation causes a loss of energy from the spins, transverse relaxation occurs by a mutual swapping of energy *between spins,* for example, one spin being excited to the β state while another simultaneously drops to the α state; a so called flip-flop process. This gives rise to the original term of *spin–spin relaxation* which is still in widespread use. Longitudinal relaxation is thus an enthalpic process whereas transverse relaxation is entropic. Although the measurement of T_2 has far less significance in routine spectroscopy, methods for this are described below for completeness and an alternative practical use of these is also presented.

2.4.4. Measuring T_2 with a spin-echo sequence

The measurement of the natural transverse relaxation time T_2 could in principle be obtained if the contribution from magnetic field inhomogeneity was removed. This can be achieved, as has been suggested already, by use of a spin-echo sequence. Consider again a sample of like spins and imagine the sample to be composed of microscopically small regions such that within each region the field is perfectly homogeneous. Magnetisation vectors within

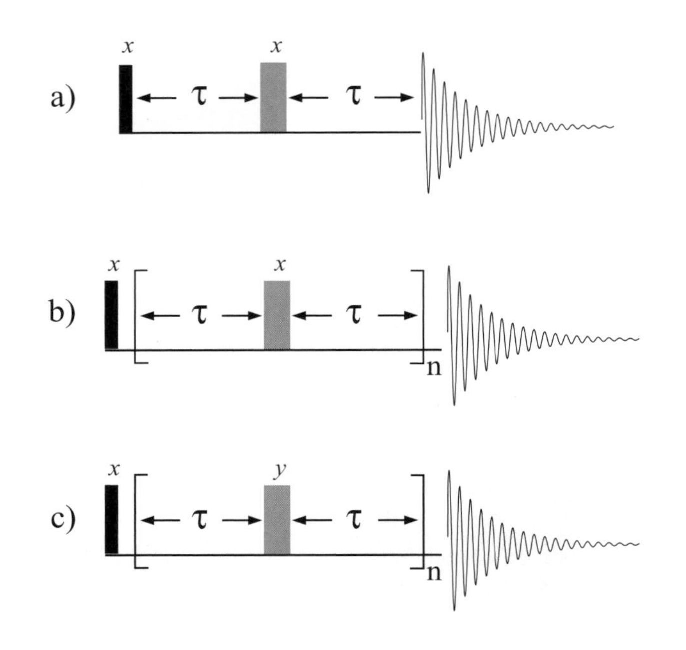

Figure 2.27. Spin-echo sequences for measuring T_2 relaxation times. (a) A basic spin-echo, (b) the Carr–Purcell sequence and (c) the Carr–Purcell–Meiboom–Gill (CPMG) sequence.

any given region will precess at the same frequency and these are sometimes referred to as *isochromats* (meaning 'of the same colour' or frequency). In the basic two-pulse echo sequence (Fig. 2.27a) some isochromats move ahead of the mean whilst others lag behind during the time period, τ (Fig. 2.28). The 180° pulse rotates the vectors toward the $-y$-axis and following a further period τ the faster moving vectors coincide with the slower ones along the $-y$-axis. Thus, the echo has refocused the blurring in the x–y plane caused by field inhomogeneities. If one were to start acquiring data immediately after the 90° pulse, one would see the FID decay away initially but then reappear after a time 2τ as the echo forms (Fig. 2.29a). However, during the 2τ time period, some loss of phase coherence by natural transverse relaxation also occurs, and this is *not* refocused by the spin-echo since, in effect, there is no phase memory associated with this process to be undone. This means that at the time of the echo the intensity of the observed magnetisation will have decayed according to the *natural* T_2 time constant, independent of field inhomogeneity. This can clearly be seen in a train of spin-echoes applied during the acquisition of an FID (Fig. 2.29b).

A logical experiment for determinig T_2 would be to repeat the sequence with increasing τ and measure the amplitude of the echo versus time, by analogy with the inversion–recovery method above. However, some care is required when using such an approach as the formation of the echo depends on the isochromats experiencing exactly the same field throughout the duration of the pulse sequence. If any given spin diffuses into a neighbouring region during the sequence it will experience a slightly different field from that where it began, and thus will not be fully refocused. As τ increases such diffusion losses become more severe and the experimental relaxation data less reliable (although this method does provide the basis for measuring molecular diffusion in solution by NMR; see Chapter 9).

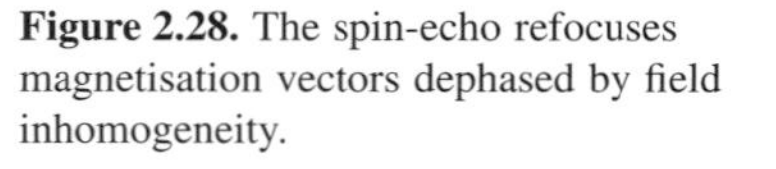

Figure 2.28. The spin-echo refocuses magnetisation vectors dephased by field inhomogeneity.

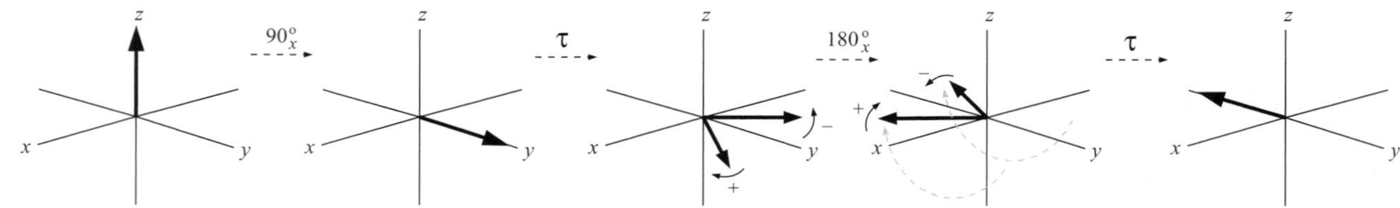

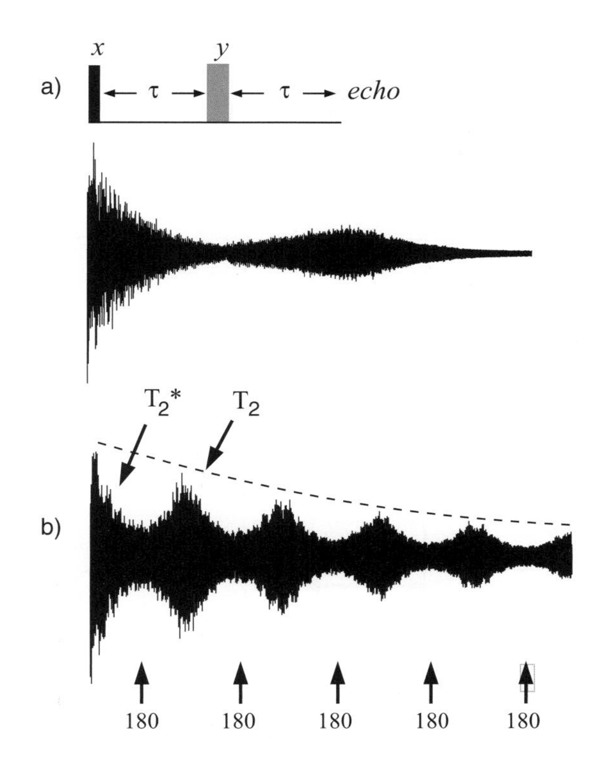

Figure 2.29. Experimental observation of spin-echoes. (a) Signal acquisition was started immediately after a 90° excitation pulse and a 180° pulse applied to refocus field inhomogeneity losses and produce the observed echo. (b) A train of spin-echoes reveals the true T_2 relaxation of magnetisation (dashed line).

A better approach to determining T_2, which minimises the effect of diffusion, is to repeat the echo sequence within a single experiment using a short τ to form multiple echoes, the decay of which follow the time constant T_2. This is the Carr–Purcell sequence (Fig. 2.27b) which causes echoes to form alternately along the $-y$ and $+y$ axes following each refocusing pulse. Losses occur from diffusion between the echo peaks, or in other words in the time period 2τ, so if this is kept short relative to the rate of diffusion (typically $\tau < 100$ ms) such losses become negligible. The intensity of the echo at longer time periods is attenuated by repeating the $-\tau$–180–τ– sequence many times prior to acquisition. The problem with this method is the fact that any errors in the length of the 180° pulse will be cumulative leading to imperfect refocusing as the experiment proceeds. A better implementation of this scheme is the Carr–Purcell–Meiboom–Gill (CPMG) sequence (Fig. 2.27c) in which $180°_y$ (as opposed to $180°_x$) pulses cause refocusing to take place in the $+y$ hemisphere for every echo. Here errors in pulse lengths are not cumulative but are small and constant on every odd-numbered echo but will cancel on each even-numbered echo (Fig. 2.30).

T_2 may then be extracted by performing a series of experiments with increasing $2\tau n$ (by increasing n) and acquiring data following the last even echo peak in each case. Application of the CPMG sequence is shown in Fig. 2.31 for a sample with differing resonance linewidths and illustrates the faster disappearance of the broader resonances i.e. those with shorter T_2. In reality, the determination of T_2 by any of these methods is still not straightforward. The most significant problem is likely to be from homonuclear couplings which are not refocused by the spin-echo and hence will impose additional phase modulations on the detected signals. As a result, studies involving T_2 measurements are even less widespread than those involving T_1. Fortunately, from the point of view of performing practical day to day spectroscopy, exact T_2 values are not important and the value of T_2^* (which may be calculated from linewidths as described above) has far greater significance. It is this value that determines the rate of decay of transverse magnetisation, so it effectively defines how long a multipulse experiment can be before the

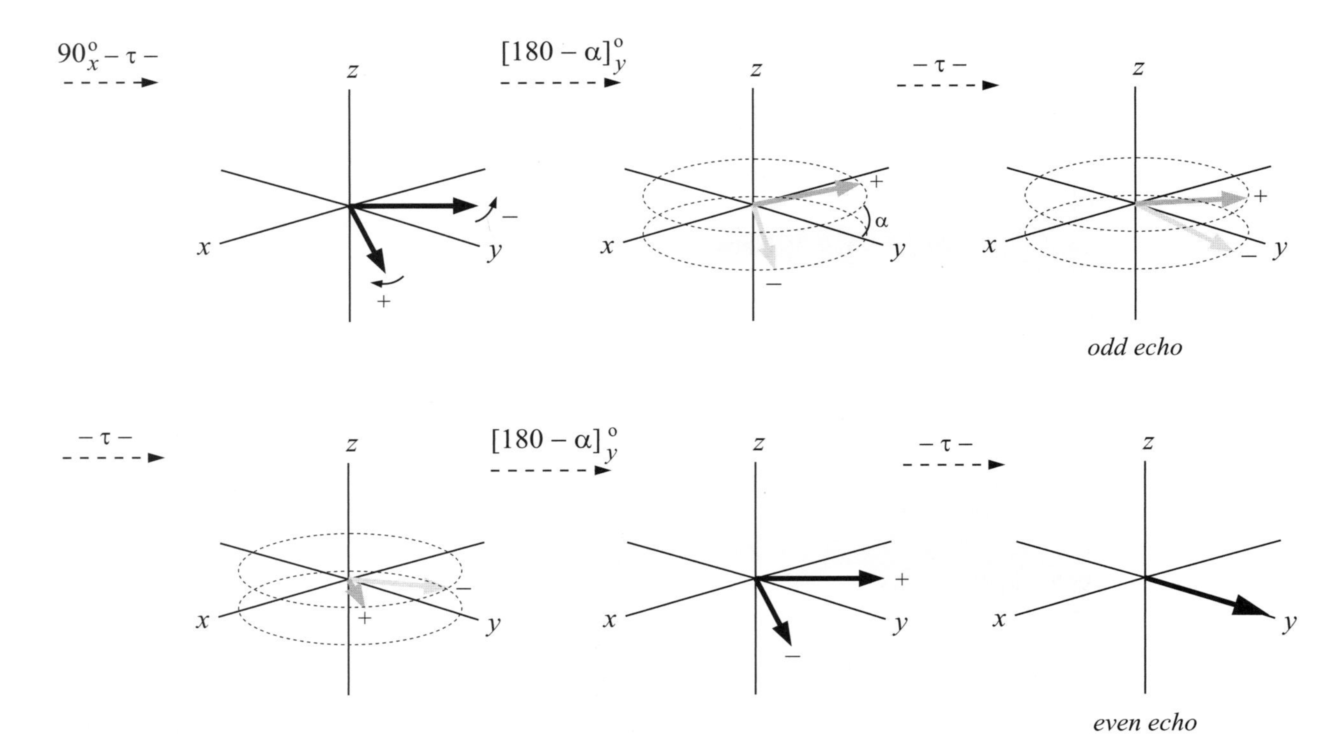

Figure 2.30. The operation of the CPMG sequence in the presence of pulse imperfections. The 180° pulse is assumed to be too short by α° meaning vectors will fall above (dark grey) or below (light grey) the x–y plane following a single 180° pulse and so reduce the intensity of 'odd' echoes. By repeating the sequence the errors are cancelled by the imperfect second 180° pulse so 'even' echoes can be used to accurately map T_2 relaxation.

system has decayed to such an extent that there is no longer any signal left to detect.

T_2 spectrum editing

One interesting use of these echo techniques lies in the exploitation of gross differences in transverse relaxation times of different species. Larger molecules typically display broader resonances than smaller ones since they posses shorter T_2 spin relaxation times. If the differences in these times are sufficiently large, resonances of the faster relaxing species can be preferentially reduced in intensity with the CPMG echo sequence, whilst the resonances of smaller, slower relaxing molecules decrease by a lesser amount (Fig. 2.32). This therefore provides a means, albeit a rather crude one, of editing a spectrum according to molecular size, retaining the resonances of the smaller components at the expense of the more massive ones. This approach has been widely used in the study of biofluids to suppress background contributions from very large macromolecules such as lipids and proteins.

The selective reduction of a solvent water resonance [3] can also be achieved in a similar way if the transverse relaxation time of the water protons can be reduced (i.e. the resonance broadened) such that this becomes very much shorter than that of the solutes under investigation. This can be achieved by the addition of suitable paramagnetic relaxation agents (about which the water molecules form a hydration sphere) or by reagents that promote chemical exchange. Ammonium chloride and hydroxylamine have been used to great effect in this way [4,5], as illustrated for the proton spectrum of the reduced arginine vasopressin peptide in 90% H_2O [6] (Fig. 2.33). This method of solvent suppression has been termed WATR (water attenuation by transverse relaxation). Whilst capable of providing impressive results it does have limited application; more general solvent suppression procedures are described in Chapter 9.

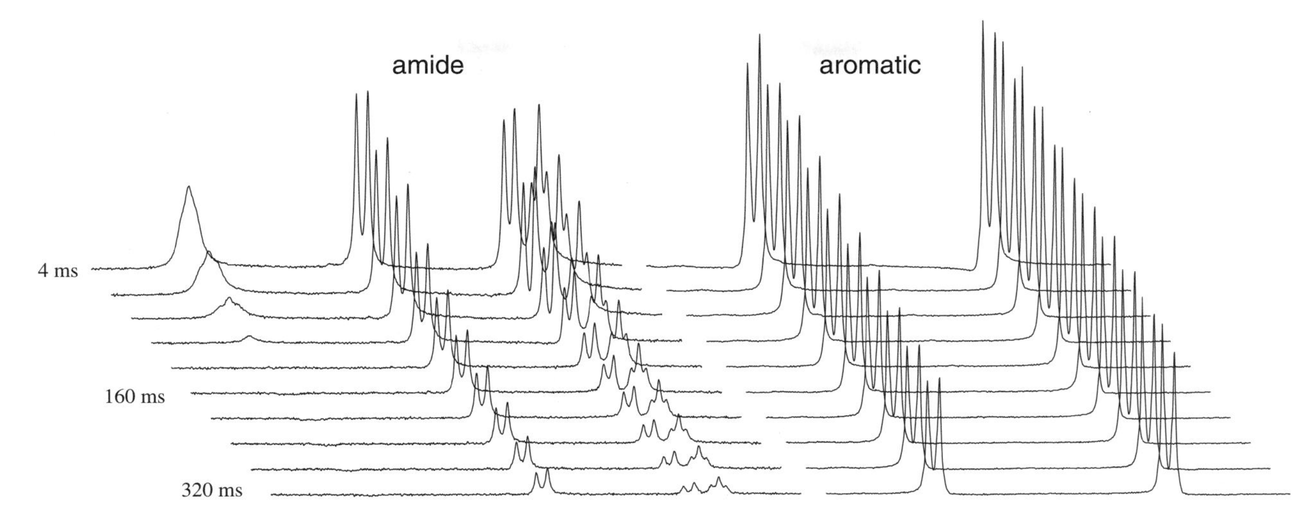

Figure 2.31. The CPMG sequence performed on the pentapeptide Leu-enkephalin **2.2** in DMSO. The very fast decay of the highest frequency amide proton occurs because this is in rapid chemical exchange with dissolved water, broadening the resonance significantly. The numbers show the total T_2 relaxation period $2\tau n$.

Tyr-Gly-Gly-Phe-Leu

2.2

2.5. MECHANISMS FOR RELAXATION

Nuclear spin relaxation is not a spontaneous process, it requires stimulation by a suitable fluctuating field to induce the necessary spin transitions and there are four principal mechanisms that are able to do this, the dipole–dipole, chemical shift anisotropy, spin rotation and quadrupolar mechanisms. Which of these is the dominant process can directly influence the appearance of an NMR spectrum and it is these factors we consider here. The emphasis is not so much on the explicit details of the underlying mechanisms, which can be found in physical NMR texts [7], but on the manner in which the spectra are affected by these mechanisms and how, as a result, different experimental conditions influence the observed spectrum.

2.5.1. The path to relaxation

The fundamental requirement for longitudinal relaxation of a spin-$^1/_2$ nucleus is a time-dependent magnetic field fluctuating at the Larmor frequency of the nuclear spin. Only through this can a change of spin state be induced or, in other words, can relaxation occur. Local magnetic fields arise from a

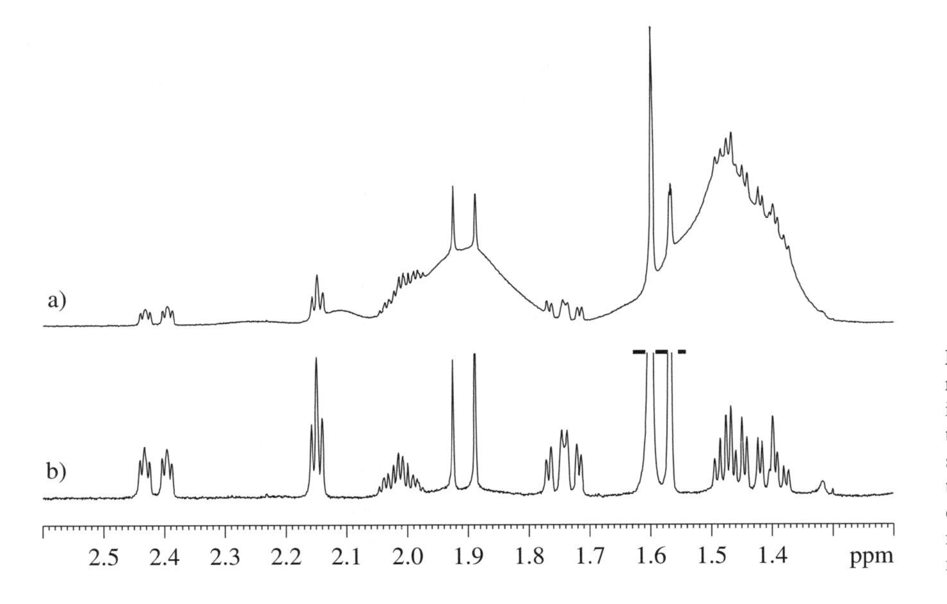

Figure 2.32. The T_2 filter. The broad resonances of polystyrene (Mr = 50,000) in (a) have been suppressed in (b) through T_2-based editing with the CPMG sequence, leaving only the resonances of the smaller camphor molecule. The τ delay was 1.5 ms and the echo was repeated 150 times to produce a total relaxation delay period $2\tau n$ of 450 ms.

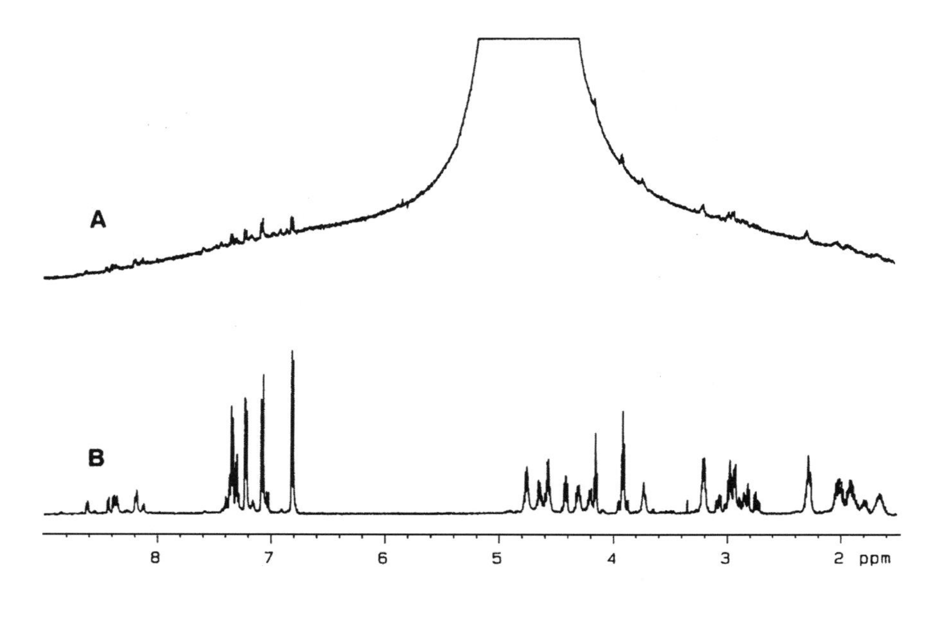

Figure 2.33. Solvent attenuation with the WATR method. (a) The 1D proton spectrum of 8 mM reduced arginine vasopressin in 90% H_2O/10% D_2O, pH = 2.75 containing 0.2 M NH_2OH. (b) The same sample recorded with the CPMG sequence using a total relaxation delay period of 235 ms (reproduced with permission from [6]).

number of sources, described below, whilst their time-dependence originates in the motions of the molecule (vibration, rotation, diffusion etc). In fact, only the chaotic tumbling of a molecule occurs at a rate that is appropriate for nuclear spin relaxation, others being either too fast or too slow. This random motion occurs with a spread of frequencies according to the molecular collisions, associations and so on experienced by the molecule, but is characterised by a rotational *correlation time*, τ_c, the average time taken for the molecule to rotate through one radian. Short correlation times therefore correspond to rapid tumbling and vice versa. The frequency distribution of the fluctuating magnetic fields associated with this motion is termed the *spectral density*, $J(\omega)$, and may be viewed as being proportional to the probability of finding a component of the motion at a given frequency, ω (in rad s^{-1}). Only when a suitable component exists at the spin Larmor frequency can longitudinal relaxation occur. The spectral density function has the general form

$$J(\omega) = \frac{2\tau_c}{1 + \omega^2\tau_c^2} \tag{2.14}$$

and is represented schematically in Fig. 2.34a for fast, intermediate and slow molecular tumbling rates (note the conventional use of the logarithmic scale). As each curve represents a probability, the area under each remains constant. For the Larmor frequency ω_0 indicated in Fig. 2.34a the corresponding graph of T_1 against molecular tumbling rates is also given (Fig. 2.34b). Fast molecular motion has only a relatively small component at the Larmor frequency so relaxation is slow (T_1 is long). This is the region occupied by small molecules in low-viscosity solvents, know as the *extreme narrowing limit*. As the tumbling rates decrease the spectral density at ω_0 initially increases but then falls away once more for slow tumbling so the T_1 curve has a minimum at intermediate rates. Thus, *for small rapidly tumbling molecules, faster motion corresponds to slower relaxation and hence narrower linewidths*, since longitudinal and transverse relaxation rates are identical ($T_2 = T_1$) under these conditions. A reduction in tumbling rate, such as by an increase in solvent viscosity or reduction in sample temperature, reduces the relaxation times and broadens the NMR resonance. The point at which the minimum is encountered and the slow motion regime approached is field dependent because ω_0 itself is field dependent (Fig. 2.34b). Behaviour in the slow motion regime is slightly more complex. The energy-conserving flip-flop processes that lead to transverse

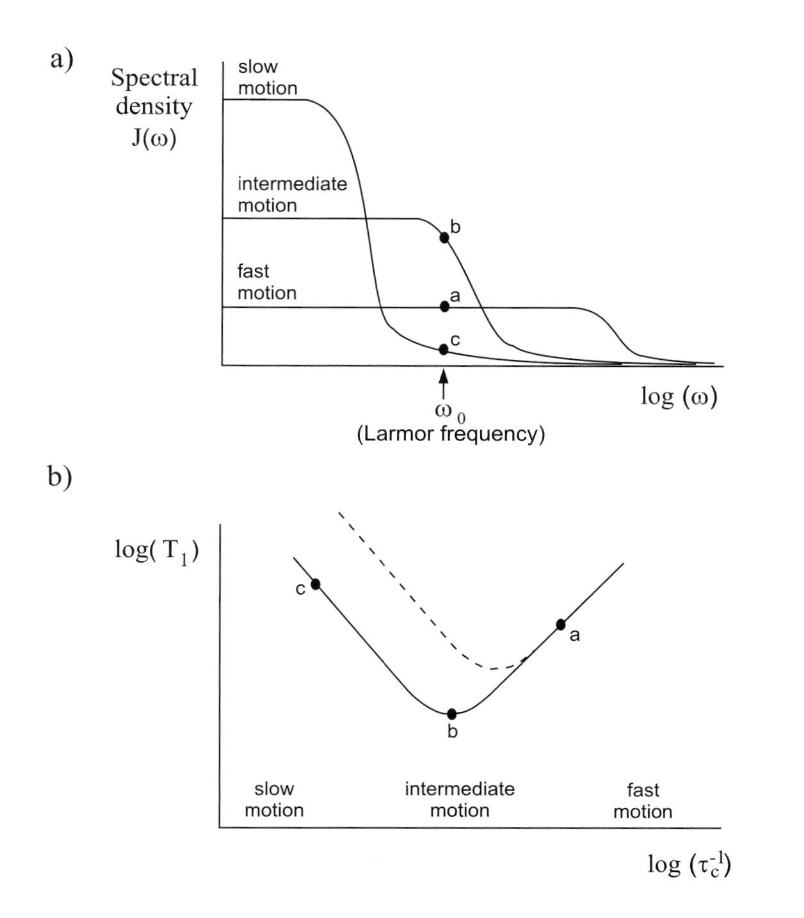

Figure 2.34. (a) A schematic representation of the spectral density as a function of frequency shown for molecules undergoing fast, intermediate and slow tumbling. For spins with a Larmor frequency ω_0, the corresponding T_1 curve is shown in (b) as a function of molecular tumbling rates (inverse correlation times, τ_c). The T_1 curve is field dependent because ω_0 is field dependent and the minimum occurs for faster motion at higher fields (dashed curve in (b)).

relaxation are also stimulated by very low frequency fluctuations and the T_2 curve differs markedly from that for T_1 (Fig. 2.35). Thus, for slowly tumbling molecules such as polymers and biological macromolecules, T_1 relaxation times can again be quite long but linewidths become rather broad as a result of short T_2s.

Molecular motion is therefore fundamental to the process of relaxation, but it remains to be seen how the fields required for this arise and how these mechanisms influence observed spectra.

2.5.2. Dipole–dipole relaxation

The most important relaxation mechanism for many spin-$^1/_2$ nuclei arises from the dipolar interaction between spins. This is also the source of the tremendously important nuclear Overhauser effect and further discussions on

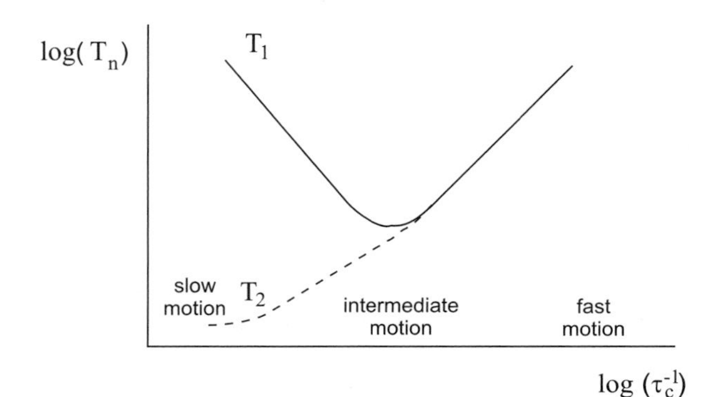

Figure 2.35. A schematic illustration of the dependence of T_1 and T_2 on molecular tumbling rates. T_1 relaxation is insensitive to very slow motions whilst T_2 relaxation may still be stimulated by them.

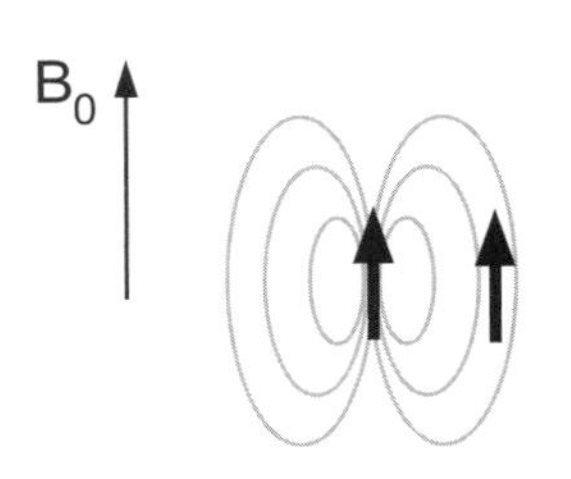

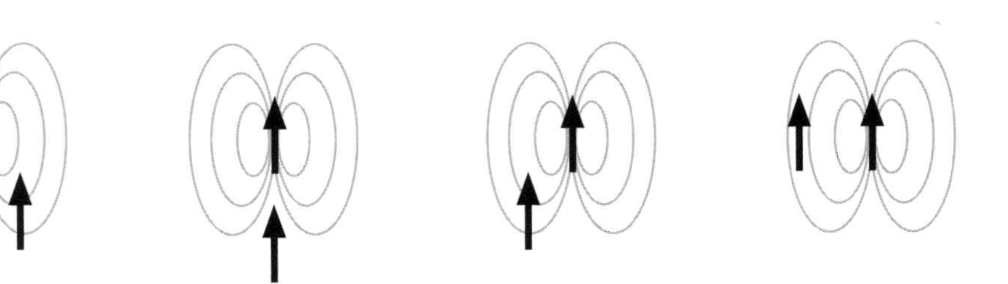

Figure 2.36. Dipole–dipole relaxation. The magnitude of the direct through space magnetic interaction between two spins is modulated by molecular tumbling and so induces spin transitions and hence relaxation.

this mechanism can be found in Chapter 8, so are kept deliberately brief here. Dipolar interactions can be visualised using the 'bar magnet' analogy for a spin-$^1/_2$ nucleus in which each is said to possess a magnetic North and South pole. As two such dipoles approach their associated magnetic fields interact; they attract or repel depending on their relative orientations. Now suppose these dipoles were two neighbouring nuclei in a molecule that is tumbling in solution. The orientation of each nucleus with respect to the static magnetic field does not vary as the molecule tumbles just as a compass needle maintains its direction as a compass is turned. However, their relative positions in space will alter and the local field experienced at one nucleus as a result of its neighbour will fluctuate as the molecule tumbles (Fig. 2.36). Tumbling at an appropriate rate can therefore induce relaxation.

This mechanism is often the dominant relaxation process for protons which rely on their neighbours as a source of magnetic dipoles. As such, protons which lack near-neighbours relax more slowly (notice how the methine protons in α-pinene (Fig. 2.23) all have longer T_1s than the methylene groups). The most obvious consequence of this is lower than expected integrals in routine proton spectra due to the partial saturation of the slower relaxing spins which are unable to recover sufficiently between each pulse-acquire sequence. If T_1 data are available, then protons with long relaxation times can be predicted to be remote from others in the molecule. Carbon-13 nuclei are also relaxed primarily by dipolar interactions, either with their directly bound protons or, in the absence of these, by more distant ones. In very large molecules and at high field, the chemical shift anisotropy mechanism described below can also play a role, especially for sp^2 centres, as it can for spin-$^1/_2$ nuclei which exhibit large chemical shift ranges. Dipolar relaxation can also arise from the interaction of a nuclear spin with an unpaired electron, the magnetic moment of which is over 600 times that of the proton and so provides a very efficient relaxation source. This is sometimes referred to as the *paramagnetic relaxation* mechanism. Even the presence of dissolved oxygen, which is itself paramagnetic, can contribute to spin relaxation and the deliberate addition of relaxation agents containing paramagnetic species, the most common being chromium(III) acetylacetonate, $Cr(acac)_3$, for organic solvents or manganese(II) chloride for water, are sometimes used to reduce relaxation times and so speed data acquisition (Chapter 4).

2.5.3. Chemical shift anisotropy relaxation

The electron distribution in chemical bonds is inherently unsymmetrical or *anisotropic* and as a result, the local field experienced by a nucleus, and hence its chemical shift, will depend on the orientation of the bond relative to the applied static field. In solution, the rapid tumbling of a molecule averages this *chemical shift anisotropy* (CSA) such that one observes only a single frequency for each chemically distinct site, sometimes referred to as the isotropic chemical shift. Nevertheless, this fluctuating field can stimulate relaxation if sufficiently strong. This is generally the case for nuclei which exhibit a large chemical shift

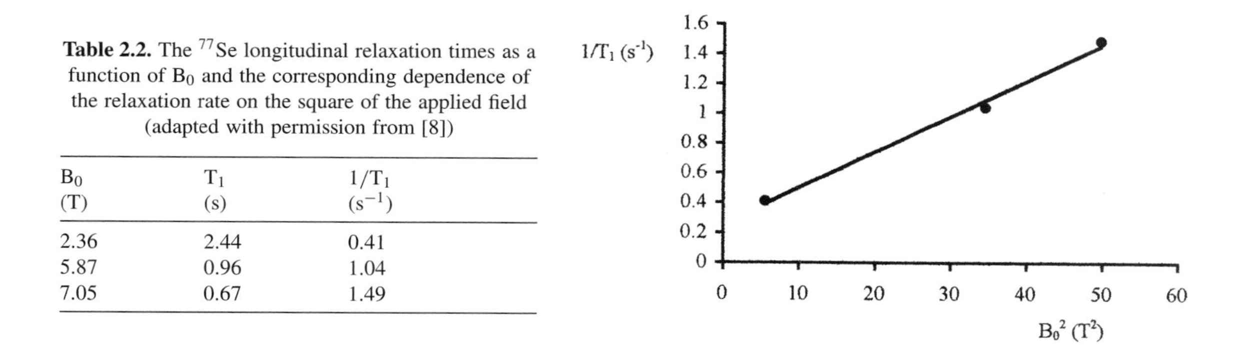

Table 2.2. The ^{77}Se longitudinal relaxation times as a function of B_0 and the corresponding dependence of the relaxation rate on the square of the applied field (adapted with permission from [8])

B_0 (T)	T_1 (s)	$1/T_1$ (s^{-1})
2.36	2.44	0.41
5.87	0.96	1.04
7.05	0.67	1.49

range since these posses the greatest shift anisotropy, for example ^{19}F, ^{31}P and, in particular, many metals.

The characteristic feature of CSA relaxation is its dependence on the *square* of the applied field, meaning it has greater significance at higher B_0. For example, the selenium-77 longitudinal relaxation rate in the selone **2.3** was shown to be linearly dependent on B_0^2, indicating CSA to be a significant relaxation mechanism [8] (Table 2.2), whilst no proton-selenium nuclear Overhauser effect (NOE) could be detected, demonstrating the ^{1}H–^{77}Se dipole–dipole mechanism to be ineffectual. Nuclei whose relaxation is dominated by the CSA mechanism may show significantly larger linewidths at higher fields if they posses a large shift anisotropy, and any potential benefits of greater dispersion and sensitivity may be lost by line broadening. For this reason, the study of some metal nuclei may be more successful at lower fields. Reducing the correlation time, by warming the sample for example, may attenuate the broadening effect, although this approach clearly has rather limited application. In some cases, enhanced CSA relaxation at higher fields can be advantageous. A moderately enhanced relaxation rate, such as in the ^{77}Se example above, allows for more rapid data collection (Section 4.1), thus providing an improvement in sensitivity (per unit time) above that expected from the increase in magnetic field alone.

Se

2.3

The CSA mechanism can also have a perhaps unexpected influence on the spectra of nuclei that are scalar spin-coupled to the CSA-relaxed spin. If CSA causes very rapid relaxation, the satellites arising from coupling to this spin will broaden and may even disappear altogether. Thus, whilst the coupling may be apparent at low fields, it may vanish at higher ones. This effect can be seen in the proton spectra of the platinum complex **2.4** recorded at 80 and 400 MHz [9] (Fig. 2.37). The increased linewidth of the satellites relative to the parent line at higher field scales with the square of the applied field, as expected for the CSA mechanism. To understand why this occurs, recall the origin of the Pt satellites themselves. These doublet components arise from the spin-½ ^{195}Pt nuclei existing in one of two states, α and β, which result in a frequency difference of J Hz for the corresponding proton signals. The CSA relaxation induces rapid transitions of the platinum spins between these states causing the doublet components to repeatedly switch positions. As this exchange rate (i.e. relaxation rate) increases the satellites first broaden and will eventually merge into the parent line, as for any dynamic process. This rapid and repeated change of spin states has a direct analogy with conventional spin decoupling (Chapter 4).

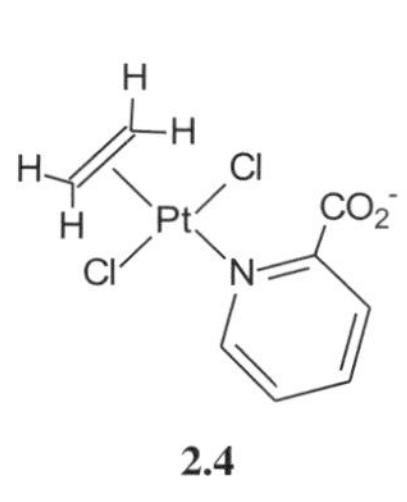

2.4

2.5.4. Spin-rotation relaxation

Molecules or groups which rotate very rapidly have associated with them a *molecular* magnetic moment generated by the rotating electronic and nuclear

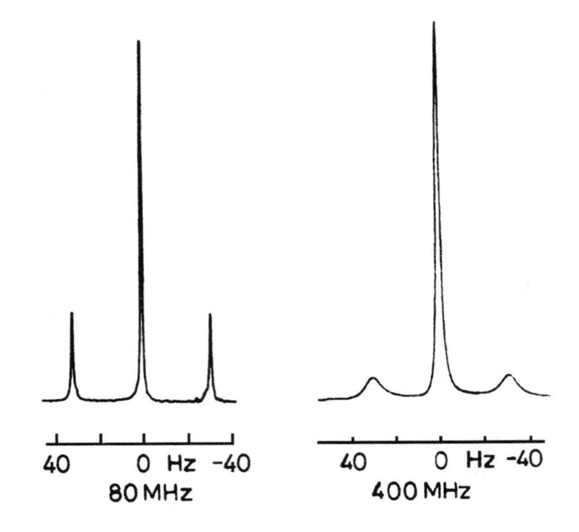

Figure 2.37. The ethene proton resonance of the platinum complex **2.4** in $CDCl_3$ at 80 and 400 MHz showing broadening of the ^{195}Pt satellites at higher field (reproduced with permission from [9]).

charges. The field due to this fluctuates as the molecule or group changes its rotational state as a result of, for example, molecular collisions and this provides a further mechanism for nuclear relaxation. This is most effective for small, symmetrical molecules or for freely rotating methyl groups and its efficiency *increases* as tumbling rates *increase*. This is in contrast to the previously described mechanisms. Thus, heating a sample enhances spin-rotation relaxation, this temperature dependence being characteristic of this mechanism and allowing its presence to be established.

2.5.5. Quadrupolar relaxation

The quadrupolar relaxation mechanism is only directly relevant for those nuclei that have a nuclear spin quantum number, I, greater than $^1/_2$ (quadrupolar nuclei) and is often the dominant relaxation process for these. This can also be a very efficient mechanism and the linewidths of many such nuclei can be hundreds or even thousands of hertz wide. The properties of selected nuclei with I > $^1/_2$ are summarised in Table 2.3. Whilst the direct observation of these

Table 2.3. Properties of selected quadrupolar nuclei

Isotope	Spin (I)	Natural abundance (%)	Quadrupole moment (10^{-28} m^2)	NMR frequency (MHz)	Relative sensitivity
^{2}H	1	0.015	2.8×10^{-3}	61.4	1.45×10^{-6}
^{6}Li	1	7.42	-8.0×10^{-4}	58.9	6.31×10^{-4}
^{7}Li	*3/2*	*92.58*	*-4×10^{-2}*	*155.5*	*0.27*
^{10}B	*3*	*19.58*	*8.5×10^{-2}*	*43.0*	*3.93×10^{-3}*
^{11}B	3/2	80.42	4.1×10^{-2}	128.3	0.13
^{14}N	1	99.63	1.0×10^{-2}	28.9	1.01×10^{-3}
^{17}O	5/2	0.037	-2.6×10^{-2}	54.2	1.08×10^{-5}
^{23}Na	3/2	100	0.10	105.8	9.27×10^{-2}
^{27}Al	5/2	100	0.15	104.2	0.21
^{33}S	3/2	0.76	-5.5×10^{-2}	30.7	1.72×10^{-5}
^{35}Cl	3/2	75.73	−0.1	39.2	3.55×10^{-3}
^{37}Cl	3/2	24.47	-7.9×10^{-2}	32.6	6.63×10^{-4}
^{59}Co	7/2	100	0.38	94.5	0.28

The observation of such nuclei is generally most favourable for those of low quadrupole moment and high natural abundance which exist in more highly symmetric environments; those listed in italics are considered the least favoured of the available isotopes. The NMR frequencies are quoted for a 400 MHz instrument (9.4 T magnet) and sensitivities are relative to ^{1}H and take account of both the intrinsic sensitivity of the nucleus and its natural abundance.

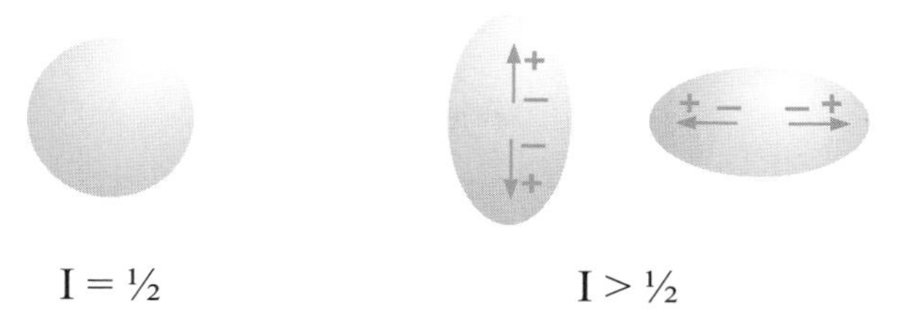

Figure 2.38. Quadrupolar nuclei lack the spherical charge distribution of spin $^{1}/_{2}$ nuclei, having an ellipsoidal shape which may be viewed as arising from pairs of electric dipoles. Thus quadrupolar nuclei interact with electric field gradients.

nuclei may not be routine for many organic chemists, their observation can, at times, prove very enlightening for specific problems and the indirect effects they have on the spectra of spin-$^{1}/_{2}$ nuclei should not be overlooked.

Quadrupolar nuclei possess an electric quadrupole moment in addition to a magnetic dipole moment. This results from the charge distribution of the nucleus deviating from the usual spherical symmetry associated with spin-$^{1}/_{2}$ nuclei and becoming ellipsoidal in shape. This can be viewed as arising from two back-to-back *electric* dipoles (Fig. 2.38). As such, the quadrupole moment is influenced by electric field *gradients* about the nucleus, but not by symmetric electric fields. The gradient is modulated as the molecule tumbles in solution and again if this occurs at the appropriate frequency it can induce flipping of nuclear spin states and thus stimulate relaxation. This is analogous to the relaxation of nuclear dipoles by time-dependent local magnetic fields, but the quadrupolar relaxation mechanism is the only one that depends on electric rather than magnetic interactions.

The relaxation rates of a quadrupolar nucleus are dictated by two new factors not previously considered. The first is the magnitude of the quadrupole moment itself (Table 2.3). Larger values contribute to more efficient spin relaxation and hence broader linewidths, whereas smaller values typically produce sharper lines. Thus, those nuclei with smaller quadrupole moments are usually more favoured for NMR observation. As before, for the mechanism to be effective, molecular tumbling must occur at an appropriate frequency, so again fast molecular tumbling reduces the effectiveness, leading to longer relaxation times and sharper lines. High temperatures or lower-viscosity solvents are thus more likely to produce narrow linewidths. The ultimate in low viscosity solvents are supercritical fluids which have viscosities more like those of a gas yet solubilising properties more like liquids. These have indeed been used in the study of quadrupolar nuclei, [10] but since they are only supercritical at very high pressures they demand the use of single-crystal sapphire NMR tubes so their use cannot be considered routine!

The second new factor is the magnitude of the electric field gradient. In highly symmetrical environments, such as tetrahedral or octahedral symmetries, the field gradient is, in principle, zero and the quadrupolar mechanism is suppressed. In reality, local distortions still arise, if only momentarily, introducing an element of asymmetry and hence enhanced relaxation and line broadening. Nevertheless, a higher degree of electrical symmetry can be correlated with narrower resonances. Thus, for example, the ^{14}N linewidth of $N(Me)_4^+$ is less than 1 Hz whereas that for NMe_3 is nearer to 80 Hz. Linewidth changes in ^{11}B spectra (I = $^{3}/_{2}$) have been used in the identification of tetrahedral boronic acid complexes at the active site of β-lactamases [11], enzymes responsible for the destruction of β-lactam antibiotics such as penicillins, and part of the defence mechanism of bacteria. Boronic acids, such as 3-dansylamidophenylboronic acid **2.5**, are known to be reversible inhibitors of active site serine β-lactamases and the complexes so formed display significant changes in the ^{11}B chemical shift of the boronic acid together with a reduction in linewidth relative to the free acid (Fig. 2.39). This reduction is attributed to the boron nucleus taking up a more symmetrical tetrahedral environment as this becomes bound by the enzyme's active site serine. This shift and line-narrowing can be mimicked

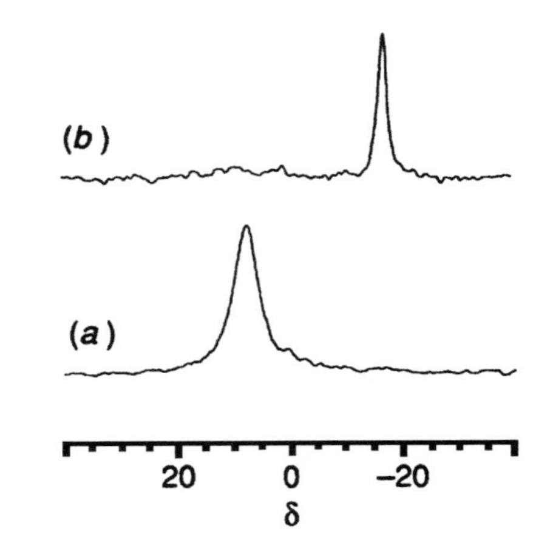

Figure 2.39. The ^{11}B NMR spectrum (128 MHz) of dansylamidophenylboronic acid **2.5** (a) as the free trigonal boronic acid ($\Delta\nu_{1/2}$ 580 Hz) and (b) as the tetrahedral complex with the active site serine of the P99 β-lactamase from *Enterobacter cloacae* ($\Delta\nu_{1/2}$ 160 Hz). Spectra are referenced to external trimethylborate (adapted with permission from [11]).

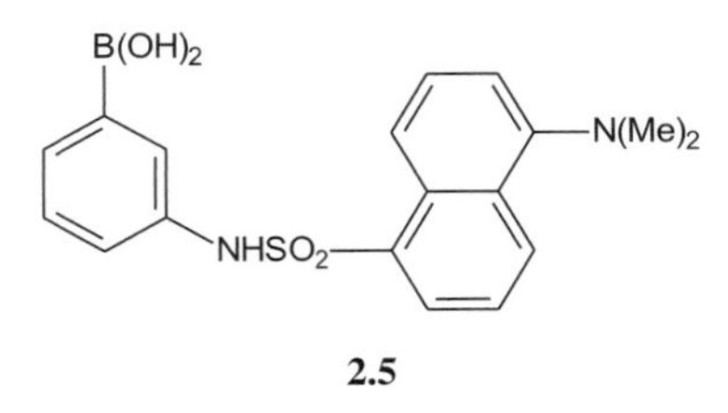

2.5

by placing the boronic acid in alkaline solution in which the R–B(OH)$_3^-$ ion predominates.

The broad resonances of many quadrupolar nuclei means field inhomogeneity makes a negligible contribution to linewidths so the methods described previously for measuring relaxation times are no longer necessary. For small molecules at least, T_2 and T_1 are identical and can be determined directly from the half-height linewidth. The broad resonances together with the sometimes low intrinsic sensitivity and low natural abundance of quadrupolar nuclei are the principal reasons for their relatively low popularity for NMR studies relative to spin-½ nuclei. The very fast relaxation of certain quadrupolar nuclei can also make their direct observation difficult with conventional high-resolution spectrometers; see Section 4.5.

Scalar coupling to quadrupolar nuclei

Probably of more relevance to the practising organic chemist is the influence quadrupolar nuclei have on the spectra of spin-½ nuclei, by virtue of their mutual scalar coupling. Coupling to a quadrupolar nucleus of spin I produces, in theory, 2I + 1 lines so, for example, the carbon resonance of $CDCl_3$ appears as a 1:1:1 triplet (^{2}H has I = 1) by virtue of the 32 Hz ^{13}C–^{2}H coupling. However, more generally, if the relaxation of the quadrupolar nucleus is rapid relative to the magnitude of the coupling, the splitting can be lost, in much the same way that coupling to a nucleus experiencing rapid CSA relaxation is lost. The carbon resonance of $CDCl_3$ is only a triplet because deuterium has a relatively small quadrupole moment making its coupling apparent whereas all coupling to the chlorine nuclei (^{35}Cl and ^{37}Cl have I = 3/2) is quenched by the very rapid relaxation of these spins (Fig. 2.40). Similarly, the proton resonance

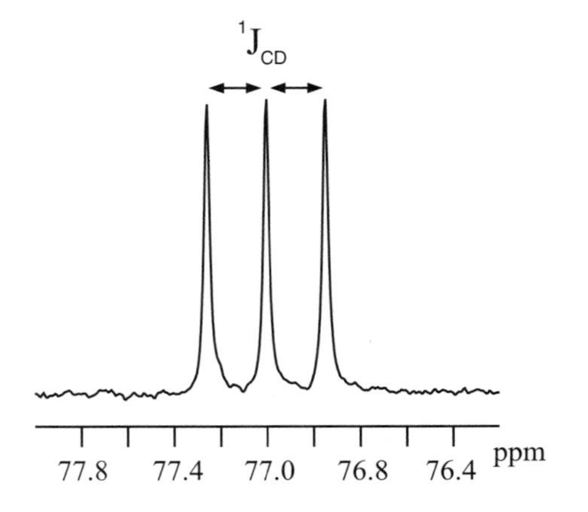

Figure 2.40. The carbon-13 spectrum of $CDCl_3$ reveals coupling to deuterium but not to chlorine-35 or chlorine-37.

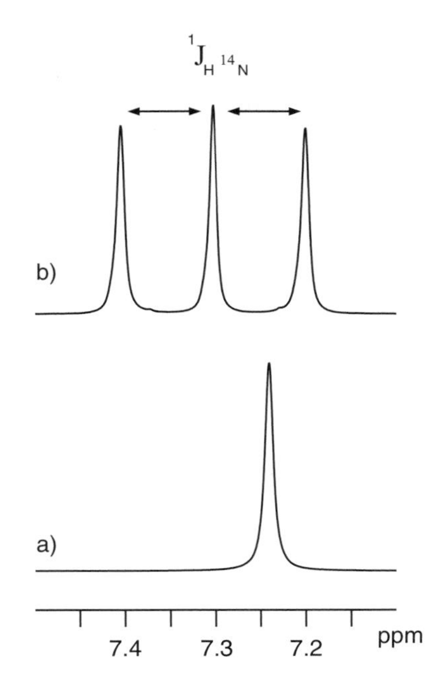

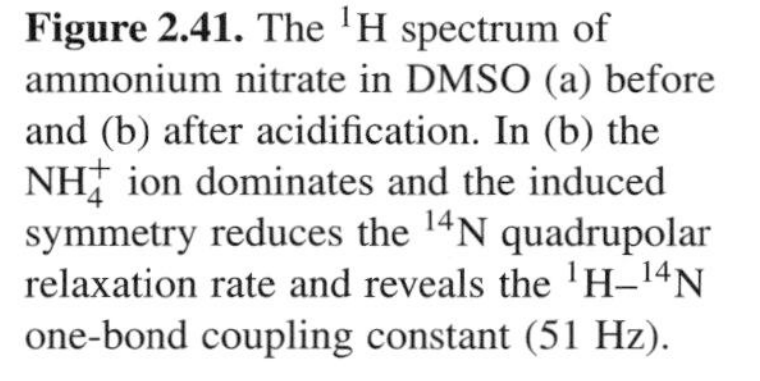

Figure 2.41. The ^1H spectrum of ammonium nitrate in DMSO (a) before and (b) after acidification. In (b) the NH_4^+ ion dominates and the induced symmetry reduces the ^{14}N quadrupolar relaxation rate and reveals the ^1H–^{14}N one-bond coupling constant (51 Hz).

of $CHCl_3$ is a sharp singlet despite the presence of the neighbouring chlorine atoms.

The appearance of the spin-$^1/_2$ nucleus spectrum is therefore also influenced by the factors described above which dictate the rate of quadrupolar relaxation. Couplings to quadrupolar nuclei that exist in a highly symmetrical environment are likely to be seen because of the slower relaxation the nuclei experience. For this reason the proton spectrum of $^{14}NH_4^+$ is an unusually sharp 1:1:1 triplet (Fig. 2.41, ^{14}N has $I = 1$) and the fluorine spectrum of $^{11}BF_4^-$ is a sharp 1:1:1:1 quartet (Fig. 2.42, ^{11}B has $I = ^3/_2$) . Increasing sample temperature results in slower relaxation of the quadrupolar nucleus so there is also a greater chance of the coupling being observable. In contrast, reducing the temperature increases relaxation rates and collapses coupling fine structure. This is contrary to the usual behaviour associated with dynamic systems where heating typically leads to simplification of spectra by virtue of resonance coalescence. The likelihood of coupling fine-structure being lost is also increased as the magnitude of the coupling constant decreases. In general then, the observation of scalar coupling to a quadrupolar nucleus is the exception rather than the rule.

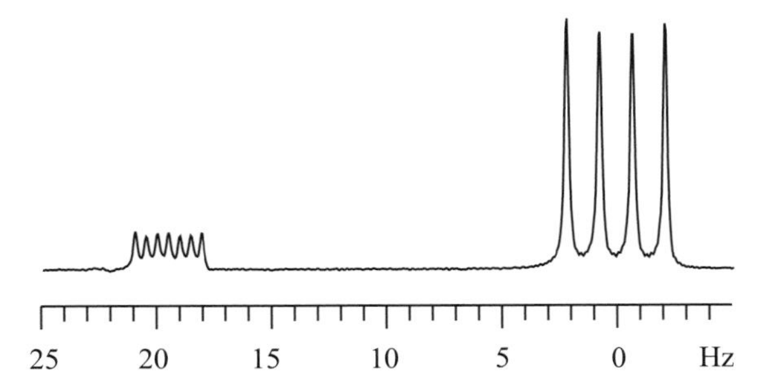

Figure 2.42. The ^{19}F spectrum of sodium borofluoride in D_2O. The $^{11}BF_4^-$ ion produces the even quartet (^{11}B $I = ^3/_2$, 80% abundance, J = 1.4 Hz) and the $^{10}BF_4^-$ ion produces the even septet (^{10}B $I = 3$, 20% abundance, J = 0.5 Hz). The frequency difference between the two is due to the $^{10}B/^{11}B$ isotope shift.

REFERENCES

[1] R.K. Harris, J. Kowalewski and S.C.D. Menezes, *Magn. Reson. Chem.*, 1998, **36**, 145–149.
[2] M.L. Martin, J.-J. Delpeuch and G.J. Martin, Practical NMR Spectroscopy, Heydon, London, 1980.

[3] P.J. Hore, *Method. Enzymol.*, 1989, **176**, 64–77.
[4] D.L. Rabenstein, S. Fan and T.K. Nakashima, *J. Magn. Reson.*, 1985, **64**, 541–546.
[5] D.L. Rabenstein, G.S. Srivasta and R.W.K. Lee, *J. Magn. Reson.*, 1987, **71**, 175–179.
[6] C.K. Larive and D.L. Rabenstein, *Magn. Reson. Chem.*, 1991, **29**, 409–417.
[7] R.K. Harris, Nuclear Magnetic Resonance Spectroscopy, Longman, Harlow, 1986.
[8] T.C. Wong, T.T. Ang, F.S. Guziec Jr. and C.A. Moustakis, *J. Magn. Reson.*, 1984, **57**, 463–470.
[9] I.M. Ismail, S.J.S. Kerrison and P.J. Sadler, *Polyhedron*, 1982, **1**, 57–59.
[10] M.P. Waugh and G.A. Lawless, in: Advanced Applications of NMR to Organometallic Chemistry, eds. M. Gielen, R. Willem, B. Wrackmeyer, Wiley, Chichester, 1996.
[11] J.E. Baldwin, T.D.W. Claridge, A.E. Derome, B.D. Smith, M. Twyman and S.G. Waley, *J. C. S. Chem. Commun.*, 1991, 573–574.

Chapter 3

Practical aspects of high-resolution NMR

NMR spectrometers are expensive instruments, representing one of the largest financial investments a chemical laboratory is likely to make, and to get the best results from these they must be operated and maintained in the appropriate manner. This chapter explores some of the fundamental experimental aspects of relevance to high-resolution, solution-state NMR spectroscopy, from instrumental procedures through to the preparation of samples for analysis. Later sections also deal with the basics of calibrating a spectrometer and assessing its performance.

3.1. AN OVERVIEW OF THE NMR SPECTROMETER

A schematic illustration of a modern NMR spectrometer is presented in Fig. 3.1. The fundamental requirement for high-resolution NMR spectroscopy is an intense static magnetic field which is provided, nowadays exclusively, by superconducting solenoid magnets manufactured from niobium-alloy wire. These are able to produce the stable and persistent magnetic fields demanded by NMR spectroscopy of up to 18.8 tesla with current technology, corresponding to proton frequencies of 800 MHz. The drive for ever increasing magnetic

Figure 3.1. A schematic illustration of the modern NMR spectrometer. The greyed sections are essential components of the spectrometer, whereas the others may be considered optional features (although pulsed field gradients are becoming standard on new instruments).

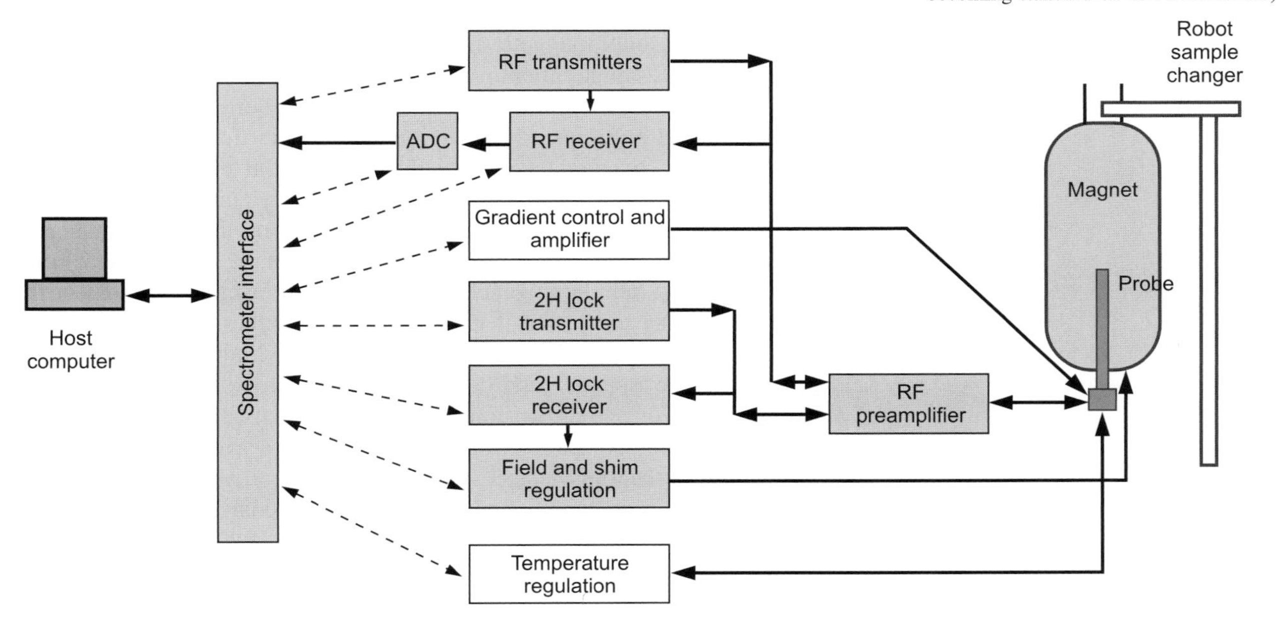

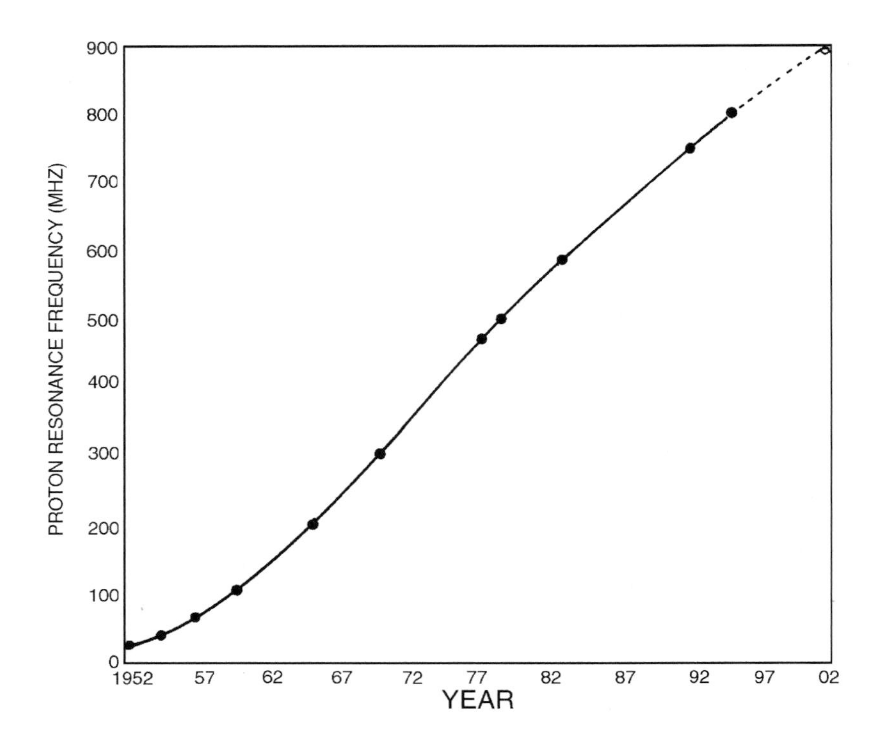

Figure 3.2. The increase in proton resonance frequency since the introduction of NMR spectroscopy as an analytical method (adapted with permission from [1]).

fields, encouraged by the demands for greater signal dispersion and instrument sensitivity, has continued since the potential of NMR spectroscopy as an analytical tool was first realised over 50 year years ago (Fig. 3.2) and still represents an area of intense competition between magnet manufacturers.

The magnet solenoid operates in a bath of liquid helium (at 4 K), surrounded by a radiation shield and cooled by a bath of liquid nitrogen (at 77 K), itself surrounded by a high vacuum. This whole assembly is an extremely efficient system and once energised the magnet operates for many years free from any external power source. It requires only periodic refilling of the liquid cryogens, typically on a weekly or bi-weekly basis for the nitrogen but only every 2–12 months for helium depending on magnet age and construction. The central bore of the magnet dewar is itself at ambient temperature, and this houses a collection of electrical coils, known as the shim coils, which generate their own small magnetic fields and are used to trim the main static field and remove residual inhomogeneities. This process of optimising the magnetic field homogeneity, known as *shimming*, is required for each sample to be analysed, and is discussed in Section 3.4. Within the shim coils at the exact centre of the magnetic field sits the head of the probe, the heart of any NMR spectrometer. This houses the radio-frequency coil(s) and associated circuitry that act as antennae, transmitting and receiving the electromagnetic radiation. These coils may be further surrounded by pulsed field gradient coils that serve to *destroy* field homogeneity in a controlled fashion (this may seem a rather bizarre thing to do, but it turns out to have rather favourable effects in numerous experiments). The sample, sitting in a cylindrical glass tube, is held in a turbine or 'spinner' and descends into the probe head on a column of air or nitrogen. For one-dimensional experiments it is still common practice to spin the sample at 10–20 Hz whilst in the probe to average to zero field inhomogeneities in the transverse (x–y) plane and so improve signal resolution. Sample spinning is now rarely used for multidimensional NMR experiments on modern instruments, as it can induce additional signal modulations and associated undesirable artefacts. This requires that acceptable resolution can be obtained on a static sample and while this is perfectly feasible with

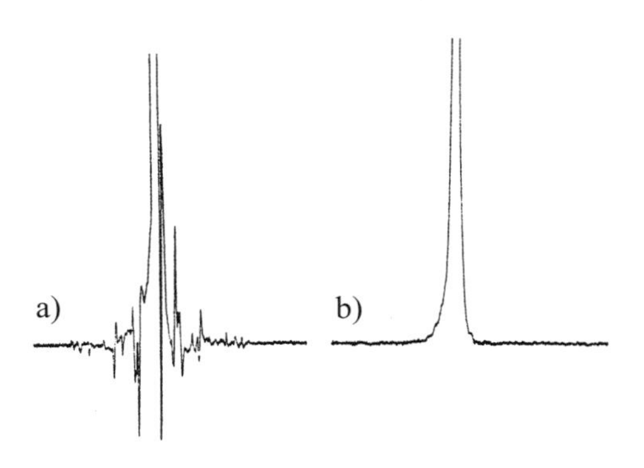

Figure 3.3. Floor vibrations can introduce unwelcome artefacts around the base of a resonance (a) which can be largely suppressed by mounting the magnet assembly on an anti-vibration stand (b) (figure courtesy of Bruker).

modern shim technology, older instruments may still demand spinning for all work.

Probes come in various sizes of diameter and length, depending on the magnet construction, but are more commonly referred to by the diameter of the sample tube they are designed to hold. The most widely used tube diameter is now 5 mm, other common sizes being 3, 8 and 10 mm (Section 3.4). Probes may be dedicated to observing one frequency (*selective* probes), may be tuneable over a very wide frequency range (*broadband* probes) or may tune to predefined frequency ranges, for example 4- or quad-nucleus probes. In all cases they will also be capable of observing the deuterium frequency simultaneously to provide a signal for field regulation (the 'lock' signal). A second (outer) coil is often incorporated to allow the simultaneous application of pulses on one or more additional nuclei.

In many locations it is advantageous to mount the whole of the magnet assembly on a vibration damping system as floor vibrations (which may arise from a whole host of sources including natural floor resonances, air conditioners, movement in the laboratory and so on) can have deleterious effects on spectra, notably around the base of resonances (Fig. 3.3). Whilst such artefacts have lesser significance to routine 1D observations, they may severely interfere with the detection of signals present at low levels, for example those in heteronuclear correlation or nuclear Overhauser effect experiments.

Within the spectrometer cabinet sit the radio-frequency transmitters and the detection system for the observation channel, additional transmitter channels [1], the lock channel and the pulsed field gradient transmitter. Most spectrometers come in either a two-channel or three-channel configuration, plus the lock channel. The spectrometer is controlled via the host computer, the brains of the spectrometer, which in recent years has generally been a UNIX based system linked to the spectrometer via a suitable interface but the trend for some manufacturers seems to be toward PC workstations taking on this role. The electrical analogue NMR signals are converted to the digital format required by the host computer via the analogue-to-digital converter (ADC), the characteristics of which can have important implications for the acquisition of NMR data (Section 3.2). The computer also processes the acquired data, although this may also be performed 'off-line' with one of the many available NMR software packages.

Various optional peripherals may also be added to the instrument, such as

[1] Reference to the 'decoupler channel(s)' is often used when referring to these additional channels but this should not be taken too literally as they may only be used for the application of only a few pulses rather than a true decoupling sequence. This nomenclature stems from the early developments of NMR spectrometers when the additional channel was only capable of providing 'noise decoupling', usually of protons.

variable temperature units which allow sample temperature regulation within the probe, robotic sample changers and so on. Recently, the coupling of NMR with other analytical techniques such as HPLC, has gained popularity especially within the pharmaceutical industry. The need for these will obviously depend on the type of samples handled and the nature of the experiments employed.

With the hype surrounding the competition between instrument manufacturers to produce ever increasing magnetic fields, it is all too easy for one to become convinced that an instrument operating at the highest available field strength is essential in the modern laboratory. Whilst the study of biological macromolecules no doubt benefits from the greater sensitivity and dispersion available, problematic small or mid-sized molecules are often better tackled through the use of the appropriate modern techniques. Signal dispersion limitations are generally less severe, and may often be overcome by using suitably chosen higher-dimensional experiments. Sensitivity limitations, which are usually due to a lack of material rather than solubility or aggregation problems, may be tackled by utilising smaller probe geometries (Section 3.3.3), after all, a new probe will cost at least an order of magnitude less than a new spectrometer. So for example, if one has insufficient material to collect a carbon-13 spectrum, rather than seeking time on a higher-field instrument, one could consider employing a proton-detected heteronuclear correlation experiment to determine these shifts indirectly. Beyond such considerations, there are genuine physical reasons largely relating to the nature of nuclear spin relaxation which mean that certain experiments on small molecules are likely to work *less well* at very high magnetic fields. In particular this relates to the nuclear Overhauser effect (Chapter 8), one of the principal NMR methods in structure elucidation. For many cases commonly encountered in the chemical laboratory a lower field instrument of modern specification is sufficient to enable the chemist to unleash a myriad of modern pulse NMR experiments on the samples of interest and subsequently solve the problem in hand. A better understanding of these modern NMR methods should aid in the selection of the appropriate experiments.

3.2. DATA ACQUISITION AND PROCESSING

This section examines some of the spectrometer procedures that relate to the collection, digitisation and computational manipulation of NMR data, including some of the fundamental parameters that define the way in which data is acquired. Such technicalities may not seem relevant to anyone who does not consider themselves a spectroscopist, but the importance of understanding a few basic relationships between experimental parameters comes from the need to recognise spectrum artefacts or corrupted data that can result from inappropriate parameter settings and to appreciate the limitations inherent in NMR measurements. Only then can one make full and appropriate use of the spectroscopic information at hand.

3.2.1. Pulse excitation

It is widely appreciated that modern NMR spectrometers use a ‘short pulse’ of radiofrequency energy to excite nuclear resonances over a range of frequencies. This pulse is supplied as *monochromatic* radiation from the transmitter, yet the nuclear spin transitions giving rise to our spectra vary in energy according to their differing Larmor frequencies and so it would appear that the pulse will be unable to excite all resonances in the spectrum simultaneously. However, Heisenberg’s Uncertainty principle tells us that an excitation pulse of duration Δt has associated with it a frequency uncertainty or spread of around $1/\Delta t$ Hz

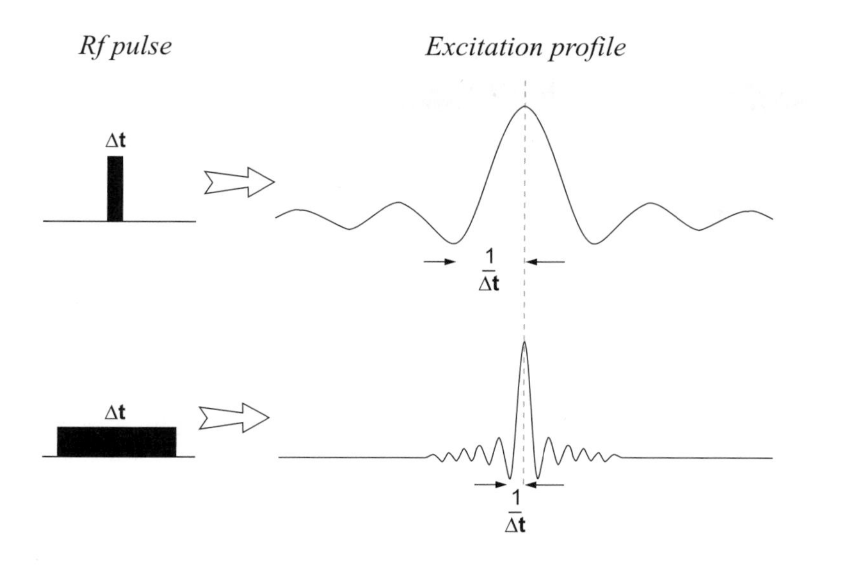

Figure 3.4. A single monochromatic radiofrequency pulse has an effective excitation bandwidth that depends inversely on the duration of the pulse. A short intense pulse is therefore able to excite over a wide frequency window (a), whereas a longer weaker pulse provides a more selective excitation profile (b).

and so effectively behaves as if it were polychromatic. The duration of the rf pulse Δt is usually referred to as the *pulse width*. A short, high-power pulse provides excitation over a wide frequency window whilst a longer, low-power pulse (which will provide the same net tip angle) is effective over a much smaller region (Fig. 3.4).

Consider a proton spectrum recorded at 400 MHz with the transmitter frequency placed in the centre of the spectral region of interest. A 10 ppm spectral window corresponds to 4000 Hz, therefore we need to excite over ± 2000 Hz, meaning the pulse duration must be 0.5 ms or less. The observation frequency for carbon-13 on the same instrument would be 100 MHz so a typical 200 ppm spectral width corresponds to ± 10 kHz, requiring a pulse of only 100 μs but with considerably higher power as its energy is now spread over a wider area. If the pulse were to be made very long (say tens of milliseconds) and weak, it would only excite over a rather small frequency range, giving rise to *selective excitation* of only part of the spectrum. As will become apparent in later chapters, the use of selective excitation methods are now commonplace in numerous NMR experiments. Long, low-power pulses which excite only a selected region of a spectrum are commonly referred to as 'soft pulses' whereas those that are of short duration and of high power are termed 'hard pulses'. Unless stated explicitly, all pulses in this book will refer to non-selective hard pulses.

Off-resonance effects

In practice, modern NMR instruments are designed to deliver high-power 90° pulses closer to 10 μs, rather than the hundreds predicted from the above arguments. This is to suppress the undesirable effects that arise when the pulse rf frequency is *off-resonance*, that is, when the transmitter frequency does not exactly match the nuclear Larmor frequency, a situation of considerable practical significance that has been ignored thus far.

As shown in the previous chapter, spins that are on-resonance have their magnetisation vector driven about the rf B_1 field toward the x–y plane during the pulse. Those spins that are off-resonance will, in addition to this B_1 field, experience a *residual* component ΔB of the *static* B_0 field along the z-axis of the rotating frame for which:

$$\frac{\gamma \Delta B}{2\pi} = \gamma \Delta B = \Delta \nu \,\text{Hz} \tag{3.1}$$

where $\Delta\nu$ represents the offset from the reference frequency. The vector sum of B_1 and ΔB is the *effective field* B_{eff} experienced by an off-resonance spin about

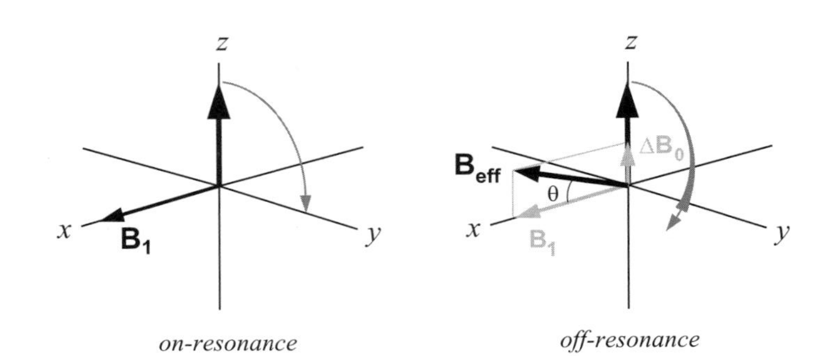

Figure 3.5. Excitation of magnetisation for which the rf is on-resonance (a) results in the rotation of the bulk vector about the applied rf field, B_1. Those spins which experience off-resonance excitation (b) are instead driven about an effective rf field, B_{eff}, which is tipped out of the x–y plane by an angle θ, which increases as the offset increases.

which it rotates (or more correctly, nutates) (Fig. 3.5). This is greater than B_1 itself and is tilted away from the x–y plane by an angle θ, where

$$\tan\theta = \frac{\Delta B}{B_1} \tag{3.2}$$

For those spins further from resonance, the angle θ becomes greater and the net rotation toward the x–y plane diminishes until, in the limit, θ becomes 90°. In this case the bulk magnetisation vector simply remains along the $+z$-axis and thus experiences no excitation at all. In other words, the nuclei resonate outside the *excitation bandwidth* of the pulse. Since an off-resonance vector is driven away from the y-axis during the pulse it also acquires a (frequency dependent) phase difference relative to the on-resonance vector (Fig. 3.6). This is usually small and an approximately linear function of frequency so can be corrected by phase adjustment of the final spectrum (Section 3.2.8).

How deleterious off-resonance effects are in a pulse experiment depends to some extent on the pulse tip-angle employed. A 90° excitation pulse ideally transfers magnetisation from the $+z$ axis into the x–y plane, but when off-resonance the tilt of the effective field will act to place the vector above this plane. However, the greater B_{eff} will mean the magnetisation vector follows a longer trajectory and this increased net flip angle offers some compensation, and hence a 90° pulse is fairly tolerant to off-resonance effects as judged by the elimination of z-magnetisation (Fig. 3.7a). In contrast, a 180° inversion pulse ideally generates pure $-z$ magnetisation leaving none in the transverse plane, but now the increased effective flip angle is detrimental, tending to move the vector further from the South Pole. Thus, 180° pulses do not perform well when applied off-resonance (Fig. 3.7b), and can be a source of poor experimental performance. In practice it is relatively easy to provide sufficiently short pulses (ca. 10 μs) to ensure that excitation and inversion is reasonably uniform over the relatively small frequency ranges encountered in ^{1}H NMR spectroscopy.

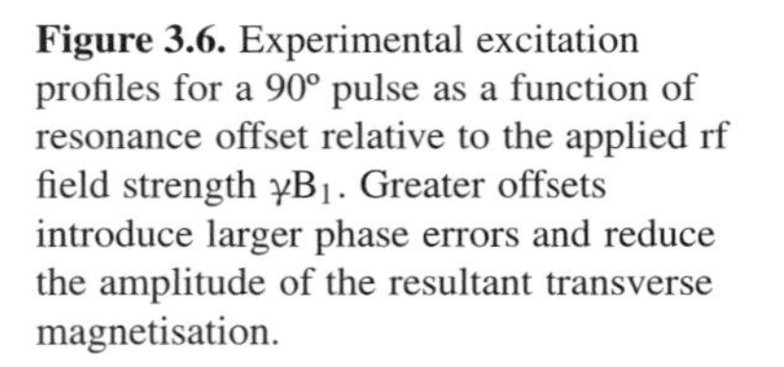

Figure 3.6. Experimental excitation profiles for a 90° pulse as a function of resonance offset relative to the applied rf field strength γB_1. Greater offsets introduce larger phase errors and reduce the amplitude of the resultant transverse magnetisation.

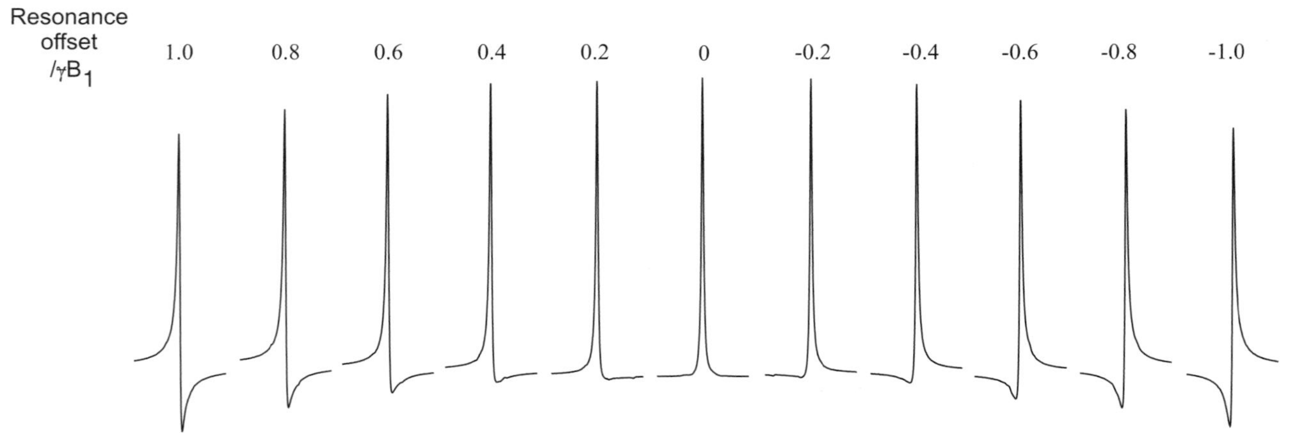

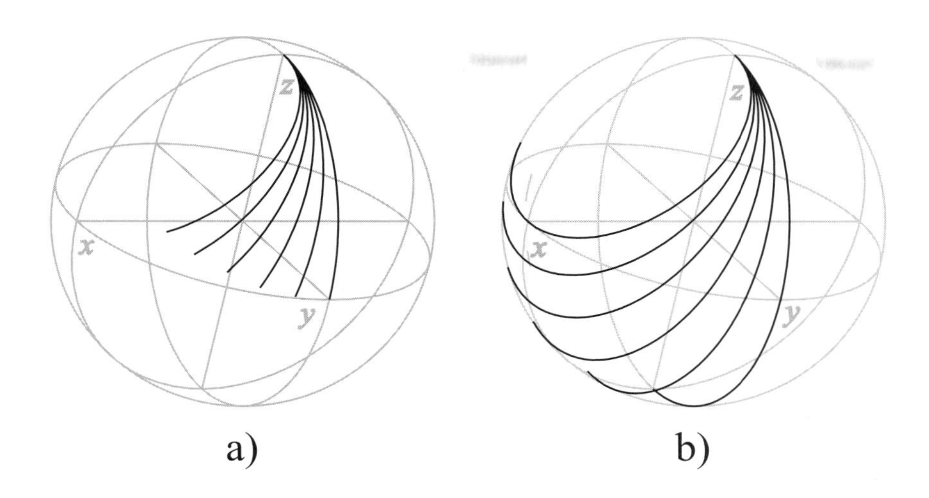

Figure 3.7. Excitation trajectories as a function of resonance offset for a) a 90° pulse and b) a 180° pulse. The offset moves from zero (on-resonance) to $+\gamma B_1$ Hz in steps of $0.2\gamma B_1$ (as in Fig. 3.6). The 90° pulse has a degree of 'offset-compensation' as judged by its ability to generate transverse magnetisation over a wide frequency bandwidth. In contrast the 180° pulse performs rather poorly away from resonance, leaving the vector far from the target South Pole and with a considerable transverse component.

However, for nuclei that display a much greater frequency dispersion, such as ^{13}C or ^{19}F, this is often not the case, and resonance distortion and/or attenuation can occur, and spurious signals may arise in multipulse experiments as a result. One approach to overcoming these limitations is the use of clusters of pulses known as *composite-pulses* which aim to compensate for these (and other) defects; see Chapter 9.

3.2.2. Signal detection

Before proceeding to consider how one collects the weak NMR signals, we briefly consider the detection process that occurs within the NMR receiver. The energy emitted by the excited spins produces tiny analogue electrical signals that must be converted into a series of binary numbers to be handled by the computer. This *digitisation* process (Section 3.2.3) must occur for all the frequencies in the spectrum. As chemists we are really only interested in knowing the chemical shifts *differences* between nuclei rather than their *absolute* Larmor frequencies since it is from these differences that one infers differences in chemical environments. As we already know, Larmor frequencies are typically tens or hundreds of megahertz whilst chemical shift ranges only cover some kilohertz. For example, a proton spectrum recorded at 200 MHz corresponds to a frequency range of only 2–3 kHz, and even for carbon (50 MHz on the same instrument) the frequency range is only around 10 kHz. Therefore, rather than digitising signals over many megahertz and retaining only the few kilohertz of interest, a more sensible approach is to subtract a reference frequency from the detected signal *before* digitisation, leaving only the frequency window of interest (Fig. 3.8). The resultant signals are now in the *audio* range which, for humans at least, corresponds to frequencies less than 20 kHz (so it is also possible to hear NMR resonances!).

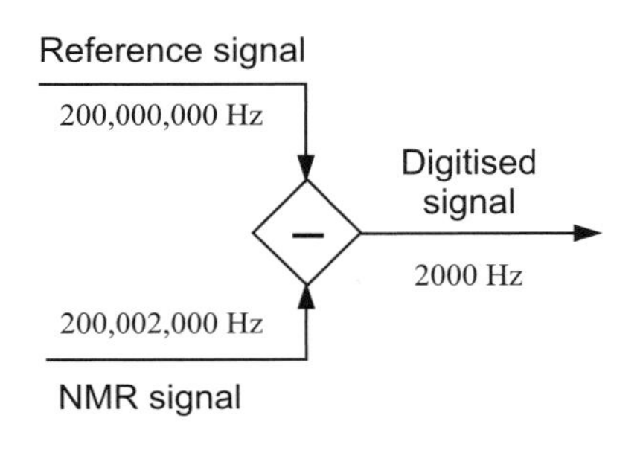

Figure 3.8. The NMR detection process. A fixed reference frequency is subtracted from the detected NMR signal so that only the frequency *differences* between resonances are digitised and recorded.

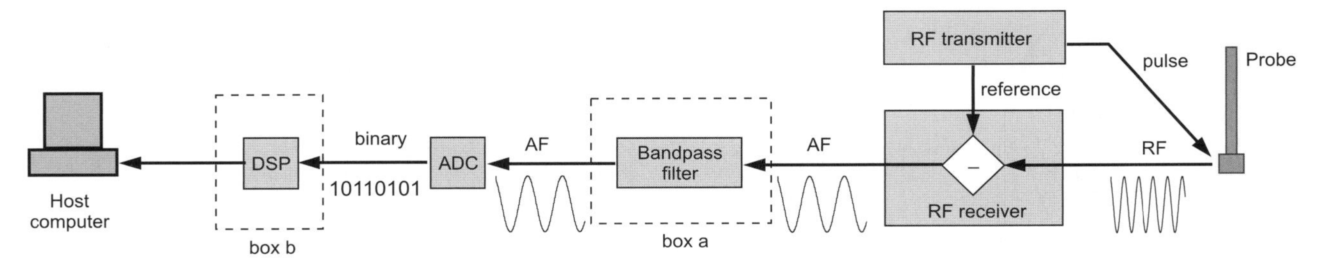

Figure 3.9. Schematic illustration of NMR data collection. Pulsed rf excitation stimulates the NMR response in the probe head which is then amplified and detected in the receiver. The receiver reference frequency (that of the pulse rf) is subtracted to leave only audio frequencies (AF) which are digitised by the analogue-to-digital converter (ADC) and subsequently stored within the computer. The functions of the boxed sections are described later in the text (DSP: digital signal processor).

The reference frequency is usually chosen to be that of the original pulse used to excite the spin system, and is supplied as a continuous reference signal from the transmitter (Fig. 3.9). This detection process is exactly analogous to the use of the rotating frame representation introduced in the previous chapter where the rotating frame reference frequency is also that of the rf pulse. If you felt a little uneasy about the seemingly unjustified use of the rotating frame formalism previously then perhaps the realisation that there is a genuine experimental parallel within all spectrometers will help ease your concerns. The digitised FID you see therefore contains only the audio frequencies that remain after subtracting the reference and it is these that produce the resonances observed in the final spectrum following Fourier transformation.

3.2.3. Sampling the FID

The Nyquist condition

To determine the frequency of an NMR signal correctly, it must be digitised at the appropriate rate. The *Sampling* or *Nyquist Theorem* tells us that to characterise a regular, oscillating signal correctly it must be defined by at least two data points per wavelength. In other words, to characterise a signal of frequency F Hz, we must sample at a rate of at least 2F; this is also known as the *Nyquist condition*. In NMR parlance the highest recognised frequency is termed the *spectral width*, SW (Fig. 3.10). The time interval between sampled data points is referred to as the *dwell time*, DW, as given by:

$$\mathrm{DW} = \frac{1}{2\mathrm{SW}} \tag{3.3}$$

Signals with frequencies less than or equal to the spectral width will be characterised correctly as they will be sampled at two or more points per wavelength, whereas those with higher frequencies will be incorrectly determined and in fact will appear in the spectrum at frequencies which are lower than their true values. To understand why this occurs, consider the

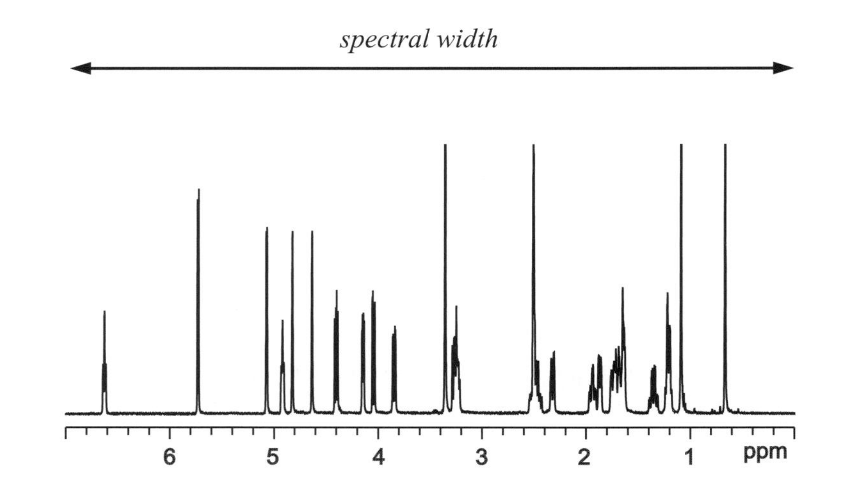

Figure 3.10. The spectral width defines the size of the observed frequency window. Only within this window are the line frequencies correctly characterised.

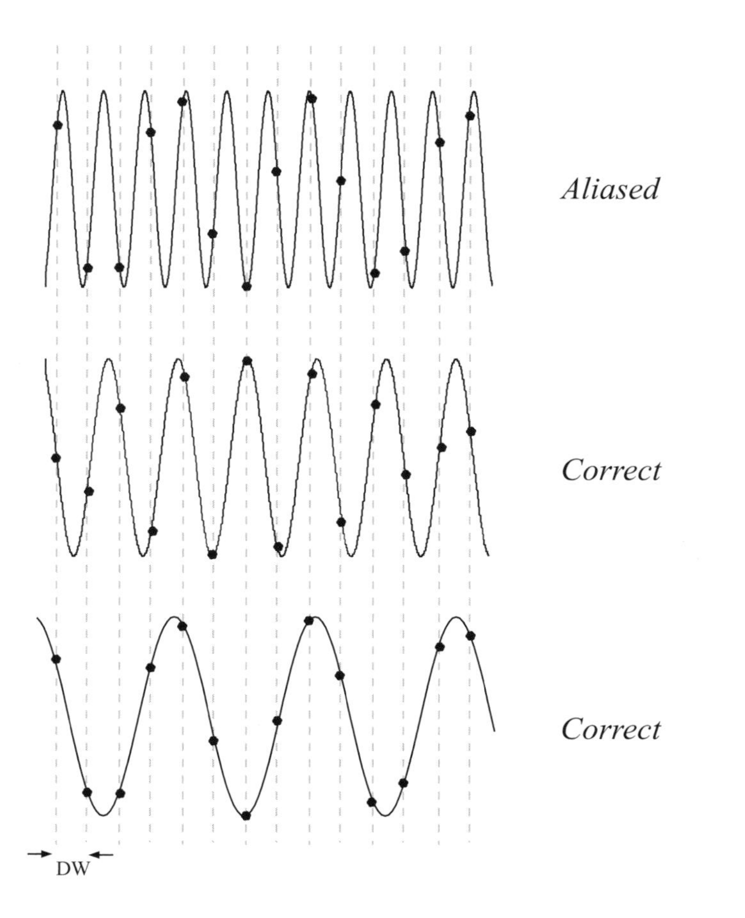

Figure 3.11. The Nyquist condition. To correctly characterise the frequency of an NMR signal it must be sampled at least twice per wavelength, it is then said to fall within the spectral width. The sampling of signals (a) and (b) meet this criteria. Signals with frequencies too high to meet this condition are aliased back within the spectral width and so appear with the wrong frequency. The sampling pattern of signal (c) matches that of (a) and hence it is incorrectly recorded as having the lower frequency.

sampling process for three frequencies, two within and the third outside the spectral width (Fig. 3.11). The sampled data points of the highest frequency clearly match those of the lower of the three frequencies and this signal will therefore appear with an incorrect frequency within the spectral window. Resonances that appear at incorrect frequencies, because in reality they exist outside the spectral width, are said to be *aliased* or *folded* back into the spectrum. Their location when corrupted in this manner is dependent upon which of the two commonly employed *quadrature detection* schemes are in use, as explained in Section 3.2.4.

Filtering noise

One particularly insidious effect of aliasing or folding is that not only will NMR resonances fold into the spectral window but noise will as well. This can seriously compromise sensitivity, since noise can extend over an essentially infinite frequency range (so-called white noise) all of which, potentially, can fold into the spectrum and swamp the NMR resonances. It is therefore essential to prevent this by filtering out all signals above a certain frequency threshold with an audio *bandpass filter* after detection but prior to digitisation (Fig. 3.9, box a). The effect of the noise filter is demonstrated in Fig. 3.12 and the gain in signal-to-noise in the spectrum employing this is clearly evident in Fig. 3.13. The cut-off bandwidth of the filter is variable, to permit the use of different spectral widths, and is usually automatically set by the spectrometer software to be a little greater than the spectral window. This is necessary because analogue filters are not perfect and do not provide an ideal sharp frequency cut-off, but instead tend to fall away steadily as a function of frequency, as is apparent in Fig. 3.12. A practical consequence is that these filters can attenuate the intensity of signals falling at the extreme ends of the spectrum so may

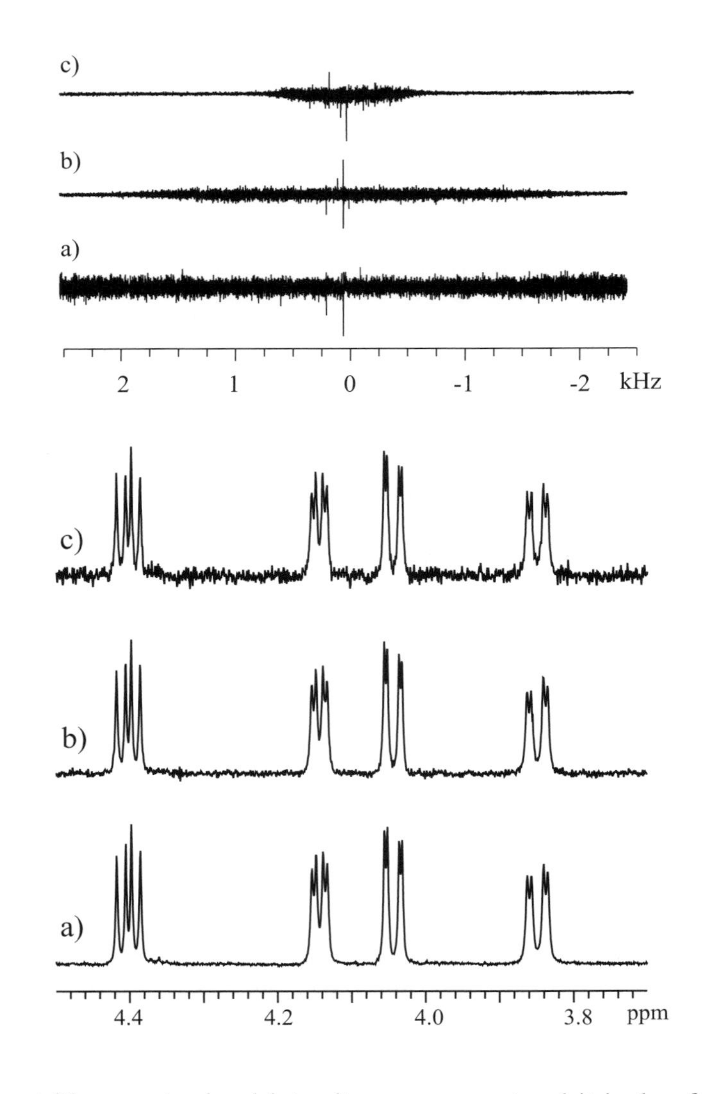

Figure 3.12. Eliminating noise with analogue filters. Spectra were recorded over a spectral width of 5 kHz with no sample in the probe. The analogue filter window was set to (a) 6 kHz, (b) 3 kHz and (c) 1 kHz and the associated noise attenuation is apparent, although the cut-off point is ill defined.

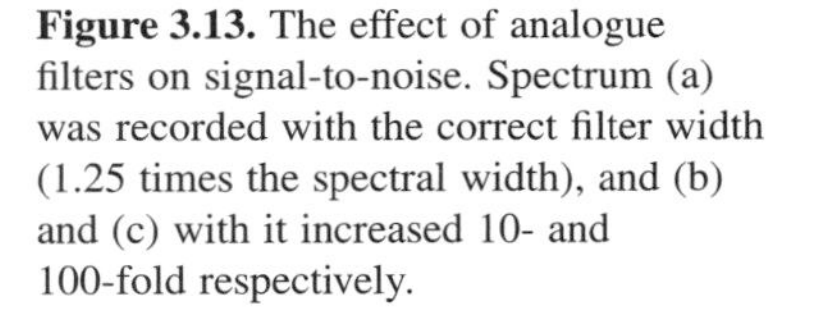
Figure 3.13. The effect of analogue filters on signal-to-noise. Spectrum (a) was recorded with the correct filter width (1.25 times the spectral width), and (b) and (c) with it increased 10- and 100-fold respectively.

interfere with accurate signal intensity measurement and it is therefore wise to ensure that resonances do not fall at the edges. Analogue filters can also introduce a variety of distortions to the spectrum, and in recent years the use of digital filtration methods having superior characteristics have become a standard feature in modern NMR spectrometers. Digital signal processing is considered in Section 3.2.6.

Acquisition times and digital resolution

The total sampling period of the FID, known as the *acquisition time*, is ultimately dictated by the frequency resolution required in the final spectrum. From the Uncertainty principle, the resolution of two lines separated by $\Delta\nu$ Hz requires data collection for at least $1/\Delta\nu$ seconds. If one samples the FID for a time that is too short then the frequency differences cannot be resolved and fine structure is lost (the minimum linewidth that can be resolved is given approximately by 0.6/AQ). Likewise, if the signal decays rapidly then one is unable to sample it for a long period of time and again can resolve no fine structure. In this case, the rapid decay implies a large linewidth arising from natural (transverse) relaxation processes, or from poor field inhomogeneity, and one will be unable to recover resolution by extending the acquisition time. Thus, when selecting the appropriate acquisition time one needs to consider the likely frequency differences that need to be resolved and the relaxation times of the observed spins.

The acquisition time is defined by the digitisation rate (which is dictated by the spectral width and defines the sampling dwell time DW) and on how many data points are sampled in total. If the FID contains TD time-domain data points then:

$$AQ = DW \cdot TD = \frac{TD}{2 \cdot SW} \tag{3.4}$$

When dealing with the final spectrum we are concerned with frequency and not time, and what one really needs is some measure of how well the resonances in the spectrum are digitised. The figure of interest is the frequency between adjacent data points in the spectrum, the *digital resolution*, DR (quoted in hertz per data point or simply Hz/point). It should be stressed that this is not what spectroscopists speak of when they refer to the 'resolution' achieved on a spectrometer as this relates to the homogeneity of the magnetic field. Digital resolution relates only to the frequency window to data point ratio, which is small for a well digitised spectrum but large when poorly digitised.

Following the Fourier transform, two data sets are generated representing the 'real' and 'imaginary' spectra (Section 3.3) so the real part with which one usually deals contains half the data points of the original FID (in the absence of further manipulation), and its data size, SI, is therefore TD/2. Digital resolution is then:

$$DR = \frac{\text{total frequency window}}{\text{total number of data points}} = \frac{SW}{SI} = \frac{2 \cdot SW}{TD} = \frac{1}{AQ} \tag{3.5}$$

Thus, digital resolution is simply the reciprocal of the acquisition time, so to collect a well digitised spectrum one must sample the data for a long period of time; clearly this is the same argument as presented above.

The effect of inappropriate digital resolution is demonstrated in Fig. 3.14. Clearly with a high value, that is short AQ, the fine structure cannot be resolved. Only with the extended acquisition times can the genuine spectrum be recognised. For proton spectroscopy, one needs to resolve frequency differences of somewhat less than 1 Hz to be able to recognise small couplings, so acquisition times of around 2–4 seconds are routinely used, corresponding to digital resolutions of around 0.5 to 0.25 Hz/point. This then limits the accuracy with which frequency measurements can be made, including shifts and couplings, (even though peak listings tend to quote resonance frequencies to many decimal places).

The situation is somewhat different for the study of nuclei other than protons however, since they often exhibit rather few couplings (especially in the presence of proton decoupling), and because one is usually more concerned with optimising sensitivity, so does not wish to sample the FID for extended periods and thus sample more noise. Since one does not need to define lineshapes with high accuracy, lower digital resolution suffices to resolve chemical shift differences and 1–2 Hz/point (0.5–1 s AQ) is adequate in ^{13}C NMR. Exceptions occur in the case of nuclei with high natural abundance which may show homonuclear couplings such as ^{31}P, or when the spectrum is being used to estimate the relative ratios of compounds in solution. This is true, for example, in the use of 'Mosher's acid' derivatives for the determination of enantiomeric excess from ^{19}F NMR spectra, for which the resonances must be well digitised to represent the true intensity of each species. In such cases the parameters that may be used for the 'routine' observation of ^{19}F are unlikely to be suitable for quantitative measurements. Similarly, the use of limited digital resolution in routine carbon-13 spectra is one reason why signal intensities do not provide a reliable indication of relative concentrations. In contrast, the spectra of rapidly relaxing quadrupolar nuclei contain broad resonances that require only low digital resolution and hence short acquisitions times.

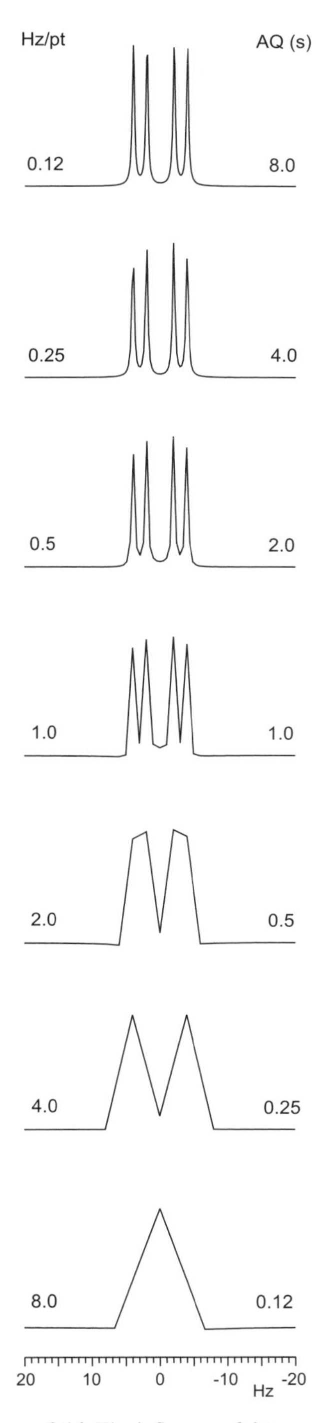

Figure 3.14. The influence of data acquisition times on the ability to resolve fine structure. Longer acquisition times correspond to higher digitisation levels (smaller Hz/pt) which here enable characterisation of the coupling structure within the double-doublet (J = 6 and 2 Hz).

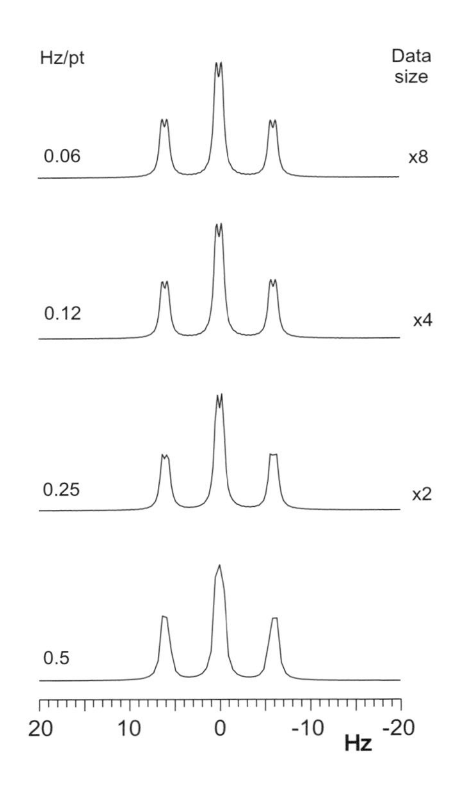

Figure 3.15. Zero-filling can be used to enhance fine structure and improve lineshape definition.

Zero-filling and truncation artefacts

Whilst the NMR response decays throughout the FID, the noise component remains essentially constant and will eventually dominate the tail of the FID. At this point there is little advantage in continuing acquisition since this only adds noise to the final spectrum. *Provided the FID has fallen to zero* when acquisition stops, one can artificially improve the digital resolution by appending zeros to the end of the FID. This process is known as *zero-filling* and it interpolates these added data points in the frequency-domain and so enhances the definition of resonance lineshapes (Fig. 3.15).

It has been shown [2] that by doubling the number of data points in the time-domain by appending zeros (a single 'zero-fill') it is possible to improve the frequency resolution in the spectrum. The reason for this gain stems from the fact that information in the FID is split into two parts (real and imaginary) after the transformation so one effectively looses information when considering only the real (absorption) part. Doubling the data size regains this lost information and is therefore a useful tool in the routine analysis of proton spectra (one could of course simply double the acquired data points but this leads to reduced sensitivity and requires more spectrometer time). Further zero-filling simply increases the *digital* resolution of the spectrum by interpolating data points, so no new information can be gained and the improvement is purely cosmetic. However, because this leads to a better definition of each line it can still be extremely useful when analysing multiplet fine-structure in detail or when measuring accurate resonance intensities; in the determination of enantiomeric excess via chiral solvating reagents for example. In 2D experiments acquisition times are necessarily kept rather short and zero-filling is routinely applied to improve the appearance of the final spectrum. A more sophisticated approach to extending the time-domain signal known as *linear prediction*, is described below.

If the FID has not decayed to zero at the end of the acquisition time, the data set is said to be *truncated* and this leads to distortions in the spectrum after zero-filling and Fourier transformation. The distortions arise from the

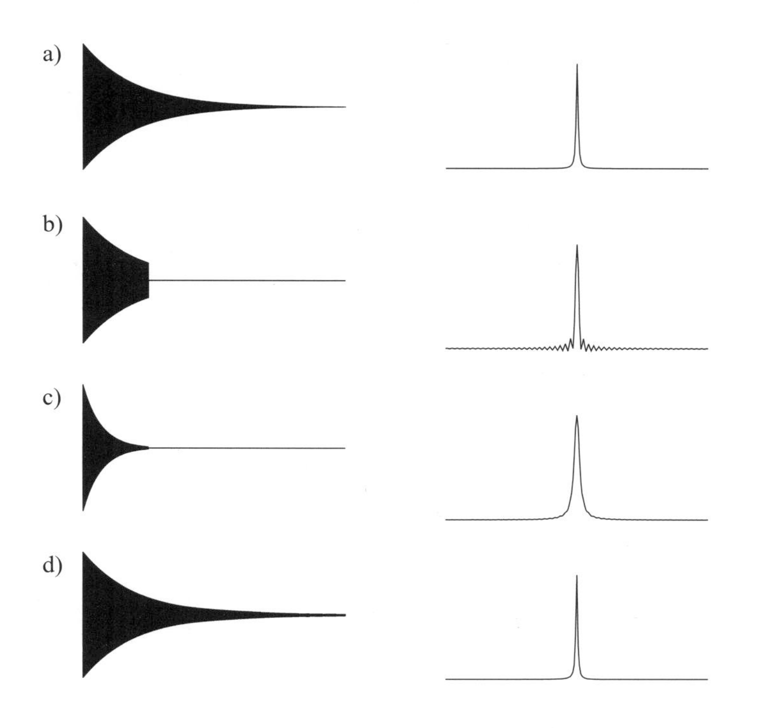

Figure 3.16. Processing a truncated FID. (a) A complete FID and the corresponding resonance, (b) a truncated FID which has been extended by zero-filling produces sinc wiggles in the spectrum, (c) apodisation of the FID in (b) together with zero filling and (d) linear prediction of the FID in (b).

FT of the sudden step within the FID, the result of which is described by the function (sin $x)/x$, also known as sinc x. Fig. 3.16 shows that this produces undesirable ringing patterns that are symmetrical about the base of the resonance, often referred to as 'sinc wiggles'. To avoid this problem it is essential to either ensure the acquisition time is sufficiently long, to force the FID to decay smoothly to zero with a suitable shaping function (Section 3.2.7) or to artificially extend the FID by linear prediction.

In proton spectroscopy, acquisitions times are sufficiently long that truncation artefacts are rarely seen in routine spectra, although they may be apparent around the resonances of small molecules with long relaxation times, solvent lines for example. As stated above, acquisition times for other nuclei are typically kept short and FIDs are usually truncated. This, however, is rarely a problem as it is routine practice to apply a window function to enhance sensitivity of such spectra, which itself also forces the FID to zero so eliminating the truncation effects. Try processing a standard ^{13}C spectrum directly with an FT but without the use of a window function; for some resonances you are likely to observe negative responses in the spectra that appear to be phasing errors. In fact these distortions arise purely from the low digital resolution used, i.e. the short acquisition times, and the corresponding truncation of the FID [3] (Fig. 3.17a). Use of a simple decaying function (Section 3.2.7) removes these effects (Fig. 3.17b) and when used in this way is often referred to as *apodisation*, literally meaning 'removing the feet'. Similar considerations apply to truncated 2D data sets where shaping functions play an essential role.

Linear prediction

The method of linear prediction (LP) can play many roles in processing of NMR data [4,5], from the rectification of corrupted or distorted data, through to the complete generation of frequency-domain data from an FID; an alternative to the FT. Here we consider its most popular usage, known as forward linear prediction, which extends a truncated FID. Rather than simply appending zeros, this method, as the name suggests, predicts the values of the missing data

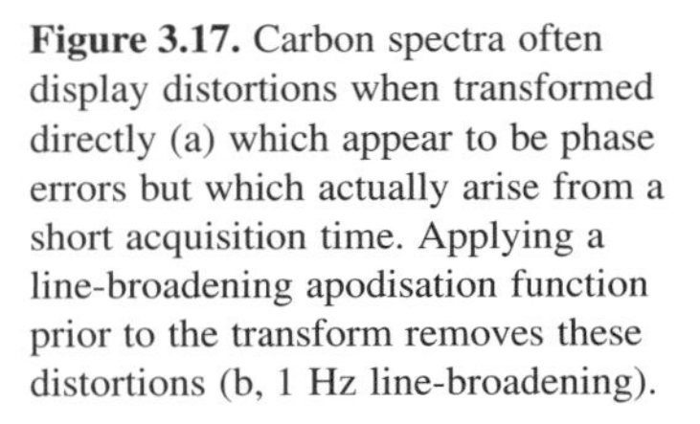

Figure 3.17. Carbon spectra often display distortions when transformed directly (a) which appear to be phase errors but which actually arise from a short acquisition time. Applying a line-broadening apodisation function prior to the transform removes these distortions (b, 1 Hz line-broadening).

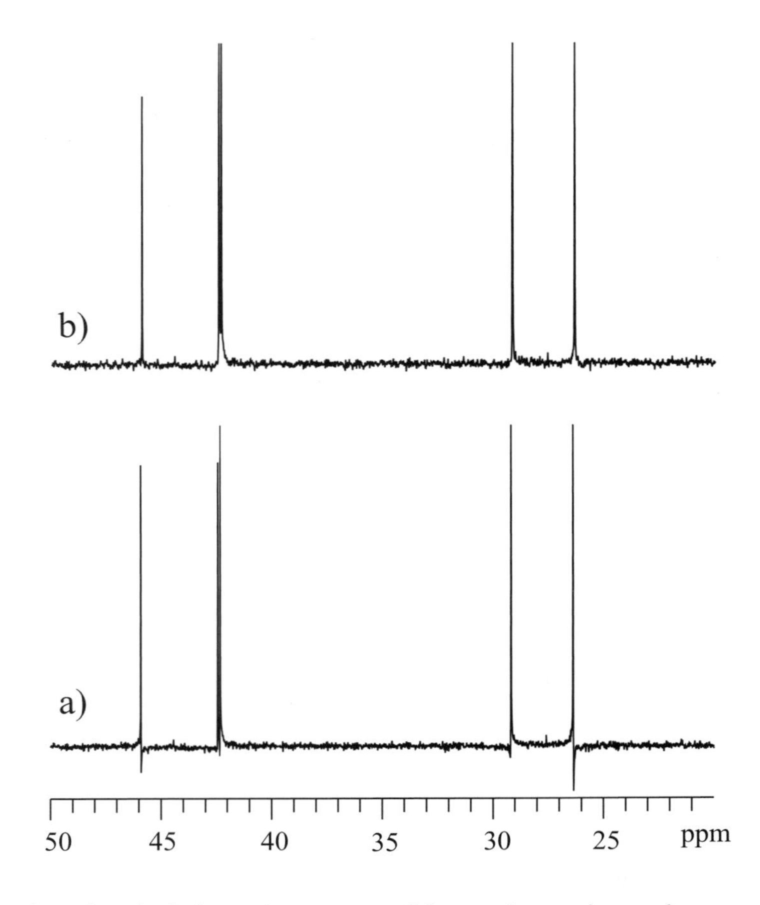

points by using the information content of the previous points and so genuinely extends the FID (Fig. 3.16d).

In a time sequence of data points the value of a single point, d_n, can be estimated from a linear combination of the immediately preceding values:

$$d_n = a_1 d_{n-1} + a_2 d_{n-2} + a_3 d_{n-3} + a_4 d_{n-4} \ldots \quad (3.6)$$

where a_1, a_2,... and so on represent the LP coefficients. The number of coefficients (referred to as the *order* of the prediction) corresponds to the number of data points used to predict the next value in the series. Provided the coefficients can be determined from the known data, it is then possible to extrapolate beyond the acquired data points. Repetition of this process incorporating the newly predicted points ultimately leads to the extended FID.

This process is clearly superior to zero-filling and produces a much better approximation to the true data than does simply appending zeros. It improves resolution, avoids the need for strong apodisation functions and greatly attenuates truncation errors. Naturally, the method has its limitations, the most severe being the requirement for high signal-to-noise in the FID for accurate estimation of the LP coefficients. Successful execution also requires that the number of points used for the prediction is very much greater than the number of lines that comprise the FID. This may be a problem for 1D data sets with many component signals and LP of such data is less widely used. The method is far more valuable for the extension of truncated data sets in the indirectly detected dimensions of a 2 or 3D experiment (you may wish to return to these discussions when you are familiar with the 2D approach, see Chapter 5). Here individual interferograms contain rather few lines (look at a column from a 2D data set) and are thus better suited to prediction, providing sensitivity is adequate. The use of LP in the routine collection of 2D spectra of organic molecules has been subject to detailed investigation [6]. Typically

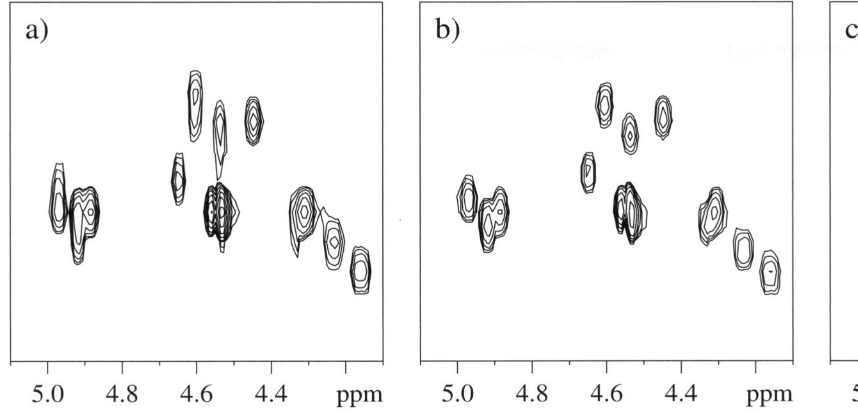

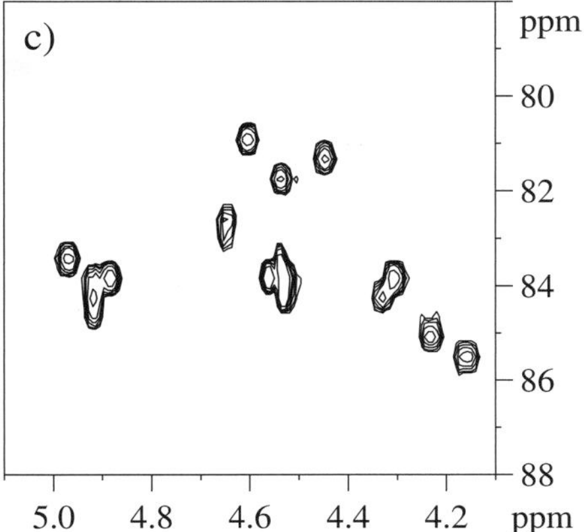

Figure 3.18. Improved peak resolution in a two-dimensional heteronuclear proton-carbon correlation experiment (Chapter 6) through linear prediction. The same raw data was used in each spectrum, with the F_1 (carbon) dimension processed with (a) no data extension, (b) one zero-fill and (c) linear prediction in place of zero-filling.

a two to four fold prediction is used, so for example, 128 t_1 data points may be readily extended to 256 or 512. Apodisation of the data is still likely to be required, although not as severely as for untreated data, and zero-filling may also be applied to further improve digitisation. Fig. 3.18 clearly demonstrates the improved resolution attainable through the use of linear prediction in the indirect dimension of a two-dimensional heteronuclear correlation experiment. The same principles can be used in *backward* linear prediction. Data points at the start of an FID are sometimes corrupted by ringing in receiver circuitry when detection starts resulting in baseline distortion of the spectrum. Replacing these points with uncorrupted predicted points eliminates the distortion; an example of this is found in Section 3.5.

3.2.4. Quadrature detection

It was described above how during the NMR signal detection process a reference frequency equal to that of the excitation pulse is subtracted from the NMR signal to produce an audio-frequency signal that is digitised and later subject to Fourier transformation. The problem with analysing the data produced by this *single-channel detection* is that the FT is intrinsically unable to distinguish frequencies above the reference from those below it, that is, it cannot differentiate positive from negative frequencies. This results in a magnetisation vector moving at $+\nu$ Hz in the rotating frame producing two resonances in the spectrum at $+\nu$ and $-\nu$ Hz after Fourier transformation (Fig. 3.19). The inevitable confusing overlap in a spectrum acquired with the reference positioned in the centre of the spectral width can be avoided by placing the reference at one edge of the spectrum to ensure all rotating-frame frequencies have the same sign. Whilst this will solve potential resonance overlap problems, it introduces a number of other undesirable factors. Firstly, although there will be no mirror image NMR resonances remaining within the

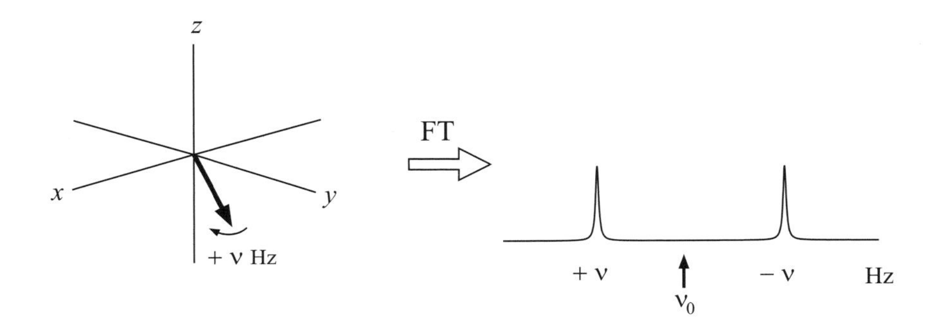

Figure 3.19. A single-channel detection scheme is unable to differentiate positive and negative frequencies in the rotating frame. This results in a mirror image spectrum being superimposed on the true one if the transmitter is placed in the centre of the spectrum.

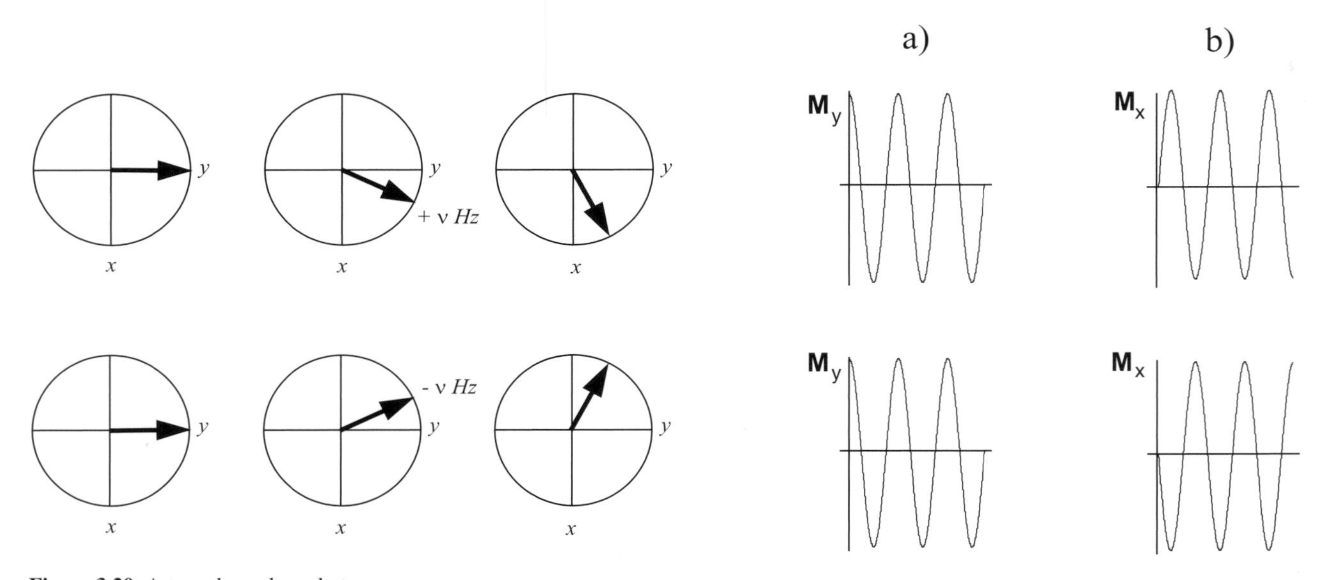

Figure 3.20. A two-channel quadrature detection system monitors magnetisation on two orthogonal axes, providing both (a) cosine and (b) sine modulated data which ultimately allow the sense of precession to be determined (see text).

spectral window, noise will still be mirrored about the reference frequency and added to that already present. This leads to a decrease in signal-to-noise by a factor of $\sqrt{2}$, or about 1.4 (not by a factor of 2 because noise is random and does not add coherently). Furthermore, this produces the greatest possible frequency separation between the pulse and the highest frequency resonance, so enhancing undesirable off-resonance effects. The favoured position for the reference is therefore in the centre of the spectrum and two-channel (*quadrature*) detection is then required to distinguish sign.

To help visualise why single-channel detection cannot discriminate positive and negative frequencies, and how the quadrature method can, consider again a single magnetisation vector in the rotating frame. The use of the single-channel detector equates to being able to observe the precessing magnetisation along *only one axis*, say the *y*-axis. Fig. 3.20 shows that the resultant signal along this axis for a vector moving at $+\nu$ Hz is identical to that moving at $-\nu$ Hz, both giving rise to a cosinusoidal signal, so the two are indistinguishable (Fig. 3.20a). If, however, one were able to make use of two (phase-sensitive) detection channels whose reference signals differ in phase by 90° (hence the term *quadrature*), one would then be able to observe magnetisation simultaneously along both the *x* and the *y* axes such that one channel monitors the cosine signal, the other the sine (Fig. 3.20a and b). With the additional information provided by the sinusoidal response of the second channel, the sense of rotation can be determined and vectors moving at $\pm\nu$ Hz can be distinguished (Fig. 3.21). Technically, the Fourier transform is then *complex*, with the *x* and *y* components being handled separately as the real and imaginary inputs to the transform, following which the positive and negative frequencies are correctly determined. In the case of the single channel, the data are used as input to a *real* FT.

Figure 3.21. A two-channel detection scheme is able to differentiate positive and negative frequencies in the rotating frame, allowing the transmitter to be placed in the centre of the spectrum without the appearance of mirror image signals.

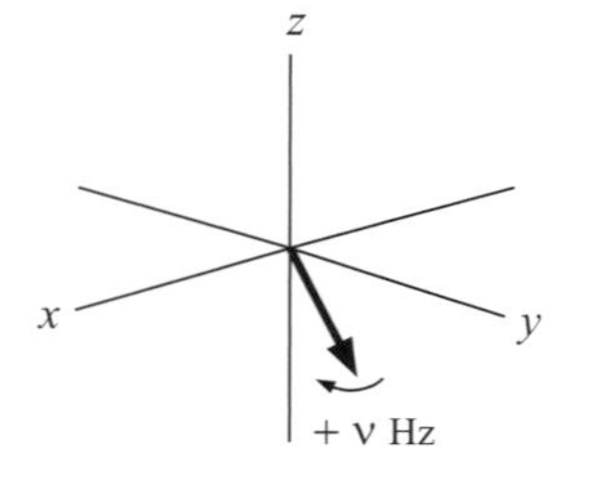

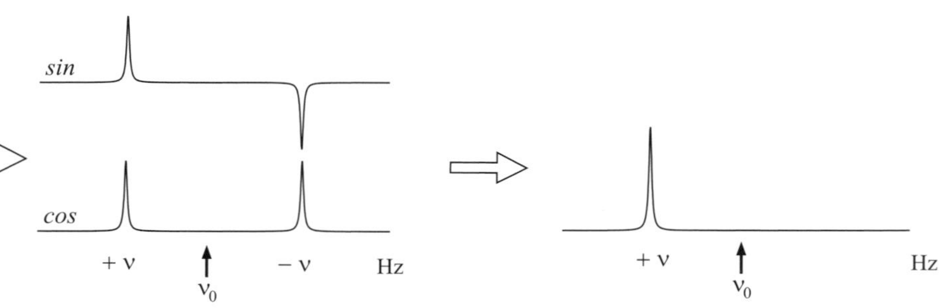

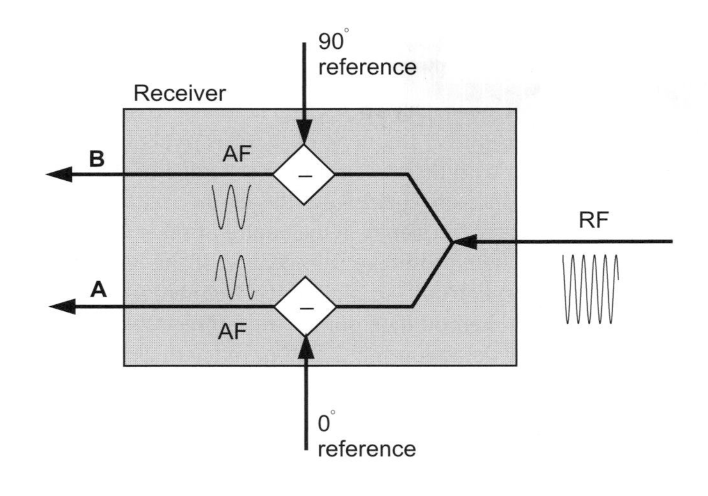

Figure 3.22. Schematic illustration of the experimental implementation of quadrature detection in the NMR receiver. The incoming rf signal is split in two and the reference signal, differing in phase by 90° in the two channels, subtracted. Channels A and B therefore provide the required sine and cosine components.

Simultaneous and sequential sampling

Quadrature detection is universally employed in all modern spectrometers. However, there exist two experimental schemes for implementing this and which of these you are likely to use will be dictated by the spectrometer hardware and perhaps by the age of the instrument (on some modern instruments the operator can choose between the two methods).

In both implementations the incoming rf signal from the probe and preamplifier is split in two and each fed to separate phase-sensitive detectors whose reference frequencies are identical but differ in phase by 90° (Fig. 3.22). The resulting audio signals are then sampled, digitised and stored for subsequent analysis, and it is in the execution of these that the two methods differ. The first relies on *simultaneous* sampling of the two channels (i.e. the channels are sampled at precisely the same point in time) and the data points for each are stored in separate memory regions (Fig. 3.23a). The two sets of data so generated represent the cosine and sine components required for the sign discrimination and are used as the real and imaginary input to a complex FT routine. As this enables positioning of the transmitter frequency in the centre of the spectrum, a frequency range of only $\pm\mathrm{SW}/2$ need be digitised. According to the Nyquist criterion the sampling rate now becomes equal to SW i.e. (2*SW/2) or half of that required for single channel detection, so the equations given in the previous section are modified slightly for *simultaneous quadrature detection* to give:

$$\mathrm{DW} = \frac{1}{\mathrm{SW}} \text{ and } \mathrm{AQ} = \mathrm{DW} \cdot \frac{\mathrm{TD}}{2} = \frac{\mathrm{TD}}{2 \cdot \mathrm{SW}} \tag{3.7}$$

also

$$\mathrm{DR} = \frac{\mathrm{SW}}{\mathrm{SI}} = \frac{2 \cdot \mathrm{SW}}{\mathrm{TD}} \text{ and so again } \mathrm{DR} = \frac{1}{\mathrm{AQ}} \tag{3.8}$$

Notice that in the calculation of the acquisition time we need only consider

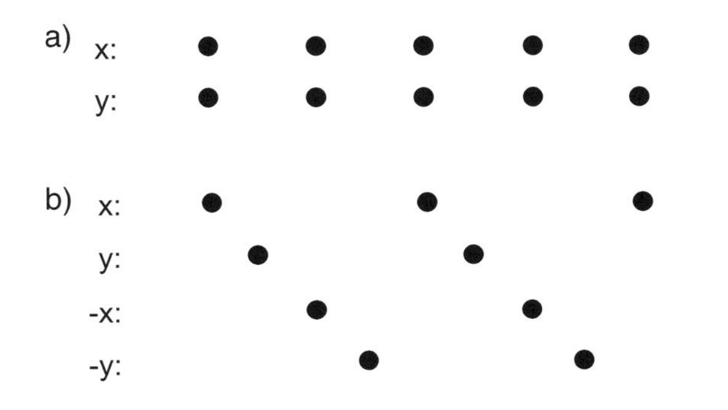

Figure 3.23. Data sampling schemes for the two common quadrature detection methods. (a) Simultaneous sampling: the two quadrature channels (representing x and y magnetisation) are sampled at the same point in time. (b) Sequential sampling: the two channels are sampled alternately at twice the rate of method (a), and the phase inverted for alternate pairs of data points (see text).

half the total time-domain data points since now two points are sampled at the same time.

The second method for quadrature detection actually strives to eliminate the problems of positive and negative frequencies by effectively mimicking a single channel detection scheme with the transmitter at the edge of the spectrum, so is sometimes referred to as 'pseudo-quadrature detection'. Although it allows the positioning of the transmitter frequency at the centre of the spectrum, it employs an ingenious sampling scheme to make it appear as if the reference frequency sat at one edge of the spectrum, so that only frequencies of one sign are ever characterised. In this method data points are sampled *sequentially* at a rate suitable for single channel detection (i.e. 2.SW or twice that in the simultaneous method), but for each data point sampled the reference phase is incremented by 90° (Fig. 3.23b). The effect on a magnetisation vector is that it appears to have advanced in the rotating frame by 90° more between each sampling period than it actually has, so the signal appears to be precessing faster than it really is. As digitisation occurs at a rate of 2SW and each 90° phase shift corresponds to an advance of 1/4 of a cycle, the frequency appears to have increased by 2.SW/4 or SW/2. Since the transmitter is positioned at the centre of the spectrum, the genuine frequency range runs from +SW/2 to −SW/2, meaning the artificial increase of +SW/2 moves the frequency window to +SW to 0. Thus, there are no longer negative frequencies to distinguish. Experimentally, the required 90° phase shifts are achieved by alternating between the two phase-sensitive detectors to give the 0° and 90° shifts, and simply inverting the signals from these channels to provide the 180° and 270° shifts respectively (Fig. 3.23b). The sampled data is handled by a single memory region as *real* numbers which are used as the input to a *real* FT. In this *sequential quadrature detection* scheme, the sampling considerations in Section 3.2.3 apply as they would for single-channel detection. In Chapter 5 these ideas are extended to quadrature detection in two-dimensional spectroscopy.

For either scheme, the total number of data points digitised, acquisition times and spectral widths are identical, so the resulting spectra are largely equivalent. The most obvious difference is in the appearance of aliased signals, that is, those that violate the Nyquist condition, as described below. Experimentally, flatter baselines are observed for the simultaneous method as a result of the *symmetrical* sampling of the initial data points in the FID and this is the recommended protocol.

Aliased signals

In Section 3.2.3 it was shown that a resonance falling outside the spectral window (because it violates the Nyquist condition) will still be detected but will appear at an incorrect frequency and is said to be *aliased* or *folded* back into the spectrum. This can be confusing if one is unable to tell whether the resonance exhibits the correct chemical shift or not. The precise location of the aliased signal in the spectrum depends on the quadrature detection scheme in use and on how far outside the window it truly resonates. With the simultaneous (complex FT) scheme, signals appear to be 'wrapped around' the spectral window and appear at the *opposite* end of the spectrum (Fig. 3.24b), whereas with the sequential (real FT) scheme signals are 'folded back' at the *same* end of the spectrum (Fig. 3.24c). If you are interested to know why this difference occurs see reference [7].

Fortunately, in proton spectroscopy it is generally possible to detect the presence of an aliased peak because of its esoteric phase which remains distorted when all others are correct. In heteronuclear spectra that display only a single resonance such phase characteristics cannot provide this information as there is no other signal with which to make the comparison. In such cases it is necessary to widen the spectral window and record any movement of the

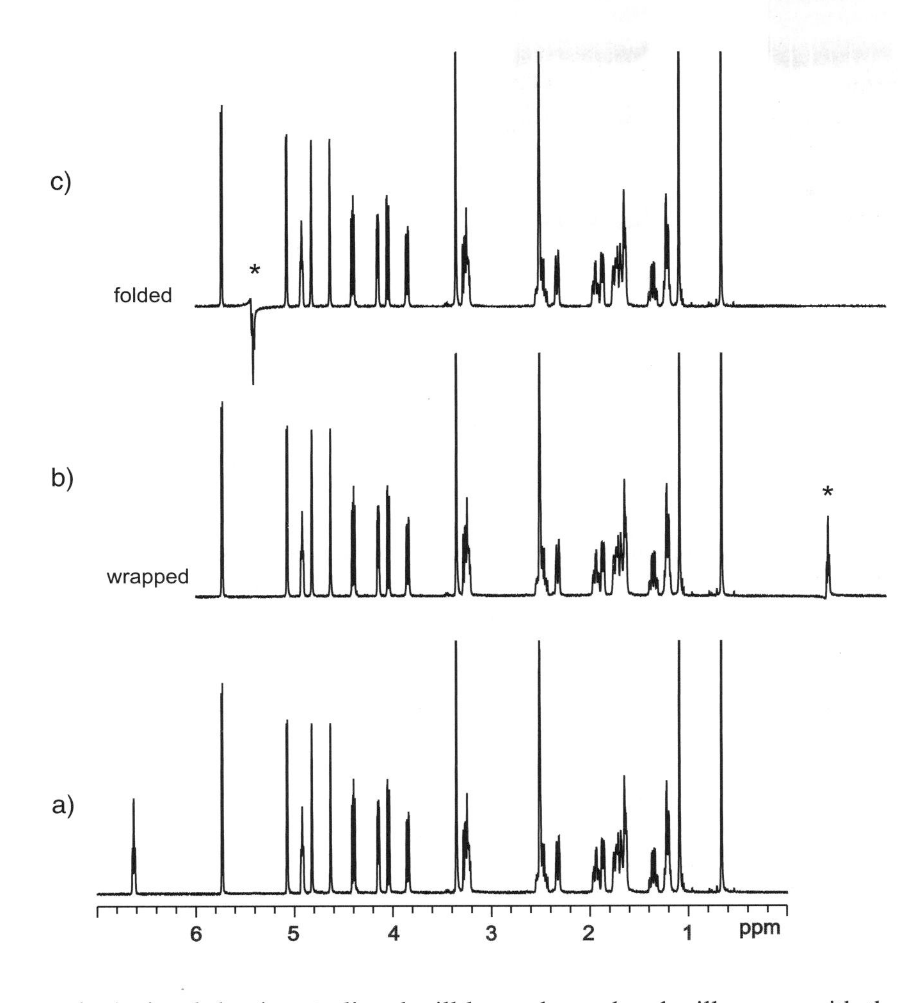

Figure 3.24. Aliasing of resonances. Spectrum (a) displays all resonances at their correct shifts, whilst (b) and (c) result from the spectral window being positioned incorrectly. Spectrum (b) shows how the resonance wraps back into the spectrum at the far end when simultaneous sampling is employed, whereas in (c) it folds back in at the near end with sequential sampling. Typically, the phase of the aliased resonance(s) is also corrupted.

peak. A signal that is not aliased will be unchanged and will appear with the same *absolute* frequency. In contrast, an aliased resonance will move toward the edge of the spectrum and will appear at a different absolute frequency.

3.2.5. Phase cycling

At this point we briefly consider one particular artefact that arises from the use of quadrature detection, not only because of the need to recognise these if and when they occur, but more importantly because one approach to their removal introduces the concept of *phase cycling*, an experimental method that lies at the very heart of almost every NMR experiment. This process involves repeating a pulse sequence with identical timings and pulse tip angles but with judicious changes to the *phases* of the rf pulse(s) and to the routing of the data to the computer memory blocks (often loosely referred to as the receiver phase cycle). The aim in all this is for the desired signals to add coherently with time averaging whereas all other signals, whether from unwanted NMR transitions or from instrumental imperfections, cancel at the end of the cycle and do not appear in the resulting spectrum. We shall see the importance of phase cycles throughout the remainder of the book, particularly in the sections on multipulse one-dimensional and two-dimensional NMR, but as an introduction we return to the idea of quadrature detection.

In the general quadrature scheme one requires two signals to be digitised that differ in phase by 90° but which are otherwise identical. The digitised data from each is then stored in two separate memory blocks, here designated 1 and 2. Experimentally it is rather difficult to ensure the phase difference is *exactly*

Figure 3.25. Quadrature images are unwanted mirror-image artefacts that arise from spectrometer imperfections. Here an image of $CHCl_3$ can be seen at about one-half the height of the carbon satellites in a single scan spectrum. Phase cycling is typically employed to suppress these, although recent hardware designs can eliminate them completely.

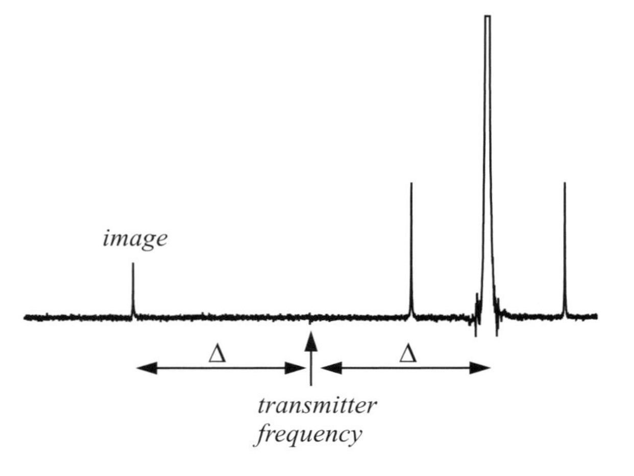

90° and the signal amplitude identical in each channel. Such channel imbalance in phase and amplitude leads to spurious images of resonances mirrored about the transmitter frequency known as *quadrature images* or colloquially as *quad images* (Fig. 3.25). Their origin can be realised if one imagines the signal amplitude in one channel to fall to zero. The scheme is then one of single-channel detection so again positive and negative frequencies cannot be distinguished and thus mirror images appear about the reference (see Fig. 3.19 again). In a correctly balanced receiver these images are usually less than 1% on a single scan, but even these can be significant if you wish to detect weak resonances in the presence of very strong ones, so some means of compensating the imbalance is required.

One solution is to ensure that data from each channel, A and B, contributes equally to the two memory blocks 1 and 2. This can be achieved by performing two experiments, the first with a $90°_x$ pulse the second with $90°_y$ and adding the data. To keep the cosine and sine components in separate memory regions for use in the FT routine, appropriate data routing must also be used (Fig. 3.26a and b). This is generally handled internally by the spectrometer, being defined as the 'receiver phase' from the operators point of view, and taking values of 0, 90, 180 or 270° or x, y, $-x$ and $-y$. Note, however, that the phase of the receiver *reference rf* does not alter on sequential scans, only the data routing is changed. For any pulse NMR sequence it is necessary to define the rf phases *and* the receiver phase to ensure retention of the desired signals and cancellation of artefacts.

By use of the two-step phase cycle it is therefore possible to compensate for the effects of imbalance in the two receiver channels. It is also possible to remove extraneous signals that may occur, such as from DC offsets in the receiver, by simultaneously inverting the phase of the rf pulse and the receiver, thus Fig. 3.26a steps to 26c and likewise 26b becomes 26d. The NMR signals will follow the phase of the pulse so will add in the memory whereas offsets or spurious signals will be independent of this so will cancel. This gives us

Table 3.1. The four-step CYCLOPS phase cycle illustrated in Fig. 3.26

Scan number	Pulse phase	Receiver phase
1	x	x
2	y	y
3	$-x$	$-x$
4	$-y$	$-y$

This shorthand notation is conventionally used to describe all pulse sequence phase cycles.

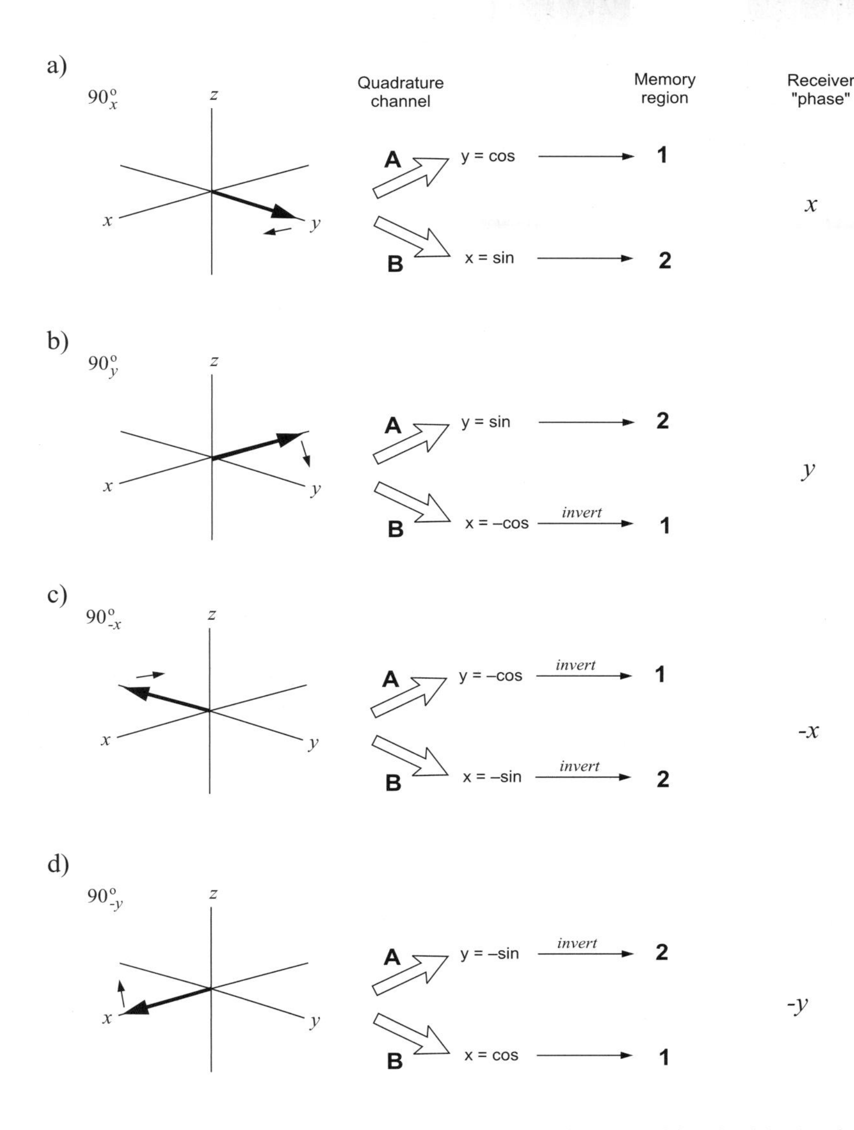

Figure 3.26. Phase cycling. The CYCLOPS scheme cancels unwanted artefacts whilst retaining the desired NMR signals. This four-step phase cycle is explained in the text.

a second possible two-step phase cycle which, when combined with the first, produces an overall four-step cycle known as CYCLOPS [8] (CYCLically Ordered PhaSe cycle, Table 3.1). This is the standard phase cycle used for one-pulse acquisitions on all spectrometers, and is often nested within the phase cycles of two-dimensional experiments again with the aim of removing receiver artefacts.

3.2.6. Dynamic range and signal averaging

In sampling the FID, the analogue-to-digital converter (ADC or digitiser) limits the frequency range one is able to characterise (i.e. the spectral width) according to how fast it can digitise the incoming signal. In addition to limiting the *frequencies*, ADC performance also limits the *amplitudes* of signals that can be measured. The digitisation process converts the electrical NMR signal into a binary number proportional to the magnitude of the signal. This digital value is defined as a series of computer bits, the number of which describes the *ADC resolution*. Typical digitiser resolutions on modern spectrometers operate with 14 or 16 bits. The 16-bit digitiser is able to represent values in the range ±32767 i.e. $2^{15}-1$ with one bit reserved to represent the sign of the signal. The ratio between the largest and smallest detectable value (the most- and least-significant bits), 32767:1, is the *dynamic range* of the digitiser. If we

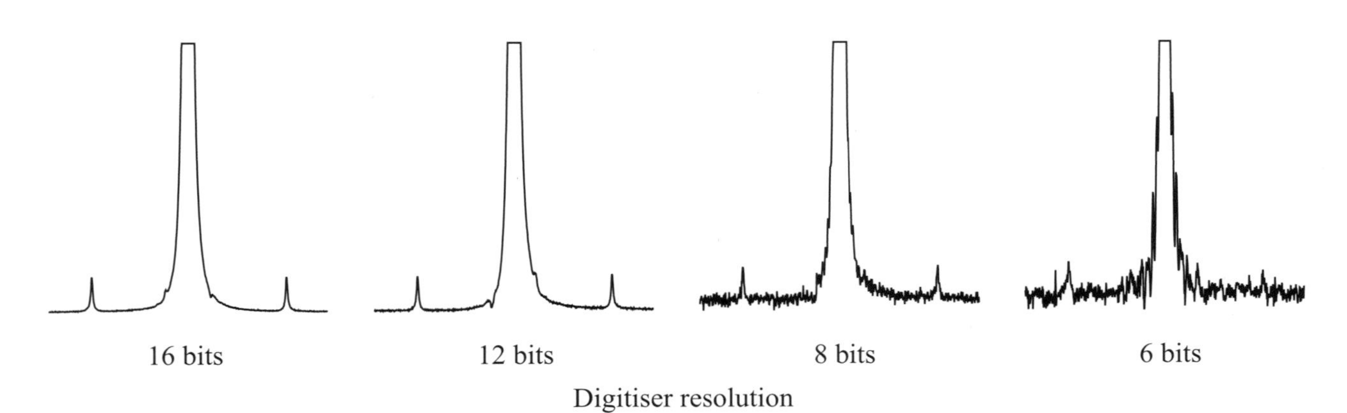

Figure 3.27. Dynamic range and the detection of small signals in the presence of large ones. As the digitiser resolution and hence its dynamic range are reduced, the carbon-13 satellites of the parent proton resonance become masked by noise until they are barely discernible with only 6-bit resolution (all other acquisition parameters were identical for each spectrum). The increased noise is *digitisation* or *quantisation noise* (see text below).

assume the receiver amplification (or *gain*) is set such that the largest signal in the FID on each scan fills the digitiser, then the smallest signal that can be recorded has the value 1. Any signal whose amplitude is less than this will not trigger the ADC; the available dynamic range is insufficient. However, noise will also contribute to the detected signal and this may be sufficiently intense to trigger the least significant bit of the digitiser. In this case the small NMR signal *will* be recorded as it rides on the noise and signal averaging therefore leads to summation of this weak signal, meaning even those whose amplitude is below that of the noise may still be detected. However, the digitiser may still limit the detection of smaller signals in the presence of very large ones when the signal-to-noise ratio is high. Fig. 3.27 illustrates how a reduction in the available dynamic range limits the observation of smaller signals when thermal noise in the spectrum is low. This situation is most commonly encountered in proton studies, particularly of protonated aqueous solutions where the water resonance may be many thousand times that of the solute. Such intensity differences impede solute signal detection, so many procedures have been developed to selectively reduce the intensity of the H_2O resonance and ease the dynamic range requirements; some of these *solvent suppression* schemes are described in Chapter 9.

If any signal is so large that it cannot fit into the greatest possible value the ADC can record, its intensity will not be measured correctly and this results in severe distortion of the spectrum (Fig. 3.28a). The effect can be recognised in

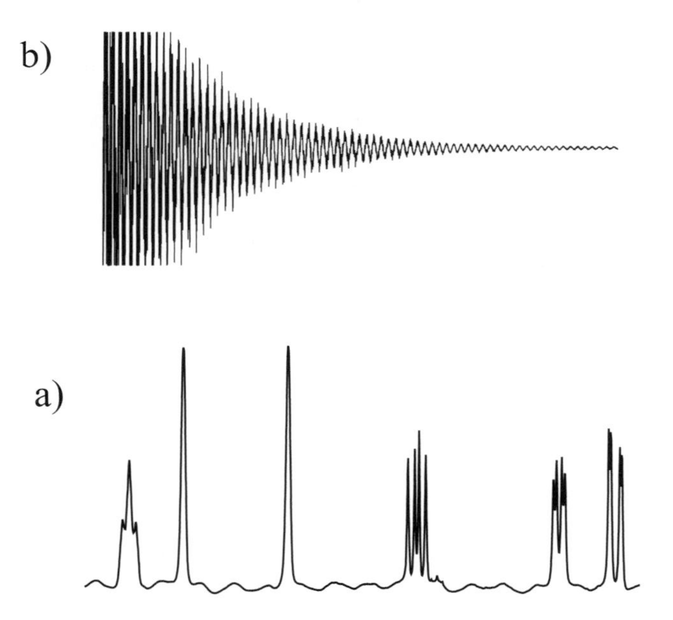

Figure 3.28. Receiver or digitiser overload distorts the spectrum baseline (a). This can also be recognised as a 'clipping' of the FID (b).

extra 3 bits. The degree of oversampling that can be achieved is limited by the digitisation speed of the ADC and typical values of N_{os} for proton observation are 16–32, meaning spectral widths become a few hundred kHz and the ADC resolution may be increased by 2–3 bits in favourable circumstances (when thermal noise can be considered insignificant relative to digitisation noise).

To maintain the desired digital resolution in the spectral region of interest when oversampling, data sets would have to be enlarged according to the oversampling factor also, which would demand greater storage capacity and slower data processing. To overcome these limitations, modern spectrometers generally combine oversampling with digital signal processing methods. Since one is really only interested in a relatively small part of our oversampled data set ($1/N_{os}$ of it) the FID is reduced after digitisation to the conventional number of data points by the process of *decimation* (literally 'removing one-tenth of') prior to storage. This, in effect, takes a running average of the oversampled data points, leaving one point for every N_{os} sampled, at intervals defined by the Nyquist condition for the desired spectral window (the peculiar distortions seen at the beginning of a digitally processed FID arise from this decimation process). The resulting FID then has the same number of data points as it would have if it had been sampled normally. These digital signal processing steps are generally performed with a dedicated processor after the ADC (Fig. 3.9, box b) and are typically left invisible to the user (short of setting a few software flags perhaps) although they can also be achieved by separate post-processing of the original data or even included in the FT routine itself [10]. The use of fast dedicated processors means the calculations can readily be achieved as the data are acquired thus not limit data collection.

One further advantage of digital processing of the FID is the ability to mathematically define frequency filters that have a far steeper and more complete cut-off than can be achieved by analogue filtration alone, meaning signal aliasing can be eliminated. Spectral widths may then be set to encompass only a sub-section of the whole spectrum, allowing *selective detection* of NMR resonances [11]. The ability to reduce spectral windows in this way without complications from signal aliasing has considerable benefit when acquiring two-dimensional NMR data in particular. The principle behind the filtration is as follows (Fig. 3.33). The digital time domain filter function is given by the inverse FT of the desired frequency domain window. This function may then be convoluted with the digitised FID such that, following Fourier transformation, only those signals that fell within the originally defined window remain in the spectrum. One would usually like this window to be rectangular in shape although in practice such a sharp cut-off profile cannot be achieved without introducing distortions. Various alternative functions have been used which

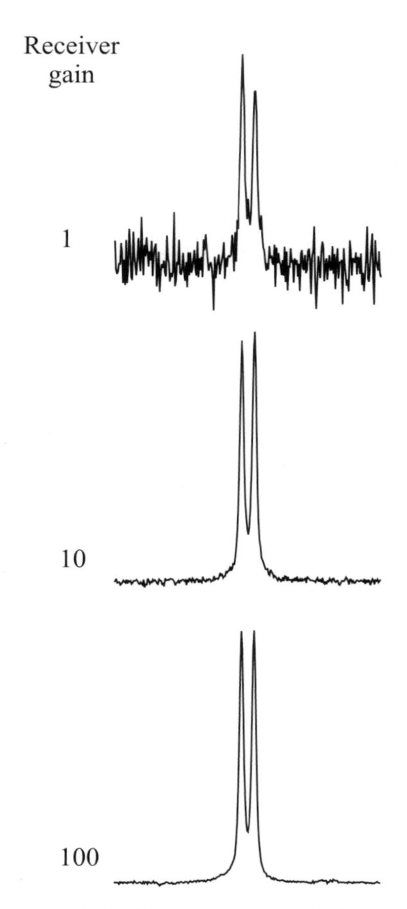

Figure 3.31. Digitisation noise. At high receiver gain settings the noise in this proton spectrum is vanishingly small, meaning system thermal noise is low. As the gain is reduced, the amplitude of the NMR signal (and thermal noise) is reduced and digitisation noise becomes significant relative to the NMR signal.

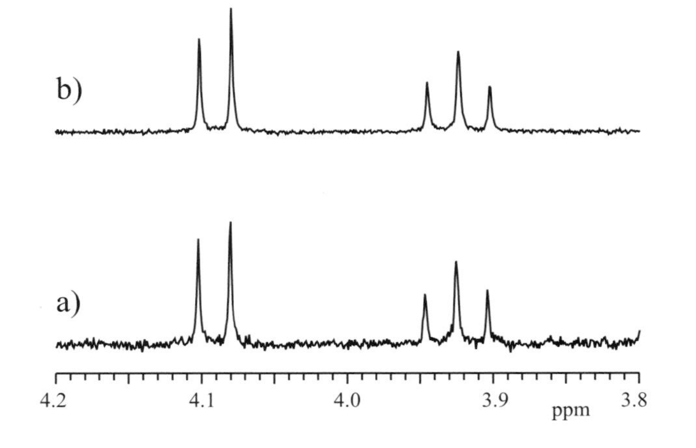

Figure 3.32. Enhanced sensitivity can be realised in favourable cases by oversampling the data and hence reducing digitisation noise. Spectrum (a) shows part of a conventional proton spectrum sampled according to the Nyquist criterion. Oversampling the data by a factor of 24 as in (b) provides a sensitivity gain (all other conditions as for (a)).

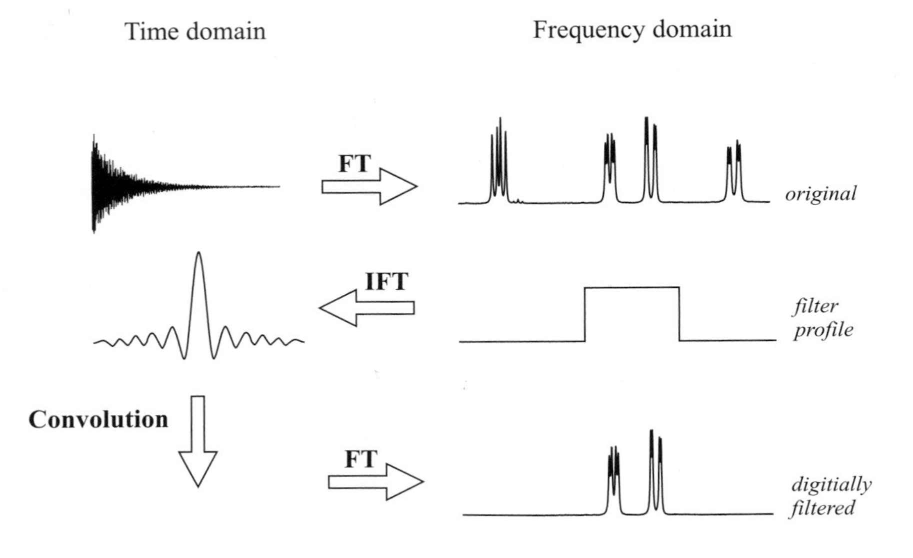

Figure 3.33. The use of a digital filter to observe a selected region of a spectrum. The desired frequency window profile is subject to an inverse FT and the resulting time domain function convoluted with the raw FID. Transformation of the modified data produces a spectrum containing only a subset of all resonances as defined by the digital filter.

approach this ideal, generally with the property that the more coefficients used in the function, the steeper is their frequency cut-off. Since these operate only *after* the ADC, the analogue filter (Fig. 3.9, box a) is still required to reject broadband noise from outside the oversampled spectral width, but its frequency cut-off is now far removed from the normal spectral window and its performance less critical.

3.2.7. Window functions

There exist various ways in which the acquired data can be manipulated prior to Fourier transformation in an attempt to enhance its appearance or information content. Most commonly one would either like to improve the signal-to-noise ratio of the spectrum to help reveal resonances above the noise level, or to improve the resolution of the spectrum to reveal hidden fine structure. Many mathematical weighting functions, known as *window functions*, have been proposed to achieve the desired result, but a relatively small number have come into widespread use, some of which are illustrated in Fig. 3.34. The general philosophy behind all these functions is the same. In any FID the NMR signal intensity declines throughout the acquisition time whereas the noise amplitude remains constant, meaning the relative noise level is greater in the latter part. Decreasing the tail of the FID will therefore help reduce the noise amplitude in the transformed spectrum, so enhancing *sensitivity*. Conversely, increasing the middle and latter part of the FID equates to retarding the decay of the signal, thus narrowing the resonance and so enhancing *resolution*.

One word of warning here before proceeding. The application of window functions has also been referred to in the past as 'digital processing' or 'digital filtering' of the data. However, this should not be confused with the digital signal processing terminology introduced in previous sections and in widespread use nowadays. The adoption of the terms 'window' or 'weighting' function is therefore recommended in the context of sensitivity or resolution enhancement.

Sensitivity enhancement

As suggested above, the noise amplitude in a spectrum can be attenuated by de-emphasising the latter part of the FID, and the most common procedure for achieving this is to multiply the raw data by a decaying exponential function (Fig. 3.34a). This process is therefore also referred to as *exponential multiplication*. Because this forces the tail of the FID towards zero, it is also suitable for the apodisation of truncated data sets. However, it also increases the apparent decay rate of the NMR signal, causing lines to broaden, meaning one compromises resolution for the gain in sensitivity (Fig. 3.35). Using too strong a function, that is one that causes the signal to decay too rapidly, can actually lead to a decrease in signal-to-noise ratio in the resulting spectrum because the broadening of the lines causes a reduction of their peak heights. Spectrometer software usually allows one to define the amount of line-broadening directly in hertz (here parameter lb) so exponential multiplication is rather straightforward to use and allows the chemist to experiment with different degrees of broadening to attain a suitable result. The optimum balance between reducing noise and excessive line broadening is reached when the decay of the window function matches the natural decay of the NMR signal, in which case it is known as the *matched filter* [12] which results in a doubling of the resonance linewidth. For the exponential function to truly match the FID, the decay of the NMR signal must also be an exponential (meaning the NMR resonances have a Lorentzian lineshape) which on a correctly adjusted spectrometer is usually assumed to be the case (at least for spin-$^1/_2$ nuclei). Despite providing the maximum gain in sensitivity, the matched condition is often not well suited to routine use in proton spectroscopy as the resulting line-broadening and loss of resolution may preclude the separation of closely space lines, so less line-broadening than this may be more appropriate. Furthermore, different resonances in the spectrum often display different unweighted linewidths so the matched condition cannot be met for all resonances simultaneously. For routine proton work it turns out to be convenient to broaden the line by an amount equal to the digital resolution in the spectrum, as this leads to some sensitivity enhancement but to a minimal increase in linewidth. For heteronuclear spectra resolution is usually a lesser concern and line-broadening of a few hertz is commonly employed (comparable to or slightly greater than the digital resolution). This can be considered essential to attenuate the distortions about the base of resonances arising, in part, from truncation of the data set (Section 3.2.3 and Fig. 3.17). For spectra that display resonances that are tens or even hundreds of hertz wide (most notably those of quadrupolar nuclei) the amount of line-broadening must be increased accordingly to achieve any appreciable sensitivity gain.

Resolution enhancement

One might suppose that to improve the resolution in spectra we should apply a function that enhances the latter part of the FID, so increasing the decay time. The problem with this approach is that by doing this one also increases the noise amplitude in the tail of the FID and emphasises any truncation of the signal that may be present, so increasing the potential for undesirable 'truncation wiggles'. A better approach is to apply a function which initially counteracts the early decay but then forces the tail of the FID to zero to provide the necessary apodisation. The most popular function for achieving this has been the *Lorentz–Gauss transformation* [13] (Fig. 3.34 b and c), sometimes loosely referred to as *Gaussian multiplication* (although this strictly refers to yet another mathematical weighting function) and also known as double-exponential multiplication. This transforms the usual Lorentzian lineshape into a Gaussian lineshape which has a somewhat narrower profile, especially around the base, (Fig. 3.36) and it is this feature that allows the resolution of closely spaced lines.

The shape of the function is altered by two variable parameters which define the degree of line-narrowing in the resulting spectra and the point during the acquisition time at which the function reaches its maximum value. These are usually presented to the operator as negative line-broadening in hertz (here parameter lb) and as a fraction of the total acquisition time (here parameter gb), respectively. The choice of suitable values for these usually comes down

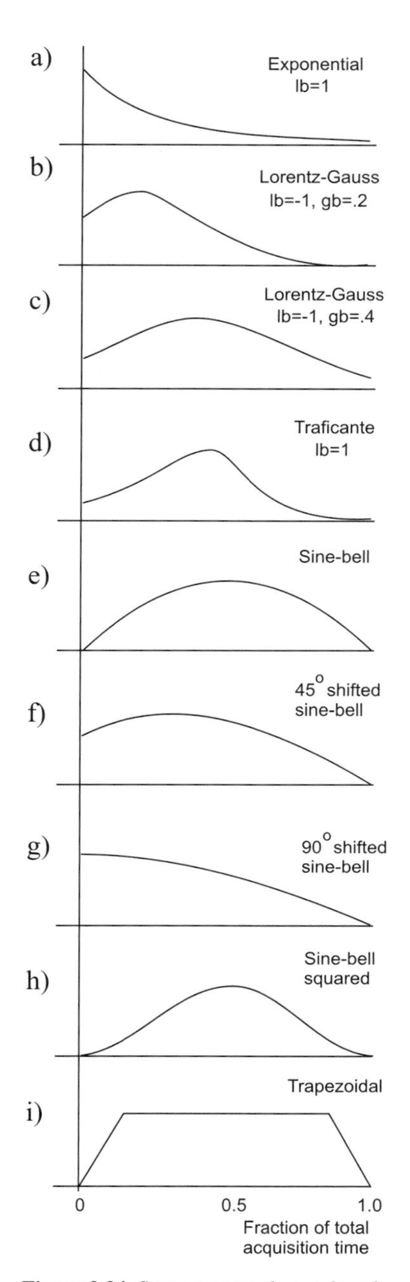

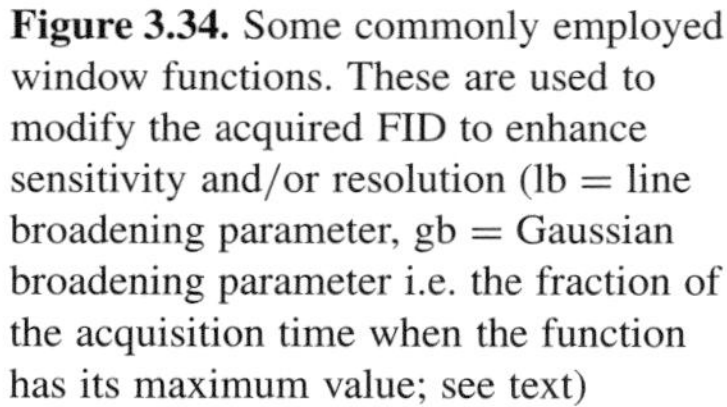

Figure 3.34. Some commonly employed window functions. These are used to modify the acquired FID to enhance sensitivity and/or resolution (lb = line broadening parameter, gb = Gaussian broadening parameter i.e. the fraction of the acquisition time when the function has its maximum value; see text)

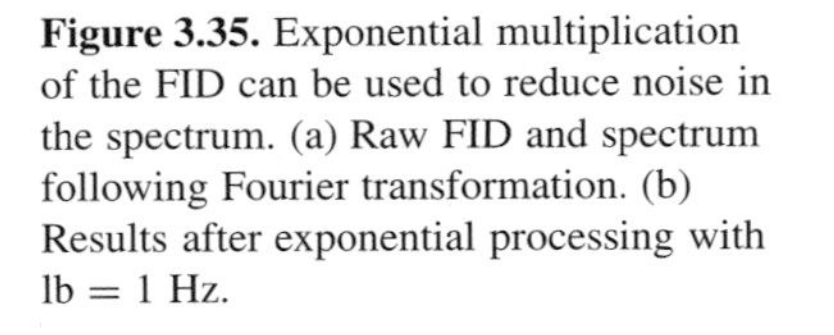
Figure 3.35. Exponential multiplication of the FID can be used to reduce noise in the spectrum. (a) Raw FID and spectrum following Fourier transformation. (b) Results after exponential processing with lb = 1 Hz.

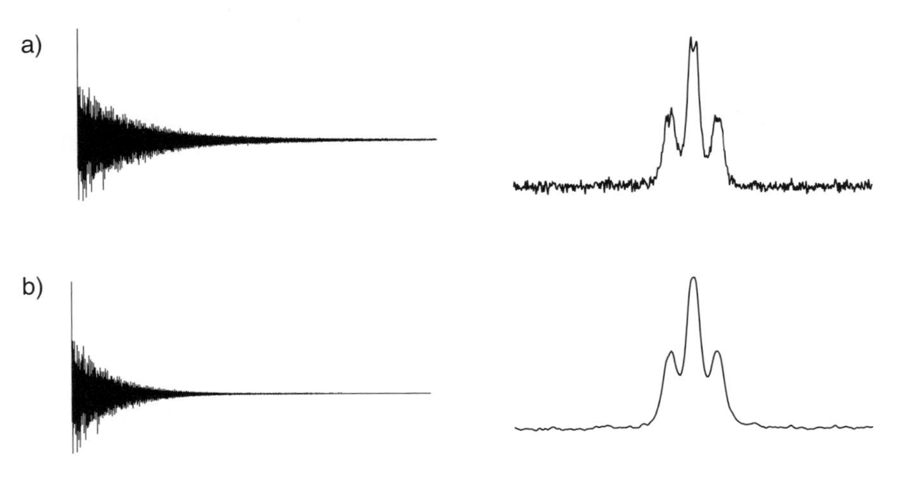

to a case of trial-and-error, and different optimum values may be found for different groups of peaks within the same spectrum. Modern NMR processing packages usually allow an interactive variation of the parameters during which one can observe both the window function itself and the resulting spectrum so an optimum can rapidly be determined. A more negative line-broadening will produce narrower lines whilst positioning the maximum further along the decay will enhance this effect. This will also lead to greater distortion of the resonances about the base and to degradation of sensitivity as the early signals are sacrificed relative to the noise (Fig. 3.37). In any case, some reduction in sensitivity will invariably result and it is necessary to have reasonable signal-to-noise for acceptable results. It has been suggested [14] that the optimum Gaussian resolution enhancement for routine use on spectra with narrow lines aims for a reduction in linewidth by a factor of 0.66 for which the function maximum should occur at a time of $1/(\Delta\nu_{1/2})$ seconds, that is, the inverse of the resonance linewidth in the absence of window functions. The appropriate figures for setting in the processing software can then be arrived at by simple arithmetic. For example, a starting linewidth of 1 Hz will require a line narrowing of 0.66 Hz and the maximum to occur after 1 second. With a typical proton acquisition time of around 3 s, the function should therefore reach a maximum at 1/3 of the total acquisition. Gaussian multiplication can be used to good effect when combined with zero-filling, thus ensuring the digital resolution is sufficiently fine to define any newly resolved fine-structure.

In more recent years new window functions have been introduced [15,16] that are similar to the Lorentz–Gauss window but which aim to improve resolution without a discernible reduction in sensitivity. These so called *TRAF*

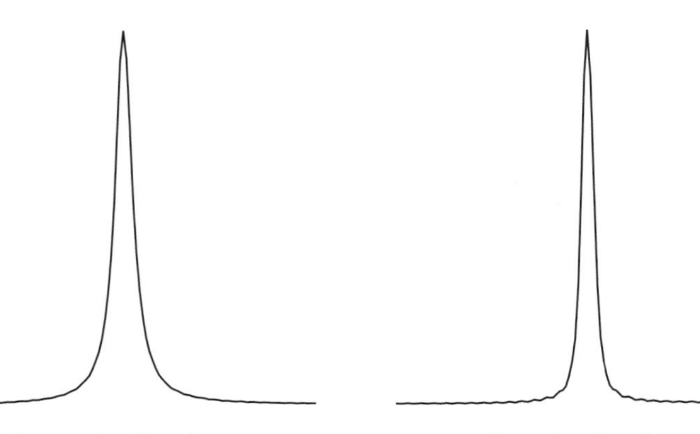

Figure 3.36. A comparison of the Lorentzian and Gaussian lineshapes.

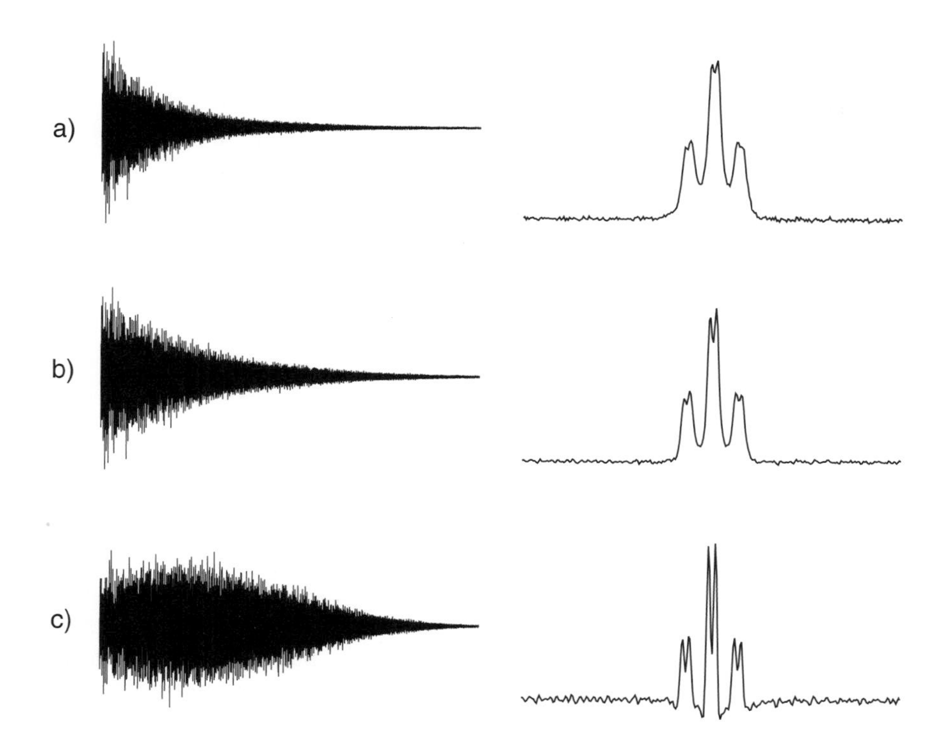

Figure 3.37. The Lorentz–Gauss transformation ('Gaussian multiplication') can be used to improve resolution. (a) Raw FID and spectrum following Fourier transformation and results after the L–G transformation with (b) lb = −1Hz, gb = 0.2 and (c) lb = −3 Hz and gb = 0.2.

functions (Fig. 3.34d) generally aim to enhance the middle part of the FID but also use matched filtration of the later part to attenuate noise. They are appearing in NMR software packages and provide an alternative to the established functions. Another function, more commonly used in the processing of two-dimensional data sets, is the *sine-bell* window, which has no adjustable parameters. This comprises one half of a sine wave, starting at zero at the beginning of the FID, reaching a maximum half way through the acquisition and falling back to zero by the end of the decay (Fig. 3.34e). The function always has zero intensity at the end, so eliminates the truncation often encountered in 2D data sets. This tends to be a rather severe resolution enhancement function which can introduce undesirable lineshape distortions and also produces severe degradation in sensitivity because the early part of the NMR signal is heavily attenuated, and hence this is rarely used in one-dimensional spectra. A variation on this is the *phase-shifted sine-bell* for which the point at which the maximum occurs is a variable (Fig. 3.34f and g) and can be moved toward the start of the FID. Again, this is commonly applied in 2D processing and has the advantage that the user has the opportunity to balance the gain in resolution against the lineshape distortion and sensitivity degradation. A final variant is the (optionally shifted) *squared sine-bell* (Fig. 3.34h), which has similar properties to the sine-bell but tails more gently at the edges, which can invoke subtle differences in 2D spectra. Likewise, the *trapezoidal* function (Fig. 3.34i) is sometimes used in the processing of 2D data.

3.2.8. Phase correction

It has already been mentioned in Section 3.2 that the phase of a spectrum needs correcting following Fourier transformation because the receiver reference phase does not exactly match the initial phase of the magnetisation vectors. This error is constant for all vectors and since it is independent of resonance frequencies it is referred to as the 'zero-order' phase correction (Fig. 3.38). Practical limitations also impose the need for a frequency-dependent or 'first-order' phase correction. Consider events immediately after the

Figure 3.38. Zero-order (frequency independent) phase errors arise when the phase of the detected NMR signals does not match the phase of the receiver reference rf. All resonances in the spectrum are affected to the same extent.

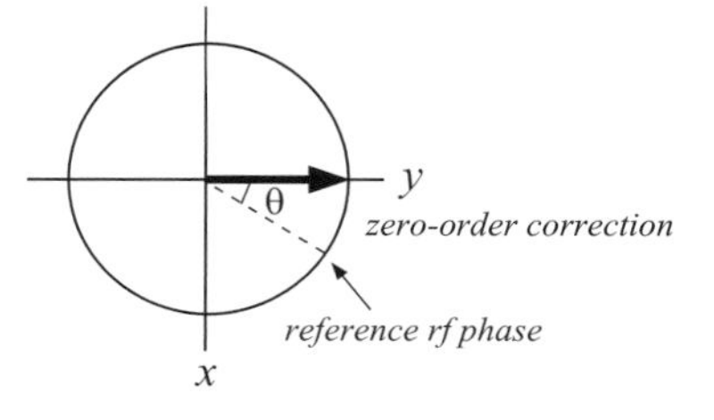

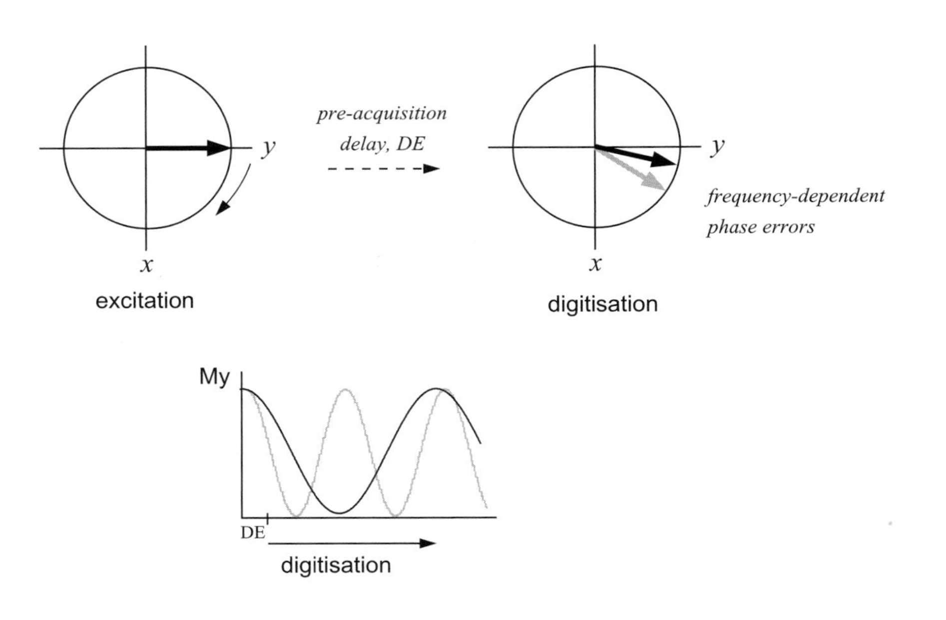

Figure 3.39. First-order (frequency dependent) phase errors arise from a dephasing of magnetisation vectors during the pre-acquisition delay which follows the excitation pulse. When data collection begins, vectors with different frequencies have developed a significant phase difference which varies across the spectrum.

pulse is applied to the sample. A short period of time, the pre-acquisition delay, DE, is required for the spectrometer electronics to recover from the effect of the pulse before an undistorted FID can be collected. This delay is typically tens of microseconds, during which magnetisation vectors will evolve a little according to their chemical shifts so that at the point digitisation begins they no longer have the same phase (Fig. 3.39). Clearly those resonances with the greatest shifts require the largest corrections. If the pre-acquisition delay is small relative to the frequency offsets, the phase errors have an approximately linear offset dependency and can be removed. If the delay becomes large, the correction cannot be made without introducing a rolling spectrum baseline. The appearance of these zero- and first-order phase errors is illustrated in the spectra of Fig. 3.40.

Typically both forms of error occur in a spectrum directly after the FT. The procedure for phase correction is essentially the same on all spectrometers. The zero-order correction is used to adjust the phase of one signal in the spectrum to pure absorption mode, as judged 'by eye' and the first-order correction is then

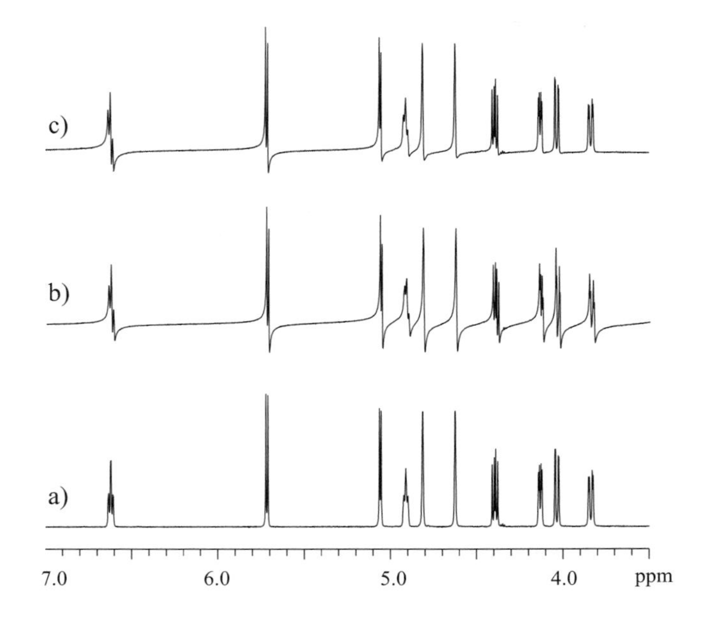

Figure 3.40. Zero- and first-order phase errors in a ^{1}H spectrum. (a) The correctly phased spectrum, (b) the spectrum in the presence of frequency-independent (zero-order) phase errors and (c) the spectrum in the presence of frequency-dependent (first-order) phase errors.

used to adjust the phase of a signal far away from the first in a similar manner. Ideally the two chosen resonances should be as far apart in the spectrum as possible to maximise the frequency dependent effect.

3.3. PREPARING THE SAMPLE

This section describes some of the most important aspects of sample preparation for NMR analysis, a topic that is all too frequently given insufficient consideration in the research laboratory. Even for routine applications of NMR it is wise for the chemist to adopt a systematic strategy for sample preparation that will give consistently good results, so saving instrument time and eliminating frustration.

3.3.1. Selecting the solvent

Since all modern NMR spectrometers rely on a deuterium lock system to maintain field stability, a deuterated solvent is invariably required for NMR, and these are now available from many commercial suppliers. In a busy organic chemistry laboratory the volume of NMR solvents used can be surprisingly large, and the cost of the solvent may play some part in its selection. Generally, chloroform and water are the most widely used solvents and are amongst the cheapest, although DMSO is becoming increasingly favoured, notably in the industrial sector, because of its useful solubilising properties. Acetone, methanol and dichloromethane are also in widespread routine use.

The selection of a suitable solvent for the analyte is based on a number of criteria. The most obvious requirement is that the analyte be soluble in it *at the concentration required for the study*. This will be dependent on a diverse range of factors including the fundamental sensitivity of the nucleus, the overall sensitivity of the instrument, the nature of the experiment and so on. If experiments are likely to be performed at very low temperatures then it is also important to know that the solute remains in solution when the temperature is reduced so that a precipitate does not form in the NMR tube, and degrade resolution. For experiments at temperatures other than ambient probe temperatures (which may be a few degrees greater than the ambient room temperature), the melting and boiling points of the solvents must also be considered. Table 3.2 summarises some important properties of the most commonly encountered solvents. For work at very high temperatures, dimethylsulphoxide or toluene are generally the solvents of choice, whilst for very low temperatures dichloromethane, methanol or tetrahydrofuran are most appropriate. Even for experiments performed at ambient temperatures solvent melting points may be limiting; on a cold day DMSO solutions can freeze in the laboratory. The viscosity of the solvent will also influence the resolution that can be obtained with the best performance provided by the least viscous solvents, such as acetone (which, for this reason, is used as the solvent for spectrometer resolution tests). The sharpest possible lock signal is also beneficial in experiments based on difference spectroscopy, the most important of these being the NOE difference experiment (Chapter 8).

For proton and carbon spectroscopy, the chemical shifts of the solvent resonances need to be anticipated as these may occur in a particularly unfortunate place and interfere with resonances of interest. In proton spectroscopy, the observed solvent resonance arises from the residual *protonated* species (NMR solvents are typically supplied with deuteration levels in excess of 99.5%). For routine studies, where a few milligrams of material may be available, the lower specification solvents should suffice for which the residual protonated resonance is often comparable in magnitude to that of the solute. Solvents with

Table 3.2. Properties of the common deuterated solvents

Solvent	δ_H (ppm)	$\delta_{(HDO)}$ (ppm)	δ_C (ppm)	Melting point (°C)	Boiling point (°C)
Acetic acid-d_4	11.65, 2.04	11.5	179.0, 20.0	16	116
Acetone-d_6	2.05	2.0	206.7, 29.9	−94	57
Acetonitrile-d_3	1.94	2.1	118.7, 1.4	−45	82
Benzene-d_6	7.16	0.4	128.4	5	80
Chloroform-d_1	7.27	1.5	77.2	−64	62
Deuterium oxide-d_2	4.80	4.8	–	4	101
Dichloromethane-d_2	5.32	1.5	54.0	−95	40
N,N-dimethyl formamide-d_7	8.03, 2.92, 2.75	3.5	163.2, 34.9, 29.8	−61	153
Dimethylsulfoxide-d_6	2.50	3.3	39.5	18	189
Methanol-d_4	4.87, 3.31	4.9	49.2	−98	65
Pyridine-d_5	8.74, 7.58, 7.22	5.0	150.4, 135.9, 123.9	−42	114
Tetrahydrofuran-d_8	3.58, 1.73	2.4	67.6, 25.4	−109	66
Toluene-d_8	7.09, 7.00, 6.98, 2.09	0.4	137.9, 129.2, 128.3, 125.5, 20.4	−95	111
Trifluoroacetic acid-d_1	11.30	11.5	164.2, 116.6	−15	75
Trifluoroethanol-d_3	5.02, 3.88	5.0	126.3, 61.5	−44	75

Proton shifts, δH, and carbon shifts, δC, are quoted relative to TMS (proton shifts are those of the residual partially ***protonated*** solvent). The proton shifts of residual HDO/H_2O vary depending on solution conditions.

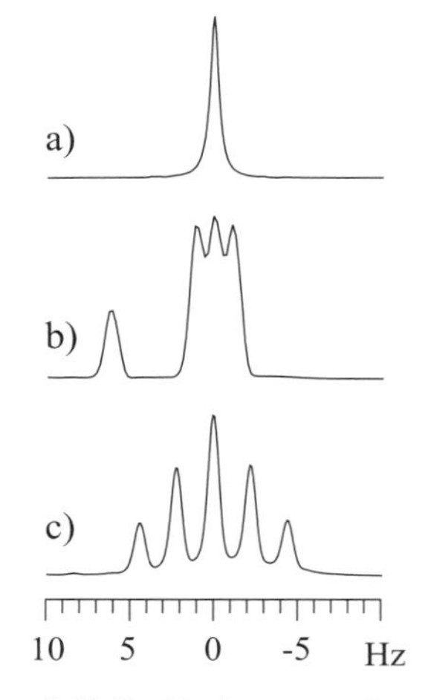

Figure 3.41. Residual protonated resonances of deuterated solvents (a) $CHCl_3$ in $CDCl_3$, (b) $CHDCl_2$ in CD_2Cl_2 and (c) CHD_2COCD_3 in $(CD_3)_2CO$. The multiplicity seen in (b) and (c) arises from two-bond (geminal) couplings to spin−1 deuterium producing a 1:1:1 triplet and a 1:2:3:2:1 quintet respectively. The left-hand singlet in (b) is residual CH_2Cl_2 in the solvent, the shift difference arises from the H–D isotope shift of 6 Hz.

higher levels of deuteration are beneficial for proton spectroscopy when sample quantities are of the order of tens of micrograms or less and are likely to be used in conjunction with micro-sample techniques (Section 3.3.3). The solvent proton resonance, with the exception of chloroform and water, will comprise a multiplet from coupling to spin−1 deuterium. For example dichloromethane displays the triplet of $CDHCl_2$ whilst acetone, dimethylsulphoxide or methanol show a quintuplet from CD_2H (Fig. 3.41). In carbon spectroscopy, the dominant resonance is often that of the deuterated solvent which again will be a multiplet owing to coupling with deuterium. For other common nuclei, interference from the solvent is rarely a consideration, the one notable exception to this being deuterium itself. Unless very large sample quantities are being used (many tens of milligrams) and one can be sure the solvent resonance will not overlap those of the solute, it is usually necessary to record ^2H spectra in *protonated* solvent since the huge signals of the deuterated solvent are likely to swamp those of interest. This then precludes the use of the lock system for maintaining field stability. The lack of a suitable lock reference requires a small quantity of the deuterated solvent to be added to the solution to provide an internal chemical shift reference (see below).

Aside from the solvent resonance itself, the other significant interference seen in proton spectra is water, which is present to varying degrees in all solvents and is also often associated with solutes. The water resonance is often rather broad and its chemical shift can vary according to solution conditions. All solvents are hygroscopic to some degree (including deuterated water) and should be exposed to the atmosphere as little as possible to prevent them from becoming wet. Very hygroscopic solvents such as dimethylsulphoxide, methanol, and water are best kept under a dry inert-gas atmosphere. Solvents can also be obtained in smaller glass ampoules which are particularly suitable for those which are used only infrequently. Molecular sieves may also be used to keep solvents dry, although some care is required in filtering the prepared solution before analysis to remove fine particulates that may be present. Alternatively the NMR solution may be passed through activated alumina to remove water, as part of the filtration process.

The solvent may also result in the loss of the resonances of exchangeable protons since these will become replaced by deuterons. In particular this is likely to occur with deuterated water and methanol. To avoid deuterium

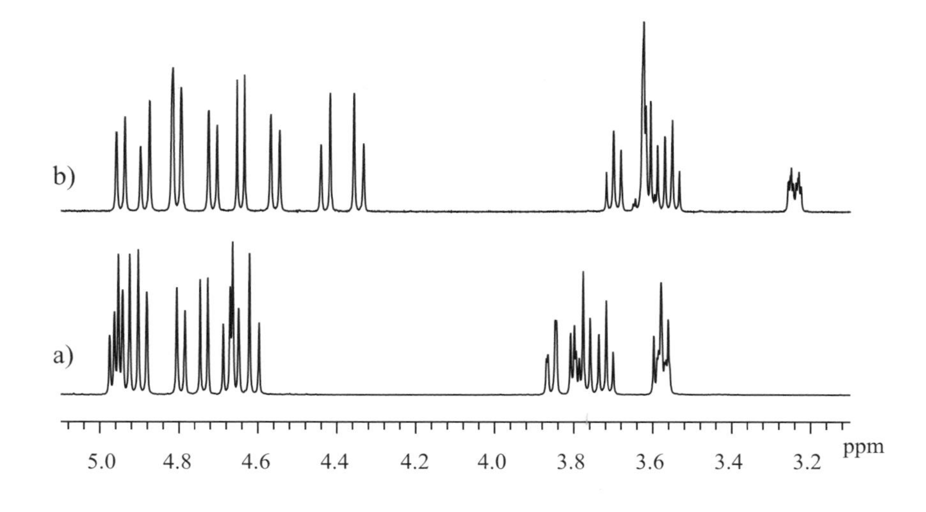

Figure 3.42. Changes in solvent can be used to improve resonance dispersion. The proton spectrum of the sugar **3.1** is shown in (a) $CDCl_3$ and (b) C_6D_6. Notice the appearance of the resonance at 3.2 ppm in (b) that was hidden at 3.6 ppm by another resonance in (a).

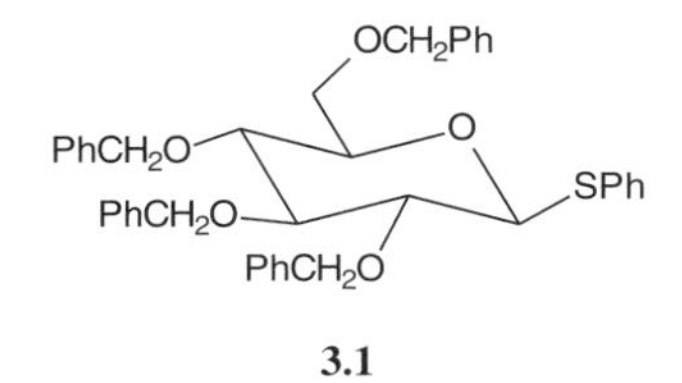

3.1

replacement altogether one must consider using H_2O containing 10–20% D_2O (to provide a lock signal) or d_3-methanol (CD_3OH). In these cases, a suitable solvent suppression scheme (Chapter 9) is required to attenuate the large solvent resonance (which may still lead to the loss of the exchangeable protons of the solute by the process known as saturation transfer). Studies in 90% H_2O dominate the NMR of biological macromolecules because solvent exchangeable protons, such as the backbone amide NH protons in peptides and proteins or the imido protons of DNA and RNA base-pairs, often play a key role in the structure determination of these molecules [17,18].

The nature of the solvent can also have a significant influence on the appearance of the spectrum, and substantial changes can sometimes be observed on changing solvents. Whether these changes are beneficial is usually difficult to predict and a degree of trial- and error is required. A useful switch of solvent is from a non-aromatic solvent into an aromatic one, for example chloroform to benzene (Fig. 3.42), making use of the magnetic anisotropy exhibited by the latter. In cases of particular difficulty, where the change from one solvent to another simply produces a different but equally unsuitable spectrum, then titration of the second solvent into the first may provide a suitable compromise between the two extremes. Such changes may prove useful in 1D experiments that use selective excitation of a specific resonance, by revealing a target proton when previously it was hidden, as in the example in Fig. 3.42. The selective removal of a resonance by the addition of another solvent can also prove useful in spectrum interpretation. Adding a drop of D_2O to an organic solution in the NMR tube, mixing thoroughly and leaving the mix to settle removes (or attenuates) acidic exchangeable protons. Acidic protons that are protected from the solvent, such as those in hydrogen bonds, may not fully exchange immediately, but may require many hours to disappear. This can provide a useful probe of H-bonding interactions.

3.3.2. Reference compounds

When preparing a sample it is common practice to add a suitable compound to act as an *internal* chemical shift reference in the spectrum, and the selection of this must be suitable for the analyte and solvent. In proton and carbon NMR, the reference used in organic solvents is tetramethylsilane (TMS, 0.0 ppm) which has a number of favourable properties; it has a sharp 12-proton singlet resonance that falls conveniently to one end of the spectrum, it is volatile so can be readily removed and it is chemically inert. In a few cases this material may be unsuitable such as in the study of silanes or cyclopropanes. For routine

Table 3.3. Spectrum reference materials (0 ppm) and reference frequencies (Ξ values) for selected nuclides

Nuclide	Primary reference	Ξ value (MHz)	Alternative reference
^{1}H	Me_4Si	100.000 000	TSP-d_4 (aq)
^{2}H	Me_4Si	15.351	trace deuterated solvent
^{6}Li	LiCl (aq)	14.717	
^{7}Li	LiCl (aq)	38.866	
^{10}B	$BF_3O(Et)_2$	10.746	H_2BO_3 (aq)
^{11}B	$BF_3O(Et)_2$	32.089	H_2BO_3 (aq)
^{13}C	Me_4Si	25.145 004	1,4-dioxan @ 67.5 ppm, TSP-d_4 (aq)
^{14}N	CH_3NO_2	7.224	
^{15}N	CH_3NO_2	10.136 767	NH_4NO_2 (aq), NH_3 (liq) [a]
^{17}O	H_2O	13.557	
^{19}F	$CFCl_3$	94.094 003	
^{29}Si	Me_4Si	19.867 184	
^{31}P	H_3PO_4 (85%)	40.480 747	
^{119}Sn	Me_4Sn	37.290 665	
^{195}Pt	$K_2[Pt(CN)_6]$	21.414 376	
^{207}Pb	Me_4Pb	20.920 597	

[a] ^{15}N shifts of biomolecular materials are more often quoted relative to external liquid ammonia.

work it is often not necessary to add any internal reference as the residual lines of the solvent itself can serve this purpose (Table 3.2). For aqueous solutions, the water soluble equivalent of TMS is partially deuterated sodium 3-(trimethylsilyl)propionate-d_4 (TSP-d_4) which is also referenced to 0.0 ppm. A volatile alternative is 1,4-dioxane (^{1}H 3.75 ppm, ^{13}C 67.5 ppm) which can be removed by lyophilisation. The standard reference materials for some other common nuclides are summarised in Table 3.3. Often these are not added into the solution being studied but are held in an outer, concentric jacket or within a separate axial capillary inside the solution, in which case the reference material may also be in a different solvent to that of the sample.

An alternative to adding additional reference materials is to use a so-called external reference. Here the spectrum of a separate reference substance is acquired before and/or after the sample of interest and the spectrum reference value carried over. Identical field settings should be used for both which, on some older instruments, requires the same lock solvent, or an additional correction to the spectrum reference frequency must be used to compensate any differences. This restriction does not arise on instruments that use shifting of the lock transmitter frequency to establish the lock condition.

The referencing of more 'exotic' nuclei is generally less clear-cut than for those in common use and in many cases it is impractical to add reference materials to precious samples, and it is sometimes even difficult to identify what substance is the 'accepted' reference standard. In such cases the Ξ-scale can be used, which does not require use of a specific reference material. Instead this scheme defines the *reference frequency* for the reference material of each nuclide at a field strength at which the proton signal of TMS resonates at exactly 100.000 MHz. The reference frequencies are scaled appropriately for the magnetic field in use and this then defines the absolute frequency at 0.0 ppm for the nuclide in question. The Ξ values for selected nuclei are also summarised in Table 3.3, whilst more extensive tables are available [19].

3.3.3. Tubes and sample volumes

Newcomers to the world of practical NMR often find the cost of NMR tubes surprisingly high. The prices reflect the need to produce tubes that conform to strict requirements of straightness (camber) and concentricity, as deviation from these can produce undesirable artefacts in spectra, usually in the form of 'spinning-sidebands'. Generally, the higher quality (and more expensive) tubes are required on higher-field instruments and this is particularly so for proton work. Some experimenting may be required to decide on a suitable balance between cost and performance for your particular instruments.

The diameter of the tube is generally determined by the dimensions of the probe to be used. Common tube diameters in chemical laboratories are 2.5 or 3 mm (for use with the so-called microprobes), 5 mm (the most widely used), or 10 mm (typically used for the observation of low-sensitivity nuclei where the solubility of the material to be studied is limiting). Increasingly, 8 mm tubes are finding use in biological macromolecular work, where solubility and aggregation considerations preclude the use of concentrated samples. As sensitivity is usually of prime importance when performing any NMR experiment, it is also prudent to use the correct sample volume for the probe/tube configuration and not to dilute the sample any more than is necessary. The important factor here is the length of the detection coil within the probe. The sample volume should be sufficient to leave a little solution above and below the coil since magnetic susceptibility differences at the solution/air and solution/glass interfaces lead to local distortions of the magnetic field. For a standard 5 mm tube, the required volume will be around 400–600 μl whereas for the micro tubes this reduces to about 100–150 μl. For larger tubes it may be necessary to use a 'vortex-suppresser' to prevent whirlpool formation in the tube if the sample is spun. These are usually PTFE plugs designed to fit tightly within the NMR tube which are pushed down to sit on the surface of the sample and thus hold the solution in place. Some care is required when using these plugs in variable temperature work due to the thermal expansion or contraction of the plug, and at low temperatures the plug may even fall into the solution so its use may be inappropriate.

For the handling of smaller sample quantities, microcells can be obtained that fit within standard 5 mm NMR tubes but which require only tens of microlitres of solution allowing one to concentrate the available material within the detection coil. When using such cells it is advisable to fill the volume around the outside of the cell with solvent, as this minimises susceptibility differences and improves resolution. More recently, special tubes and plugs have become available that allow smaller sample volumes to be held within the coils sensitive region (Fig. 3.43). These use glass that has a magnetic susceptibility matched to that of the solvent and so allows shorter sample heights without the introduction of lineshape defects [20]. Presently, tubes are available individually matched to water and to a small selection of organic solvents.

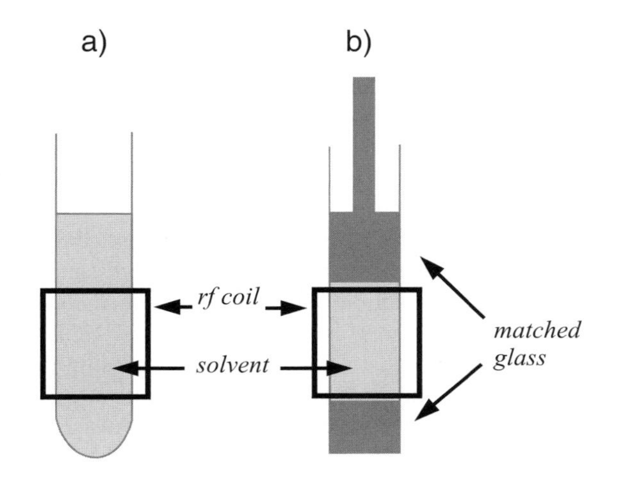

Figure 3.43. Schematic comparison of (a) a conventional NMR tube and (b) a matched tube and plug in which the magnetic susceptibility of the glass is matched to that of the solvent. This allows a smaller sample volume and so concentrates more of the analyte within the detection coil.

NMR tubes should be kept clean, dry, free from dust, and must also be free from scratches on the glass as this distorts the cylinder in which the sample is confined. New tubes are not always clean when delivered, although this is, unfortunately, often assumed to be the case. Organic lubricants used in their manufacture may remain on the glass and become all too apparent when the tube is first used. For routine washing of tubes, rinsing with a suitable organic solvent, such as acetone, or with distilled water a few times is usually sufficient. More stubborn soiling is best tackled by chemical rather than mechanical means and either soaking in detergent or strong mineral acids is recommended. The use of chromic acid must be avoided as this can leave traces of paramagnetic chromium(VI) which will degrade resolution. Tubes should not be dried by subjecting them to high temperatures for extended periods of time as this tends to distort them. A better approach is to keep the tubes under vacuum, at slightly elevated temperatures if possible, or to blow filtered nitrogen into the tubes. If oven drying is used, the tubes must be laid on a flat tray and heated for only 30 minutes, not placed in a beaker or tube rack as this allows them to bend. If the removal of all traces of protiated water from the tube is important, then it is necessary to rinse the tube with D_2O prior to drying to ensure the exchange of water adsorbed onto the surface of the glass. Unused tubes should be stored with their caps on to prevent dust from entering. Finally, it is also important to avoid contamination of the *outside* of the NMR tube (which includes fingerprint oils) as this leads to the transfer of the contaminants into the probe head. The accumulation of these contributes to degradation of the instrument's performance over time.

3.3.4. Filtering and degassing

For any NMR experiment it is necessary to achieve a uniform magnetic field throughout the whole of the sample to obtain a high-resolution spectrum and it is easy to imagine a range of circumstances that detract from this condition. One of the most detrimental is the presence of particulates in the sample since these can distort of the local magnetic field, so it is of utmost importance to remove these from an NMR sample prior to analysis. Even very slight changes in the field within the sample can produce a noticeable reduction in resolution. This should not come as a surprise if one recalls that one often wishes to resolve lines whose resonant frequencies are many hundreds of MHz but which are separated by less than 1 Hz. This corresponds to a resolution of 1 part in 10^9, roughly equivalent to measuring the distance from the Earth to the Sun to an accuracy of 1 km! Commercial sintered filters can be obtained but require constant washing between samples. A convenient and cheap alternative is to pass the solution through a small cotton-wool plug in a Pasteur pipette directly into the NMR tube (it is good practise to rinse the cotton-wool with a little of the NMR solvent before this). If cotton wool is not suitable for the sample, a glass wool plug may be used although this does not provide such a fine mesh. If metal ions are likely to be present in solution (paramagnetic ions being the most unwelcome), either from preparation of the sample or from decomposition of the sample itself, then passing the solution through a small column of a metal chelating resin should remove these.

In situations where very high resolution is demanded or if relaxation studies are to be performed then it may also be necessary to remove all traces of oxygen from the solution. This may include the measurement of nuclear Overhauser enhancements, although for routine NOE measurements it is usually not necessary to go to the trouble of degassing the sample (see Chapter 8). The need to remove oxygen arises because O_2 is paramagnetic and its presence provides an efficient relaxation pathway which leads to line-broadening. Essentially there are two approaches to degassing a sample.

The first involves bubbling an inert gas through the solution to displace oxygen. This is usually oxygen-free nitrogen or argon which need be passed through the solution in the NMR tube for about one minute for organic solvents and double this for aqueous solutions. Great care is required when attempting this and a fine capillary must be used to introduce the gas *slowly*. It is all too easy to blow the sample clean out of the tube in an instant by introducing gas too quickly so it is probably wise to get some feel for this before attempting the procedure on your most precious sample! Note that with volatile solvents a significant loss in volume is likely to occur and TMS, if used, may well also be lost.

The second and more thorough approach is the 'freeze-pump-thaw' technique. The NMR tube containing the solution is frozen with liquid nitrogen or dry ice and placed under vacuum on a suitable vacuum line. Commercial tube manufacturers produce a variety of specialised tubes and adapters for this purpose but a simple alternative is to connect a standard tube to a vacuum line via a needle through a rubber septum cap on the tube. The tube is then isolated from the vacuum by means of a stop-cock and allowed to thaw, during which the dissolved gases leave the solution. The procedure is then repeated typically at least twice more, this usually being sufficient to fully degas the sample. When using an ordinary vacuum pump it may be necessary to place a liquid nitrogen trap between the pump and the sample to avoid the possibility of vacuum oils condensing in the sample tube. Furthermore, when freezing aqueous samples it is easy to crack the tube if this is carried out too fast; holding the sample tube just above the freezing medium whilst tilting it is usually sufficient to avoid such disasters. An alternative approach for aqueous samples containing involatile solutes is to carefully place the sample directly under vacuum without freezing, for example in a vacuum jar or schlenk, and allowing dissolved gas to bubble out of solution.

Following the degassing procedure the tube should be sealed. If the sample is likely to be subject to short-term analysis, say over a few hours, then a standard tight-fitting NMR cap wrapped with a *small* amount of paraffin wax film or, better still, use of a rubber septum, is usually sufficient since diffusion into solution of gases in the tube will be rather slow. For longer duration studies, specialised adapters or screw-top tubes can be purchased or standard tubes can be flame-sealed, for which NMR tubes with restrictions toward the top are available to make the whole process very much easier. When flame-sealing it is highly advisable to cool or freeze the solution and then seal whilst pulling a gentle vacuum. **Failure to cool the solution or evacuate the tube may cause the pressure within in it to become dangerously high as the sample warms**. The seal should be symmetric otherwise the sample will spin poorly, so practice on discarded tubes is likely to pay dividends.

3.4. PREPARING THE SPECTROMETER

Whenever a new sample is placed within the NMR spectrometer, the instrument must be optimised for this. The precise nature of the adjustments required and the amount of time spent making these will depend on the sample, the spectrometer and the nature of the experiment but in all cases the aim will be to achieve optimum resolution and sensitivity and to ensure system reproducibility. The details of the approaches required to make these adjustments depend on the design of spectrometer in question so no attempt to describe such detail is made here. They all share the same general procedures however, which are summarised in Scheme 3.1 and described below.

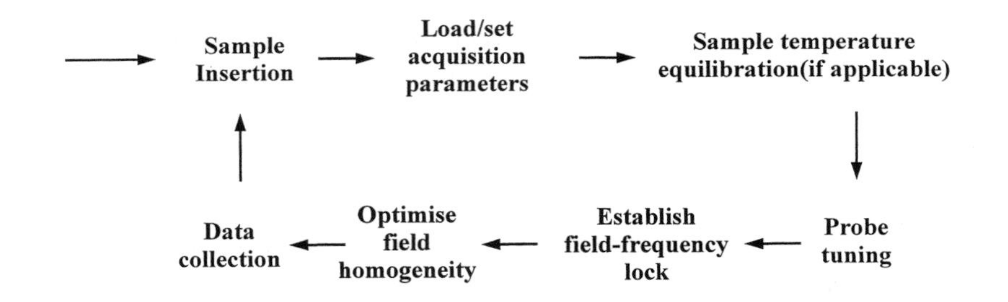

Scheme 3.1. The typical procedure followed in preparing a spectrometer for data collection. All these steps can be automated, although at the time of writing there is still only a very limited selection of fully auto-tuneable probeheads.

3.4.1. The probe

If you are a chemist making use of the NMR facilities available to you, then choosing the appropriate probe for the study in question may not be relevant as you will probably be forced to use that which is available. If, however, you are involved with instrument purchasing or upgrading or you are fortunate enough to have available a variety of probes for a given instrument, then it is important to be able to make the appropriate selection. Since it is the probe that must receive the very weak NMR signals it is perhaps the most critical part of the NMR spectrometer and its particular design and construction will influence not only the types of experiments it is able to perform but also its overall performance. There has been considerable competition over the years in the design of probes, resulting in ever increasing performance and whereas previously one was restricted to choosing a probe from the spectrometer manufacturer, this is no longer the case as there now exist a small number of companies that produce probes that are compatible with the instruments of the main NMR vendors.

There are two main factors that have a significant bearing on improved probe performance. The first is the material used in the construction of the receiver coil itself. Since this coil sits in very close proximity to the sample, this material may distort the magnetic field within this, compromising homogeneity. Modern composite metals are designed so that they do not lead to distortions of the field (they are said to have *zero magnetic susceptibility*) so allowing better lineshapes to be obtained which ultimately leads to improved signal-to-noise figures. The second factor lies in the coil design and dimensions, with modern coils tending to be longer than previously so that a greater sample volume sits within them. This demands a greater volume in which the magnetic field is uniform so these changes in probe construction have largely followed improvements in room-temperature shim systems (see below).

There exists a great variety of probe designs of various dimensions that can be obtained for any given instrument. The probe size is usually described by the sample tube diameter it is designed to hold, and as mentioned previously, common sizes are 3, 5, 8 and 10 mm, with the 5 mm configuration being that in standard use at present. Purely on the grounds of the inherent sensitivity of a given coil construction, the narrower 3 mm coils found in micro-probes will have the better performance and so are favoured when very small sample quantities, perhaps in the microgram region, are to be analysed. The use of these microprobes has largely been in the area of natural product chemistry [21] and in the study of biosynthetic materials, where these small sample quantities are commonplace. For extreme cases of limited material, so-called nano-probes have been designed which allow the available sample to be concentrated into even smaller volumes of around 40 μl, but these are not in widespread use. Wider diameter 8 or 10 mm tubes allow greater sample volumes to be held within the active region of the coil, so these would generally be used when sample solubility is the factor limiting sensitivity.

The other important consideration when selecting a probe is the range of nuclei it is able to observe and for which of these the coil configuration has been optimised. The simplest design is a probe containing a single coil which is designed for the observation of only one nucleus (in fact, it would be 'doubly-tuned' to enable the simultaneous observation of deuterium for the field-frequency regulation via the lock system, although the presence of a deuterium channel is usually implicit when discussing probe configurations). However, many modern NMR experiments require pulses to be applied to two (or more) different nuclides, of which one is most often proton, for which two coils are necessary. In this case, two possible configurations are in widespread use depending on whether one wishes to observe the proton or another nuclide (referred to as the X-nucleus). The traditional two-coil design is optimised for the observation of the X-nucleus, with the X-coil as the inner of the two allowing it to sit closer to the sample so offering the best possible sensitivity for X-observation (it is said to have the greatest 'filling factor'). This configuration can be described by the shorthand notation X{^{1}H}. Nowadays multipulse experiments tend to utilise the higher sensitivity offered by proton observation wherever possible, and benefit from probes in which the proton coil sits closest to the sample with the X-coil now the outer most; ^{1}H{X}. It is this design of probe that is widely referred to as having the *inverse* configuration because of this switch in geometry. In either case, the X-coil circuitry can be designed to operate at only a single frequency or can be tuneable over a wide frequency range, such probes being known as *broadband observe* or *broadband inverse* probes. An alternative popular configuration in organic chemistry is the *quad-nucleus* probe that allows observation of the four most commonly encountered nuclei ^{1}H, ^{13}C, ^{19}F, and ^{31}P. In studies of biological macromolecules, and more recently in organic chemistry, *triple-resonance* probes are employed, allowing proton observation and pulsing or decoupling of two other nuclei; ^{1}H{X,Y}. A further feature offered is the addition of magnetic field gradient coils to the probe head. These surround the usual rf coils and are designed to destroy the static magnetic field homogeneity throughout the sample for short periods of time in a very reproducible manner; gradient-selected techniques are introduced in Chapter 5. The most recent probe designs aim to reduce thermal noise by utilising superconducting materials for the detection coil (requiring this to be cooled with liquid helium) and, although proposed some years ago [22,23], have only recently become available [24]. So far these are available only in certain configurations, limited by a number of engineering factors, not least of which is the need to maintain the detection coil at around 4 K whilst the adjacent sample temperature remains at ambient! These typically offer sensitivity gains of a factor of 4–8 over conventional probes (under optimum conditions) and currently represent a promising area for future developments.

3.4.2. Tuning the probe

The NMR probe is a rather specialised (and expensive) piece of instrumentation whose primary purpose is to hold the transmit and receive coils as close as possible to the sample to enable the detection of the weak NMR signals. For the coils to be able to transmit the rf pulses to the sample and to pick up the NMR signals efficiently, the electrical properties of the coil circuit should be optimised for each sample. The adjustments are made via variable capacitors which sit in the probe head a short distance from the coil(s) and comprise the tuning circuitry. There are two aspects to this optimisation procedure known as *tuning* and *matching*, although the whole process is more usually referred to as 'tuning the probe'. The first of these, as the name implies, tunes the coil to the radiofrequency of the relevant nucleus and is analogous to the tuning of a radio-receiver to the desired radio station. A poorly tuned probe will lead to a severe degradation in sensitivity, just as a radio broadcast becomes swamped with hissing noise. The second aspect aims to equalise (or match) the impedance (the total effective resistance to alternating current) of the coil/sample combination with that of the transmitter and receiver so that the maximum possible rf energy can pass from the transmitter into the sample and subsequently from the sample into the receiver. As electrical properties differ between samples, the optimum tune and match conditions will also vary and so require checking for each new sample. These differences can be largely attributed to differences between solvents with the most significant changes occurring between non-polar organic solvents and ionic aqueous solutions.

Probe tuning is necessary for a number of reasons. Other than the fundamental requirement for maximising sensitivity, it ensures pulse-widths can be kept short which in turn reduces off-resonance effects and minimises the power required for broadband decoupling. A properly tuned probe is also required if previously calibrated pulse-widths are to be reproducible, an essential feature for the successful execution of multipulse experiments.

Tuning and matching

The process of probe tuning involves applying rf to the probe, monitoring the response from it by some suitable means and making adjustments to the capacitors in the head of the probe (via long rods that pass through its base) to achieve the desired response. In the case of broadband probes, which may be tuneable over very wide frequency ranges, the capacitors may even need to be physically exchanged, either by removing them from the probe altogether or by means of a switching mechanism held within the probe. Various procedures exist for monitoring the response of the probe but the most useful and the one supplied widely on modern instruments, uses a frequency sweep back and forth over a narrow region (of typically a few MHz) about the target frequency during which the probe response is compared to that of an internal 50 Ω reference load (all NMR spectrometers are built to this standard impedance). The display then provides a simultaneous measure of the tune and match errors (Fig. 3.44a),

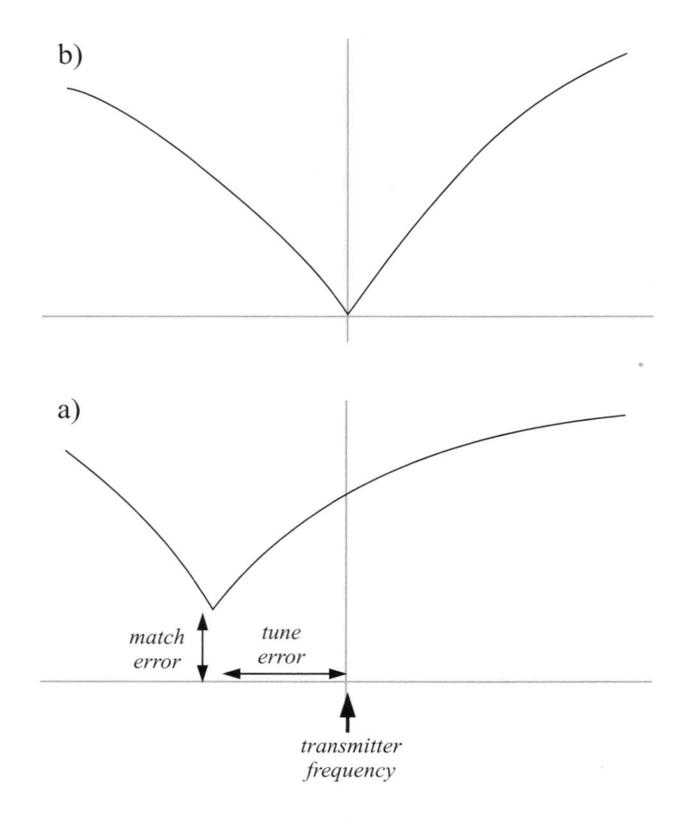

Figure 3.44. Tuning and matching an NMR probehead. The dark line represents the probe response seen for (a) a mis-tuned and (b) a correctly tuned probe head.

allowing one to make interactive adjustments to the tuning capacitors to achieve the desired result (Fig. 3.44b); the response reaches a minimum when the probe is matched to the 50 Ω load. With this form of display it is possible to see the direction in which adjustments should be made to arrive at the correct tune and match condition. It is important to note that the tune and match controls are not mutually exclusive and adjustments made to one are likely the alter the other, so a cyclic process of tune, match, tune etc. will be required to reach the optimum.

The method for probe tuning on older spectrometers that are unable to produce the frequency sweep display is to place a *directional coupler* between the transmitter/receiver and the probe and to apply rf as a series of very rapid pulses. The directional coupler provides some form of display, usually a simple meter, which represents the total power being reflected back from the probe. The aim is to *minimise* this response by the tuning and matching process so that the maximum power is able to enter the sample. Unfortunately with this process, unlike the method described above, there is no display showing errors in tune and match separately, and there is no indication of the direction in which changes need be made, one simply has an indication of the overall response of the system. This method is clearly the inferior of the two, but may be the only option available.

When more than one nuclear frequency is of interest, the correct approach is to make adjustments at the *lowest* frequency first and work up to the highest. It is also prudent to remove rf filters, couplers and so on from the rf path since it is the probe itself one wishes to match. Tuning will also be influenced by sample and probe temperature and must be checked whenever changes are made. If large temperature changes are required it is wise to quickly recheck the tuning every ten or twenty degrees so that one never becomes too far from the optimum; this is especially important if using the directional coupler method. Where the spectrometer is used in an open access environment, where interaction with the spectrometer is kept to a minimum or where the instrument runs automatically, probe tuning for each sample is generally not viable (although probes with complete auto-tuning capabilities are now appearing), in which case it is appropriate to tune the probe for the most frequently used solvent, and accept some degradation in performance for the others. Different solvents may require different pulse width calibrations under these conditions.

Finally in this section we consider one particular situation in which it is beneficial to deliberately detune a probe. When performing studies in *protonated* water the linewidth of the solvent resonance is broadened significantly in a well tuned probe, because of the phenomenon of *radiation damping* [25]. This is where the FID of the solvent decays at an accelerated rate because the relatively high current generated in the coil by the intense NMR signal itself produces a secondary rf field which drives the water magnetisation back to the $+z$ axis at a faster rate than would be expected from natural relaxation processes alone. The rapid decay of the FID in turn results in a broadened water resonance. This only occurs for very intense resonances and has greatest effect when the sample couples efficiently with the coil, that is, when the probe is well tuned. Detuning the probe a little provides a sharper (and weaker) water resonance whose lineshape gives a better indication of the field homogeneity. Retuning the probe is essential for subsequent NMR observations employing solvent suppression schemes.

3.4.3. The field-frequency lock

Despite the impressive field stability provided by superconducting magnets they still have a tendency to drift significantly over a period of hours, causing NMR resonances to drift in frequency leading to a loss of resolution. To overcome this problem some measure of this drift is required so that corrections may be applied. On all modern spectrometers the measurement is provided by monitoring the frequency of the deuterium resonance of the solvent. The deuterium signal is collected by a dedicated ^{2}H observe spectrometer within the instrument that operates in parallel with the principal channels, referred to as the *lock channel* or simply the *lock*.

The lock system

The lock channel regulates the field by monitoring the *dispersion* mode deuterium resonance rather than the absorption mode signal that is usually considered in NMR, and aims to maintain the centre of this resonance at a constant frequency (Fig. 3.45). A drift in the magnetic field alters the

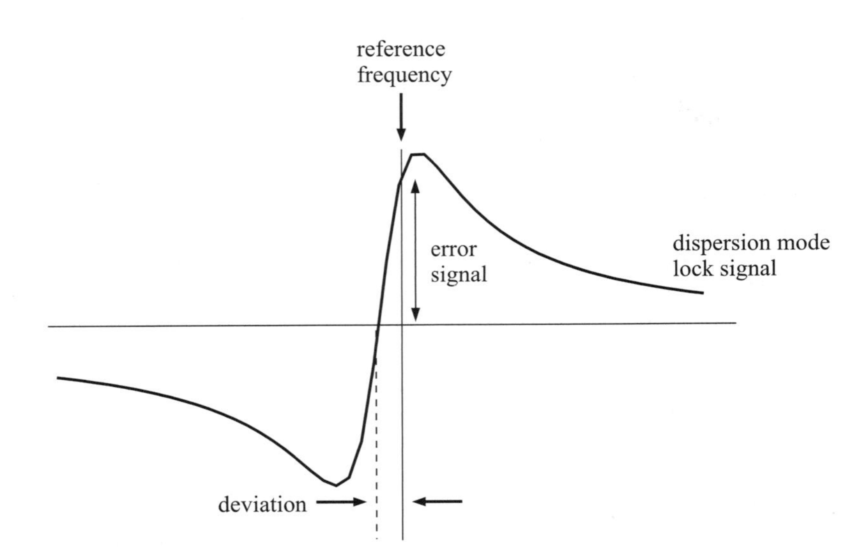

Figure 3.45. The spectrometer lock system monitors the dispersion mode signal of the solvent deuterium resonance. A shift of the resonance frequency due to drift in the static field generates an error signal that indicates the magnitude and direction of the drift, enabling a feedback system to compensate this.

resonance frequency and therefore produces an error signal that has both magnitude and sign (unlike the absorption mode resonance which varies only in magnitude). This then controls a feedback system which adjusts the field setting. The dispersion signal also has the advantage of having a rather steep profile, providing the greatest sensitivity to change. Monitoring the deuterium resonance also provides a measure of the magnetic field homogeneity within the sample since only a homogeneous field produces a sharp, intense resonance. In this manner the lock signal may be used as a guide when optimising (shimming) the magnetic field (Section 3.4.4) where the absorption mode signal is presented to the operator as the lock display. Since the best field-frequency regulation is provided by the strongest and sharpest lock resonance, the more highly deuterated non-viscous solvents provide the best regulation. Water is especially poor since it tends to have a rather broad resonance because of exchange processes which are very dependent on solvent temperature and pH.

Optimising the lock

The first procedure for locking is to establish the resonance condition for the deuterium signal, which involves altering either the field or the frequency of the lock transmitter. Of these two options the latter is preferred since it avoids the need for changing transmitter frequencies and is now standard on modern instruments. Beyond this, there are three fundamental probe-dependent parameters that need to be considered for optimal lock performance. The first of these is the lock transmitter *power* used to excite the deuterium resonance. This needs to be set to the highest usable level to maximise the ^{2}H signal-to-noise ratio but must not be set so high that it leads to lock *saturation*. This is the condition in which more energy is applied to the deuterium spins than can be dissipated through spin relaxation processes, and is evidenced by the lock level drifting up and down erratically. The saturation level can be determined by making small changes to the lock power and observing the effect on the lock signal. If the power is increased the lock signal should also increase, but if saturated it will then drift back down. Conversely, if the power is decreased, the lock level will drop but will then tend to creep up a little. The correct approach is to determine the point of saturation and then to be sure you are operating below this before proceeding. The second feature is the lock *gain*. This is the amplification that is applied to the detected lock signal, and is generally less critical than the power although it should not be so high as to introduce excessive lock noise. The final feature is the lock *phase*. We have already seen that the field-frequency regulation relies on monitoring the pure dispersion-mode lineshape and this is present only when the resonance has been phased correctly. This is the case when the *observed* lock signal produces the maximum intensity as the lock phase is adjusted since only then does this have the required pure-absorption lineshape.

Acquiring data unlocked

In cases where deuterium is not available for locking, for example if one wishes to observe or decouple deuterium itself, then it may be necessary to acquire data 'unlocked' (although recent hardware developments allow the decoupling of deuterium during acquisition by appropriate blanking of the lock channel). Running unlocked is perfectly feasible over limited time periods which depend on the drift rate of the magnet. For experiments that must run for very many hours for reasons of sensitivity, a useful approach is to acquire a series of FIDs each being collected over a shorter time period, say one hour. Drift within each data set will then be essentially negligible whereas any drift occurring over the whole duration of the experiment can be corrected by adding *frequency domain spectra* for which any frequency drift has been compensated manually by internal spectrum referencing. The addition then provides the required enhancement of signal-to-noise without the deleterious effects of field drift.

3.4.4. Optimising the field homogeneity: shimming

NMR experiments require a uniform magnetic field over the whole of the NMR sample volume that sits within the detection coil. Deviation from this ideal introduces various lineshape distortions, compromising both sensitivity and resolution. Thus, each time a sample is introduced into the magnet it is necessary to 'fine-tune' the magnetic field and a few minutes spent achieving good resolution and lineshape is time well spent. For anyone actually using an NMR spectrometer, competence in the basic level of field optimisation is essential, but even if you only need to *interpret* NMR spectra, perhaps because someone else has acquired the data or if the whole process is performed through automation, then some understanding of the most common defects arising from remaining field inhomogeneities can be invaluable.

The shim system

Maintaining a stable magnetic field that is uniform to 1 part in 10^9 over the active volume within modern NMR probes (typically 0.1 to 1.5 cm^3) is extremely demanding. This amazing feat is achieved through three levels of field optimisation. The first lies in the careful construction of the superconducting solenoid magnet itself, although the field homogeneity produced by these is rather crude when judged by NMR criteria. This basic field is then modified at two levels by sets of 'shim' coils. These coils carry electrical currents that

Table 3.4. Shim gradients found on high-resolution spectrometers

Shim gradient	Gradient order	Interacting shim gradients
Z^0 (the main field)	0	–
Z^1 (Z)	1	–
Z^2	2	Z
Z^3	3	Z, (Z^2)
Z^4	4	Z^2, Z^0,(Z, Z^3)
Z^5	5	Z^3, Z, (Z^2, Z^4)
Z^6	6	Z^4, Z^2, Z^0, (Z, Z^3, Z^5)
X	1	Y, (Z)
Y	1	X, (Z)
XZ	2	X, (Z)
YZ	2	Y, (Z)
XY	2	X, Y
X^2–Y^2	2	XY, (X, Y)
XZ^2	3	XZ, (X, Z)
YZ^2	3	YZ, (Y, Z)
ZXY	3	XY, (X, Y, Z)
Z(X^2–Y^2)	3	X^2–Y^2, (X, Y, Z)
X^3	3	X
Y^3	3	Y
XZ^3	4	XZ^2, (XZ, X, Z)
YZ^3	4	YZ^2, (YZ, Y, Z)

Not all these may be present on lower-field instruments, whilst on very high-field spectrometers additional high-order shims may be found. Those shown in brackets interact less strongly, and thus may not require subsequent readjustment. Those that interact with Z^0 (the main field) may cause momentary disruption of the lock signal when adjusted.

generate small magnetic fields of their own which are employed to cancel remaining field gradients within the sample. In fact, shims are small wedges of metal used in engineering to make parts fit together, and were originally used in the construction of iron magnets to modify the position of the poles to adjust the field. Still in the present day where superconducting magnets dominate, this name permeates NMR, as does the term 'shimming', referring to the process of field homogeneity optimisation[2]. The superconducting shim coils sit within the magnet cryostat, and remove gross impurities in the magnet's field. The currents are set when the magnet is first installed and do not usually require altering beyond this. The room-temperature shims are set in a former which houses the NMR probe itself, the whole assembly being placed within the bore of the magnet such that the probe coil sits at the exact centre of the static field. These shims (of which there are typically around twenty to thirty on a modern instrument) remove any remaining field gradients by adjusting the currents through them, although in practice only a small fraction of the total number need be altered on a regular basis (see below).

The static field in vertical bore superconducting magnets also sits vertically and this defines, by convention, the z-axis. Shims that affect the field along this axis are referred to as axial or Z-shims, whereas those that act in the horizontal plane are known as radial or X/Y shims (Table 3.4). When acquiring high-resolution spectra it is traditional practice to spin the sample (at about 10–20 Hz) about the vertical axis. This has the effect of averaging field inhomogeneities in the X–Y plane, so improving resolution. The averaging means that adjustments to shims containing an X or Y term must be made when the sample is static, hence these shims are also commonly referred to as the 'non-spinning shims'.

[2] Note that some (older) texts may refer to this process as 'tuning', which is now exclusively reserved for processes involving radio-frequencies; for example, one may *shim* a magnet but will *tune* a probe.

Modern shim sets are capable of delivering non-spinning lineshapes that almost match those when spinning, and it is becoming increasing common *not* to spin samples. For multidimensional studies this is certainly the case, since sample spinning can introduce modulation effects to the acquired data, leading to unwanted artefacts particularly in the form of so-called t_1-noise (see Chapter 5).

Shimming

In order to achieve optimum field homogeneity, high-quality samples are essential. The depth of a sample also has a considerable bearing on the amount of Z shimming required, which can be kept to a minimum by using solutions of similar depth each time. Most spectrometers possess software that is capable of carrying out the shimming process automatically, and clearly this is essential if an automatic sample changer is used. However, such systems are not infallible and can produce spectacularly bad results in some instances. Here, reproducible sample depths are vitally important for the auto-shimming procedures to be successful and to reach an optimum rapidly. It is also crucial for the whole of the sample to be at thermal equilibrium so that convection currents do not exist, which, for aqueous solution in particular, may demand 10–20 minutes equilibration in the probe.

To provide an indication of progress when shimming, one requires a suitable indicator of field homogeneity. Essentially, there are three schemes that are in widespread use, all of which have their various advantages and disadvantages; 1) the lock level, 2) the shape of the FID, and 3) the shape of the NMR resonance. The ultimate measure of homogeneity is the NMR resonance itself, since defects apparent in the spectrum can often be related directly to deficiencies in specific shim currents, as described below. Most often field homogeneity is monitored by the height of the deuterium lock resonance which one aims to maximise. Whilst conceptually this is a simple task, in reality it is complicated by the fact that most shims interact with others. In other words, having made changes to one it will then be necessary to re-optimise those with which it interacts. Fortunately, shims do not influence all others, but can be sub-divided into smaller groups which are dealt with sequentially during the shimming process. A detailed account of the shimming procedure has been described [26] and the fundamental physics behind field-gradient shims has also been presented [27], but here we shall be concerned more with addressing the lineshape defects that are commonly encountered in the daily operation of an NMR spectrometer.

When shimming, it is not always sufficient to take the simplest possible approach and maximise the lock level by adjusting each shim in turn, as this is may lead to a 'false maximum', in which the lock level appears optimum yet lineshape distortions remain. Instead, shims must be adjusted interactively. As an example of the procedure that should be adopted, the process for adjusting the Z and Z^2 shims (as is most often required) should be:

(1) Adjust the Z shim to maximise the lock level, and note the new level
(2) Alter Z^2 so that there is a noticeable change in the lock level, which may be up or down, and remember the direction in which Z^2 has been altered
(3) Readjust Z for maximum lock level.
(4) Check whether the lock level is greater than the starting level. If it is, repeat the whole procedure, adjusting Z^2 in the same sense, until no further gain can be made. If the resulting level is lower, the procedure should be repeated but Z^2 altered in the opposite sense.

If the magnetic field happens to be close to the optimum for the sample when it is initially placed in the magnet, then simply maximising the lock level with each shim directly will achieve the optimum since you will be close to this already. Here again a reproducible sample depth makes life very much easier.

Shimming is performed by concentrating on one interacting group at a time, always starting and finishing with the lowest order shim of the group. Principal interactions are summarised in Table 3.4. Whenever it is necessary to make changes to a high order shim, it will be necessary to readjust all the low order shims within the same interacting group, using a similar cyclic approach to that described for the adjustment of Z and Z^2 above. Generally, the order of optimisation to be followed will be:

(1) Optimise Z and Z^2 interactively, as described above. If this is the first pass through Z and Z^2, then adjust the lock phase for maximum lock level.
(2) Optimise Z^3: make a known change, then repeat step 1. If the result is better than previously, repeat this procedure, if not, alter Z^3 in the opposite sense and repeat step 1.
(3) Optimise Z^4 interactively with Z^3, Z^2 and Z.
(4) Stop the sample spinning (if applicable) and adjust Z to give the maximum response (this is likely to have changed a little as the position of the sample relative to the field will change). Adjust X and Y in turn to give the maximum response.
(5) Optimise X and XZ interactively. Adjust Z to give the maximum response.
(6) Optimise Y and YZ interactively. Adjust Z to give the maximum response.
(7) Optimise XY interactively with X and Y.
(8) Optimise X^2-Y^2 interactively with XY, X and Y.
(9) Repeat step 1.

The higher the shim order, the greater the changes will be required and when far from the optimum shim settings, large changes to the shim currents may have only a small effect on the lock level and the shim response will feel rather 'sluggish'. When close to the optimum the response becomes very sensitive and small changes can have a dramatic effect. The above procedure should be sufficient for most circumstances and any field strength, unless the basic shims set has become grossly misset. If lineshape distortions remain then it may be possible to identify the offending shim(s) from the nature of the distortion (see below), allowing the appropriate corrections to be applied.

Common lineshape defects

The NMR resonance lineshape gives the ultimate test of field homogeneity, and it is a useful skill to be able to recognise the common distortions that are caused by errors in shim settings (Fig. 3.46). Thus, the Z-shims all influence the width of the NMR resonance, but in subtly different ways; impurities in the even order shims (Z^2, Z^4 and Z^6) will produce unsymmetrical distortions to the lines whereas those in the odd orders (Z, Z^3, Z^5) will result in symmetrical broadening of the resonance. In any case, the general rule is that the higher the order of the shim, the lower down the resonance the distortions will be seen. Errors in Z^3 usually give rise to a broadening of the base of a resonance and, since a broad resonance corresponds to a rapid decay of the FID, such errors are sometimes seen as a sharp decay in the early part of the FID. Another commonly observed distortion is that of a shoulder on one side of a peak, arising from poorly optimised Z and Z^2 shims (this is often associated with reaching a 'false maximum' simply by maximising the lock level with each shim and is usually overcome by making a significant adjustment to Z^2, and following the procedure described above).

Errors in the low order X/Y shims give rise to the infamous 'spinning sidebands' (for a spinning sample!). These are images of the main resonance displaced from it by multiples of the spinning frequency. Shims containing a single X or Y term produce '1st order sidebands' at the spinning frequency from the main line whereas XY and X^2-Y^2 give second order sidebands at double the spinning frequency. However, unless something has gone seriously

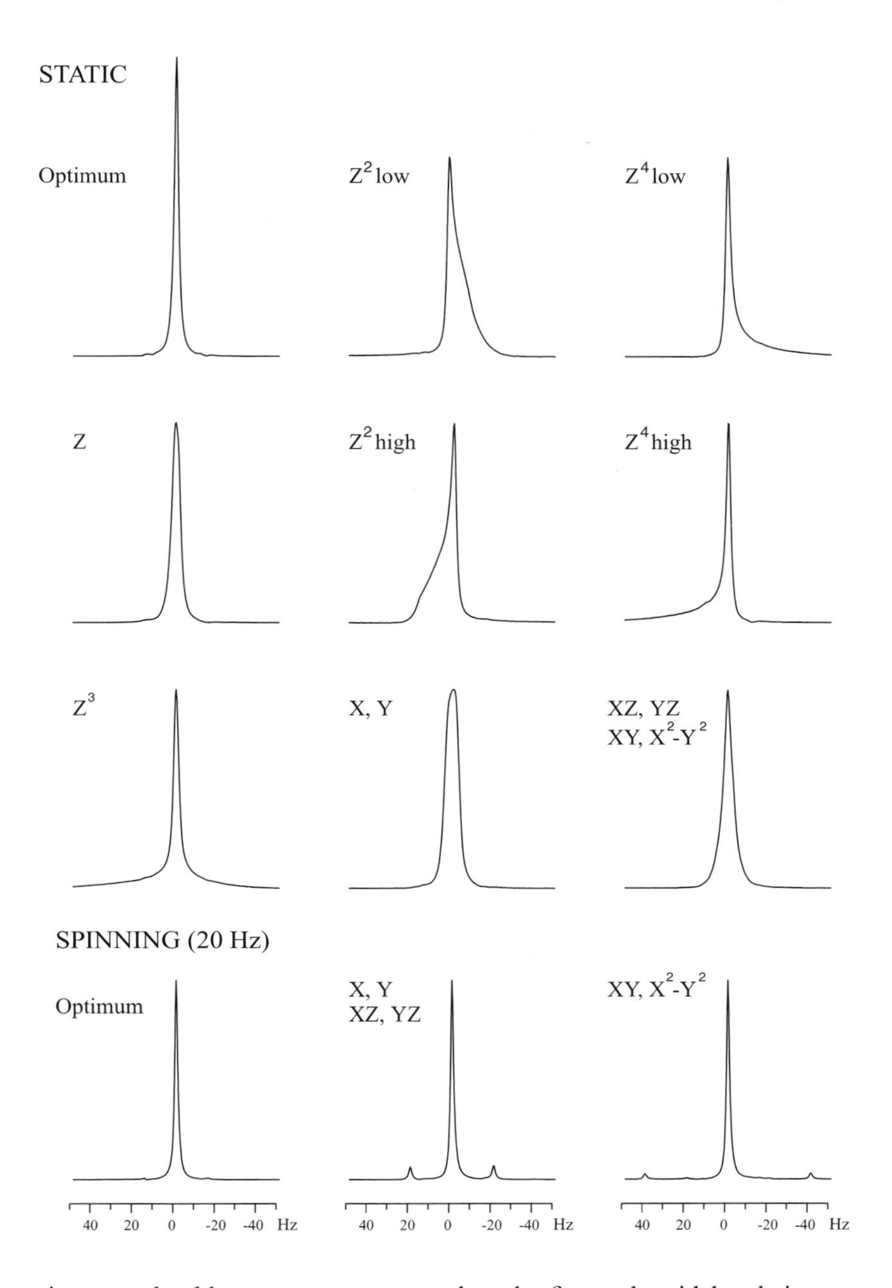

Figure 3.46. Lineshape defects that arise from inappropriate settings of various shims. These effects have been exaggerated for the purpose of illustration.

amiss, you should not encounter more than the first order sidebands in every day NMR at most, and these should certainly be no greater than one or two percent of the main resonance. If there is any doubt as to the presence of sidebands, a simple test is to alter the spinning speed by say 5 Hz and re-acquire the data; only the sidebands will have moved. If the sample is not spinning, errors in the low order X/Y shims contribute to a general broadening of the resonance.

Shimming using the FID or spectrum

Although the lock level is used as the primary indicator of field homogeneity, it is not always the most accurate one. The lock level is dependent upon only one parameter; the height of the deuterium resonance. This, whilst being rather sensitive to the width of the main part of the resonance, is less sensitive to changes in the broad base of the peak. The presence of such low level lumps can be readily observed in the spectrum (particularly in the case of protons) but for this to be of use when shimming, the spectrometer must be able to supply a 'real-time display' of a single-scan spectrum so that changes to the shim currents can be assessed rapidly. With modern host computers, the Fourier transform and phase correction of a spectrum can be performed very rapidly, allowing one to correct for lineshape distortions as one shims. Alternatively, the shape and the duration of the FID may be used as a more immediate indicator of homogeneity and can often be used simultaneously with the lock display. This approach works best when a singlet resonance dominates the spectrum (such as for aqueous solutions) for which the shape of the FID should be a smooth exponential decay. Since with this method of shimming it is likely that changes to the shim currents will be made *during* the acquisition of the spectrum (which will certainly lead to a peculiar lineshape) it is essential that one assesses a later spectrum for which there have been no adjustments during acquisition to decide whether improvements have been made.

Gradient shimming

The most recently introduced approach to field optimisation comes from the world of magnetic resonance imaging and makes use of field gradients to map B_0 inhomogeneity within a sample. This can then be cancelled by calculated changes to the shim settings [28]. The results that can be attained by this approach are little short of astonishing when seen for the first time, especially for anyone who has had to endure the tedium of extensive manual shimming of a magnet, and this relatively new method will undoubtedly enjoy greater popularity in the future.

The discussions below assume some understanding of the action of pulsed field gradients (PFGs) and the reader not familiar with these may wish to return to this section after they are introduced in Chapter 5. In any case, an appreciation of the capabilities of this method should be readily achieved from what follows. Here we shall consider the basis of the method with reference to the optimisation of z-shims which requires z-axis pulsed field gradients (although recent work [29] has demonstrated the use of conventional shim assemblies to generate the appropriate gradients with so-called homospoil pulses, which has the advantage of not requiring specialised gradient hardware). The underlying principle is that all spins throughout a sample contributing to a singlet resonance will posses the same precession (Larmor) frequency *only* if the static field is homogeneous throughout (of course this is what we aim for when shimming!). Any deviation from this condition will cause spins in physically different locations within the sample to precess at differing rates according to their *local* static field. If the excited spins are allowed to precess in the transverse plane for a fixed time period prior to detection, these differing rates simply correspond to different phases of their observed signals (Fig. 3.47). By detecting these signals in the presence of an applied field gradient, the *spatial distribution* of the spins becomes encoded as the *frequency distribution* in the spectrum allowing the inhomogeneity (encoded as phase

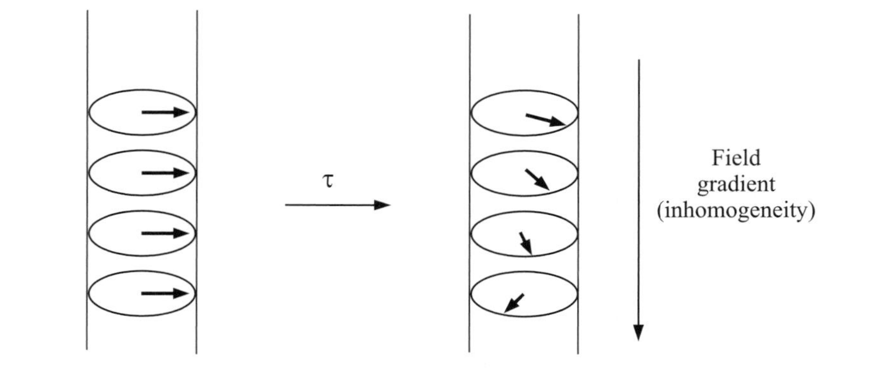

Figure 3.47. Errors in the local static field along the length of a sample are encoded as signal phases when magnetisation is allowed to precess in a inhomogeneous static field for a time τ. The resulting spatially dependent phase differences are used as the basis of gradient shimming.

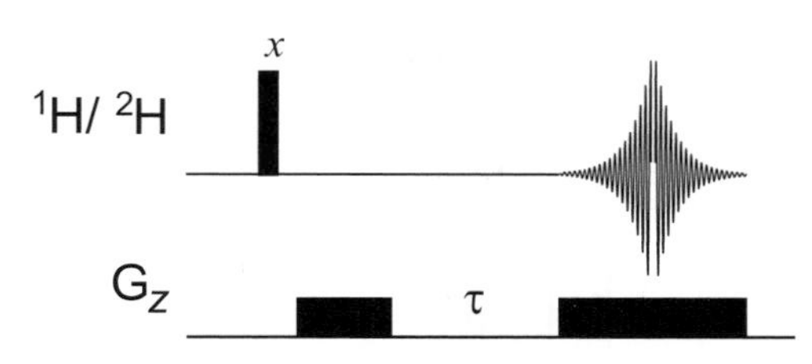

Figure 3.48. A gradient echo sequence suitable for *z*-axis field gradient shimming. The pulsed field gradients provide the spatial encoding whilst the delay τ encodes the static field inhomogeneity as signal phase.

differences) to be mapped along the length of the sample in the case of *z*-axis gradients (or across the sample for *x*- and *y*-axis gradients).

A suitable scheme for recording this is the gradient echo of Fig. 3.48 in which spins are first dephased by a PFG and later rephased after a period of precession τ_1 to allow detection. The resulting spectrum is the one-dimensional spatial profile (or image) of the sample (Fig. 3.49). Recording a second echo with delay τ_2 and taking the difference yields the *phase map* in which only free precession during the period ($\tau_2-\tau_1$) is encoded. The phase distribution in this profile therefore directly maps the inhomogeneity along the sample. The necessary corrections to shims currents to remove these inhomogeneities are calculated from a series of reference phase maps recorded with known offsets in each of the *z*-shims. Once these reference maps have been recorded for a given probe, they can be used for the gradient shimming of all subsequent samples, with the whole process operating automatically.

The primary experimental requirement for gradient shimming is a sample containing a dominant strong, singlet resonance. A good candidate for proton observation is 90% H_2O, but although this is ideal for biomolecular studies, it

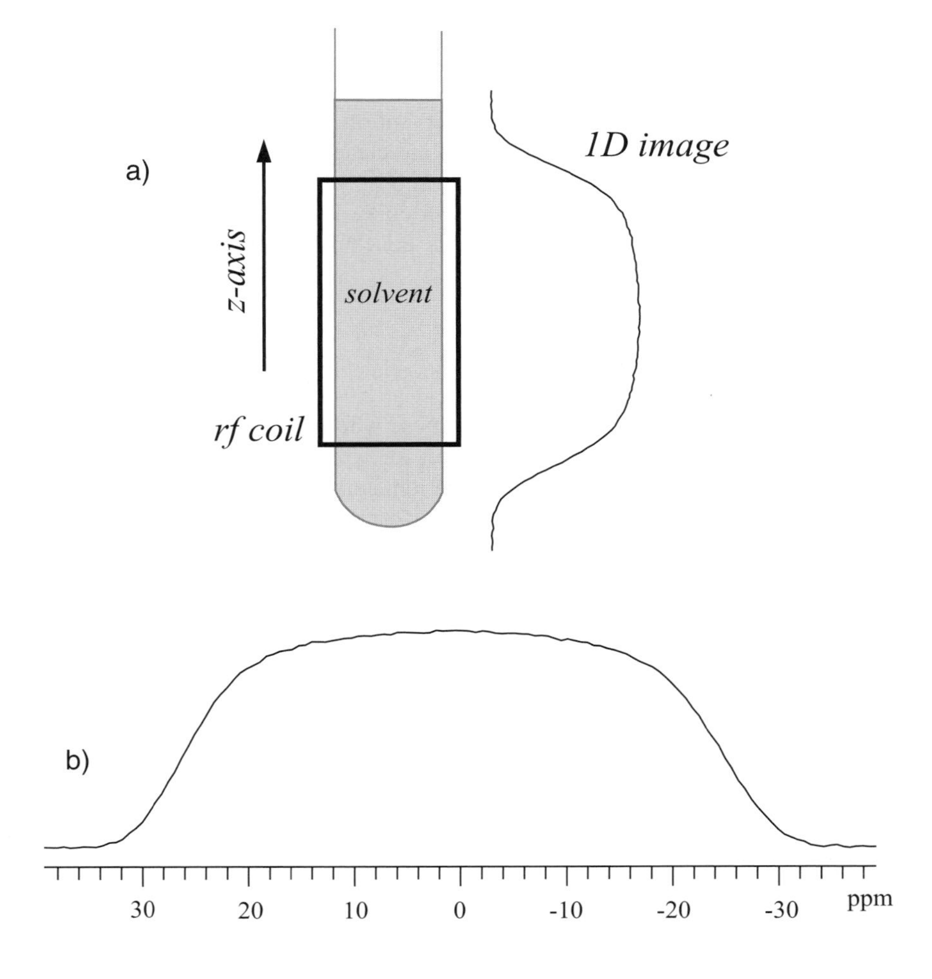

Figure 3.49. (a) The 1D *z*-axis deuterium image of DMSO (magnitude mode). (b) The frequency axis of this image encodes the spatial dimension along the length of the sample.

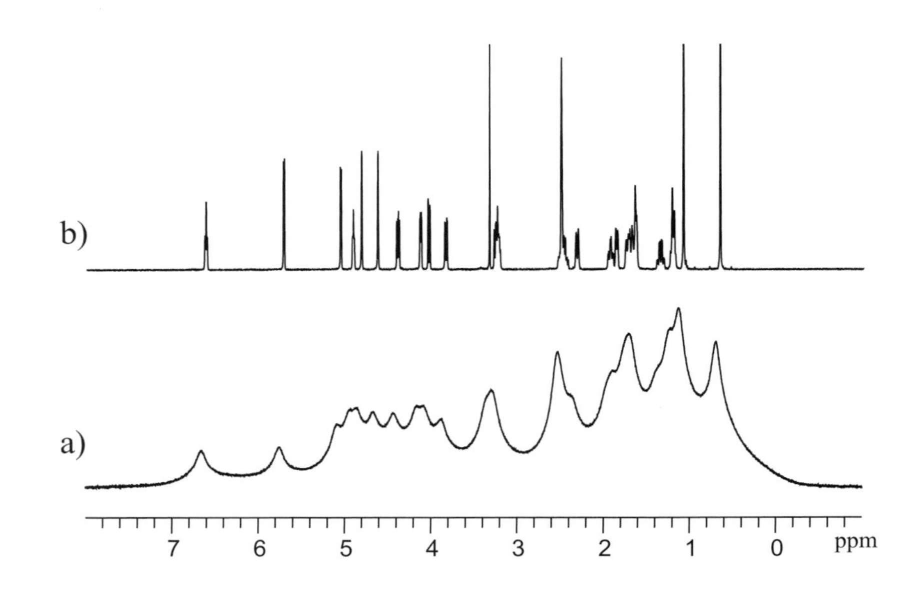

Figure 3.50. Automatic deuterium gradient shimming. Spectrum (a) was acquired with all *z*-axis shims set to zero. After less than two minutes of gradient shimming of the z–z^5 shims (3 iterations) spectrum (b) was obtained. The solvent was DMSO and one scan was acquired for each deuterium gradient echo collected via the probe lock coil. Automatic switching for deuterium observation was achieved with a home-built switching device.

is clearly of little use for the majority of solvents used in organic spectroscopy. An alternative in this case is to observe the deuterium resonance of the solvent [30] (which in most cases is also a singlet) using the lock channel of the probe. The potential problem is then one of sensitivity and the need for appropriate hardware to allow deuterium observation on the lock coil without manually recabling the instrument each time a sample is shimmed. The necessary lock channel switching devices are now commercially available.

The remarkable power of gradient shimming is illustrated in Fig. 3.50. The lower proton spectrum was recorded with the z–z^5 shims all set to zero whilst the upper trace was the result of only 3 iterations of deuterium gradient shimming using the DMSO solvent resonance. The whole process took less than 2 minutes without operator intervention. Although a rather extreme example, the capabilities of this approach are clearly evident and it is likely to play a valuable role in automated spectroscopy, where irreproducible sample depths can lead to rather poor results with conventional simplex optimisation shim routines. The individual mapping of field errors within each and every sample overcomes these problems.

3.5. SPECTROMETER CALIBRATIONS

This section is primarily intended for those who need to set-up experiments or those who have new hardware to install for which new calibrations are required. As with any analytical instrumentation, correct calibrations are required for optimal and reproducible instrument performance. All the experiments encountered in this book are critically dependent on the application of rf and gradient pulses of precise amplitude, shape and duration and the calibrations described below are therefore fundamental to the correct execution of these sequences. Periodic checking of these calibrations, along with the performance tests described in the following section, also provides an indication of the overall health of the spectrometer.

3.5.1. Radiofrequency pulses

Modern multipulse NMR experiments are critically dependent on the application of rf pulses of known duration (the *pulse width*) that correspond to precise magnetisation tip angles, most frequently 90° and 180°. Pulse width calibrations are normally defined for the 90° pulse (PW_{90}), from which all other tip angles may be derived. Those with PW_{90} in the microsecond range generally excite over a rather wide bandwidth and are termed *non-selective* or *hard* pulses whilst those in the millisecond range are of lower-power and are effective over a much smaller frequency window and are thus termed *selective* or *soft* pulses. The implementation of selective pulses and their calibration is considered separately in Chapter 9, whilst here we shall concentrate on the more widely used hard pulses, and on weaker rf fields used for decoupling purposes.

Rf field strengths

The calibration of a pulse width is equivalent to determining the radiofrequency (B_1) field strength of the pulse, and, in fact, it is often more useful to think in terms of field strengths than pulse widths when setting up experiments. For example, excitation bandwidths and off-resonance effects are best considered with reference to field strengths, as are decoupling bandwidths. The relationship between pulse width and the rf field strength γB_1 is straightforward:

$$\gamma B_1 = \frac{1}{PW_{360}} \equiv \frac{1}{4PW_{90}}\ \text{Hz} \tag{3.9}$$

Thus, a 90° pulse width of 10 μs corresponds to a field strength of 25 kHz, a typical value for pulse excitation on a modern spectrometer.

Although spectrometer transmitters are frequently used at full power when applying single pulses, there are many instances when lower transmitter powers are required, for example the application of decoupling sequences or of selective pulses. These lower powers are derived by attenuating the transmitter output, and the units used for defining the level of attenuation are the deciBel or dB. This is, in fact, a measure of the ratio between two power levels, P_1 and P_2, as defined by:

$$dB = 10\log_{10}\left(\frac{P_1}{P_2}\right) \tag{3.10}$$

When one speaks of attenuating the transmitter output by so many dBs one must think in terms of a change in power *relative to the original output*. In fact, it is more convenient to think in terms of changes in output voltage rather than power, since the rf field strength and hence pulse widths (our values of interest) are proportional to voltage. Since power is proportional to the *square* of the voltage ($P = V^2/R$, where R is resistance), we may rewrite Eq. 3.10 as:

$$dB = 20\log_{10}\left(\frac{V_1}{V_2}\right) \equiv 20\log_{10}\left(\frac{PW_1}{PW_2}\right) \tag{3.11}$$

where PW_1 and PW_2 are the pulse widths for the same net tip angle at the two attenuations. Thus, if one wished to double a pulse width, an additional 20 log 2 dB (6 dB) attenuation is required. An alternative expression has the form:

$$+n\,dB \Rightarrow PW \times 10^{n/20} \tag{3.12}$$

or in other words, addition of *n* dB attenuation increases the pulse width by a factor of $10^{n/20}$. Some example attenuation values are presented in Table 3.5.

Table 3.5. Pulse width vs attenuation

Attenuation/dB	1	3	6	10	12	18	20	24
Pulse width factor	1.1	1.4	2.0	3.2	4.0	7.9	10	15.8

The additional attenuation of transmitter output levels increases the pulse width by the factors shown. Thus, adding 6 dB will double the pulse width whereas removing 6 dB will halve it.

An illuminating experiment to perform, if your spectrometer is able to alter the output attenuation internally, is to determine the pulse width over a wide range of attenuations. A plot of pulse widths vs $\log_{10}$(attenuation) should yield a single straight line over the full range. Discontinuities in the plot may arise when different power amplifiers come into use or when large attenuators are switched in place of many smaller ones.

Observe pulses: high sensitivity

We begin with the most basic NMR pulse calibration, that for the observed nucleus. When the spectrum of the analyte can be obtained in a single scan, or only a few scans, it is straightforward to calibrate the pulse width simply by following the behaviour of the magnetisation as the excitation pulse is increased, with nulls in the signal intensity occurring with a 180° or 360° pulse. To perform any pulse calibration it is essential that the probe is first tuned for the sample so that the results are reproducible. The transmitter frequency should be placed on-resonance for the signal to be monitored, to eliminate potential inaccuracies arising from off-resonance effects, and a spectrum recorded with a pulse width less than 90°, say 2–3 μs. The resonance is used to define the phase correction for all subsequent experiments, and is phased to produce the conventional positive absorption signal. The experiment is repeated with identical phase correction but with a progressively larger pulse width (Fig. 3.51). Passing through the 180° null then provides the PW_{180} calibration, whilst going beyond this yields an inverted resonance, so it becomes clear when you have gone too far. Detecting the null precisely can sometimes be tricky, as there is often some slight residual signal remaining, the exact appearance depending on the probe used, so in practice one aims for the minimum residual signal.

Between acquisitions there must be a delay sufficient for complete relaxation of the spins, and if signal averaging is used for each experiment, this must also be applied between each scan to obtain reliable calibrations. In proton spectroscopy, such delays are unlikely to be too much of a burden, but can be tediously long for slower-relaxing spins. If you have some feel for what the pulse width is likely to be, perhaps from similar samples or a 'reference' calibration, then a better approach in this case is to search for the 360° null. Since magnetisation remains close to the $+z$ axis, its recovery demands less time and the whole process can be performed more quickly. The process can

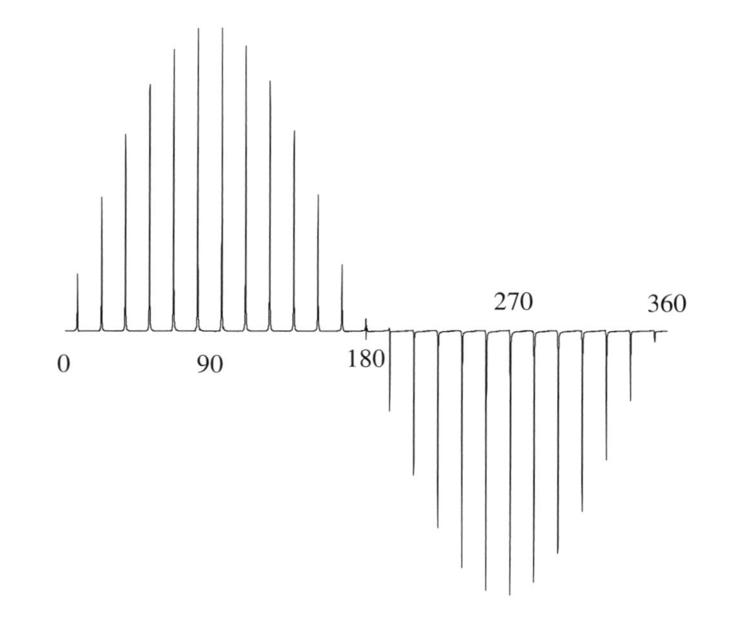

Figure 3.51. Pulse width calibration for the observe channel. A sequence of experiments is recorded with a progressively incremented excitation pulse. The maximum signal is produced by a 90° pulse and the first null with a 180° pulse. Either the 180° or 360° condition can be used for the calibration (but be sure to know which of these you are observing!).

be further simplified by monitoring the FID alone, which also has minimum amplitude with the 360° pulse. In any case, it is wise to check the null obtained is in fact the one you believe it to be, as all multiples of 180° give rise to nulls.

Observe pulses: low sensitivity

When the sample is too weak to allow its observation within a few scans, one possible approach is to use the method of signal averaging and to search for the 360° null as described above [31], although this could become a rather laborious affair. An alternative approach [32] requires that only two spectra are collected with the pulse width of the second being exactly double that of the first ($PW_2 = 2PW_1$). Once again, a sufficient delay between scans is required to avoid saturation effects and the spectrometer must be operating in an 'absolute intensity' scaling mode so that the relative signal intensities from both experiments can be compared. Heteronucleur spectra should also be acquired without enhancement by the NOE, and the use of the inverse-gated decoupling scheme (Section 4.2.3) allows such spectra to be collected with broadband decoupling during acquisition to aid sensitivity. Following some simple arithmetic [32], it can be shown that:

$$\theta_1 = \cos^{-1}\left(\frac{I_2}{2I_1}\right) \tag{3.13}$$

where I_1 and I_2 correspond to the measured signal intensities in the first and second experiment respectively. The duration of the pulse in the first experiment corresponds to the pulse angle θ_1, from which the 90° pulse width can be derived. For optimum results, θ_1 should be greater than 30° and θ_2, the tip angle in the second experiment, should be less than 180°, so again some rough feel for the likely answer is required for this approach to be efficient.

A more convenient approach in many instances is to perform the calibrations not on the sample of interest itself but on a suitable sample made up for the purposes of calibration that is strong enough to permit the observation of the 180/360° null directly and quickly. The calibration can then be recorded and used in future experiments, as this should be quite reproducible if similar solvents are used and the probe correctly tuned. The largest discrepancies occur between aqueous and organic solvents, especially when the aqueous solutions are highly ionic, and separate calibrations may be required.

'Indirect' pulses

In contrast to above, the calibration of pulses for nuclei other than that being observed must be performed 'indirectly'. This is often referred to as the calibration of the 'decoupler' pulses although this is something of a misnomer nowadays since the pulses may have nothing to do with decoupling itself but may be an integral part of the pulse sequence. Most commonly, the indirect channel is used in two modes; the first is to apply short, hard pulses typically at maximum power, the second is to apply a pulsed decoupling sequence for longer periods of time, at significantly less power. It is therefore commonplace to record indirect calibrations at these two power levels, for which the following method is equally applicable, and having performed the calibration at high power, the low power attenuation can be estimated from Eq. 3.11 prior to its accurate determination. With more recent spectrometers that make use of linear amplifiers together with internal power corrections, low power calibrations become superfluous since all pulse widths can be calculated from a single high power calibration with this equation, so it is worthwhile checking if your spectrometer is equipped for this.

To relate the behaviour of the indirect spin, A, to that being observed, X, one exploits the mutual heteronuclear scalar coupling which must exist between them, J_{AX}, for the calibration to be possible. The simplest calibration sequence

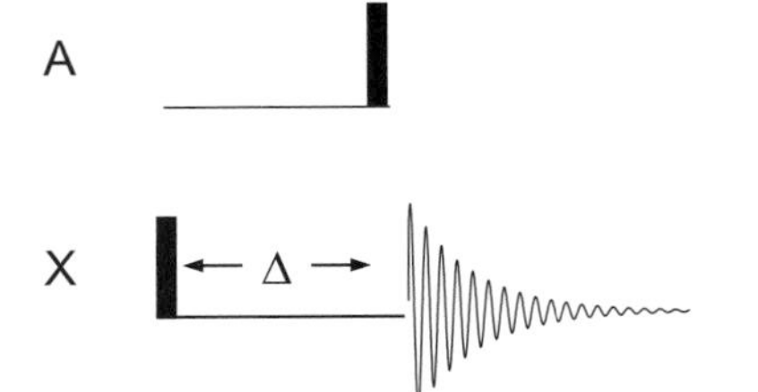

Figure 3.52. The sequence for the calibration of 'indirect' transmitter pulses (those on nuclei other than that being observed). The period Δ is set to $1/2J_{AX}$.

[33,34] suitable for spin-½ nuclei is shown in Fig. 3.52. The A and X pulses should be on resonance, the X-pulse calibration must already be known, and the value of J_{AX} measured directly from the (coupled) X spectrum. The delay Δ is set to $1/2J_{AX}$ seconds for an AX group so that the X-spin doublet vectors are antiphase after this evolution period. The subsequent A pulse renders the X magnetisation *unobservable* when $\theta = 90°$ as this generates pure heteronuclear multiple-quantum coherence which is unable to generate a detectable signal in the probe (see Chapter 5 for an explanation of this effect).

The approach is to begin with θ very small and to phase the spectrum so that the doublet lines are in antiphase. As θ increases the doublet intensity will decrease and become zero when θ is 90°, whilst beyond this the doublet reappears but with inverted phase (Fig. 3.53). If it is necessary to perform the calibration with an A_2X group, the delay Δ should be 1/4J and it is the outer lines of the triplet that behave as described above whilst the centre line remains unaffected. When calibrating lower powers for the purpose of broadband decoupling, it is usually more convenient to set the *duration* of the θ pulse according to the decoupler bandwidth required (Section 9.2) and to vary the output attenuation to achieve the null condition.

A readily available alternative that is particularly suitable for (although not restricted to) proton calibration in ^1H–X systems is the DEPT sequence (Chapter 4). This improves observation sensitivity by making use of polarisation transfer from ^1H to X and is dependent upon the faster relaxing protons for repetition rates in signal averaging. The 90° proton pulse is achieved when XH_2 or XH_3 groups pass through a null whereas XH groups have maximum intensity at this point (Fig. 3.54). One caveat here is that for the purposes of pulse calibration, *all* the proton pulses in the sequence need to be altered on each experiment, not just the final 'θ' proton pulse (see Chapter 4 for an explanation of this sequence).

As is often the way in experimental procedures, simplicity is best and the indirect pulse calibrations can most conveniently be performed on strong calibration samples using the sequence of Fig. 3.52. For ^1H–^{13}C systems, $CHCl_3$ or (^{13}C-labelled) methanoic acid are convenient materials for organic and aqueous calibrations respectively. When directly observing protons (inverse configuration), the major ^{12}C line of $CHCl_3$ will be apparent as a large dispersive signal between the ^{13}C satellites of interests (Fig. 3.55), but this should not pose a problem so long as the sensitivity is sufficient for the satellites to be observed. For the indirect calibration of natural abundance ^{15}N with ^1H observation, the dominant ^{14}N-bound proton resonance is rather broad owing to the quadrupolar ^{14}N nucleus and may mask the small ^{15}N satellites. In such cases, the ^{14}N line can be preferentially suppressed, owing to its faster transverse relaxation rate, by the application of a spin-echo prior to the sequence of Fig. 3.52; see reference [35] for further details. A more convenient alternative is to use ^{15}N labelled materials for calibration purposes.

Homonuclear decoupling field strength

Homonuclear decoupling requires the decoupling (B_2) field to be on during the acquisition of the FID and, as described in Chapter 4, this demands the use of rapid gating of the decoupler during this period, so the methods presented above are therefore no longer applicable. The field strengths used in homonuclear decoupling are considerably less than those required for pulsing and it is sometimes useful to have some measure of these when setting up decoupling experiments, to achieve the desired selectivity. A direct measure of the mean field strength produced during homonuclear decoupling can be arrived at through measurement of the Bloch–Siegert shift [36], which is the change in the resonance frequency of a line on the application of nearby irradiation. Provided the offset of the decoupler from the unperturbed resonance is greater than the decoupler field strength ($\nu_o - \nu_{dec} >> \gamma B_2$), then:

$$\gamma B_2 = ((\nu_{BS} - \nu_o) \cdot 2(\nu_o - \nu_{dec}))^{1/2}\,\text{Hz} \qquad (3.14)$$

where ν_{BS} is the signal frequency in the presence of the irradiating field, ν_o is that in the absence of the irradiating field and ν_{dec} the frequency of the applied rf. An example of such a calibration is shown in Fig. 3.56.

Heteronuclear decoupling field strength

Complementary to the pulse methods described above for the measurement of the 'indirect' pulse widths, is the measurement of the heteronuclear decoupling field strength. This method is applicable to the measurement of medium- to low-power outputs, such as those used in broadband decoupling, soft pulses and selective decoupling, and makes use of continuous, rather than pulsed, irradiation. Once again the method exploits the heteronuclear coupling between spins and observes the changes to this on the application of off-resonance heteronuclear decoupling [37]. Consider again the AX pair in which continuous A-spin decoupling is applied so far from the A resonance that it has no effect; the X spectrum displays a doublet. As the decoupler frequency moves towards the A resonance the observed coupling begins to collapse until eventually the X spectrum displays only a singlet when decoupling is on-resonance. Measurement of the reduced splitting in the intermediate stages provides a measure of the field strength according to [37]:

$$\gamma B_2 = \left[\left(\frac{J \cdot \nu_{off}}{J_r}\right)^2 + \frac{(J_r^2 - J^2)}{4} - \nu_{off}^2\right]^{1/2}\,\text{Hz} \qquad (3.15)$$

where J is the AX coupling constant, J_r the reduced splitting in the presence of the decoupling field and ν_{off} the decoupler offset from the A resonance. When the decoupler field strength is much larger than the decoupler resonance offset, this simplifies to:

$$\gamma B_2 \approx \frac{J \cdot \nu_{off}}{J_r}\,\text{Hz} \qquad (3.16)$$

Thus, by measuring the resonance splitting in the absence and presence of the irradiation, the field strength is readily calculated. For ^1H–^{13}C spectroscopy, the one-bond coupling is suitable for the medium to strong powers, whilst for very weak decoupler fields (<50Hz) it may be necessary to utilise long-range ^1H–^{13}C couplings to obtain significant changes in splitting. The calibration of proton field strength with this method is illustrated in Fig. 3.57.

3.5.2. Pulsed field gradients

Numerous modern NMR experiments utilise field gradient pulses for signal selection or rejection (see Chapter 5), requiring these to be applied for well defined durations at known gradient strengths. Experiments presented in the literature will (or should) state the gradient strengths required to achieve the desired results, so it is necessary to have some knowledge of the strengths provided by the instrument if wishing to implement these techniques. Practical procedures for determining gradient strengths and gradient recovery periods are described here, as it is these two parameters that must be defined by the operator when preparing a gradient selected experiment. Discussions are restricted to static B_0 gradients rather than the far less widely used rf B_1 gradients [38,39]. However, the use of B_1 gradients has a number of significant advantages [40], and it may be that these become more widely used as the theoretical and technical limitations are overcome.

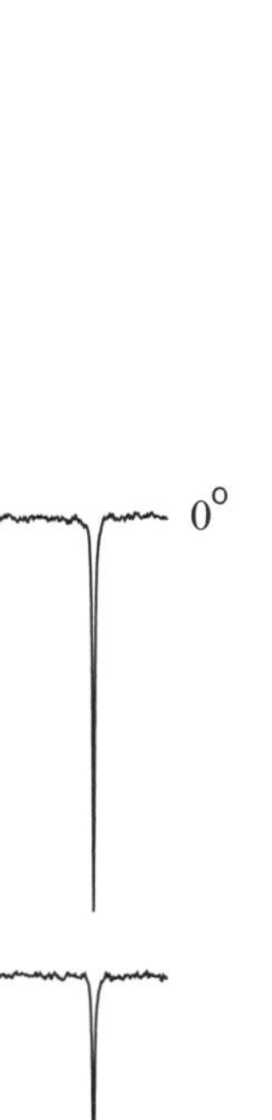

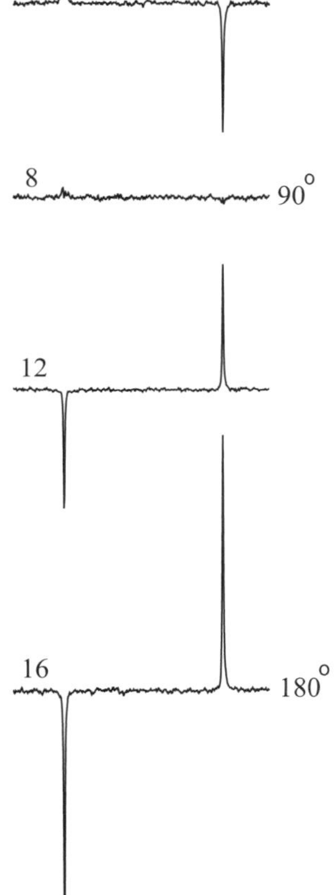

Figure 3.53. Indirect calibration of the proton pulse width with carbon observation using the sequence of Fig. 3.52. As the proton pulse width increases the carbon signal diminishes until it disappears at the $^1H(90^o)$ condition. Going beyond this causes the signal to reappear but with inverted phase (the same phase correction is used for all spectra). The sample is ^{13}C-labelled methanoic acid in D_2O.

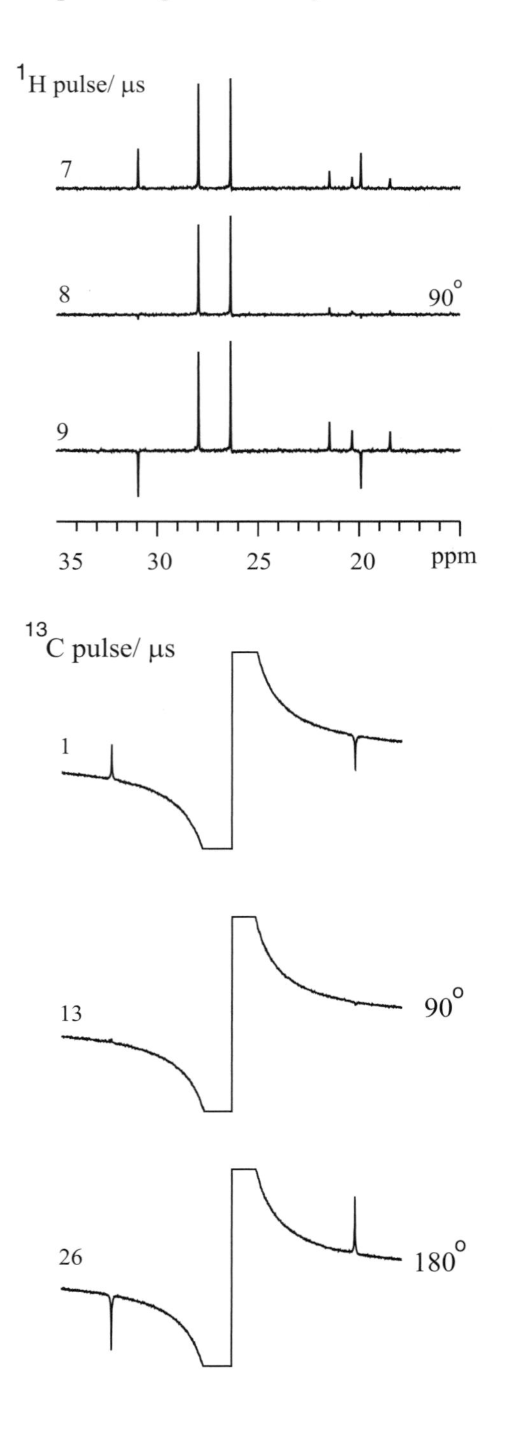

Figure 3.54. Calibration of indirect proton pulses using the DEPT sequence. When the proton editing pulse in the sequence is exactly 90°, only methine resonances are apparent (see Chapter 4 also). The sample is menthol in $CDCl_3$.

Figure 3.55. Calibration of indirect pulses (here ^{13}C) with proton observation using the sequence of Fig. 3.52. The parent ^{1}H(^{12}C) resonance appears as a large dispersive signal, but the behaviour of the satellites is clear, again disappearing with a 90° pulse on the indirect channel. The sample is $CHCl_3$.

Gradient strengths

For the application of pulsed field gradients in high-resolution NMR, knowledge of the exact gradient strengths produced is not, in fact, crucial to the success of the experiments. This is in contrast to the application of rf pulses, where precise pulse widths *are* required. Provided the gradients are *sufficiently strong* for the experiment, the critical factors are pulse reproducibility and the ability to produce gradients pulses of precise amplitude *ratios*. The procedure described below can be used to calibrate gradient strengths and to check the linearity of the gradient amplifier output.

Essentially there are two approaches to gradient calibration, one based on the measurement of molecular diffusion, the other using an image of a suit-

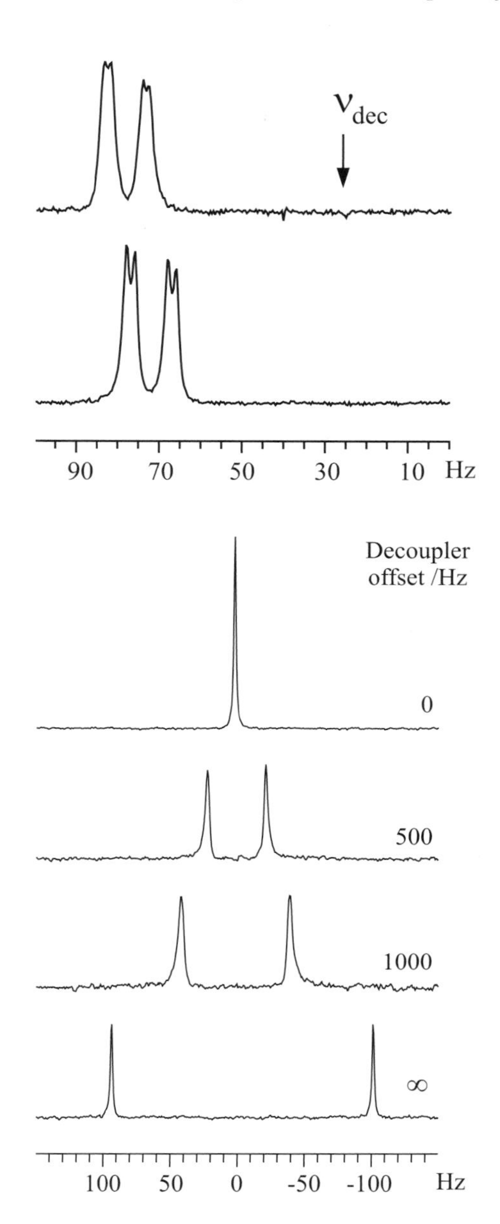

Figure 3.56. Calibration of the homonuclear decoupling field strength via the Bloch–Siegert shift. The decoupler offset from the unperturbed resonance was 47.5 Hz causing a shift of 5.5 Hz, indicating the mean rf field to be 23 Hz.

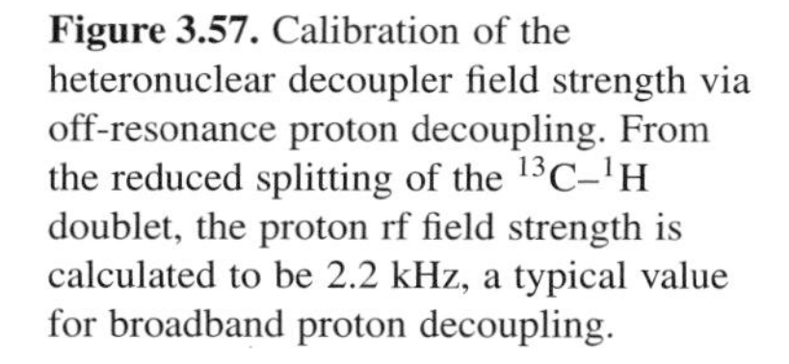

Figure 3.57. Calibration of the heteronuclear decoupler field strength via off-resonance proton decoupling. From the reduced splitting of the ^{13}C–^{1}H doublet, the proton rf field strength is calculated to be 2.2 kHz, a typical value for broadband proton decoupling.

able calibration sample. With the relatively low gradient field strengths used in high-resolution NMR, the second approach is rather more straightforward and does not depend upon precise experimental conditions or the need for a reference solute with well defined diffusion properties. The calibration of *z*-gradients is based on the measurement of a one-dimensional image profile of a phantom of known spatial dimensions placed in a standard NMR tube containing H_2O (Fig. 3.58). The phantom is typically a disk 2–4 mm in length of rubber or teflon (a cut-down vortex suppresser for example), which is positioned in the tube such that it sits at the centre of the rf coil. The sample is not spun and need not be locked, in which case the lock field sweep must be stopped. The proton spectrum is first acquired in the absence of the field gradient and the transmitter placed on resonance. The spectrum is then acquired with a very large spectral width, say 50 kHz, in the presence of the field gradient. This can, in principle, be achieved by a simple pulse-acquire scheme with the gradient turned on throughout (Fig. 3.59a), although this approach has its limitations because the application of rf pulses in the presence of the gradient requires excitation over a very wide bandwidth. Thus, it is preferable

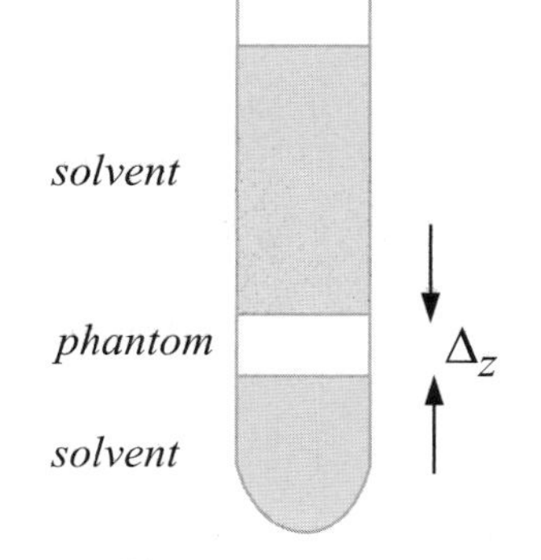

Figure 3.58. A phantom sample suitable for calibrating *z*-axis gradient field strength. The teflon or rubber phantom has precise dimensions (Δ_z, typically 2–4 mm in height) and excludes solvent from a slice of the tube. It is positioned in the tube such that it sits at the centre of the rf coil when placed in the probe.

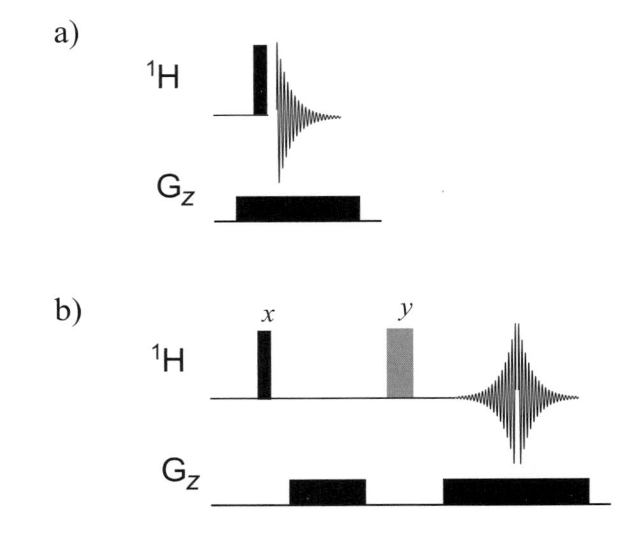

Figure 3.59. Sequences for collecting gradient profiles for the calibration of the *z*-axis gradient field strength. Sequence (a) collects the FID directly in the presence of the applied gradient whilst sequence (b) makes use of a gradient echo (see text). In (b) the second gradient has the same strength but twice the duration of the first.

to record the spectrum with a gradient spin-echo sequence (Fig. 3.59b), in which the second of the two gradients has the same strength but twice the duration of the first, and is applied throughout the acquisition period during which the spin-echo refocuses. In either case, the acquisition time must be kept short (10 ms or less) and relatively low gradient strengths applied. The resulting data are processed with a *non-shifted* sine-bell window function (this matches the shape of the echo FID) and magnitude calculation. The spectrum then displays a dip in the profile corresponding to the region of the sample that contains the phantom, and hence no solution, (Fig. 3.60) from which the gradient strength may be calculated as:

$$G_z = \frac{2\pi\Delta\nu}{\gamma\Delta_z} \equiv \frac{\Delta\nu}{4358 \times \Delta_z}\,\mathrm{G\,cm^{-1}} \text{ or } \frac{\Delta\nu}{4.358 \times 10^5 \times \Delta_z}\,\mathrm{T\,m^{-1}} \quad (3.17)$$

where $\Delta\nu$ is the width of the dip in hertz and Δ_z is the height of the phantom in cm. The SI units for a field gradient are T m^{-1}, although it is common for values quoted in the literature to be given in G cm^{-1} (1 G cm^{-1} $\equiv$ 0.01 T m^{-1}). For calibrations of *x* and *y* gradients a similar method may be used, in which case the phantom is not required and the internal diameter of the NMR tube defines the sample width, with the areas outside the tube providing the empty region, and the width of the resulting 'peak' used in the above expression.

Repeating the above procedure with increasing gradient field strengths indicates the linearity of the amplifier, with larger gradients producing a proportionally wider dip in the profile. Fig. 3.61 shows the result obtained for a 5 mm inverse *z*-gradient probe. Assuming linearity over the whole amplifier range, the maximum gradient strength for this system is 0.51 T m^{-1} (51 G cm^{-1}).

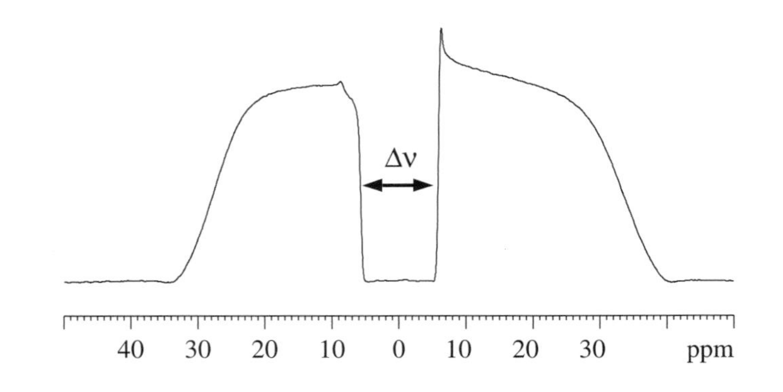

Figure 3.60. A typical gradient profile collected with the sequence of Fig. 3.59b. Here the dip is 5896 Hz wide, corresponding to a gradient strength of 0.051 T m^{-1} (10% of maximum).

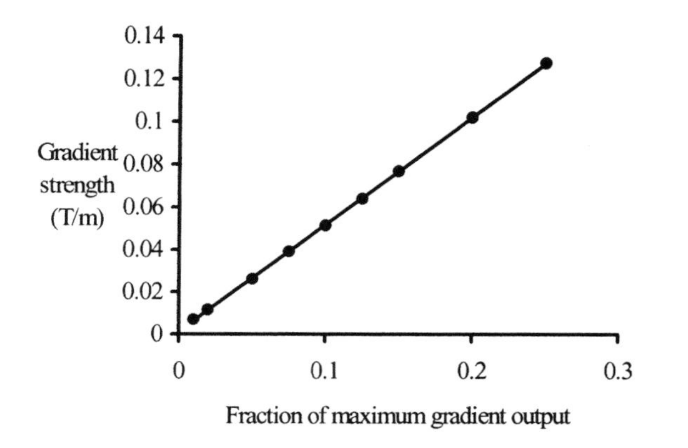

Figure 3.61. Calibration of the gradient amplifier linearity using the sequence of Fig. 3.59b. Assuming system linearity over the full output range and extrapolating the calibration, the maximum gradient strength provided by this system is 0.51 T m^{-1} (51 G cm^{-1}).

Gradient recovery times

As is discussed in Chapter 5, significant distortions in the magnetic field surrounding a sample will result following the application of a gradient pulse and to acquire a high-resolution spectrum the eddy currents responsible for these field distortions must be allowed to decay before data is collected. A simple scheme for the measurement of the gradient recovery period is shown in Fig. 3.62, in which the initial gradient pulse is followed by a variable recovery period, after which the data is acquired. Distortions of the spectrum are readily seen by observing a sharp proton singlet, such as that of chloroform. Initially, the recovery period, τ, is set to a large value (1 s) to establish the reference spectrum, and the value progressively decreased until distortions appear in the spectrum (Fig. 3.63). This provides some indication of the recovery period required following a gradient pulse, during which data should not be collected or further gradient or rf pulses applied. The success of high-resolution gradient-selected NMR is critically dependent on being able to use very short recovery periods, which are typically around 100 μs on modern instruments that have actively-shielded probeheads. Whilst NMR experiments generally utilise *shaped* gradient pulses (as these help reduce eddy current generation), a more demanding test of hardware performance can be achieved by using

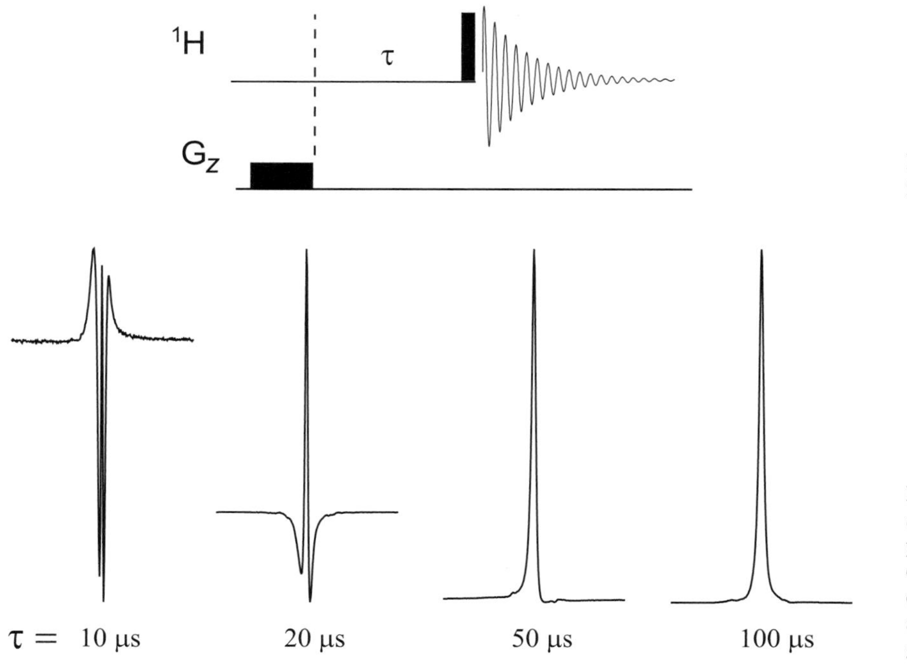

Figure 3.62. A simple scheme for investigating gradient recovery times.

Figure 3.63. Gradient recovery after application of a 1 ms square gradient pulse of 0.25 Tm^{-1} (50% of maximum output). The sample is chloroform and data were acquired with an actively-shielded 5 mm inverse *z*-axis gradient probehead.

rectangular gradients, as in Fig. 3.63. Performing this test with a half-sine shaped gradient pulse showed complete recovery of the resonance within 10 μs.

3.5.3. Sample temperature

Many spectrometers are equipped with facilities to monitor and regulate the temperature within a probe head. Usually the sensor takes the form of a thermocouple whose tip is placed close to the sample in the gas flow used to provide temperature regulation. However, the readings provided by these systems may not reflect the true temperature of the sample unless they have been subject to appropriate calibration. One approach to such calibration is to measure a specific NMR parameter that has a known temperature dependence to provide a more direct reading of *sample* temperature. Whilst numerous possibilities have been proposed as reference materials [41], two have become accepted as the standard temperature calibration samples for solution spectroscopy. These are methanol for the range 175–310 K and 1,2-ethanediol (ethylene glycol) for 300–400 K.

For the low temperature calibration, neat MeOH is used, with a trace of HCl to sharpen the resonances, for which the following equation holds [42,43]:

$$T(K) = 403 - 29.53\Delta\delta - 23.87(\Delta\delta)^2 \tag{3.18}$$

where $\Delta\delta$ is the OH–CH_3 chemical shift difference (ppm). This expression may be approximated to three linear equations of the form:

$$175\text{–}220\,\text{K}: \quad T(K) = 537.4 - 143.1\Delta\delta \tag{3.19}$$

$$220\text{–}270\,\text{K}: \quad T(K) = 498.4 - 125.3\Delta\delta \tag{3.20}$$

$$270\text{–}310\,\text{K}: \quad T(K) = 468.1 - 108.9\Delta\delta \tag{3.21}$$

which are presented as calibration charts in Fig. 3.64. For calibrations with neat ethane-1,2-diol, the appropriate equation is [43]:

$$T(K) = 466.0 - 101.6\Delta\delta \tag{3.22}$$

where $\Delta\delta$ now represents the OH–CH_2 chemical shift difference (ppm) (Fig. 3.65). For both samples, the peak shift difference decreases as the sample

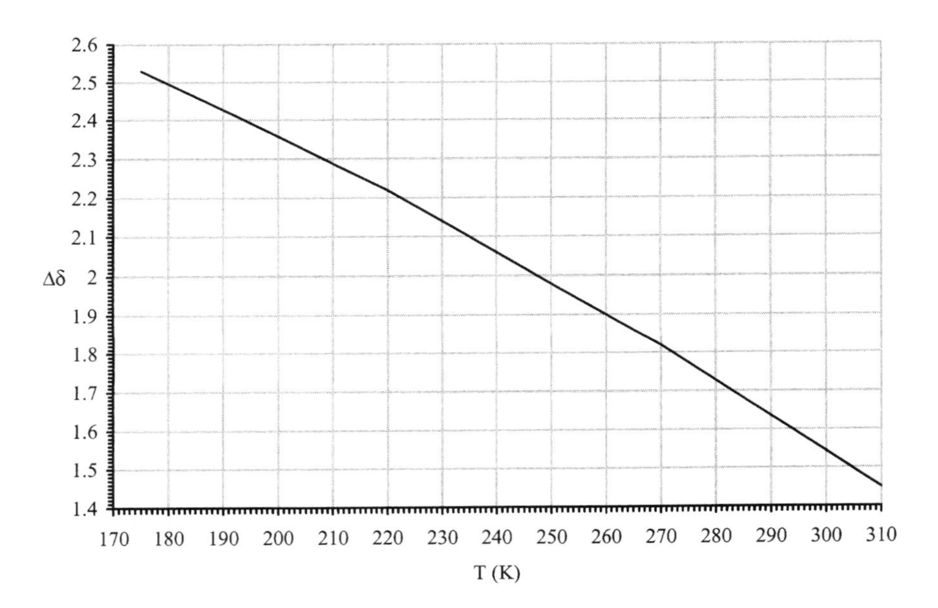

Figure 3.64. Temperature calibration chart for neat methanol. The shift difference $\Delta\delta$ (ppm) is measured between the CH_3 and OH resonances.

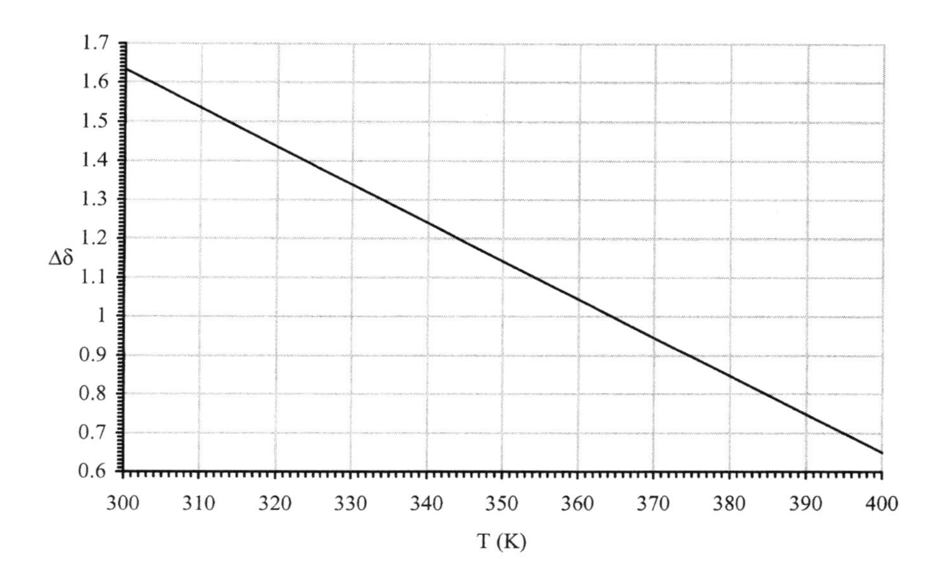

Figure 3.65. Temperature calibration chart for neat Ethane-1,2-diol (ethylene glycol). The shift difference $\Delta\delta$ (ppm) is measured between the CH_2 and OH resonances.

temperature increases due to the reduced intermolecular hydrogen-bonding, which in turn causes the hydroxyl resonance to move to lower frequency.

Before making temperature measurements, the sample should be allowed to equilibrate for at least 5–10 minutes at each new temperature, and it is also good practice to take a number of measurements to ensure they do not vary before proceeding with the calculation.

3.6. SPECTROMETER PERFORMANCE TESTS

There exist a large number of procedures designed to test various aspects of instrument performance and it would certainly be inappropriate, not to mention rather tedious, to document them in a book of this sort. Instead I want to introduce briefly only those that are likely to be of use in a routine spectrometer 'health check' and as such this section will be of most interest to anyone with responsibility for the upkeep of an instrument. These tests are generally accepted in the NMR community and by instrument manufacturers as providing a quantitative measure of instrument performance and, in theory at least, should also provide a comparison between the capabilities of different spectrometers. Naturally, this leads to aggressive claims by instrument vendors regarding their results and some caution is required when comparing advertised figures. This demands attention to the details of procedures used, for example, whether a lineshape test was performed with or without sample rotation or how wide a region of the spectrum was used for estimating the noise level in a sensitivity test. When utilising these tests for in-house checks, consistency in your methodology is the important factor if the results are to be compared with previous measurements. Ideally these tests should be performed at regular intervals, say every six months or once a year, and appropriate records kept. Additionally more frequent updating of the 'master' shim settings may also be required for optimal performance, although this tends to be very dependent on instrument stability (as well as that of the surrounding environment). One should also be aware that, perhaps not surprisingly, the installation test results achievable on a modern state-of-the-art instrument will certainly be substantially better than the initial test data of an instrument installed some years ago (see, for example, the discussions on probes in Section 3.4.1). The idea of what is an acceptable test result for an older instrument should therefore be scaled accordingly and reference to previous records becomes essential if one wishes to gauge instrument performance.

3.6.1. Lineshape and resolution

The most important test, in may be argued, is that for the NMR resonance lineshape, since only when this is optimised can other quantities also be optimum. The test measures the properties of a suitable singlet resonance which, in a perfectly homogeneous field and in the absence of other distortions, produces a precise Lorentzian lineshape, or in other words, its FID possesses an exact exponential decay. Naturally the only way to achieve this is through careful optimisation of the magnetic field via shimming, which can take many hours to achieve good test results, depending on the initial state of the system. Nowadays both the lineshape and resolution measurements are often taken from a single set of test data. The *lineshape* is defined by the width of the resonance at 0.55% and 0.11% of the peak height (these numbers having evolved from measurement of proton linewidths at the height of the ^{13}C satellites and at one-fifth thereof). The *resolution* is defined by the half-height linewidth, $\Delta\nu_{1/2}$ of the resonance. In fact, by definition, this cannot be a true measure of resolution (how can one measure resolving power with only a single line?), rather the ability of the instrument to separate neighbouring lines is implied by the narrowness of the singlet. For a genuine Lorentzian line, the widths at 0.55 and 0.11% should be 13.5 and 30 times that at half-height respectively, although a check for this is often never performed as attempts are often made (erroneously) to simply minimise all measurements. The tests may be performed on both spinning and static samples (be sure you know which if comparing results), the first being of relevance principally for highest-resolution proton measurements whereas the second indicates suitability for non-spinning one- and two-dimensional experiments.

The proton lineshape test uses chloroform in deuteroacetone typically at concentrations of 3% at or below 400 MHz, and 1% at or above 500 MHz. Older instruments and/or probes of lower sensitivity or observations via outer 'decoupler coils', may require 10% at 200 MHz and 3% at 500 MHz to prevent noise interfering with measurements close to the baseline. A single scan is collected and the data recorded under conditions of high digital resolution (acquisition time of 16 s ensuring the FID has decayed to zero) and processed without window functions. Don't be tempted to make measurements at the height of the satellites themselves unless these are confirmed by measurement to be 0.55%. Since these arise from protons bound to ^{13}C, which relax faster than those of the parent line, they may be relatively enhanced should full equilibrium not be established after previous pulses. The test results for a 400 MHz instrument is shown in Fig. 3.66. The traditional test for proton resolution which dates back to the CW era (*o*-dichlorobenzene in deuteroacetone) is becoming less used nowadays, certainly by instrument manufacturers, and seems destined to pass into NMR history.

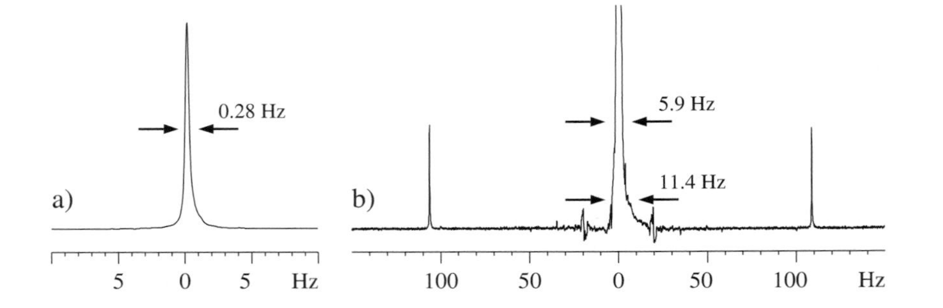

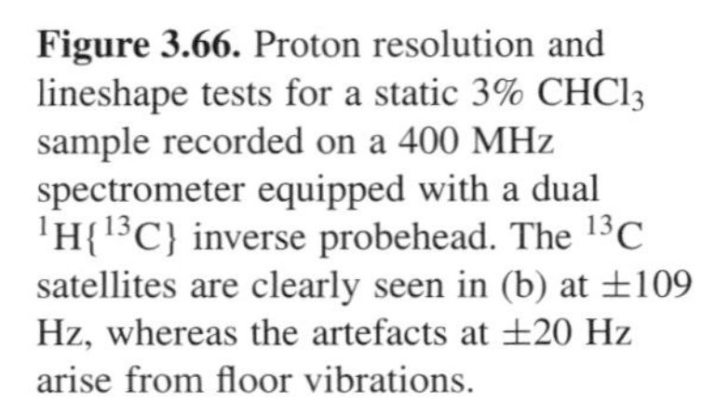

Figure 3.66. Proton resolution and lineshape tests for a static 3% $CHCl_3$ sample recorded on a 400 MHz spectrometer equipped with a dual $^{1}H\{^{13}C\}$ inverse probehead. The ^{13}C satellites are clearly seen in (b) at $\pm$109 Hz, whereas the artefacts at $\pm$20 Hz arise from floor vibrations.

Lineshape and resolution tests on other nuclei follow a similar procedure to that above. Not all nuclei available with a given probe need be tested and typically only tests for 'inner' and 'outer' coil observations on multinuclear probes are required. This means the second test will often involve carbon-13 for which two samples are in widespread use; the ASTM (American Society for Testing and Materials) test sample (40% *p*-dioxane in deuterobenzene; also used for the sensitivity test) or 80% benzene in deuteroacetone. In either case on-resonance continuous-wave (CW) decoupling of protons should be used as this provides improved results for a single resonance relative to broadband decoupling. Rather long (30–40 s) acquisition times will be required for a well shimmed system.

3.6.2. Sensitivity

A great disadvantage of NMR spectroscopy relative to many other analytical techniques is the intrinsically low sensitivity from which it suffers. This, of course, is greatly outweighed by its numerous benefits, yet is still one of the more likely causes of experiment 'failure' and so deserves serious attention. The term 'sensitivity' strictly defines a minimum amount of material that is detectable under defined conditions, but is used rather loosely throughout NMR and often interchangeably with 'signal-to-noise-ratio'. The instrument sensitivity test is indeed a signal-to-noise measurement in which the peak height of the analyte is compared with the noise level in the spectrum. For maximum peak height, optimum lineshape is essential and generally the lineshape and resolution test should be performed first and the shim settings re-optimised on the sensitivity sample (for similar reasons, sensitivity tests are most often performed with sample spinning). Likewise the probe must be tuned for the test sample and the 90° pulse determined accurately. The definition of noise intensity is of prime importance for this measurement since different approaches may lead to differing test results. This has particular significance nowadays as the older method of 'manual' noise estimation is being superseded by automated computational routines provided by instrument vendors. The traditional method compares peak intensity, P, to twice the root-mean-square noise level, N_{rms}, in the spectrum. Although not directly measurable, N_{rms} may be estimated from the peak-to-peak noise of the baseline, N_{pp}, whereby N_{rms} is one-fifth N_{pp} as measured directly on paper plots. The signal-to-noise is thus given by

$$\frac{S}{N} = \frac{2.5P}{N_{pp}} \qquad (3.23)$$

The peak-to-peak measurement must include all noise bands within the defined frequency window but is somewhat susceptible to what one might call 'operator optimism' in the exclusion of 'spikes' or 'glitches'. Whilst computational methods for determining noise have the advantage of removing such operator bias, the algorithms used may leave an air of mystery surrounding the measurement which may cloud comparisons between vendors' published results. Whilst this may be of significance from the point of view of instrument purchasing, for routine instrument maintenance it is consistency of approach that is important for performance monitoring and the computerised approach is likely to be more reliable.

The proton sensitivity test uses 0.1% ethylbenzene in deuterochloroform. As for all sensitivity tests, a single scan spectrum is recorded following a 90° pulse on a fully equilibrated sample (relaxation delay of 60 s in this case). The spectrum is processed with matched filtration corresponding to a line-broadening of 1 Hz for the methylene group on which the signal intensity is measured (this resonance is broadened slightly by unresolved long-range

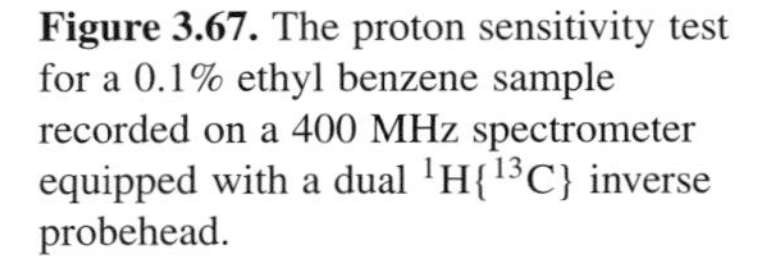

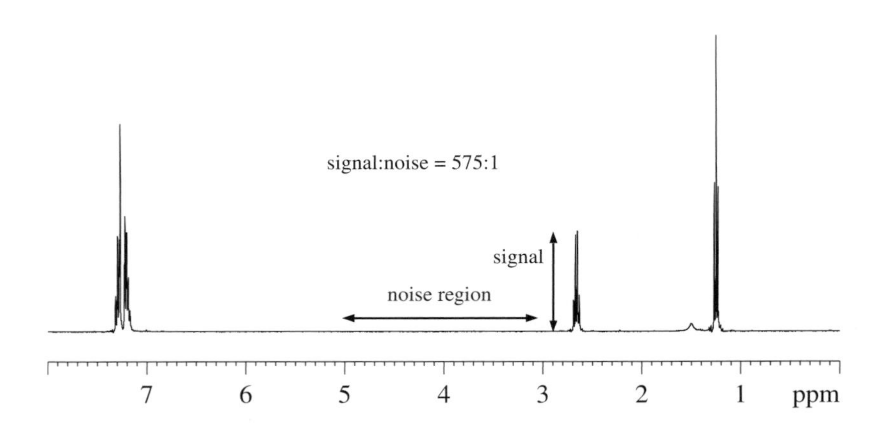

Figure 3.67. The proton sensitivity test for a 0.1% ethyl benzene sample recorded on a 400 MHz spectrometer equipped with a dual $^{1}H\{^{13}C\}$ inverse probehead.

couplings to the aromatic protons). Traditionally the noise is measured over the region between 3 and 5 ppm although there is an increasing tendency for instrument manufacturers to limit this to a 200 Hz window within this region, which invariably produces more impressive signal-to-noise figures (be sure you know which method is used if comparing results!). Those for a 400 MHz instrument are presented in Fig. 3.67.

The carbon sensitivity test comes in two guises, without and with proton decoupling. These give rise to quite different results so its is again important to be aware of which approach has been used. Whilst the first provides an indication of absolute instrument sensitivity, the second represents a more realistic test of overall performance since it also takes into account the efficiency of proton decoupling and is thus more akin to how one performs genuine experiments. Both approaches acquire carbon spectra under atypical conditions of high digital resolution to correctly define peak shapes, demanding acquisition times of around 4 s. The proton coupled version makes use of the ASTM sample again (40% *p*-dioxane in deuterobenzene) but this time measures peak height of the deuterobenzene triplet. A line-broadening of 3.5 Hz is applied and the noise measured over the 80 to 120 ppm window. The proton decoupled version uses 10% ethyl benzene in deuterochloroform and employs *broadband* composite-pulse decoupling usually via the WALTZ-16 sequence (Chapter 9). A 0.3 Hz line-broadening is used and the noise recorded over the same region, with the peak height determined for the tallest aromatic resonance. Tuning of the proton coil and appropriate calibration of the proton decoupling pulses are required in this case for optimum results. Test samples for other common nuclei are summarised in Table 3.6. Should you have frequent interests in other nuclei, a suitable standard should be decided upon and used for future measurements.

Table 3.6. Standard sensitivity test samples of some common nuclei

Nucleus	Sensitivity test sample	Notes
^{1}H	0.1% Ethylbenzene in $CDCl_3$	
^{13}C	10% Ethylbenzene in $CDCl_3$	Use broadband proton decoupling
^{13}C	40% dioxane in C_6D_6 (ASTM sample)	No decoupling, C_6D_6 used for measurement
^{31}P	0.0485 M triphenylphosphate in d_6-acetone	No decoupling
^{19}F	0.05% trifluorotoluene in $CDCl_3$	No decoupling
^{15}N	90% formamide in d_6-DMSO	Use inverse gated decoupling to suppress negative NOE
^{29}Si	85% hexamethyldisiloxane in d_6-benzene	No decoupling

3.6.3. Solvent presaturation

Interest in biologically and medicinally important materials frequently demands NMR analysis to be undertaken on samples in 90% H_2O if solvent-exchangeable protons are also to be observed. The various tests presented so far have all made use of organic solvents yet the dielectric properties of these and of water are substantially different and a probe that performs well for, say, chloroform may not be optimum for an ionic aqueous solution, particularly with regard to sensitivity. Accurate tuning of the probe and pulse width calibration play an important role here, but protonated aqueous solutions also demand effective suppression of the solvent resonance if the analyte is to be observed (see Chapter 9 for discussions on suppression methods). The solvent suppression test makes use of solvent presaturation and can be used to measure a number of performance characteristics including the suppression capability, resolution and sensitivity. Good suppression performance places high demands on both static (B_0) and rf (B_1) field homogeneity. A narrow lineshape down to the baseline, and hence good B_0 homogeneity, is again a pre-requisite for good results and ensures all solvent nuclei resonate within a small frequency window. Good B_1 homogeneity means solvent nuclei in all regions of the sample experience similar rf power and are thus suppressed to the same degree. Much of the residual solvent signal that is observed following presaturation arises from peripheral regions of the sample that experience a reduced B_1 field.

The test sample is 2 mM sucrose with 0.5 mM sodium 2,2-dimethyl-2-silapentane-5-sulphonate (DSS) in 90% $H_20/10\%$ D_2O, plus a trace of sodium azide to suppress bacterial growth. In the absence of suppression only the solvent resonance is observed. The test involves on-resonance presaturation of the solvent for a two second period, followed by acquisition with a 90° pulse. The presaturation power is selected to attenuate the resonance significantly ($\gamma B_1 \approx 25$ Hz) and this setting used in future comparative tests. Naturally, higher powers will produce greater suppression but these should not be so high as to reduce the neighbouring sucrose resonances. Typically 8 transients are collected following two 'dummy' scans, which achieve a steady-state. The suppression performance is judged by measuring the linewidth of the residual solvent signal at 50% and 10% of the height of the DSS resonance. With probes of recent design that have appropriate screening of the rf coil leads, the 50% linewidth should be somewhat less than 100 Hz and reasonably symmetrical. Much older probes may show poorer performance and may be plagued by wide unsymmetrical humps (despite careful attention to shim optimisation) which result from signal pick-up in unscreened coils leads [44]. Resolution is judged by the splitting of the anomeric proton doublet at around 5.4 ppm, which should be resolved at least down to 40% of the anomeric peak height. Sensitivity may also be measured from this resonance with the noise determined for the 5.5 to 7.0 ppm region using either of the methods described above (baseline

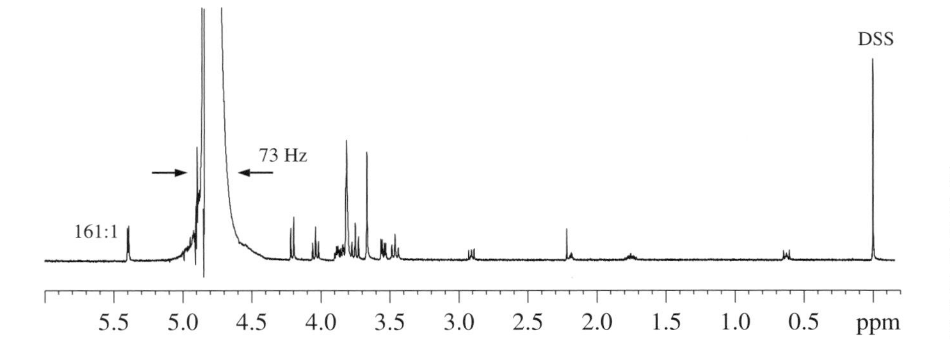

Figure 3.68. The proton solvent presaturation test. The data were acquired on a 400 MHz spectrometer equipped with a dual $^1H\{^{13}C\}$ inverse probe and the sample was 2 mM sucrose and 0.5 mM DSS in 90% $H_2O/10\%$ D_2O.

correction may be required). Test results for a 400 MHz $^{1}H\{^{13}C\}$ probe using a moderate presaturation power are shown in Fig. 3.68; further reductions in the water resonance could be achieved through the use of greater rf power.

REFERENCES

[1] J.N. Shoolery, *Prog. Nucl. Magn. Reson. Spectrosc.*, 1995, **28**, 37–52.
[2] E. Bartholdi and R.R. Ernst, *J. Magn. Reson.*, 1973, **11**, 9–19.
[3] M.B. Comisarow, *J. Magn. Reson.*, 1984, **58**, 209–218.
[4] J.J. Led and H. Gesmar, *Chem. Rev.*, 1991, **91**, 1413–1426.
[5] J.C. Hoch and A.S. Stern, NMR Data Processing, Wiley-Liss, New York, 1996.
[6] W.F. Reynolds, M. Yu, R.G. Enriquez and I. Leon, *Magn. Reson. Chem.*, 1997, **35**, 505–519.
[7] C.J. Turner and H.D.W. Hill, *J. Magn. Reson.*, 1986, **66**, 410–421.
[8] D.I. Hoult and R.E. Richards, *Proc. Roy. Soc. (Lond.)*, 1975, **A344**, 311.
[9] M.A. Delsuc and J.Y. Lallemand, *J. Magn. Reson.*, 1986, **69**, 504–507.
[10] E. Kupce, J. Boyd and I. Campbell, *J. Magn. Reson. (A)*, 1994, **109**, 260–262.
[11] M.E. Rosen, *J. Magn. Reson. (A)*, 1994, **107**, 119–125.
[12] R.R. Ernst, *Adv. Mag. Reson*, 1966, **2**, 1–135.
[13] A.G. Ferrige and J.C. Lindon, *J. Magn. Reson.*, 1978, **31**, 337–340.
[14] G.A. Pearson, *J. Magn. Reson.*, 1987, **74**, 541–545.
[15] D.D. Traficante and G.A. Nemeth, *J. Magn. Reson.*, 1987, **71**, 237–245.
[16] C.R. Pacheco and D.D. Traficante, *J. Magn. Reson. (A)*, 1996, **120**, 116–120.
[17] J.N.S. Evans, Biomolecular NMR Spectroscopy, Oxford University Press, Oxford, 1995.
[18] J. Cavanagh, W.J. Fairbrother, A.G. Palmer and N.J. Skelton, Protein NMR Spectroscopy. Principles and Practice, Academic Press, San Diego, 1996.
[19] R.K. Harris and B.E. Mann, eds. NMR and the Periodic Table, Academic Press, London, 1978.
[20] R.W. Dykstra, *J. Magn. Reson. (A)*, 1995, **112**, 255–257.
[21] R.C. Crouch and G.E. Martin, *Magn. Reson. Chem.*, 1992, **30**, S66–S70.
[22] P. Styles, N.F. Soffe, C.A. Scott, D.A. Cragg, F. Row, D.J. White and P.C.J. White, *J. Magn. Reson.*, 1984, **60**, 397–404.
[23] P. Styles, N.F. Soffe and C.A. Scott, *J. Magn. Reson.*, 1989, **84**, 376–378.
[24] H.D.W. Hill, *Magnetic Moments (Varian Newsletter)*, 1996, **8**, 4–6.
[25] N. Bloembergen and R.V. Pound, *Phys. Rev.*, 1954, **95**, 8–12.
[26] W.W. Conover, in Topics in Carbon-13 NMR Spectrosocopy, ed. G.C. Levy, Wiley, New York, 1983.
[27] G.N. Chmurny and D.I. Hoult, *Concept. Magn. Reson.*, 1990, **2**, 131–159.
[28] P.C.M. Van Zijl, S. Sukumar, M.O. Johnson, P. Webb and R.E. Hurd, *J. Magn. Reson. (A)*, 1994, **111**, 203–207.
[29] H. Barjat, P.B. Chilvers, B.K. Fetler, T.J. Horne and G.A. Morris, *J. Magn. Reson.*, 1997, **125**, 197–201.
[30] S. Sukumar, M.O. Johnson, R.E. Hurd and P.C.M. van Zijl, *J. Magn. Reson.*, 1997, **125**, 159–162.
[31] J.R. Wesener and H. Günther, *J. Magn. Reson.*, 1985, **62**, 158–162.
[32] E. Haupt, *J. Magn. Reson.*, 1982, **49**, 358–364.
[33] D.M. Thomas, M.R. Bendall, D.T. Pegg, D.M. Doddrell and J. Field, *J. Magn. Reson.*, 1981, **42**, 298–306.
[34] A. Bax, *J. Magn. Reson.*, 1983, **52**, 76–80.
[35] E. Kupce and B. Wrackmeyer, *J. Magn. Reson.*, 1991, **94**, 170–173.
[36] F. Bloch and A. Siegert, *Phys. Rev.*, 1940, **57**, 522–527.
[37] S.D. Simova, *J. Magn. Reson.*, 1985, **63**, 583–586.
[38] W.S. Price, *Ann. Rep. NMR Spectrosc.*, 1996, **32**, 51–142.
[39] D.G. Cory, F.H. Laukien and W.E. Maas, *J. Magn. Reson. (A)*, 1993, **105**, 223–229.
[40] W.E. Maas, F. Laukien and D.G. Cory, *J. Magn. Reson. (A)*, 1993, **103**, 115–117.
[41] M.L. Martin, J.-J. Delpeuch and G.J. Martin, Practical NMR Spectroscopy, Heydon, London, 1980.
[42] A.L. Van Geet, *Anal. Chem.*, 1970, **42**, 679–680.
[43] D.S. Raidford, C.L. Fisk and E.D. Becker, *Anal. Chem.*, 1979, **51**, 2050–2051.
[44] R.W. Dykstra, *J. Magn. Reson.*, 1987, **72**, 162–167.

Chapter 4

One-dimensional techniques

The approach to any structural or mechanistic problem will invariably start with the acquisition of one-dimensional spectra. Since these provide the foundations for further work, it is important that these are executed correctly and full use is made of the data they provide before more extensive and potentially time-consuming experiments are undertaken. This chapter describes the most widely used one-dimensional techniques in the chemistry laboratory, beginning with the simple single-pulse experiment and progressing to consider the various multipulse methods that enhance the information content of our spectra. The key characteristics of these are summarised briefly in Table 4.1. This chapter does not cover the wide selection of techniques that are strictly one-dimensional analogues of two-dimensional experiments, as these are more appropriately described in association with the parent experiment and are found throughout the following chapters.

4.1. THE SINGLE-PULSE EXPERIMENT

The previous chapter described procedures for the optimum collection of an FID, and how these were dictated by the relaxation behaviour of the nuclei and by the digital resolution required in the resulting spectrum, which in turn defines the data acquisition time. When setting up an experiment, one also need consider what is the optimum pulse excitation angle to use and how rapidly pulses can be applied to the sample for signal averaging (Fig. 4.1). These parameters also depend on the spin relaxation times. There are two extreme

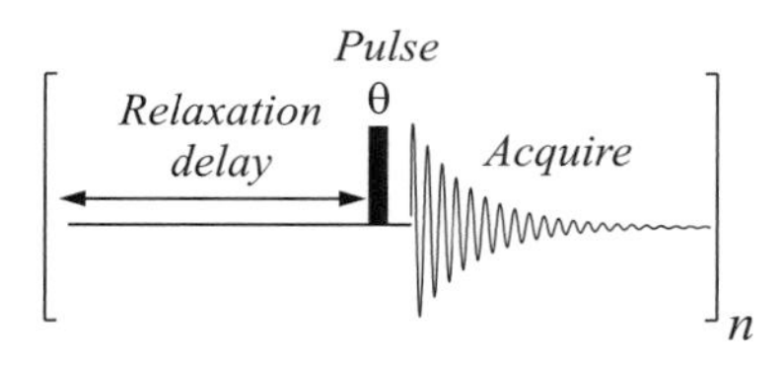

Figure 4.1. The essential elements of the single-pulse NMR experiment: the relaxation (recovery) delay, the pulse excitation angle and the data acquisition time.

Table 4.1. The principal applications of the multi-pulse techniques described in this chapter

Technique	Principal applications
J-mod. spin-echo or APT	Editing of heteronuclide spectra (notably carbon-13) according to resonance multiplicity.
INEPT	Enhancement of a low-γ nuclide through polarisation transfer from a high-γ nuclide e.g. ^{1}H, ^{31}P, ^{19}F. Editing of spectra according to resonance multiplicity.
DEPT	Enhancement of a low-γ nuclide through polarisation transfer, with editing according to resonance multiplicity. Experimentally more robust for routine carbon-13 spectra than INEPT. No responses observed for quaternary carbons.
DEPT-Q	As for DEPT but with retention of quaternary carbons.
PENDANT	Enhancement of a low-γ nuclide through polarisation transfer from a high-γ nuclide e.g. ^{1}H, ^{31}P, ^{19}F. Editing of spectra according to resonance multiplicity with retention of quaternary carbons.
RIDE or ACOUSTIC	Observation of low-frequency quadrupolar nuclei with very broad resonances. Sequence suppresses 'acoustic ringing' responses from probehead.

cases to be considered for the single-pulse experiment, in particular whether one is striving for optimum sensitivity for a given period of data accumulation or whether one requires accurate quantitative data from our sample. The experimental conditions for meeting these criteria can be quite different, so each shall be considered separately.

4.1.1. Optimising sensitivity

If one were to apply a 90° pulse to a spin system at thermal equilibrium, it is clear that the maximum possible signal intensity would result since all magnetisation is placed in the transverse plane. This may therefore appear to be the optimum pulse tip angle for maximising sensitivity. However, one is usually interested in performing signal averaging, so before applying subsequent pulses it becomes necessary to wait many times T_1 for the system to relax and recover the full signal once again; a period of $5T_1$ leads to 99.3% recovery of longitudinal magnetisation (see Fig. 2.20 in Chapter 2), complete for all practical purposes. Such slow repetition is in fact not the most efficient way of signal averaging and it turns out to be better to do away with the recovery delay altogether and to adjust pulse conditions to maximise the *steady-state* *z*-magnetisation produced.

Under conditions where there is complete decay of transverse magnetisation between scans (i.e. the FID decays to zero) the optimum tip angle for a pulse repetition time t_r, known as the *Ernst angle,* α_e, is given by [1,2]:

$$\cos \alpha_e = e^{-t_r/T_1} \tag{4.1}$$

and is illustrated in Fig. 4.2. As the repetition time decreases relative to the spin relaxation rate, that is when faster pulsing is used, smaller tip angles produce the optimum signal-to-noise ratio. Proton spectra are typically acquired with sufficient digital resolution to reveal multiplet fine structure, and acquisition times tend to be of the order of $3T_2^*$, which is sufficient to enable the almost complete decay of transverse magnetisation between scans ($3T_2^*$ corresponds to 95% decay). Furthermore, since for small to medium-sized molecules $T_2 = T_1$ and for proton observation in a well shimmed magnet $T_2^* \approx T_2$, we have $t_r \approx 3T_1$, and thus, from Fig. 4.2 the pulse angle for *maximising sensitivity* will typically be $>80°$. However, there will exist a range of T_1 values for the protons in a molecule and it will not be possible to use optimum conditions for all. In such cases, a large pulse angle is likely to lead to significant intensity differences for protons with widely differing relaxation times, with the slower relaxing spins displaying reduced signal intensity [3]. This is exemplified in

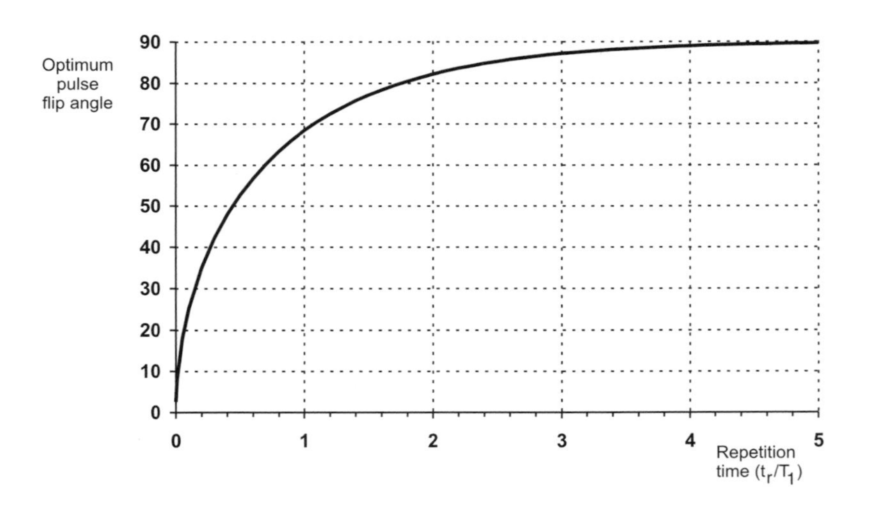

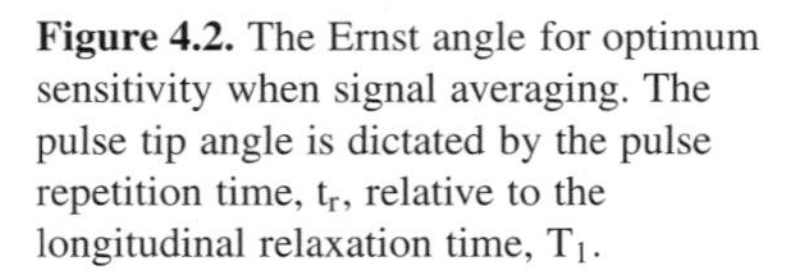
Figure 4.2. The Ernst angle for optimum sensitivity when signal averaging. The pulse tip angle is dictated by the pulse repetition time, t_r, relative to the longitudinal relaxation time, T_1.

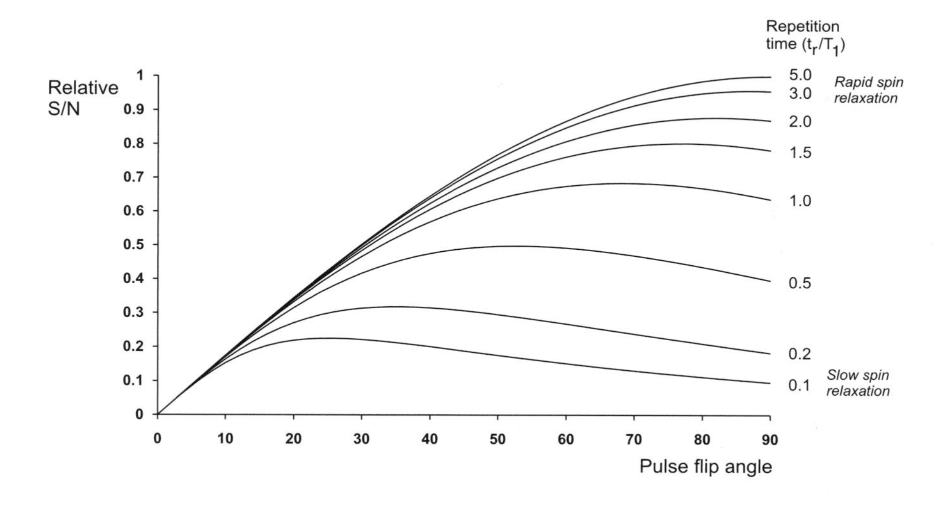

Figure 4.3. The dependence of signal-to-noise ratio on pulse tip angle for different pulse repetition times. The maximum of each curve corresponds to the Ernst angle. When a molecule contains nuclei with very different relaxation times (T_1s), pulsing with a large flip angle produces a large difference in their signal intensities and makes even semi-quantitative measurements unreliable. Optimising the tip angle for the slowest relaxing spins reduces these intensity differences. The plots have been generated from the equations given in [3].

the signal-to-noise plots of Fig. 4.3. When proton integrals are required to indicate the number of nuclei giving rise to each resonance, as for routine proton acquisitions, it is wise to optimise the pulse tip angle for the longer T_1 values and thus use shorter pulses (see the following section for details of *accurate* quantitative measurements). Assuming the longest T_1 to be around 4 s and an acquisition time of 3 s, the optimum tip angle will be around 60°, although the plots of Fig. 4.3 clearly possess rather broad maxima, meaning the precise setting of the pulse tip angle is not critical. Pulsing too rapidly, that is, using very short repetition times relative to T_1, leads to a substantial decrease in signal-to-noise until in the extreme case magnetisation has no time to recover between pulses and so no signal can be observed. This condition is known as *saturation* and causes resonances to be lost completely which can be nuisance or a bonus depending on your point of view. If you are interested in observing the signal then clearly this must be avoided, whereas if the signal is unwanted, for example a large solvent resonance, then this is to be encouraged (Chapter 9).

Conditions for the acquisition of heteronuclear spectra are usually rather different to those for proton observations. Often one does not require the spectra to be well digitised when there is little or no fine structure to be observed, therefore acquisition times are kept comparable to T_2^*, which is often less than T_2. Under such conditions transverse magnetisation does not decay completely between scans and such rapid pulsing gives rise to steady-state spin-echoes [2,4,5] where transverse magnetisation remaining from one pulse is refocused by subsequent pulses. These echoes can give rise to phase and intensity distortions in resulting spectra which are a function of resonance offset. The use of rapid pulsing, where repetition times are considerably shorter than longitudinal relaxation times as in the case of carbon-13 observations for example, further contributes to distortions of the relative intensities because of partial saturation effects. This is the principal reason for quaternary carbon centres (which relax rather slowly owing to the lack of attached protons) appearing with often characteristically weak intensities. In extreme cases these can become fully saturated and effectively lost altogether if pulsing is too rapid and/or if the pulse tip angle too large. Thus, routine carbon spectra are typically acquired with a somewhat reduced pulse angle (<45°) and with a relaxation delay of a second or so. Similar considerations apply to other heteronuclei. For routine observations there is always some compromise needed between optimum sensitivity and the undesirable intensity anomalies that arise for nuclei with widely differing relaxation times. If problematic, one approach toward reducing these anomalies is the addition of paramagnetic relaxation agents such as chromium(III) acetylacetonate, $Cr(acac)_3$ with the

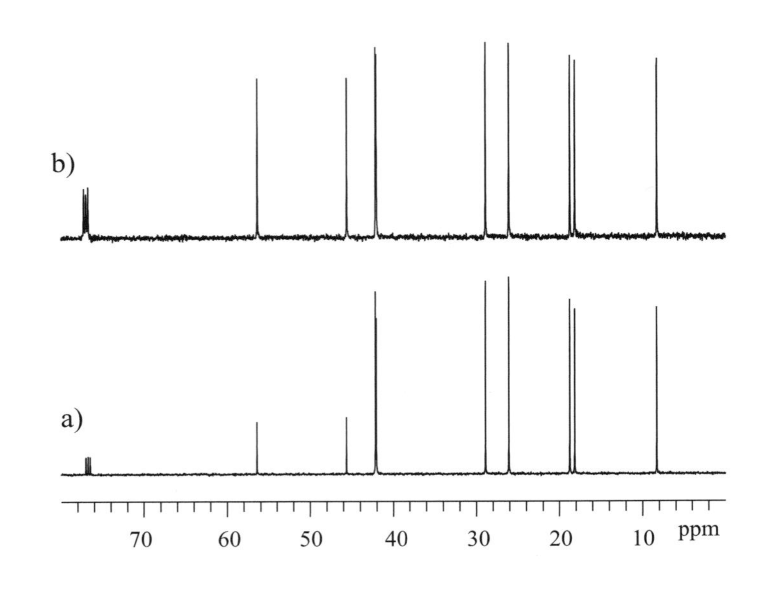

Figure 4.4. Carbon-13 spectra of camphor **4.1** acquired (a) without and (b) with the addition of $Cr(acac)_3$ under otherwise identical standard observation conditions (1 s relaxation delay, 30° excitation pulse, 0.5 s acquisition time). In (a) the two quaternary resonances display reduced signal intensities as a result of partial saturation. In (b) this difference is largely eliminated by the addition of the relaxation agent, whilst the reduced signal-to-noise ratio of the protonated resonances arises from the suppression of ^{1}H–^{13}C NOE enhancements and the line-broadening caused by the relaxation agent.

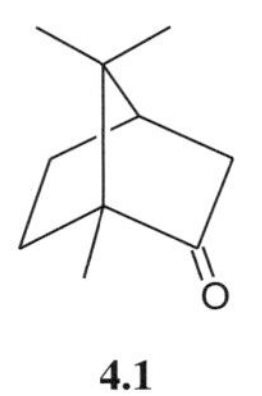

4.1

aim of reducing and, in part, equalising longitudinal relaxation times to allow faster signal averaging (Fig. 4.4; around 10–100 mM relaxation agent suffices and the solution should take on a slight pink-purple hue).

Finally we consider the situation where one is compelled by the pulse sequence to use 90° pulses in which case the above arguments no longer apply. This is in fact the case for most multipulse and multidimensional sequences. Under such conditions a compromise is required between acquiring data rapidly and the unwelcome effects of saturation, and it has been shown [6] that for steady-state magnetisation the sensitivity is maximised by setting the repetition rate equal to 1.3 T_1. This leads to a signal-to-noise ratio that is approximately 1.4 times greater than that obtained using a recycle time of 5 T_1 *for a given period of data collection* (Fig. 4.5). Thus for most sequences, a repetition time of 1.3 T_1 is the target.

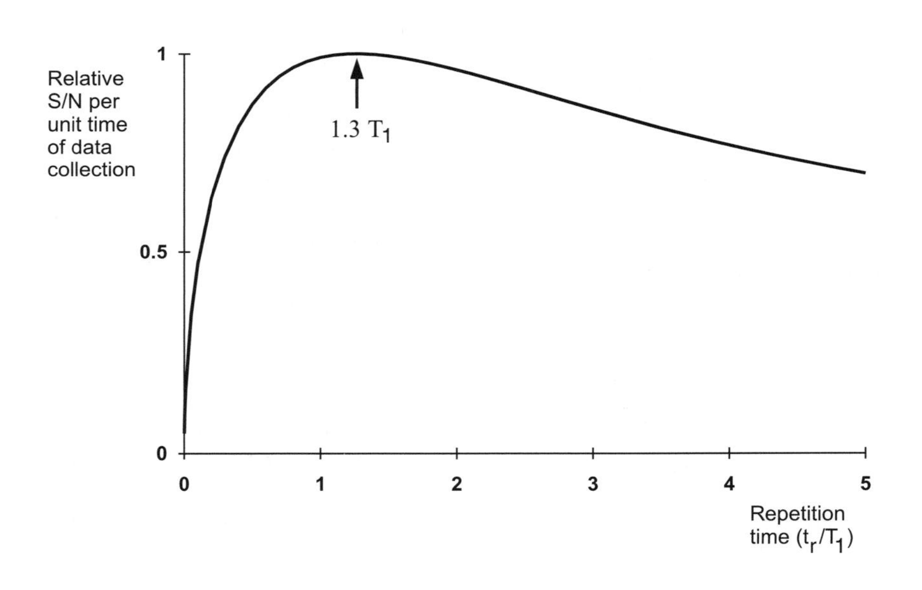

Figure 4.5. Optimum pulse repetition time when using 90° pulses. The curve has been calculated from the equation in [3].

4.1.2. Quantitative measurements and integration

The extensive use of integration in proton NMR arises from the well-used doctrine that the area of an NMR resonance is proportional to the relative number of nuclei giving rise to it. In fact this is only strictly true under well-defined experimental conditions, and in routine proton spectra integrals may only be accurate to within 10–20% or so. Whilst this level of accuracy is usually sufficient for the estimation of the relative proton count within a molecule, it is clearly inadequate for quantitative measurements where accuracy to within a few percent is required. The fact that integrals are typically reported to many decimal places by NMR software can draw one into believing such figures are significant, so inflating ones expectations of what can be derived from routine spectra. Conversely, it is widely taught in introductory NMR lectures that it is not possible to integrate carbon-13 spectra. Whilst this is indeed inappropriate for routine acquisitions, not just for carbon-13 but also for many heteronuclei, it is nevertheless quite possible to obtain meaningful integrals when the appropriate protocols are employed, as described below. In all cases it is important to be conscious of the fact that NMR measurements are always *relative* measures of signal intensities. There exist no NMR equivalent to the extinction coefficients of ultra-violet spectroscopy so *absolute* sample quantities cannot be determined directly.

Data collection

There are three features of specific importance to quantitative measurements, aside from the obvious need for adequate signal-to-noise in the spectrum. These are the avoidance of differential saturation effects, the need to characterise the NMR resonance lineshape properly and the need to avoid differential NOE enhancements. As has been mentioned in the previous section, acquiring data whilst pulsing rapidly relative to spin relaxation times leads to perturbation of the relative signal intensities in the spectrum so to avoid this it is *essential* to wait for the spins to fully relax between pulses, demanding recycle times of least $5T_1$ *of the slowest relaxing nuclei*. This allows the use of 90° observation pulses, providing the maximum possible signal per transient. Clearly, one requires some knowledge of the T_1 values for the sample of interest which may be estimated by the inversion recovery methods described in Chapter 2 or estimates made from prior knowledge of similar compounds. Whilst recycle times of the order of $5T_1$ are usually bearable for proton work, they can be tediously long in the study of heteronuclei which may demand many minutes between scans. Here relaxation reagents can be employed to reduce these periods to something more tolerable.

The second fundamental requirement is for the data be sufficiently well digitised for the lineshape to be defined properly. To minimise intensity errors it is necessary to have *at least four* acquired data points covering the resonance linewidth, although many more than this are preferable, so it is beneficial to use the minimum spectral width compatible with the sample and to adjust the acquisition times accordingly. The spectral width should not be too narrow to ensure the receiver filters do not interfere with resonance intensities at the edges of the spectrum.

A further source of intensity distortions in heteronuclear spectra recorded with broadband proton decoupling arises from the NOE produced by proton saturation (Chapter 8). Clearly, differential enhancements will prevent the collection of meaningful intensity data, so it is necessary to take measures to suppress the NOE, yet it is still desirable to collect proton-decoupled spectra for optimum signal-to-noise and minimal resonance overlap. The solution to this apparent dichotomy is to employ the inverse-gated decoupling scheme described in the following section. The lack of the NOE and the need to pulse at a slow rate means quantitative measurements can take substantially longer than the routine observation of the same sample. The addition of a relaxation agent will again speed things along by reducing recycle delays and will also aid suppression of the NOE as it will eliminate the dipolar relaxation responsible for this enhancement (see Fig. 4.4 above).

Data processing

Having taking the necessary precautions to ensure the acquired data genuinely reflects the relative ratios within the sample, appropriate processing can further enhance results. The spectra of heteronuclei in particular benefit from the application of an exponential function that broadens the lines. This helps to ensure the data are sufficiently well digitised in addition to improving the signal-to-noise ratio. For proton spectroscopy, achieving adequate sensitivity is not such a demanding problem, although the use of a matched exponential window will again help to ensure sufficient digitisation. The use of zero-filling will further assist with the definition of the lineshape and is highly recommended, although this *must not* be used as a substitute for correct digitisation of the acquired FID. Careful phasing of the spectrum is also essential. Deviations from pure absorption-mode lineshapes will reduce integrated intensities with contributions from the negative-going components; in the extreme case of a purely dispersive lineshape, the integrated intensity is zero! Another potential source of error arises from distortions of the spectrum baseline, which have their origins in the spectrometer receiver stages. These errors mean the regions of the spectrum that should have zero intensity, that is, those that are free from signals, have a non-zero value, and make a positive or negative contribution to measurements. NMR software packages incorporate suitable correction routines for this.

The final consideration when integrating is where the integral should start and finish. For a Lorentzian line, the tails extend a considerable distance from the centre and the integral should, ideally, cover 20 times the linewidth each side of the peak if it is to include 99% of it. For proton observation this is likely to be 10–20 Hz each side. In practice it may not be possible to extend the integral over such distances before various other signals are met. These may arise from experimental imperfections (such as spinning sidebands), satellites from coupling to other nuclei or from other resonances in the sample. Satellites can be particularly troublesome in some cases as they may constitute a large fraction of a total signal, owing to high natural abundance of the second nuclide, and one must decide on whether to include them or exclude them *for all measurements*. In some instances satellites can be used to one's advantage in quantitative measurements. One example is in the estimation of enantiomeric or diastereomeric excesses by proton NMR where the minor isomer is present at only a few percent of the major. In such cases, the *ee* or *de* measurement demands the comparison of a very large integral verses a very small one, a situation prone to error. Comparison of similar size integrals can be made if one considers only the carbon-13 satellites of the major species (each present at 0.55%), and scales the calculation accordingly.

4.2. SPIN DECOUPLING METHODS

The use of spin–spin decoupling is no doubt familiar to most regular users of solution NMR spectroscopy, and it has constituted an important tool for the chemist since the early applications of NMR to problems of chemical structure [7]. It is now so widely used in the observation of heteronuclear spectra (when did you last see a fully proton-coupled carbon spectrum?) that one can almost become oblivious to the fact that it is an integral part of numerous pulse experiments. Nowadays, scalar spin–spin decoupling is most often applied with one of two goals in mind; either the *selective* decoupling of a single resonance in an attempt to identify its coupling partner(s), or non-selective (*broadband*) decoupling of one nuclide to enhance and simplify the spectrum of another. Decoupling is usually classified as being *homonuclear*, in which the decoupled and observed nuclides are the same, or *heteronuclear* in which they differ. That which experiences the decoupling rf is conventionally distinguished by placing it in curly brackets so, for example, $^{13}C\{^{1}H\}$ would indicate carbon observation in the presence of proton decoupling, whilst $^{1}H\{^{13}C,^{31}P\}$ would represent proton observation in the presence of simultaneous carbon and phosphorus decoupling (or more generally carbon and phosphorus pulsing). One may also find spin decoupling being referred to as one example of a 'double-resonance' experiment, a term originating from CW experiments in which the observe and decouple rf were applied simultaneously. Likewise, the H, C, P example above may be termed a 'triple-resonance' experiment as it requires three rf channels.

4.2.1. The basis of spin decoupling

A simplified [8] yet convenient description of how spin decoupling operates considers two spin-$^1/_2$ nuclei, A and X, that share a mutual scalar coupling of J Hz. The resonant frequency of the X-spins will depend on whether their coupled partners are oriented parallel (α orientation) or anti-parallel (β orientation) to the applied static field. For a spin ensemble we can assume there exist an equal number of A nuclei in the α and β states, owing to the very small energy difference between the two orientations, and thus the X spectrum displays the familiar doublet pattern. Application of an rf field, designated the B_2 field (recall the transmitter field is termed B_1), at the frequency of the A spins causes these to undergo continuous, rapid transitions between the α and β orientations by continually inverting these spins. If this reorientation is fast relative to the coupling constant, the X-spin doublet coalesces into a singlet since the lifetimes of the α and β orientations are no longer sufficient for the coupling to be distinguished. Thus, if the A spins are irradiated during data acquisition with a sufficiently strong field such that $\gamma B_2 > J$ Hz, the X resonance displays no coupling to A and the spins are said to be decoupled.

Whilst the removal of scalar spin–spin coupling is the usual goal of such experiments, a number of additional effects can arise from the application of the additional rf field, which may be beneficial or detrimental depending on the circumstances. Incomplete decoupling can introduce residual line-broadening or, even worse, leave rather esoteric partially decoupled multiplets, whilst the non-uniform irradiation of a resonance can introduce population transfer effects which cause intensity distortions *within* multiplets (Section 4.4). Population disturbances caused by the rf may also produce intensity changes that arise from the nuclear Overhauser effect (Chapter 8) which operates quite independently from J coupling. Finally, changes in the resonant positions of signals close in frequency to the applied rf may also be observed, so-called Bloch–Siegert shifts. In many cases it is possible to have some control over these factors, according to the experimental protocol used, as described in later sections.

4.2.2. Homonuclear decoupling

Homonuclear decoupling involves the selective application of a coherent decoupling field to a target resonance with the aim of identifying scalar spin-coupled partners, and is most often applied in proton spectroscopy (Fig. 4.6). Although the use of this method for identifying coupled $^{1}H–^{1}H$ spins has been largely superseded by two-dimensional correlation methods (Chapter 5), selective decoupling can still be a very useful and convenient tool in the NMR armoury. It is very simple to set-up, providing rapid answers to relatively simple questions and can be particularly useful in identifying spins that share

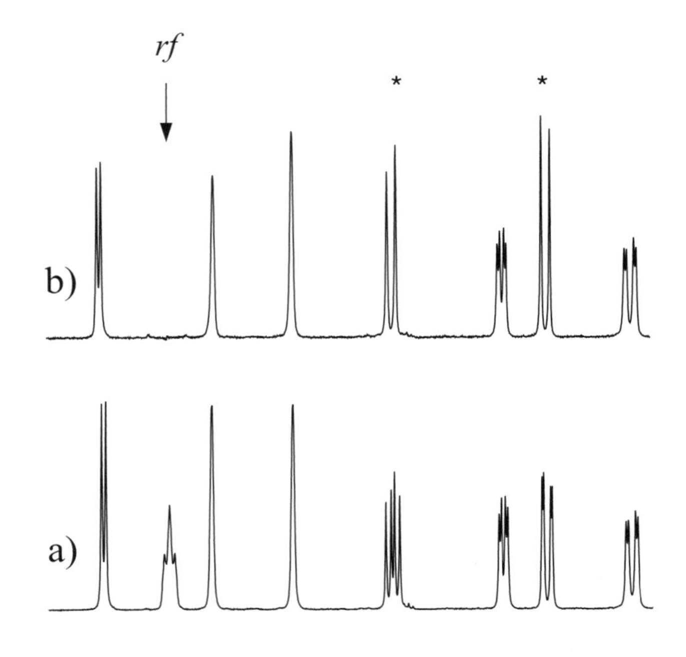

Figure 4.6. Homonuclear decoupling allows the rapid identification of coupled partners by removing couplings to the irradiated spin. (a) Control spectrum and (b) decoupled spectrum.

very small couplings which do not always reveal themselves in 2D correlation experiments. In very crowded spectra where only a specific interaction is to be investigated, the affected resonances may not be obvious and the use of *difference spectroscopy* can aid interpretation [9,10]. Here, a control spectrum recorded in the absence of decoupling [1] is subtracted from that collected in the presence of on-resonance decoupling to reveal any changes that arise. In such cases a two-dimensional correlation experiment may, however, be more appropriate.

One of the limitations to the use of selective decoupling lies in the need to irradiate only a single resonance to identify the coupling partners of the desired target. Any saturation spill-over onto neighbouring resonances introduces a degree of ambiguity into the interpretation. Where the target multiplet is free from other resonances this posses little problem and the decoupler power can be set sufficiently high to ensure complete decoupling ($\gamma B_2 > J$ Hz). When other resonances are close in frequency to the target, it may be necessary to reduce the decoupler power and thus reduce the frequency spread to avoid disturbing neighbouring resonances. The penalty for reducing the decoupler power may be incomplete decoupling of the target spins, although changes within the fine-structure of other resonances should be sufficient to identify coupled partners.

Bloch–Siegert shifts

The application of an rf field during the acquisition of the FID may also move signals that resonate close to the decoupler frequency. This effect is known as the *Bloch–Siegert* shift [11,12] and, more formally, it occurs when γB_2 becomes comparable to the shift difference in hertz between the decoupling frequency and the resonance. It arises because the decoupling field acts on the neighbouring spins such that they experience a modified effective field that is inversely dependent upon their resonance offsets from the decoupling frequency but proportional to $(\gamma B_2)^2$, which causes resonances to move *away* from this decoupling frequency (Fig. 4.7). The effect is principally limited to homonuclear decoupling experiments, where resonances may be

[1] For any form of difference spectroscopy it is better practice to collect the control FID with the decoupler frequency applied far away from all resonances rather than being turned off completely.

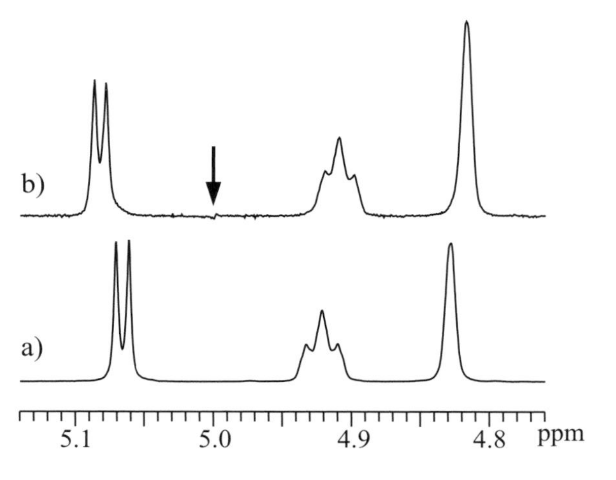

Figure 4.7. The Bloch–Siegert shift causes resonances near to an applied rf to move away its point of application. (a) is the conventional spectrum and (b) is that acquired with a decoupling field applied at the position of the arrow.

very close to the decoupling frequency but is of no concern in most pulse NMR experiments since the rf pulse is turned off prior to data collection. A notable exception is when solvent presaturation is applied during the evolution time (t_1) of proton homonuclear 2D experiments. This can lead to shifts of f_1 frequencies of resonances close to the solvent, but since this is not present in f_2, it introduces asymmetry in the shifts of crosspeaks associated with these resonances. In homonuclear decoupling experiments it is rarely a major problem since the requirement for selectivity limits the B_2 field and thus keeps the shifts small. However, such small shifts may still introduce subtraction artefacts into decoupling difference spectra since the reference will not contain Bloch–Siegert shifts, so one should be cautious not to interpret these as evidence of coupling. Caution is also required should one need to measure accurate chemical shifts from decoupled spectra. In Section 3.5.1 the Bloch–Siegert shift is used quantitatively as a means of calibrating decoupler powers.

Experimental implementation

The application of homonuclear irradiation whilst acquiring the FID poses some challenging instrumental problems. Whilst needing to detect the responses of the excited spins, one must not have the receiver open when the decoupler is on since this will simply swamp the NMR signal. The solution lies in the discrete sampling of the FID, and the application of homonuclear irradiation *only* when data points are *not* being sampled, that is, during the FID dwell time [13]. Spectrometers have purpose-built homonuclear decoupling modes to handle the necessary gating internally. The time in which the decoupler is gated on is thus only a small fraction of the total acquisition time, this so-called *duty cycle* being typically 20% or less. The low duty-cycle means the effective mean decoupler power is somewhat less than the instantaneous B_2 field, hence rf powers are usually greater than those required for presaturation of a resonance where such receiver conflict does not arise. If the gating-off of the decoupler is not perfect during data collection, a number of spectrum artefacts may be introduced [14,15]. Most notable are a reduction in signal-to-noise, owing to 'leakage' of the decoupler rf into the receiver, and a significant 'spike' occurring at the decoupler frequency, although this is really only a question of aesthetics. The leakage problem varies greatly, it seems, from one instrument to another and if a serious problem may be cured by instrumental modification [15]. The second may be eliminated simply by setting the *transmitter* frequency to match that of the decoupler [16] so that the usual phase-cycling routines employed in 1D acquisitions remove the unsightly 'zero-frequency' spike (Fig. 4.8).

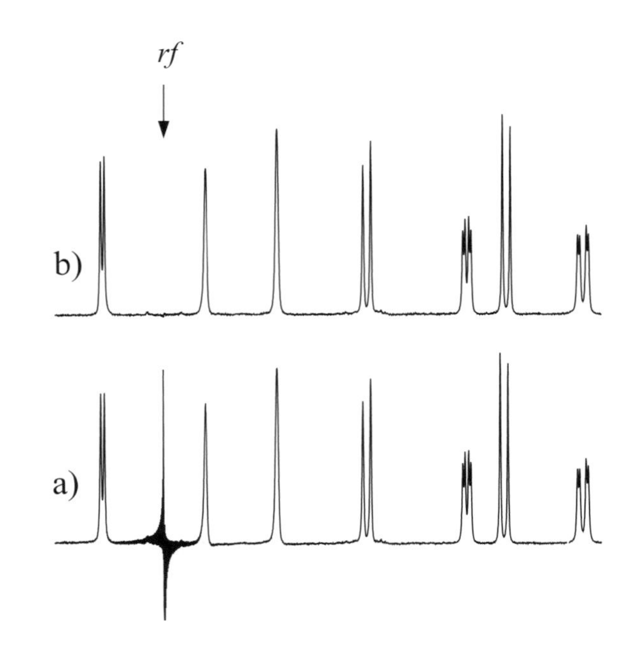

Figure 4.8. The aesthetically unappealing decoupler frequency 'spike' sometimes observed in homonuclear decoupling experiments as in (a) can be readily removed by setting the transmitter and decoupler frequencies to be the same, as in (b).

4.2.3. Heteronuclear decoupling

X{^{1}H} decoupling

The application of broadband proton decoupling during the acquisition of carbon-13 spectra is now universally applied. The removal of all ^{1}H–^{13}C couplings concentrates all the carbon resonance intensity into a single line providing a significant increase in signal intensity and simplification of the spectrum. As an added bonus, the continuous saturation of the proton spins provides further enhancement of the signal (by as much as 200% in the case of carbon-13) due to the nuclear Overhauser effect (Chapter 8). For these reasons, the use of broadband proton decoupling is standard practice for the routine observation of most commonly encountered nuclides.

The principles behind heteronuclear decoupling are no different from the homonuclear case, the major difference being the very much larger frequency separation between the decoupling and observed frequencies, meaning Bloch–Siegert shifts and receiver interference are no longer a problem. The use of broadband heteronuclear decoupling is itself associated with a number of other practical problems. Firstly, there is the need to decouple uniformly over the whole of the proton spectrum and this requires the applied rf to be effective over a far greater frequency window. Decoupling bandwidths are typically of the order of many kilohertz and are greater at higher field strengths. A 10 ppm proton spectrum covers 4 kHz at 400 MHz but 6 kHz at 600 MHz, for example. Generally, a wider decoupling bandwidth can be achieved by increasing the decoupler power, although the continuous application of high power rf is more likely to destroy the probe and sample than achieve the desired results. High powers also give rise to the problem of rf sample heating, which is most pronounced in ionic samples, and can be disastrous for heat sensitive compounds. The effect is insidious as the heating occurs within the sample itself, so the usual method of sample temperature monitoring (a thermocouple placed within the airflow surrounding the sample) is largely insensitive to its presence, the usual indicator being a drift of the lock level when the pulse sequence is started. To overcome the problem of decoupling over wide bandwidths without the need for excessive powers, specially designed modulated decoupling schemes are employed (so-called *composite pulse decoupling sequences*, described in Chapter 9). The careful

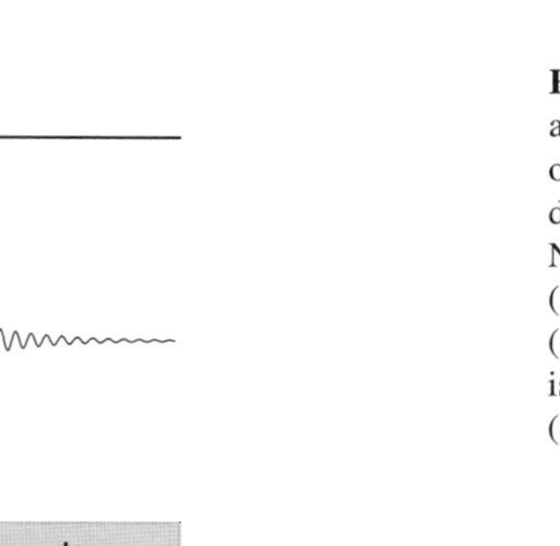

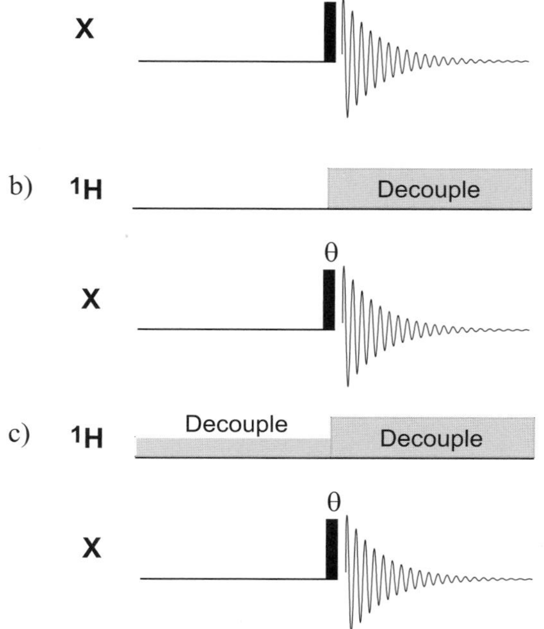

Figure 4.9. Possible schemes for applying proton decoupling when observing a heteronucleus. (a) Gated decoupling (coupled spectrum with NOE), (b) inverse-gated decoupling (decoupled spectrum without NOE) and (c) power-gated decoupling where the rf is applied at two different powers (decoupled spectrum with NOE).

design of probes can also assist by reducing the sample heating which is caused by the *electric* component of the rf. As with pulse excitation, the required frequency spread can be kept to a minimum by placing the decoupler frequency in the *centre* of the region to be decoupled. If your heteronuclear spectra display unexpectedly broad or even split resonances, this may be due to the decoupler frequency being incorrect, and/or the decoupler channel being poorly tuned or wrongly calibrated (Chapter 3).

Sample heating can also be reduced by gating-off the rf when not essential to the experiment (Fig. 4.9). Gating the decoupler off during the recovery delay (*inverse-gated decoupling*, Fig. 4.9b) also removes the NOE and provides a decoupled spectrum *without* NOE enhancements. This is because any NOE that builds up during the acquisition time affects only longitudinal magnetisation so does not influence the detected (transverse) signals, and is then allowed to decay during the recovery delay, so again has no influence. This method has particular value in the observation of nuclei with negative magnetogyric ratios, since the NOE causes a *decrease* in signal intensity for such nuclei (Chapter 8). It is also employed for accurate quantitative measurements, as discussed in the previous section. Conversely, running the decoupler during the relaxation delay but gating it off during the acquisition period (*gated decoupling*, Fig. 4.9a) results in a spectrum that retains spin coupling and is enhanced by the NOE. A further method for reducing sample heating provides a decoupled spectrum with enhancement by the NOE and is known as *power-gated decoupling* (Fig. 4.9c). Here, a high decoupler power is applied during acquisition to achieve complete decoupling, but a reduced power is applied between scans to maintain a degree of saturation and hence develop the NOE. This approach is particularly applicable to studies at high-field where larger bandwidths demand relatively high decoupler powers. A comparison of these various decoupling schemes is illustrated in Fig. 4.10 for the terpene α-pinene **4.2**.

4.2

By analogy with homonuclear decoupling above, *selective* heteronuclear decoupling is also possible. The decoupling of a single proton resonance could in

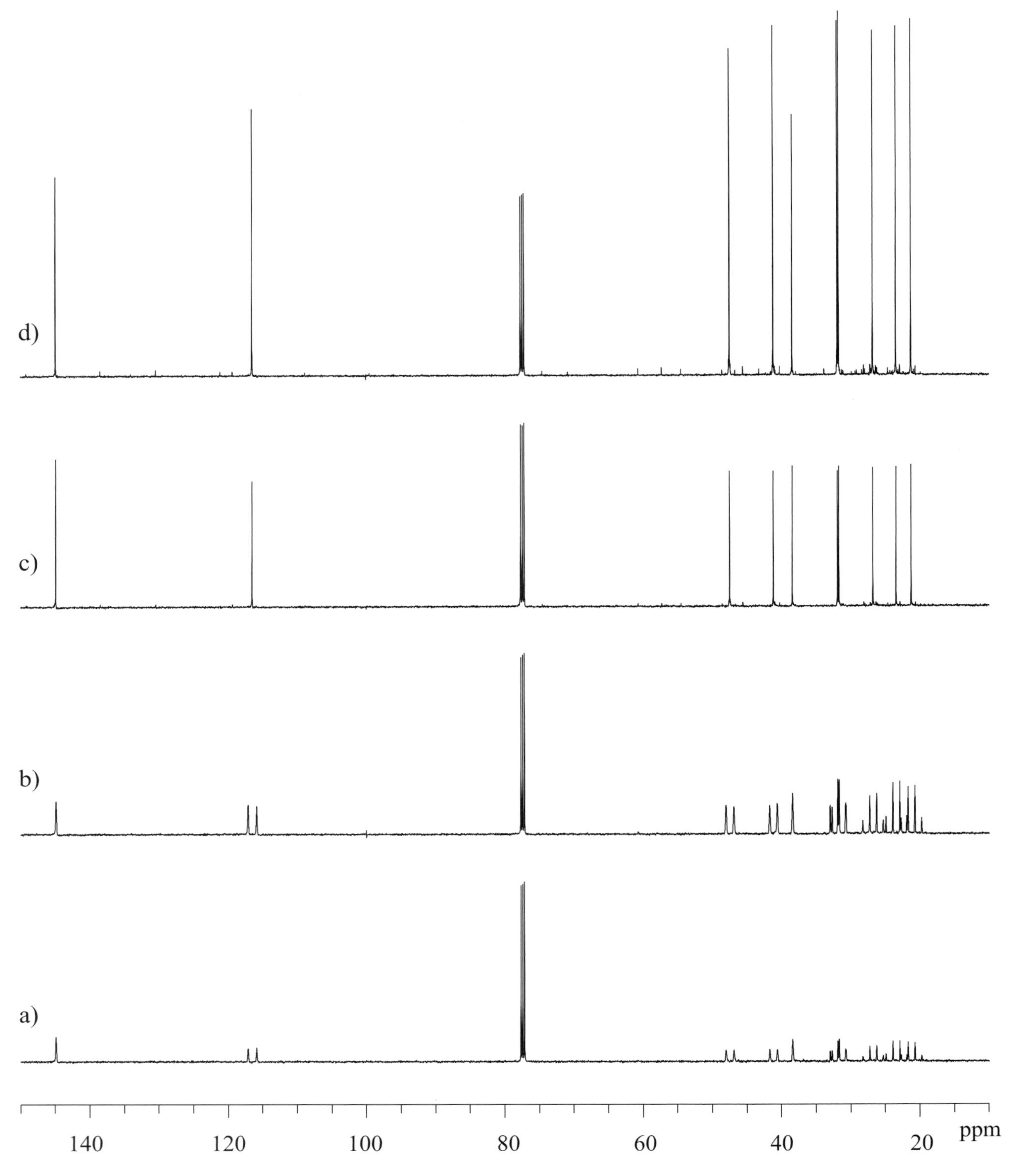

Figure 4.10. The carbon-13 spectrum of α-pinene acquired (a) without proton irradiation at any stage (coupled spectrum without NOE), (b) with gated decoupling (coupled spectrum with NOE), (c) with inverse-gated decoupling (decoupled spectrum without NOE) and (d) power-gated decoupling (decoupled spectrum with NOE). All other experimental conditions were identical and the same absolute scaling was used for each plot.

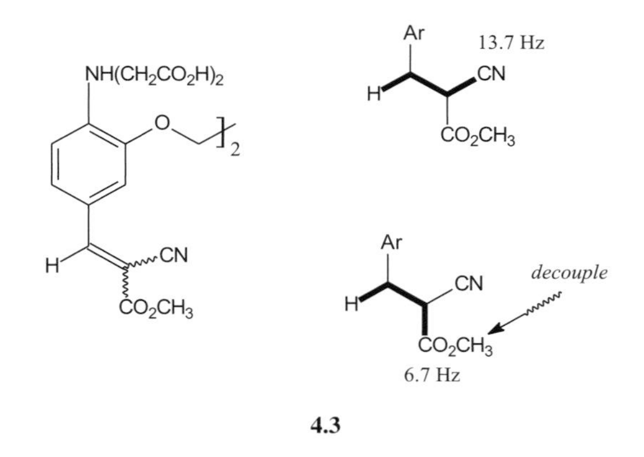

principle identify coupled partners through the observation of collapsed multiplet structure in the heterospin spectrum. This however requires the decoupling rf to be effective over the *satellites* of the parent proton resonance, which may be far apart if $^1J_{XH}$ is large. One-bond proton–carbon couplings exceed 100 Hz and may be hard to remove whilst retaining sufficient selectivity, so here a two-dimensional heteronuclear correlation spectrum is likely to be a more efficient approach. The selective removal of long-range (2 or 3-bond) proton–carbon couplings is more readily achieved since these are typically less than 10 Hz. An example of this is illustrated by the identification of configurational isomers of **4.3** through the measurement of three-bond proton–carbon couplings (Fig. 4.11) which share a Karplus-type dependence on dihedral bond

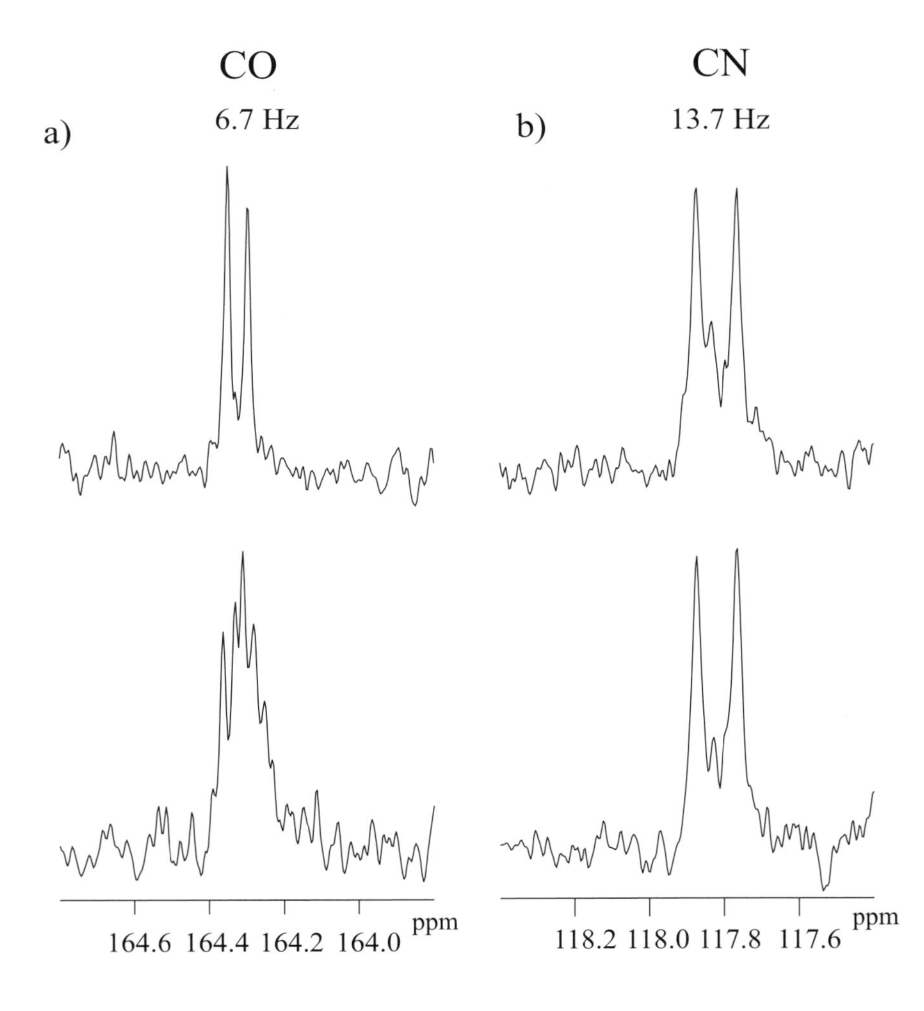

Figure 4.11. The application of selective proton decoupling in the measurement of heteronuclear long-range proton-carbon coupling constants. Lower traces are from the fully proton-coupled carbon-13 spectrum and the upper traces from that in which the methyl ester protons of **4.3** were selectively decoupled to reveal the three-bond coupling of the carbonyl carbon across the alkene.

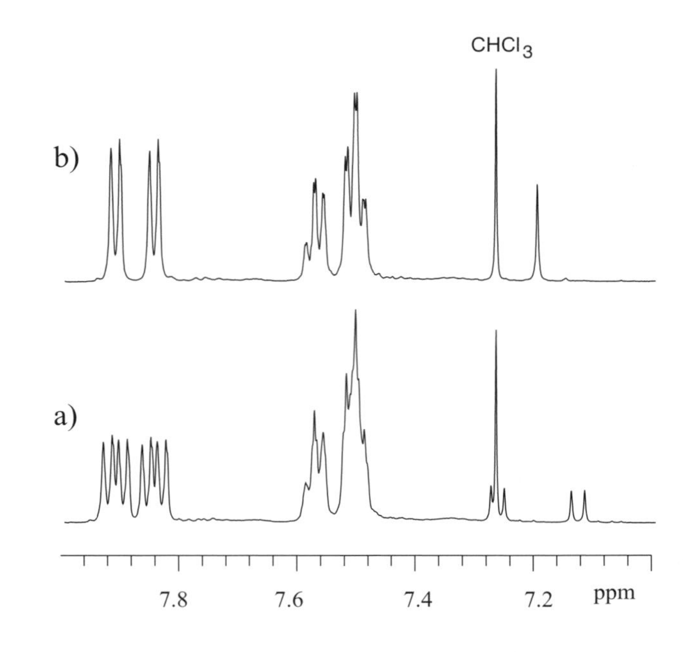

Figure 4.12. Simplification of the conventional proton spectrum (a) of the palladium phosphine **4.4** in $CDCl_3$ by the application of broadband phosphorus decoupling (b). All long-range ^1H–^{31}P couplings are removed, as is most apparent for the alkene proton (7.2 ppm) and the *ortho* protons of the phenyl rings (above 7.8 ppm).

angles. The magnitude of the proton couplings to the carbon atoms of the nitrile and the ester across the alkene would therefore indicate their relative stereochemistry, but whilst the nitrile coupling presented a clear doublet in the proton-coupled ^{13}C spectrum, the carbonyl coupling was masked by additional three-bond couplings to the ester methyl protons. Selective decoupling of these methyl protons eliminated this interference resulting in a carbonyl doublet, sufficient to identify the nitrile as being trans to the alkene proton.

$^1H\{X\}$ decoupling

Traditionally, FT spectrometers were built with the ability to decouple only protons whilst observing the heteronucleus, as described above. Modern instruments now have the capability of providing X-nucleus pulsing and decoupling whilst observing protons, $^1H\{X\}$ (the 'inverse' configuration). The bandwidths required for the broadband decoupling of, for example, carbon-13, far exceed those needed for proton decoupling, and the composite pulse decoupling methods described in Chapter 9 become essential for success. Thus, carbon-13 decoupling over 150 ppm (a typical range for proton-bearing carbons) at 400 MHz-^1H requires a bandwidth of 15 kHz. Broadband X-nucleus decoupling is most frequently applied as part of a multidimensional pulse sequence to simplify crosspeak structures, although it can be a useful tool in interpreting 1D proton spectra when a heteronucleus has high natural abundance and thus makes a significant contribution to coupling fine structure, for example ^{19}F or ^{31}P. Fig. 4.12 demonstrates the simplification of the ^1H spectrum of the palladium phosphine **4.4** on application of broadband ^{31}P decoupling, and this procedure often aids the interpretation of multiplet structures or the extraction of homonuclear proton couplings. Likewise, if the X-nucleus chemical shifts are known, *selective* X-decoupling can be used to identify the coupled partners in the proton spectrum, and in simple cases this may serve as a ready alternative to a two-dimensional heteronuclear correlation experiment[2].

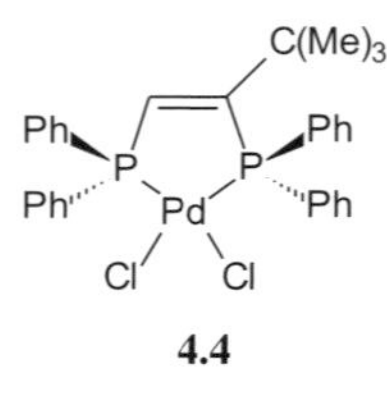

4.4

[2] In fact, this approach was used to identify the resonant frequencies of heteronuclei prior to the advent of the pulse-FT methods [7] which made their direct observation possible. Thus, the proton spectrum was recorded whilst applying a second rf field to the heteronucleus. Successively stepping the decoupler frequency and repeating the measurement would ultimately indicate the X resonance position when the X–^1H coupling was seen to disappear from the proton spectrum.

4.3. SPECTRUM EDITING WITH SPIN-ECHOES

The principal reason behind the application of broadband proton decoupling of heteronuclei is the removal of the coupling structure to concentrate signal intensity, thereby improving the signal-to-noise ratio and reducing resonance overlap (Fig. 4.10 above). Additional gains such as signal enhancements from the NOE and the clarification of any remaining homonuclear couplings may also arise. Set against these obvious benefits is the loss of multiplicity information present in the proton-coupled spectrum (Fig. 4.10a), meaning there is no way, *a priori*, of distinguishing, for example, a methine from a methylene carbon resonance. It is therefore desirable to be able to record fully proton-decoupled spectra yet still retain the valuable multiplicity data. Some of the earliest multipulse sequences were designed to achieve this aim. These were based on simple spin-echoes and provided spectra in which the multiplicities were encoded as signal intensities and signs. Despite competition from the methods based on polarisation transfer described shortly, these techniques still find widespread use in organic chemistry laboratories. As they are also rather easy to understand, they provide a suitable introduction to the idea of spectrum multiplicity editing. The sections that follow utilise the rotating-frame vector model to explain pictorially the operation of the experiments and familiarity with the introduction in Chapter 2 is assumed.

4.3.1. The J-modulated spin-echo

One of the simplest approaches to editing is the J-modulated spin-echo sequence [17] (Fig. 4.13a, also referred to as SEFT, Spin-Echo Fourier Transform, [18]) which can be readily appreciated with reference to the vector model. The key to understanding this sequence is the realisation that the evolution of carbon magnetisation vectors under the influence of the $^1J_{CH}$ coupling only occurs when the proton decoupler is gated off whilst at all other times only carbon chemical shifts are effective. Further simplification comes from ignoring carbon chemical shifts altogether since shift evolution during the first Δ period is precisely refocused during the second by the 180°(C) refocusing pulse (see Section 2.2). Thus, to understand this sequence one only need consider the influence of heteronuclear coupling during the first Δ period.

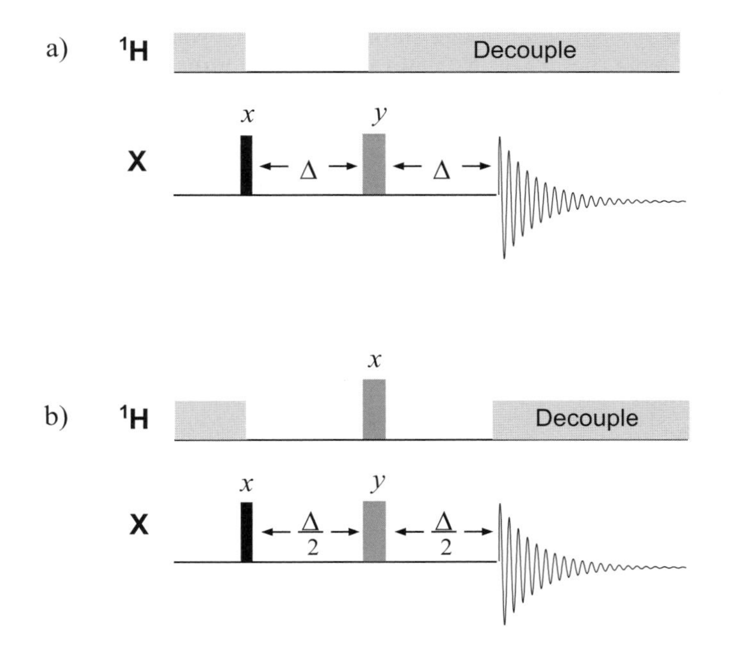

Figure 4.13. J-modulated spin-echo sequences. (a) The decoupler-gated variant and (b) the pulsed variant.

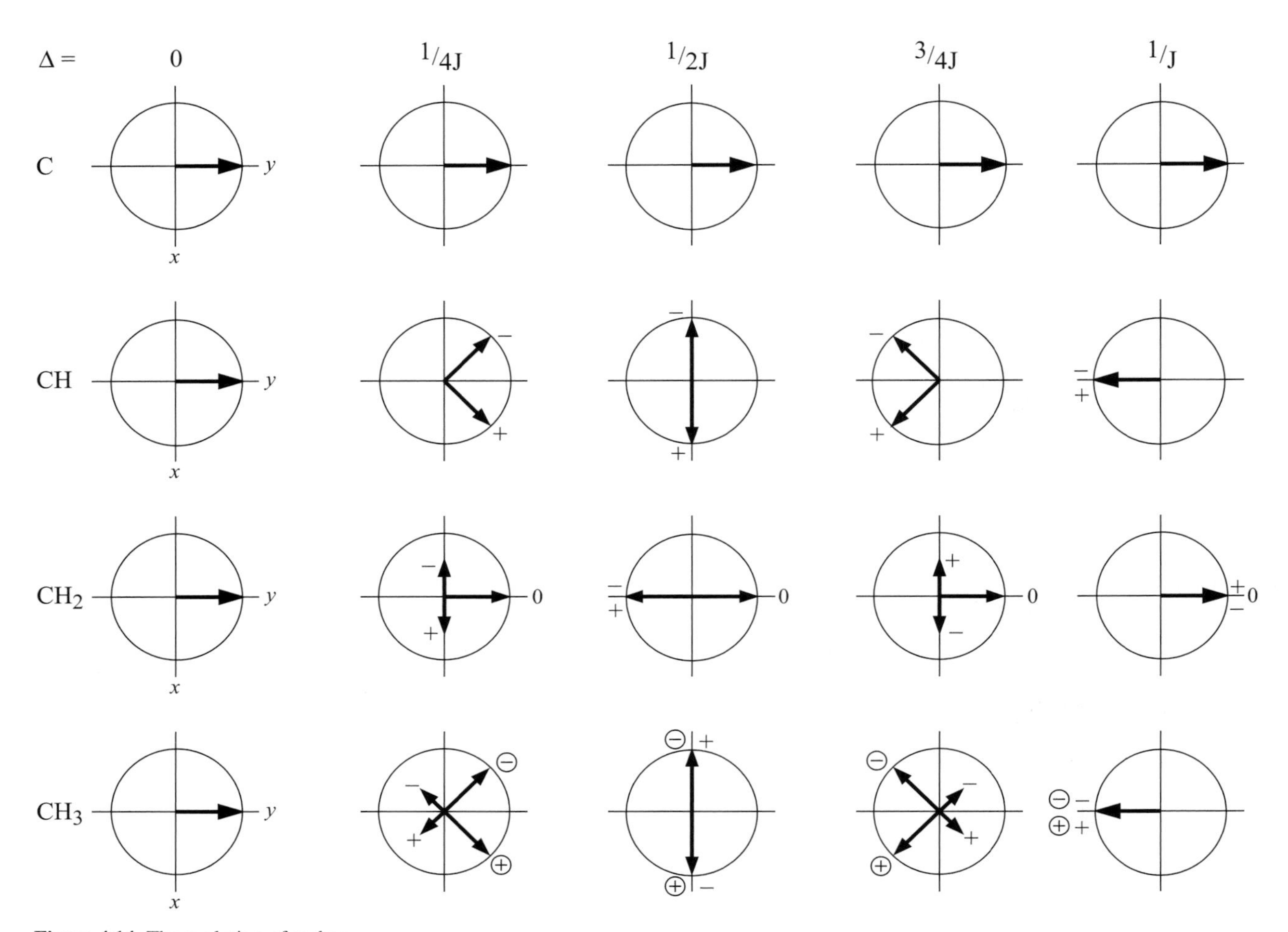

Figure 4.14. The evolution of carbon magnetisation vectors under the influence of proton-carbon couplings.

Consider events with $\Delta = 1/J$ s, for which the relevant vector evolution for different multiplicities is illustrated in Fig. 4.14. Since chemical shifts play no part, quaternary carbons remain stationary along $+y$ during Δ (long-range C–H coupling that may exist will be far smaller than the one-bond coupling and may be considered negligible). Doublet vectors for a methine pair evolve at $\pm J/2$ Hz, so will each rotate through one-half cycle in $1/J$ s and hence meet once more along $-y$. As these now have a 180° phase difference with respect to the quaternary signals, they will ultimately appear inverted in the final spectrum. Applying these arguments to the other multiplicities shows that methylene vectors evolving at $\pm J$ Hz will align with $+y$ whilst methyl vectors evolving at $\pm J/2$ and $\pm 3J/2$ Hz will terminate along $-y$. More generally, by defining an angle θ such that $\theta = 180J\Delta$ degrees, the signal intensities of carbon multiplicities, I, vary according to:

$$\begin{aligned} &\text{C:} && I = 1 \\ &\text{CH:} && I \propto \cos\theta \\ &\text{CH}_2\text{:} && I \propto \cos^2\theta \\ &\text{CH}_3\text{:} && I \propto \cos^3\theta \end{aligned}$$

as illustrated in Fig. 4.15. The spectrum for $\Delta = 1/J$ s ($\theta = 180°$) will therefore display quaternaries and methylenes positive with methines and methyls negative if phased as for the one-pulse carbon spectrum (Fig. 4.16b), although edited spectra are often presented with methine resonances positive and methylene negative, so check local convention. The carbon multiplicities therefore become encoded as signal intensities and at least some of the

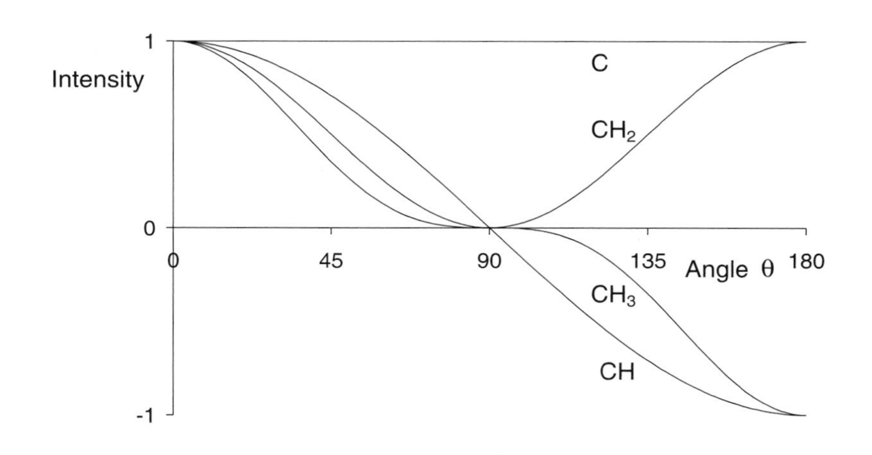

Figure 4.15. The variation of carbon signal intensities in the J-modulated spin-echo as a function of the evolution time Δ ($\theta = 180J\Delta^{o}$).

multiplicity information lost in the single-pulse carbon experiment has been recovered, whilst still benefiting from decoupled resonances. The application of proton decoupling for all but a short time period Δ also ensures spectra are acquired with enhancement from the NOE. Setting $\Delta = 1/2J$ s corresponds to a null for all protonated carbons (Fig. 4.15), so producing a quaternary-only spectrum (Fig. 4.16c). The accuracy of editing is clearly critically dependent on the correct setting of Δ, which is in turn dependent upon J, and as there are likely to be a wide variety of J values within a sample one is forced to make some compromise setting for Δ. One-bond proton–carbon couplings range between 125 and 250 Hz, although they are more commonly between 130 and 170 Hz (Table 4.2), so a typical value for Δ would be ≈7 ms ($^{1}J_{CH} \approx 140$ Hz) when aromatics and carbon centres bearing electronegative heteroatoms are anticipated. If couplings fall far from the chosen value the corresponding resonance can display unexpected and potentially confusing behaviour. Alkyne carbons are particularly prone to this, owing to exceptionally large $^{1}J_{CH}$ values.

The spin-echo experiment is particularly simple to set-up as it does not require proton pulses or their calibration, a desirable property when the experiment was first introduced but of little consequence nowadays. The same results can, in fact, be obtained by the use of proton 180° pulses rather than

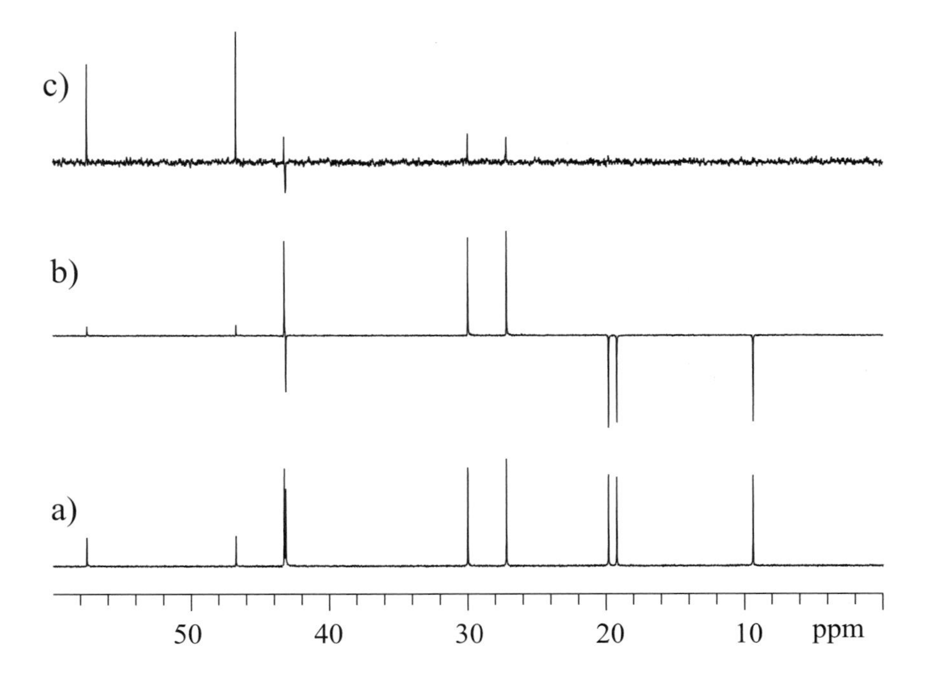

Figure 4.16. Carbon spectrum of camphor **4.1** edited with the J-modulated spin-echo sequence. (a) Conventional carbon spectrum (carbonyl not shown), and edited spectra with (b) $\Delta = 1/J$ ($\theta = 180^{o}$) and (c) $\Delta = 1/2J$ ($\theta = 90^{o}$) with J assumed to be 130 Hz. Some breakthrough of protonated carbons is observed in (c) due to variations in coupling constants within the molecule.

Table 4.2. Typically ranges for one-bond proton-carbon coupling constants

Carbon environment	Typical $^{1}J_{CH}$ range (Hz)
Aliphatic, CH_n-	125–135
Aliphatic, CH_nX (X = N, O, S)	135–155
Alkene	155–170
Alkyne	240–250
Aromatic	155–165

gating of the decoupler [19] (Fig. 4.13b). In this case the Δ period is broken into two periods of 1/2J separated by the simultaneous application of proton and carbon 180° pulses. These serve to refocus carbon chemical shifts but at the same time allow couplings to continue to evolve during the second Δ/2 period (Section 2.2). Hence, the total evolution period in which coupling is active is 1/J, as in the decoupler-gating experiment above, and identical modulation patterns are produced. It is this shorter, pulsed form of the heteronuclear spin-echo that is widely used in numerous pulse sequences to refocus shift evolution yet leave couplings to evolve.

4.3.2. APT

The principal disadvantage of the J-modulated spin-echo described above is the use of a 90° carbon excitation pulse which, as discussed in Section 4.1, is not optimum for signal averaging and may lead to signal saturation, notably of quaternary centres. The preferred approach using an excitation pulse width somewhat less than 90° requires a slight modification of the J-modulated experiment, giving rise to the APT (Attached Proton Test) sequence [20] (Fig. 4.17). Use of a small tip-angle excitation pulse leaves a component of magnetisation along the $+z$ axis, which is inverted by the 180°(C) pulse that follows. In systems with slowly relaxing spins, this inverted component may cancel magnetisation arising from the relaxation of the transverse components, leaving little or no net signal to observe on subsequent cycles. It is therefore necessary to add a further 180°(C) which returns the problematic $-z$ component to $+z$ prior to acquisition, thereby eliminating possible cancellation. Transverse components also experience a 180° rotation prior to detection, but are otherwise unaffected beyond this phase inversion. The APT experiment is commonly used with excitation angles of 45° or less and is thus better suited to signal averaging than the basic echo sequence above, but otherwise gives similar editing results.

The poor editing accuracy of spin-echoes in the presence of a wide range of J values and the inability to fully characterise all carbon multiplicities are the major limitations of these techniques. More complex variations on the pulsed J-modulated spin-echo are to be found that do allow a complete decomposition of the carbon spectra into C, CH, CH_2 and CH_3 sub-spectra [21] and which also show greater tolerances to variations in $^{1}J_{CH}$ [22]. Likewise,

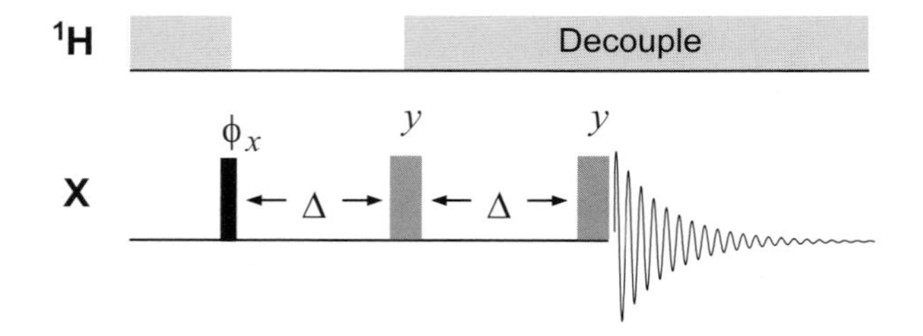

Figure 4.17. The Attached Proton Test (APT) sequence.

J-compensated APT sequences have been developed for greater tolerance to a spread of J values [23], and for the direct generation of complete sub-spectra [24]. Invariably the simplest sequences still find widest use.

4.4. SENSITIVITY ENHANCEMENT AND SPECTRUM EDITING

One of the principal concerns for the chemist using NMR spectroscopy is the relatively poor sensitivity of the technique when compared to other analytical methods, which stems from the small energy differences and hence small population differences between spin states. Developments that provide increased signal intensity relative to background noise are a constant goal of research within the NMR community, be it in the design of the NMR instrumentation itself or in novel pulse techniques and data processing methods. The continued development of higher field magnets clearly contributes to gains in sensitivity, and is paralleled by other instrumental advances. Aside from these developments, optimum sensitivity is provided by observing the nuclear species giving rise to the strongest signal, as judged by the intrinsic sensitivity of the nucleus and its natural abundance. The intrinsic sensitivity depends upon the magnetogyric ratio, γ, of the spin in three ways, with a greater γ contributing to:

- a high resonant frequency, which in turn implies a large transition energy difference and hence a greater Boltzmann population difference,
- a high magnetic moment and therefore a stronger signal, and
- a high *rate* of precession which induces a greater signal in the detection coil (just as cycling faster causes a bicycle dynamo lamp to glow brighter).

Thus, in general, the strength of an NMR signal is proportional to γ^3 for a single nuclide, but as noise itself increases with the square-root of the observation frequency [25] the signal-to-noise ratio scales as $\gamma^{5/2}$. Notice also that as two of the above terms depend upon resonant frequency, the signal-to-noise ratio for a single nuclide scales with the static field according to $B_0^{3/2}$ (all other things being equal). When more than one nuclide is involved in a sequence, a general expression for the signal-to-noise ratio of a one dimensional experiment may be derived:

$$\frac{S}{N} \propto N\,A\,T^{-1}B_0^{3/2}\gamma_{exc}\gamma_{obs}^{3/2}T_2^*(NS)^{1/2} \tag{4.2}$$

where N is the number of molecules in the observed sample volume, A is a term that represents the abundance of the NMR active spins involved in the experiment, T is temperature, B_0 is the static magnetic field, γ_{exc} and γ_{obs} represent the magnetogyric ratios of the initially excited and the observed spins respectively, T_2^* is the effective transverse relaxation time and NS is the total number of accumulated scans. The high magnetogyric ratio of the proton, in addition to its 100% abundance (and ubiquity), explains why proton observation is favoured in high-resolution NMR. Indeed, many of the newly developed multipulse heteronuclear experiments utilise direct proton observation to achieve greater sensitivity, whilst the heteronuclear spin is observed indirectly, as described in Chapter 6. This section deals with those one-dimensional pulse methods in widespread use that assist in the *direct* observation of relatively low-γ nuclei, here termed X-nuclei, for example, ^{13}C, ^{15}N, ^{29}Si and so on. These methods are characterised by initial excitation of a high-γ spin, typically ^{1}H, ^{19}F or ^{31}P, followed by *polarisation transfer* onto the low-γ nucleus to which it is scalar spin–spin coupled. This process introduces γ_{high} as the γ_{exc} term in Eq. 4.2 in place of γ_{low}, and so provides sensitivity gains by a factor of $\gamma_{high}/\gamma_{low}$.

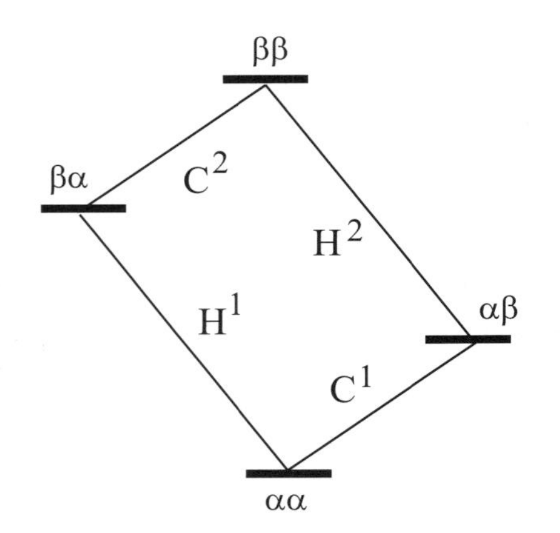

Figure 4.18. The schematic energy level diagram for a two-spin ^1H–^{13}C system. The two transitions for each nucleus correspond to the two lines in each doublet.

It has been alluded to in the previous section, and will be considered in full in Chapter 8, that significant enhancements of low-γ nuclei may be obtained on saturation of, for example, protons, owing to the NOE. This effect, however, can be ill-suited to nuclei that posses a negative γ, since the NOE then causes a *reduction* in signal intensity and may, in the worst case, lead to a complete loss of signal. Polarisation transfer methods do not suffer such disadvantages and, in many cases, are able to provide considerably greater enhancements than the NOE alone. In addition, these experiments provide another means for the multiplicity editing of spectra and enable the differentiation of groups that possess differing numbers of attached nuclei. The ability to edit carbon spectra and hence to distinguish quaternary, methine, methylene and methyl groups offers an alternative to the spin-echo methods of the previous section and these polarisation transfer techniques, the DEPT experiment in particular, have become routine experiments in the organic chemist's repertoire. Methods that relate to the multiplicity editing of proton spectra are principally dealt with in Chapter 6.

4.4.1. Polarisation transfer

Polarisation transfer methods enhance signal intensity by transferring the greater population differences of high-γ spins onto their spin-coupled low-γ partners. Through this they replace one of the three γ_{low} signal intensity dependencies described above with γ_{high}, leading to a signal enhancement by a factor of $(\gamma_{high}/\gamma_{low})$. The principles behind all polarisation transfer methods can be understood by considering a spin-$\frac{1}{2}$ scalar-coupled pair which for illustrative purposes is here taken to be a ^1H–^{13}C pair *both with 100% abundance*, whose energy level diagram is shown in Fig. 4.18. At thermal equilibrium the populations for each of the four spin transitions are governed by the Boltzmann law as represented in Fig. 4.19a. From this the population *differences* across the transitions are $2\Delta H$ and $2\Delta C$ for the proton and carbon spins respectively, and since these scale linearly with γ, the ratio $2\Delta H/2\Delta C$ equates to $\gamma H/\gamma C$ or ≈ 4. The resonance intensities for proton are thus four times greater than those of carbon when judged by these populations (Fig. 4.19b). Now consider the result of selectively inverting *one* line of the H-spin doublet, say H^1, (with a weak selective pulse for example) which equates to inverting the population difference across the corresponding transition (Fig. 4.19c). As H^2 remains untouched, the population difference across the H^2 transition is no different to that before the inversion. The salient point is that both the C-spin population differences have now been altered by this process. Thus, the C^1 population difference becomes $-2\Delta H+2\Delta C$ whilst that of C^2 is $2\Delta H+2\Delta C$ and the population differences previously associated with the protons have been transferred onto the carbons. Since $2\Delta H$

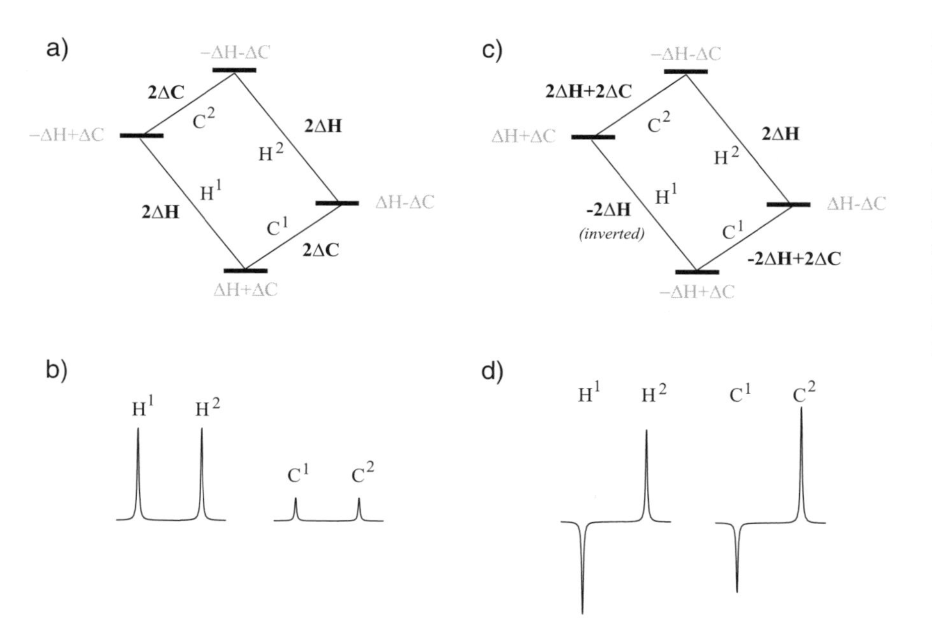

Figure 4.19. Polarisation transfer in a two-spin system. (a) Populations (in grey) and population differences (in bold) for each transition at equilibrium and (b) the corresponding spectra obtained following pulse excitation illustrating the four-fold population difference (see text). (c) The situation after selective inversion of one-half of the proton doublet and (d) the corresponding spectra showing the enhanced intensity of the carbon resonances.

is four-times greater than 2ΔC, the C-transitions will display relative intensities of −3:5 so that sampling the carbon magnetisation at this point would produce a spectrum with one half of the doublet inverted and with signal intensities greater than in the single pulse carbon spectrum (Fig. 4.19d). For a $^{1}H–^{15}N$ pair, signal intensities following polarisation transfer from protons are +11 and −9, ($\gamma_H/\gamma_N \approx 10$) and the signal enhancement is even more impressive. Since one half of the H-spin populations have been inverted but the line intensities are otherwise unaffected, there has been no *net* transfer of magnetisation from proton to carbon. The *integrated intensity* of the whole carbon doublet is the same as that in the absence of polarisation transfer, so we say there has been a *differential* transfer of polarisation.

The experiment described above is termed *selective population transfer* (SPT), or more precisely in this case with proton spin inversion, *selective population inversion*, (SPI). It is important to note, however, that the complete inversion of spin populations is not a requirement for the SPT effect to manifest itself. Any *unequal perturbation* of the lines within a multiplet will suffice, so, for example, saturation of one proton line would also have altered the intensities of the carbon resonance. In heteronuclear polarisation (population) transfer experiments, it is the heterospin-coupled satellites of the parent proton resonance that must be subject to the perturbation to induce SPT. The effect is not restricted to heteronuclear systems and can appear in proton spectra when homonuclear-coupled multiplets are subject to unsymmetrical saturation. Fig. 4.20 illustrates the effect of selectively but unevenly saturating a double doublet and shows the resulting intensity distortions in the multiplet structure of its coupled partner, which are most apparent in a difference spectrum. Despite these distortions, the integrated intensity of the proton multiplet is unaffected by the presence of the SPT because of the equal positive and negative contributions (see Fig. 4.19d). Distortions of this sort have particular relevance to the NOE difference experiment described in Chapter 8.

The greatest limitation of the SPI experiment is its lack of generality. Although it achieves the desired polarisation transfer, it is only able to produce this for one resonance in a spectrum at a time. To accomplish this for all one would have to repeatedly step through the spectrum, inverting satellites one-by-one and performing a separate experiment at each step. Clearly a more efficient approach would be to invert one half of each proton doublet

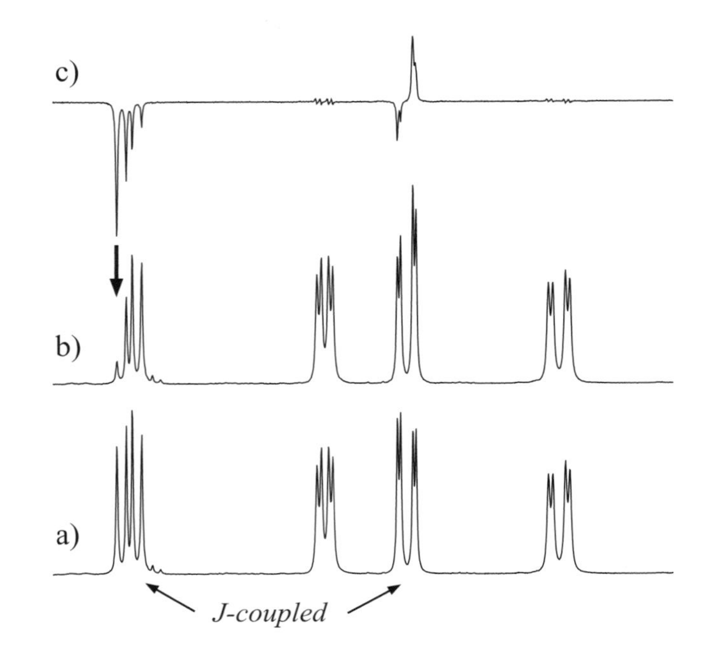

Figure 4.20. Selective population transfer (SPT) distorts proton multiplet intensities when the resonance of a coupled partner is unevenly saturated (b). These perturbations are more apparent in the difference spectrum (c), when the unperturbed spectrum (a) is subtracted.

simultaneously for all resonances in a single experiment and this is precisely what the INEPT sequence achieves.

4.4.2. INEPT

The INEPT experiment [26] (Insensitive Nuclei Enhanced by Polarisation Transfer) was one of the forerunners of many of the pulse NMR experiments developed over subsequent years and still constitutes a feature of some of the most widely used multidimensional experiments in modern pulse NMR. Its purpose is to enable *non-selective* polarisation transfer between spins, and its operation may be readily understood with reference to the vector model. Most often it is the proton that is used as the source nucleus and these discussion will relate to XH spin systems throughout, although it should be remembered that any high-γ spin-$^1/_2$ nucleus constitutes a suitable source.

The INEPT sequence (Fig. 4.21a) provides a method for inverting one-half of each XH doublet in a manner that is independent of its chemical shift requiring the use of non-selective pulses only. The sequence begins with excitation of all protons which then evolve under the effects of chemical shift and heteronuclear coupling to the X-spin. After a period $\Delta/2$, the proton vectors experience a 180° pulse which serves to refocus chemical shift evolution (and field inhomogeneity) during the second $\Delta/2$ period. The simultaneous application of a 180°(X) pulse ensures the heteronuclear coupling continues to evolve by inverting the proton vector's sense of precession. This is once again a spin-echo sequence in which only coupling evolution requires consideration. For an XH pair, a total Δ period of 1/2J ($\Delta/2 = 1/4$J) leaves the two proton vectors opposed or *antiphase* along $\pm x$, so that the subsequent $90°_y$(H) pulse aligns these along the $\pm z$-axis (Fig. 4.22). This therefore corresponds to the desired inversion of one-half of the proton doublet, as for SPI, but for all spin pairs simultaneously. The 90°(X) pulse samples the newly created population differences to produce the enhanced spectrum. In practice, these last two pulses are applied simultaneously, although in this pictorial representation it proves convenient to consider the proton pulse to occur first.

The asymmetrical peak intensities produced by the SPI effect, -3 and $+5$ for a CH pair, arise from the contribution of the natural X-spin magnetisation, which may be removed by the use of a simple phase cycle. Thus, repeating the

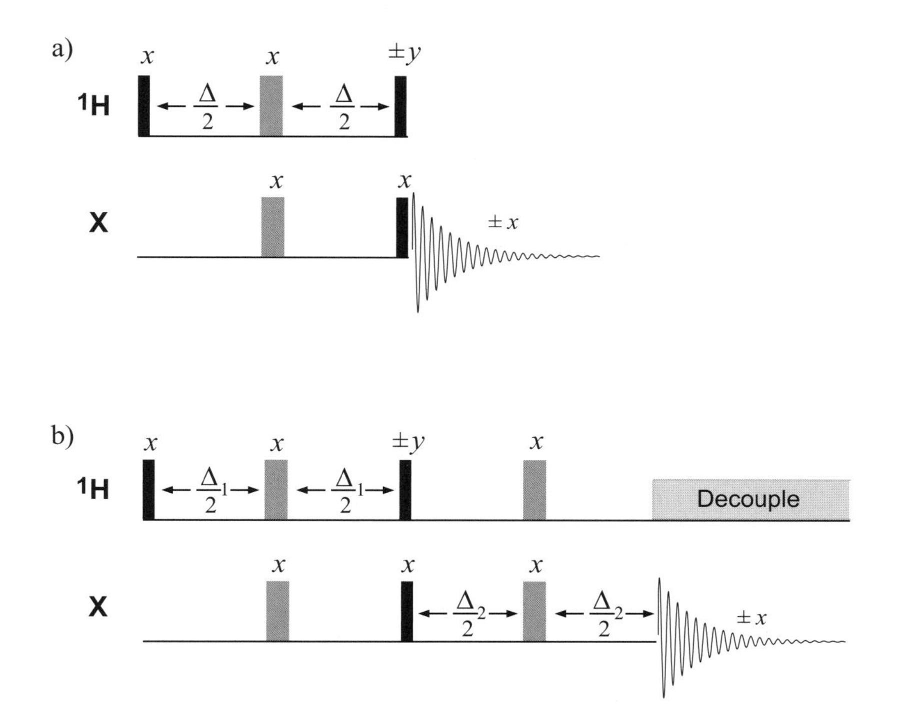

Figure 4.21. (a) the INEPT sequence and (b) the refocused INEPT sequence.

experiment with the phase of the last proton pulse at $-y$ causes the antiphase lines to adopt the inverted disposition relative to the $+y$ experiment, that is, what was $-z$ now becomes $+z$ and vice versa (Fig. 4.23). The resulting X-spin doublets are likewise inverted, although the natural X magnetisation, being oblivious to the proton pulse, is unchanged. Subtraction of the two experiments by inverting the receiver phase causes the polarisation transfer contribution to add but cancels the natural magnetisation (Fig. 4.24). This two-step phase-cycle is the basic cycle required for INEPT. As only the polarisation transfer component is retained by this process, a feature of pure polarisation transfer experiments is the lack of responses from nuclei without significant proton coupling.

Refocused INEPT

One problem with the basic INEPT sequence described above is that it precludes the application of proton spin decoupling during the acquisition of the X-spin FID. Since this removes the J-splitting it will cause the antiphase

Figure 4.22. The evolution of proton vectors during the INEPT sequence. Following initial evolution under J_{XH} the 180°(H) pulse flips the vectors about the x-axis and the 180°(C) pulse inverts their sense of precession. After a total evolution period of $1/2J_{HX}$ the vectors are antiphase and are subsequently aligned along $\pm z$ by the $90°_y$(H) pulse. This produces the desired inversion of one half of each H–X doublet for all resonances.

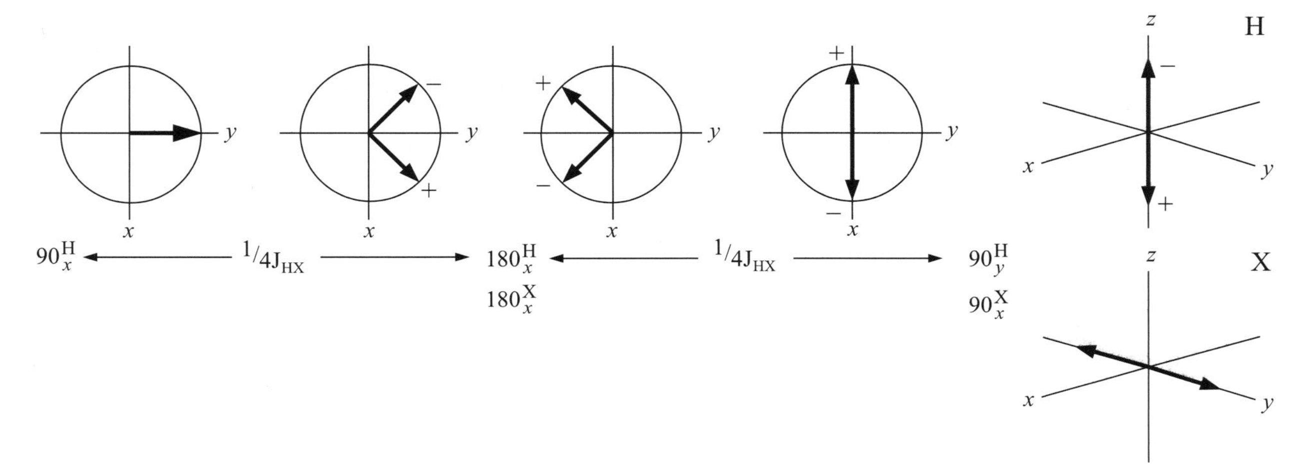

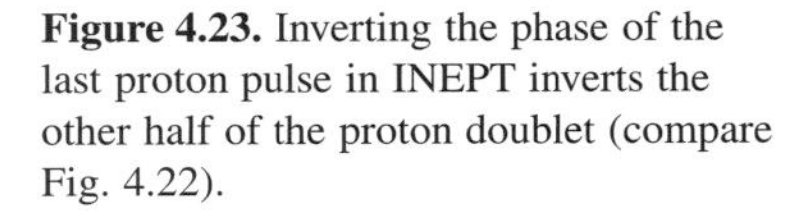

Figure 4.23. Inverting the phase of the last proton pulse in INEPT inverts the other half of the proton doublet (compare Fig. 4.22).

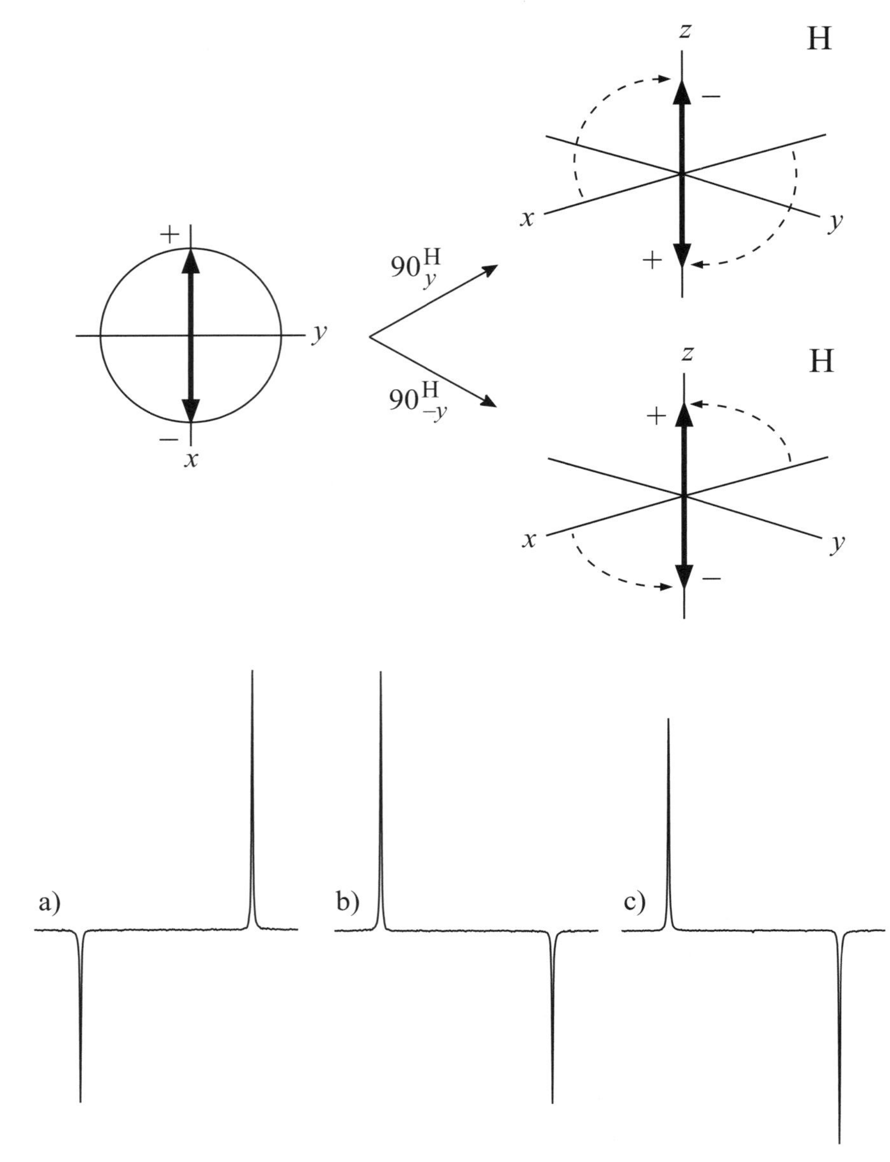

Figure 4.24. Experimental carbon-13 INEPT spectra of methanoic acid recorded with the phase of the last proton pulse set to (a) y and (b) $-y$. Subtracting the two data sets (c) cancels natural magnetisation that has not been generated by polarisation transfer and equalises the intensity of the two lines.

lines to become coincident and cancel to leave no observable signal. The addition of a further period, Δ_2, after the polarisation transfer allows the X-spin magnetisation to refocus under the influence of the XH coupling, giving the *refocused-INEPT* sequence [27] (Fig. 4.21b), the results of which are illustrated in Fig. 4.25. Once again, the refocusing period is applied in the form of a spin-echo to remove chemical shift dependencies, and for the XH system to fully refocus the appropriate period is again 1/2J. Since it is now the X-spin magnetisation that refocuses, it is necessary to consider coupling to all attached protons. Choosing $\Delta_2 = 1/2J$ is not appropriate for those nuclei coupled to more than one proton and does not lead to the correct refocusing of vectors, rather they remain antiphase and produce no signal on decoupling (this provides a clue as to how one might use INEPT for spectrum editing). It has been shown [28] that for optimum sensitivity a refocusing period for an XH_n group requires:

$$\Delta_2 = \frac{1}{\pi J}\sin^{-1}\left(\frac{1}{\sqrt{n}}\right) \tag{4.3}$$

where the angular term must be calculated in radians. For carbon spectroscopy this corresponds to 1/2J, 1/4J and ≈1/5J for CH, CH_2 and CH_3 groups re-

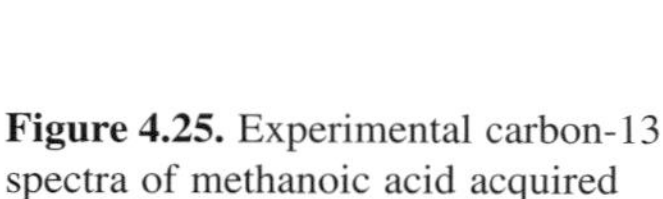

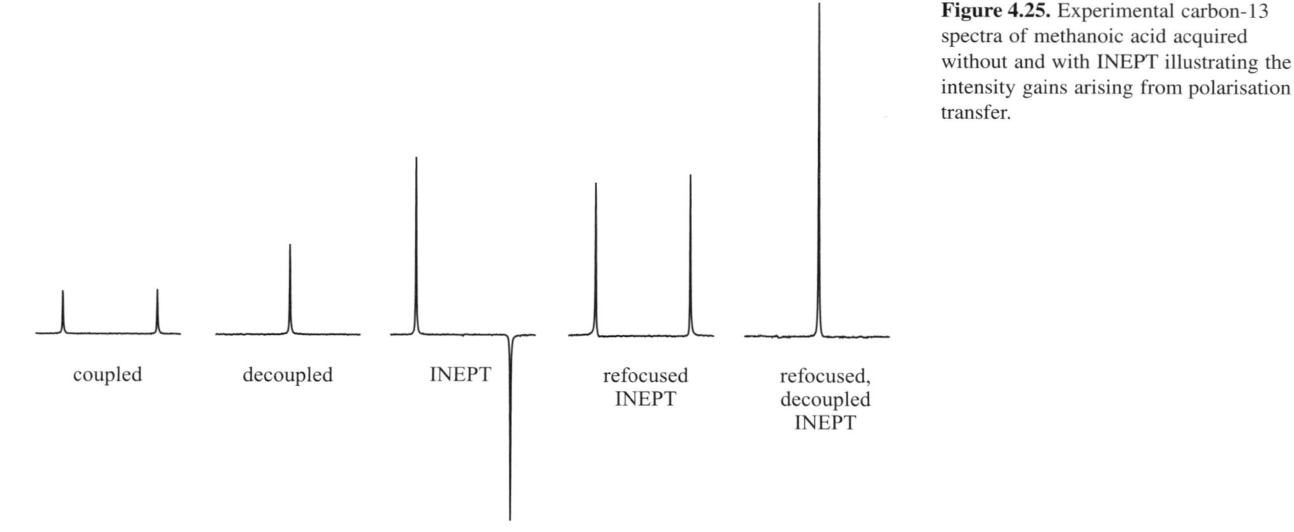

Figure 4.25. Experimental carbon-13 spectra of methanoic acid acquired without and with INEPT illustrating the intensity gains arising from polarisation transfer.

spectively. If all multiplicities are to be observed simultaneously, a compromise setting of $\approx 1/3.3J$ is appropriate, which for an assumed 140 Hz coupling gives $\Delta_2 = 2.2$ ms. A spectrum recorded under these conditions but without proton decoupling will display significant phase errors since the refocusing period will not be optimal for any multiplicity; such errors are removed with proton decoupling however.

In cases where the XH coupling fine-structure is of interest and proton decoupling is not applied, the relative line intensities within multiplets will be distorted relative to the coupled spectrum without polarisation transfer, and this is one potential disadvantage of the above INEPT sequences. Even with suppression of the contribution from natural X-spin magnetisation, intensity anomalies remain for XH_n groups with $n > 1$ (Fig. 4.26). A potentially confusing feature is the disappearance of the central line of XH_2 multiplets. One sequence proposed to generate the usual intensities within multiplets is INEPT$^+$, comprising the refocused-INEPT with an additional proton 'purge' pulse to remove the magnetisation terms responsible for the intensity anomalies [29]. In practice, the DEPT-based sequences described below are preferred as these show greater tolerance to experimental mis-settings.

Finally note that these INEPT discussions assume the evolution of magnetisation is dominated by the coupling between the X and H spins, with all other couplings being negligibly small. In situations where homonuclear proton coupling becomes significant ($J_{HH} \geq \approx 1/3\ J_{HX}$), it is necessary to modify the Δ_1 period for optimum sensitivity, and analytical expressions for this have been derived [30]. Such considerations are most likely to be significant with polarisation transfer from protons via long-range couplings, that is, when the protons are not directly bound to the heteronucleus. Examples may include coupling to tertiary nitrogens, quaternary carbons or to phosphorus. In some instances, the situation may be considerably simplified if the homonuclear couplings are removed by the application of *selective* proton decoupling during the proton Δ_1 evolution periods [31]. Fig. 4.27 illustrates the acquisition of ^{15}N

	(a)	(b)	(c)
CH	1 1	-3 +5	-1 1
CH_2	1 2 1	-7 2 +9	-1 0 1
CH_3	1 3 3 1	-11 -9 15 13	-1 -1 1 1

Figure 4.26. Relative multiplet line intensities in coupled INEPT spectra. (a) conventional multiplet intensities and those from INEPT (b) without and (c) with suppression of natural magnetisation.

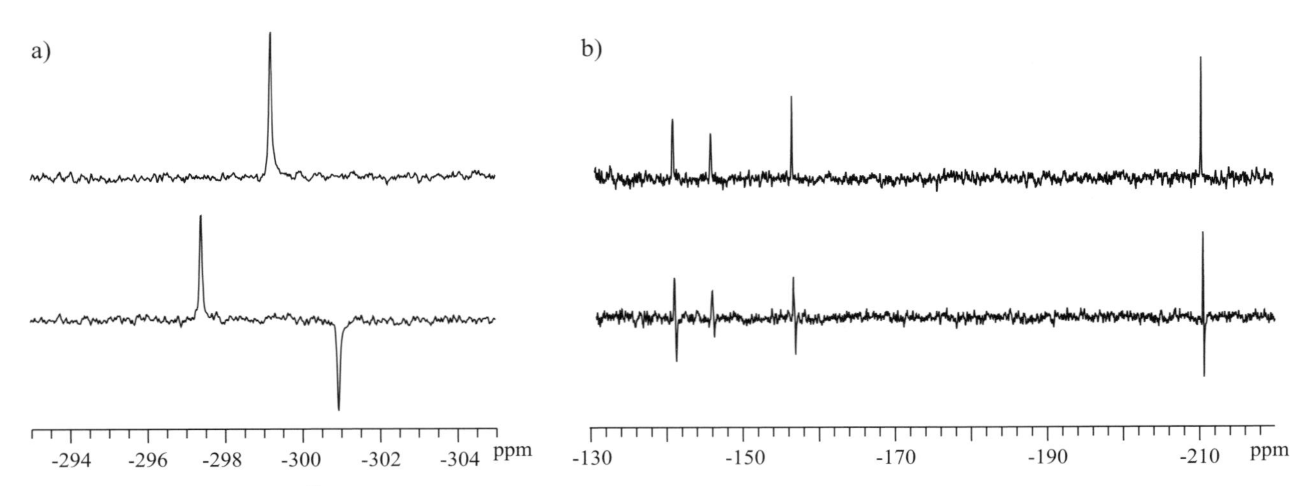

Figure 4.27. Natural abundance ^{15}N INEPT spectra of adenosine-5′-monosulphate **4.5**. Lower traces show results with the INEPT sequence and upper traces with the refocused-INEPT sequence. Delays were calculated assuming (a) J = 90 Hz and (b) J = 10 Hz. For (a) the lower trace displays the 1:0:−1 pattern of the NH_2 group whilst in (b) all resonances display antiphase two-bond ^1H–^{15}N couplings. Spectra are referenced to nitromethane.

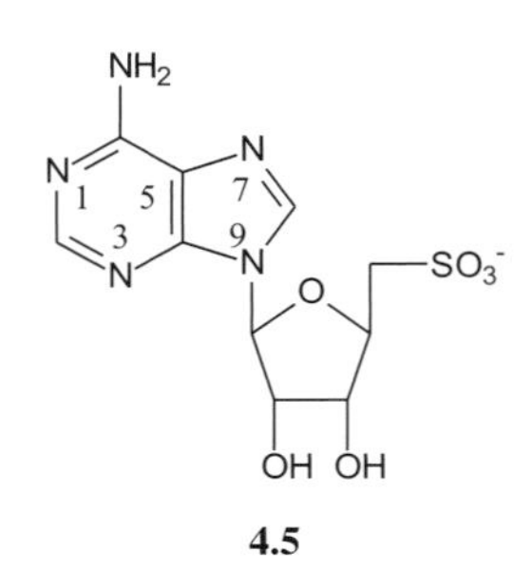

spectra of adenosine-5′-sulphate **4.5** by polarisation transfer through one-bond and long-range (two-bond) couplings. All nitrogen centres in the purine unit are observed in two experiments in which delays were optimised for J = 90 Hz and 10 Hz, which were assumed values for one- and two-bond couplings, respectively.

Sensitivity gains

To appreciate the sensitivity gains from the application of the INEPT sequence, one should compare the results with those obtained from the usual direct observation of the low-γ species. This invariably means the spectrum obtained in the presence of proton broadband decoupling for which signal enhancement will occur by virtue of the ^{1}H-X NOE (Chapter 8). Thus, to make a true comparison, we need to consider the signal arising from polarisation transfer versus that from observation with the NOE, which *for an XH pair* is given by:

$$I_{INEPT} = I_0 \left|\frac{\gamma_H}{\gamma_X}\right| \qquad I_{NOE} = I_0\left(1 + \frac{\gamma_H}{2\gamma_X}\right) \tag{4.4}$$

where I_{INEPT} is the signal intensity following polarisation transfer, I_{NOE} is that in the presence of the NOE and I_0 represents the signal intensity in the absence of any enhancement. Notice firstly that the NOE makes a contribution that *adds* to the natural magnetisation and secondly, because of this, the resulting signal intensity is also dependent upon the *sign* of the magnetogyric ratios, whereas polarisation transfer depends only on their magnitudes. The NOE therefore causes a *decrease* in signal intensity for those nuclei with a *negative* magnetogyric ratio, and may cause the observed signal to be inverted if the NOE is greater than the natural magnetisation or to become close to zero if comparable to it. Table 4.3 compares the theoretical *maximum* signal intensities that can be expected for polarisation transfer and the NOE from protons to heteronuclear spins in XH pairs. The degree of signal enhancement does not

Table 4.3. Signal intensities for the X-spin in ^{1}H–X pairs arising from polarisation transfer (I_{INEPT}) and from direct observation with the maximum NOE (I_{NOE})

X	^{13}C	^{15}N	^{29}SI	^{31}P	^{57}Fe	^{103}Rh	^{109}Ag	^{119}Sn	^{183}W	^{195}Pt	^{207}Pb
I_{INEPT}	3.98	9.87	5.03	2.47	30.95	31.77	21.50	2.81	24.04	4.65	4.78
I_{NOE}	2.99	−3.94	−1.52	2.24	16.48	−14.89	−9.75	−0.41	13.02	3.33	3.39

Intensities are given relative to those obtained by direct observation in the absence of the NOE (I_0).

scale linearly with the number of attached source nuclei, n, used for polarisation transfer. In other words, the enhancement expected for an XH_2 group is not twice that for an XH group. Although greater enhancements do arise with more attached protons when the optimum refocusing delay is used, the gains over that for an XH group are only modest [29].

In practice, owing to experimental imperfections or to other relaxation processes reducing the magnitude of the NOE (chemical shift anisotropy for the metals in particular), the figures in Table 4.3 may not be met, although they provide some guidance as to which method would be the most appropriate. The results for the higher-γ heteronuclei such as ^{31}P and ^{13}C are clearly comparable, whereas polarisation transfer provides far greater gains for the lower-γ species. A further important benefit of the polarisation transfer approach not reflected in the figures of Table 4.3 is that the repetition rate of the experiment depends on the longitudinal relaxation times of the *protons*, since the populations of interest originate only from these spins. In contrast, the direct observation experiment, with or without the NOE, depends on the relaxation times of the X-spins, which are typically very much longer. The opportunity for faster signal averaging provides another significant gain in sensitivity per unit time when employing polarisation transfer and in practice this feature can be as important or sometimes more important than the direct sensitivity gains from the transfer itself.

In short, polarisation transfer methods provide greatest benefits for those nuclei that have low magnetogyric ratios and are slow to relax. It is also the preferred approach to direct observation for those nuclei with negative γs where the NOE may lead to an overall signal reduction. Nitrogen-15 [32] (Fig. 4.28) and silicon-29 routinely benefit from PT methods, as does the observation of metals, in particular transition metals which often posses very low γs and are extremely insensitive, despite their sometimes high natural abundance, some examples being ^{57}Fe, ^{103}Rh, ^{109}Ag, and ^{183}W.

Despite these impressive sensitivity gains when directly observing the X-spin, the more modern approach is to observe the X-spin *indirectly* through the coupled proton when possible, which can be achieved through a number of heteronuclear correlation experiments. As these methods additionally employ proton observation, they benefit from a further theoretical gain of $(\gamma_{high}/\gamma_{low})^{3/2}$ over X-observe schemes (see Eq. 4.2). These topics are pursued in Chapter 6, and should be considered as potentially faster routes to X-nucleus data.

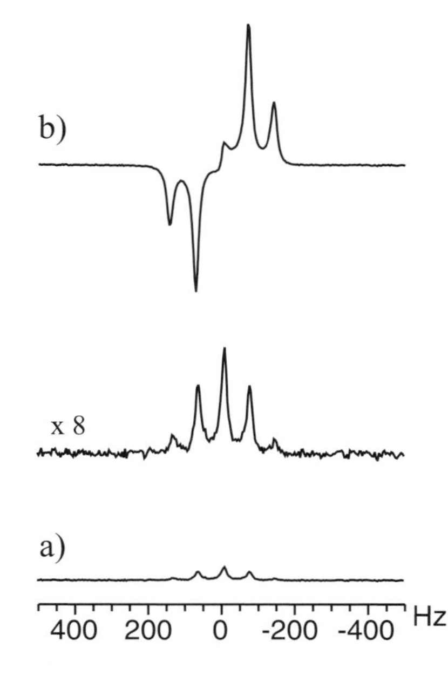

Figure 4.28. Signal enhancement of the ^{15}N spectrum of ammonium nitrate with INEPT. Direct observation using (a) the Ernst angle optimised for the nitrogen T_1 and (b) INEPT optimised for the proton T_1. Both spectra were collected in the same total accumulation time.

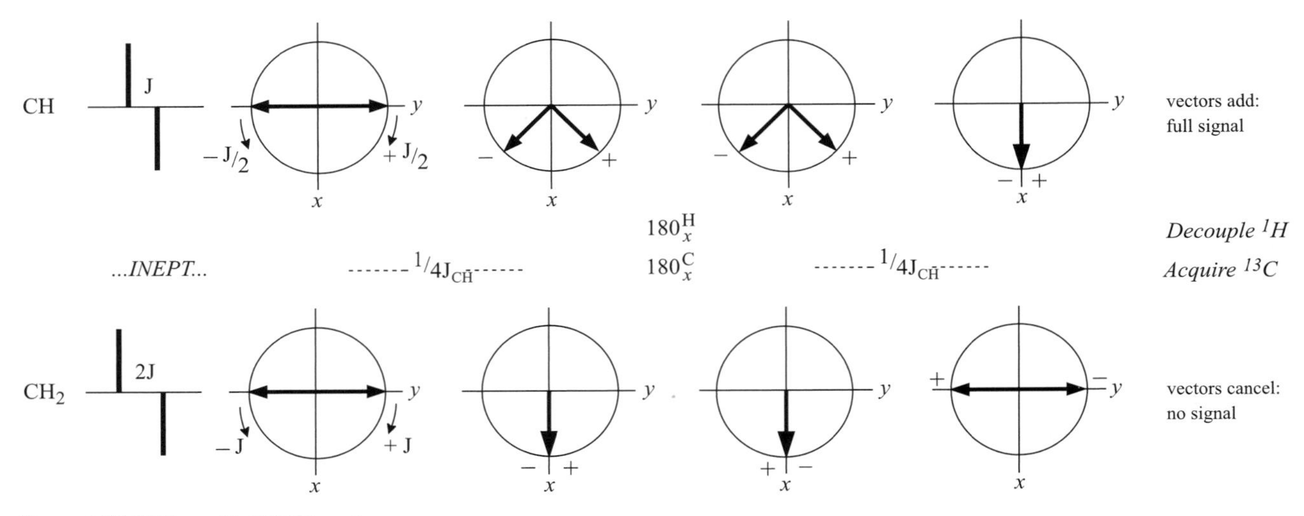

Figure 4.29. Editing with INEPT can be achieved with a judicious choice of refocusing delay, Δ_2. A total delay of $1/2J_{CH}$ retains CH resonances but eliminates CH_2 (and CH_3) resonances.

Editing with INEPT

Returning to the selection of the Δ_2 period in the refocused INEPT sequence, it is apparent that whilst a period of 1/2J produces complete refocusing of doublets, the triplets (and quartets) remain antiphase and will be absent from the spectrum recorded with proton decoupling (Fig. 4.29). Choosing $\Delta_2 = 1/2J$ therefore yields a sub-spectrum containing only methine resonances. This idea of editing heteronuclear spectra according to multiplicities is closely related to the editing with spin-echoes described in Section 4.3. Extending this idea to the selection of other carbon multiplicities it is again convenient to define an angle $\theta = 180J\Delta_2$ degrees from which the signal intensities in the decoupled experiment are:

$$C{:}\quad I \propto \sin\theta$$
$$CH_2{:}\quad I \propto 2\sin\theta\cos\theta$$
$$CH_3{:}\quad I \propto 3\sin\theta\cos^2\theta$$

as presented graphically in Fig. 4.30. To differentiate all protonated carbons it is sufficient to record three spectra with Δ_2 adjusted suitably to give θ = 45, 90 and 135 degrees. The 90° experiment corresponds to $\Delta_2 = 1/2J$ mentioned above and hence displays methine groups only, $\theta = 45°$ produces all responses whilst $\theta = 135°$ has all responses again, but with methylene groups inverted. This process combines signal enhancement by polarisation transfer with multiplicity determination through spectrum editing. The editing

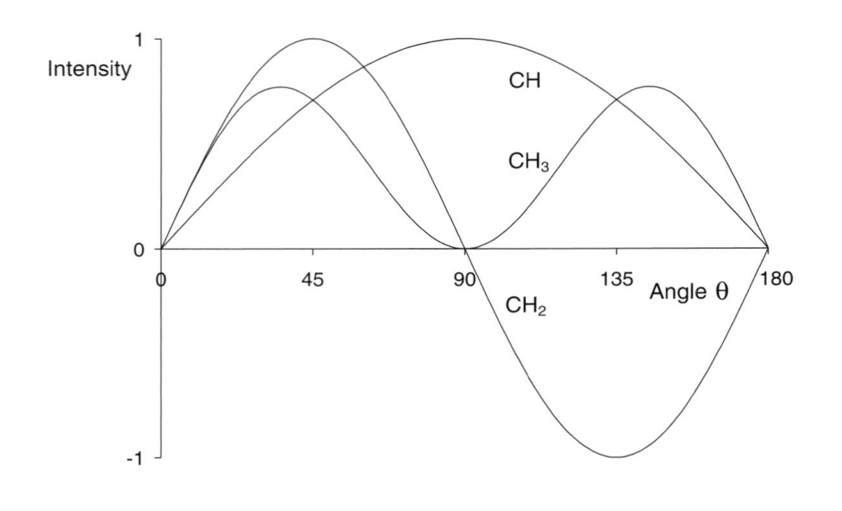

Figure 4.30. The variation of carbon signal intensities in the refocused INEPT experiment as a function of the evolution time Δ ($\theta = 180J\Delta_2$°). Identical results are obtained for the DEPT experiment for which the angle θ represents the tip angle of the last proton pulse.

attainable here is superior to that provided by the basic spin-echo methods of Section 4.3 since comparison of the three INEPT spectra allows one to determine the multiplicity of all resonances in the spectrum. However, in routine carbon-13 spectroscopy, this information tends not to be derived from INEPT but from the related DEPT sequence, which has better tolerance to experimental imperfections. Alas, DEPT, cannot be described fully by recourse to the vector model, and an understanding of INEPT goes a long way to help one appreciate the features of DEPT.

In summary, the INEPT sequence provides signal enhancement for all proton-coupled resonances in a single experiment by virtue of polarisation transfer. This also enables more rapid data collection since the repetition rate is now dictated by the faster relaxing protons and not the heteronucleus. By judicious choice of refocusing delays, refocused INEPT can also provide multiplicity editing of, for example, carbon-13 spectra. However, those nuclei that do not share a proton coupling cannot experience polarisation transfer and are therefore missing from INEPT spectra. The INEPT sequence is also used extensively as a building-block in heteronuclear two-dimensional sequences, as will become apparent in later chapters.

4.4.3. DEPT

The DEPT experiment [33] (Distortionless Enhancement by Polarisation Transfer) is the most widely used polarisation transfer editing experiment in carbon-13 spectroscopy, although its application is certainly not limited to the proton–carbon combination. It enables the complete determination of all carbon multiplicities, as does the refocused INEPT discussed above, but has a number of distinct advantages. One of these is that it directly produces multiplet patterns in proton-coupled carbon spectra that match those obtained from direct observation, meaning methylene carbons display the familiar 1:2:1 and methyl carbons the 1:3:3:1 intensity patterns; this is the origin of the term 'distortionless'. However, for most applications proton decoupling is applied during acquisition and multiplet structure is of no consequence, so the benefits of DEPT must lie elsewhere.

The DEPT sequence

The DEPT pulse sequence is illustrated in Fig. 4.31. To follow events during this, consider once more a ^{1}H–^{13}C pair and note the action of the two 180° pulses is again to refocus chemical shifts where necessary. The sequence begins in a similar manner to INEPT with a 90°(H) pulse after which proton magnetisation evolves under the influence of proton–carbon coupling such that after a period 1/2J the two vectors of the proton satellites are antiphase. The application of a 90°(C) pulse at this point produces a new state of affairs that has not been previously encountered, in which both transverse proton and carbon magnetisation evolve coherently. This new state is termed *heteronuclear multiple quantum coherence* (hmqc) which, in general, cannot be visualised with the vector model, and without recourse to mathematical formalisms it is

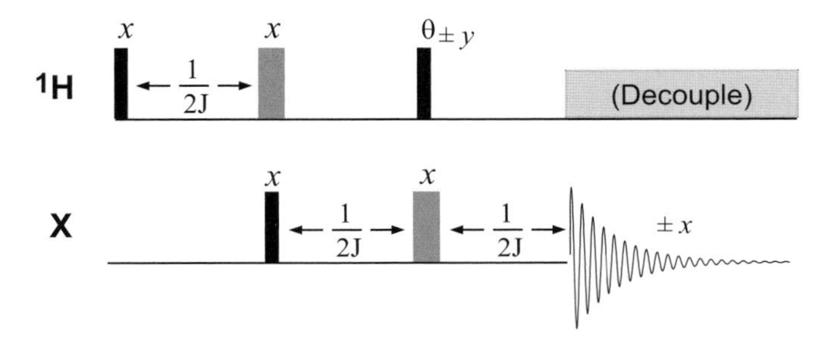

Figure 4.31. The DEPT sequence.

not possible to rigorously describe its behaviour. However, for these purposes we can imagine this multiple quantum coherence to be a pooling of both proton and carbon magnetisation, to be separated again at some later point [3].

The hmqc now evolves for a further time period under the influence of the proton chemical shift but also, simultaneously, under the influence of the carbon chemical shifts until application of the proton θ pulse. However, it will *not* be influenced by the proton–carbon coupling (this is another interesting property of multiple-quantum coherence, which need not concern us further at this time, see Chapter 5). To remove the effects of proton chemical shifts, a second evolution period is also set to 1/2J and a proton 180° pulse is applied between the two delays, coincident with the 90°(C) pulse. The action of the θ pulse is to transfer the hmqc into antiphase transverse carbon magnetisation, that is, to regenerate observable magnetisation. The details of the outcome of this transfer process depend upon the multiplicity of the carbon resonance. In other words, methine, methylene and methyl groups respond differently to this pulse and this provides the basis of editing with DEPT. In the final 1/2J delay, the carbon magnetisation refocuses under the influence of proton coupling but also evolves according to carbon chemical shift, as it did in the second 1/2J period. Thus, the simultaneous application of a 180°(C) pulse with the proton θ pulse leads to an overall refocusing of the carbon shifts during the final 1/2J period. Carbon magnetisation is therefore detected without chemical shift dependent phase errors, with or without proton decoupling. Phase alternation of the θ pulse, combined with data addition/subtraction by the receiver, leads to the cancellation of natural carbon magnetisation, as for INEPT. The net result is again polarisation transfer from protons to carbon, combined with the potential for spectrum editing.

Editing with DEPT

As was hinted at above, the key to spectrum editing with DEPT is the realisation that the phase and intensity of carbon resonances depends upon the proton tip angle, θ. In fact, the editing results for angle θ are analogous to those produced by refocused INEPT with $\Delta_2 = \theta/180J$, so the previous discussions relating to θ apply equally to DEPT, as does the graph of Fig. 4.30. However, with INEPT the editing delay Δ_2 must be chosen according to the spin coupling constant whereas with DEPT the editing is achieved through the variable pulse angle θ which has no J dependence. This means the editing efficiency of DEPT tends to be superior to INEPT when a range of J values are encountered, and this is the principal advantage of DEPT in routine analysis. Evolution delays in DEPT do of course depend on J, although it turns out that the experiment is quite tolerant of errors in these settings.

The starting point for determining resonance multiplicities with DEPT is the collection of three spectra with $\theta = 45$, 90 and 135°, noting that the 90° experiment requires twice as many scans to attain the same signal-to-noise ratio as the others. The signs of the responses are then as summarised in Table 4.4. Knowing these patterns it is a trivial matter to determine multiplicities by direct comparison, whilst quaternaries can be distinguished by their appearance only in the direct carbon spectrum. Example spectra are shown for the bicyclic terpene andrographolide **4.6** (Fig. 4.32). An extended approach is to combine these spectra appropriately to produce 'sub-spectra' that display separately CH, CH_2 and CH_3 resonances. Personally, I think it is always better to examine

[3] The transverse magnetisation we observe directly in an NMR experiment is known as *single quantum coherence*. *Multiple quantum coherence*, however, cannot be directly observed because it induces no signal in the detection coil. For multiple quantum coherence to be of use to us, it must be transferred back into signal quantum coherence by the action of rf pulses. The concept of coherence is developed further in Chapter 5.

Table 4.4. Signs of multiplet resonances in DEPT spectra

	DEPT-45	DEPT-90	DEPT-135
XH	+	+	+
XH_2	+	0	−
XH_3	+	0	+

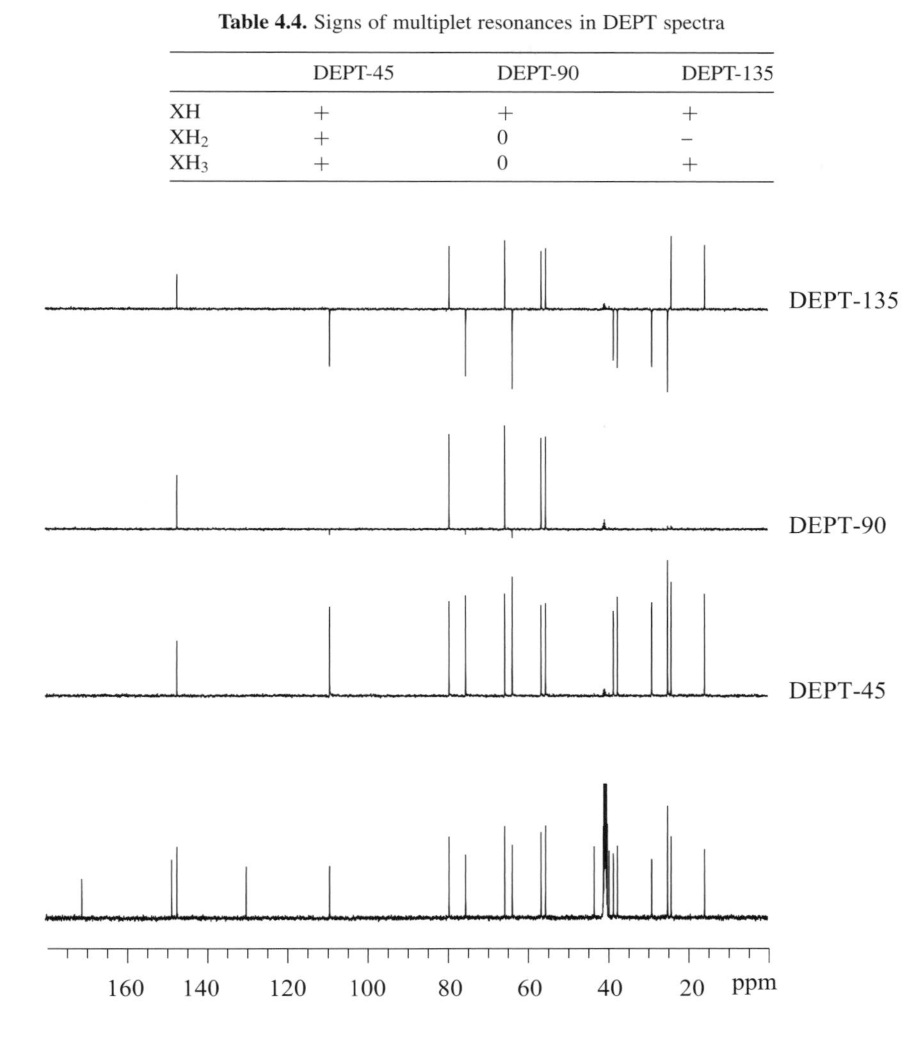

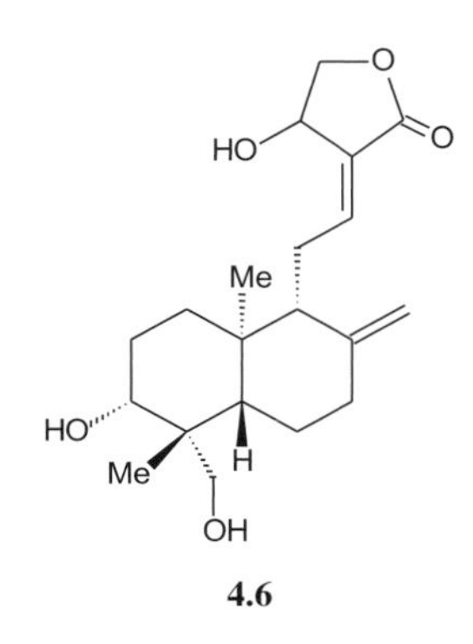

Figure 4.32. The conventional carbon and DEPT-edited spectra of the terpene andrographolide **4.6**.

the original DEPT spectra directly since this is such a simple matter. The combining of spectra introduces greater potential for 'artefact' generation, and it also becomes more difficult to identify what is going on if editing fails owing to the presence of an unusually high coupling constant or to instrument miscalibration, for example. In practice, the recording of a single DEPT spectrum with $\theta = 135°$ is often sufficient to provide the desired information when methyls are absent or are easily recognised on account of their chemical shifts. Recently, a modified DEPT sequence, termed DEPTQ [34], has been introduced which includes the detection of quaternary resonances and retains the advantages and editing capabilities of the original sequence. Its ability to observe all multiplicities makes this a strong contender for routine laboratory use.

Errors in DEPT editing may arise from a number of sources. The most likely is incorrect setting of the proton pulses, especially the θ pulse used for editing, which may often be traced to poor tuning of the proton channel. Even small errors in θ can lead to the appearance of small unexpected peaks in the DEPT-90; if θ is too small it approaches DEPT-45 whilst too big and it approaches DEPT-135. Usually, because of their low intensity, these spurious signals are easily recognised and should cause no problems. Even with correct pulse calibrations, errors can arise when the setting for the delay period is very far from that demanded by $^1J_{CH}$ [35] and in particular CH_3 resonances may still appear weakly in the CH sub-spectrum. A typical compromise value for $^1J_{CH}$ for determining 1/2J is ≈140 Hz, for which 1/2J = 3.6 ms. Alkyne carbons in particular exhibit very high values of $^1J_{CH}$ (usually >200 Hz), and may appear as weak signals displaying bizarre sign behaviour. Even more unexpected can be the appearance of the *quaternary* carbon of an alkyne group which experiences polarisation transfer from the alkyne proton two bonds away because of the exceptionally large $^2J_{CH}$ value.

Optimising sensitivity

It was shown above that for the refocused INEPT experiment optimum sensitivity was obtained for a given multiplicity when the refocusing period was set according to the number of coupled protons and the associated coupling constant. When DEPT is used primarily as a means of enhancing the sensitivity of the heteronucleus rather than for spectrum editing, similar considerations apply. The difference here is that it is the θ proton pulse that must be optimised according to the number of spins in the XH_n group according to:

$$\theta_{opt} = \sin^{-1}\left(\frac{1}{\sqrt{n}}\right) \tag{4.5}$$

Notice the similarity with Eq. 4.3 except that the dependence on J is again not relevant for DEPT. For an XH group, θ_{opt} is 90°, whilst this angle decreases with higher multiplicities.

4.4.4. PENDANT

The polarisation transfer sequences presented above provide signal enhancement of insensitive nuclei but suffer from a lack of responses from those that do not posses directly-bound protons (an exception is the DEPTQ sequence [34]). Conversely, the spin-echo based editing sequences do display such responses but gain enhancement only from the NOE and require signal averaging that is dictated by the typically slower-relaxing insensitive spins. One sequence that attempts to combine the best of both approaches is PENDANT [36] (Polarisation Enhancement Nurtured During Attached Nucleus Testing) (Fig. 4.33). This relies on polarisation transfer from protons as for INEPT and indeed the sequence bears close similarity with the refocused-INEPT of Fig. 4.21b. From the point of view of protonated nuclei, PENDANT can, in essence, be understood by reference to this experiment. However, it also contains an additional 90°(X) pulse to elicit responses from non-protonated nuclei and lacks the usual phase cycling employed which suppresses such signals. Spectrum editing is achieved by judicious choice of the refocusing period, with 5/8J suggested as optimum for $^1/_2\Delta_2$. This should allow the simultaneous observation of both protonated and non-protonated carbons and provide similar editing to the DEPT-135 experiment (Fig. 4.34), although in practice quaternary carbons can still appear with rather low intensity and may not be readily apparent, particularly with weaker samples. This is undoubtedly because of their slower relaxation coupled

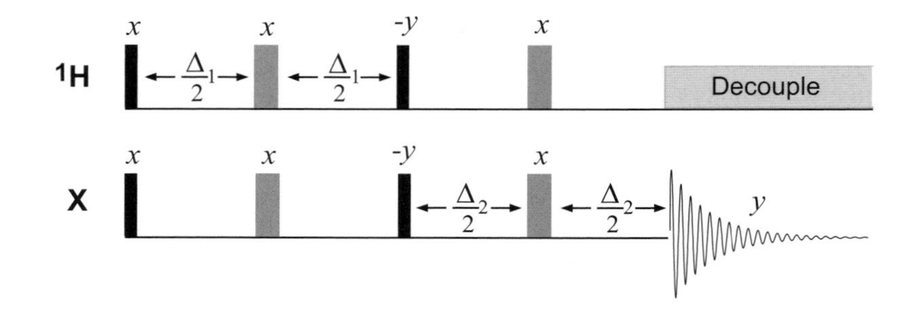

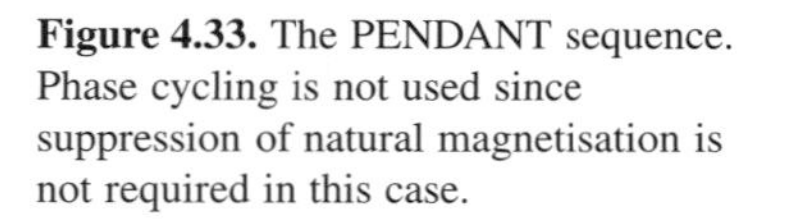
Figure 4.33. The PENDANT sequence. Phase cycling is not used since suppression of natural magnetisation is not required in this case.

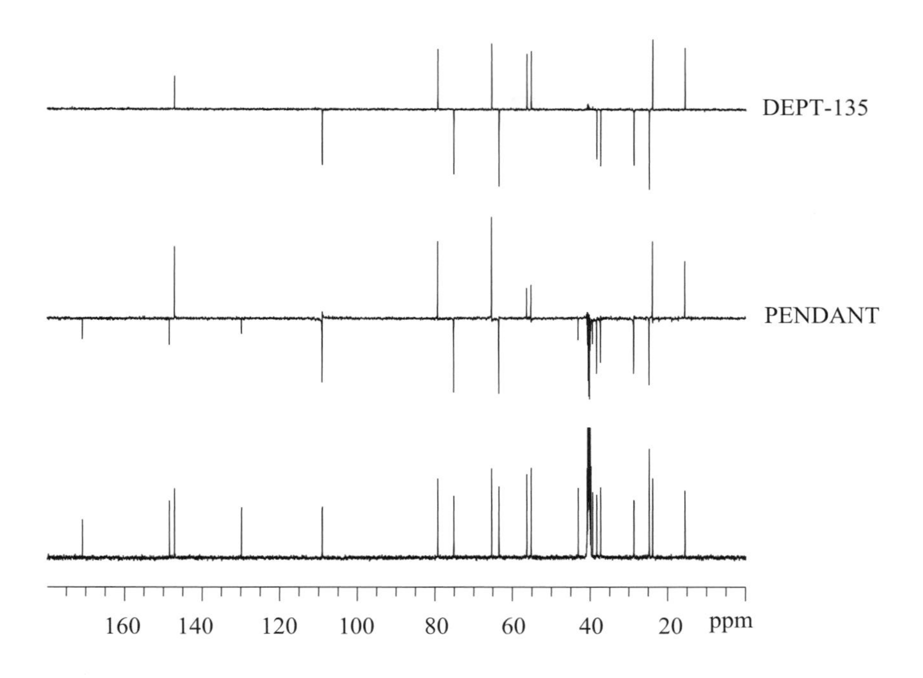

Figure 4.34. The carbon PENDANT and DEPT-135 spectra of andrographolide **4.6**. Similar results are obtained for both with additional responses appearing from quaternary carbons in the PENDANT spectrum.

with the use of 90° ^{13}C excitation pulses, analogous to the problems associated with the simpler spin-echo editing experiments of Section 4.3.1, and because pulse repetition rates in polarisation transfer sequences would normally be optimised according to proton T_1s. For the routine editing of carbon spectra, the recent DEPTQ experiment appears to be a more robust technique [34]. The PENDANT sequence would appear to offer greater benefit in the transfer of polarisation through *long-range* couplings, particularly in X–H_n systems where n is large ($3 < n < 9$), for which the 5/8J delays in the refocusing period becomes 1/16J for optimum results [37].

Before leaving these sections on the editing of X-nucleus spectra, a final comment on the utility of such experiments in the modern NMR laboratory. As has already been mentioned, techniques based on proton observation are becoming dominant in the world of organic structure elucidation, with the chemical shifts of the X-nuclei determined indirectly from 2D heteronuclear correlation spectra. Recent developments in this area have evolved techniques in which both 1D and 2D spectra are edited in such a way as to indicate X-nucleus multiplicities (Chapter 6). The higher sensitivity and faster data collection possible with these methods compared with direct X-nucleus observation, together with the additional correlation data provided by 2D techniques, will inevitably lead to reduced dependence on the traditional, more time-consuming X-nucleus observation and editing experiments. The principal driving force behind this change of emphasis is the application of pulsed field gradients and as these become increasingly common on routine instruments, the chemist's approach to structure elucidation can be expected evolve accordingly.

4.5. OBSERVING QUADRUPOLAR NUCLEI

A characteristic feature of many quadrupolar nuclei is the broad lines they produce, due to rapid quadrupolar relaxation (Section 2.5.5). The rapid recovery of the spins following excitation means they can often be acquired under conditions of very fast pulsing with full excitation by a 90° pulse, which is clearly beneficial for signal averaging purposes. However, the corresponding rapid decay of an FID can make the direct observation of nuclei with linewidths

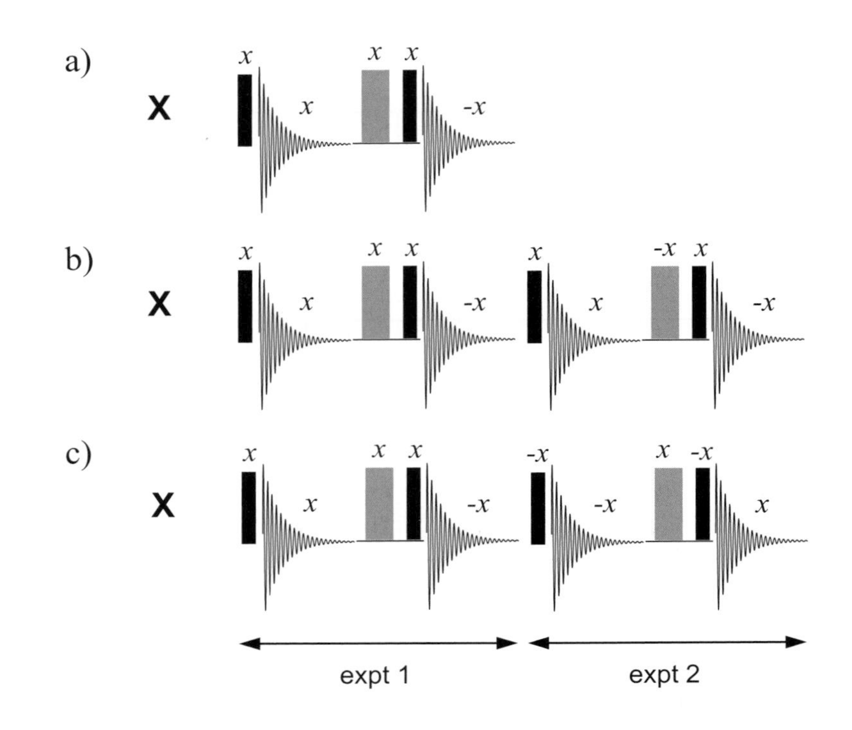

Figure 4.35. Sequences for the observation of quadrupolar nuclei with very broad lines for which acoustic ringing is a problem. Sequence (a) eliminates ringing associated with the 90° pulse whilst ACOUSTIC (b) and RIDE (c) further eliminate that associated with the 180° pulses.

of hundreds or thousands of hertz experimentally challenging. Following an rf pulse, a delay is required for the probe circuits and receiver electronics to recover before the NMR signal can be digitised. One particular concern is so-called *acoustic ringing* in the probehead [38] which is most severe for low-abundance, low-frequency nuclei at lower fields and which may take many tens of microseconds to decay. In the observation of spin-$^1/_2$ nuclei a pre-acquisition delay of this order can be used without problem but in cases of fast relaxing nuclei, such as ^{17}O or ^{33}S, this can lead to a loss of much of the FID before detection begins (resonance linewidths of 1 and 5 kHz correspond to relaxation times of only ca. 320 and 60 μs, respectively). Not only does this compromise sensitivity it also introduces substantial phase errors to the spectrum. The use of shorter delays is therefore essential but leads to a significant contribution to the spectrum in the form of broad baseline distortions from spectrometer transient responses. Numerous sequences to suppress this ring-down contribution have been proposed, [38] two of which are illustrated in Fig. 4.35b and c. In both cases suppression of the acoustic response from the 90° excitation pulses is achieved in two scans by inverting the NMR signal phase on the second transient with an additional 180° pulse, together with simultaneous inversion of the receiver phase [39] (Fig. 4.35a). Whilst the NMR signals are added by this process, the acoustic response remains unchanged in the two experiments and therefore cancels with receiver inversion. The remaining unwanted feature is now the acoustic response from the additional 180° pulse, which is suppressed in a slightly different manner in the two sequences. With ACOUSTIC [40] (Alternate Compound One-eighties USed to Suppress Transients In the Coil, Fig. 4.35b) the whole experiment is repeated with inversion of the 180° pulse. This, in effect, inverts the associated acoustic response relative to the first experiment so adding the data from the two experiments cancels this component. The alternative RIDE sequence [41] (RIng-down DElay, Fig. 4.35c) instead repeats the whole process but with inversion of the NMR response by inverting the phase of both 90° pulses, and subtracts this second experiment from the first. Both sequences suffer from potential problems with off-resonance effects because of the use of 180° pulses, but otherwise provide significant reductions in baseline distortions. Fig. 4.36

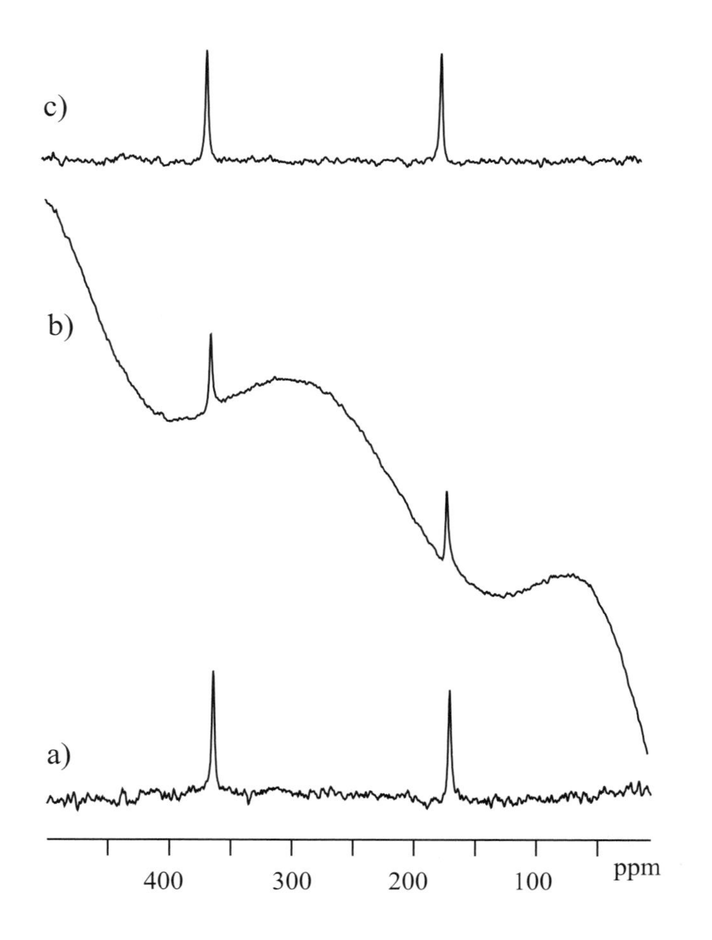

Figure 4.36. ^{17}O spectra of ethyl acetate recorded (a) with and (b) without the RIDE sequence. The severe baseline distortion in (b) arises from acoustic ringing in the probehead. Spectrum (c) was from the same FID of (b) but this had the first 10 data points replaced with backward linear predicted points, computed from 256 uncorrupted points. The spectra are referenced to D_2O and processed with 100 Hz line-broadening.

illustrates the use of RIDE in the collection of ^{17}O spectrum of ethyl acetate, where the suppression of the baseline distortion is clear.

An alternative approach now available with modern processing software is to collect the distorted FID with the simple one-pulse-acquire sequence and to replace the early distorted data points with uncorrupted points generated through backward linear prediction. Fig. 4.36c was produced from the same raw data as 4.36b, but with the first 10 data points of the FID replaced with predicted points. The baseline distortion is completely removed and there is no signal-to-noise loss through experimental imperfections, as occurs for the RIDE data set.

REFERENCES

[1] R.R. Ernst and W.A. Anderson, *Rev. Sci. Instr.*, 1966, **37**, 93–102.
[2] D.E. Jones and H. Sternlicht, *J. Magn. Reson.*, 1972, **6**, 167–182.
[3] E.D. Becker, J.A. Ferretti and P.N. Gambhir, *Anal. Chem.*, 1979, **51**, 1413–1420.
[4] R. Freeman and H.D.W. Hill, *J. Magn. Reson.*, 1971, **4**, 366–383.
[5] P. Waldstein and W.E. Wallace, *Rev. Sci. Instr.*, 1971, **42**, 437–440.
[6] J.S. Waugh, *J. Mol. Spectrosc.*, 1970, **35**, 298–305.
[7] R.A. Hoffman and S. Forsén, *Prog. Nucl. Magn. Reson. Spectrosc.*, 1966, **1**, 15–204.
[8] W.A. Anderson and R. Freeman, *J. Chem. Phys.*, 1962, **37**, 85–103.
[9] G. Massiot, S.K. Kan, P. Gonard and C. Duret, *J. Am. Chem. Soc.*, 1975, **97**, 3277–3278.
[10] J.K.M. Sanders and J.D. Mersh, *Prog. Nucl. Magn. Reson. Spectrosc.*, 1982, **15**, 353–400.
[11] F. Bloch and A. Siegert, *Phys. Rev.*, 1940, **57**, 522–527.
[12] N.F. Ramsey, *Phys. Rev.*, 1955, **100**, 1191–1194.
[13] J.P. Jesson, P. Meakin and G. Kneissel, *J. Am. Chem. Soc.*, 1973, **95**, 618–620.
[14] R.W. Dykstra, *J. Magn. Reson.*, 1990, **88**, 388–392.
[15] R.W. Dykstra, *J. Magn. Reson.*, 1992, **100**, 571–574.

[16] R.W. Dykstra, *J. Magn. Reson. (A)*, 1993, **102**, 114–115.
[17] D.W. Brown, T.T. Nakashima and D.L. Rabenstein, *J. Magn. Reson.*, 1981, **45**, 302–314.
[18] C.L. Cocq and J.Y. Lallemand, *J. C. S. Chem. Commun.*, 1981, 150–152.
[19] F.-K. Pei and R. Freeman, *J. Magn. Reson.*, 1982, **48**, 318–322.
[20] S.L. Patt and J.N. Shoolery, *J. Magn. Reson.*, 1982, **46**, 535–539.
[21] H. Bildsøe, S. Dønstrup, H.J. Jakobsen and O.W. Sørensen, *J. Magn. Reson.*, 1983, **53**, 154–162.
[22] O.W. Sørensen, S. Dønstrup, H. Bildsøe and H.J. Jakobsen, *J. Magn. Reson.*, 1983, **55**, 347–354.
[23] A.M. Torres, T.T. Nakashima and R.E.D. McClung, *J. Magn. Reson. (A)*, 1993, **101**, 285–294.
[24] J. Ollerenshaw, T.T. Nakashima and R.E.D. McClung, *Magn. Reson. Chem.*, 1998, **36**, 445–448.
[25] R. Freeman, A Handbook of Nuclear Magnetic Resonance, 2nd ed., Longman, Harlow, 1997.
[26] G.A. Morris and R. Freeman, *J. Am. Chem. Soc.*, 1979, **101**, 760–762.
[27] D.P. Burum and R.R. Ernst, *J. Magn. Reson.*, 1980, **39**, 163–168.
[28] D.T. Pegg, D.M. Doddrell, W.M. Brooks and M.R. Bendall, *J. Magn. Reson.*, 1981, **44**, 32–40.
[29] O.W. Sørensen and R.R. Ernst, *J. Magn. Reson.*, 1983, **51**, 477–489.
[30] K.V. Schenker and W. von Philipsborn, *J. Magn. Reson.*, 1985, **61**, 294–305.
[31] A. Mohebbi and O. Gonen, *J. Magn. Reson. (A)*, 1996, **123**, 237–241.
[32] W. von Philipsborn and R. Müller, *Angew. Chem. Int. Ed. Engl.*, 1986, **25**, 383–413.
[33] D.M. Doddrell, D.T. Pegg and M.R. Bendall, *J. Magn. Reson.*, 1982, **48**, 323–327.
[34] R. Burger and P. Bigler, *J. Magn. Reson.*, 1998, **135**, 529–534.
[35] M.R. Bendall and D.T. Pegg, *J. Magn. Reson.*, 1983, **53**, 272–296.
[36] J. Homer and M.C. Perry, *J. C. S. Chem. Commun.*, 1994, 373–374.
[37] J. Homer and M.C. Perry, *J. C. S. Perkin Trans.*, 1995, **2**, 533–536.
[38] I.P. Gerothanassis, *Prog. Nucl. Magn. Reson. Spectrosc.*, 1987, **19**, 267–329.
[39] D. Canet, J. Brondeau, J.P. Marchal and B. Robin-Lherbier, *Org. Magn. Reson.*, 1982, **20**, 51–53.
[40] S.L. Patt, *J. Magn. Reson.*, 1982, **49**, 161–163.
[41] P.S. Belton, I.J. Cox and R.K. Harris, *J. C. S. Faraday Trans.*, 1985, **81** (2), 63–75.

Chapter 5

Correlations through the chemical bond I: Homonuclear shift correlation

---

Having discussed a number of methods in the previous chapter that are considered routine one-dimensional techniques, this chapter leads into the world of two-dimensional NMR. Whereas many of the methods in Chapter 4 used through-bond interactions as a tool toward a specific goal, such as sensitivity enhancement through polarisation transfer, the two-dimensional methods in this and the following chapter exploit spin coupling to map specific through-bond interactions. It is quite likely that you have already made some use of two-dimensional methods in your research, most probably in the form of the COSY experiment, and it is with this technique that the chapter begins. Not only is this one of the most widely used 2D experiments, its simple form provides a convenient introduction to multidimensional experiments. Whilst the methods in this chapter are aimed at determining correlations between like spins (*homonuclear correlations*), a large part of the discussion is equally relevant to all two-dimensional NMR experiments, and many parts of this chapter provide the foundations for those that follow.

The principles underlying the generation of a two-dimensional spectrum were first presented in a lecture in 1971 [1], although it was a number of years later that the approach found wider application [2]. During the 1980s the world of NMR, and consequently the chemist's approach to structure determination, was revolutionised by the development of numerous two-dimensional techniques, and nowadays many higher-dimensionality methods (3D and 4D) also exist. Those methods utilising three or more dimensions have found greatest application in the hands of biological NMR spectroscopists studying macromolecules (a specialist area in its own right) and, at present, tend to be used rather less frequently within what may be considered main-stream organic chemistry. These methods therefore fall beyond the realms of this book, but are covered in texts dedicated to biomolecular NMR [3,4].

This chapter begins by introducing the principles that lie at the heart of all two-dimensional methods. The techniques themselves, summarised briefly in Table 5.1, begin with basic Correlation Spectroscopy (COSY) which maps nuclei sharing a mutual scalar coupling within a molecule, making the proton–proton COSY one of the workhorse techniques in organic structure elucidation. Later in the chapter some variants of the basic COSY experiment are presented that display a number of beneficial characteristics and therefore also find widespread application. The last two sections cover somewhat different techniques. The first of these, Total Correlation Spectroscopy (TOCSY), provides an alternative to the COSY approach for establishing correlations within a molecule. This provides efficient transfer of information along a network of coupled spins, a feature that can be extremely powerful in the analysis of more

Table 5.1. The principal applications of the main techniques described in this chapter

Technique	Principal applications
COSY-90	Correlating coupled homonuclear spins. Typically used for correlating protons coupled over 2- or 3-bonds, but may be used for any high-abundance nuclide. The basic COSY experiment.
DQF-COSY	Correlating coupled homonuclear spins. Typically used for correlating protons coupled over 2- or 3-bonds. Higher-resolution display than basic COSY. Additional information on magnitudes of coupling constants may be extracted from 2D peak fine-structure. Singlets suppressed.
COSY-β	Correlating coupled homonuclear spins. Typically used for correlating protons coupled over 2- or 3-bonds, but may be used for any high-abundance nuclide. Reduced 2D peak structure over basic COSY. Vicinal and geminal coupling relationships can be differentiated in some cases.
Delayed-COSY	Correlating coupled homonuclear spins through small couplings. Often used to identify proton correlations over many bonds (>3) hence also known as long-range COSY.
TOCSY	Correlating coupled homonuclear spins and those that reside within the same spin-system but which may not share mutual couplings. Employs the propagation of magnetisation along a continuous chain of spins. Powerful technique for analysing complex proton spectra.
INADEQUATE	Correlating coupled homonuclear spins of low natural abundance ($<20\%$). Typically used for correlating adjacent carbon centres at natural abundance, but has extremely low sensitivity.

complex spectra. The final method, INADEQUATE, establishes correlations between like spins of low natural abundance. This may be used, for example, to directly identify neighbouring carbon centres and in this form is perhaps the ultimate experiment for defining a molecular skeleton although, alas, it is also about the least sensitive.

The approach in this chapter is again a pictorial one. Whilst this avoids the need for becoming embroiled in formalisms that rigorously describe the behaviour of magnetisation during multipulse experiments, it has its limitations in that many of the techniques cannot be described completely by such a simplified approach. We have already come up against these limitations when attempting to understand the DEPT experiment in the previous chapter, where it was not possible to describe the behaviour of heteronuclear multiple quantum coherence in terms of the classical vector model. An explanation of this technique was thus derived largely from the more readily understood INEPT sequence. Despite these limitations it is still possible to develop some physical insight into how the experiments operate without resorting to esoteric mathematical descriptions. Section 5.4 introduces graphical formalisms, known as *coherence transfer pathways*, that provide a simple representation of the 'flow' of magnetisation during pulse experiments, which proves particularly enlightening for two-dimensional sequences and for those experiments that utilise pulsed field gradients for signal selection. Before any of this however, we must first develop some feel for how two-dimensional spectra are generated.

5.1. INTRODUCING TWO-DIMENSIONAL METHODS

The first point to clarify when discussing two-dimensional techniques is the fact that the two dimensions refer to two *frequency* dimensions, where as so-called one-dimensional methods have, of course, only one. In either case there will also be a dimension representing signal *intensity*, although this is never usually included when describing the dimensionality of an experiment. The two frequency dimensions may represent any combination of chemical shifts or scalar couplings. More recently, methods have been developed where the idea of a two-dimensional representation has been adapted to include one frequency and one 'other' dimension. For example, hybrid HPLC-NMR methods are capable of producing 2D plots with the usual 1D NMR spectrum comprising one dimension and the chromatographic retention time in the other. Likewise, it is possible to disperse the spectra of solutes according to their diffusion properties, in which case the second dimension represents the diffusion coefficients (Chapter 9).

Conventional two-dimensional spectra find such wide utility in chemical research because they map out interactions within, or sometimes between, our molecules of interest. The interactions that can be probed may be broken down into three distinct categories which relate to quite different physical phenomena; through-bond coupling, through-space coupling and chemical exchange. The first of these is considered in this and the following chapter under the division of homonuclear (between like nuclei, e.g. ^{1}H–^{1}H) and heteronuclear (between unlike nuclei e.g. ^{1}H–^{13}C) couplings, and the experiments considered are principally aimed at the identification and subsequent piecing together of structural fragments within a molecule. Through-space coupling provides the basis for the nuclear Overhauser effect which is most often employed to deduce molecular stereochemistry and conformation, as described in Chapter 8. Lastly, the mapping of chemical exchange pathways, which may be intra- or inter-molecular, will also be considered in Chapter 8 since the methods for studying both the NOE and exchange are essentially identical.

5.1.1. Generating a second dimension

No matter what the nature of the interaction to be mapped, all two dimensional sequences have the same basic format and can be subdivided into four well defined units termed the *preparation*, *evolution*, *mixing* and *detection* periods (Fig. 5.1). The preparation and mixing periods typically comprise a

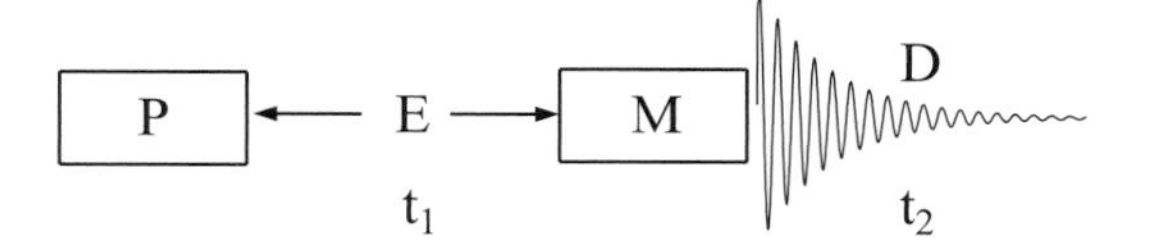

Figure 5.1. The general scheme for any two-dimensional experiment. P: Preparation, E: Evolution, M: Mixing and D: Detection.

pulse or a cluster of pulses and/or fixed time periods, the exact details of which vary depending on the nature of the experiment. The detection period is entirely analogous to the detection period of any one-dimensional experiment during which the spectrometer collects the FID of the excited spins. The evolution period provides the key to the generation of the second dimension. Before proceeding, briefly recall what happens during the acquisition of a one-dimensional FID and the processes that generate a single frequency domain spectrum. The detection process involves sampling (digitising) the oscillating free induction decay at regular time intervals dictated by the Nyquist condition for an appropriate acquisition period. The collected data are then subject to Fourier transformation to produced the required frequency domain spectrum. In other words, the general requirement for creating a frequency dimension is the regular sampling of magnetisation as it varies in some way as a function of time. Extending this idea one can conclude that to generate a spectrum with *two* frequency domains, f_1 and f_2, it is necessary to sample data as a function of *two* separate time variables, t_1 and t_2. Clearly one time domain and hence one frequency domain originates from the usual detection period (t_2 in Fig. 5.1), but how does the other arise?

Figure 5.2. An illustrative two-dimensional sequence in which P and M are 90° pulses. This is also the basic COSY sequence.

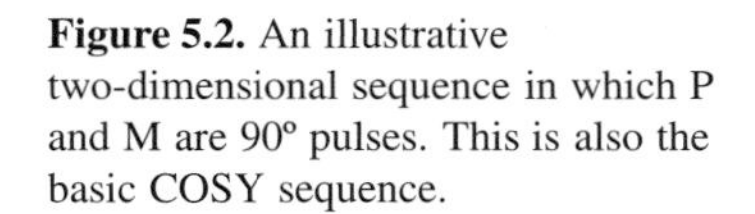

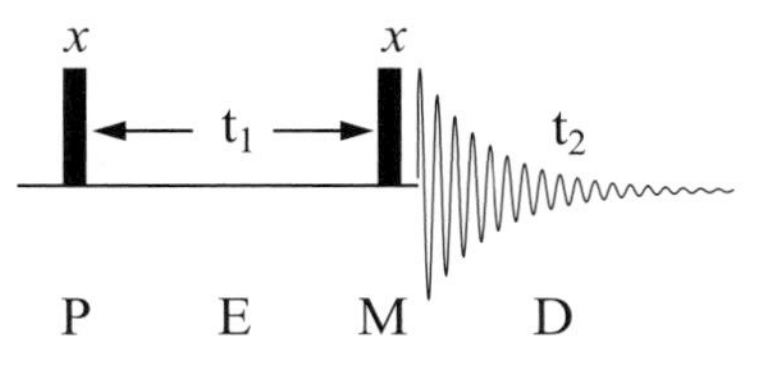

To illustrate the required procedure consider a simple pulse sequence in which both the preparation and mixing units of Fig. 5.1 are each single 90_x pulses (Fig. 5.2) acting on a sample that contains only a single, uncoupled proton resonance, say chloroform, with a chemical shift offset of ν Hz. Viewing events with the vector model (Fig. 5.3) the initial 90_x pulse places the equilibrium magnetisation in the x–y plane along the $+y$-axis, after which

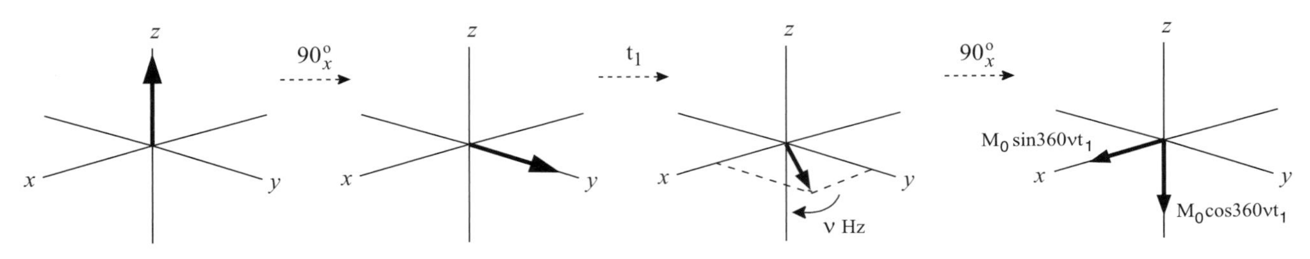

Figure 5.3. The action of the COSY sequence of Fig. 5.2 on a single, uncoupled proton resonance.

it will precess (or *evolve*) according to this chemical shift offset. After a time period, t_1, the vector has moved through an angle of $360\nu t_1$ degrees, and is then subject to the second pulse. To appreciate the influence of this pulse, it is convenient to consider the vector as comprising two orthogonal components, one along the original *y*-axis ($M_0.\cos 360\nu t_1$) and the other along the *x*-axis ($M_0.\sin 360\nu t_1$). The second 90_x pulse places the *y*-component along the $-z$ axis whilst the *x*-component is unaffected and remains precessing in the transverse plane and produces the detected FID. Fourier transformation of this FID will therefore produce a spectrum containing a single resonance whose intensity depends upon the factor $\sin 360\nu t_1$.

Performing this experiment with t_1 set to zero means the two 90_x pulses will add to produce a net 180_x pulse which simply corresponds to inversion of the equilibrium vector. As there exists no transverse magnetisation there is no signal to detect and the spectrum contains only noise. Now imagine repeating the experiment a number of times with the t_1 interval increased by a uniform amount each time, and the resulting FIDs stored separately. As t_1 increases from zero, the resulting signal intensity also increases as *x*-magnetisation has time to develop during the evolution period, reaching a maximum when it has evolved through an angle of 90°. Further increases in t_1 cause the *x*-component to diminish, pass through a null, then become negative and so on according to the sine modulation. Subjecting each of the acquired t_2 FIDs to Fourier transformation produces a series of spectra containing a single resonance whose intensity (or *amplitude*) varies as a function of time according to $\sin 360\nu t_1$ (Fig. 5.4). With longer values of t_1, relaxation effects diminish the intensity of the transverse magnetisation according to T_2^*, so the signal intensities show a steady (exponential) decay in addition to the *amplitude modulation*. The intensity of the resonance as a function of time therefore represents another free induction decay for the t_1 time domain (referred to as an interferogram) that has been generated artificially or *indirectly* (Fig. 5.5). The frequency of the amplitude modulation corresponds to the chemical shift offset of the resonance in the rotating frame and we say the magnetisation has been *frequency labelled* as a function of t_1. Subjecting this time domain data to Fourier transformation, one would again expect this to produce a single

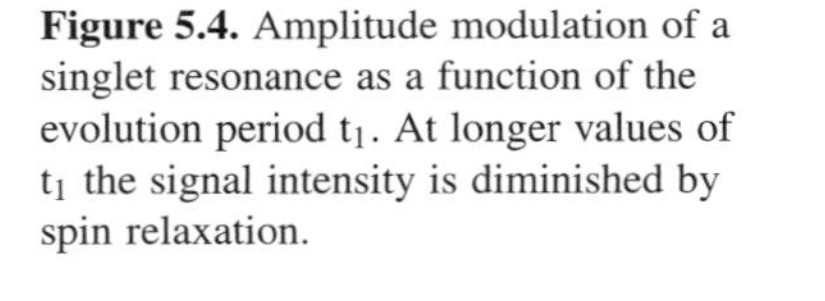

Figure 5.4. Amplitude modulation of a singlet resonance as a function of the evolution period t_1. At longer values of t_1 the signal intensity is diminished by spin relaxation.

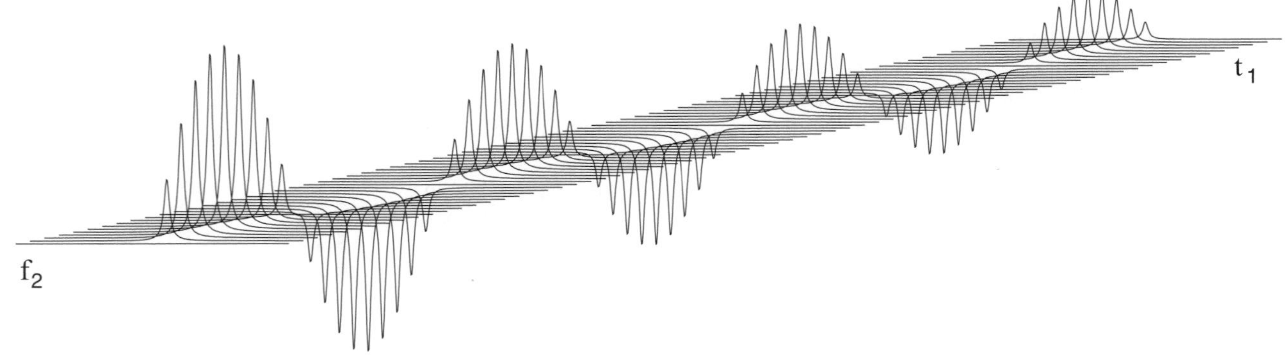

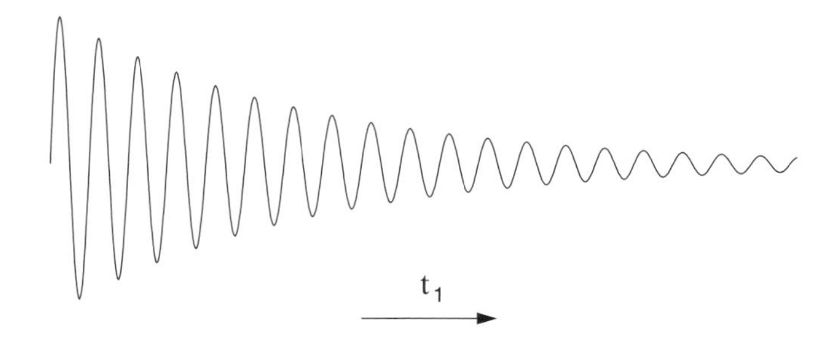

Figure 5.5. The variation in peak intensity of the amplitude modulated resonance of Fig. 5.4 produces an FID (interferogram) for the t_1 domain.

resonance with a chemical shift of ν Hz *in the f_1 frequency domain*, just as for the f_2 domain. Generalising this process requires monitoring the intensity of *every data point* in the f_2 domain as a function of t_1, to produce a complete two-dimensional data set. Following Fourier transformation with respect to t_1 another complete frequency domain is generated which, in combination with the conventional domain, produces a two-dimensional spectrum displaying a single resonance at ν Hz in both dimensions (Fig. 5.6). Note that the labelling of the frequency axes as f_1 and f_2 follows from the ordering of the corresponding time domains in the pulse sequence, that is, t_1 followed by t_2. The new artificial (t_1) time domain has been sampled discretely by recording a FID for each t_1 time point and storing each separately, in analogy with the sampling of the t_2 data points described earlier. The repeated acquisition of FIDs with systematically incremented t_1 time periods (Fig. 5.7) is fundamental to the generation of *all* two-dimensional data sets. Indeed, this simple idea can be extended to produce a three-dimensional spectrum simply by having

Figure 5.6. The two-dimensional spectrum resulting from the sequence of Fig. 5.2 for a sample containing a single uncoupled spin. One peak results with a shift of ν Hz in both dimensions.

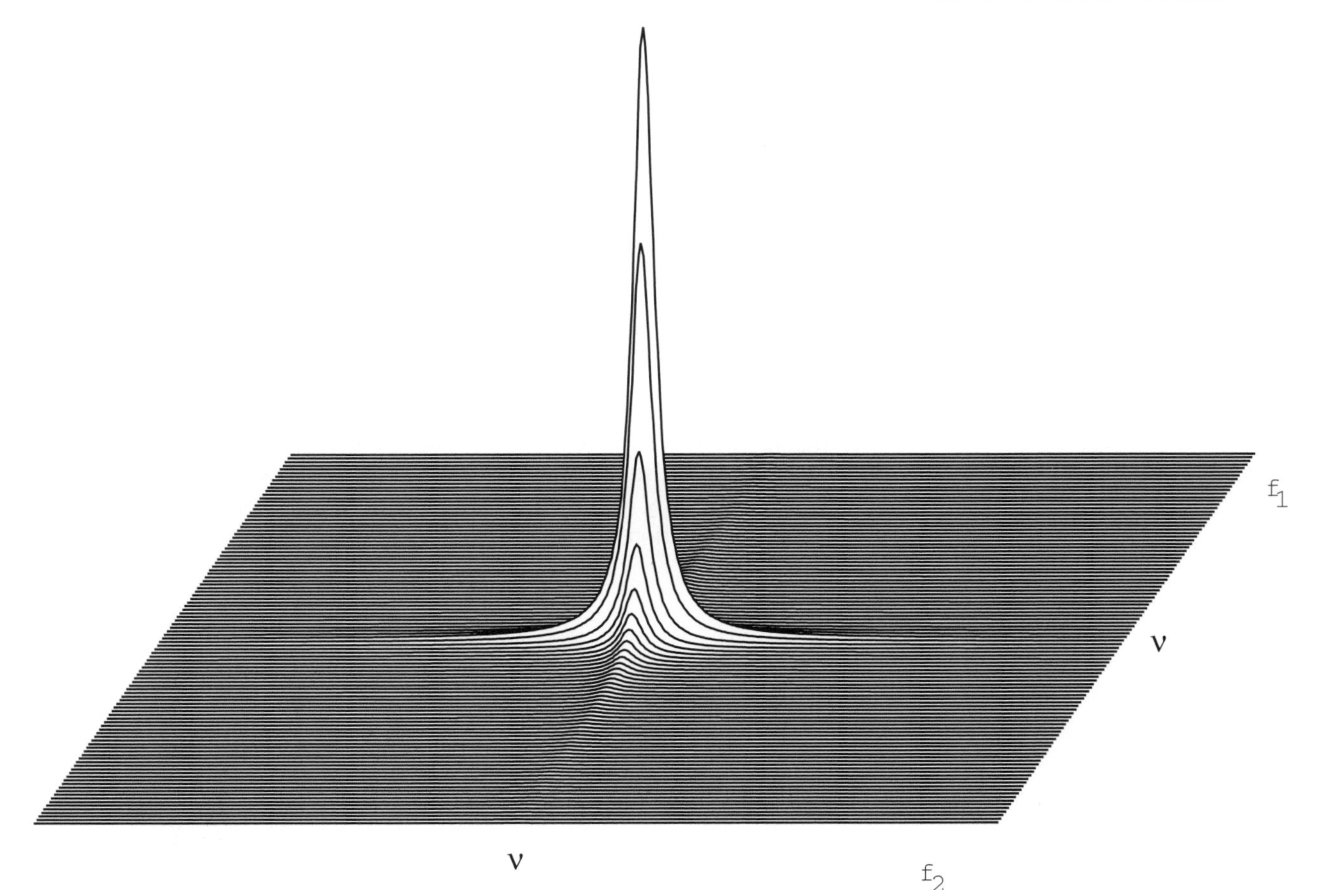

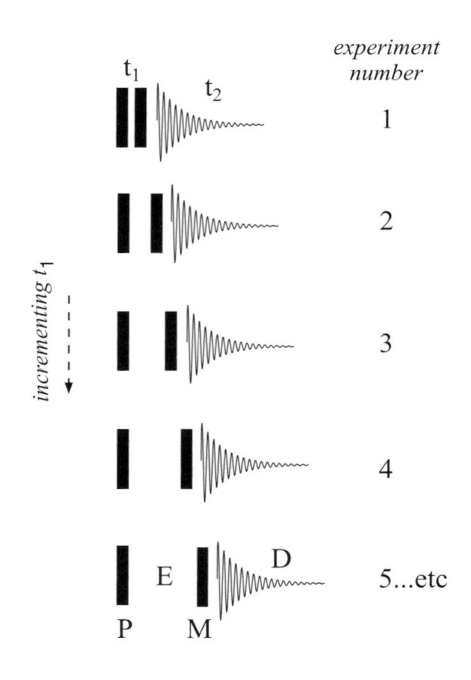

Figure 5.7. A generalised scheme for the collection of a two dimensional data set. The experiment is repeated many times with the t_1-period incremented at each stage and the resulting FIDs stored separately. Following double Fourier transformation with respect to first t_2 and then t_1, the two-dimensional spectrum results.

three independently incremented time periods, one detected 'directly' and two 'indirectly' (Fig. 5.8); whilst the technical details may be a little complex, the principle is quite straightforward. The philosophy behind the multidimensional approach is that one signal modulated as a function of one time variable (say t_1) is detected at some later point as a function of another (t_2 in the case of a 2D experiment). These general descriptions apply equally to spectra that contain many resonances. Repeating the above experiment for a sample containing just two uncoupled spins with offsets of ν_A and ν_X Hz results in a spectrum with two 2D singlets whose frequencies in both dimensions correspond simply to the respective chemical shift offsets of the resonances (Fig. 5.9).

Having gone to the trouble of generating a two-dimensional data set, the spectra of Figs. 5.6 and 5.9 tell us nothing that we could not derive from the corresponding one-dimensional spectra, simply the chemical shifts of the participating nuclei. This provides no new information because the 2D peaks map (or correlate) exactly the same information in both dimensions ($\nu_1 = \nu_2$), in this case the chemical shifts. Two-dimensional spectra become useful when the peaks they contain correlate *different* information on the two axes, that is when $\nu_1 \neq \nu_2$. This requires that magnetisation evolving at frequency ν_1 during the t_1 time period then evolves at a different frequency ν_2 in the detection (t_2) period. Clearly for this to happen there must be some mechanism by which the magnetisation precession frequency changes during the sequence, and more specifically within the mixing period. The details of this period dictate the information content of the resulting spectrum according to the

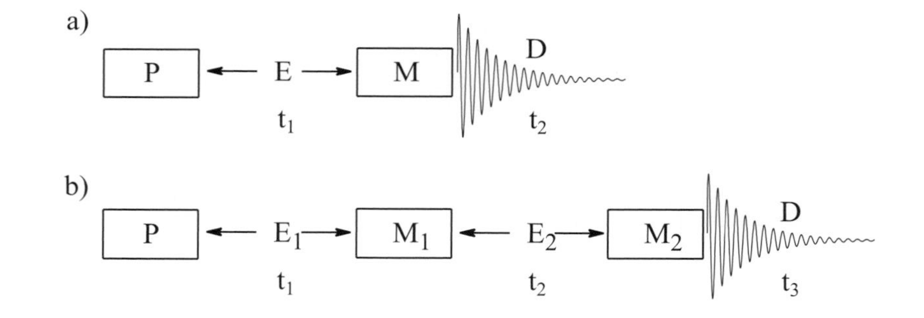

Figure 5.8. Schematic representations of (a) two-dimensional and (b) three-dimensional experiments. A 3D experiment is collected by independently varying the t_1 and t_2 intervals to generate the f_1 and f_2 domains respectively, by analogy with a 2D experiment.

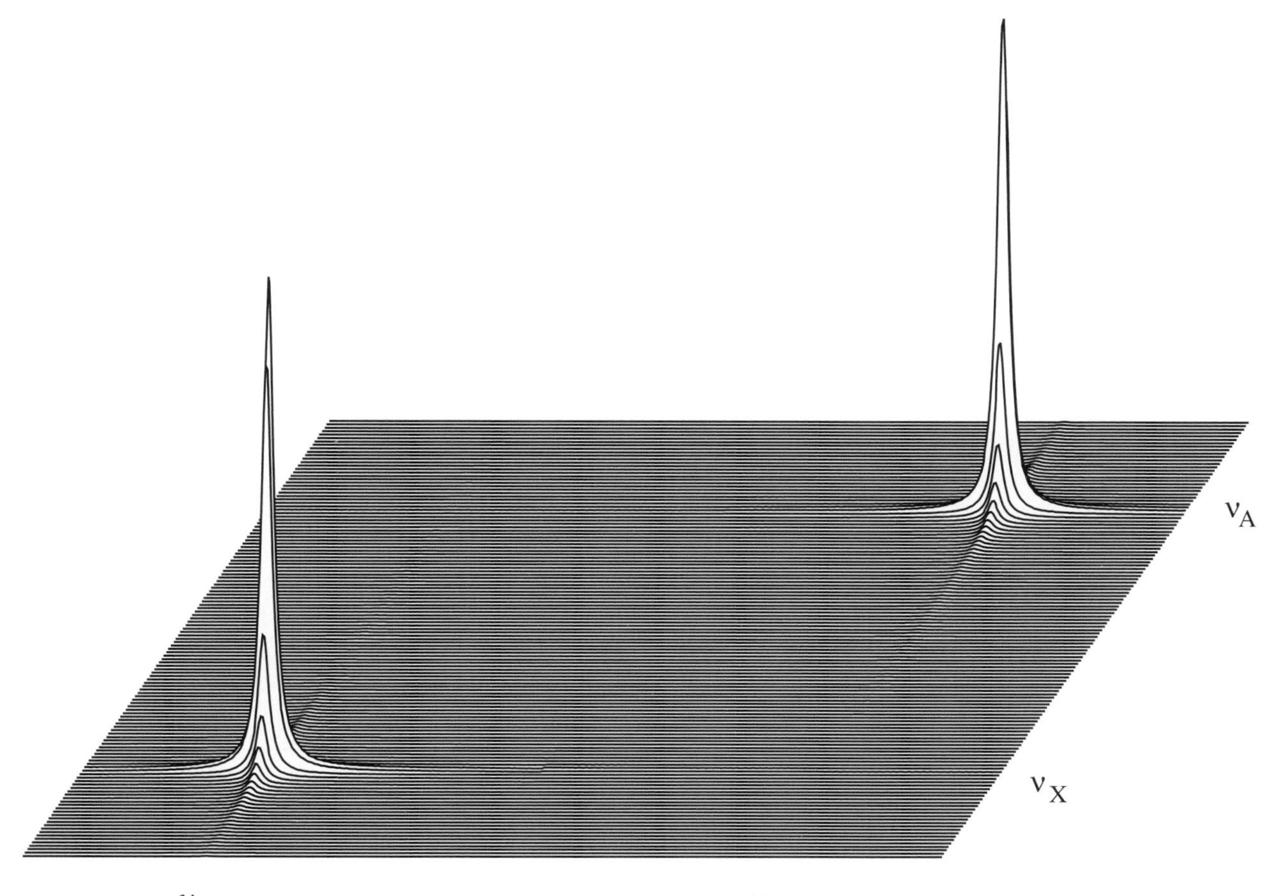

Figure 5.9. The two-dimensional spectrum resulting from the sequence of Fig. 5.2 for a sample containing two uncoupled spins, A and X, of offsets ν_A and ν_X. Each produces a 2D peak at its corresponding chemical shift offset in both dimensions.

interaction that is exploited, i.e. coupling, NOE, or chemical exchange. Just how the informative correlation peaks are produced will become apparent as various sequences are examined in the sections that follow.

Before moving on, a comment on the presentation of 2D spectra is required. The spectra of Figs. 5.6 and 5.9 have been presented in the 'stacked-plot' mode to emphasise the similarity with one-dimensional spectra, and the presence of two frequency axes and one intensity axis. Although these may look aesthetically impressive, this form of presentation is of little use in practice. The usual way to present 2D spectra is via 'contour plots', in which peak intensities are represented by contours, as a mountain range would be represented on a map. Fig. 5.10 shows the equivalent contour presentation of Figs. 5.6 and 5.9 and unless stated otherwise, all 2D spectra from now on will make use of this contour mode.

5.2. CORRELATION SPECTROSCOPY (COSY)

It was stated in the opening remarks of this chapter that COSY was the first two-dimensional sequence proposed, and it is in fact that given in Fig. 5.2 above, utilising just two 90° pulses. The sequence, which correlates the chemical shifts of spins that share a mutual J-coupling, is most often applied in proton spectroscopy although is equally applicable to any high-abundance nuclide. It is without doubt the mostly widely used of all two-dimensional methods, and is thus considered first. This section provides an introduction to

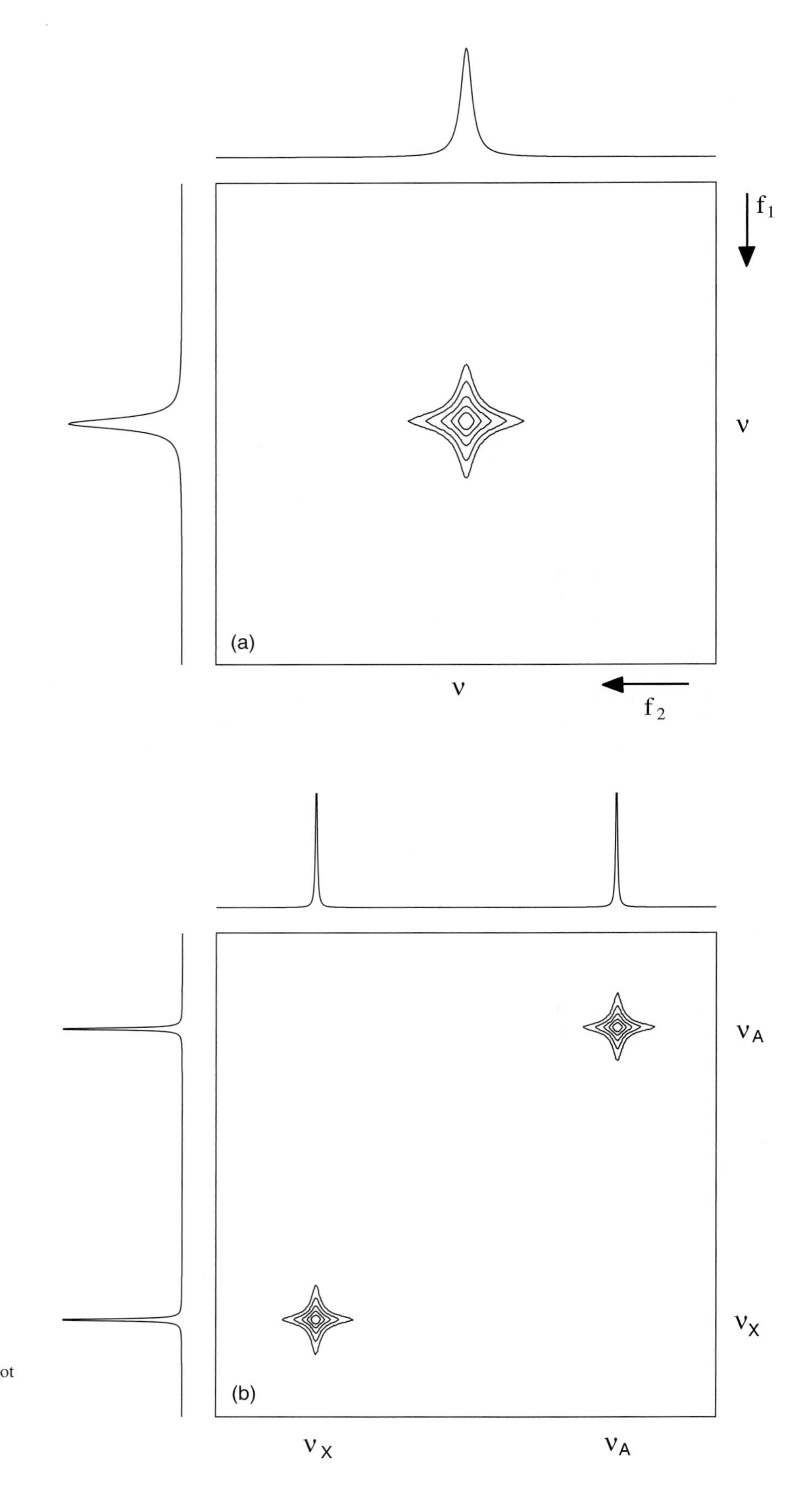

Figure 5.10. The equivalent contour plot representations of Figs. 5.6 and 5.9. Alongside the f_1 and f_2 axes are the conventional 1D spectra for reference purposes.

the operation of the experiment and its principal features, after which you should be in a position to understand and interpret routine COSY spectra. Later sections look at more specific and useful features of the basic experiment and those of its cousins.

5.2.1. Correlating coupled spins

Moving on from the previous discussions concerning uncoupled spins, we now consider the more general case of spins that are scalar coupled. We first note that the frequency labelling component of the experiment described above is in operation for all excited spins so it is unnecessary to consider this further. Of interest now is the mixing part of the sequence, in this case the action of the pulse following t_1. In short, this pulse transfers 'magnetisation' originally associated with one spin onto neighbouring spins to which it is coupled, this process being known formally as *coherence transfer* (Section 5.4), which is only possible because of the J-coupling. Coherence transfer has, in fact, already been described in Chapter 4 where it was presented in a slightly different guise as polarisation transfer from proton to carbon, which is nothing more than *heteronuclear coherence transfer*. In this, the original spin populations ('magnetisation') associated with the proton were ultimately transferred to carbon. In the heteronuclear case events could be readily pictured by reference to population diagrams (SPI) and to the vector model (the INEPT sequence) because one could consider the action of pulses on the source proton and the target carbon quite independently. For the homonuclear case things are rather more complicated because all spins experience the same pulses simultaneously and the simple picture of events tends to break down. Nevertheless, similar transfer processes are operative and with COSY we simply have the analogous *homonuclear coherence transfer* taking place between spins. The similarity can be further exemplified if one compares the COSY sequence of Fig. 5.2 with the basic INEPT sequence of Fig. 4.21a. Ignoring the refocusing 180° pulses in INEPT, the two sequences are largely identical, both with initial excitation being followed by a period of evolution, after which *both* coupled spins experience a 90° pulse which elicits the transfer. The concept of coherence and coherence transfer is central to most NMR experiments and this is further considered in Section 5.4.

The key point in all this is that magnetisation transfer occurs between coupled spins. To appreciate the outcome of this in the final COSY spectrum, consider the case of two J-coupled spins, A and X, with a coupling constant of J_{AX} and chemical shift offsets ν_A and ν_X. The magnetisation associated with spin A will, after the initial 90° pulse, precess during t_1 according to its chemical shift offset, ν_A. The second 90° pulse then transfers some part of this magnetisation to the coupled X spin, whilst some remains associated with the original spin A (the reason for this segregation is described in Section 5.2.3). That which remains with A will then precess in the detection period at a frequency ν_A, just as it did during t_1, so in the final spectrum it will produce a peak at ν_A in both dimensions, denoted (ν_A, ν_A). This peak is therefore equivalent to that observed for the uncoupled AX system of Fig. 5.9 and because it represents the same frequency in both dimensions it sits on the diagonal of the 2D spectrum and is therefore referred to as a *diagonal peak*. In contrast, the transferred magnetisation will precess in t_2 at the frequency of the new 'host' spin X, and will thus produce a peak corresponding to two different chemical shifts in the two dimensions (ν_A, ν_X). This peak sits away from the diagonal and is therefore referred to as an off-diagonal or, more commonly, a *crosspeak* (Fig. 5.11). This is the peak of interest as it provides direct evidence of coupling between spins A and X. The whole process also operates in the reverse direction, that is the same arguments apply for magnetisation originally

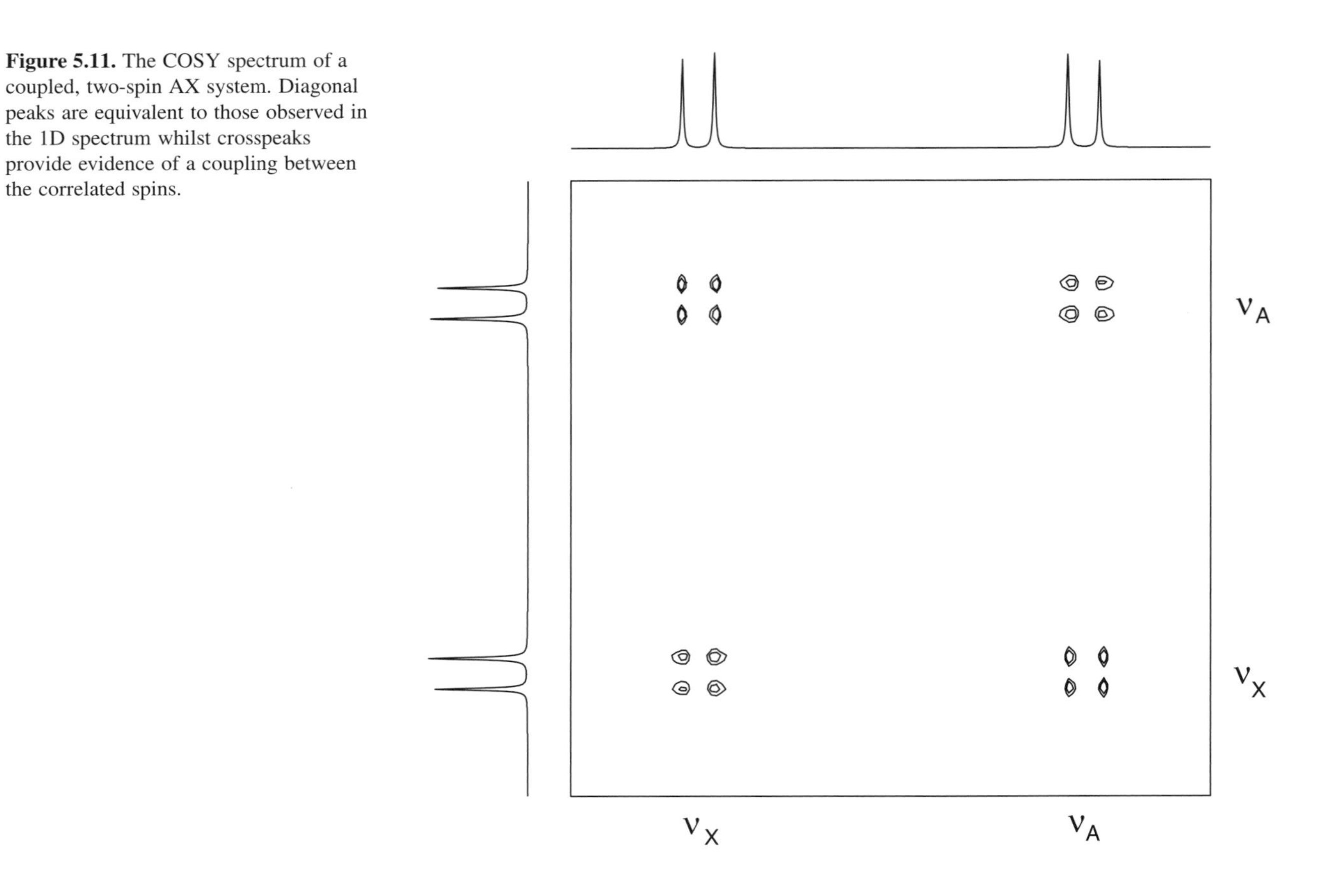

Figure 5.11. The COSY spectrum of a coupled, two-spin AX system. Diagonal peaks are equivalent to those observed in the 1D spectrum whilst crosspeaks provide evidence of a coupling between the correlated spins.

associated with the X spin, giving rise to a diagonal peak at (ν_X, ν_X) and a crosspeak at (ν_X, ν_A). Thus, the COSY spectrum is symmetrical about the diagonal, with crosspeaks on either side of it mapping the same interaction (Fig. 5.11).

5.2.2. Interpreting COSY

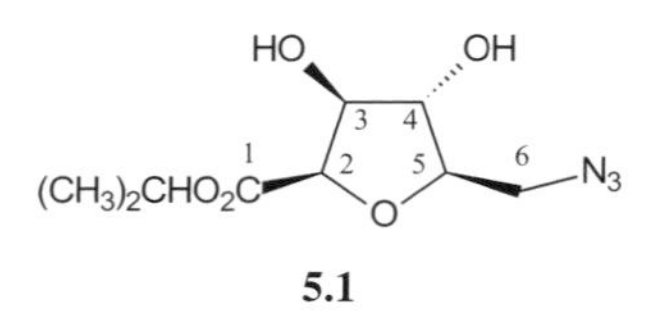

The application of the COSY technique in spectral assignment is best illustrated by considering some real examples. The COSY spectrum of the azo-sugar **5.1**, a monomer for the synthesis of oligomeric carbopeptoids, demonstrates both the power of the technique as well as some of its limitations or potential pitfalls (Fig. 5.12). The goal is to assign all proton resonances in the spectrum of **5.1**. In this simple example, this could be accomplished through considered analysis of the multiplet fine-structure of each resonance. However, this process fails when spectra display significant resonance overlap and the greater dispersion available in two-dimensions can often bypass such limitations. Furthermore, COSY has the added advantage of speed, by avoiding the need to employ time-consuming analysis of the 1D spectrum. Naturally, one may return to such levels of detail when assignments have been made and the gross structure identified to glean additional structural data, from coupling constants for example, but COSY can very rapidly provide evidence in support of a structure without the need for this. The first step in analysis is the matching of diagonal peaks to the equivalent resonances in the 1D spectrum, as this provides orientation for all subsequent assignments. Coupled spins are then correlated by stepwise movements, starting and finishing at a diagonal peak via an intermediate crosspeak. Assignments for **5.1** begin with the characteristic doublet of H2. Correlation to its vicinal partner in the sugar is made via either of the symmetrically related crosspeaks which thus assigns H3, and

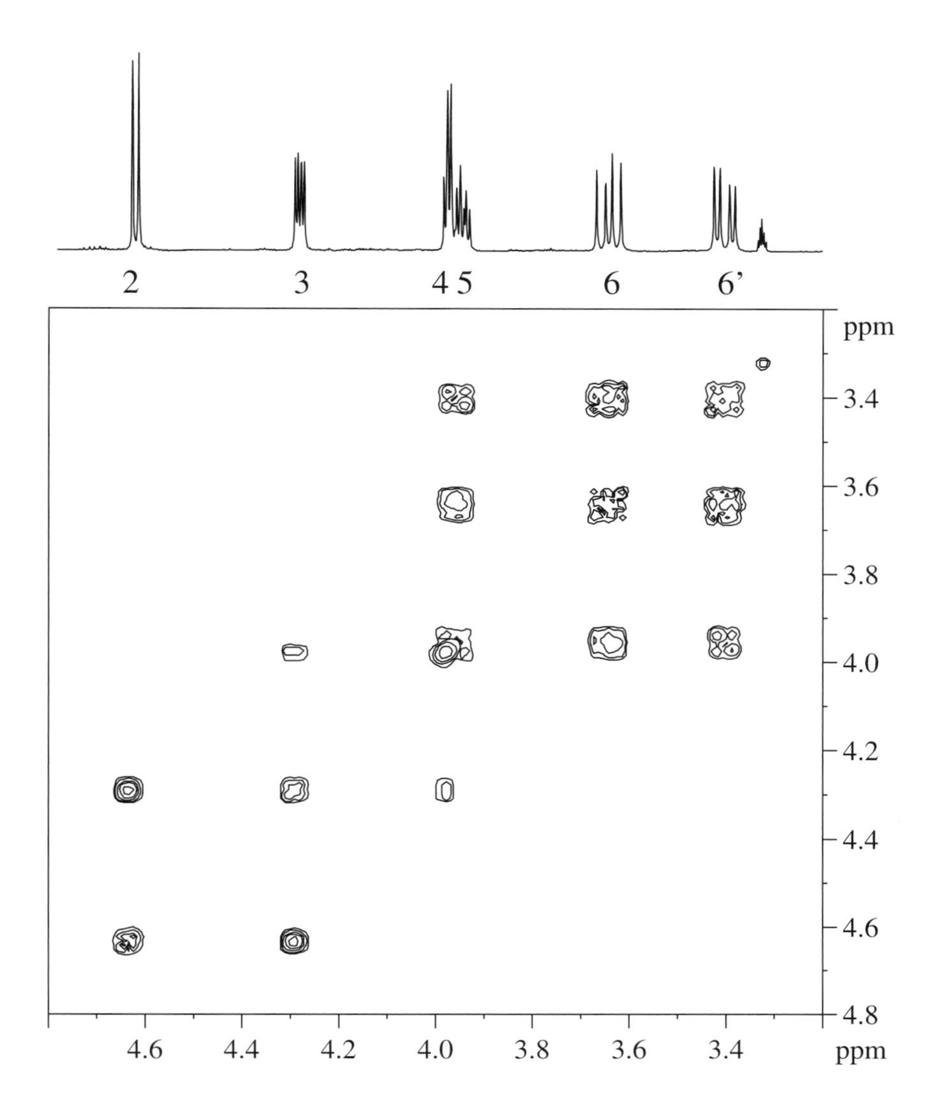

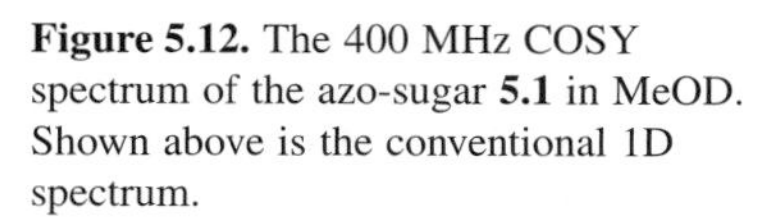

Figure 5.12. The 400 MHz COSY spectrum of the azo-sugar **5.1** in MeOD. Shown above is the conventional 1D spectrum.

so on. This trivial procedure provides the basis on which COSY spectra are analysed. The power and simplicity of the experiment is further illustrated by a second example, a carbopeptoid dimer **5.2** (Fig. 5.13). In this case, two discrete spin systems exist within the molecule on adjacent sugar rings, and the differentiation and assignment of these follows readily from the characteristic doublets of H2 in each. The two ring systems are traced in Fig. 5.13, above and below the diagonal in red (ring A) and black (ring B) for clarity. Further correlations from the H6 protons of ring B to the adjacent amide NH are not shown in this figure, but their presence readily differentiates rings A and B. Indeed, correlations from the lone amide resonance would also provide a suitable starting point to map ring B.

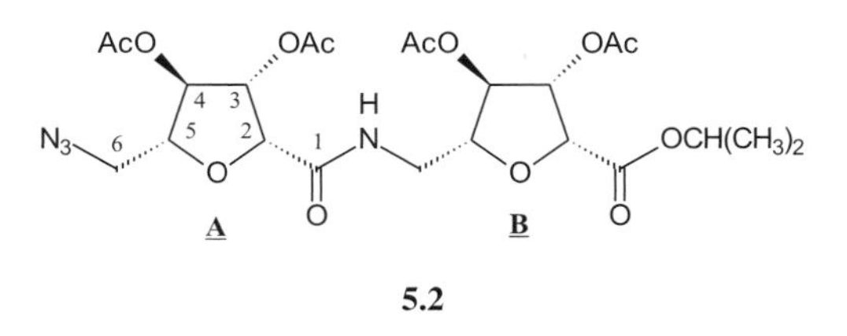

5.2

The COSY experiment, despite its simplicity, is, however, subject to a number of limitations and caveats that one should also be aware of. The correlation

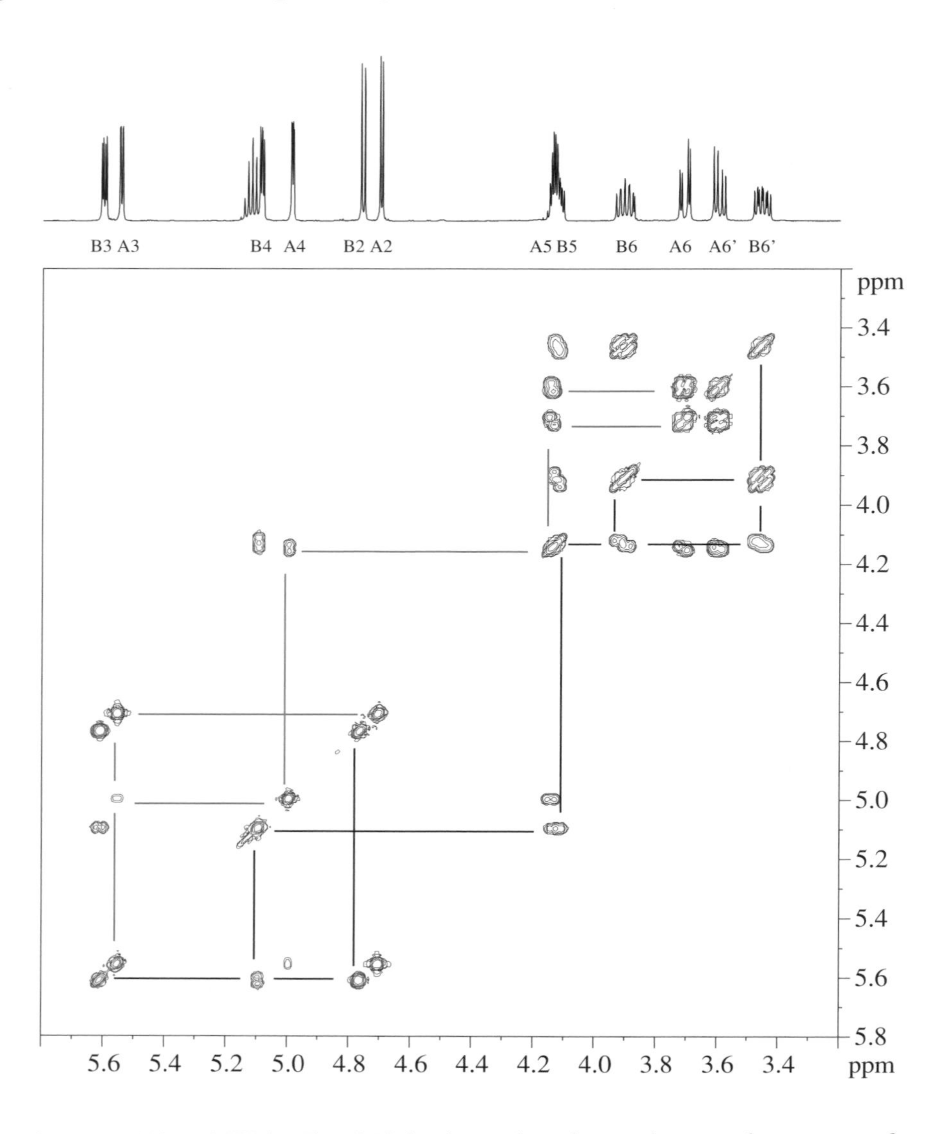

Figure 5.13. The 500 MHz COSY spectrum of the carbopeptoid dimer **5.2** in $CDCl_3$ shown with the 1D spectrum above. Traced in red above the diagonal are the assignments for sugar ring A and in black below the diagonal are those for sugar ring B (this plot excludes correlations to the amide proton of residue B). Both can be assigned starting from the H2 proton of each sugar. The rings are labelled from the N-terminus of the carbopeptoid and the sugar ring numbering follows IUPAC recommendations for carbohydrate nomenclature.

between H4 and H5 in Fig. 5.12 is almost lost due to the near degeneracy of these resonances, and when coupled spins overlap in this manner the stepwise assignment process can breakdown since it may not be clear from which resonance subsequent correlations should be traced. Some caution is required in such instances to avoid incorrect assignments, and more sophisticated methods may ultimately be required to provide unambiguous results. With complex structures, one may be left with various groups of coupled spins identified from COSY spectra which cannot be linked through homonuclear correlations. These pieces of the molecular 'jigsaw' require additional experimental data if they are to be joined together, such as from heteronuclear correlation experiments or NOE measurements. Most proton–proton couplings operate over two or three bonds so the proton COSY spectrum typically identifies vicinal and geminal relationships. A further general limitation of COSY is the inability, *a priori*, to differentiate between these two types of correlation; in Fig. 5.12 the vicinal H5–H6 crosspeaks appear similar to the geminal H6–H6′ crosspeak. When analysing COSY these two possible relationships should be borne in mind, particularly when diastereotopic geminal protons may be anticipated, notably in cyclic systems or those with stereogenic centres. A variant on the basic COSY sequence (COSY-β, Section 5.6.3) can, in favourable circumstances, differentiate between vicinal and geminal relationships and this method also finds widespread use. Structural features, such as unsaturation or the well

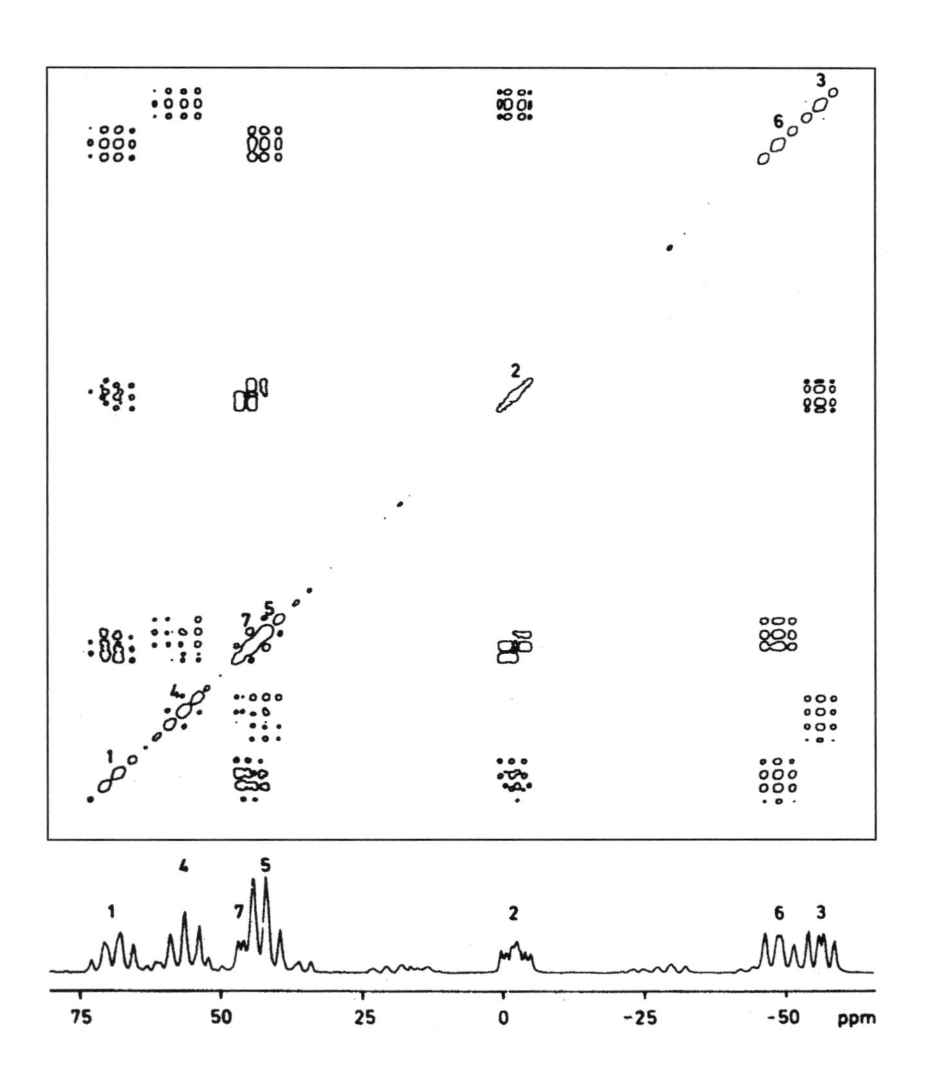

Figure 5.14. The ^{31}P–^{31}P COSY spectrum of the P_7Me_5 cluster **5.3** with the ^{31}P 1D spectrum below (reproduced with permission from [5]).

known *w*-geometry, can also enhance long-range coupling constants and COSY can be surprisingly effective at revealing these correlations. As such, it is *not* safe to automatically assume *all* crosspeaks correspond to either vicinal or geminal coupling relationships. Geometrical factors can also cause vicinal couplings to be close to zero, in which case no COSY crosspeak is produced despite the presence of adjacent protons. Thus, the *absence* of a crosspeak between protons does *not* always exclude them from being adjacent in a structure.

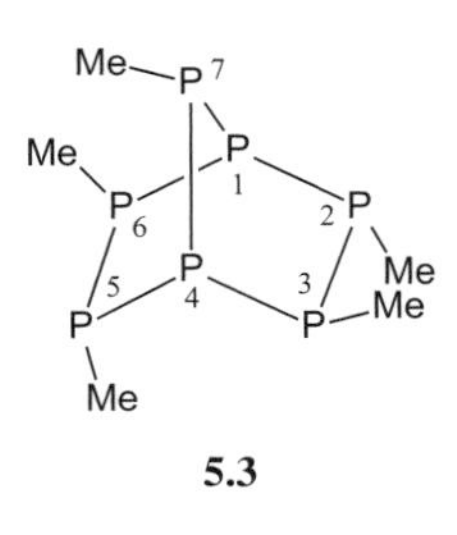

5.3

The COSY experiment is not limited to proton spectroscopy, but is suitable for establishing homonuclear correlations for any high abundance nuclide, the most common examples being ^{19}F, ^{31}P or ^{11}B. Fig. 5.14 illustrates the ^{31}P–^{31}P COSY spectrum of the P_7Me_5 cluster **5.3** for which the assignment procedure is exactly as described for the proton spectra above. In this case, single-bond P–P couplings are identified directly, along with a lone two-bond coupling between P^2 and P^7 which arises from the sizeable $^2J_{PP}$ coupling.

5.2.3. Peak fine structure

The spectrum of Fig. 5.11 also reveals within each diagonal and crosspeak fine structure that is equivalent to the structure seen within the one-dimensional multiplets. Whether such fine structure is always resolved depends on the experimental settings used in the acquisition of the data. Whilst in the illustrative

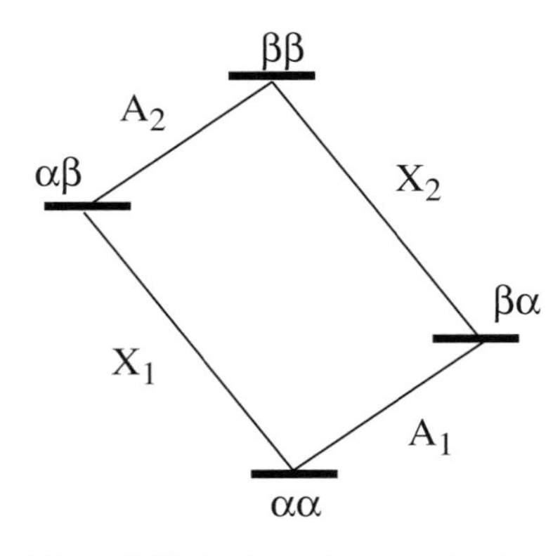

Figure 5.15. A schematic energy level diagram for the coupled two-spin AX system.

example of Fig. 5.11 the data were recorded so as to reveal these details, the fine structure in Figs. 5.12 and 5.13 is not resolved, as is often the case in routine analysis. Interpretation of multiplet fine structure can, under the appropriate experimental conditions, provide valuable information, a topic addressed in Section 5.5.2 where we encounter a version of the COSY experiment better suited to this purpose. To realise the origin of this fine structure in the 2D spectrum, it is necessary to modify slightly the description of magnetisation transfer given above. It was stated that this transfer occurs between coupled spins but more precisely one should say it occurs between the *transitions* associated with these spins. Reference to the energy level diagram for a coupled 2 spin system (Fig. 5.15) should clarify this. Magnetisation associated with, say, transition A_1, which gives rise to one half of the A spin doublet, is redistributed to the other three transitions A_2, X_1 and X_2 by the second 90° pulse of the COSY sequence. That associated with A_2 during the detection period gives rise to the fine structure *within* the diagonal peak, whilst that now present as X_1 and X_2 provides the structure of the crosspeak. The result is that, in addition to the modulation due to chemical shifts during t_1 and t_2, there is additional modulation arising from J-couplings, and this produces the coupling fine structure in both dimensions.

Recall also that following the second pulse, 'some magnetisation remains associated with the original spin'. Thinking back to the discussions of polarisation transfer in the INEPT experiment, it was shown that the basic requirement for the transfer of polarisation was an antiphase disposition of the doublet vectors of the source spin, which for INEPT was generated by a spin-echo sequence. Magnetisation components that were in-phase just before the second 90° pulse would not contribute to the transfer, hence the Δ period was optimised to maximise the antiphase component. The same condition applies for magnetisation transfer between two protons as in the COSY experiment. This requires that the proton–proton coupling be allowed to evolve to give a degree of antiphase magnetisation that may be transferred by the second pulse, whilst the in-phase component remains associated with the original spin (Fig. 5.16). The coupling evolution period for COSY is the t_1 period so that the amount of transferred magnetisation detected in t_2 is also modulated as a function of t_1 (sin $180Jt_1$); this is the modulation mentioned at the end of the last paragraph that ultimately characterises the crosspeak coupling fine structure in f_1. Likewise, the amplitude of the in-phase, non-transferred component is also modulated in t_1 by the coupling (cos $180Jt_1$) and this produces the coupling fine structure of the diagonal peak in f_1. Multiplet structure is further considered in Section 5.5.2 with the aim of extracting coupling constants from 2D crosspeaks.

5.3. PRACTICAL ASPECTS OF 2D NMR

This section introduces the most important experimental aspects relating to two-dimensional data sets, and again uses the COSY experiment to illustrate

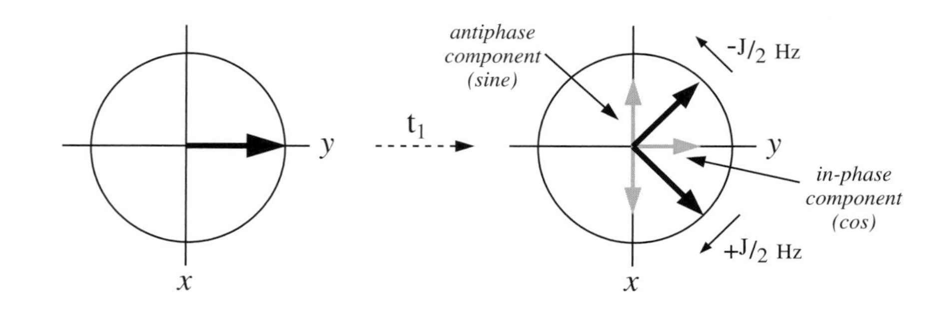

Figure 5.16. Coupling evolution during t_1 produces in-phase and antiphase magnetisation components. Only the antiphase component contributes to magnetisation transfer and hence to crosspeaks in the 2D spectrum.

these. Many of the discussions are simply extensions of what has already been discussed in Chapter 3 for the one-dimensional experiment and familiarity with this is assumed. No new concepts are introduced here, although a modified approach to experimental set-up and to data processing is essential if 2D experiments are to be successful.

5.3.1. 2D lineshapes and quadrature detection

All the two-dimensional spectra presented so far have made use of quadrature detection in both dimensions, enabling the reference frequencies for each to be placed at the centre of the spectrum. Quadrature detection for the f_2 data is achieved by either the simultaneous or sequential sampling schemes described in Section 3.2.4 and is therefore entirely analogous to that used for one-dimensional acquisitions. Quadrature detection in f_1 is also necessary for the same reasons. As with the 1D case, this demands some means of distinguishing frequencies that are higher than that of the reference from those that are lower when evolving during t_1. In other words, it is again necessary to distinguish positive and negative frequencies in the rotating frame so that mirror image signals do not appear either side of $f_1 = 0$. There are two general approaches to this in widespread use, one providing so-called *phase-sensitive* data displays, whilst the other provides data that it conventionally displayed in the *absolute-value* mode in which all phase information has been discarded. The first of these displays lineshapes in which absorption- and dispersion-modes are separated, meaning the preferred absorption-mode signal is available for a high-resolution display. The second approach is inferior in that it produces lineshapes in which the absorptive and dispersive parts are inextricably mixed making it poorly suited to high-resolution work. However, since absolute value spectra are rather easy to process and manipulate, they still find use in some routine work and in fully automated processes, so are also considered here.

Phase-sensitive presentations

As for 1D data, f_1 quadrature detection requires two data sets to be collected which differ in phase by 90°, thus providing the necessary sine and cosine amplitude-modulated data. Since the f_1 dimension is generated artificially, there is strictly no reference rf to define signal phases so it is the phase of the pulses that bracket t_1 that dictate the phase of the detected signal. Thus, *for each t_1 increment* two data sets are collected, one with a 90_x preparation pulse (t_1 sine modulation), the other with 90_y (t_1 cosine modulation), and both stored separately (Fig. 5.17). These two sets are then equivalent to the

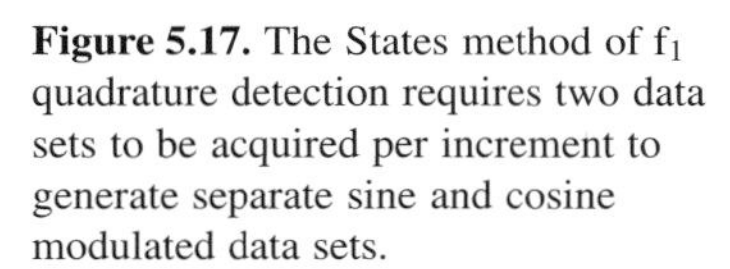
Figure 5.17. The States method of f_1 quadrature detection requires two data sets to be acquired per increment to generate separate sine and cosine modulated data sets.

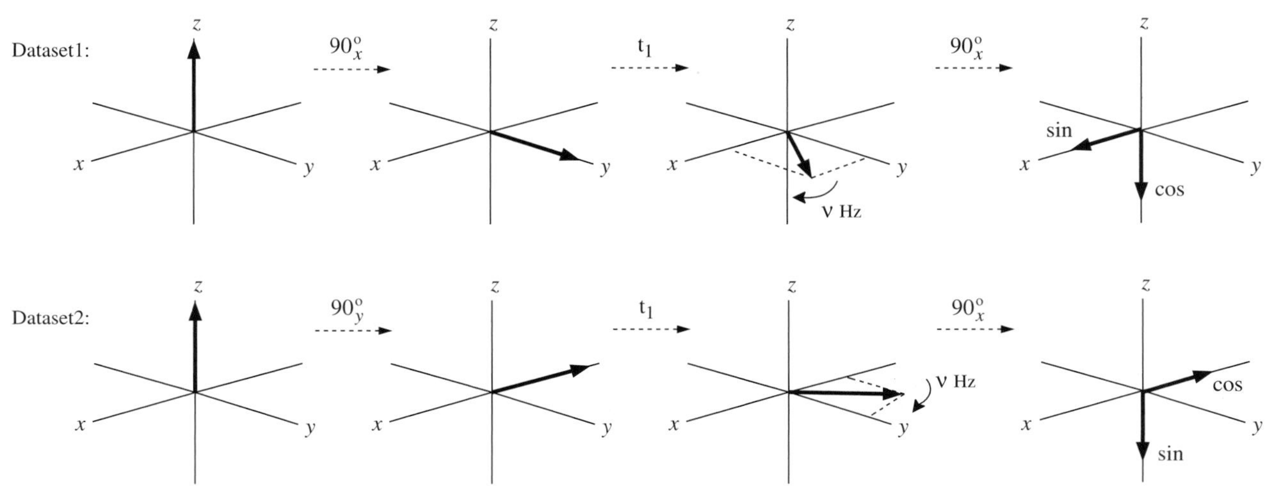

two channel data collected with *simultaneous* acquisition which produces the desired frequency discrimination when subject to a complex Fourier transform (also referred to as a hypercomplex transform in relation to 2D data). The rate of sampling in t_1 or, in other words, the size of the t_1 time increment, is dictated by the f_1 spectral width and is subject to the same rules as for the simultaneous sampling of one-dimensional data. This method is derived from the original work of States, Haberkorn and Ruben [6,7] and is therefore referred to in the literature as the 'States' method of f_1 quad-detection.

An alternative approach is analogous to the method of *sequential* sampling introduced in Section 3.2.4. As in the 1D approach the aim is to avoid both positive and negative frequencies arising by shifting the *apparent* frequency range from $\pm\frac{1}{2}SW_1$ to 0 to $+SW_1$ Hz by making the evolution frequencies during t_1 appear faster than they actually are. As for the States method above, there is no reference rf phase to shift for this artificial domain, so the equivalent effect is achieved by incrementing the phase of the preparation pulse by 90° for each t_1 increment, and sampling the data twice as fast as for the States method (by halving the t_1 increment). Only one data set is acquired for each t_1 period, but twice as many t_1 increments are collected, so the total t_1 acquisition time, and hence the digital resolution, is equal for both methods. This approach to f_1 quad-detection [8] is now referred to as Time Proportional Phase Incrementation or TPPI.

The States and TPPI methods produce equivalent data sets [9] (although they differ subtly in the appearance of aliased signals and of the artefacts known as axial peaks; see below) and the one you choose is likely to depend on the scheme favoured by your instrument vendor, although both options may be found on modern instruments. The most significant point from all this is the two-dimensional lineshapes they produce. Both methods involve the detection of a signal that is *amplitude modulated* as a function of the t_1 evolution period (Fig. 5.4), and this is the fundamental requirement for producing spectra that have absorption and dispersion parts separated (*pure-phase* spectra) so allowing a *phase-sensitive* presentation. Separate real and imaginary parts of the data exist for both the f_1 and f_2 dimensions, again analogous to the real and imaginary parts of a 1D spectrum. This gives rise to four data quadrants (Fig. 5.18) with only the (real, real) data set being presented to the user as the final 2D spectrum, the others being retained for phase correction. It is usual for the displayed spectrum to contain absorption-mode lineshapes in both dimensions (Fig. 5.6) wherever possible since the double-absorption lineshape affords the highest possible resolution (Fig. 5.18, RR quadrant). The phase information contained within crosspeaks can also provide additional information in some circumstances and this is especially true for the phase-sensitive COSY experiment.

Signal phases in phase-sensitive COSY

When discussing two-dimensional spectra it is important to consider the relative phases and lineshapes for all resonances in the 2D plot. For the COSY spectrum, as explained above, the diagonal peaks arise from the in-phase component of magnetisation produced by evolution under spin–spin coupling. This is not transferred to the coupled spin by the mixing pulse, and diagonal peaks therefore have in-phase multiplet fine-structure in both dimensions. The crosspeaks, in contrast, arise from the transferred antiphase component so have antiphase multiplet structure in both dimensions. The initial phase of these two sets of signals also differs by 90° since the diagonals arise from magnetisation that was cosine modulated whilst the crosspeaks arise from that which was sine modulated. Thus, in the final spectrum, the phases of the diagonal and crosspeaks will also differ by 90° and since the crosspeaks are of most interest these are phased so as to have (antiphase) absorption-mode

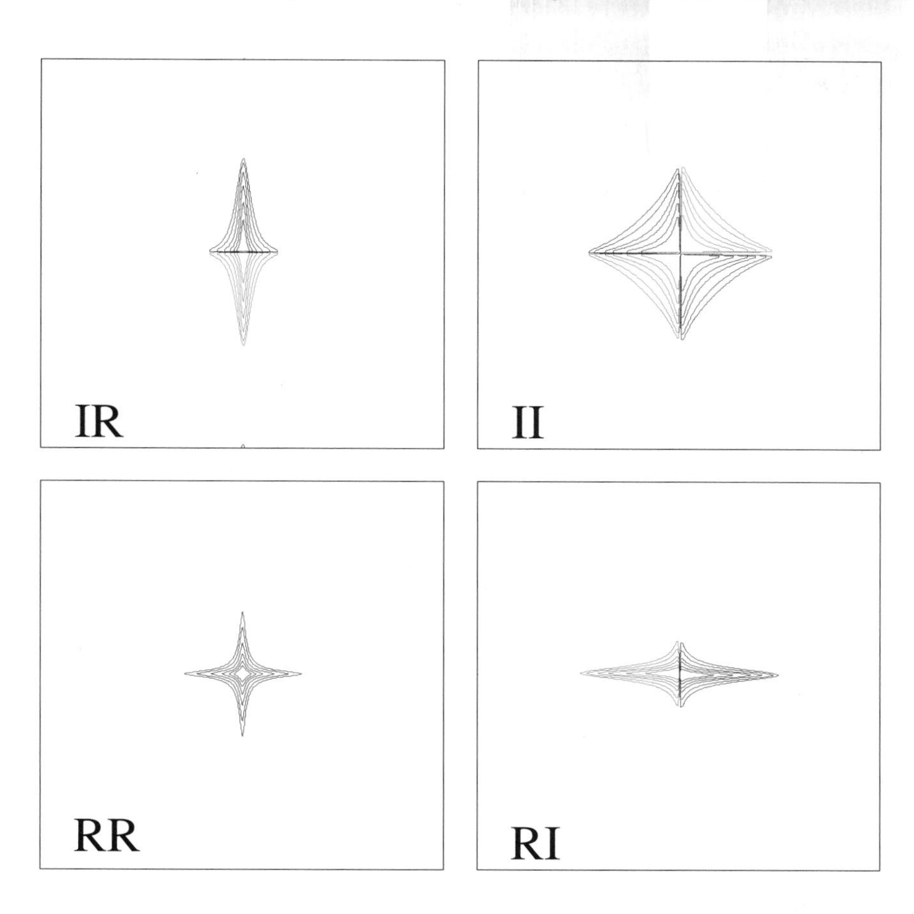

Figure 5.18. The four quadrants of a phase-sensitive data set. Only the RR quadrant is presented as the 2D spectrum and this is phased to contain absorption-mode lineshapes in both dimensions to provide the highest resolution (R = real, I = imaginary). Positive contours are in black and negative in red.

lineshapes in both dimensions, meaning the diagonal peaks posses (in-phase) double-dispersion lineshapes (Fig. 5.19). The unfortunate presence of the dispersion mode lineshapes is an unavoidable consequence of the COSY sequence with the wide tails of the diagonals potentially masking peaks that sit close by. One popular variant of the experiment (DQF-COSY) removes the dispersive contribution so is favoured for high-resolution work. The antiphase character of COSY crosspeaks can also be troublesome when couplings are small relative to resonance linewidths since cancellation of the lines occurs and the crosspeak disappears. This is an important factor in determining whether correlations due to small couplings can be detected with this experiment, a topic addressed in Section 5.6.2.

Aliasing in two dimensions

Resonances that fall outside the chosen spectral width will be characterised with incorrect frequencies and so will appear aliased in the 2D spectrum. For symmetrical data sets such as COSY, the two spectral widths are chosen to be the same, so aliasing will occur in both dimensions, and the position of the aliased signal can usually be predicted from the quadrature detection scheme used for each dimension. Thus, in analogy with one-dimensional spectra, simultaneous or States sampling causes signals to be *wrapped* around, appearing at the far end of the spectrum, whilst those from sequential or TPPI sampling appear as *folded* signals at the near end. Some confusion can be introduced if different sampling schemes are used for the two dimensions, for example, simultaneous sampling in f_2 but TPPI in f_1 [10]. Deliberately aliasing signals can be a useful trick in acquiring 2D data since reduced spectral widths imply reduced acquisition times and smaller data sets, or alternatively, greater resolution. For example, it is often feasible to eliminate lone phenyl groups

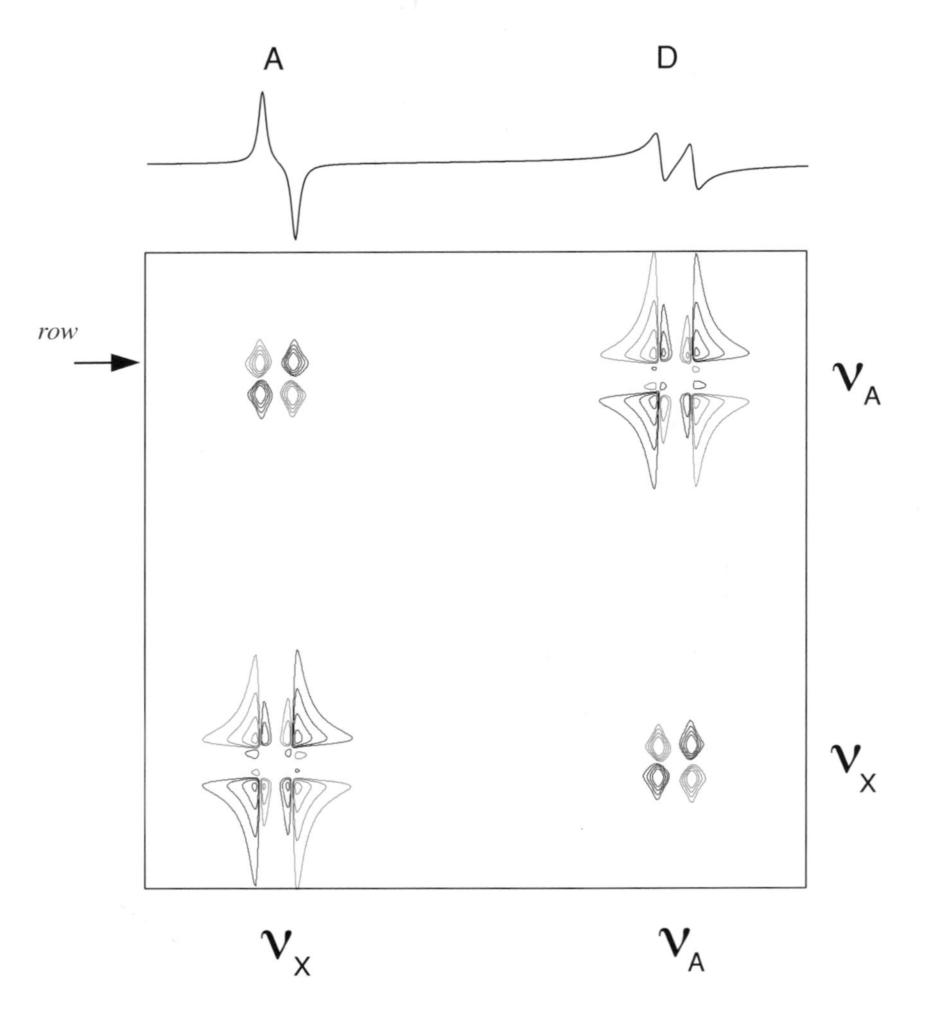

Figure 5.19. The phase-sensitive COSY for a coupled two-spin AX system. Diagonal peaks have broad, in-phase double-*dispersion* lineshapes (**D**) whereas crosspeaks have narrow, antiphase double-*absorption* lineshapes (**A**), as further illustrated in the row extracted from the spectrum.

from COSY spectra and concentrate on the aliphatic region only as this is where the interesting correlations most often lie. With modern digital filters aliased signals in f_2 can be eliminated, although those aliased in f_1 remain since there is no equivalent filtration in the indirect dimension.

Absolute-value presentations

Prior to the advent of the above methods that allowed the presentation of phase-sensitive displays, 2D data sets were collected that were *phase-modulated* as a function of t_1 rather than amplitude-modulated. Phase-modulation arises when the sine and cosine modulated data sets collected for each t_1 increment are combined (added or subtracted) by the steps of the phase cycle, meaning each FID per t_1 increment contains a mixture of both parts. Here it is the sense of phase precession that allows the differentiation of positive and negative frequencies. This method is inferior to the phase-sensitive approach because of the unavoidable mixing of absorptive and dispersive lineshapes, so is generally only suitable for routine, low-resolution work.

The selection of only one of the two possible mirror image data sets in f_1 is now achieved by suitable combination of the sine and cosine data sets. *Addition* of the two data sets within the phase-cycle (Table 5.2) selects those signals that have the *same* sense of precession in t_1 as they have during t_2 (say, both positive) and these are referred to as *P-type* signals. *Subtraction* within the phase cycle selects those signals that have the *opposite* sense of precession during t_1 and t_2 (say, negative in t_1, positive in t_2) and are referred to as *N-type* signals. To clarify this point, remember the two senses of precession we speak

Table 5.2. COSY phase cycles to select the N-type (echo) or P-type (anti-echo) signals in f_1

	N-type			**P-type**		
	Pulse 1	Pulse 2	Acquire	Pulse 1	Pulse 2	Acquire
Cycle 1	x	x	x	x	x	x
Cycle 2	y	x	$-x$	y	x	x

of here simply represent the two possible signals that would be detected either side of $f_1 = 0$ if one were *not* using quadrature detection. By employing this it becomes possible to select only one of these to appear in the final spectrum whilst cancelling the mirror image. The information content of the P-type or N-type spectra are equivalent, only their appearance differs by virtue of reflection about $f_1 = 0$. One significant difference arises from the fact that with N-type selection signals are chosen that have opposite senses of precession in the two time periods. This may be thought of as being analogous to a spin-echo where vectors move in opposite senses either side of the refocusing pulse. A similar effect arises during t_2 with N-type selection whereby echoes also occur, known as *coherence transfer echoes*, (Fig. 5.20). For this reason, N-type selection is also referred to as *echo selection* and P-type signals as *anti-echo selection*, since the refocusing effect does not arise for signals that precess in the same sense. Refocusing of field inhomogeneity with echo selection together with the fact that P-type signals are subject to more severe distortions, means that N-type selection is preferred when this method of quad-detection is used. Conventionally, these spectra are then presented with the diagonal running from bottom left to top right, as in earlier figures.

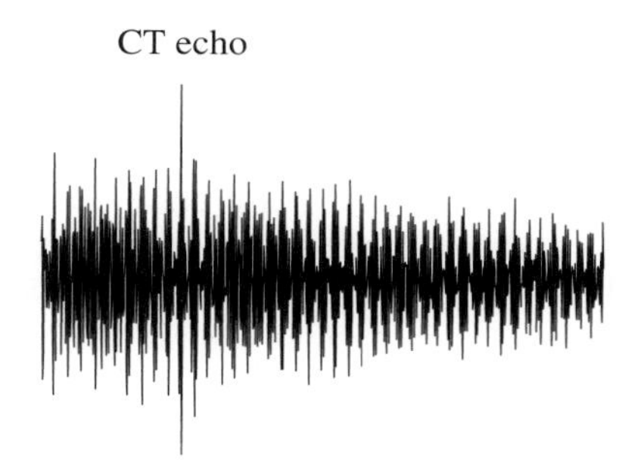

Figure 5.20. The coherence transfer echo is apparent in an FID taken from an N-type COSY data set.

The greatest drawback with data collected with phase-modulation is the inextricable mixing of absorption and dispersion-mode lineshapes. The resonances are said to posses a *phase-twisted* lineshape (Fig. 5.21a), which has two principal disadvantages. Firstly the undesirable and complex mix of both positive and negative intensities and secondly, the presence of dispersive contributions and the associated broad tails that are unsuitable for high-resolution spectroscopy. To remove confusion from the mixed positive and negative intensities, spectra are routinely presented in absolute-value mode, usually after a *magnitude calculation* (Fig. 5.22):

$$\mathbf{M} = (\text{real}^2 + \text{imaginary}^2)^{1/2}$$

where **M** represents the resulting spectrum. To improve resolution, severe window functions are also employed to eliminate the dispersive tail from the lineshape. Although a number of shaping functions are suitable [11], the most frequently used are the *sine-bell*, or the *squared-sine bell* (Section 3.2.7), which are simple to use as they have only one variable, the position of the

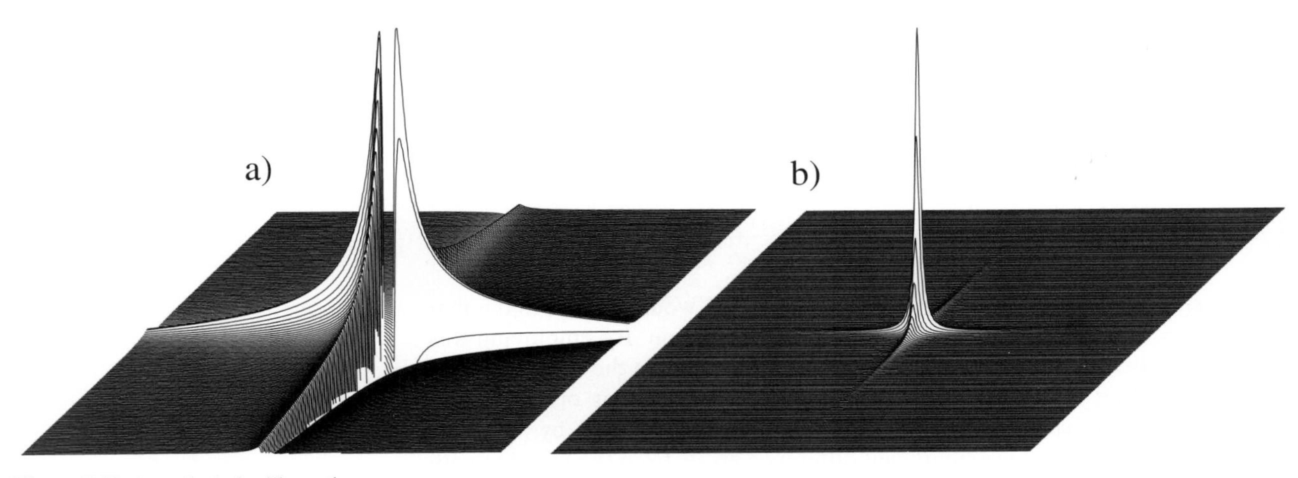

Figure 5.21. A stacked plot illustration of (a) the phase-twisted line shape and (b) the double-absorption lineshape. Clearly the resolution in (b) is far superior and for this reason phase-sensitive methods are preferred.

maximum. When used unshifted i.e. when the half sine-wave has a maximum at the centre of the acquisition time, the resulting peaks possess no dispersive component, thus reproducing the desired absorption lineshape (Fig. 5.22c). The sine-bell shape is also beneficial in enhancing the crosspeak intensities relative to the diagonal. Since the crosspeaks arise from sine modulations, as described above, they initially have zero intensity which builds within the acquisition time. Diagonal peaks are cosine modulated so begin with maximum intensity and are therefore attenuated by the window function. However, the attenuation presents a problem for signals with differing relaxation times (or, in other words, linewidths) as these will be attenuated to different extents and, in particular, broader lines will be notably reduced in intensity by the early part of the sine-bell. The moral when interpreting absolute value data processed in this way is to be wary of differing crosspeak intensities and not simply to associate these with smaller coupling constants. The final problem with the use of these extreme resolution enhancing window functions is that much of the early signal is attenuated and the later noise enhanced, leading to a potential loss of sensitivity. It is possible to shift the function maximum to earlier in the FID to help improve this, with the inevitable re-introduction of some dispersive component, although a better approach is acquire phase-sensitive data which does not demand the application of such severe window functions. Despite this, when sensitivity is not a limiting factor, the magnitude experiment does have the advantage of not requiring phase corrections to be made, so is well suited to routine acquisitions and to automated methods, especially when implemented as the COSY-β variant described in Section 5.6.3.

Figure 5.22. Contour plots of (a) the phase-twist lineshape, (b) the same following magnitude calculation, and (c) the same following resolution enhancement with an unshifted sine-bell window and magnitude calculation.

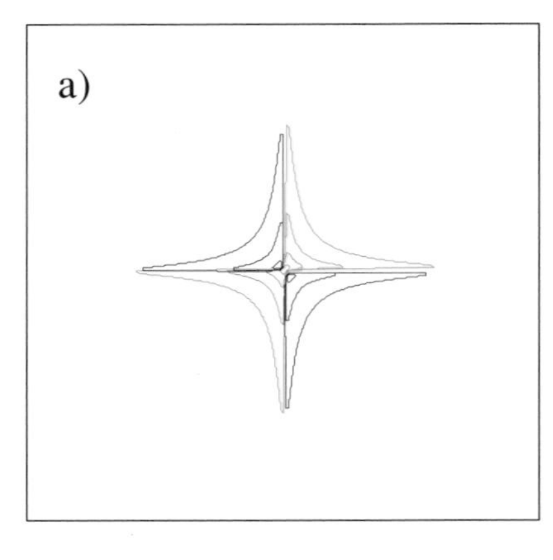

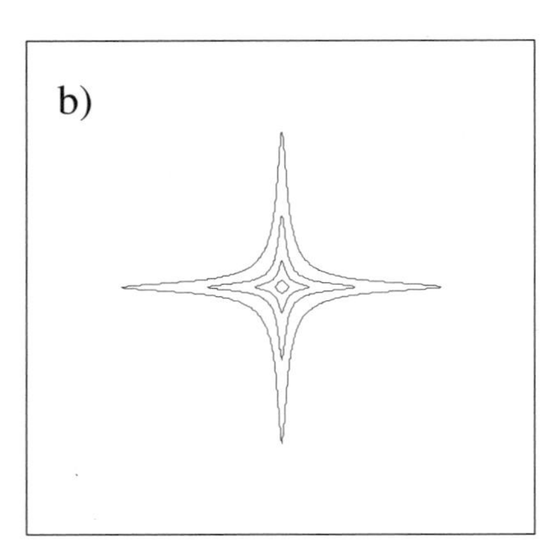

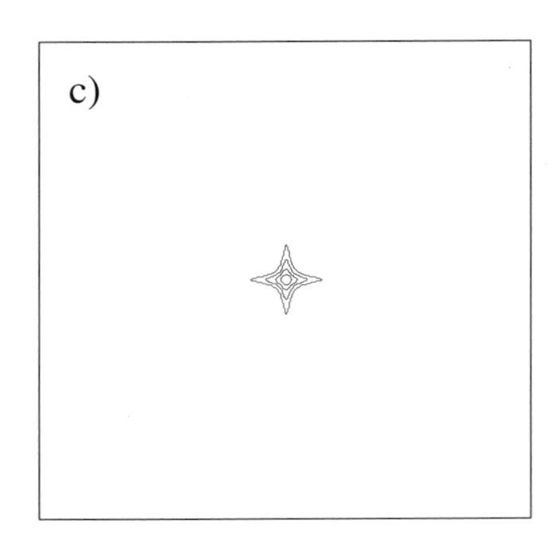

5.3.2. Axial peaks

Axial peaks are experimental artefacts that arise from magnetisation that was longitudinal (aligned along the *z*-axis) during the t_1 evolution period which is subsequently returned to the transverse plane by the mixing pulse and detected in t_2. Since this magnetisation does not evolve during t_1 it is not frequency labelled and therefore has zero f_1 frequency. The peaks can be associated with all resonances in the spectrum and therefore appear as a band of signals across the spectrum at $f_1 = 0$ Hz. For the States or absolute-value methods, this occurs at the centre of the f_1 dimension (Fig. 5.23) and, unless eliminated, may interfere with genuine crosspeaks. If TPPI is used for quadrature detection in f_1, the peaks are shifted toward one edge of the spectrum which, as described above, is where $f_1 = 0$ appears to be, and although aesthetically undesirable, are less of an interference.

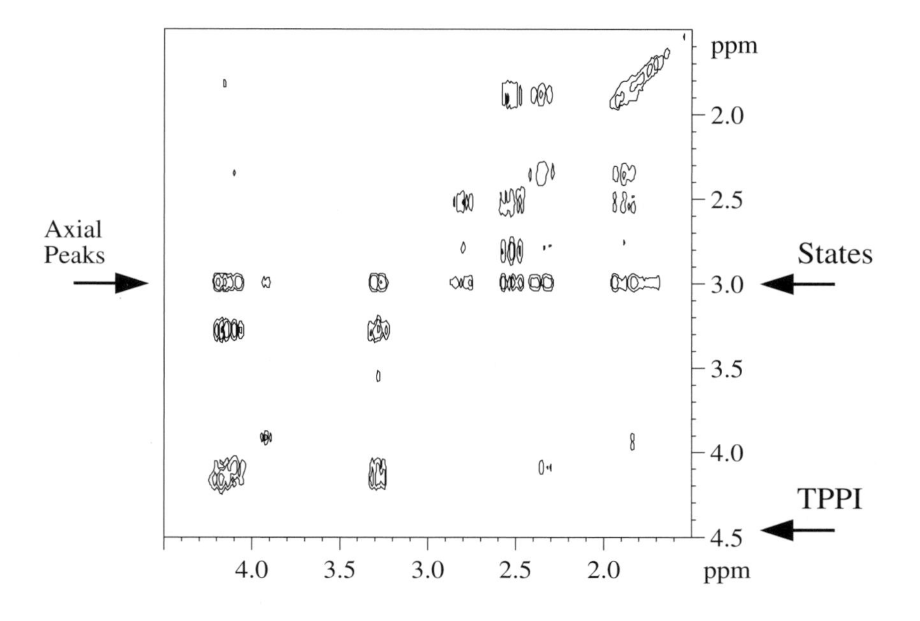

Figure 5.23. Axial peaks in a COSY spectrum form a band of signals along f_2 at (a) the midpoint of f_1 (States or absolute-value data sets) or (b) at the high-frequency edge of the spectrum (TPPI data sets).

The longitudinal magnetisation responsible for axial peaks arises from both experimental imperfection and from relaxation during t_1. Thus, if the preparation sequence does not place all magnetisation in the transverse plane, which for COSY means the initial pulse is not exactly 90°, some residual *z*-component remains. Elimination of these unwanted peaks is achieved simply by repeating the experiment with the receiver phase inverted. The phase of the axial peaks will be unchanged, so will cancel when the two data sets are subtracted by the phase cycle. This procedure also has the benefit of cancelling any offset of the FIDs that arises from receiver imperfections. To avoid simultaneous cancellation of the desired signals the phase of the preparation pulse is also inverted to ensure that it matches the receiver phase, leading to the two-step phase cycle of Table 5.3.

Table 5.3. The basic two-step phase cycle for the elimination of axial peaks

	Pulse 1	Pulse 2	Acquire
Scan 1	x	x	x
Scan 2	$-x$	x	$-x$

5.3.3. Instrumental artefacts

Whilst there exist a whole host of artefacts that can arise in 2D spectra according to the details of the experiment, the two most likely to be encountered are briefly considered here. Both arise from instrumental imperfections, and how significant they are to you will be somewhat dependent on your instrument and its performance, but in any case it is useful to be able to recognise these artefacts if and when they appear.

f_2 quadrature artefacts

Artefacts appearing along f_2 rows are in many instances the exact parallel of those that may be observed in 1D acquisitions, and, not surprisingly, the same solutions apply. Mirror-image quadrature artefacts arising from imbalance of the receiver quadrature channels may again be suppressed by integration of the 4-step CYCLOPS routine (Section 3.2.5) into the existing phase cycle. This involves incrementing the phase of all pulses in the sequence together with the receiver phase in steps of 90°, and leads to a four-fold increase in the duration of the phase cycle. On modern instruments, the intensity of quadrature artefacts are so small (or even non-existent with digital, frequency-shifted detection schemes) that the addition of CYCLOPS can often be avoided and time savings made when sensitivity is not limiting.

t_1-noise

Digitisation of data during a 2D experiment is subject to the same thermal noise arising from the probe head and preamplifier as in a 1D experiment, and this contributes to the noise baseplane observed in the 2D spectrum. There also exists a particularly objectionable artefact associated with 2D experiments, referred to as t_1-noise (note the t_1 here refers to the evolution period and should not be confused with the T_1 relaxation time constant). This appears as bands of noise parallel to the f_1 axis where an NMR resonance exists and it is sometimes this that limits the observation of peaks in the spectrum rather than the true thermal noise. Indeed, it appears that the very earliest work on 2D NMR was unpublished due to excessive t_1-noise present in the spectra. Generally speaking, this is caused by instrument instabilities which lead to random fluctuations in signal intensities and phase from one FID to the next over the course of the 2D experiment. Since these fluctuations relate to perturbations of the NMR resonance, the bands of t_1-noise characteristically appear only at the f_2 shifts of resonances and have intensities proportional to the corresponding resonance amplitude (Fig. 5.24). Thus, stronger bands of noise are associated with intense, sharp peaks, most notably singlets.

The instrumental instabilities that contribute to t_1-noise are numerous [12, 13], including irreproducible rf pulse phase and amplitude, instability in the field-frequency lock and field homogeneity, and so on. These factors are in the hands of the instrument manufacturers, who have made steady progress in reducing various sources of instrumental instabilities over the years, although the problems are by no means completely eliminated. In preparing a 2D acquisition, there are a number of steps that will also help minimise these artefacts. The same arguments apply to any experiment that uses difference spectroscopy to retain selected signals and cancel others (for example the NOE difference experiment in Chapter 8), so are not exclusive to the world of 2D NMR. Use of a strong, sharp lock signal will provide optimum field-frequency regulation, and the use of auto-shim routines on the lower-order shims for long-term acquisitions should also help maintain lineshape reproducibility. Stability of the sample environment is also important. Magnetic field stability is, of course, essential, so movement of any magnetic materials anywhere in the vicinity of the instrument is highly undesirable; this could even include motion

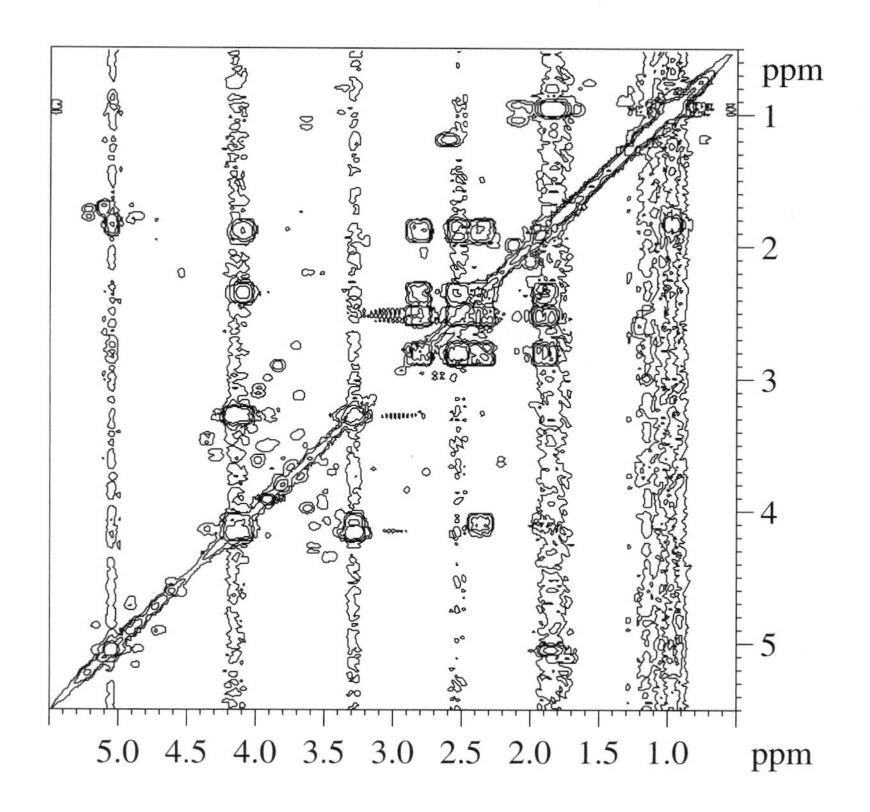

Figure 5.24. Bands of noise parallel to the f_1 axis (t_1-noise) often appear in 2D spectra.

(human or otherwise) above or below the magnet or in the lab. next door. Temperature stability should also be maintained with a suitable temperature control unit, and equilibration of the sample prior to running the experiment is essential, especially for exchanging solutes or solvents, notably aqueous solutions. Finally, it is recommended practice ***not*** to spin samples during 2D acquisitions. This typically causes additional modulation of the detected signal as the sample rotates, notably from an imperfect tube or a 'wobbling' sample spinner, further contributing to t_1 noise.

Symmetrisation

Unwanted t_1-noise in spectra can also be reduced by a number of post-processing strategies. One of the most common procedures for absolute-value COSY experiments is symmetrisation [14]. This involves replacing symmetrically related data points either side of the diagonal with the lesser of the two values so retaining crosspeaks but suppressing unsymmetrical artefacts. This must be applied with caution however, as it may enhance artefacts that have coincident symmetry, such as regions of t_1-noise associated with intense singlets, giving them the appearance of genuine correlation peaks; confusion can usually be avoided by comparison with the non-symmetrised spectrum. This method is not generally suitable for phase-sensitive data as it may introduce distortion of crosspeak structure and can lead to reduced sensitivity, so this symmetrisation finds rather limited use in modern NMR.

Significant reductions in the level of t_1-noise can also be obtained in certain experiments where signal selection is achieved by pulsed field gradients, although homonuclear experiments such as COSY show little benefit in this respect, more notable gains being apparent for proton-detected heteronuclear correlation experiments. More recent post-processing methodologies such as reference deconvolution [15,16] have also provided impressive reductions in t_1-noise [17,18], and the algorithms required for this are gradually finding their way into commercial processing software.

5.3.4. 2D data acquisition

The success of any NMR experiment is, of course, crucially dependent on the correct setting of the acquisition parameters. In the case of 2D experiments one has to consider the parameters for each dimension separately, and we shall see that the most appropriate parameter settings for f_2 are rarely optimum for f_1. Likewise, one has to give rather more thought to the setting up of a 2D experiment than is usually required for 1D acquisitions to make optimum use of the instrument time available and data storage space. Once again, the general considerations below will be applicable to all 2D experiments, although we restrict our discussion at this stage to COSY.

At the most basic level one has to address three fundamental questions when setting up an experiment; 1) will the defined parameters enable (or limit) the detection of the desired information and exclude the unwanted? 2) how long will it take and do I have enough time? and 3) how much storage space is required for the data and is this available? Ideally we would wish to acquire data sets rapidly and we would like these to be as small as possible for speed of processing and ease of handling yet still able to provide the information we require, so we set up our experiment with these goals in mind.

Firstly, spectral widths, which should be the same in both dimensions of the COSY experiment, should be kept to minimum values with transmitter offsets adjusted so as to retain only the regions of the spectrum that will provide useful correlations. It is usually possible to reduced spectral widths to well below the 10 or so ppm proton window observed in 1D experiments. The use of excessively large windows leads to poorer digital resolution in the final spectrum, or requires greater data sizes, neither of which is desirable. The spectral widths in turn define the sampling rates for data in t_2, in exact analogy with 1D acquisitions, and the size of the t_1 increment, again according to the Nyquist criteria. The acquisition time (AQ_t), and hence the digital resolution ($1/AQ_t$), for each dimension is then dictated by the number of data points collected in each. For t_2, this is the number of data points digitised in each FID, whilst for t_1 this is the number of FIDs collected over the course of the experiment. The appropriate setting of these parameters is a most important aspect to setting up a 2D experiment, and the way in which one thinks about acquisition times and digital resolution in a 2D data set is, necessarily, quite different from that in a 1D experiment. As an illustration, imagine transferring the typical parameters used in a 1D proton acquisition into the two dimensions of COSY. The acquisition time might be 4 s, corresponding to a digital resolution of 0.25 Hz/pt, with no relaxation delay between scans. On, for example, a 400 MHz instrument, with a 10 ppm spectral width, this digital resolution would require 32K words to be collected per FID. The 2D equivalent, with States quad-detection in f_1 and with axial peak suppression, requires 4 scans to be collected for each t_1 increment. The mean acquisition time for each would be 6 s (t_2 plus the mean t_1 value), corresponding to 24 s of data collection per FID. If 16K t_1-increments were to be made for the f_1 dimension (two data sets are collected for each t_1 increment remember) this would correspond to a total experiment time of about $4^1/_2$ days. Furthermore, the size of the resulting data matrix would be a little over 1000 million words and, with a typical 32 bit-per-word computer system this requires some 4 gigabytes of disk space! I trust you will agree that four days for a basic COSY acquisition is quite unacceptable, let alone the need for a separate hard disk per experiment so acquiring data with such high levels of digitisation in both dimensions is clearly not possible.

The key lies in deciding on what level of digitisation is required for the experiment in hand. The first point to notice is that adding data points to extend the t_2 dimension leads to a relatively small increase in the overall length of the experiment, so we may be quite profligate with these (although they will lead to a corresponding increase in the size of the data matrix). Adding t_1 data points on the other hand requires that a complete FID of potentially many scans is required *per increment*, which makes a far greater increase to the total data collection time. Thus, one generally aims to keep the number of t_1 increments to a minimum that is consistent with resolving the correlations of interest, and increasing t_2 as required when higher resolution is necessary. For this reason, the digital resolution in f_2 is often better than that in f_1, particularly in the case of phase-sensitive data sets. The use of smaller t_1 acquisition times (AQ_{t1}) is, in general, also preferred for reasons of sensitivity since FIDs recorded for longer values of t_1 will be attenuated by relaxation and so will contribute less to the overall signal intensity. The use of small AQ_{t1} is likely to lead to truncation of the t_1 data, and it is then necessary to apply suitable window functions that force the end of the data to zero to reduce the appearance of truncation artefacts.

For COSY in particular, one of the factors that limits the level of digitisation that can be used is the presence of intrinsically antiphase crosspeaks, since too low a digitisation will cause these to cancel and the correlation to disappear (see Section 5.6.2 for further discussions). The level of digitisation will also depend on the type of experiment and the data one expects to extract from it. For absolute-value COSY one is usually interested in establishing where correlations exist, with little interest in the fine-structure within these crosspeaks. In this case it is possible to use a low level of digitisation consistent with identifying correlations. As a rule of thumb, a digital resolution of J to 2J Hz/pt (AQ of 1/J to 1/2J s) should enable the detection of most correlations arising from couplings of J Hz or greater. Thus for a lower limit of, say, 3 Hz a digital resolution of 3–6 Hz/pt (AQ of ca. 300–150 ms) will suffice. The acquisition time for t_1 is typically half that for t_2 in this experiment, with one level of zero-filling applied in t_1 so that the final digital resolution is the same in both dimensions of the spectrum (as required for symmetrisation).

For phase-sensitive data acquisitions one is likely to be interested in using the information contained within the crosspeak multiplet structures, and a higher degree of digitisation is required to adequately reflect this, a more appropriate target being around J/2 Hz/pt or better (AQ of 2/J s or greater). Again, digitisation in t_2 is usually 2 or even 4 or 8 times greater than that in t_1. In either dimension, but most often in t_1, this may be improved by zero filling, although one must always remember that it is the length of the time domain acquisition that places a fundamental limit on peak resolution and the effective linewidths after digitisation, regardless of zero-filling. The alternative approach for extending the time-domain data is to use forward linear prediction when processing the data (Section 3.2.3). The rule as ever is that high resolution requires long data sampling periods.

Having decided on suitable digitisation levels and data sizes, one is left to choose the number of scans or transients to be collected per FID and the repetition rates and hence relaxation delays to employ. The minimum number of transients is dictated by the minimum number of steps in the phase cycle used to select the desired signals. Further scans may include additional steps in the cycle to suppress artefacts arising from imperfections. Beyond this, further transients should only be required for signal averaging when sensitivity becomes a limiting factor. Since most experiments are acquired under 'steady-state' conditions, it is also necessary to include 'dummy' scans prior to data acquisition to allow the steady-state to establish. On modern instruments that utilise double-buffering of the acquisition memory, dummy scans are required only at the very beginning of each experiment so make a negligible increase to the total time required. On older instruments which lack this feature it is necessary to add dummy scans for each t_1 increment, and these

Table 5.4. Illustrative data tables for COSY experiments

Experiment	Spectral width (ppm)	$N(t_2)$	$N(t_1)$	Hz/pt (t_2)	Hz/pt (t_1)	Experiment time	Raw data-set size
a) Phase-Sensitive	10 × 10	32K	32K	0.25	0.25	4.5 days	1000 Mwords
b) Phase-Sensitive	6 × 6	2K	1K	2.3	2.3	55 mins	2 Mwords
c) Absolute-Value	6 × 6	1K	256	4.6	4.6	22 mins	0.25 Mwords

Scenario (a) transplants acquisition parameters from a typical 1D proton spectrum into the second dimension leading to unacceptable time requirements, whereas (b) and (c) use parameters more appropriate to 2D acquisitions. All calculations use phase cycles for f_1 quad-detection and axial peak suppression only and, for (b) and (c), a recovery delay of 1s between scans. A single zero-filling in f_1 was also employed for (b) and (c).

may then make a significant contribution to the total duration of the experiment. The repetition rate will depend upon the proton T_1s in the molecule and since the sequence uses 90° pulses, the optimum sensitivity is achieved by repeating every 1.3 T_1 s.

Returning to the example 400 MHz acquisition discussed above, we can apply more appropriate criteria to the selection of parameters. Table 5.4 compares the result from above with more realistic data, and it is clear that under these conditions COSY becomes a viable experiment, requiring only hours or even minutes to collect, rather than many days. The introduction of pulse field gradients to high-resolution spectroscopy (Section 5.6) allows experiments to be acquired with only 1 transient per FID where sensitivity is not limiting so further reducing the total time required for data collection. The data storage requirements in these realistic examples are also well within the capabilities of modern computing hardware, and are likely to become increasingly less significant as this develops further. Although illustrated for COSY spectra, the general line of reasoning presented here is applicable to the set-up of any 2D experiment. These issues are briefly considered with reference to different classes of techniques in the following chapters describing other 2D methods.

5.3.5. 2D data processing

The general procedure for processing any 2D data set is the same and essentially follows that outlined in the flow chart of Fig. 5.25. The details of the parameters selected for each stage will depend upon the technique used, the acquisition conditions and the nature of the sample being studied, so only the general principles that are to be considered during data processing are considered here. All of these are extensions of the ideas already introduced in Chapter 3 for the handling of 1D data and this section assumes familiarity with these (Sections 3.2.3 and 3.2.8). Processing techniques have even greater importance for two-dimensional work where the operator is able to tailor the appearance of the final spectrum to a large degree according to the chosen parameters. The choice of window functions has a major impact on the final spectrum and this selection differs markedly for phase-sensitive and absolute-value data sets, so initially we restrict discussion to the phase-sensitive case. These functions are applied typically with three goals in mind:

- To ensure truncation of the time-domain data and thus attenuate truncation artefacts,
- To enhance resolution in the spectrum, and
- To provide optimum signal-to-noise.

Owing to the short acquisition times used, data truncation typically occurs in both dimensions. Particularly when zero-filling is applied, as is nearly always

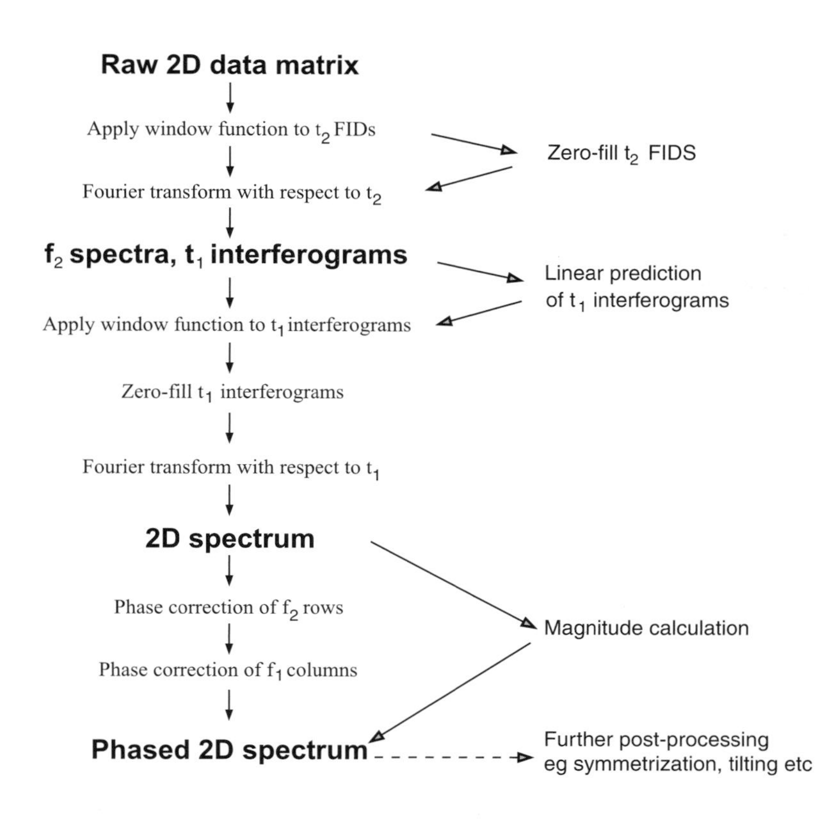

Figure 5.25. The typical scheme followed in the processing of a two-dimensional data set. The items shown to the right are additional or alternative procedures that may be executed.

the case in t_1, the window function must force the FID to zero if truncation wiggles are to be avoided. The longer t_2 acquisition times often means zero-filling is not required in this dimension and similar considerations apply to those for handing a 1D FID. Generally some enhancement of resolution without degradation of signal-to-noise is of primary concern for which any of the resolution enhancements functions of Section 3.2.8 can be used, the most popular being the Lorentz–Gauss transformation, or the shifted sine-bell or squared-sine-bell. The later two guarantee forcing the decay to zero, so are also suitable for the shaping of t_1 data prior to zero-filling, which is typically applied at least once. In either dimension, excessively strong enhancement leads to the appearance of negative-going excursions about resonances and trial and error is generally the best approach to achieving optimum results. Often the first FID of a 2D data matrix can be used as a convenient test 1D spectrum to visualise the effects different window functions will have on the final f_2 data. Modern software now allows interactive changes of shaping parameters along with a display of the final spectrum making such experimenting rather straightforward. An alternative to the zero-filling of t_1 data for high signal-to-noise data sets is the extension of t_1 interferograms by linear prediction to typically two or four times their original size. Not only does this extend the decay, it reduces the need for apodisation of highly truncated data, both of which contribute to improved resolution in f_1. Further zero-filling beyond this may also be applied to improve the digitisation of the final spectrum if desired.

Absolute-value data sets (typically presented following a magnitude calculation) demand stronger resolution enhancement functions in both dimensions to suppress the undesirable dispersion-mode contributions to the phase-twisted lineshapes. This is most often achieved with the sine-bell or squared-sine-bell functions. These are well suited to the anti-phase peaks that give rise to COSY correlations since these start with zero intensity (being initially antiphase they have no net magnetisation) and grow in during the course of the FID. However, most 2D experiments produce in-phase resonances which provide full intensity at the start the FID and if used non-shifted these functions lead to attenuation of much of the signal and can severely degrade sensitivity (try processing a 1D proton spectrum with a non-shifted sine-bell!).

Phase correction

Phase correction of a 2D spectrum is undertaken independently for the two frequency domains. Where the first FID of a 2D matrix contains signals, this can be processed and phased as for a 1D spectrum and the resulting zero- and first-order phase constants applied to the f_2 data. The frequency independent phase correction for f_1 will depend on the details of the phase cycling used in the pulse sequence, but should be close to 0° or 90°. In theory, a frequency-dependent (1st order) phase correction in f_1 should not be necessary since the time delays prior to data sampling that produce these errors do not arise for the artificially generated time domain. In practice some small correction is often required to both domains for optimum results and this is undertaken interactively as for a 1D spectrum. Thus, one aims to observe signals at either end of the spectrum on which to make the corrections. This involves independently selecting f_2 rows (or f_1 columns) from the 2D spectrum. To provide the necessary frequency distribution of resonances, 2 or more from each dimension are usually required, and the resulting phase constants applied to all other rows (or columns).

Presentation

It has already been stated that the most informative and convenient way to present 2D data is as a contour plot, in which the horizontal and vertical axes represent the frequency dimensions and the peak intensity is indicated with appropriate contours. The lower contour setting defines the cut-off below which no signals are presented. Typically this will be set some way above the baseplane noise level but if set too high may eliminate potentially useful information such as weak correlations arising from small couplings. In some instances a cut-off level suitable for revealing weak signals will be too low for more intense ones and may cause the appearance of artefacts such as t_1-noise, which does not make for an aesthetically appealing plot. In such situations it may be worthwhile producing multiple plots with differing settings. Producing expansions of different regions can also be beneficial, especially in crowded situations or where the details of crosspeak structures are to be analysed. Such analysis is best performed interactively on screen if possible, where the use of cursors and cross-hairs is more accurate than that of pencil and ruler. In situations where both positive and negative contours must be displayed and differentiated, different colours can be employed or different line styles if monochrome output is demanded.

5.4. COHERENCE AND COHERENCE TRANSFER

The notion of coherence has already been touched upon in the introduction to COSY in this chapter and has been briefly mentioned in Chapter 4. This lies at the foundation of every NMR experiment, and to appreciate many of the topics that follow in this book it is useful to develop some feel for this and the notion of coherence transfer. Coherence is a generalisation of the notion of transverse magnetisation yet it is important to appreciate some specific features of coherence, which may be illustrated with reference to the energy level diagram of a coupled 2-spin AX system (Fig. 5.26a). Following the application of a 90° pulse on this system at equilibrium, transverse magnetisation exists which, according to the vector model, may be represented as four vectors in the rotating frame, one for each of the transitions in Fig. 5.26a. These give

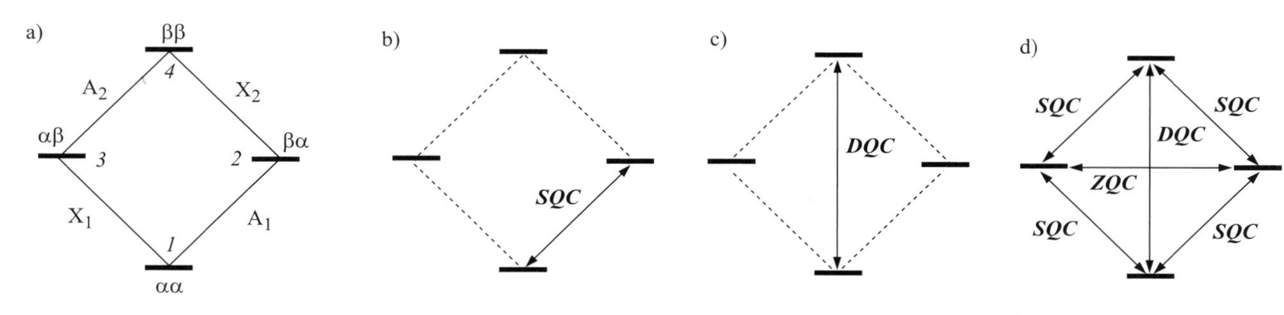

Figure 5.26. Schematic energy level diagrams for a coupled two-spin AX system (see text). SQC = single-quantum coherence, DQC = double-quantum coherence and ZQC = zero-quantum coherence.

rise to the four lines of the two doublets in the resulting spectrum. These transitions are termed *single-quantum* transitions because they are associated with a change in magnetic quantum number, M, of one ($\Delta M = 1$). The magnetisation associated with each is therefore referred to as a *single-quantum coherence* which is said to possess a *coherence order*, p, of 1. Single-quantum coherence is the conventional transverse magnetisation detected in any NMR experiment as this is able to induce a voltage in the NMR detector. In this respect it is useful to consider the difference between a transition excited by a 90° pulse, as above, and one that is saturated by weak rf irradiation. Both situations correspond to equal α and β populations and hence to zero population differences across the transitions yet only the first produces a detectable signal. The saturated transition has no *net* magnetisation to detect because the spins are distributed about the *x–y* plane with random phase. The spins excited by the pulse, however, are bunched together along one axis in the transverse plane and so produce a net component that is detectable; they have *phase coherence* imposed on them by the pulse (Fig. 5.27). In general we may distinguish a single-quantum coherence as a phase coherence between two states that have $\Delta M = \pm 1$.

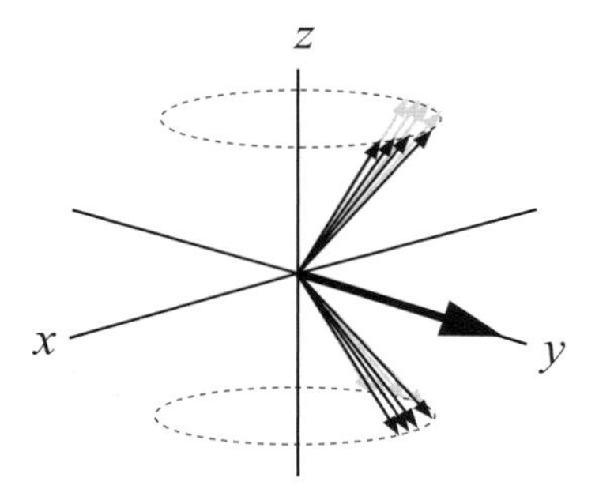

Figure 5.27. Net transverse magnetisation is produced from the bunching of individual magnetic moments which gives rise to an observable NMR signal. These spins are said to posses *phase coherence*, and because only single-quantum spin transitions ($\alpha \leftrightarrow \beta$) are involved in generating this state, it is termed *single-quantum coherence*.

Now imagine the application of a selective 90° pulse to the 2-spin system at equilibrium but to just one of the transitions, say A_1. This then creates a single-quantum coherence for this transition alone (Fig. 5.26b). Next imagine applying a selective 90° pulse to the transition X_2 and hence equalising the populations across this transition. This process transfers some of the spins associated with the βα state to the ββ state and some part of the phase coherence originally associated with A_1 is now associated with the 1–4 transition (αα–ββ) (Fig. 5.26c). This is a new form of coherence not previously considered in any detail. It corresponds to a change in magnetic quantum number of 2 ($\Delta M = 2$) so is therefore termed two-quantum or more usually, *double-quantum coherence* (p = 2) which cannot readily be represented with the vector model notation used thus far. In the COSY experiment a similar state of affairs arises through the use of non-selective 90° pulses applied simultaneously to all transitions. Thus, during t_1 the single-quantum coherences (transverse magnetisation) generated by the first 90° pulse evolve under the influence of spin–spin coupling to give the antiphase magnetisation components required for *coherence transfer*, as described in Section 5.2.3. On application of the second 90° pulse, the coherence originally associated with the A_1 single-quantum transition is distributed amongst the other transitions within the spin system (X_1, X_2, A_2, 1–4 and 2–3) whilst some part remains with A_1 (Fig. 5.26d). The first two again correspond to single-quantum coherence but now of the X spin, so this coherence transfer from A to X (Fig. 5.28) gives rise to the COSY crosspeak. Coherence remaining with A_1 or that transferred to A_2 ultimately produces the diagonal peak multiplet that we have already seen. The 1–4 coherence is the double-quantum coherence described above whilst the 2–3 coherence corresponds to a change in magnetic quantum number of zero, so is termed *zero-quantum coherence* (p = 0). Longitudinal magnetisation, although

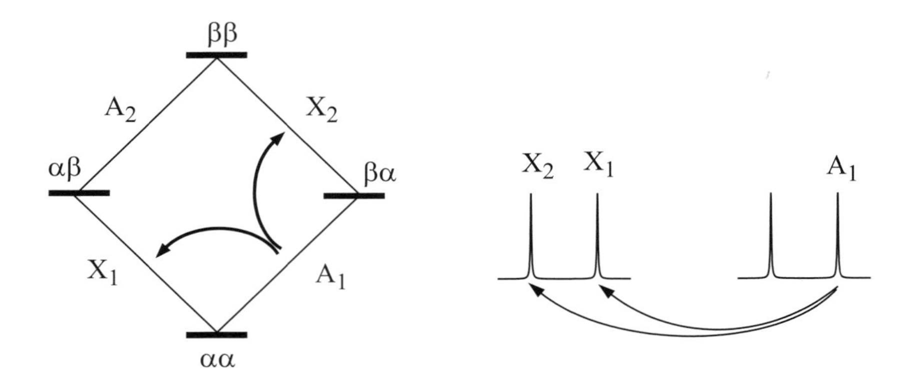

Figure 5.28. The coherence transfer process responsible for generating crosspeaks in correlation spectroscopy, illustrated for the A_1 transition. Coherence of spin-A during t_1 of a 2D sequence becomes coherence of spin-X during t_2, thus correlating the two spins in the resulting spectrum.

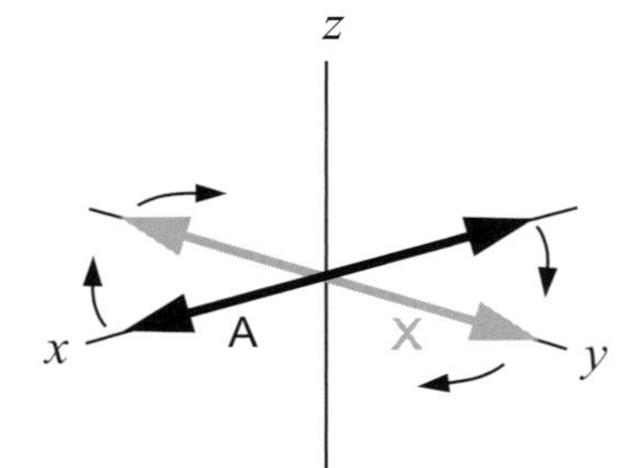

Figure 5.29. A simplified picture of multiple-quantum coherence in an AX system views this as being composed of groups of evolving antiphase vectors which have zero net magnetisation, and hence can never be directly observed.

not a state of coherence, also has p = 0 since it does not correspond to transverse magnetisation. All coherences for which $p \neq \pm 1$ are referred to as *multiple-quantum coherences*. Quantum mechanical selection rules dictate that only those coherences corresponding to $\Delta M = \pm 1$ are able to induce a signal in the detection coil or, in other words, *only single-quantum coherences may be observed directly* since all others have no *net* magnetisation associated with them. Unfortunately, simply stating this fact leaves a slightly uncomfortable air of mystery surrounding the 'invisible' multiple-quantum coherences. A simple physical picture of these states views them as combinations of pairs of evolving vectors that are *always* antiphase and hence never produce observable *net* magnetisation (Fig. 5.29). The evolution frequency of these states is dictated by the chemical shifts of the participating nuclei and the coherence order. Whereas single-quantum coherences evolve in the rotating frame according to the chemical shift offset of the spin from the transmitter, double-quantum coherence evolves according to the *sums* of the chemical shift offsets of the two-spins ($\nu_A + \nu_X$) and zero-quantum coherences according to their offset *differences* ($\nu_A - \nu_X$). In a system containing more than two coupled spins, higher orders of multiple-quantum coherence may also be excited, so, for example, at least three coupled spins are required for the generation of triple-quantum coherence, and so on. Although again these higher orders cannot be observed *directly* their presence may be detected *indirectly* if they are reconverted to single-quantum coherences prior to detection. For example, adding a third pulse to COSY would regenerate single-quantum coherence from the otherwise invisible multiple-quantum coherences that were generated; an example of this is found in the double-quantum filtered COSY described shortly. Just as spin–spin relaxation produces a loss of observable signals by destroying the phase coherence of transverse (single-quantum) magnetisation, multiple-quantum coherences are also dephased by such relaxation processes and similarly have a finite lifetime in which they can be manipulated.

It should also be noted that a coherence order has sign associated with it, so single-quantum coherence may have an order of p = +1 or p = −1, whereas double-quantum coherences will have $p = \pm 2$ and so on. The positive and negative signs represent hypothetical vectors that precess at the same rate but in opposite senses in the rotating frame. For single-quantum coherence, these simply correspond to the two sets of mirror-image signals that would be observed either side of the reference frequency in the absence of quadrature detection, so have a realistic physical manifestation. By using quadrature detection when collecting data we choose to keep one of these sets of signals and eliminate the other, to avoid possible confusion.

5.4.1. Coherence-transfer pathways

The existence of the different levels of coherence present during any pulse sequence may be illustrated by means of coherence level diagrams. These were originally introduced to NMR for the purpose of designing pulse sequence phases cycles [19–21] but can also provide an extremely effective yet simple graphical means of following the flow of magnetisation through a pulse sequence. This formalism proves to be especially powerful when considering the pathway taken by magnetisation from the start of a 2D pulse sequence to its arrival at the detector. It also provides a means of following multiple-quantum coherences which cannot be represented by the vector model, so overcoming some of the limitations of this without resorting to mathematical formalisms. The coherence level diagram for a single-pulse 1D acquisition is presented in Fig. 5.30. A single pulse generates only coherences with $p = \pm 1$, and these are represented in this figure as solid and dashed lines. As stated above, one of these pathways is selected by the hardware quadrature detection whilst the other is rejected, and purely by convention the $p = -1$ pathway is retained (note that in the original paper [19] the opposite convention was chosen). Coherence level diagrams are used to indicate the pathway followed by the desired magnetisation that is selected by the phase cycle or pulsed field gradients in use. All other possible pathways (of which there could be very many) are not shown and are assumed to be eliminated, thus the dashed pathway of Fig. 5.30 would not usually be presented.

To introduce the idea of coherence-transfer pathways for 2D experiments the COSY sequence is again considered and for simplicity the absolute-value P- and N- type COSY experiments described in Section 5.3.1 are described. The desired pathways for both these experiments are shown in Fig. 5.31 with the only difference between these two being the relative sense of precession of (single-quantum) coherence in the t_1 and t_2 time periods, this being the same for the P-type and opposite for the N-type signals. This difference is then clearly illustrated by the coherence transfer pathways followed, both of which begin with equilibrium longitudinal magnetisation which has coherence order zero. Since coherence order $p = -1$ is selected in t_2, the corresponding coherence in t_1 is therefore $p = -1$ for the P-type signal but $p = +1$ for the N-type. These pathways are selected by the phase cycle described previously and illustrated in Table 5.2. During the t_1 period, and in general any period of free precession under the influence of chemical shifts and/or couplings, the coherence orders remain unchanged; only rf pulses are able to alter coherence levels (although relaxation will ultimately regenerate longitudinal magnetisation and hence coherence level zero).

For the phase-sensitive COSY experiment both $p = \pm 1$ pathways must be preserved during t_1. Recall that for a pure-phase 2D spectrum the signal detected in t_2 *must* be amplitude-modulated as a function of t_1. Such a signal can only be obtained if both of the counter-rotating coherences are retained since the retention of only one of these unavoidably leads to a phase-modulated

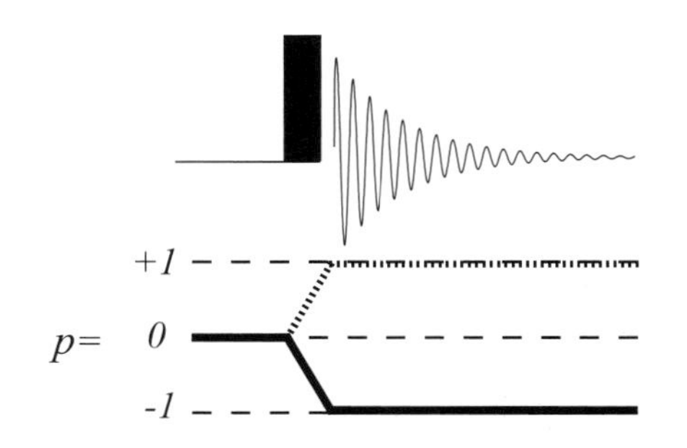

Figure 5.30. The coherence level diagram for a single-pulse 1D experiment. The solid line represents the coherence transfer pathway of the detected magnetisation. The dotted line represents the mirror image pathway that is rejected by the quadrature detection scheme.

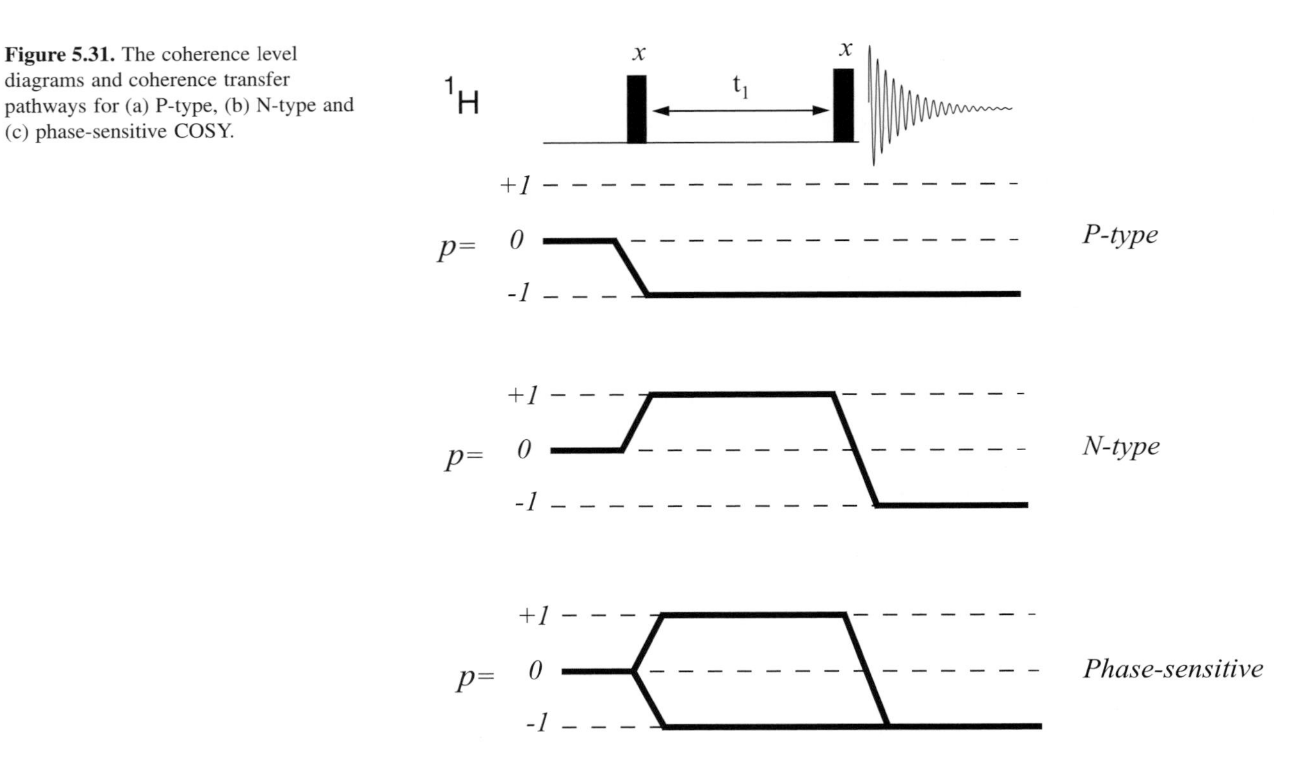

Figure 5.31. The coherence level diagrams and coherence transfer pathways for (a) P-type, (b) N-type and (c) phase-sensitive COSY.

signal (Fig. 5.32) and hence the phase-twist lineshape, as for the P- or N-type COSY. However, the presence of both pathways implies the detection of the NMR resonance plus its mirror (quadrature) image in f_1, so the use of a suitable quadrature detection scheme (States or TPPI) to eliminate one set of these responses at the data processing stage is essential. The need to retain both pathways during t_1 for the acquisition of pure phase spectra is fundamental to all 2D experiments. This has prime importance when utilising pulsed field gradients for coherence pathway selection.

5.5. GRADIENT-SELECTED SPECTROSCOPY

It may be argued that the most significant fundamental development in experimental methodology in high-resolution NMR during the 1990s has been the introduction of pulsed field gradients, which extend the scope of numerous NMR techniques [22–24]. The concept of applying magnetic field *gradients* across a specimen or sample has long been used in (nuclear) magnetic resonance imaging [25] and although the potential for their use in high-resolution NMR was realised some time ago [26,27], technical limitations precluded their widespread use for some years [28]. Nowadays, pulsed field gradients find ap-

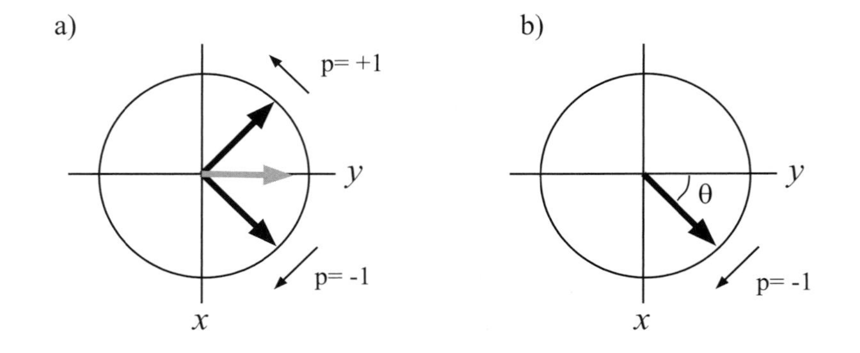

Figure 5.32. (a) Amplitude modulation requires both counter–rotating vectors (p $= \pm 1$ coherences) be retained during t_1. The resultant signal (greyed) is modulated in amplitude along the y–axis as the vectors precess. (b) Phase modulation results if only one vector (p $= 1$ or p $= -1$) is retained as the phase angle θ varies as the vector precesses.

plication in numerous one and multidimensional NMR techniques as a means of selecting those signals deemed interesting and suppressing those which are not, and have proved hugely beneficial to many of the routine methods used within structural organic chemistry. This section introduces the general principles behind the use of pulsed field gradients (PFGs), and although nestled in this introduction to two-dimensional NMR techniques, the discussions are quite general and applicable to all gradient-selected experiments. Gradient-based methods have not been collected together within a separate chapter, but are presented alongside the more conventional equivalents throughout the book. This more properly reflects the way in which the gradient methods have steadily permeated the NMR world and may now be considered standard techniques as more instruments are supplied with the appropriate hardware incorporated. Furthermore, the majority of methods that use PFGs are often simple variations on the original non-gradient equivalent (hence the use of terms such as *gradient-enhanced* or *gradient-accelerated* experiments in the literature) and thus require no greater understanding to enable one to apply the experiment and interpret the data. In fact the advantages endowed by PFGs over traditional methods largely relate to the practicalities of data collection, often allowing the collection of superior quality data in shorter times.

The term 'traditional' in this context, refers to those experiments that make use of phase-cycling to select the desired signals and suppress all others. The notion of phase-cycling has already been encountered in previous sections; the point to recall at this stage is that this procedure involves the repetition of a pulse sequence with the phases of the rf pulse(s) adjusted on each transient and the data from each combined such that the desired signals add constructively whilst those that are unwanted cancel. Phase-cycling procedures are thus all based on difference spectroscopy and as a result are subject to problems of imperfect subtraction of the undesired signals which leads to artefacts in the final spectrum. The dependence on a difference procedure also means that on each transient data is collected which is ultimately discarded which clearly does not make optimum use of the receiver's dynamic range. These methods also require that a full phase-cycle be completed to achieve the required selection regardless of signal strength, so place a fundamental limit on how quickly an experiment can be collected. Experiments which make use of PFGs for signal selection do not suffer from these limitations because *only* the *desired signal* is collected *on each transient*, so leading to cleaner spectra that are free from difference artefacts and, when sensitivity is not limiting, to faster data collection because one is not limited by the number of steps in a phase-cycle.

5.5.1. Signal selection with pulsed field gradients

Defocusing and refocusing with PFGs

Before subjecting any sample to NMR analysis, one of the operations that must be performed is that of shimming the magnetic field to remove inhomogeneities. This ensures that all spins in the same chemical environment experience the same applied field and hence possess the same Larmor frequency regardless of their location within the sample. If, however, one were to deliberately impose a linear magnetic field *gradient* (known as a B_0 gradient) across the sample, say along the z-axis, chemically equivalent spins would now experience a different applied field according to their position in the sample, in this case, along its length. This may be understood by considering the sample to be composed of microscopically thin disks in the x–y plane with each experiencing a different local magnetic field according to their physical location (Fig. 5.33). If the field gradient were imposed for a discrete period of time as a *gradient pulse* (Fig. 5.34a), then during this the spins, having been

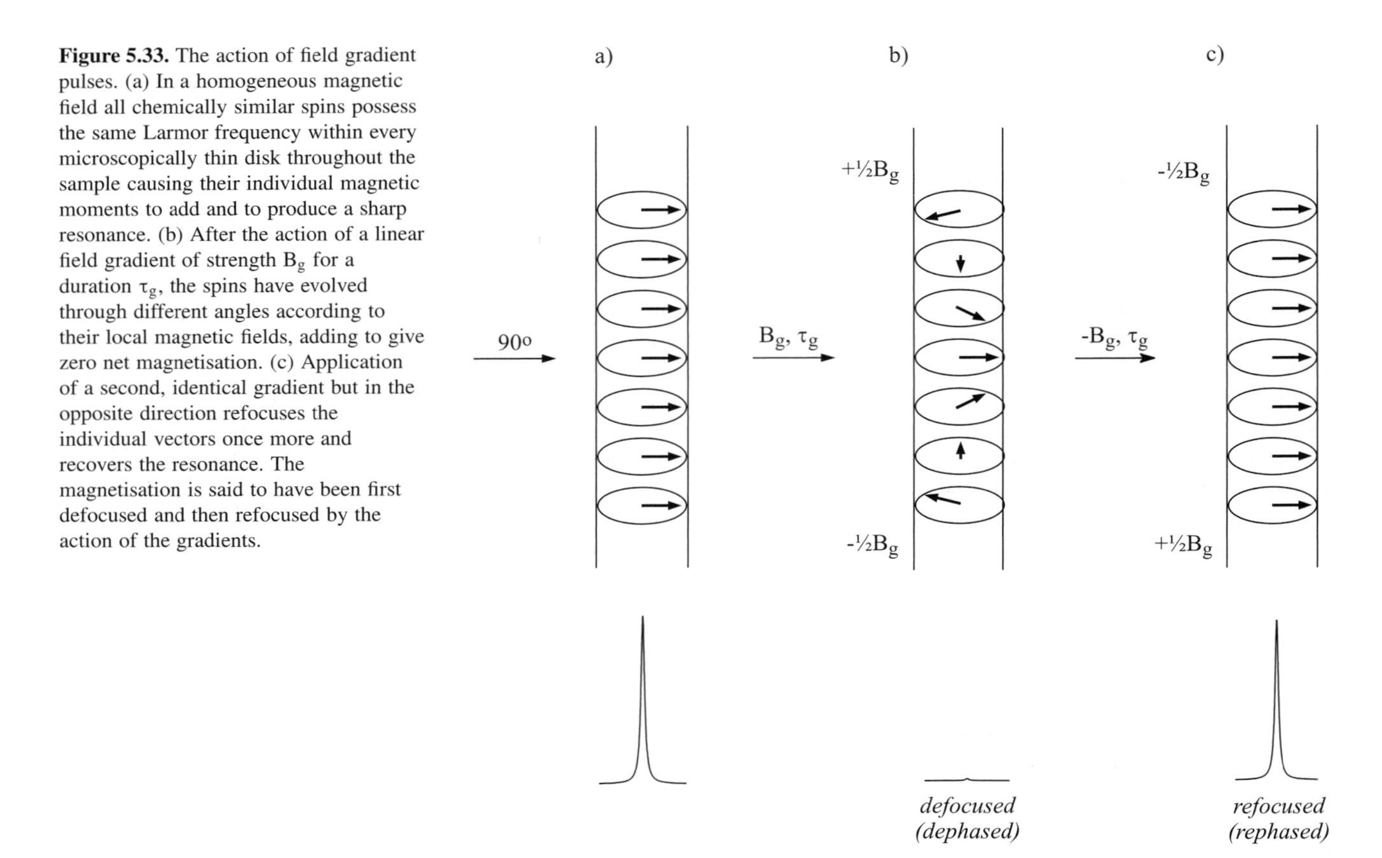

Figure 5.33. The action of field gradient pulses. (a) In a homogeneous magnetic field all chemically similar spins possess the same Larmor frequency within every microscopically thin disk throughout the sample causing their individual magnetic moments to add and to produce a sharp resonance. (b) After the action of a linear field gradient of strength B_g for a duration τ_g, the spins have evolved through different angles according to their local magnetic fields, adding to give zero net magnetisation. (c) Application of a second, identical gradient but in the opposite direction refocuses the individual vectors once more and recovers the resonance. The magnetisation is said to have been first defocused and then refocused by the action of the gradients.

previously excited with an rf pulse, precess with differing frequencies and will thus rotate through differing angles (Fig. 5.33b). Provided the gradient pulse is of sufficient strength and duration, the *net* magnetisation remaining, which is the sum of the transverse magnetisation from all disks in the sample, will be zero and there would be no detectable NMR signal. The gradient pulse is said to have *defocused* (or *dephased*) the magnetisation. The effect of applying such a sequence with increasing gradient strengths is shown in Fig. 5.35.

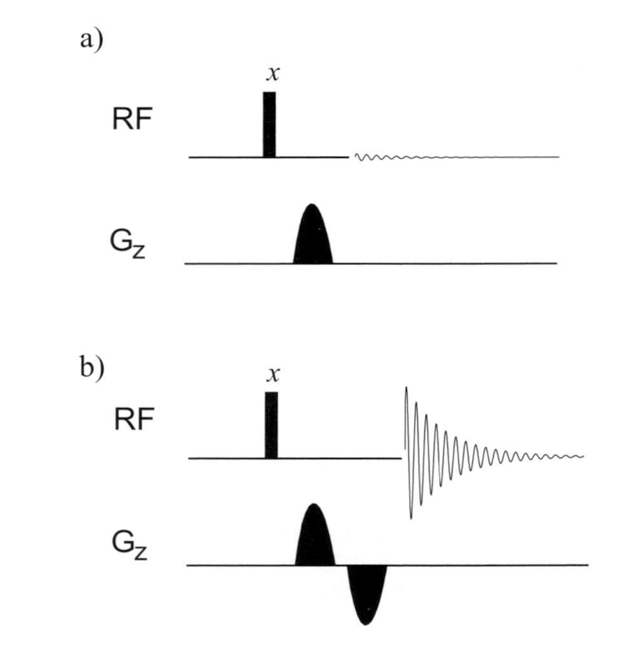

Figure 5.34. The pulse sequence representation of (a) a single *z*-axis field gradient pulse and (b) two *z*-axis gradient pulses applied in opposite directions. Gradient pulses typically have a shaped rather than square profile, such as the sine profile illustrated here (see Section 5.5.4). RF identifies the radiofrequency pulse channel.

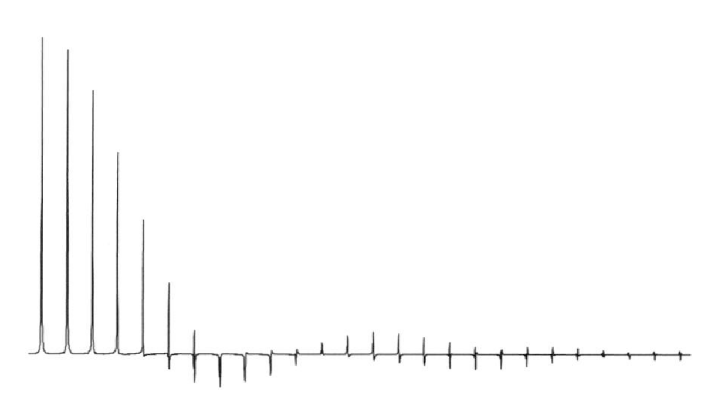

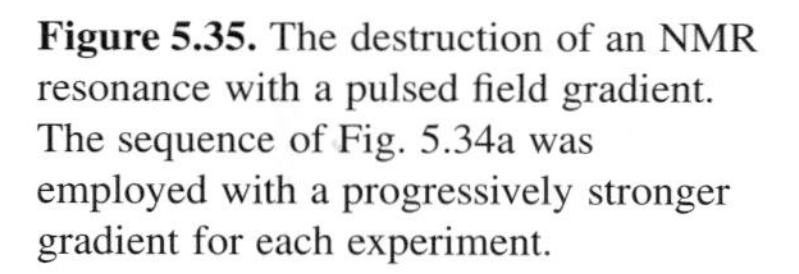

Figure 5.35. The destruction of an NMR resonance with a pulsed field gradient. The sequence of Fig. 5.34a was employed with a progressively stronger gradient for each experiment.

Although there is now no *detectable* signal, magnetisation has not been lost from the transverse plane and may be recovered. Applying a second gradient pulse immediately after the first, of equal intensity and duration but in the opposite sense (now along the $-z$-axis, Fig. 5.34b), causes the dephasing of the first gradient to be exactly 'unwound' by the action of the second (Fig. 5.33c) and the magnetisation vectors from each disk would *refocus* (or *rephase*) to produce a *gradient echo* and an observable net signal (Fig. 5.36). The concept of defocusing all unwanted signals and *selectively* refocusing only those that one desires lies at the heart of every gradient selected experiment but to understand how this refocusing can be made selective, we must consider the action of the gradient pulse on transverse magnetisation in a slightly more formal context.

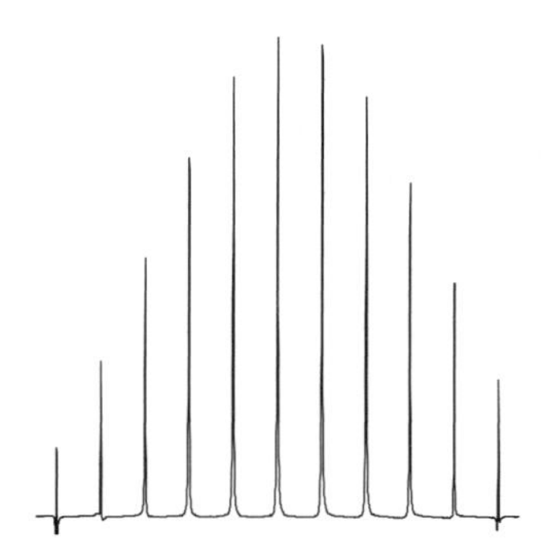

Figure 5.36. The recovery of an NMR resonance in a gradient echo. The sequence of Fig. 5.34b was employed with the strength of the first gradient fixed and that of the second varied from 90 to 110% of the first.

Selective refocusing

In the absence of a field gradient pulse, the Larmor frequency, ω, of a spin depends upon the applied static field, B_0, such that $\omega = \gamma B_0$. When the gradient pulse is applied there is an additional spatially dependent field, $B_{g(z)}$ associated with the gradient giving rise to a spatially dependent Larmor frequency, $\omega_{(z)}$:

$$\omega_{(z)} = \gamma(B_0 + B_{g(z)}) \text{ rad s}^{-1} \tag{5.1}$$

If the gradient is applied for a duration τ_g seconds, the magnetisation vector rotates through a spatially dependent phase angle, $\Phi_{(z)}$ of

$$\Phi_{(z)} = \gamma(B_0 + B_{g(z)})\tau_g = \gamma B_0 \tau_g + \gamma B_{g(z)} \tau_g \text{ rad} \tag{5.2}$$

The first term in the final expression simply represents the Larmor precession in the absence of the field gradient which is constant across the whole sample and as such shall not be considered further at this point, except to note that this precession has bearings on the practical implementation of field gradients, as described later. The second term represents the spatially dependent phase caused by the gradient pulse itself, and it is this quantity that is of interest. Generalising the above expression which is only applicable to transverse single-quantum magnetisation, it is necessary to include a term that represents the coherence order, *p*, of the magnetisation, so that:

$$\Phi_{(z)} = p\gamma B_{g(z)} \tau_g \tag{5.3}$$

This modification reflects the fact that a p-quantum coherence dephases at a rate proportional to p. Thus, double-quantum coherences dephase twice as fast as single-quantum coherences yet zero-quantum coherences are insensitive to field gradients. When the coherence involves different nuclear species, allowance must be made for the magnetogyric ratio and coherence order for each, such that:

$$\Phi_{(z)} = B_{g(z)} \tau_g \sum p_I \gamma_I \tag{5.4}$$

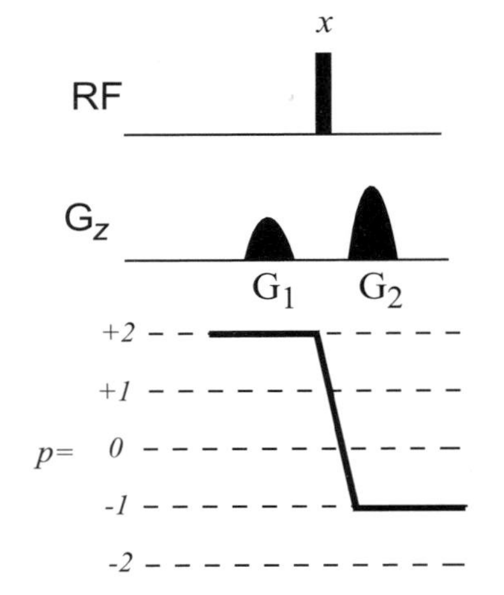

Figure 5.37. An illustration of signal selection by pulsed field gradients. Using two gradients with ratio G_1:G_2 of 1:2 selects only the coherence transfer pathway shown, leaving all others defocused and unobservable.

It is therefore apparent that the degree of defocusing caused by a gradient is dependent upon the coherence order and the magnetogyric ratio of the spins involved, and these two features provide the key to signal selection with field gradients. The purpose of gradients in the majority of experiments is the selective refocusing of magnetisation that has followed the desired coherence transfer pathway during the experiment whilst leaving all other pathways defocused and hence unobservable. Whilst PFGs themselves are unable to alter coherence orders, rf pulses can, and it is a combination of rf pulses to generate the appropriate coherence orders together with the pulsed field gradients to select these that ultimately provides the desired result.

This general process is illustrated for a homonuclear spin system in the scheme of Fig. 5.37 in which only double-quantum coherence existing prior to the pulse is to be retained. Thus, prior to the rf pulse $p = 2$ coherence exists which experiences a gradient of strength B_{g1} and duration τ_1 and thus obtains a spatially dependent phase:

$$\Phi_1 = 2\gamma B_{g1}\tau_1 \tag{5.5}$$

Following the rf pulse, the coherence is transformed into observable single-quantum magnetisation of order $p = -1$ which experiences a second pulsed gradient of strength B_{g2} and duration τ_2 which encodes a phase:

$$\Phi_2 = -\gamma B_{g2}\tau_2 \tag{5.6}$$

For coherence following a defined transfer pathway to be refocused and hence observable, its *total spatially dependent phase from all gradient pulses must be zero*:

$$\sum \Phi_i = \sum \gamma B_{gi}\tau_i = 0 \tag{5.7}$$

or, in other words, the refocusing induced by the last gradient must exactly undo the defocusing caused by all earlier gradients in the coherence transfer pathway. Thus, in this illustration Φ_2 must equal $-\Phi_1$ for this condition to be satisfied. This may be achieved by either altering the duration of the second gradient relative to the first, whilst keeping the amplitude the same, or by altering the amplitude and retaining the duration. Nowadays it is uniform practice to use gradient pulses of the same duration throughout a pulse sequence and to alter their amplitudes. Hence, setting $B_{g2} = 2B_{g1}$ in this example (remembering the coherence order has sign as well as magnitude) selects the desired pathway (Fig. 5.37). In contrast, single-quantum magnetisation originating from, for example, any triple-quantum coherence ($p = 3$) that may have originally existed, acquires a net phase of:

$$\Phi = (3\gamma B_{g1}\tau_1) + (-1\gamma 2B_{g1}\tau_1) = \gamma B_{g1}\tau_1 \tag{5.8}$$

and hence remains dephased and does not contribute to the detected signal. The gradient pair has therefore selected only the desired pathway. For *homonuclear* systems with gradients of equal duration, it is sufficient to consider only the *coherence orders* and *ratios of gradient amplitudes* involved in the selection of the coherence transfer pathway, as illustrated in the COSY examples below. For *heteronuclear* systems, *magnetogyric ratios* of the participating spins must also be included, as will be illustrated in Chapter 6. Since in all following discussions gradients within a sequence are assumed to have the same shape and duration, a gradient pulse will simply be denoted as G_i in shorthand form from now on, which is taken to represent a pulse of strength B_{gi} and duration τ_i.

Gradient selected COSY

To illustrate the incorporation of PFGs into conventional experiments, the basic COSY sequence is again used as an example. It has already been shown

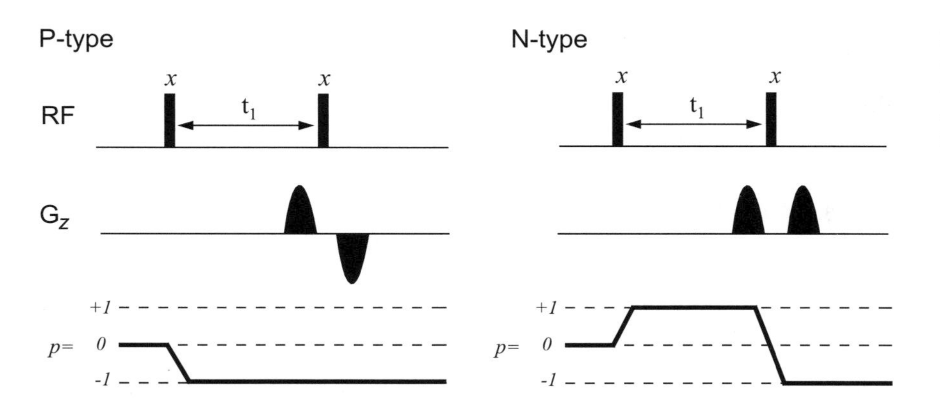

Figure 5.38. The necessary gradient combinations to select P- and N-type COSY data, shown with the corresponding selected coherence transfer pathway.

that for the absolute-value COSY experiment those signals to be retained in the final COSY spectrum follow the coherence transfer pathway illustrated in Fig. 5.38, with N-type selection being the favoured option. For signal selection it is sufficient to phase-encode coherences before the second 90° pulse and decode these after the pulse, immediately prior to detection. For N-type signals, the desired coherence orders are +1 and −1 at these points and thus the phase-encoding induced by each gradient may be represented using shorthand notation as:

$$\begin{aligned}\Phi_1 &= +1G_1\\ \Phi_2 &= -1G_2\end{aligned} \qquad (5.9)$$

Since in COSY one is dealing with a *homonuclear* spin system the magnetogyric ratios need not be included as these are constant throughout, so the only variables are the gradient strengths. Hopefully it is quite apparent that for these two expressions to cancel one another, one simply selects $G_2 = G_1$, that is, both gradients should have the same strength applied in the same direction. Conversely, for P-type selection (coherence orders −1 either side of the pulse) one must select $G_2 = -G_1$, that is, similar gradients applied in opposite directions. In either case, the desired pathway is selected *in a single scan* without the need for phase-cycling because gradients that refocus N-type signals cannot simultaneously refocus P-type; hence quadrature detection in f_1 is inherent with this scheme. Axial peaks are also suppressed by this gradient selection because, by definition, these arise from magnetisation that was along the *z*-axis during t_1 and was therefore unaffected by the first gradient. When this magnetisation is made transverse by the second 90° pulse, the final gradient dephases it and it is simply not observed. Thus, the original four steps of the phase-cycled absolute-value COSY (two for f_1 quad detection and two for axial peak suppression) can be replaced by a single step with PFGs, allowing the 2D data set to be collected with only one transient per t_1 increment, leading to a four-fold reduction in time. This assumes that sensitivity is sufficient to allow the collection of only one scan per increment, which in many routine applications is the case. As an educational exercise, try running a gradient COSY sequence with first N-type and then P-type selection as described above and compare the results. One should simply be the mirror image of the other about the midpoint of f_1.

5.5.2. Phase-sensitive experiments

One problem with the gradient approach described above for COSY is that it necessarily precludes the acquisition of high-resolution phase-sensitive data sets by selecting only one pathway in t_1. Since this leads to phase-modulated data it provides only phase-twisted lineshapes. Recall that phase-sensitive ac-

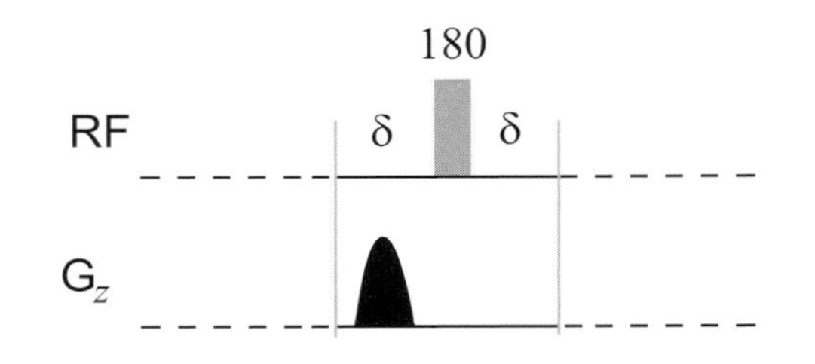

Figure 5.39. Gradient pulses are often applied within spin-echoes to refocus chemical shift evolution occurring during the pulse and thus remove phase errors from spectra.

quisitions require both $p = \pm 1$ pathways to be retained during t_1, but this is impossible when a gradient is placed within the evolution period since only one of these pathways may be refocused prior to detection. The most general solution to collecting phase-sensitive data is to avoid placing gradients within t_1 altogether, as this then allows conventional States or TPPI f_1 quadrature detection to be employed. The details of how this is done depend on the sequence in use, and examples of this approach will be illustrated for various techniques in this and the following chapters. There also exists an ingenious approach to f_1 quadrature detection [29] which provides absorption-mode phase-sensitive data and allows gradients to be placed conveniently within the evolution time. The procedure involves collecting both P- and N-type data sets alternately and storing them *separately* for each t_1 value. The two pathways are selected typically by inverting the sign of the final refocusing gradient, as for the P vs N selection above. These data sets contain mirror-image signals in the f_1 dimension when processed conventionally, so mirroring one data set about $f_1 = 0$ and adding it to the second will retain only one set of signals and hence provide the required f_1 frequency discrimination. This procedure also exactly cancels the dispersive contributions otherwise apparent within the phase-twist lineshapes of each data set and so produces the favoured absorption-mode lineshapes. The data handling required for this procedure is now found within conventional processing software, and is often referred to as the *echo-antiecho* approach.

However, this is not all that is required to produce pure-phase spectra. Following Eq. 5.2 above, the evolution of chemical shifts under the influence of the static B_0 field during the application of a gradient pulse was ignored for convenience as this played no part in determining signal selection. Such shift evolution during the gradient pulse (which is typically of the order of milliseconds) leads to appreciable frequency-dependent phase errors in spectra if not eliminated. These errors are of no consequence for absolute value presentations and hence were not considered for the COSY examples above, but prove disastrous for phase-sensitive data sets. To avoid these failings, gradient pulses are either applied within existing J-evolution (Δ) periods or must be placed within a spin-echo to refocus the chemical shift evolution (Fig. 5.39). Examples of both these approaches are to be found in the techniques that follow.

5.5.3. PFGs in high-resolution NMR

This section summarises some of the key features of pulsed field gradients when applied to high-resolution NMR spectroscopy, highlighting some of the benefits they provide and also warning of potential limitations. Many of these features will be made apparent when describing the gradient selected techniques in the remainder of the book, so this merely provides an overview of these topics. Broadly speaking, the role of PFGs in high-resolution NMR experiments may be grouped into three classes:

- Selection of a coherence transfer pathway: The examples discussed above have already provided an illustration of how gradients may be used in this manner, the aim being to select magnetisation that has followed one desired pathway and suppress that which has followed any other. This application represents the most widespread use of gradients in organic NMR spectroscopy, the resulting experiments providing alternatives to the conventional phase-cycled equivalent.

- Suppression of a solvent resonance: The ability of PFGs to completely destroy an NMR resonance makes them ideally suited for methods of solvent suppression, where the aim is to selectively remove the solvent resonance but retain all others. The most widespread use of these methods are in the studies of molecules in *protonated* water where gradient based suppression methods prove most effective (Chapter 9).
- Purging of unwanted magnetisation: Purging in this context means the elimination of responses that arise from experimental imperfections and the like, such as inaccurate pulse widths. For example, the use of an imperfect 180° inversion pulse on $+z$ magnetisation will leave an unwanted residual component in the transverse plane. This may be destroyed by the application of a single purging z-gradient (sometimes referred to as a 'homospoil' pulse), ensuring that it will not contribute to the final spectrum. The use of purging z-gradients requires all wanted magnetisation to sit along the z-axis and so be invariant to the gradient pulse. One of the earliest applications of a gradient pulse within a high-resolution experiment was as a purge pulse in a 2D exchange sequence [30].

Advantages of field gradients

The principal benefits arising from the use of field gradients for coherence pathway selection as opposed to conventional phase-cycling may be summarised thus:

- Quality: On each transient only the desired signal is detected, avoiding the need for signal addition/subtraction in the steps of a phase cycle. Difference artefacts are thus not encountered, and the t_1-noise often associated with these in 2D spectra is reduced, providing cleaner spectra.
- Speed: As there is no requirement to complete a phase-cycle, experiment times are dictated by sensitivity and resolution considerations only. When sensitivity is not limiting, 2D experiments can be acquired with a single transient per t_1 increment, leading to significant time savings.
- Dynamic range: As only the desired signal is ever detected, all others being destroyed in the probehead, optimum use can be made of the receiver's dynamic range and greater receiver gains can be employed. This compensates to some degree for the sensitivity losses sometimes associated with gradient selection (see below).
- Signal suppression: Very high signal suppression ratios can be achieved. The elimination of protons bound to carbon-12 in proton-detected ^{1}H–^{13}C correlations becomes trivial (suppression ratio of $\approx$1:100) and even the selection of natural abundance ^{1}H–^{13}C–^{13}C fragments becomes feasible [31] requiring suppression ratios of 1:10,000 (see Section 5.8.4). Likewise the suppression of large solvent resonances becomes very much easier with gradient methods [32,33] (Chapter 9).
- Ease of use: In contrast to the use of rf pulses, the accurate calibration of gradient strengths is not required and the absolute strengths used are not critical for the success of the experiment, provided they are sufficient to dephase unwanted magnetisation. More important is the fact that gradient *ratios* must be precise and the gradients reproducible. These demands can be quite readily met with appropriate instrument design, making gradient-selected techniques easy to implement and experimentally robust.

Limitations of field gradients

Despite their undoubted advantages, PFGs have a number of fundamental limitations that should also be appreciated. These have important consequences for the design of gradient experiments and although one may not be concerned with experiment design, these points help explain some of the features that are common to many of the gradient-selected sequences.

- Sensitivity: When used for coherence selection, gradients are able to refocus only one of two $\pm p$ coherence orders. In some applications this means that only one-half of the available signal is detected in contrast to phase-cycled experiments in which both pathways may be retained. Experiments which use PFGs for coherence selection may therefore exhibit lower sensitivity than the phase-cycled equivalent by a factor of typically 2 or $\sqrt{2}$ depending on the precise experimental details [24,34]. Purging gradients, however, do not cause such a sensitivity loss.
- Quadrature detection: As described above, some care is required when employing gradients in phase-sensitive experiments. This either requires that gradients are not placed within time domain evolution periods or that the echo–antiecho approach be used.
- Phase errors: Chemical shifts also evolve during the application of a gradient pulse due to the static B_0 field. Therefore, gradient pulses are generally applied within a spin-echo to refocus this evolution and thus remove phase errors when phase-sensitive displays are required.
- Diffusion losses: The detected signal can be attenuated by losses due to diffusion. The refocusing condition requires that the refocusing gradient undoes the spatially dependent phase caused by all previous gradients. If a molecule were to move along the direction of the gradient (along the length of the NMR tube for a *z*-gradient) between the dephasing and rephasing pulses the gradient strengths would not be perfectly matched and incomplete refocusing will result, leading to a loss of signal intensity (it is exactly this phenomenon that is used to study diffusion in solution by NMR [35,36]; see Chapter 9). This will be more of a problem for small, fast moving molecules in low-viscosity solvents which can be best avoided by keeping the defocusing and refocusing gradient pulses close together in the sequence. The degree of signal attenuation also depends on the *square* of the gradient strengths so the use of weaker gradients, which must still be sufficient to dephase magnetisation, will also help minimise such losses.

5.5.4. Practical implementation of PFGs

As any user of an NMR spectrometer will know, the simplest way to generate a field gradient through a sample is to offset one of the shim currents from its optimum. In principle it is possible to generate the gradient needed by momentarily driving the Z-shim at maximum current, and some spectrometers are provided with a so-called 'homospoil' capability to allow this. However, such a set-up is only suitable for providing a basic purge gradient since it produces only relatively weak fields, offers no control over amplitude and would be plagued by the dreaded eddy currents described below. For performing the full range of gradient-selected experiments a probe equipped with a dedicated gradient coil surrounding the usual rf coils is required, together with an appropriate gradient amplifier. The field gradient across the sample is then generated by applying a current to the gradient coils. Inverting the sign of a gradient corresponds to reversing the applied current. Typical maximum gradient strengths are around $0.5\ \mathrm{T\,m^{-1}}$ ($50\ \mathrm{G\,cm^{-1}}$) for routine work, although in practice the actual gradient strengths used will often be somewhat less than this. A method for calibrating gradient strengths is described in Section 3.5.2.

As simple as this scheme may sound, it is plagued by a number of technical problems that would make the collection of high-resolution data impossible if left unchecked, and it is these problems that slowed the introduction of field gradients in high-resolution NMR for so long. The worst of these are the eddy currents that are generated on application of the gradient pulse. These are currents in the conducting components that surround the sample, such as the probe case, magnet bore and even the shim coils themselves, which in turn generate further spurious magnetic fields within the sample. Since these can last for hundreds of milliseconds, they prevent the acquisition of a high-resolution spectrum. The most effective way to suppress the generation of eddy currents is the use of so-called actively-shielded gradient coils, which are now used in all commercial gradient probes. These consist of a second gradient coil surrounding the first and driven by the same current. The outer coil is designed in such a way that the field it generates *outside of the active sample region* cancels that produced by the inner coil. The net result is a field gradient within the sample, but not outside of it and hence no eddy currents are stimulated in surrounding structures. Since eddy currents are caused by the rapid change in the local magnetic field on application of the gradient, they can also be attenuated by using a shaped gradient profile with a gentle rise and fall, rather than a simple rectangular pulse that has very steep leading and trailing edges. Commonly used gradient profiles are a half sine-wave or a truncated Gaussian. With these steps it is possible to acquire an unperturbed spectrum within tens of *micro*seconds of the gradient pulse; a method for gauging gradient recovery periods is also given in Section 3.5.2.

One final point regarding field gradients is worthy of note. Since a gradient pulse induces a spatially dependent phase for all magnetisation in the sample, this also leads to the destruction of the deuterium lock signal itself, a rather alarming sight when first encountered. How then does the spectrometer retain the field-frequency stability over extended experiments? The key point stems from the fact that the regulation system integrates the lock error signal over a period of typically many seconds so is little effected by the brief disturbances caused by the gradient pulse. The signal presented to the operator is not exactly that used for the field-frequency regulation and is integrated over much shorter time periods so presents a more dramatic picture of events to the operator (which itself can be a useful indicator of whether the gradient system is operating!).

This then finishes this part of the chapter that has essentially laid the foundations for understanding 2D (or more generally *n*D) NMR. Throughout this the basic COSY sequence, either in its absolute-value or phase-sensitive forms, has been used as an illustrative 2D sequence. Not only is COSY a simple sequence, but the spectra are rather easy to interpret and require relatively little explanation. Beyond this basic sequence there are a variety of other COSY experiments which provide the chemist with new or modified information. These are, in fact, more widely used in the laboratory than the experiment described thus far, and it is these that are now addressed before progressing to consider other homonuclear correlation techniques.

5.6. ALTERNATIVE COSY SEQUENCES

There are an enormous range of experiments that are, in essence, variations on the simple COSY sequence described above. In this section we take a look at a selection of these and examine the benefits these modified sequences provide the practising chemist. Many of the COSY variants proposed in the literature offer little gain over the simpler sequences whilst being rather more complex to execute or process, or are beneficial for a rather limited class of compounds and lack wide applicability. Whilst the experiments presented below represent only a fraction of the proposed sequences, they have established themselves over the years as the most widely used and informative methods.

5.6.1. Which COSY approach?

Before moving on we briefly examine the key characteristics of the different experiments and consider why these might be of interest to the research chemist. Table 5.5 summarises the most significant attributes, some of which have already been introduced whilst others are expanded in the sections that follow. Whilst the TOCSY experiment is not strictly a member of the COSY family, its information content is so closely related to that of COSY it has also been included in the table.

Pre-empting what is to follow, some general recommendations can be made at this stage. The COSY experiments with the greatest information content are the phase-sensitive versions. For solving spectra of any complexity the most useful, by virtue of its high-resolution and informative crosspeak structure, is the phase-sensitive DQF-COSY. As described below, the theoretical loss in sensitivity is not as detrimental as it may at first sound, and this method is widely used in structure elucidation. To genuinely benefit from the information contained within crosspeak structures, it requires high digital resolution which invariably corresponds to longer experiments, and phasing of the resulting two-dimensional spectrum can be something of a black art to the uninitiated. For these reasons it is not so well suited to the rapid sample throughput demanded in busy organic laboratories. For establishing correlations in relatively well dispersed spectra, the absolute-value COSY offers faster turnover and simplicity of data handling so may be favoured when fully automated generation of spectra is required. The absolute-value COSY-β variant, where the mixing pulse β is typically 45 or 60°, provides the most compact peak structure for both diagonal and crosspeaks and is best suited for routine use. In the study of

Table 5.5. A summary of the characteristic features of the principal COSY experiments and of the related TOCSY experiment

Sequence	Advantages	Potential drawbacks
Absolute-value (magnitude-mode) COSY-90	Simple and robust, magnitude processing well suited to automated operation.	Phase-twisted lineshapes produce poor resolution which require strong resolution enhancement functions. Crosspeak fine structure not usually apparent.
Phase-sensitive COSY-90	High-resolution display due to absorptive lineshapes. Crosspeak fine structure apparent; J measurement possible.	Diagonal peaks have dispersive lineshapes which may interfere with neighbouring crosspeaks. Requires high digital resolution to reveal multiplet structures.
Phase-sensitive DQF-COSY	High-resolution display due to absorptive lineshapes. Crosspeak fine structure apparent; J measurement possible. Diagonals also have absorptive lineshapes. Singlets suppressed.	Theoretical sensitivity loss by a factor of 2 relative to the COSY-90 variant. Requires high digital resolution to reveal multiplet structures.
COSY-β	Simple and robust. Magnitude processing well suited to automated operation. Simplification of crosspeak structures reduces peak overlap. Vicinal and geminal couplings can be distinguished in some cases from tilt of peaks.	Usually requires magnitude-mode presentation as phase-sensitive variant has mixed-phase lineshapes.
Delayed-COSY	Enhances detection of small and long-range couplings (<2 Hz) such as between protons in allylic systems or those in *w*-relationships.	Requires magnitude-mode presentation. Crosspeaks due to larger couplings can be significantly attenuated.
Relayed-COSY	Provides two (or more) step transfers and can reduce ambiguities arising from crosspeak overlap.	Typically has low sensitivity and responses show mixture of lineshapes so magnitude-mode presentations may be required. TOCSY preferred.
TOCSY	Provides multistep (relayed) transfers to overcome ambiguities arising from crosspeak overlap. High sensitivity. In-phase lineshapes can provide correlations even in the presence of broad resonances.	Number of transfer steps associated with each crosspeak not known, *a priori*. In-phase lineshapes tend to mask crosspeak fine-structure and may preclude J-measurement.

complex spectra where considerable peak overlap occurs, the total correlation or TOCSY experiment can be extremely informative. It provides additional correlation information by relaying magnetisation along networks of coupled spins. This is particularly favoured in the analysis of peptides and oligosaccharides, for example, in which the molecules are comprised of discrete monomer units which themselves represent isolated spin networks. This method has all but replaced the older and inferior relayed-COSY experiment that elicits stepwise transfers between coupled spins. The delayed (or long-range) COSY experiment tends to be reserved for addressing specific questions regarding the presence of small couplings and is typically less used.

5.6.2. Double-quantum filtered COSY (DQF-COSY)

Previous sections have already made the case for acquiring COSY data such that it may be presented in the phase-sensitive mode. The pure-absorption lineshapes associated with this provide the highest possible resolution and allow one to extract information from the fine-structure within crosspeak multiplets. However, it was also pointed out that the basic COSY-90 sequence suffers from one serious drawback in that diagonal peaks possess dispersion-mode lineshapes when crosspeaks are phased into pure absorption-mode. The broad tails associated with these can mask crosspeaks that fall close to the diagonal, so there is potential for useful information to be lost. The presence of dispersive contributions to the diagonal may be (largely) overcome by the use of the double-quantum filtered variant of COSY [37], and for this reason DQF-COSY is the experiment of choice for recording phase-sensitive COSY data.

The DQF sequence

The DQF-COSY sequence (Fig. 5.40) differs from the basic COSY experiment by the addition of a third pulse and the use of a modified phase-cycle or gradient sequence to provide the desired selection. Thus, following t_1 frequency labelling, the second 90° pulse generates multiple-quantum coherence which is not observed in the COSY-90 sequence since it remains invisible to the detector. This may, however, be reconverted into single-quantum coherence by the application of the third pulse, and hence subsequently detected. The required phase-cycle or gradient combination selects only signals that existed as double-quantum coherence between the last two pulses, whilst all other routes are cancelled, hence the term double-quantum filtered COSY.

The rules for filtering multiple-quantum coherences of order p are simple; all pulses prior to the p-quantum coherences must be cycled in steps of 180/p degrees with associated alternation of the receiver phase on each step, [38] so a suitable phase-cycle (although not the only one) for a double-quantum filter

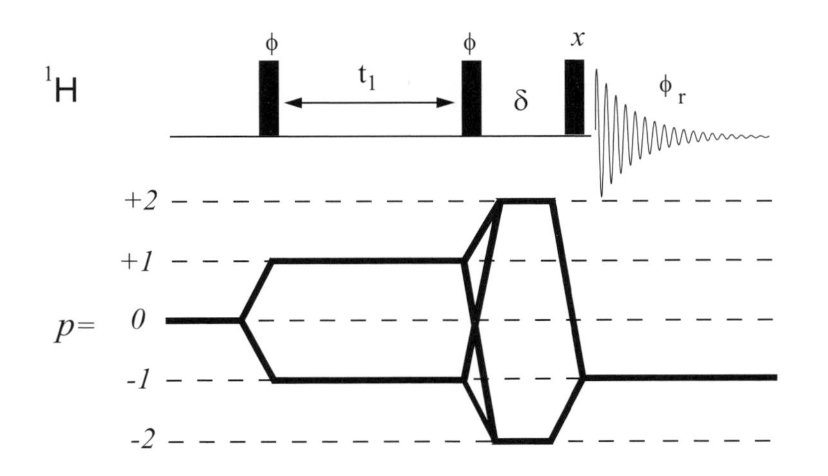

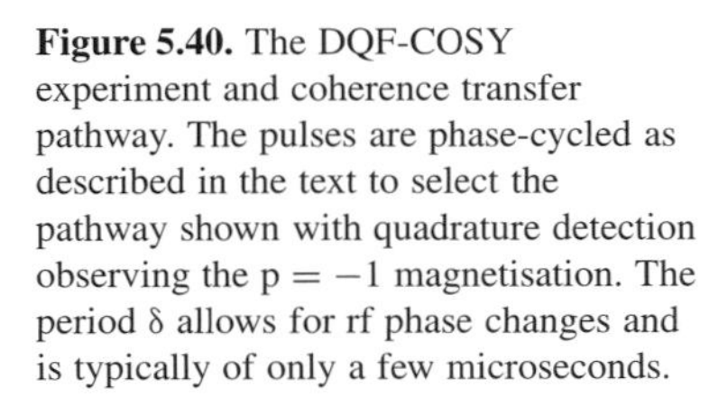
Figure 5.40. The DQF-COSY experiment and coherence transfer pathway. The pulses are phase-cycled as described in the text to select the pathway shown with quadrature detection observing the $p = -1$ magnetisation. The period δ allows for rf phase changes and is typically of only a few microseconds.

Table 5.6. A suitable phase-cycle for double-quantum filtration

ϕ	ϕ_r
x	x
y	$-x$
$-x$	x
$-y$	$-x$

Phase ϕ is used for all pulses prior to the double-quantum coherence and ϕ_r for the receiver. The phase of the final pulse remains unchanged.

would be that presented in Table 5.6. More effective filtration can be obtained through gradient selection [39]. Coherence order +2 within the filter may be selected by using gradient amplitude ratios of +1:+2 either side of the last 90° pulse, as in the example of Fig. 5.37 above, whilst order −2 is retained with ratios of +1:−2. Thus, it is possible to select either the +2 or −2 pathway, *but not both at once*. This is in contrast to the phase-cycled experiment in which both pathways are retained, so the gradient experiment detects only *one-half* of the available signal and thus has a two-fold poorer sensitivity relative to its phase-cycled cousin. This fundamental limitation of being able to refocus only one of the two possible $\pm p$ coherence order pathways on each transient is one of the disadvantages of using PFGs for coherence selection, but must be balanced against the cleaner suppression of unwanted signals. Since these gradients are not required within t_1 in this sequence, conventional quadrature detection (States or TPPI) can be employed. However, simply inserting the appropriate gradients before and after the final pulse would introduce large phase errors to the spectrum because of chemical shift evolution during the gradient pulses. This must therefore be refocused if phase-sensitive data are required so both gradients are applied within spin-echoes, producing the sequence of Fig. 5.41 for the gradient-selected, phase-sensitive DQF-COSY experiment.

The result from the filtration step, and the principal reason for its use, is that the diagonal peaks now possess antiphase absorption-mode lineshapes, as do the crosspeaks which are unaffected by the filtration. Strictly speaking, for spin systems of more than two spins the diagonal peaks still possess some dispersive contributions, but these are now antiphase so cancel and tend to be weak and rarely problematic. The severe tailing previously associated with diagonal peaks therefore is removed, providing a dramatic improvement in the quality of spectra (Fig. 5.42).

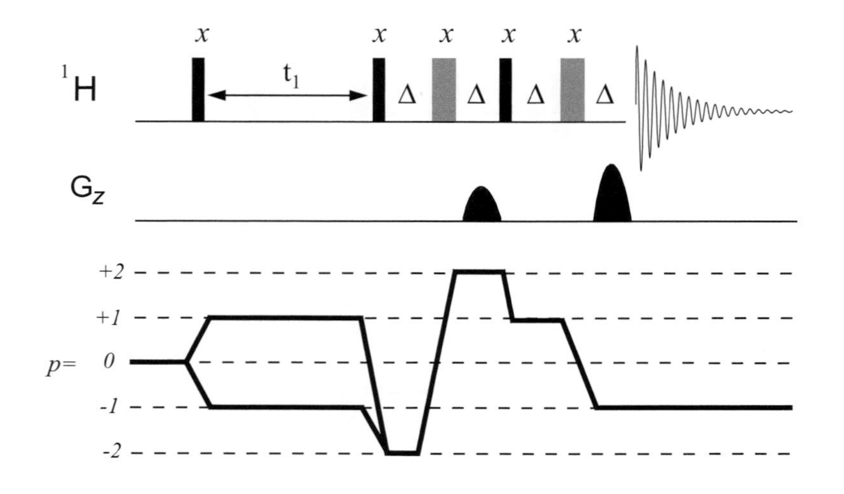

Figure 5.41. The gradient-selected DQF-COSY experiment and coherence transfer pathway. No phase-cycling is needed as the required pathway is selected with gradient ratios of 1:2. Both gradient pulses are applied within spin-echoes for phase-sensitive presentations. Note only one pathway is retained from the double-quantum filter.

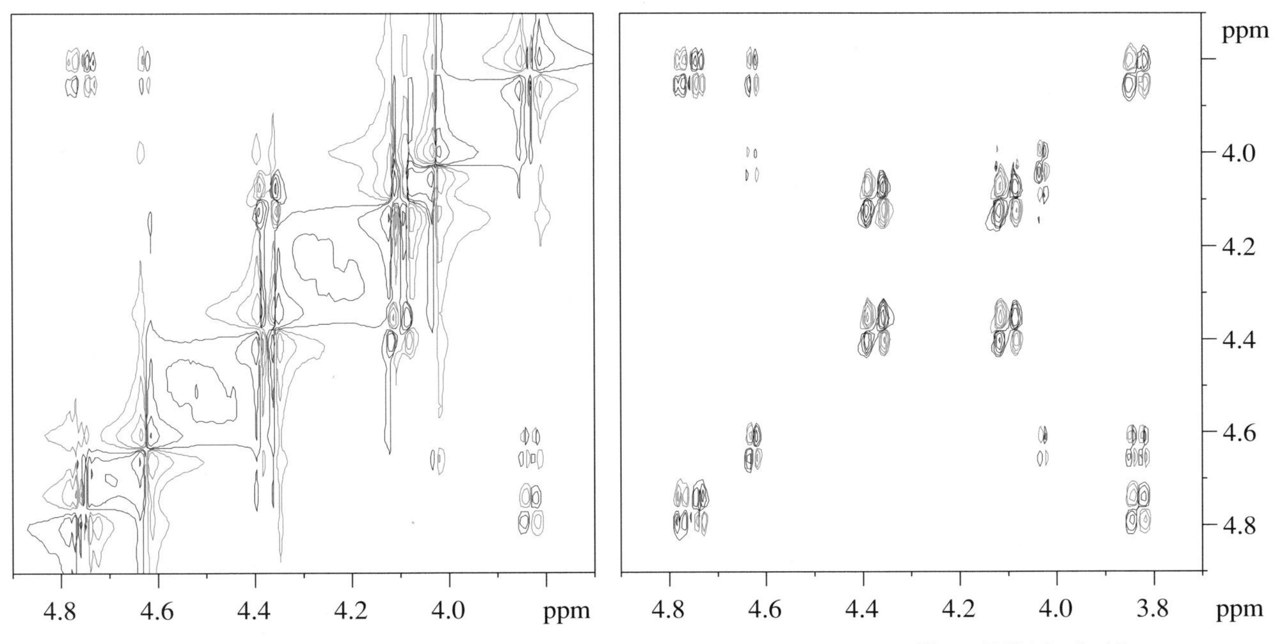

Figure 5.42. The double-quantum filtered COSY spectrum (right) provides greater clarity close to the diagonal peaks than the basic phase-sensitive COSY (left) as it does not suffer from broad, dispersive diagonal peaks.

An additional benefit from the filtration is that singlet resonances do not appear in the resulting spectrum because they are unable to create double-quantum coherence so cannot pass through the filter. Since sharp singlets produce the most intense t_1-noise bands their suppression alone can be beneficial. This filtration could also be used to suppress large solvent resonances that would otherwise dominate the spectrum. Whilst some success may be achieved through phase-cycling, with suppression ratios of the order of a few hundred, far greater suppression can be provided by gradient selection, where suppression ratios can reach 10,000:1. Editing through filtration is exemplified in the 1D double-quantum filtered spectrum of the peptide Leu-enkephalin **5.4** in CD_3OD (Fig. 5.43), produced with the 1D sequence of Fig. 5.44. The singlet solvent resonances are removed whereas those from all coupled spins are retained, and appear with the characteristic antiphase structure.

Tyr-Gly-Gly-Phe-Leu

5.4

The potential disadvantage of using double-quantum filtration is the theoret-

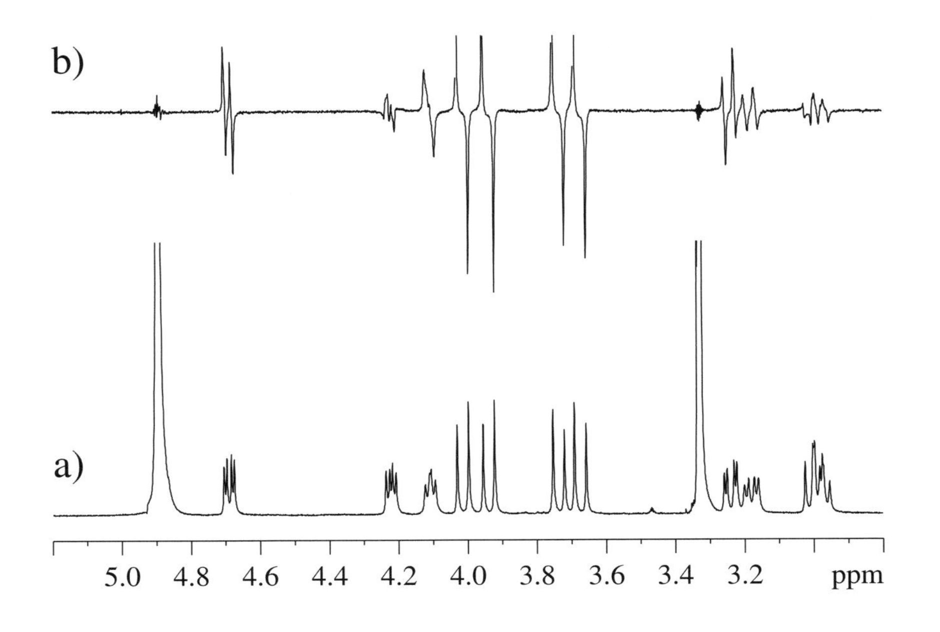

Figure 5.43. 1D double-quantum filtration of the spectrum of the peptide Leu-enkephalin **5.4** in CD_3OD. The singlet resonances of the solvent, truncated in the conventional 1D spectrum (a), have been filtered out in (b). The remaining peaks in (b) display the characteristic antiphase multiplet structure (which may be masked by magnitude calculation if desired).

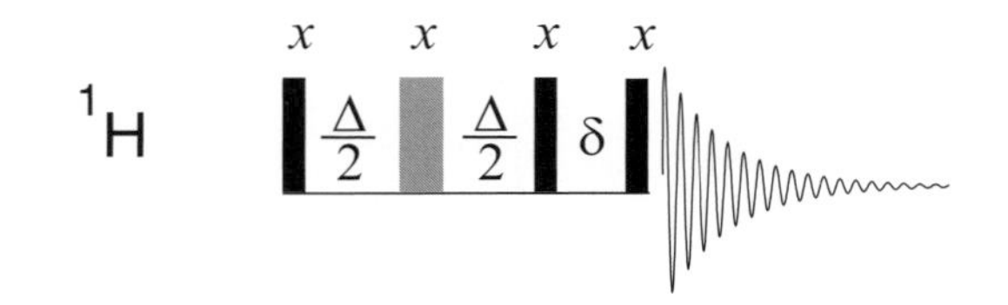

Figure 5.44. The 1D double-quantum filter. The sequence is derived from the 2D experiment by replacing the variable t_1 period with a fixed spin-echo optimised to produce antiphase vectors ($\Delta = 1/2J$) as required for the generation of double-quantum coherence. Signal selection is then as for the 2D experiment, and gradient selection may be implemented as in Fig. 5.41.

ical reduction in signal-to-noise by a factor of 2 due to losses associated with the generation of double-quantum coherence. However, the benefits arising from the removal of the dispersive contributions to the diagonal usually more than compensates the reduction in sensitivity, hence the DQF-COSY is widely employed.

Interpreting multiplet structure

A major part of 1D spectrum analysis consists of measuring coupling constants within multiplets to gain as much information as possible on relationships between spins within the molecule. Whilst this is at least plausible for well resolved multiplets the task becomes rapidly more difficult and the data obtained less reliable for resonances that overlap. In such cases it would be desirable to use the greater dispersion available in the 2D spectrum to reveal coupling patterns from otherwise intractable regions and it has already been shown that COSY crosspeak structure reflects that of the corresponding 1D multiplets. This section examines how one interprets the data within crosspeaks with the ultimate aim of extracting coupling constants.

To do this it is first is necessary to differentiate between active and passive couplings within a crosspeak. The *active* coupling is that which *gives rise to the crosspeak* that correlates the coupled spins. As described previously, the coupling responsible for the crosspeak appears as an antiphase splitting in both dimensions and so produces the basic antiphase square array from which all COSY crosspeaks are ultimately derived (Fig. 5.45). *Passive* couplings are those to *all other spins*, and appear in the spectrum as in-phase splittings. These

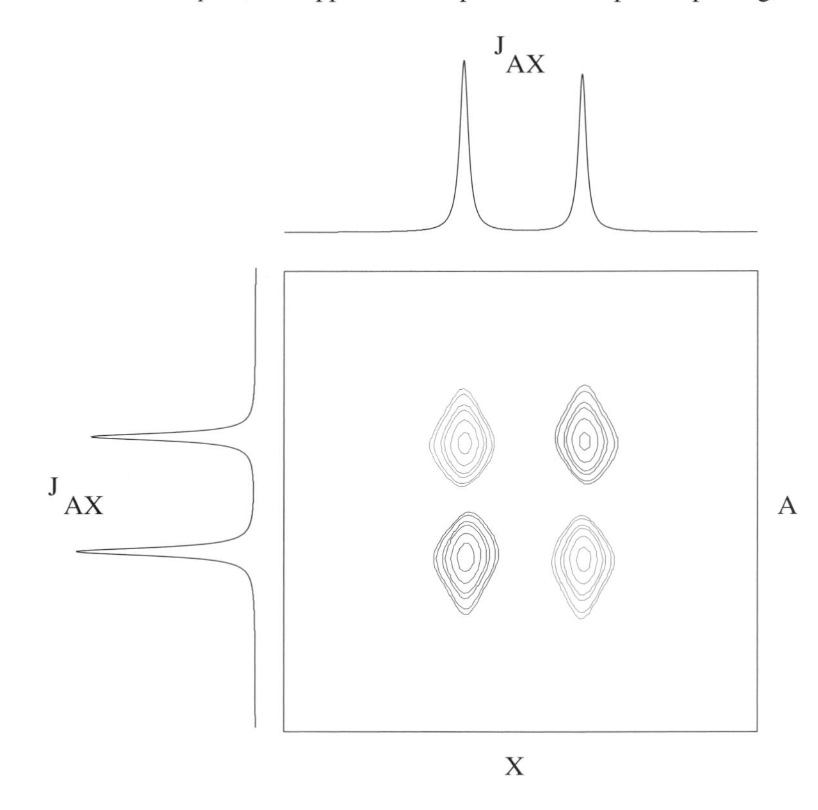

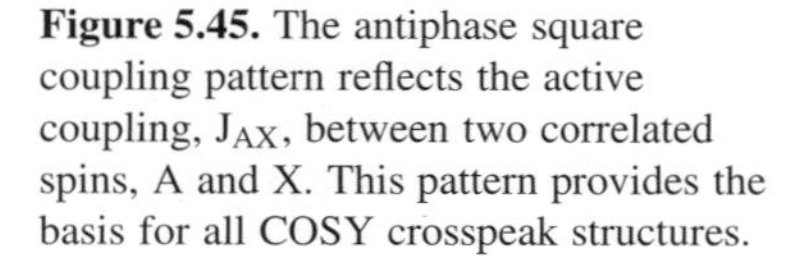
Figure 5.45. The antiphase square coupling pattern reflects the active coupling, J_{AX}, between two correlated spins, A and X. This pattern provides the basis for all COSY crosspeak structures.

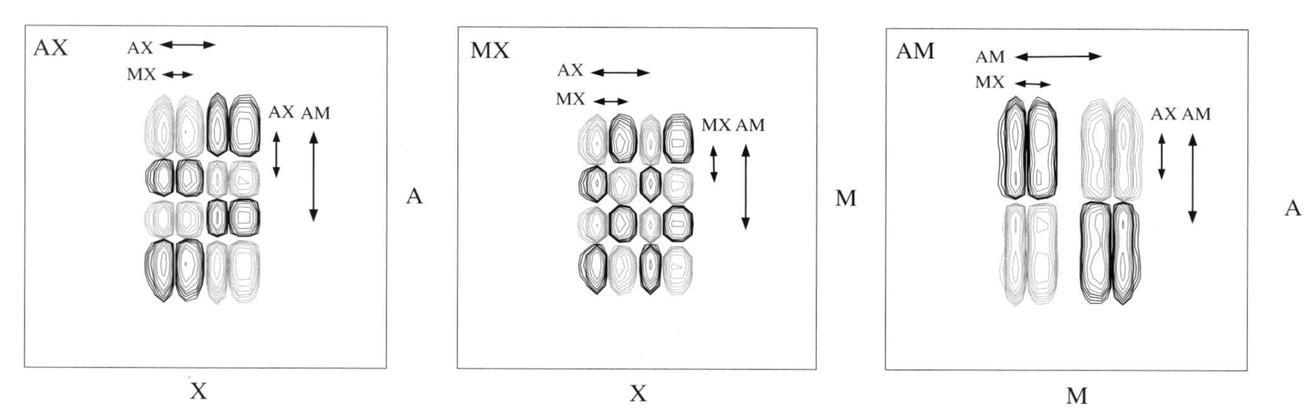

Figure 5.46. The crosspeak structures from the phase-sensitive COSY spectrum of a three-spin AMX system. The arrows indicate splittings due to the labelled couplings. The spectrum was simulated with $J_{AM} = 18$, $J_{AX} = 12$ and $J_{MX} = 6$ Hz and the final digital resolution was 2.3 Hz/pt in each dimension.

replicate the antiphase array of the active coupling without introducing further changes of sign. These features and the basic principles behind analysing multiplet structures may be illustrated by reference to the crosspeaks of a three-spin AMX system in which $J_{AM} > J_{AX} > J_{MX}$ (Fig. 5.46). Consider first the MX crosspeak, in which the antiphase square pattern of the active J_{MX} coupling is clearly apparent in the upper left corner. This whole pattern is then reproduced with the same phase by the larger passive couplings J_{AX} in f_2 and J_{AM} in f_1 to produce the final 4 x 4 structure. This stepwise splitting is the precise equivalent to that used in the analysis of 1D multiplets via coupling 'trees'. The f_2 structure of the AX multiplet arises from the large antiphase J_{AX} coupling being further split by the smaller passive J_{MX} coupling to yield the $++--$ structure. In f_1 the active coupling is now the smaller of the two so is simply repeated to yield the $+-+-$ pattern. Finally, the AM multiplet is dominated by the very large active J_{AM} coupling in both dimensions which initially yields a large square array, with each part then split by the smaller passive couplings to spin X. The rules for interpreting these structures therefore exactly parallel those for interpreting 1D multiplets, aside from the distinction between active and passive couplings. When coupling occurs between one spin and *n* equivalent spins, the resulting peak structure may be derived by imagining only one of the *n* spins to be active and all others passive. One can then construct the antiphase analogue of the familiar Pascal's triangle for coupling with equivalent spins (Fig. 5.47). Thus, for an A_3X system the crosspeak displays a doublet for A and a $+1+1-1-1$ quartet for X (Fig. 5.48).

In realistic systems multiplet structures may derive from very many couplings and resolution of all of these may not be possible. This leads to overlap within the multiplet with associated line cancellation and/or superposition. In general, crosspeak structures tend toward the simplest pattern as neighbouring lines with like-sign merge. Often in such cases smaller couplings may not be resolved, as is nearly the case for the passive MX and AX couplings within

n	(a)	(b)
0	1	1
1	1 1	1 -1
2	1 2 1	1 0 -1
3	1 3 3 1	1 1 -1 -1
4	1 4 6 4 1	1 2 0 -2 -1
5	1 5 10 10 5 1	1 3 2 -2 -3 -1
6	1 6 15 20 15 6 1	1 4 5 0 -5 -4 -1

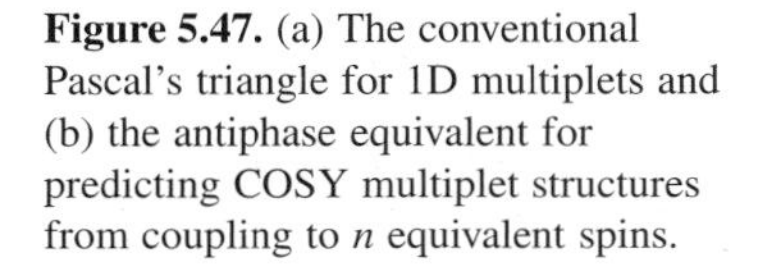

Figure 5.47. (a) The conventional Pascal's triangle for 1D multiplets and (b) the antiphase equivalent for predicting COSY multiplet structures from coupling to *n* equivalent spins.

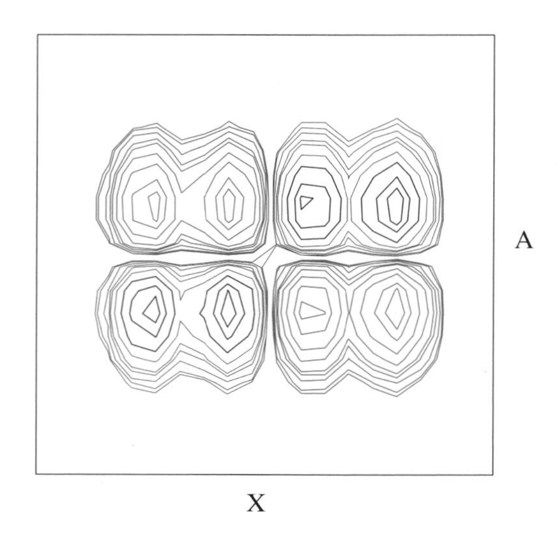

Figure 5.48. The COSY crosspeak for an A_3X group illustrating the structure predicted from Pascal's triangle.

the AM crosspeak of Fig. 5.46. When analysing fine-structure in detail it is often advantages to examine both sets of equivalent crosspeaks on either side of the diagonal since these may have differing appearances according to the different digitisation levels used, meaning a coupling may be unresolved in one dimension but quite apparent along the other.

Measuring coupling constants

Provided multiplets structures are sufficiently resolved and free from overlap with other peaks it is, in principle, possible to measure coupling constants directly from these [40]. Some caution is required however since one needs to consider whether cancellation within the multiplet could lead to an erroneous measurement. For example, if the central lines of a double-doublet were to interfere destructively and cancel, the measured splitting would then be greater than the true coupling constant. Digital resolution must also be adequate to properly characterise the coupling constant, with the highest resolution being found in the f_2 dimension. To improve the accuracy of the measurement, it is advantageous to extract 1D traces from the 2D spectrum and to subject these to *inverse* Fourier transformation, zero-filling (or better still, linear prediction) and then Fourier transformation to reproduce the 1D row with increased digital resolution. This process is illustrated in Fig. 5.49 for the αβ coupling constants in the tyrosine residue of the peptide Leu-enkephalin **5.4**. Above the 2D crosspeaks is the 1D trace through these which has been treated as described above and zero-filled to produce the same digital resolution as the conventional 1D spectrum shown on top (0.4 Hz/pt). From the antiphase and in-phase splittings the two active αβ couplings and the passive ββ coupling are readily measured as 6.1, 8.3 and 14.0 Hz.

The measurement of coupling constants in this manner is, however, limited by finite linewidths, which in 2D spectra of small to mid-sized molecules are usually dictated by the digital resolution of the spectrum rather than natural linewidths. Under such conditions, lines that are in-phase tend to merge together to produce a single maxima which may mask the coupling whereas antiphase lines remain separated but may produce apparent splittings that are greater than the true couplings. As coupling constants become smaller the splitting of antiphase lines in fact tends to a *minimum* of ≈0.6 times the linewidth [41] (Fig. 5.50), after which further reductions in the true separation act only to reduce peak intensities until complete cancellation occurs and the crosspeak is lost. One may conclude therefore that it is not possible to measure reliably coupling constants that are significantly smaller than the observed

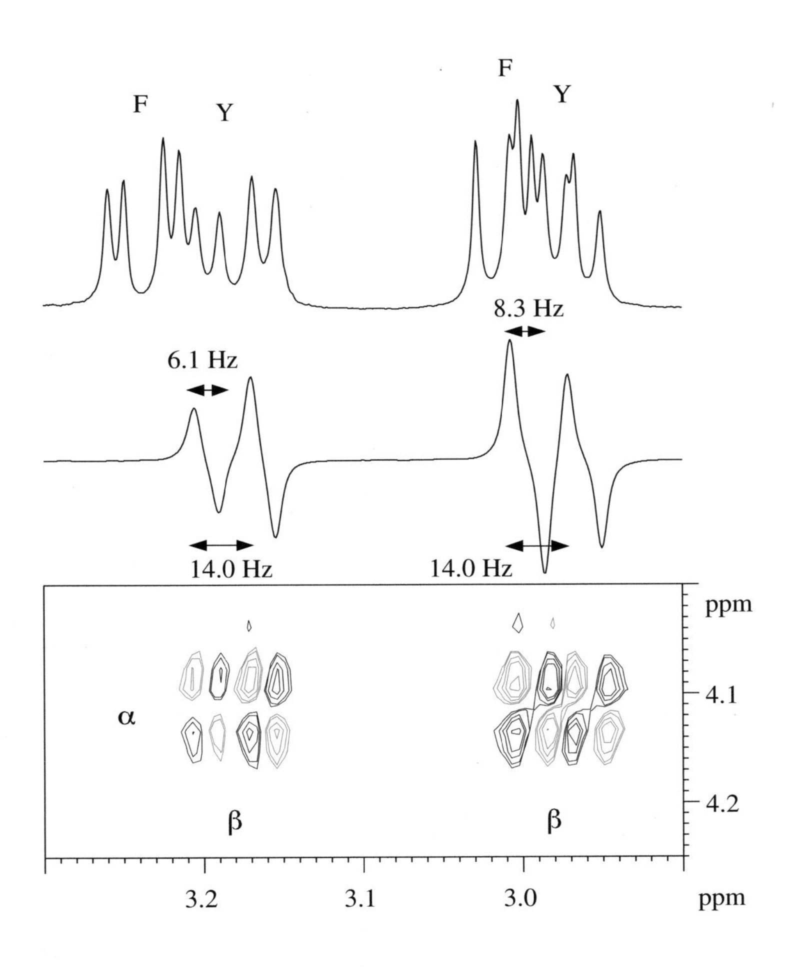

Figure 5.49. Sections from the DQF-COSY spectrum of the pentapeptide Leu-enkephalin **5.4**. The 2D crosspeaks and the f_2 1D trace taken through these correspond to correlations within the tyrosine (Y) residue. The upper trace is taken from the conventional 1D spectrum in which the β-proton resonances partially overlap with those of phenylalanine (F). The original 2D data had an f_2 resolution of 1.8 Hz/pt but the 1D trace was treated as described in the text to yield a final resolution of 0.4 Hz/pt.

linewidths after digitisation. A rule-of-thumb is that couplings should be greater than 1.5 times the *digitised* linewidth for measurements to be made and spectra recorded for the measurement of coupling constants are typically acquired with higher digital resolution in f_2 where one can afford to be profligate with data points. These considerations also indicate that crosspeaks will disappear from

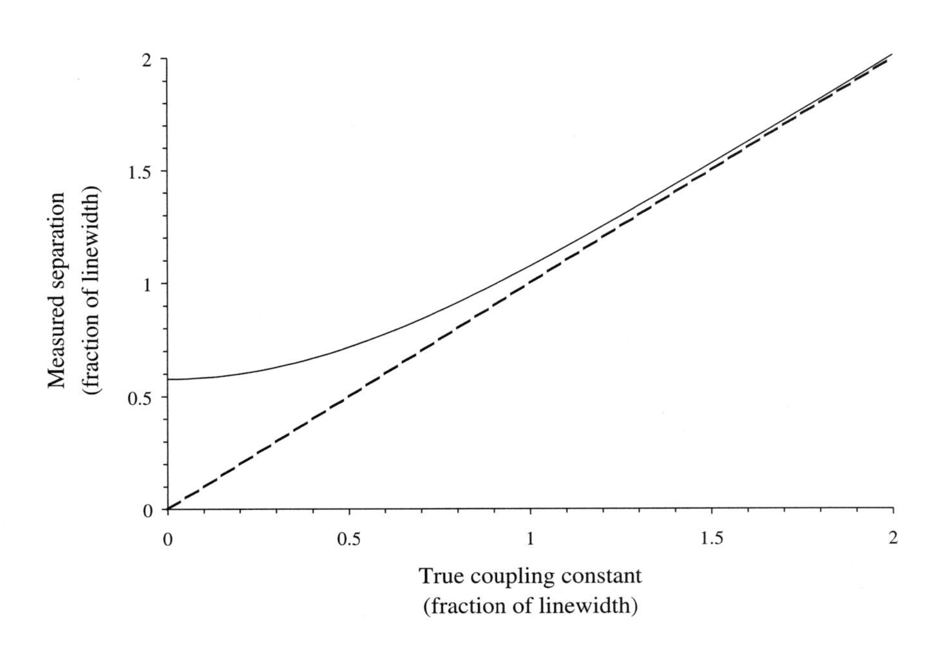

Figure 5.50. Peak separations of antiphase lines as a function of coupling constant and linewidth. As coupling constants become relatively smaller, the measured splitting (solid line) deviates from the true value (dashed line), producing measurements that are too large.

COSY spectra if linewidths are large or if digital resolution is low relative to the active coupling constant. Thus, crosspeaks due to small couplings or to broad lines are more likely to be missing. This can be a serious problem for very large molecules that have naturally broad resonances, and in such cases, the TOCSY experiment provides superior results since crosspeaks have only in-phase structure. More sophisticated methods also exist which enable the measurement of coupling constants in complex multiplets. For example, the DISCO method [42–44] of post-processing relies on the addition and subtraction of rows from COSY spectra to produce simplified traces containing fewer signals, thus making measurements more reliable. Details of the various methods available shall not be addressed here, but are discussed in a recent review [40].

Even when the data at hand precludes the measurement of coupling constants, analysis of the relative magnitudes of active and passive couplings within multiplets can prove enlightening when interpreting spectra. For example, it is possible in favourable cases to determine the relative configuration of protons within substituted cyclohexanes from consideration of the crosspeak structures. Since *axial–axial* couplings in chair conformers are typically far greater than *axial–equatorial* or *equatorial–equatorial* couplings (*ax–ax* $\approx$ 10–12 Hz, *ax–eq*/*eq–eq* $\approx$ 2–5 Hz), they appear as large antiphase splittings within crosspeak multiplets, indicative of the diaxial relationship between the correlated spins. This is illustrated in Fig. 5.51 which shows a region of the DQF-COSY spectrum of Andrographolide **5.5**, an hepato-protective agent found in traditional Indian herbal remedies. Three of the crosspeaks of proton H_C display large active couplings consistent with either geminal (H_d) or diaxial (H_a,H_e) relationships, whilst the much smaller active coupling to H_b limits these to being *ax–eq* or *eq–eq*. Such detail can be extracted from the 2D map even when multiplets are buried or are too complex for direct analysis in the 1D spectrum. For example, the correlations of H_d identify its geminal partner and indicate three vicinal *ax–eq* or *eq–eq* relationships with H_e, H_a and H_b. This type of information, which can only be determined reliably from phase-sensitive presentations, can often be used to good effect in stereochemical assignments, particularly when used in conjunction with the nuclear Overhauser effect.

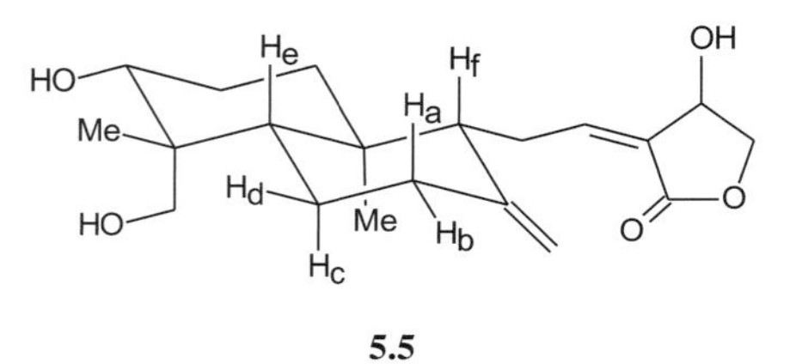

5.5

Higher order multiple-quantum filters

Using the same sequences as for DQF-COSY, but with modified phase-cycles or gradient combinations, it is possible to filter for higher orders of multiple-quantum coherence between the last two pulses, so, for example, a gradient ratio of 1:3 in the sequence of Fig. 5.41 selects triple-quantum coherence only. The triple-quantum filtered (TQF) COSY experiment eliminates responses from all singlets and two-spin systems leading to potential simplification. More generally, a p-quantum filter can be used to remove peaks arising from spin-systems with fewer than p coupled spins [38,45]. Such experiments have potential utility in the study of molecules containing well defined spin-systems that are isolated from each other in the molecule, notably amino-acids in peptides and proteins [46]. However, the great disadvantage of higher-order filtering is the reduction in sensitivity by a factor of 2^{p-1} for a p-quantum filter. Whilst a loss of a factor of 2 for the DQF-COSY relative to its unfiltered cousin

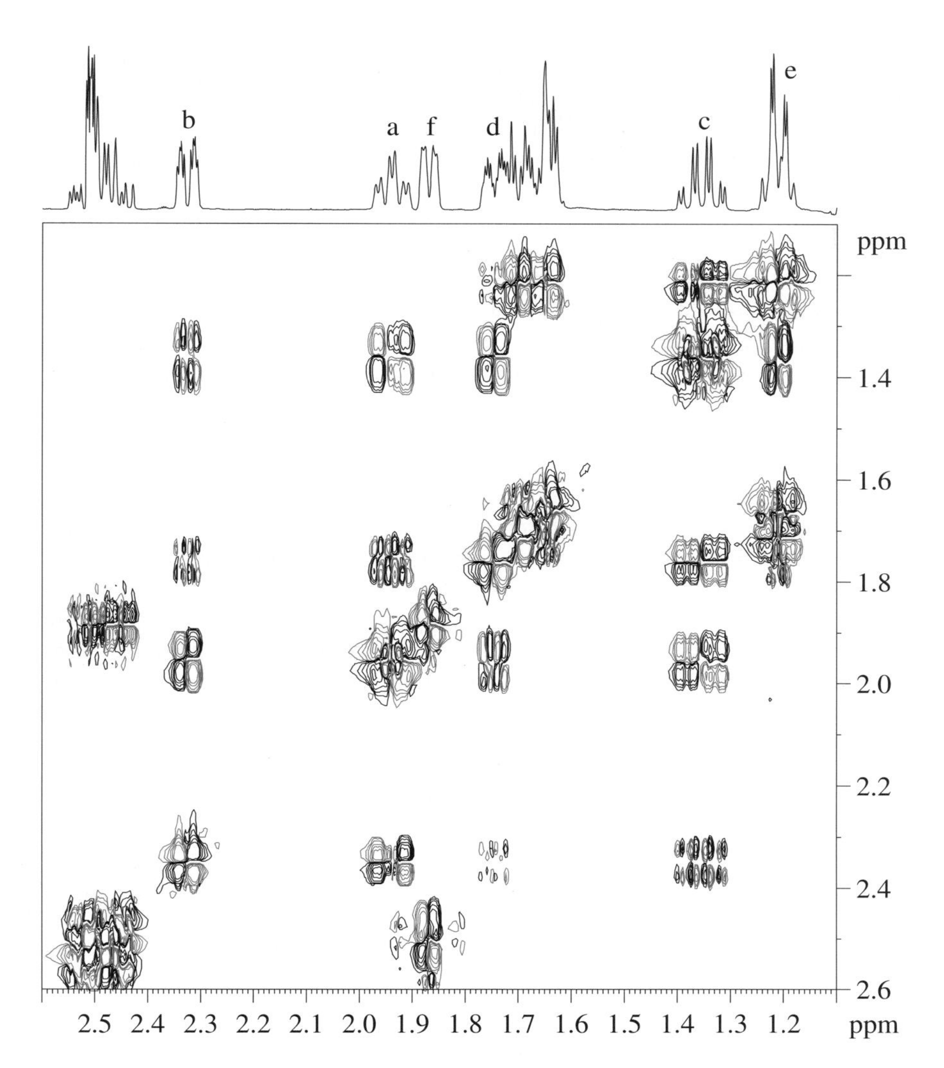

Figure 5.51. A region of the DQF-COSY spectrum of Andrographolide **5.5**. The data were collected under conditions of high f_2 resolution (1.7 Hz/pt) to reveal the coupling fine structure within the crosspeaks.

might be tolerable, a factor of 4 for the TQF-COSY may make it untenable. Largely for this reason the TQF-COSY is rarely used and higher-order filtering is essentially unheard of in routine work. The rules for interpreting multiplet fine-structure in COSY recorded with high-order filtration also require some modification [46,47].

5.6.3. COSY-β

A common modification of the basic COSY sequence is one in which the 90° mixing pulse is replaced with one of shorter tip angle, β, usually of 45 or 60 degrees (Fig. 5.52). These experiments are typically acquired and presented as absolute-value experiments since, strictly speaking, the use of a pulse angle less than 90° does not produce purely amplitude-modulated data.

The use of a reduced mixing pulse largely restricts coherence transfer between transitions that are *directly* connected, or in other words, those

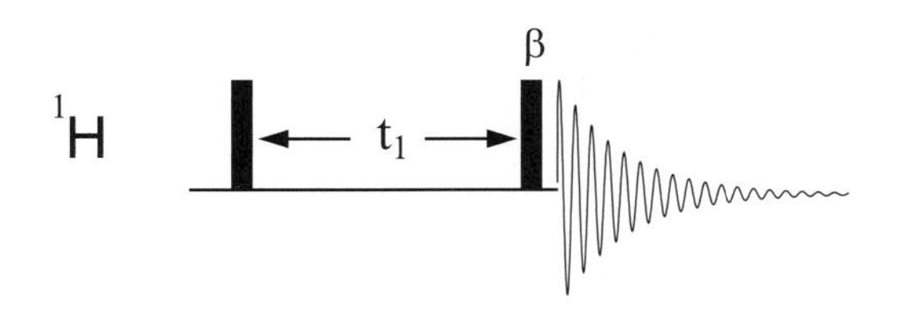

Figure 5.52. The COSY-β experiment. The β mixing pulse is usually set to 45 or 60°.

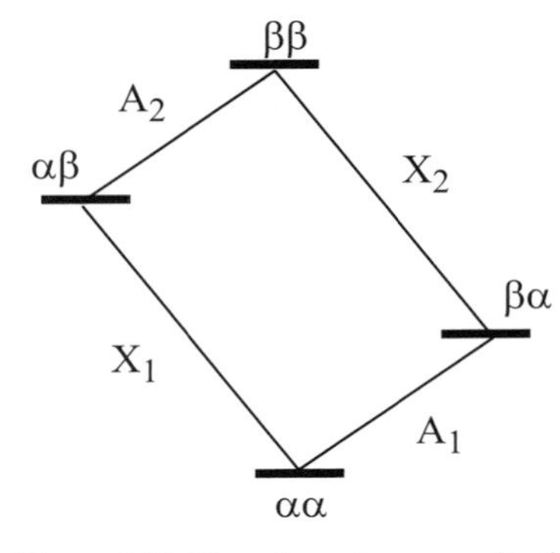

Figure 5.53. The schematic energy level diagram for a two-spin AX system.

that share an energy level, for example A_1 and X_2 in Fig. 5.53. Coherence transfer between *remotely* connected transitions, for example A_1 and A_2, is attenuated. This results in a reduction in intensity of certain lines *within* the multiplet structures of both diagonal and crosspeaks, which is particularly noticeable when spins experience more than one coupling. Diagonal peaks have a somewhat smaller 'footprint', increasing the chances of resolving crosspeaks that sit close to this, whilst crosspeaks themselves also have a simplified structure which can take on a distinctive 'tilted' appearance (Fig. 5.54). In favourable cases, this tilting effect can be used to differentiate between peaks arising from active couplings of opposite sign, such as geminal or vicinal couplings. Consider the situation for a three-spin AMX system in which all spins couple to each other. The AM crosspeak, for example, will have a positive slope and its tilt will be approximately parallel to the diagonal if the two *passive* couplings, J_{AX} and J_{MX} have the *same* sign. In contrast, the peak will display a negative slope and lie anti-parallel to the diagonal if the signs of the passive couplings differ. In proton spectroscopy, vicinal couplings are typically positive and geminal couplings negative, meaning a crosspeak between a proton and its *geminal* partner will possess a *positive slope* whereas that with the vicinal partner will display a *negative slope*. This effect is apparent in the COSY-45 spectrum of **5.1** in Fig. 5.54. Couplings to additional spins further split the crosspeak structure and in such cases the tilting may not be distinguished. Nevertheless, when this effect is observed it provides valuable additional evidence by identifying geminal pairs.

The choice between $\beta = 45°$ and $60°$ is something of personal taste. Setting $\beta = 45°$ causes the greatest intensity ratio between direct and remote transitions (a factor of about 6 for 45° versus 3 for 60°) and thus gives the greatest reduction in peak complexity, whilst setting $\beta = 60°$ provides slightly higher sensitivity. Since the primary goal of this variant is the reduction of

Figure 5.54. The absolute-value COSY-45 spectrum of **5.1** compared with the equivalent COSY-90 spectrum. The tilting apparent on some COSY-45 crosspeaks can indicate whether the active coupling arises from protons sharing geminal or vicinal relationships which produce peaks with positive and negative slopes respectively (see text).

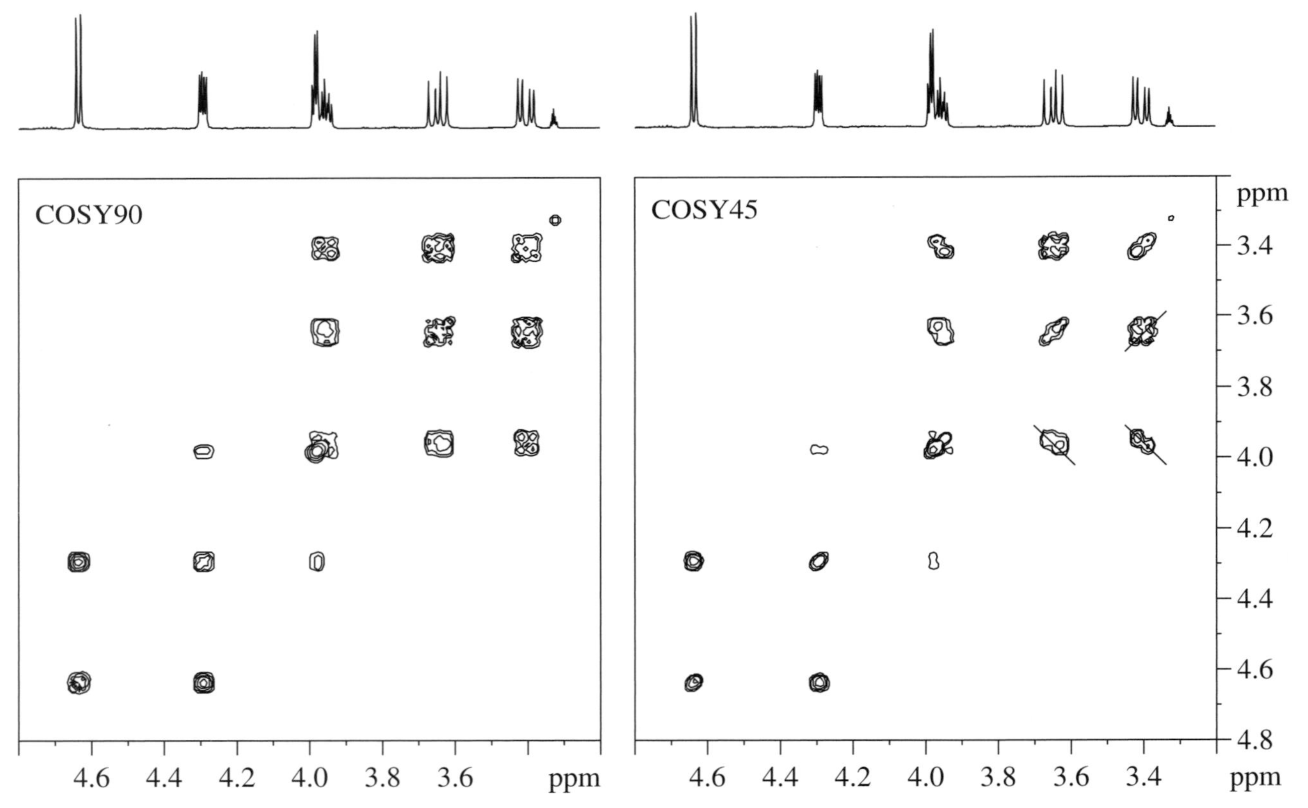

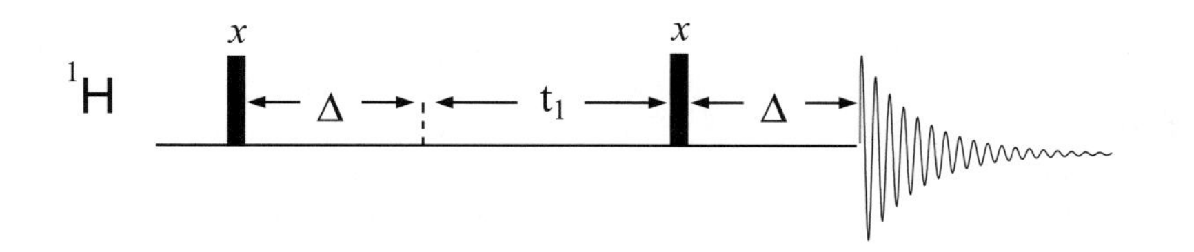

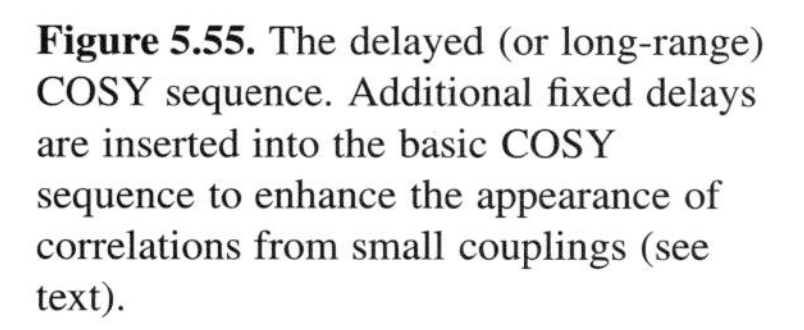

Figure 5.55. The delayed (or long-range) COSY sequence. Additional fixed delays are inserted into the basic COSY sequence to enhance the appearance of correlations from small couplings (see text).

the diagonal and crosspeak footprint the COSY-45 experiment is probably the one to choose. Due to the more compact peak structures and simplicity of processing, the absolute-value COSY-45 experiment still finds widespread use in routine analyses and automated procedures in the chemistry laboratory.

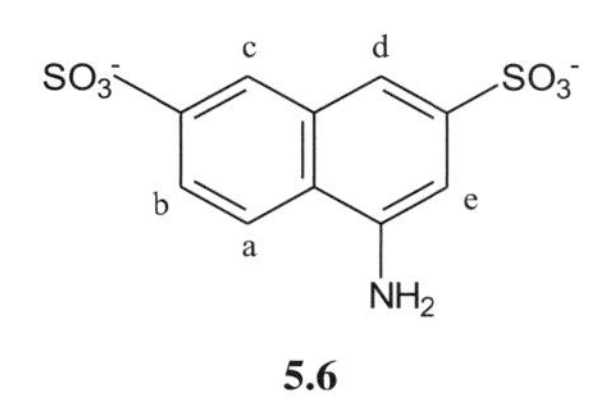

5.6.4. Delayed-COSY: detecting small couplings

The COSY sequences encountered thus far are surprisingly effective at revealing small couplings between protons. However, such crosspeaks are typically rather weak, and their identification as genuine correlations may be difficult. The delayed-COSY experiment (Fig. 5.55) enhances the intensity of correlations due to couplings that are smaller than natural linewidths and, since these are often encountered for coupling pathways over many bonds, the experiment is also referred to as *long-range-COSY*. Correlations over 4 or 5 bonds in rigid or unsaturated systems, such as those from *w*-, allylic or homoallylic couplings, can be observed readily with this experiment even though the coupling may not be resolved in the 1D spectrum (Fig. 5.56).

Previous discussions have stressed the need for an antiphase disposition between multiplet vectors for coherence transfer to occur. In COSY, this arises

Figure 5.56. (a) The conventional COSY-90 spectrum and (b) the delayed-COSY spectrum of **5.6**. Additional Δ delays of 200 ms were used in (b) whilst all other parameters were as for (a). The small (1 Hz) long-range couplings are not apparent in the conventional COSY experiment but the correlations in (b) unambiguously provide proton assignments.

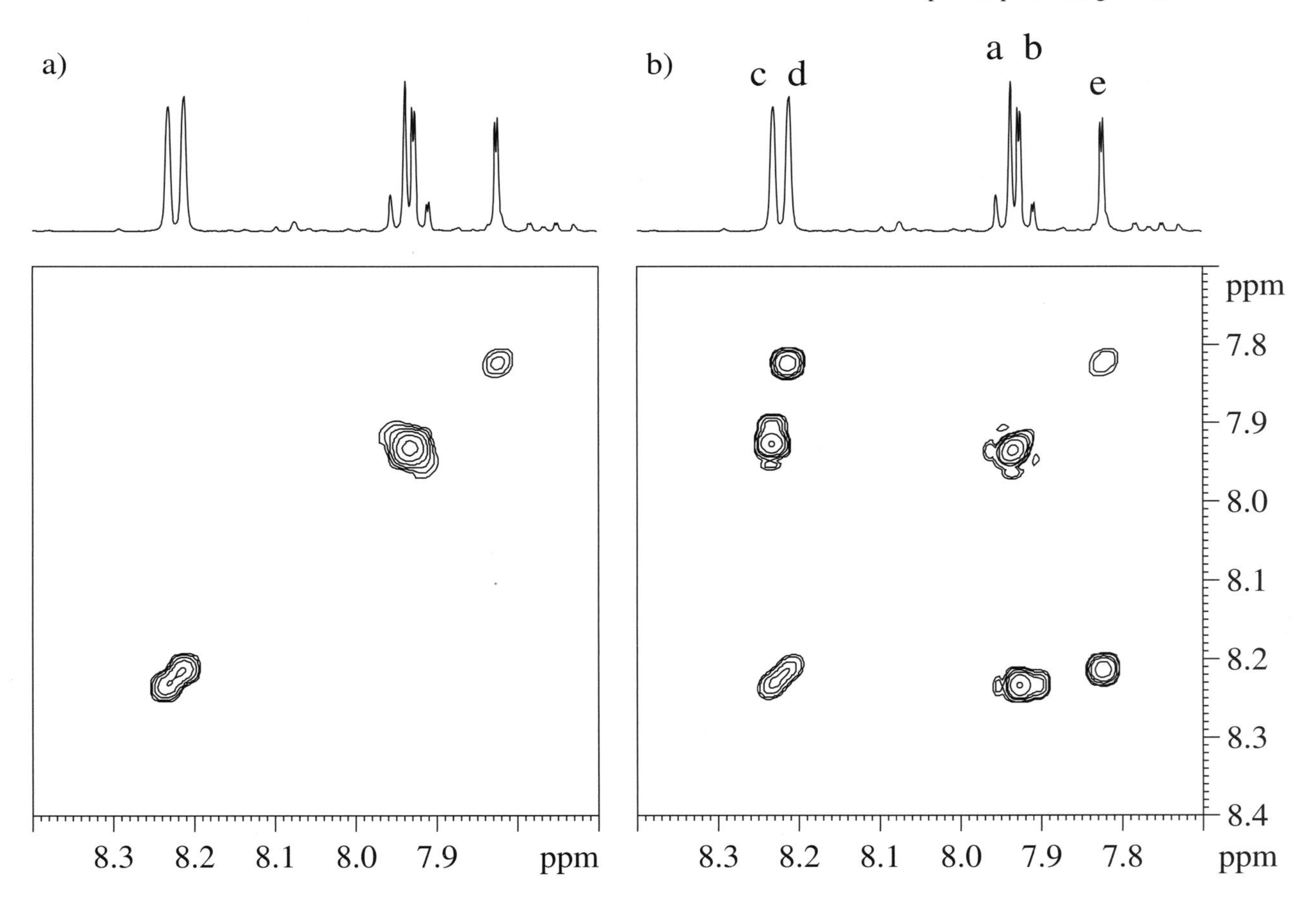

from the evolution of spin–spin couplings during t_1 and develops at a slower rate the smaller the coupling. For the case of very small coupling constants there is insufficient time for this to occur under the typical conditions employed for COSY acquisitions and hence the crosspeaks from these couplings are weak or undetectable. It is possible to enhance the intensity of these peaks by adjusting the t_1 and t_2 domains such that they are optimised for detection of the coherence transfer signal from these smaller couplings. Whilst this could in principle be achieved by extending both time periods and hence increasing the digital resolution in both dimensions, this generates enormous data sets and calls for protracted experiments. The approach used in the delayed-COSY sequence is to insert a *fixed* delay after each pulse to increase the *apparent* evolution times yet retain the same level of digitisation as in a standard COSY experiment. The use of the additional delays precludes the use of phase-sensitive presentations because of the large phase distortions that arise so absolute-value processing is necessary.

For optimum detection of small couplings, the maximum of the coherence transfer signals should occur at the *midpoint* of each time domain. It may be shown that this maximum arises at a time t, where:

$$t = \frac{\tan^{-1}(\pi J T_2)}{\pi J} \tag{5.10}$$

When J is small such that $JT_2 \ll 1$ the approximation $t = T_2$ holds and to ensure that the midpoint of each time domain coincides with this maximum, the additional delay Δ is therefore:

$$\Delta = T_2 - {}^1\!/\!{}_2 AQ \tag{5.11}$$

where AQ is the total t_1 or t_2 acquisition time. For typical COSY acquisition parameters and typical proton T_2 values, Δ falls in the range 50 to 500 ms. Since the actual values of T_2 are not usually known in advance, and there will in any case exist a spread within the molecule, a compromise setting of around 200 ms should suffice in most cases. The crosspeaks arising from couplings of greater magnitude can be severely attenuated in this experiment, so it is invariably also necessary to record the standard COSY to establish the usual coupling relationships.

5.6.5. Relayed-COSY

The relayed-COSY experiment shall be considered only very briefly because it has essentially been superseded by the far superior TOCSY experiment described below, although one may still encounter references to this experiment in older literature however. The relayed-COSY experiment (Fig. 5.57) attempts to overcome problems caused by coincidental overlap of crosspeaks in COSY that can lead to a breakdown in the stepwise tracing of coupling networks within a molecule. It incorporates an additional coherence transfer step in which that transferred from a spin, A, to its partner, M, as in the standard COSY, is subsequently relayed onto the next coupled spin X in the sequence. This produces a crosspeak in the spectrum between spins A and X *even though there exists no direct coupling between them*, by virtue of them sharing

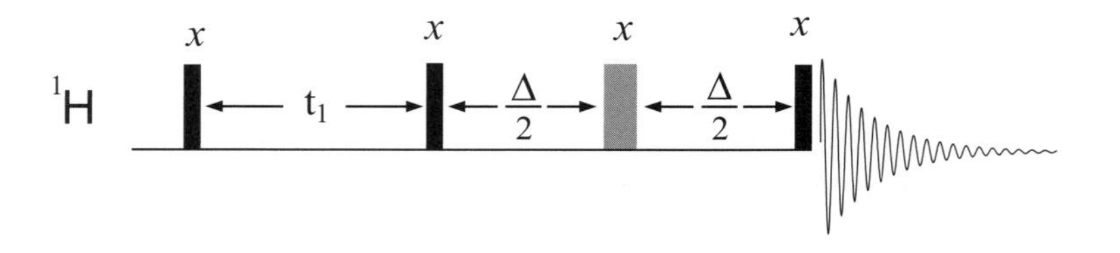

Figure 5.57. The relayed-COSY sequence in which an additional transfer step has been appended to the basic COSY experiment.

the common coupling partner M. Ambiguities from coincidental overlap of crosspeaks associated with spin M are therefore removed by providing direct evidence for A and X existing within the same spin system.

The relayed coherence transfer [48] is achieved by inserting a fixed time period, Δ, after the second pulse, again in the form of a spin-echo to remove chemical shift effects. During this the M–X coupling evolves to produce antiphase multiplet vectors (the MX coupling was passive in the initial A–M transfer so these vectors are in-phase after the second 90° pulse). The third 90° pulse then elicits the M–X coherence transfer prior to detection. The fixed delay Δ should be set optimally to 1/2J to maximise this transfer, so a compromise setting of J = 7 Hz gives Δ = 70 ms. The direct responses observed in the usual COSY experiment may also be present, but with varying intensities.

5.7. TOTAL CORRELATION SPECTROSCOPY (TOCSY)

The principal feature of all the COSY experiments described above is the direct correlation of homonuclear spins that share a scalar coupling. In proton spectroscopy this typically provides the chemist with evidence for geminal and vicinal relationships between protons within a molecule. Total Correlation Spectroscopy [49] or TOCSY, also yields homonuclear proton correlation spectra based scalar couplings, but is also able to establish correlations between protons that sit within the *same spin-system*, regardless of whether they are themselves coupled to one another. In other words, provided there is a continuous chain of spin–spin coupled protons, A–B–C–D– etc., the TOCSY sequence transfers magnetisation of spin A onto spins B, C, D etc, by relaying coherence from one proton to the next along the chain. In principle, this can correlate all protons within a spin-system and this feature earns the title *total* correlation spectroscopy. The ability to relay magnetisation in this manner provides an enormously powerful means of mapping correlations by making even greater use of the additional dispersion found in two dimensions. This is particularly advantageous in cases of severe resonance overlap, for which COSY spectra can often leave ambiguities. The TOCSY spectrum for **5.1** is shown in Fig. 5.58, and has been recorded so as to provide multistep transfers around the carbohydrate ring. This should be compared with the COSY spectrum of the same material in Fig. 5.12. The additional relayed crosspeaks seen in TOCSY provide correlations between *all* spins within the ring system and this ability to fully map spin-systems is particularly advantageous when discrete units exist within a molecule, as illustrated shortly. A second significant feature of TOCSY that contrasts with COSY is that it utilises the net transfer of *in-phase* magnetisation so does not suffer from cancellation of antiphase peaks under conditions of low digital resolution or large linewidths. In these instances, this feature makes TOCSY the more sensitive of these two methods.

An essentially identical experiment has also been referred to as Homonuclear Hartmann–Hahn spectroscopy [50,51] or HOHAHA (the two differ only in some technical details in the originally published sequences). This name arises from its similarity with methods used in solid-state NMR spectroscopy for the transfer of polarisation from proton to carbon nuclei (so-called cross-polarisation), which are based on the Hartmann–Hahn match described below. For the same reason, the transfer of magnetisation during the TOCSY sequence is sometimes referred to as *homonuclear cross-polarisation*. Throughout this text the original TOCSY terminology is used, although TOCSY and HOHAHA are now used synonymously in the chemical literature.

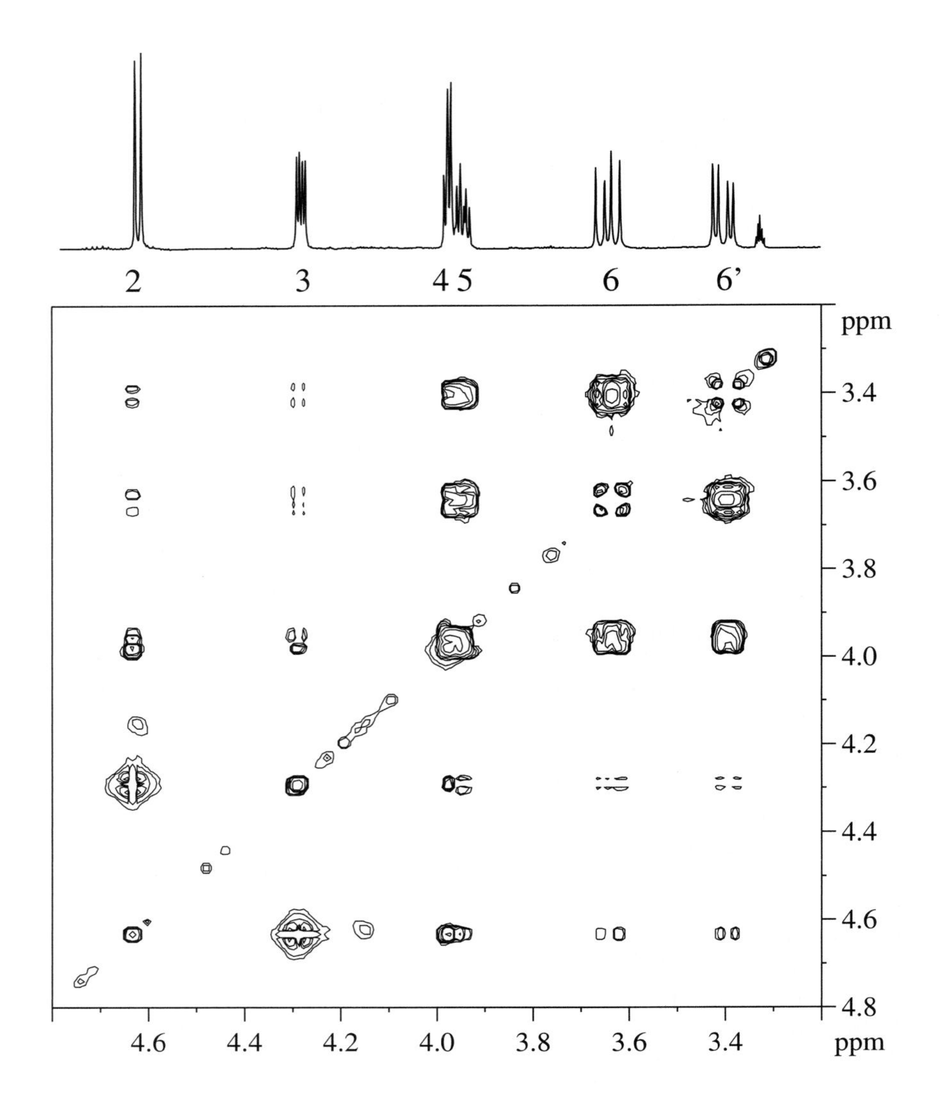

Figure 5.58. The TOCSY spectrum of **5.1**. Relayed crosspeaks are apparent between spins that lack direct scalar couplings but which exist within the same spin-system and arise from the propagation of magnetisation along the chain of coupled spins. Compare this with the COSY spectrum in Fig. 5.12.

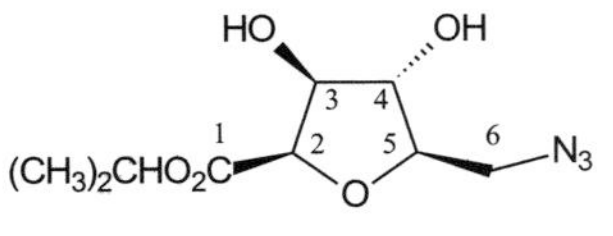

HO, OH, $(CH_3)_2CHO_2C$, N_3

5.1

5.7.1. The TOCSY sequence

The TOCSY sequence, represented in its most general form in Fig. 5.59, is rather similar to the COSY sequence already described, the only difference being the use of a mixing *sequence* in place of a single mixing pulse. This is known as a *spin-lock* or *isotropic mixing* sequence (this terminology is explained below) and its purpose is to execute the relayed magnetisation transfer mentioned above. Details such as quadrature detection and axial peak suppression parallel those for COSY, so to understand the operation of the TOCSY experiment it only becomes necessary to appreciate the influence of the spin-lock.

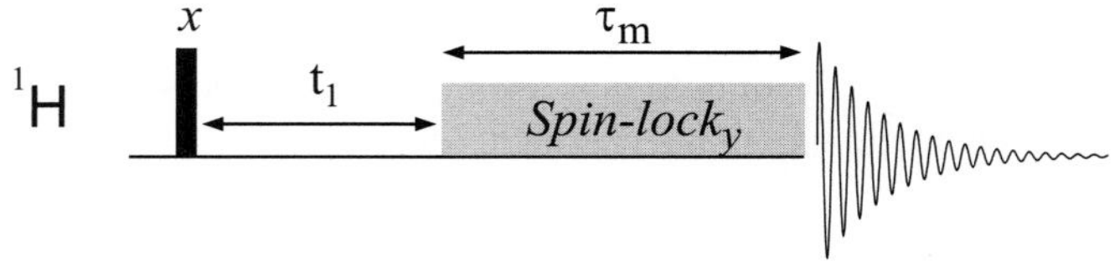

^1H — x — t_1 — τ_m — *Spin-lock$_y$*

Figure 5.59. The TOCSY sequence. The spin-lock mixing time, τ_m, replaces the single mixing pulse of the basic COSY experiment.

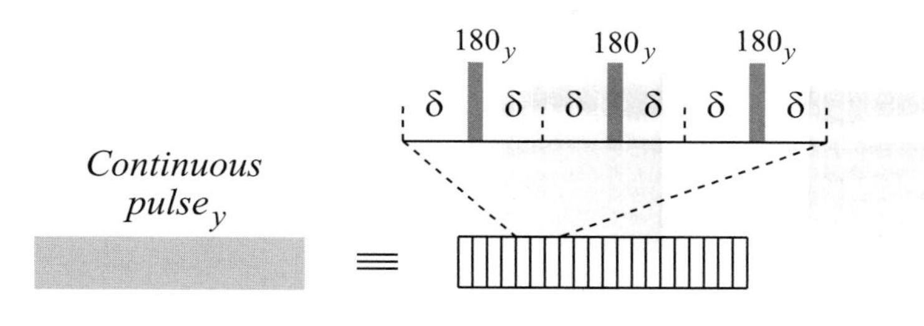

Figure 5.60. The spin-lock in its simplest form is a single, long, low-power pulse. This can be viewed as a continuous sequence of closely spaced 180° pulses bracketed by infinitely small periods, δ.

The spin-lock and coherence transfer

To understand the action of the spin-lock period in the TOCSY experiment, consider the sequence of events during the first transient of the 2D experiment (i.e. that with $t_1 = 0$). The experiment begins by exciting the spins with a $90°_x$ pulse such that they all lie along the $+y$ axis in the rotating frame. At this point the spin-lock rf field is applied, this time along the $+y$ axis, parallel to the nuclear vectors. In its simplest form the spin-lock is a continuous low-power pulse of constant phase applied for a period of typically tens of milliseconds. It is convenient to imagine this to be composed of a continuous sequence of closely-spaced $180°_y$ pulses bracketed by infinitely small periods, δ (Fig. 5.60). Each δ–180–δ period constitutes nothing more than the homonuclear spin-echo described in Chapter 2, and to picture the evolution of chemical shifts and homonuclear spin–spin couplings during the spin-lock it is sufficient to consider events within a discrete δ–180–δ period. It has already been shown in Section 2.2.4 that a 180° pulse refocuses the evolution of chemical shifts, so after each δ–180–δ period no shift evolution has accrued and all spin vectors remain along the $+y$ axis. Extending this argument for the whole of the mixing period, it is apparent that no net chemical shift evolution occurs during τ_m, and hence the nuclear vectors are said to have been *spin-locked* in the rotating frame along, in this case, the y-axis (Fig. 5.61). During the spin-lock mixing all protons have experienced the same effective field and hence the same *chemical shift offset (i.e. zero)* in the rotating frame. Section 2.2.4 has also shown that, in contrast, *homonuclear* spin–spin couplings *continue* to evolve following a 180° pulse, so that throughout the spin-lock J-couplings behave as they would for a period of free precession. In summary, during the spin-lock mixing all chemical shift differences in the rotating frame are removed yet spin–spin couplings remain active (Fig. 5.62).

Recall from your early encounters with NMR that nuclei which share a coupling and have very similar (or coincident) chemical shifts relative to their coupling constant are said to be *strongly coupled* and that under such conditions they loose their unique identity and ultimately become indistinguishable. Thus, for a system containing a coupled AB pair, it is not possible to consider the interactions of spin A independently from those of its strongly coupled neighbour B and vice versa. Such a strong coupling condition is *forced* upon all

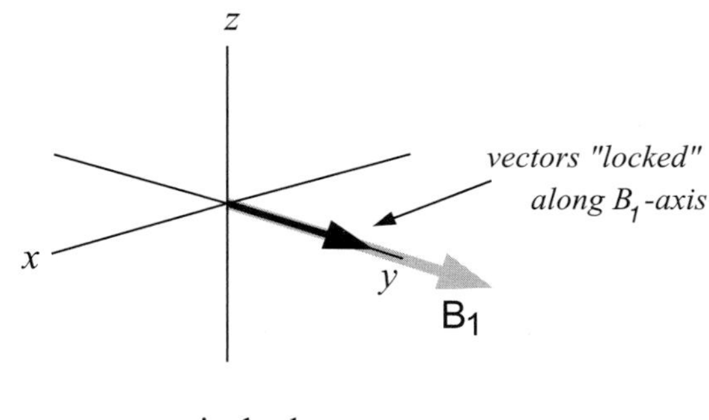

Figure 5.61. Vectors are held along the B_1 axis by the long rf pulse and are said to be *spin-locked* in the rotating frame.

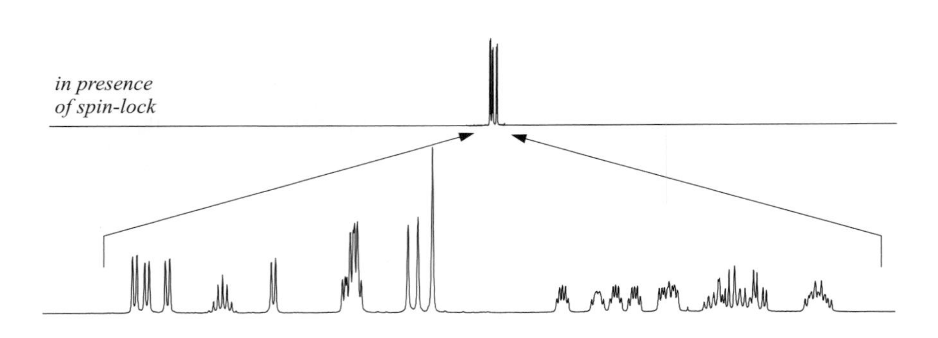

Figure 5.62. A schematic illustration of events during spin-lock mixing. All chemical shift differences between spins are eliminated yet all spin–spin couplings between them remain. This forces the *strong-coupling* condition on all spins (see text).

spin systems by the application of the spin-lock field since the protons remain coupled yet all posses the same effective chemical shift. The protons therefore loose their 'unique' identity during the mixing period and this provides the mechanism by which coherence may be shared over all spins within the same spin-system. Under these conditions there exists an oscillatory exchange of coherence between protons during the spin-lock which, for an AX system, results in complete transfer from A to X after a period of $1/2J_{AX}$ seconds [49], and a return to A after $1/J$ seconds (Fig. 5.63). Although events become more complex for larger systems, the general idea of an oscillatory exchange of coherence between spins still holds, leading to a propagation of magnetisation along the chain of coupled spins [52] (Fig. 5.64). For short mixing times in proton experiments of around 20 ms, only single-step transfers have significant intensity and the correlations seen are equivalent to those seen in COSY. Longer mixing periods enable magnetisation to propagate further along the coupled chain and relay peaks arising from multi-step transfers appear, as seen in Fig. 5.58 above. Small couplings lead to poor transfer efficiencies and can lead to a breakdown in the transfer and to loss of relay peaks. Since magnetisation may travel in either direction along a spin-chain, 2D TOCSY spectra are again symmetrical about the diagonal.

The requirement for nuclei to experience identical local fields during the mixing time for transfer to occur between them is also referred to as the *Hartmann–Hahn match*. More formally and more generally, this may be

Figure 5.63. The oscillatory transfer of magnetisation between a coupled proton pair under the influence of a spin-lock. The transfer was simulated using an MLEV-17 spin-lock (see text) of increasing duration in each experiment. In each case this was applied immediately after the selective excitation of the high frequency doublet with a pure-phase EBURP-2 selective 90° pulse (Chapter 9).

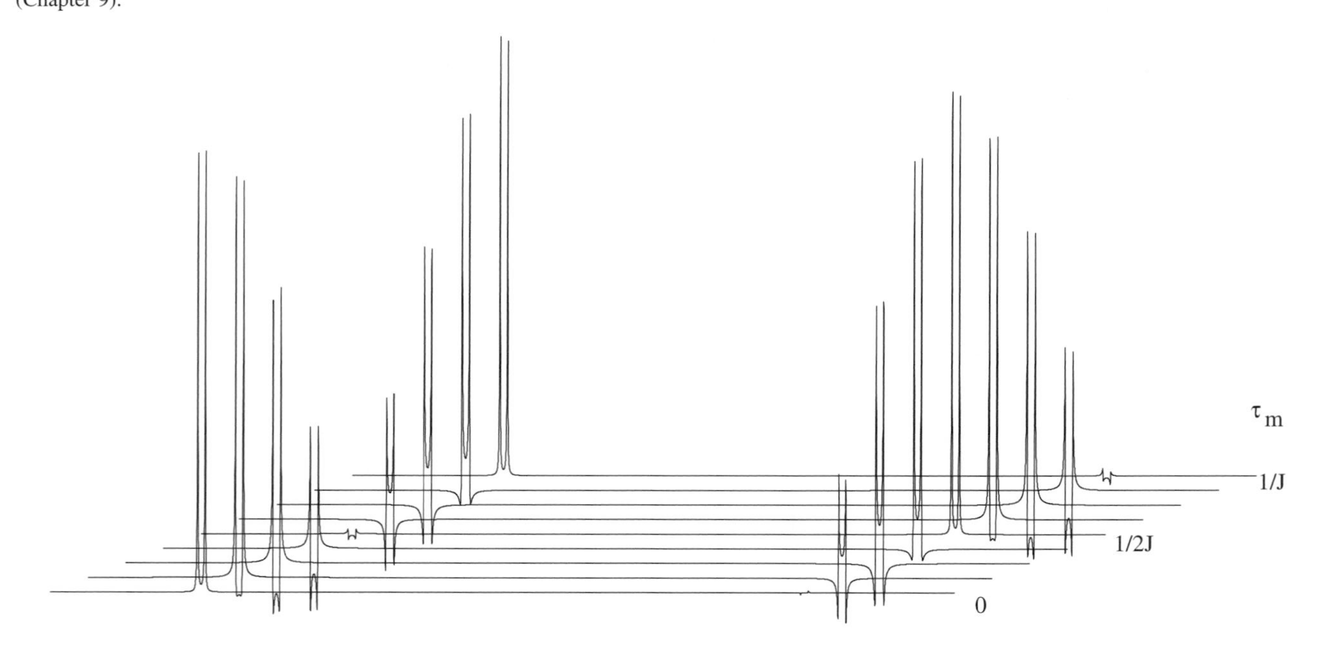

$$\tau_m \downarrow \quad \begin{array}{l} A \overset{J_{AB}}{\rightleftharpoons} B \text{ - - - } C \text{ - - - } D \text{ - - - } E \\ A \rightleftharpoons B \overset{J_{BC}}{\rightleftharpoons} C \text{ - - - } D \text{ - - - } E \\ A \rightleftharpoons B \rightleftharpoons C \overset{J_{CD}}{\rightleftharpoons} D \text{ - - - } E \\ A \rightleftharpoons B \rightleftharpoons C \rightleftharpoons D \overset{J_{DE}}{\rightleftharpoons} E \end{array}$$

Figure 5.64. Schematic illustration of the propagation of magnetisation along a chain of coupled spins as a function of the spin-lock mixing time. The arrows indicate the cyclic exchange of coherence between nuclei.

expressed as:

$$\gamma_A B_{1A} = \gamma_X B_{1X} \tag{5.12}$$

where B_{1A} and B_{1X} are the B_1 fields experienced by spins A and X and γ_A and γ_X are their magnetogyric ratios respectively. The inclusion of the γs in this means the nuclides involved in the exchange need not be the same, so, for example, transfer can be initiated between proton and carbon if simultaneous rf fields that satisfy Eq. 5.12 can be applied to each nuclide. This, by analogy, leads to hetero-TOCSY or, more amusingly, heteronuclear Hartmann–Hahn spectroscopy (HEHAHA) as a means of establishing heteronuclear correlations [53,54].

5.7.2. Using TOCSY

The ability of TOCSY to propagate magnetisation along a spin system has meant it has become a popular tool for studying molecules that are comprised of discrete and often well defined units. Obvious examples of this are proteins, peptides and oligosaccharides, for which it is often possible to trace the whole amino-acid or sugar ring system from a single resolved proton. This is illustrated in the TOCSY spectrum of Gramicidin-S, **5.7**, a cyclic decapeptide (Fig. 5.65). The aliphatic region only is shown, in which all sidechain protons of each residue can be traced from the associated α-proton (with the exception of the phenylalanine aromatic ring), as illustrated for the proline and ornithine residues. The relaying of magnetisation along a coupled chain of spins can often overcome problems due to close or coincident crosspeaks in COSY by effectively moving related correlations into regions of the spectrum that would otherwise be devoid of crosspeaks, therefore utilising redundant 'space' in the 2D spectrum. This approach was used in the sidechain assignment of **5.8**, an intermediate in the biomimetic synthesis of members of the manzamine and

Pro—Val—Orn—Leu—DPhe
DPhe—Leu—Orn—Val—Pro

5.7

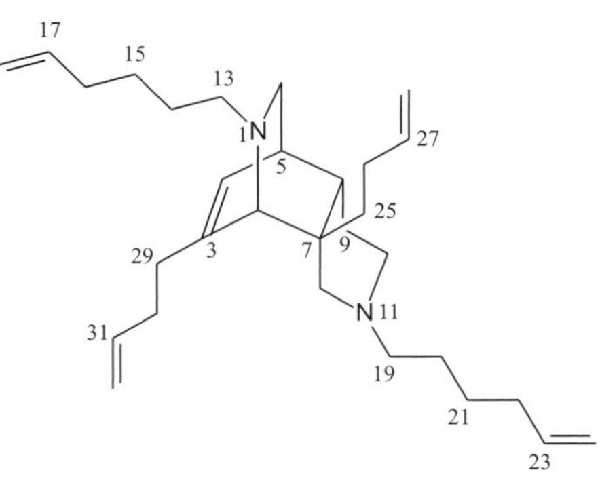

5.8

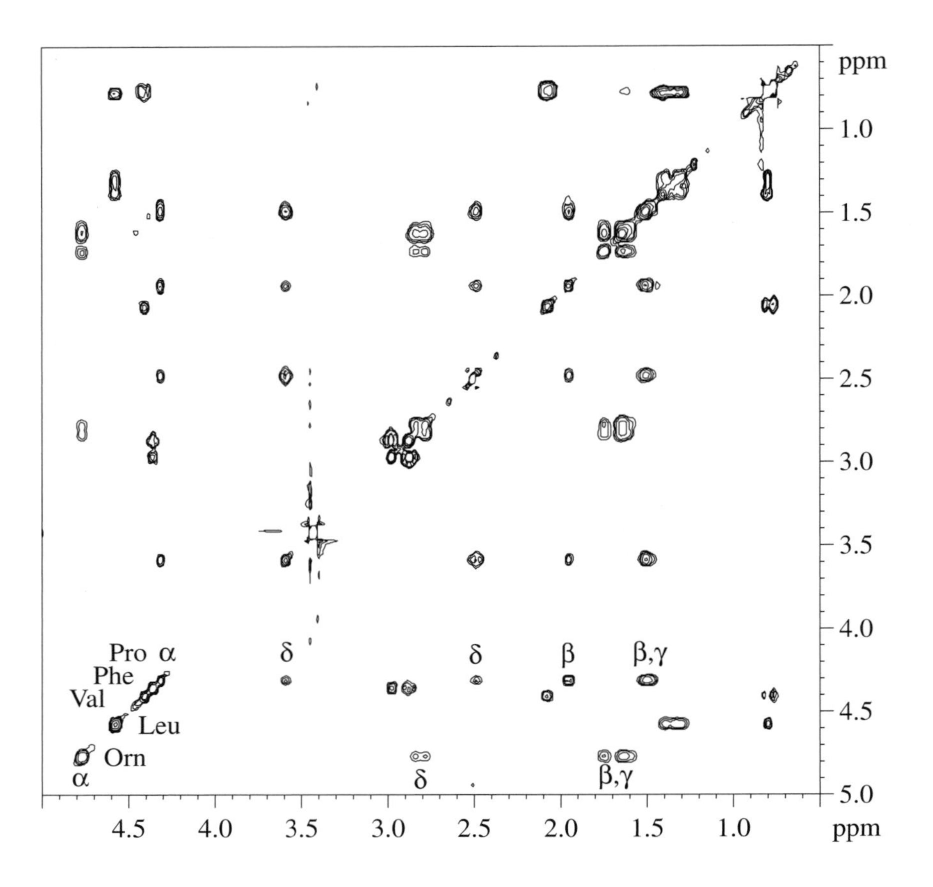

Figure 5.65. The aliphatic region of the TOCSY spectrum of the cyclic decapeptide Gramicidin-S, **5.7**. An 80 ms MLEV-17 spin-lock was used, providing complete transfer along the aliphatic sidechains of each residue. Correlations from the α-protons of proline and ornithine are labelled, with those out to the δ-protons corresponding to three-step transfers.

keramaphidine alkaloids [55,56], found naturally in marine sponges. Due to extensive crowding in some areas of the aliphatic region, sidechain assignments were more readily derived from relayed correlations involving the resolved alkene protons such as H23 (Fig. 5.66). Similarly, the numerous correlations from H4 provided confirmation of assignments derived from a DQF-COSY spectrum.

The somewhat indiscriminate propagation of magnetisation found with TOCSY can also be disadvantageous however, since it is generally not possible to state explicitly the number of relay steps involved in the production of any given correlation. A time dependence of crosspeak build-up from a number of TOCSY spectra recorded with increasing τ_m (in, for example, 20 ms steps (Fig. 5.67)) can provide an approximate indication of whether the peak arises from 1, 2 or more steps, although it is often not possible to be more precise than this. When using TOCSY on molecules that are not composed of well-characterised monomer units it is often beneficial to use this in addition to COSY spectra from which more immediate coupling neighbours can be identified. The use of very long spin-lock periods for promoting multistep transfer also leads to sensitivity losses due to relaxation (in the rotating frame) of the spin-locked magnetisation.

It has also been mentioned that TOCSY results in the net transfer of *in-phase* magnetisation, meaning the cancellation effects from *antiphase* multiplet fine-structure associated with COSY are not a feature of TOCSY. Such cancellation can be problematic for molecules that posses large natural linewidths, for example (bio)-polymers, but may also prevent the observation of COSY peaks in the spectra of small molecules that have complex multiplet structures which may cancel under conditions of poor digital resolution. In these cases the TOCSY experiment may be viewed as the more sensitive option because of the greater crosspeak intensities. The lack of antiphase structure also means spectra

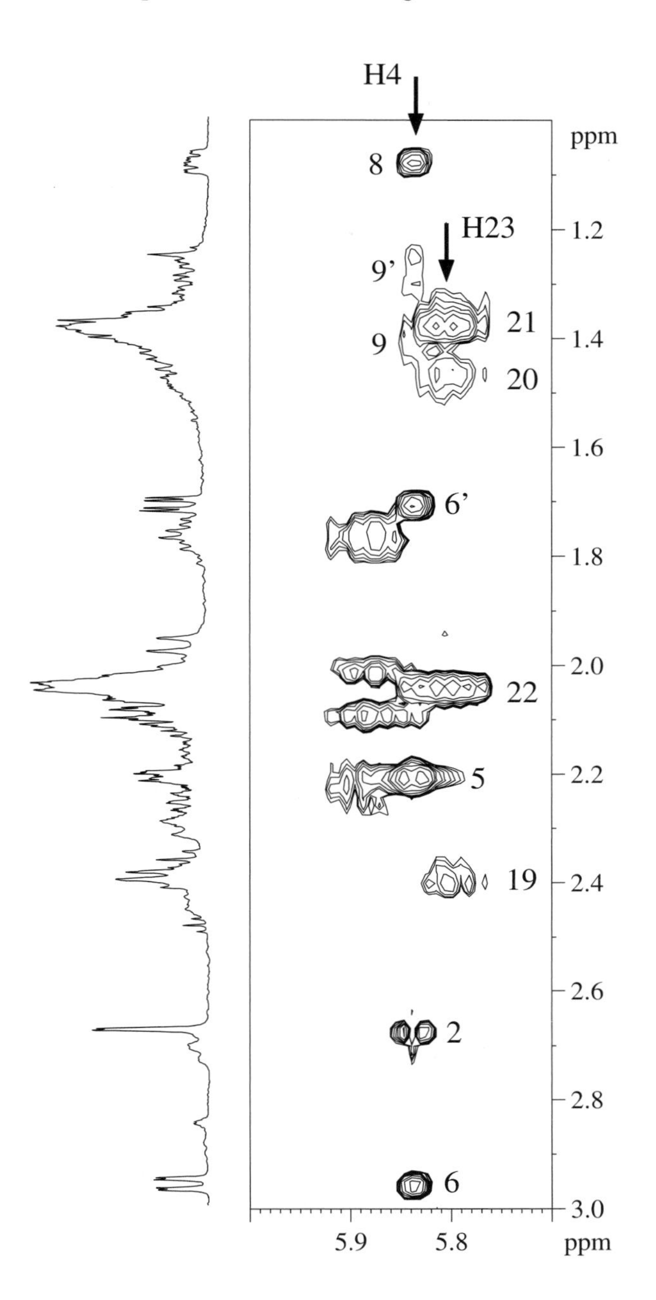

Figure 5.66. A section of the TOCSY spectrum τ_m = 80 ms) of the biomimetic intermediate **5.8** alongside the conventional 1D spectrum. Correlations from protons H4 and H23 are labelled. Despite extensive overlap in some regions of the 1D spectrum, the numerous relayed correlations in TOCSY occur in an otherwise clear region of the 2D spectrum and provide ready identification of proton shifts.

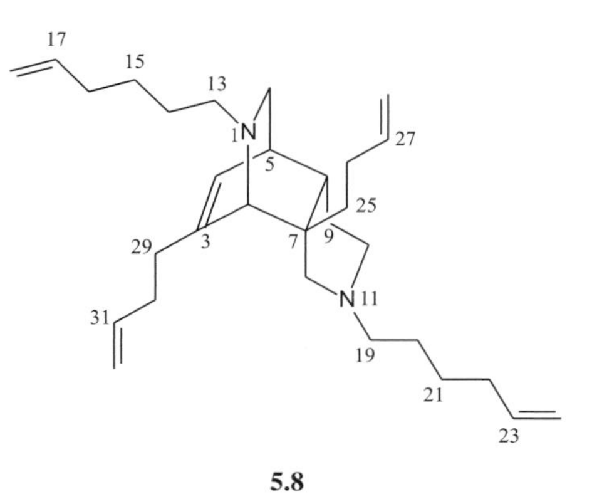

may be recorded very rapidly under conditions of rather low digital resolution without a loss of peak intensity. Finally, since both diagonal and crosspeaks possess the same phase, both produce double-absorption lineshapes meaning high-resolution phase-sensitive spectra can be obtained.

One side-effect of the spin-lock is that it also enables *incoherent* magnetisation transfer in the rotating frame. In other words, through-space effects known as rotating-frame nuclear Overhauser effects (ROEs) may, in principle, also be observed between spins. These tend to be far weaker than the TOCSY peaks particularly for small molecules and short mixing times, so are rarely problematic. They may be more of a problem for very large molecules in which they are stronger and faster to build-up. For such molecules so-called clean-TOCSY sequences have been developed [57–59] which eliminate ROE peaks by cancelling them with equal but opposite NOE peaks (Chapter 8); such methods are only applicable to very large molecules where this difference in sign exists. A far greater problem is the appearance of unwelcome TOCSY

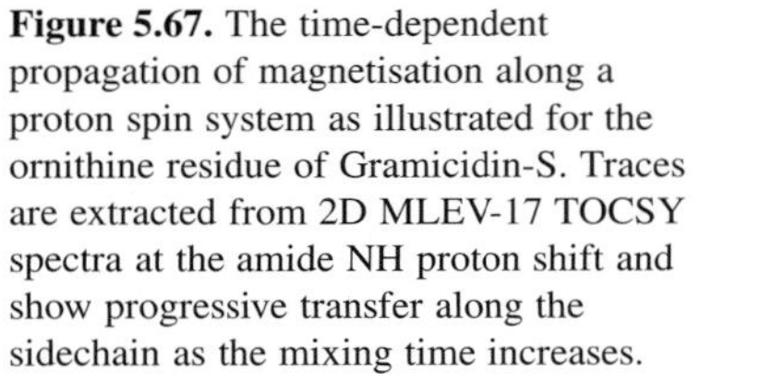

Figure 5.67. The time-dependent propagation of magnetisation along a proton spin system as illustrated for the ornithine residue of Gramicidin-S. Traces are extracted from 2D MLEV-17 TOCSY spectra at the amide NH proton shift and show progressive transfer along the sidechain as the mixing time increases.

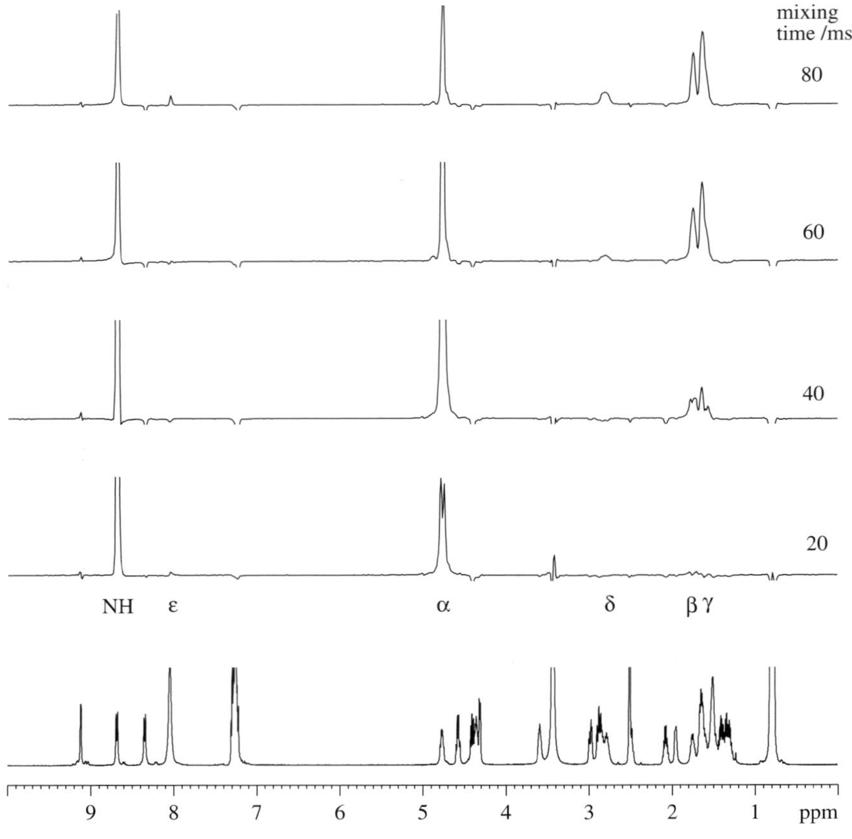

peaks in rotating-frame NOE (ROESY) spectra, a topic also addressed in Chapter 8.

5.7.3. Implementing TOCSY

Although in principle the simple scheme presented in Fig. 5.59 should provide TOCSY spectra, its suitability for practical use is limited by the effective bandwidth of the continuous-wave spin-lock. Spins which are off-resonance from the applied low-power pulse experience a reduced rf field causing the Hartmann–Hahn match to breakdown and transfer to cease. This is analogous to the poor performance of an off-resonance 180° pulse (Section 3.2.1). The solution to these problems is to replace the continuous-wave spin-lock with an extended sequence of 'composite' 180° pulses which extend the effective bandwidth without excessive power requirements. Composite pulses themselves are described in Chapter 9 alongside the common mixing schemes employed in TOCSY, so shall not be discussed here. Suffice it to say at this point that these composite pulses act as more efficient broadband 180° pulses within the general scheme of Fig. 5.60.

There are essentially two approaches based on composite-pulse methods in widespread use for the practical implementation of the TOCSY experiment (Fig. 5.68). The first of these [51] (Fig. 5.68a) is based on the so-called MLEV-17 spin-lock, in which an even number of cycles through the MLEV-17 sequence are used to produce the desired total mixing period. To ensure the collection of absorption-mode data, only magnetisation along a single axis should be retained, so it is necessary to eliminate magnetisation not parallel to this before or after the transfer sequence. In this implementation, this is achieved by the use of 'trim-pulses' applied for 2–3 ms along the chosen axis.

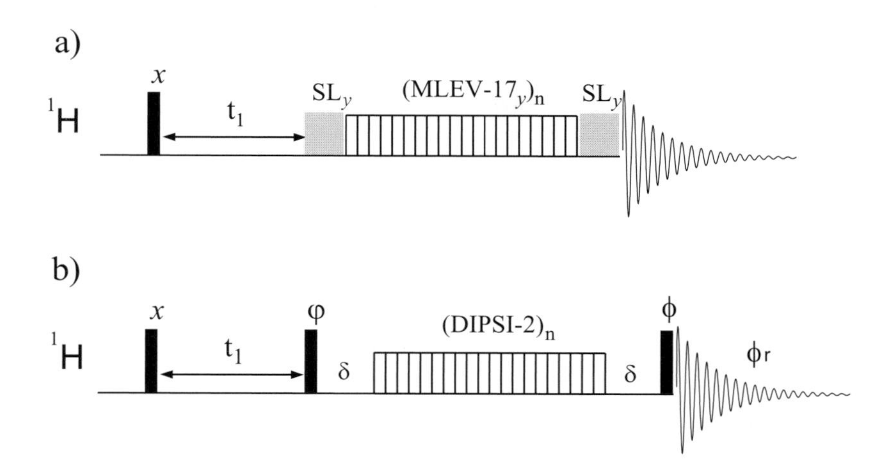

Figure 5.68. Two practical schemes for implementing TOCSY based on (a) the MLEV-17 mixing scheme and (b) the DIPSI-2 isotropic mixing scheme. The MLEV sequence is bracketed by short, continuous-wave, spin-lock trim pulses (SL) to provide pure-phase data. In scheme (b) this can be achieved by phase-cycling the 90° *z*-filter pulses that surround the mixing scheme. This demands the independent inversion of each bracketing 90° pulse with coincident receiver inversion, thus $\varphi = x, -x, x, -x$; $\phi = x, x, -x, -x$ and $\phi_r = x, -x, -x, x$. The δ periods allow for the necessary power switching.

These are periods in which a single, low-level pulse is applied, during which magnetisation not parallel to this axis will dephase due to inhomogeneity of the rf field (just as transverse magnetisation rapidly dephases in an inhomogeneous static field). This process eliminates dispersive contributions to the spectrum.

The alternative approach is to regenerate longitudinal magnetisation following the evolution period and use a suitable *isotropic mixing* scheme to transfer this *z*-magnetisation between spins [60,61] (isotropic here means that magnetisation transfer is equally effective along the *x*, *y*, or *z*-axes so the overall sequence has a similar effect to that described above). The *z*-magnetisation is then reconverted to pure-phase transverse magnetisation following the mixing, enabling the collection of absorption-mode data. The idea of generating and subsequently selecting *z*-magnetisation with the view to collecting pure-phase spectra has been termed '*z*-filtration' [62]. This also has the advantage for older instruments that the proton 'decoupler' may be used for generating the low-power mixing scheme [60] if the main proton transmitter is incapable of suitable power level control; this restriction does not arise on modern instruments. A number of isotropic mixing schemes may be used in this approach, including MLEV-16, WALTZ-16 [63,64] and DIPSI-2 [65,66], the last of these being particularly well suited for transfer in ^{1}H–^{1}H TOCSY.

In either of these two general schemes, a greater effective bandwidth is achieved with the composite-pulse mixing and significantly lower powers are required for this than for the 90° preparation pulse. For MLEV, the rf field strength need be about twice the desired bandwidth, whilst for the more efficient DIPSI-2 it may be about equal to it. Thus, a 10 ppm window at 400 MHz requires $\gamma B_1 \approx 8$ kHz (90° ≈ 30 μs) or 4 kHz (90° ≈ 60 μs) respectively, either of which can be achieved comfortably.

Gradient-selected TOCSY

For samples of sufficient strength to require only a single scan per increment, gradient versions of TOCSY may be attractive alternatives for the rapid recording of spectra. By analogy with the previous COSY discussions the absolute-value TOCSY sequence simply requires equal gradients to be placed either side of the spin-lock sequence (Fig. 5.69a) to selectively refocus the N-type pathway to provide f_1 frequency discrimination. Trim-pulses are not required since pure-phase spectra are not produced by this method. As for absolute-value COSY, this experiment may be well suited to automated acquisition schemes because of the simplicity of processing, although in general the phase-sensitive version is preferred for reasons of resolution. This may be performed by placing the gradients within spin-echoes to refocus shift evolution. Quadrature detection is then afforded by collecting P and N-type data

Figure 5.69. Gradient-selected TOCSY. Sequence (a) is suitable for absolute-value presentations with a 1:1 gradient combination selecting the N-type spectrum. Sequence (b) provides phase-sensitive data sets via the echo–antiecho method for which separate P- and N-type data are collected through inversion of the first gradient.

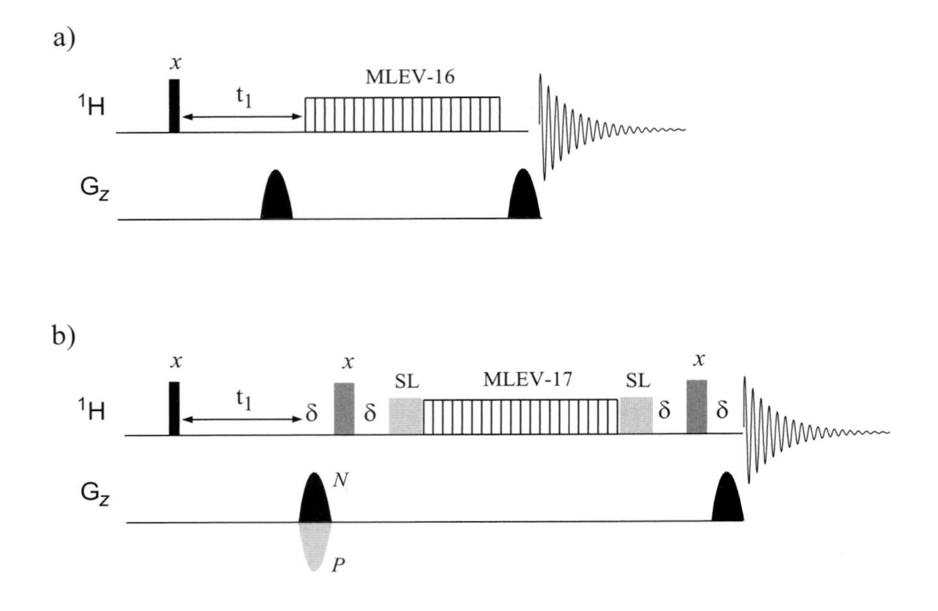

separately for each t_1 increment via gradient inversion, and processing these according to the echo–antiecho procedure of Section 5.5.2 (Fig. 5.69b).

One-dimensional TOCSY

As an alternative to the collection of a full 2D data set, the 1D analogue can prove advantageous in some circumstances. In general most 2D sequences can be adapted to produce the 1D equivalent [67,68], meaning the experiments can be quicker to acquire, require less storage space and can be recorded under conditions of higher digital resolution. These may prove particularly attractive when only specific information is required, as is often the case for small to medium-sized molecules, or when small sample quantities demand many scans for adequate signal-to-noise and so preclude the collection of 2D data sets. All 1D analogues of multidimensional experiments begin by selectively exciting one resonance in the spectrum and using this as the source for all subsequent magnetisation transfer. The resulting spectra may therefore be viewed as being equivalent to a high-resolution row through the related 2D spectrum at the shift of the excited spin. Selective excitation methods are described in Chapter 9 so are not addressed here. Numerous methods for generating 1D TOCSY spectra without [69–72] or with [73–76] gradient assistance have been presented, a generalised scheme using the MLEV-17 sequence being shown in Fig. 5.70. Following the excitation of the chosen spins, magnetisation transfer is initiated immediately with the mixing scheme since there is now no need for the evolution period. The only resonances appearing in the resulting spectra are those that have 'received' magnetisation from the source spin. The experiment produces 1D subspectra for discrete spin-systems within the molecule, potentially revealing multiplet structures that were otherwise overlapped or buried. Furthermore, the higher digital resolution afforded by the 1D analogue may resolve ambiguities arising from crosspeak overlap in 2D spectra and may also provide a more rapid method for following magnetisation transfer along a chain of spins as a function of mixing time.

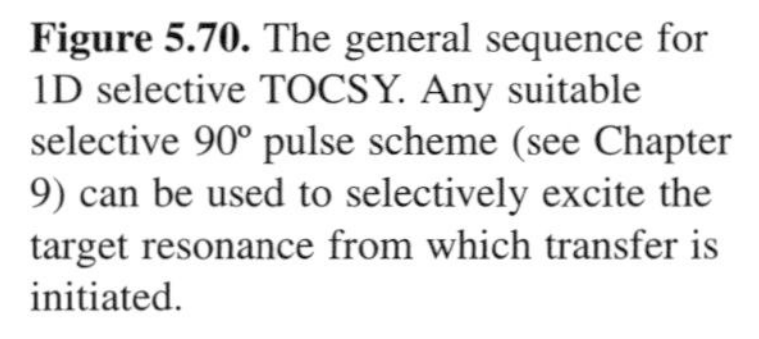
Figure 5.70. The general sequence for 1D selective TOCSY. Any suitable selective 90° pulse scheme (see Chapter 9) can be used to selectively excite the target resonance from which transfer is initiated.

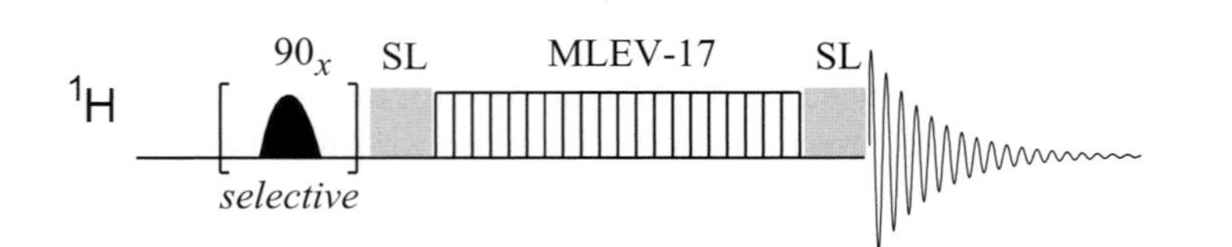

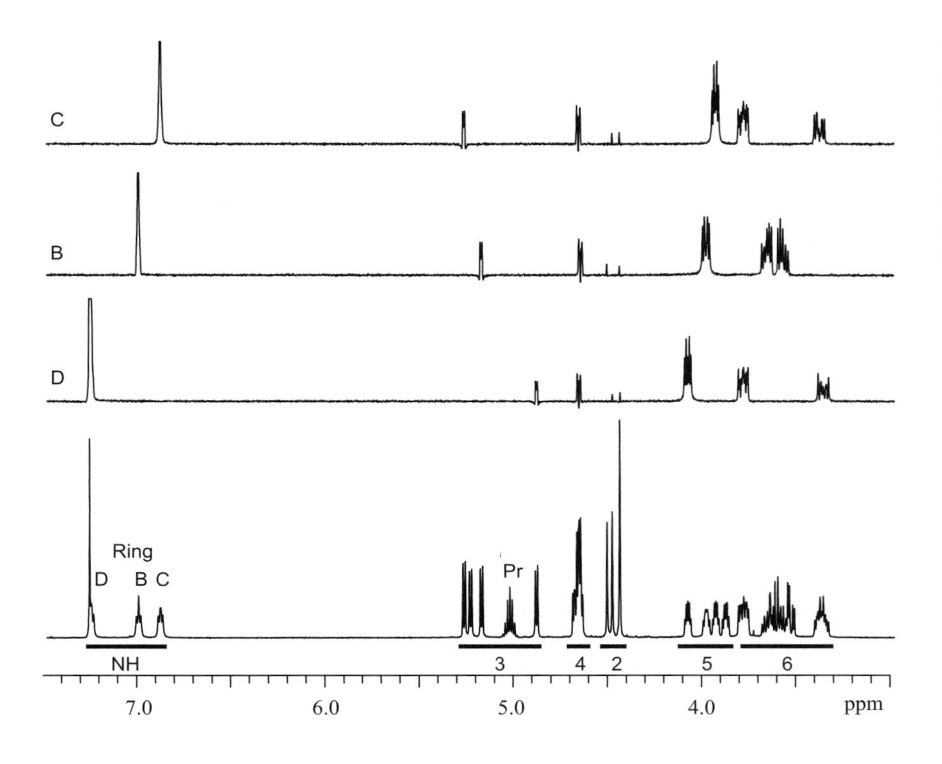

Figure 5.71. One-dimensional TOCSY spectra of the tetrameric carbopeptoid **5.9** in $CDCl_3$. Each amide proton was selectively excited and used as the starting point for coherence transfer. Selective excitation was achieved with the excitation sculpting method and mixing used a 97 ms MLEV-17 spin-lock.

Fig. 5.71 shows 1D TOCSY spectra for the carbopeptide **5.9** in which transfer is from the selected amide protons with the individual ring systems resolved in each case. This approach is often applicable to the study of oligosaccharides where the anomeric protons can provide a convenient starting point from which to establish intraresidue assignments.

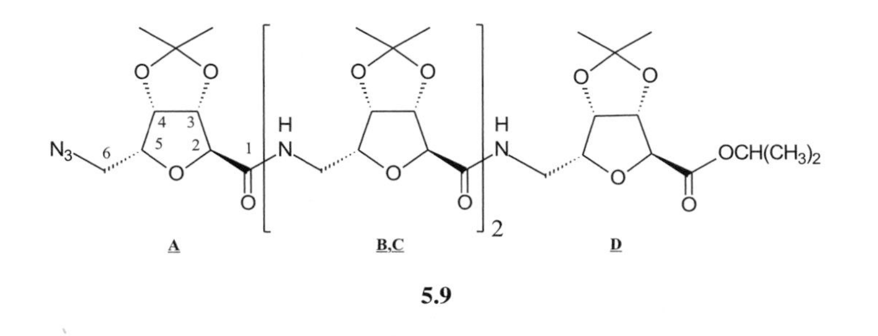

5.9

5.8. CORRELATING DILUTE SPINS: INADEQUATE

One of the major goals in determining the structure of an organic molecule is the unambiguous identification of the carbon skeleton of the compound. Carbon–carbon connectivities are, however, rarely determined directly, rather we imply their presence from the correlations observed between protons, such as in COSY-type experiments, or between proton and carbon nuclei, such as from the long-range heteronuclear correlation experiments described in the following chapter. The need for one to take this indirect approach lies in the relatively poor sensitivity and low natural abundance (1.1%) of the carbon-13 isotope. The abundance of molecules containing two adjacent ^{13}C spins is only 0.01%, so in attempting to directly identify ^{13}C–^{13}C linkages, we are forced to observe only 1 in every 10,000 molecules in our sample (in the absence of isotopic enrichment). In other words, one must look for ^{13}C satellites in the ^{13}C spectrum itself, and clearly larger sample quantities and extended acquisition periods are required for this. Despite these limitations, and because the ability to directly trace the carbon skeleton of a molecule is perhaps the ultimate approach to organic structure elucidation, considerable effort has gone into developing methods to identify C–C connectivities directly, all generally based on the INADEQUATE sequence (Incredible Natural Abundance DoublE QUAntum Transfer Experiment) [77]. These techniques are also suitable for correlating nuclides other than carbon that exist at relatively low natural abundance levels of around 1 to 20%, for example ^{29}Si, ^{119}Sn and ^{183}W (4.7, 8.6 and 14.4% respectively).

These techniques are certainly not the first to turn to when considering a structural problem because of the large sample quantities typically required, but in favourable cases may prove effective, and with molecules that posses very few or even no protons, one may have few other options. More recent techniques based on proton detection and gradient selection show some potential for reducing sample requirements, and are also presented briefly.

5.8.1. 2D INADEQUATE

The principal complication when attempting to correlate low abundance spins (referred to here as *dilute* or *rare* spins) is the interference caused by the dominant parent resonances which lack homonuclear couplings and cannot therefore provide connectivity data. The more popular correlation techniques such as COSY, whilst suitable in principle, are not well suited in these cases and the INADEQUATE sequence is more appropriate. The basis of INADEQUATE is double-quantum filtration to suppress the uninformative parent resonances [78], this filtration being analogous to that which we have already encountered in the DQF-COSY experiment (Section 5.6.2). Natural abundance carbon-13 spins are very well suited to such filtration because they are restricted to AX or AB two-spin systems only (because a molecule containing more than two ^{13}C centres will exist at such low abundance as to be non-existent) and because $^1J_{CC}$ couplings cover a limited range, meaning the sequence can be reasonably well optimised for the generation of double-quantum coherence. The two-dimensional correlation spectrum [79] is generated by allowing the double-quantum coherences associated with coupled spins to evolve during a variable t_1 period, following which they must be reconverted to single-quantum (transverse) carbon magnetisation for observation. The complete INADEQUATE sequence is given in Fig. 5.72.

The generation of double-quantum coherence requires an antiphase disposition of coupling vectors, which here develop during a period $\Delta = 1/2J_{CC}$. This is provided in the form of a homonuclear spin-echo to make the excitation independent of chemical shifts. The t_1 period represents a genuine double-quantum evolution period in which these coherences evolve at the *sums* of the rotating-frame frequencies of the two coupled spins, that is, at the sums of their *offsets*

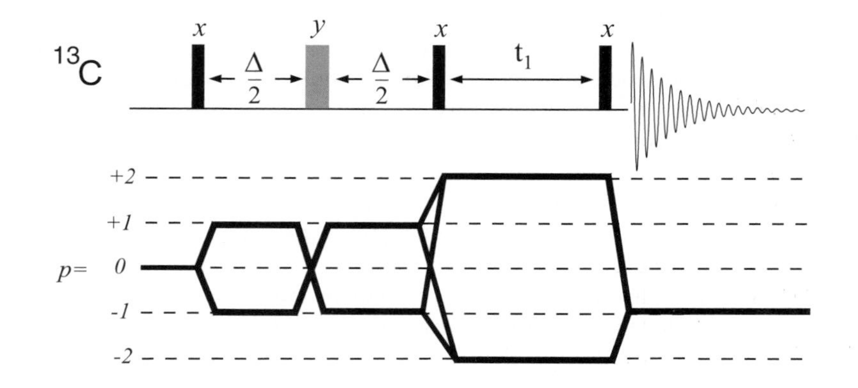

Figure 5.72. The INADEQUATE sequence and the corresponding coherence transfer pathway. The experiment selects double-quantum coherence during the evolution period with a suitable phase cycle (analogous to the selection in the DQF-COSY experiment, Section 5.6.2). In doing so it rejects all contributions from all uncoupled spins.

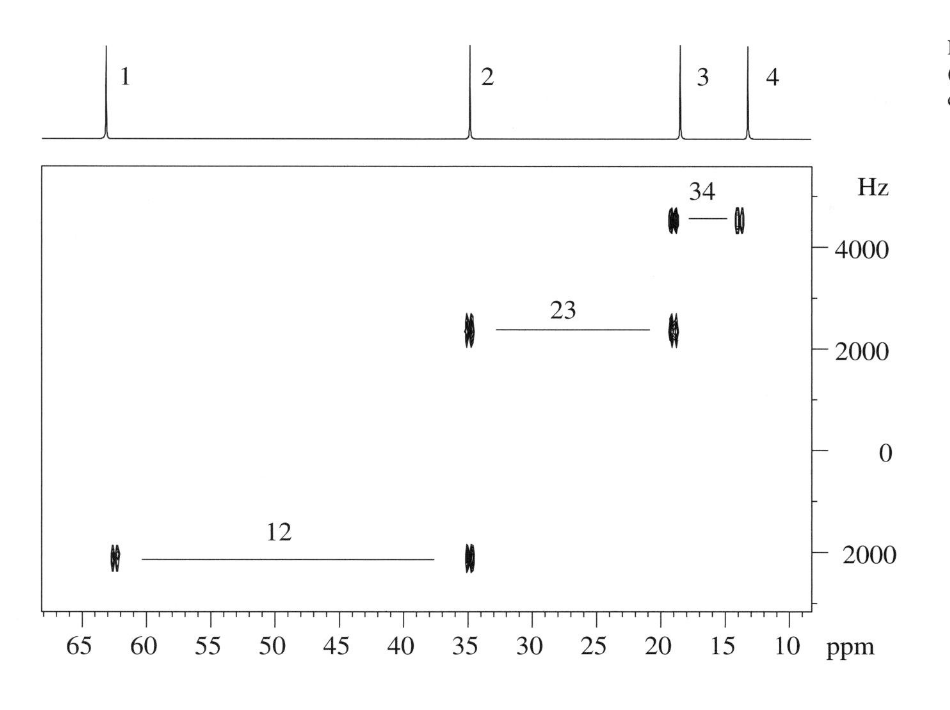

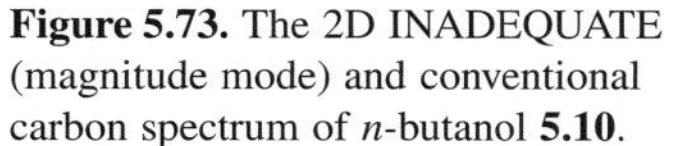

Figure 5.73. The 2D INADEQUATE (magnitude mode) and conventional carbon spectrum of *n*-butanol **5.10**.

from the transmitter frequency. Thus, for two coupled spins A and X that have offsets of ν_A and ν_X, the f_1 crosspeak frequency will be $(\nu_A + \nu_X)$. Following the reconversion of double-quantum into single-quantum coherence, crosspeaks in f_2 will be observed with the conventional chemical shift frequencies of A and X. Throughout the sequence, broadband proton decoupling can be applied to prevent evolution of heteronuclear couplings and to generate the ^{1}H to ^{13}C nuclear Overhauser enhancement to aid sensitivity.

The resulting spectrum, here illustrated for *n*-butanol **5.10** (Fig. 5.73), differs considerably in appearance from the other correlation spectra encountered in this chapter, although again the f_2 dimension represents the chemical shifts of each participating spin. Since two coupled spins share the same DQ frequency in f_1, correlations are made by following horizontal traces parallel to f_2. Carbon connectivity is therefore established by a sequence of vertical and horizontal steps. If the DQ excitation sequence is optimised for one-bond C–C couplings, then each step identifies adjacent carbon centres in the molecule. The spectra display no diagonal peaks, although the midpoints of correlated spin pairs appear along a 'pseudo' diagonal (the double-quantum diagonal) of slope 2, that is where $f_1 = 2f_2$. This characteristic symmetry can be useful in differentiating genuine responses from artefacts. Each crosspeak is composed of a doublet with a splitting of J_{CC} arising from the AX (or AB) fine structure.

CH_3 OH

5.10

An example of the power of this method in structure elucidation is shown for the sesquiterpene lactone acetylisomontanolide (Fig. 5.74). The complete carbon skeleton of the molecule can be traced directly in the spectrum, with breakdown occurring only when heteroatom linkages arise. Unfortunately, it is rare indeed to have sufficient sample quantities at hand to consider employing this technique, and still it remains a little used method in its original form. However, greater potential exits for directly tracing carbon–carbon connectivities with recently developed methods based on proton excitation and observation combined with gradient selection; see Section 5.8.4 below.

5.8.2. 1D INADEQUATE

The one-dimensional INADEQUATE experiment was in fact the original version, and was proposed as a means of measuring carbon–carbon coupling

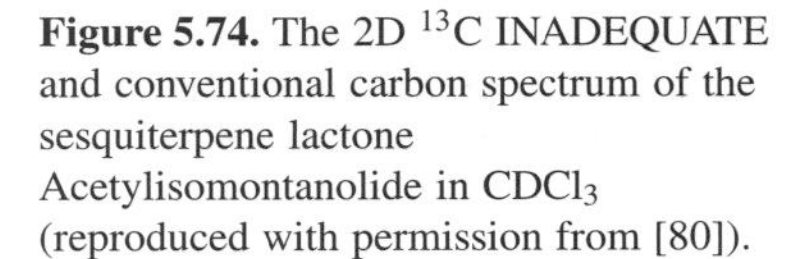

Figure 5.74. The 2D ^{13}C INADEQUATE and conventional carbon spectrum of the sesquiterpene lactone Acetylisomontanolide in $CDCl_3$ (reproduced with permission from [80]).

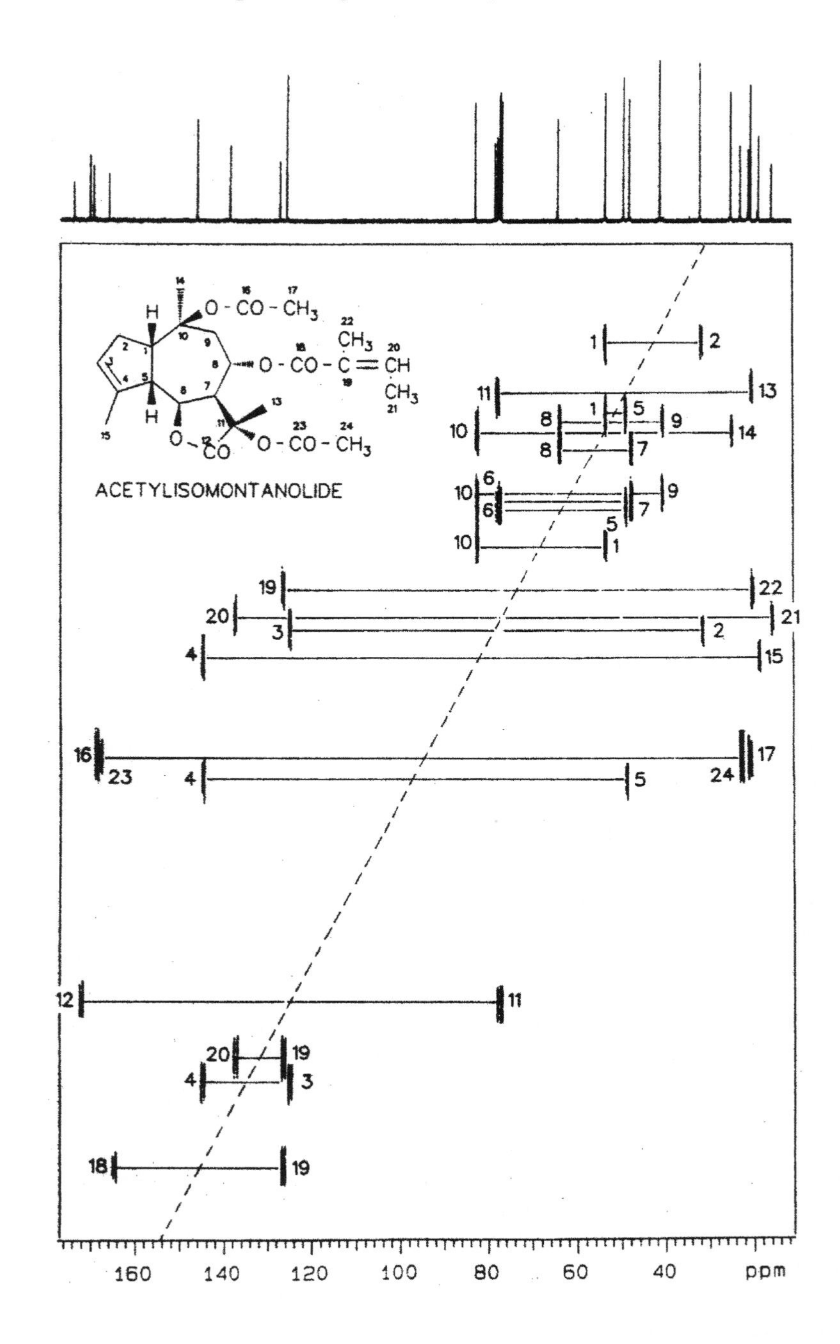

constants [78]. The use of double-quantum filtration removes the parent resonances which dominate the spectrum and mask the desired splittings, especially when smaller 2 and 3-bond couplings are to be measured. This can also suppress resonances from low-level impurities that may otherwise interfere. Replacing the t_1 evolution period by a short, fixed delay (sufficient only to allow spectrometer phase changes) results in a 1D double-quantum filtered experiment in which carbon–carbon couplings appear as antiphase doublets (Fig. 5.75). From these patterns the values of J_{CC} can be determined directly, if sufficiently resolved. An alternative application that has been used extensively is in the tracking of doubly ^{13}C-labelled precursors in the study of metabolic or biosynthetic pathways [81]. The DQ filtration is used to suppress background signals of those species that do not contain the original ^{13}C–^{13}C fragment (to the limit of natural abundance), whilst retaining only those that have survived the pathway intact.

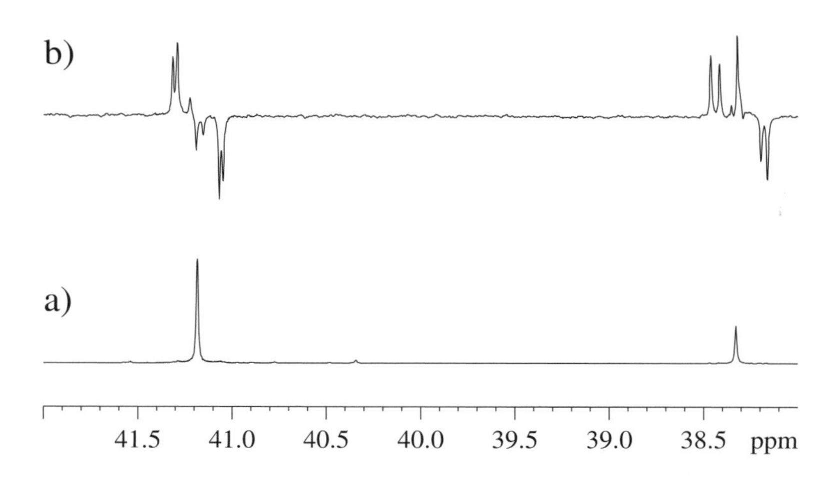

Figure 5.75. The 1D INADEQUATE experiment (b) selects for the carbon-13 satellites in the conventional carbon-13 spectra (a) and provides a means of measuring carbon–carbon coupling constants. Some residual parent resonances from lone ^{13}C centres remain in (b).

5.8.3. Implementing INADEQUATE

The basic components of the INADEQUATE phase cycle comprise double-quantum filtration and f_1 quadrature detection. The filtration may be achieved as for the DQF-COSY experiment described previously, that is, all pulses involved in the DQ excitation (those prior to t_1 in this case) are stepped x, y, $-x$, $-y$ with receiver inversion on each step (an equivalent scheme found in spectrometer pulse sequences is to step the *final* 90° pulse x, y, $-x$, $-y$ as the receiver steps in the opposite sense x, $-y$, $-x$, y; other possibilities also exist). This simple scheme may not be sufficient to fully suppress singlet contributions, which appear along $f_1 = 0$ as axial peaks and are distinct from genuine C–C correlations. Extension with the EXORCYCLE sequence (Section 7.2.2) on the 180° pulse together with CYCLOPS (Section 3.2.5) may improve this. Cleaner suppression could also be achieved by the use of pulsed field gradients, which for sensitivity reasons requires a gradient probe optimised for ^{13}C observation.

Quadrature detection requires the observed signals to exhibit 90° phase increments on each t_1 step. Since the desired signals pass through double-quantum coherence, they are *twice* as sensitive to rf phase shifts, meaning the excitation pulses prior to t_1 must themselves be incremented in 45° steps only. Such phase increments are readily achieved on modern spectrometers, so the usual States or TPPI approaches to phase-sensitive data collection can be applied. However, during the development of these sequences the available hardware was incapable of producing small phase shifts and a number of alternative schemes were developed that avoided 45° rf phase shifts altogether. One popular approach used to generate absolute-value spectra was to use pulse width dependence to enhance the intensity of the N-type signals relative to the P-type [82]. Increasing the final pulse angle to 135° produces a 6:1 intensity ratio of N:P so the P-type signal essentially disappears and frequency discrimination results, albeit with mixed-mode lineshapes which demand magnitude calculation.

The success of INADEQUATE may be compromised by deleterious off-resonance effects which are particularly troublesome for nuclei such as carbon-13 which have a relatively large spectral window. Variations employing composite-pulses to counteract offset effects have therefore been developed to minimise losses [83,84].

Experimental set-up

There is no doubt that the deciding factor when considering the use of ^{13}C INADEQUATE is sensitivity, and far greater sample quantities are required for this experiment than for other 2D sequences. As a rule-of-thumb, one should be able to obtain a conventional 1D ^{13}C spectrum *in a single scan* following a 90°

Table 5.7. Typical ranges for one-bond C–C coupling constants

C–C group	Typical $^{1}J_{CC}$ values (Hz)
C–C	35–40
C–C(O)	40–60
C=C (alkene)	70–80
C=C (aromatic)	55–70
C≡C	170–220

The principal factor in determining the magnitudes of these is carbon hybridisation.

excitation pulse, and this should display a signal-to-noise ratio of *at least* 20:1. For realistic sample quantities encountered in the chemical laboratory this is rarely achievable, and even in favourable cases, overnight acquisitions at least are typically required. The recent post-processing algorithm CCBond [85,86] (now commercially available as the FRED^{TM} routine in Varian software) has proved extremely successful at automatically analysing INADEQUATE spectra even when the signal-to-noise was too poor for the human eye to recognise connectivity patterns. Using such an approach, spectra with an order of magnitude poorer signal-to-noise than is conventionally required may be subject to reliable analysis [87,88], clearly enhancing the utility of this method (for other experimental approaches to improving sensitivity, see the following section).

Maximum sensitivity requires optimum generation of double-quantum coherence. For weakly coupled systems Δ should be equal to $1/2J_{CC}$, whilst the presence of strong coupling ($\Delta\delta/J_{CC} < 3$) requires optimum periods of $3/2J_{CC}$ [89]. As ever, the choice of Δ represents some compromise over anticipated J_{CC} values. One-bond couplings typically range from 35–70 Hz [77,90,91] (Table 5.7) but are more often less than 50 Hz in the absence of C–C unsaturation, so Δ of around 10–14 ms is a reasonable choice. Since the optimum repetition rate is 1.3 times the ^{13}C T_1s, the use of relaxation agents to speed data acquisition should also be considered, particularly when quaternary centres may be present. For organic solutions, chromium acetylacetonate is widely used for this and even dissolved oxygen has been shown to be advantageous [92]. Gadolinium triethylene-tetraamine-hexaacetate (Gd(TTHA)) [93] has been recommended for aqueous solutions and experience suggests concentrations up to 10 mM produce acceptable results without significantly affecting the carbon resonances, although this can substantially broaden the *proton* spectrum of the sample.

Due to the large spectral widths involved, spectra are collected with low resolution in both dimensions; around 10–20 Hz/pt in f_2 and 50–100 Hz/pt in f_1. Processing then requires window functions that minimise distortions arising from the truncated FIDs, such as the shifted sine-bell or squared-sine-bell. To minimise the f_1 spectral width, the transmitter frequency should be placed in the centre of the 1D ^{13}C spectrum so that double-quantum frequencies fall either side of this, that is, both positive and negative DQ frequencies appear in f_1. It is also possible to further reduce the f_1 spectral width by deliberately folding peaks in the f_1 dimension by making this equal to that in f_2. This causes no ambiguity in interpretation because both participating crosspeaks fold in together and connectivities can still be identified in horizontal traces.

5.8.4. Variations on INADEQUATE

A number of variations on the basic INADEQUATE sequence have been presented over the years, particularly with reference to ^{13}C experiments, all primar-

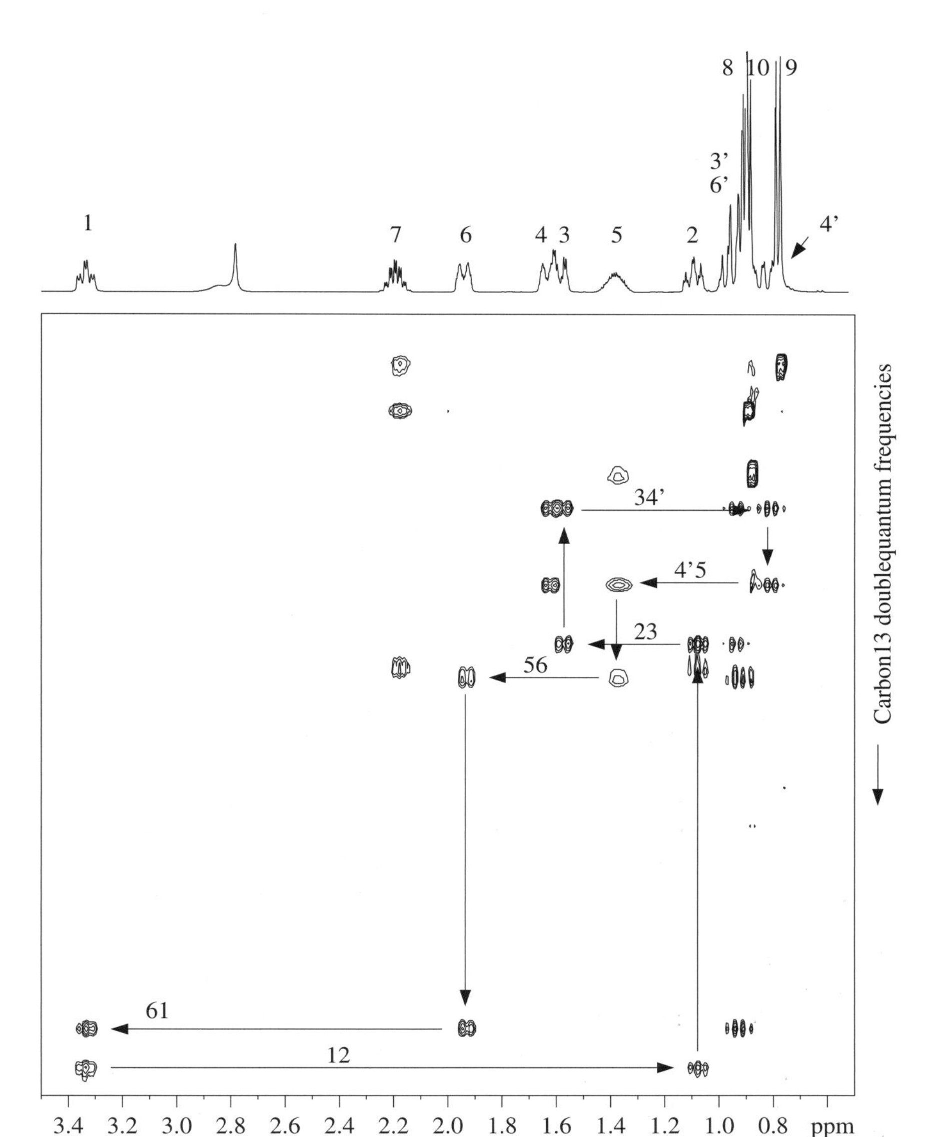

Figure 5.76. The proton-detected INEPT-INADEQUATE spectrum of menthol **5.11** (50 mg in $CDCl_3$) recorded at 400 MHz using the optimised 1,1-ADEQUATE sequence of [97]. The step-wise tracing of the carbon skeleton of the cyclohexane ring is illustrated in which carbon connectivities are identified in horizontal traces (alternative 'routes' involving the diastereotopic 3, 4 and 6 partners also exist but, for clarity, are not shown).

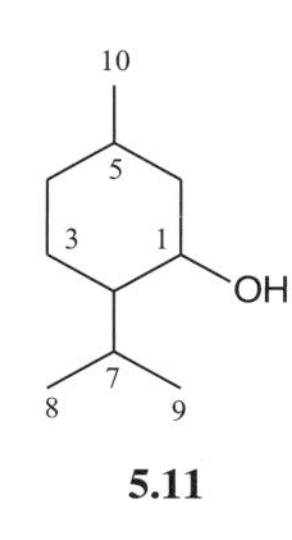

5.11

ily aimed at improving the appallingly low sensitivity of the original technique. Proton–carbon polarisation transfer prior to the INADEQUATE sequence via INEPT [94] or DEPT [95] may be used as a means of enhancing ^{13}C population differences and of allowing more rapid repetition rates since these would be dictated by faster proton spin relaxation. However, these preclude the use of

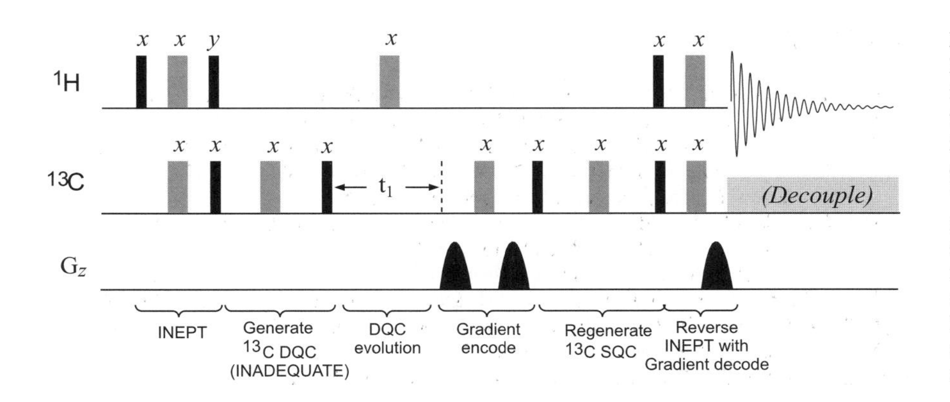

Figure 5.77. The gradient-selected INEPT-INADEQUATE sequence for identifying carbon–carbon connectivities through proton observation (time delays in the sequence have been removed for clarity). The sequence selects for $^{1}H-^{13}C-^{13}C$ fragments for which the gradient ratios are 1:−1:1. Despite the complexity of the sequence, its operation may be understood by recognising discrete, simpler segments within it that have defined roles to play, which in this case are INEPT and INADEQUATE steps.

broadband proton decoupling and hence gain no enhancement from the NOE, so only marginal improvements may obtained. Conversely, magnetisation may be transferred onto protons directly attached to ^{13}C sites by a reverse-INEPT step *after* the INADEQUATE sequence. This variant, referred to as INSIPID [96], benefits from the higher sensitivity of proton detection. Alas, it also suffers severely from the need to suppress the signals of protons attached to lone ^{13}C centres as well as those attached to ^{12}C, which possess intensities that are, respectively, 100 and 10,000 times greater than those of the desired $^{1}H–^{13}C–^{13}C$ moieties! More recent experiments, referred to as proton-detected INEPT-INADEQUATE [31] or more boldly as just ADEQUATE [97], tackle the sensitivity problem by utilising both polarisation transfer and proton observation, and overcome the proton suppression problems by exploiting pulsed field gradients for signal selection. In principle, this leads to sensitivity gains by a factor of 32 ($(\gamma_H/\gamma_C)^{5/2}$), although a factor of around 8–16 seems more realistic when allowing for experimental losses [31]. The resulting 2D spectrum presents a step-wise correlation of protons that sit on neighbouring ^{13}C centres and so maps direct carbon connectivities within the molecular framework. The spectrum in Fig. 5.76 was recorded on a 50 mg sample of unlabelled menthol **5.11** in 16 hrs, although adequate results would have be achieved in less than half this time. This powerful technique should prove valuable in the structure elucidation of unknown materials.

The original sequence is shown in Fig. 5.77 and, if you permit a small digression at this point, serves to illustrate a feature common to many modern pulse sequences. Although at first sight it may appear to be frighteningly complex and is undoubtedly the most elaborate sequence in this book, it can be broken down into more readily digestible fragments, as illustrated. Whilst the details of this sequence may be somewhat esoteric, an appreciation of its operation is rather more accessible when viewed in this stepwise manner as it becomes apparent that this is built-up from simpler sequences already encountered (the aforementioned INEPT and INADEQUATE). The sequence can then simply be viewed as a variation on the HSQC technique of Chapter 6 in which double-quantum coherence (DQC) evolves during t_1 instead of the single-quantum coherence (SQC) utilised in HSQC. When faced with trying to follow a new pulse sequence, a good approach is to try to recognise smaller fragments within it that are derived from shorter, well established sequences. We shall encounter more examples of this in the following chapter.

Naturally, the above methods require proton-bearing carbons to be present, and when applicable should prove to be far more useful in structural analysis than their ^{13}C detected counterparts. However, even these are likely to be employed only after more sensitive methods, such as those based on proton–proton and proton–carbon correlations, have failed to solve the problem at hand.

REFERENCES

[1] J. Jeener, Ampère International Summer School, Basko Polje, Yugoslavia, 1971.
[2] W.P. Aue, E. Bartholdi and R.R. Ernst, *J. Chem. Phys.*, 1976, **64**, 2229–2246.
[3] J.N.S. Evans, Biomolecular NMR Spectroscopy, Oxford University Press, Oxford, 1995.
[4] J. Cavanagh, W.J. Fairbrother, A.G. Palmer and N.J. Skelton, Protein NMR Spectroscopy. Principles and Practice, Academic Press, San Diego, 1996.
[5] J. Hahn, in J.G. Verkade, L.D. Quin, eds., Phosphorus-31 NMR Spectroscopy in Stereochemical Analysis, VCH Publishers, Florida, 1987.
[6] D.J. States, R.A. Haberkorn and D.J. Ruben, *J. Magn. Reson.*, 1982, **48**, 286–292.
[7] M. Ohuchi, M. Hosono, K. Furihata and H. Seto, *J. Magn. Reson.*, 1987, **72**, 279–297.
[8] D. Marion and K. Wüthrich, *Biochem. Biophys. Res. Commun.*, 1983, **113**, 967–974.
[9] J. Keeler and D. Neuhaus, *J. Magn. Reson.*, 1985, **63**, 454–472.
[10] C.J. Turner and H.D.W. Hill, *J. Magn. Reson.*, 1986, **66**, 410–421.

[11] A. Bax, R. Freeman and G.A. Morris, *J. Magn. Reson.*, 1981, **43**, 333–338.
[12] A.F. Mehlkopf, D. Korbee, T.A. Tiggleman and R. Freeman, *J. Magn. Reson.*, 1984, **58**, 315–323.
[13] G.A. Morris, *J. Magn. Reson.*, 1992, **100**, 316–328.
[14] R. Baumann, G. Wider, R.R. Ernst and K. Wüthrich, *J. Magn. Reson.*, 1981, **44**, 402–406.
[15] G.A. Morris, *J. Magn. Reson.*, 1988, **80**, 547–552.
[16] G.A. Morris and D. Cowburn, *Magn. Reson. Chem.*, 1989, **27**, 1085–1089.
[17] T.J. Horne and G.A. Morris, *J. Magn. Reson. (A)*, 1996, **123**, 246–252.
[18] A. Gibbs, G.A. Morris, A.G. Swanson and D. Cowburn, *J. Magn. Reson. (A)*, 1993, **101**, 351–356.
[19] A.D. Bain, *J. Magn. Reson.*, 1984, **56**, 418–427.
[20] G. Bodenhausen, H. Kogler and R.R. Ernst, *J. Magn. Reson.*, 1984, **58**, 370–388.
[21] H. Kessler, M. Gehrke and C. Griesinger, *Angew. Chem., Int. Ed. Engl.*, 1988, **27**, 490–536.
[22] J. Keeler, R.T. Clowes, A.L. Davis and E.D. Laue, *Method. Enzymol.*, 1994, **239**, 145–207.
[23] S. Berger, *Prog. Nucl. Magn. Reson. Spectrosc.*, 1997, **30**, 137–156.
[24] T. Parella, *Magn. Reson. Chem.*, 1998, **36**, 467–495.
[25] P.C. Lauterbur, *Nature*, 1973, **242**, 190–191.
[26] A. Bax, P.G. de Long, A.F. Mehlkopf and J. Smidt, *Chem. Phys. Lett.*, 1980, **69**, 567–570.
[27] P. Barker and R. Freeman, *J. Magn. Reson.*, 1985, **64**, 334–338.
[28] R.E. Hurd, *J. Magn. Reson.*, 1990, **87**, 422–428.
[29] A.L. Davis, J. Keeler, E.D. Laue and D. Moskau, *J. Magn. Reson.*, 1992, **98**, 207–216.
[30] J. Jeener, B.H. Meier, P. Bachmann and R.R. Ernst, *J. Chem. Phys.*, 1979, **71**, 4546–4553.
[31] J. Weigelt and G. Otting, *J. Magn. Reson. (A)*, 1995, **113**, 128–130.
[32] V. Sklenár, M. Pioto, R. Leppik and V. Saudek, *J. Magn. Reson. (A)*, 1993, **102**, 241–245.
[33] T.L. Hwang and A.J. Shaka, *J. Magn. Reson. (A)*, 1995, **112**, 275–279.
[34] G. Kontaxis, J. Stonehouse, E.D. Laue and J. Keeler, *J. Magn. Reson. (A)*, 1994, **111**, 70–76.
[35] E.O. Stejskal and J.E. Tanner, *J. Chem. Phys.*, 1965, **42**, 288–292.
[36] W.S. Price, *Ann. Rep. NMR. Spectrosc.*, 1996, **32**, 51–142.
[37] M. Rance, O.W. Sørensen, G. Bodenhausen, G. Wagner, R.R. Ernst and K. Wüthrich, *Biochem. Biophys. Res. Commun.*, 1983, **117**, 479–485.
[38] U. Piantini, O.W. Sørensen and R.R. Ernst, *J. Am. Chem. Soc.*, 1982, **104**, 6800–6801.
[39] A.L. Davis, E.D. Laue, J. Keeler, D. Moskau and J. Lohman, *J. Magn. Reson.*, 1991, **94**, 637–644.
[40] M. Eberstadt, G. Gemmecker, D.F. Mierke and H. Kessler, *Angew. Chem. Int. Ed. Engl.*, 1995, **34**, 1671–1695.
[41] D. Neuhaus, G. Wagner, M. Vasák, J.H.R. Kági and K. Wüthrich, *Eur. J. Biochem.*, 1985, **151**, 257–273.
[42] H. Kessler and H. Oschkinat, *Angew. Chem. Int. Ed. Engl.*, 1985, **24**, 690–692.
[43] H. Kessler, A. Müller and H. Oschkinat, *Magn. Reson. Chem.*, 1985, **23**, 844–852.
[44] H. Oschkinat and R. Freeman, *J. Magn. Reson.*, 1984, **60**, 164–169.
[45] A.J. Shaka and R. Freeman, *J. Magn. Reson.*, 1983, **51**, 169–173.
[46] N. Müller, R.R. Ernst and K. Wüthrich, *J. Am. Chem. Soc.*, 1986, **108**, 6482–6492.
[47] J. Boyd and C. Redfield, *J. Magn. Reson.*, 1986, **68**, 67–84.
[48] G. Eich, G. Bodenhausen and R.R. Ernst, *J. Am. Chem. Soc.*, 1982, **104**, 3731–3732.
[49] L. Braunschweiler and R.R. Ernst, *J. Magn. Reson.*, 1983, **53**, 521–528.
[50] D.G. Davis and A. Bax, *J. Am. Chem. Soc.*, 1985, **107**, 2820–2821.
[51] A. Bax and D.G. Davis, *J. Magn. Reson.*, 1985, **65**, 355–360.
[52] J. Cavanagh, W.J. Chazin and M. Rance, *J. Magn. Reson.*, 1990, **87**, 110–131.
[53] G.A. Morris and A. Gibbs, *J. Magn. Reson.*, 1991, **91**, 444–449.
[54] A. Gibbs and G.A. Morris, *Magn. Reson. Chem.*, 1992, **30**, 662–665.
[55] J.E. Baldwin, L. Bischoff, T.D.W. Claridge, F.A. Heupel, D.R. Spring and R.C. Whitehead, *Tetrahedron*, 1997, **53**, 2271–2290.
[56] J.E. Baldwin, T.D.W. Claridge, A.J. Culshaw, F.A. Heupel, V. Lee, D.R. Spring, R.C. Whitehead, R.J. Boughtflower, I.M. Mutton and R.J. Upton, *Angew. Chem. Int. Ed. Engl.*, 1998, **37**, 2661–2663.
[57] C. Griesinger, G. Otting, K. Wüthrich and R.R. Ernst, *J. Am. Chem. Soc.*, 1988, **110**, 7870–7872.
[58] J. Briand and R.R. Ernst, *Chem. Phys. Lett.*, 1991, **185**, 276–285.
[59] J. Cavanagh and M. Rance, *J. Magn. Reson.*, 1992, **96**, 670–678.
[60] M. Rance, *J. Magn. Reson.*, 1987, **74**, 557–564.
[61] R. Bazzo and I.D. Campbell, *J. Magn. Reson.*, 1988, **76**, 358–361.
[62] O.W. Sørensen, M. Rance and R.R. Ernst, *J. Magn. Reson.*, 1984, **56**, 527–534.
[63] A.J. Shaka, J. Keeler and R. Freeman, *J. Magn. Reson.*, 1983, **53**, 313–340.
[64] A.J. Shaka, J. Keeler, T. Frenkiel and R. Freeman, *J. Magn. Reson.*, 1983, **52**, 335–338.
[65] S.P. Rucker and A.J. Shaka, *Mol. Phys.*, 1989, **68**, 509–517.
[66] A.J. Shaka, C.J. Lee and A. Pines, *J. Magn. Reson.*, 1988, **77**, 274–293.

[67] H. Kessler, S. Mronga and G. Gemmecker, *Magn. Reson. Chem.*, 1991, **29**, 527–557.
[68] T. Parella, *Magn. Reson. Chem.*, 1996, **34**, 329–347.
[69] D.G. Davis and A. Bax, *J. Am. Chem. Soc.*, 1985, **107**, 7197–7198.
[70] H. Kessler, U. Anders, G. Gemmecker and S. Steuernagel, *J. Magn. Reson.*, 1989, **85**, 1–14.
[71] H. Kessler, H. Oschkinat, C. Griesinger and W. Bermel, *J. Magn. Reson.*, 1986, **70**, 106–133.
[72] S. Subramanian and A. Bax, *J. Magn. Reson.*, 1987, **71**, 325–330.
[73] P. Adell, T. Parella, F. Sanchez-Ferrando and A. Virgili, *J. Magn. Reson. (B)*, 1995, **108**, 77–80.
[74] T. Fäcke and S. Berger, *J. Magn. Reson. (A)*, 1995, **113**, 257–259.
[75] C. Dalvit and G. Bovermann, *Magn. Reson. Chem.*, 1995, **33**, 156–159.
[76] G.Z. Xu and J.S. Evans, *J. Magn. Reson. (B)*, 1996, **111**, 183–185.
[77] J. Buddrus and H. Bauer, *Angew. Chem. Int. Ed. Engl.*, 1987, **26**, 625–642.
[78] A. Bax, R. Freeman and S.P. Kempsell, *J. Am. Chem. Soc.*, 1980, **102**, 4849–4851.
[79] A. Bax, R. Freeman and T.H. Frenkiel, *J. Am. Chem. Soc.*, 1981, **103**, 2102–2104.
[80] M. Budesinsky and D. Saman, *Ann. Rep. NMR. Spectrosc.*, 1995, **30**, 231–475.
[81] J.C. Vederas, *Nat. Prod. Rep.*, 1987, **4**, 277–337.
[82] T.H. Mareci and R. Freeman, *J. Magn. Reson.*, 1982, **48**, 158–163.
[83] J. Lambert, H.J. Kuhn and J. Buddrus, *Angew. Chem. Int. Ed. Engl.*, 1989, **28**, 738–740.
[84] A.M. Torres, T.T. Nakashima, R.E.D. McClung and D.R. Muhandiram, *J. Magn. Reson.*, 1992, **99**, 99–117.
[85] R. Dunkel, C.L. Mayne, J. Curtis, R.J. Pugmire and D.M. Grant, *J. Magn. Reson.*, 1990, **90**, 290–302.
[86] R. Dunkel, C.L. Mayne, R.J. Pugmire and D.M. Grant, *Anal. Chem.*, 1992, **64**, 3133–3149.
[87] R. Dunkel, C.L. Mayne, M.P. Foster, C.M. Ireland, D. Li, N.L. Owen, R.J. Pugmire and D.M. Grant, *Anal. Chem.*, 1992, **64**, 3150–3160.
[88] J.K. Harper, R. Dunkel, S.G. Wood, N.L. Owen, D. Li, R.G. Cates and D.M. Grant, *J.C.S., Perkin Trans.*, 1996, **2**, 191–200.
[89] A. Bax and R. Freeman, *J. Magn. Reson.*, 1980, **41**, 507–511.
[90] E. Breitmaier, W. Voelter, Carbon-13 NMR Spectroscopy, VCH Publishers, Weinheim, 1987.
[91] K. Kamienska-Trela, *Ann. Rep. NMR. Spectrosc.*, 1995, **30**, 131–230.
[92] D.L. Mattiello and R. Freeman, *J. Magn. Reson.*, 1998, **135**, 514–521.
[93] J. Lettvin and A.D. Sherry, *J. Magn. Reson.*, 1977, **28**, 459–461.
[94] O.W. Sørensen, R. Freeman, T.A. Frenkiel, T.H. Mareci and R. Schuck, *J. Magn. Reson.*, 1982, **46**, 180–184.
[95] S.W. Sparks and P.D. Ellis, *J. Magn. Reson.*, 1985, **62**, 1–11.
[96] P.J. Keller and K.E. Vogele, *J. Magn. Reson.*, 1986, **68**, 389–392.
[97] B. Reif, M. Kock, R. Kerssebaum, H. Kang, W. Fenical and C. Griesinger, *J. Magn. Reson., Series (A)*, 1996, **118**, 282–285.

Chapter 6

Correlations through the chemical bond II: Heteronuclear shift correlation

6.1. INTRODUCTION

This second chapter on establishing correlations through the chemical bond concentrates on techniques which correlate different nuclides, so-called *heteronuclear* shift correlations. For an organic chemist this means, in the vast majority of cases, establishing connectivities between proton and carbon nuclei and as such the techniques encountered in the sections which follow are concerned primarily with these. That is not to say that the techniques are not suitable for correlating other nuclides, for example ^{1}H with ^{15}N or ^{31}P, or even ^{31}P with ^{13}C and so on. Indeed many of the modern techniques used routinely in the chemical laboratory were originally implemented as methods for ^{1}H–^{15}N correlations in proteins and peptides. The principal techniques described in the sections that follow are summarised in Table 6.1.

A variety of methods have already been described in Chapter 4 that allow the editing of the 1D spectrum of the heteronuclear spin, for example those based on spin-echoes or polarisation transfer, so providing valuable information on the numbers of attached protons. They do not, however, provide any *direct* evidence for which protons are attached to which heteronucleus (X-spin) in the molecule and for this heteronuclear 2D correlations are widely used to transfer previously established proton assignments onto the directly bonded heteronucleus or, on occasions, vice versa. This may then provide further evidence to support or reject the proposed structure as being correct, or may provide assignments that can be used as the basis of further investigations. In

Table 6.1. The principal applications of the main techniques described in this chapter

Technique	Principal applications
HMQC	Correlating coupled heteronuclear spins across a single bond and hence identifying directly connected nuclei, most often ^{1}H–^{13}C. Employs detection of high-sensitivity nuclide e.g. ^{1}H, ^{19}F, ^{31}P (an 'inverse technique'). Experimentally robust sequence, well suited to routine structural characterisation.
HSQC	Correlating coupled heteronuclear spins across a single bond and hence identifying directly connected nuclei. Employs detection of high-sensitivity nuclide e.g. ^{1}H, ^{19}F, ^{31}P (an 'inverse technique'). Provides improved resolution over HMQC so is better suited for crowded spectra but is more sensitive to experimental imperfections.
HMBC	Correlating coupled spins across multiple bonds. Employs detection of high-sensitivity nuclide e.g. ^{1}H, ^{19}F, ^{31}P (an 'inverse technique'). Essentially HMQC tuned for the detection of small couplings. Most valuable in correlating ^{1}H–^{13}C over 2- or 3-bonds. Powerful tool for linking together structural fragments.
HETCOR	Correlating coupled heteronuclear spins across a single bond. Employs detection of the lower-γ nuclide, typically ^{13}C, so has significantly lower sensitivity than inverse techniques. Benefits from high-resolution in ^{13}C dimension, so may find use when this is critical, otherwise superseded by above methods.

addition, the experiment may be used as a means of spreading the resonances of a complex proton spectrum according to the chemical shift of the directly attached nucleus, utilising the typically greater dispersion of the X-spin chemical shifts to assist with proton interpretation. The high-sensitivity of modern correlation techniques often provides a fast method for determining indirectly chemical shifts of the X-nucleus and avoids the need for its direct observation altogether, offering considerable timesavings. Thus, the ^{1}H–^{1}H COSY and the ^{1}H–^{13}C correlation experiments (as the HMQC or HSQC sequences) represent the primary 2D techniques in structural organic chemistry.

For more complex problems there exist methods that combine the features of two, otherwise separate, techniques, which shall be referred to here as 'hybrid' experiments. Whilst a wide range of combinations have been devised, essentially limited by the imagination of the spectroscopist, two principal features are of considerable utility. The first is editing of the correlation spectrum such that it contains both shift and multiplicity data, by analogy with the 1D editing methods presented in Chapter 4. The second feature is the relaying of correlations to enhance the information content of the spectrum by providing additional neighbouring-group information. Beyond these methods that utilise one-bond heteronuclear couplings, correlations with a heteronucleus over more than one bond, so-called long-range or multiple-bond correlations, can provide a wealth of connectivity information on how molecular fragments are linked together.

6.2. SENSITIVITY

The original methods for determining heteronuclear shift correlations were based on the observation of the low-γ X-nucleus, with the proton being indirectly detected and consequently appearing along the f_1 dimension of the 2D experiment. This approach was adopted because, as we shall see, the original 2D sequences were derived from early polarisation transfer experiments, such as INEPT (Section 4.4), which were themselves designed to enhance the sensitivity of low-γ observations, and because early pulsed NMR instruments were designed with this mode of operation in mind. During the last decade or so, the approach to data collection has fundamentally changed to one in which the high-γ nucleus, most frequently the proton, is observed, with the heteronucleus now detected indirectly. This switch has given rise to a body of experiments frequently referred to as 'inverse' shift correlations, the motivation for change being improved sensitivity.

The dependence of the strength of an NMR signal on magnetogyric ratio has been discussed previously in Section 4.4 where the concept of polarisation transfer was introduced as a means of sensitivity enhancement. From the qualitative arguments presented in that section a more formal expression for the signal-to-noise ratio of a one-dimensional experiment involving spin-½ nuclei was given as:

$$\frac{S}{N} \propto N\,A\,T^{-1}B_0^{3/2}\gamma_{exc}\gamma_{obs}^{3/2}T_2^{*}(NS)^{1/2} \qquad (6.1)$$

where N is the number of molecules in the observed sample volume, A is a term that represents the abundance of the NMR active spins involved in the experiment, T is temperature, B_0 is the static magnetic field, γ_{exc} and γ_{obs} represent the magnetogyric ratios of the initially excited and the observed spins respectively, T_2^{*} is the effective transverse relaxation time and NS is the total number of accumulated scans. When choosing how to perform a heteronuclear shift correlation experiment, four general schemes may be devised, as represented in Fig. 6.1, according to which spin is used as the initial

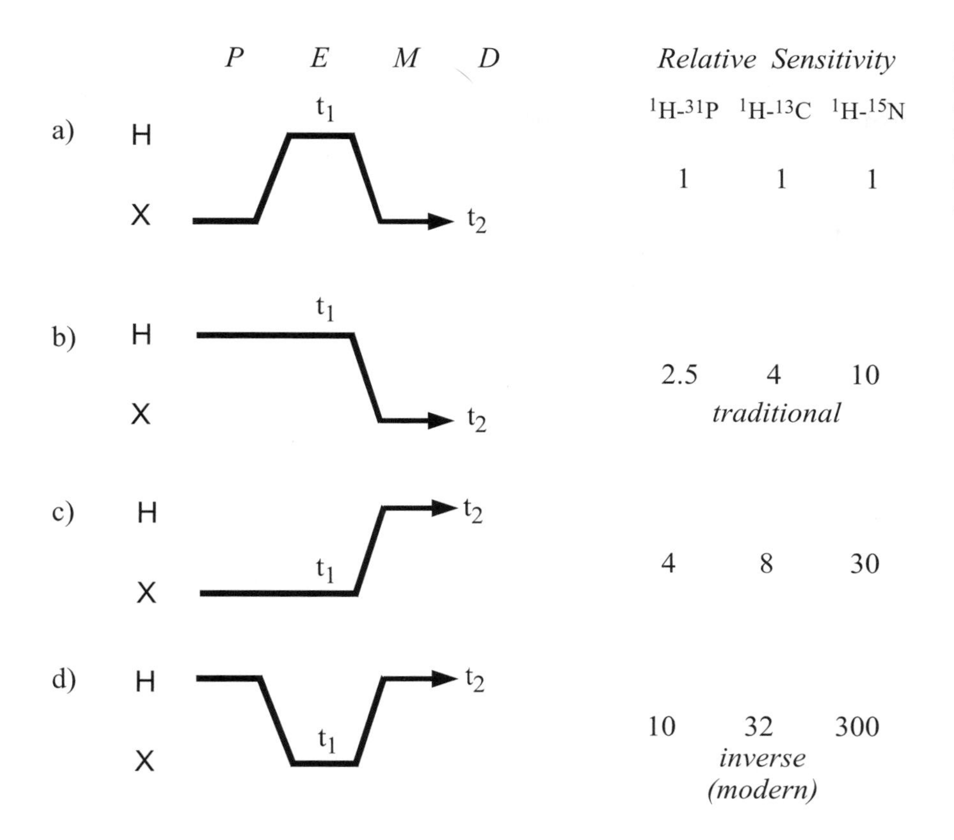

Figure 6.1. The four general schemes to produce 2D heteronuclear shift correlation spectra. The relative sensitivities of these approaches are compared for proton correlation experiments with phosphorus-31, carbon-13, and nitrogen-15.

magnetisation 'source' and which is detected (represented in the γ factors in Eq. 6.1). Alongside each schematic is summarised the relative theoretical sensitivities expected for an H–X pair based on the participating γ values. Scheme 6.1b represents the approach adopted in the traditional shift correlation experiments, in which initial proton magnetisation is frequency-labelled in t_1 and then transferred onto the X-spin for detection. Scheme 6.1d is the modern 'inverse' approach in which the proton nucleus is used both as the source and observed spin. Clearly the predicted sensitivity is significantly greater than that of the traditional method, and the dependence on the magnetogyric ratio of the low-γ spin has been completely removed, although the natural abundance of this spin is still an important factor for all approaches (term A in the above equation). The lower the γ of the X-spin, the greater the gains arising from proton detection; compare the figures for ^{31}P and ^{15}N in Fig. 6.1 for example. Additional theoretical gains by a factor of 2 or 3 can be anticipated for XH_2 or XH_3 groups respectively with the proton detected experiments.

The impressive gains illustrated in Fig. 6.1 for inverse vs traditional methods may not be met in practice when the details of a particular sequence are considered. Significant factors may include the different relaxation behaviour of participating spins or the presence of multiplet splittings that spread a resonance and so reduce the signal-to-noise ratio. However, even a realistic gain of a factor of four in ^{1}H–^{13}C correlations corresponds to a time saving of a factor of sixteen, meaning experiments that once required an overnight acquisition with ^{13}C detection can be collected in about an hour, whilst those traditionally requiring a couple of hours can be completed within a matter of minutes. Furthermore, studies on very dilute samples that would have been considered intractable become viable targets with proton detection. Such considerations have led to the universal adoption of proton-detected inverse-correlation methods whenever possible in both chemical and biological spectroscopy, and it is these techniques that are focused upon below, along with only a relatively brief consideration of the more traditional X-detected methods that may still have utility in specific circumstances.

The adoption of the inverse approach also has implications for the design of the NMR instrument. Conventional probes have been constructed so as to optimise the sensitivity for observation of the low-γ X-nucleus, which entails placing the X-nucleus coil closest to the sample and positioning the proton coil outside this. Inverse probes have this configuration switched such that the proton coil sits closest to the sample for optimum sensitivity, thus providing a greater *filling factor*. However, even with conventional probes, the proton detected experiments can still be performed, albeit with less than optimum sensitivity, and may still provide a faster approach than the former X-observe experiments.

6.3. HETERONUCLEAR SINGLE-BOND CORRELATION SPECTROSCOPY

There are two techniques in widespread use that provide single-bond heteronuclear shift correlations, known colloquially as HMQC and HSQC. The correlation data provided by these two methods are essentially equivalent, the methods differing only in finer details which, for routine spectroscopy, are often of little consequence. As such, both methods are ubiquitous throughout the chemical literature and hence will be described in the sections that follow. Historically, the HMQC experiment has been favoured by the chemical community and HSQC by biological spectroscopists. This is, at least in part, due to the manner in which the early experiments were presented, viz HMQC for ^{1}H–^{13}C correlations in small molecules and HSQC for ^{1}H–^{15}N correlations in proteins. At a practical level, the HMQC experiment tends to be more robust with respect to experimental imperfections or miscalibrations whilst the HSQC experiment has more favourable characteristics for very high-resolution work and is more flexible with regard to modification and extension of the sequence, so is now widely used in the chemical laboratory also; these points are pursued in the following sections. Both techniques employ the optimum approach to establishing heteronuclear connectivity utilising proton detection and following the general scheme of Fig. 6.1d. However, this approach poses technical difficulties in the suppression of the parent resonance arising from protons bound to nuclides with $I \neq 1/2$, ^{1}H–^{12}C and ^{1}H–^{14}N for example. This is the dominant line observed in 1D proton spectra but is merely a source of interference in heteronuclear correlations since it is only the low-intensity *satellites* that can give rise to the desired correlations, that is, the ^{1}H–^{13}C or ^{1}H–^{15}N protons in the above examples. The necessary suppression is nowadays most effectively executed by the application of pulsed field gradients (PFGs) which have had an enormous impact on heteronuclear correlation spectroscopy in particular.

6.3.1. Heteronuclear multiple-quantum correlation (HMQC)

The heteronuclear multiple-quantum correlation (HMQC) experiment is one of the two commonly employed proton-detected single-bond correlation experiments. Although this was suggested many years ago [1,2], the experiment lacked widespread use until a scheme was presented [3] that was able to overcome the technical difficulties associated with proton observation described above. Since then, and particularly since the advent of PFGs, the technique has come to dominate organic NMR spectroscopy.

The 2D spectrum provides a simple map of connectivities in which a crosspeak correlates two attached nuclei, as seen in the ^{1}H–^{13}C HMQC spectrum of menthol **6.1** (Fig. 6.2). This illustrates three of the most significant features of the experiment when applied to routine structural problems. The

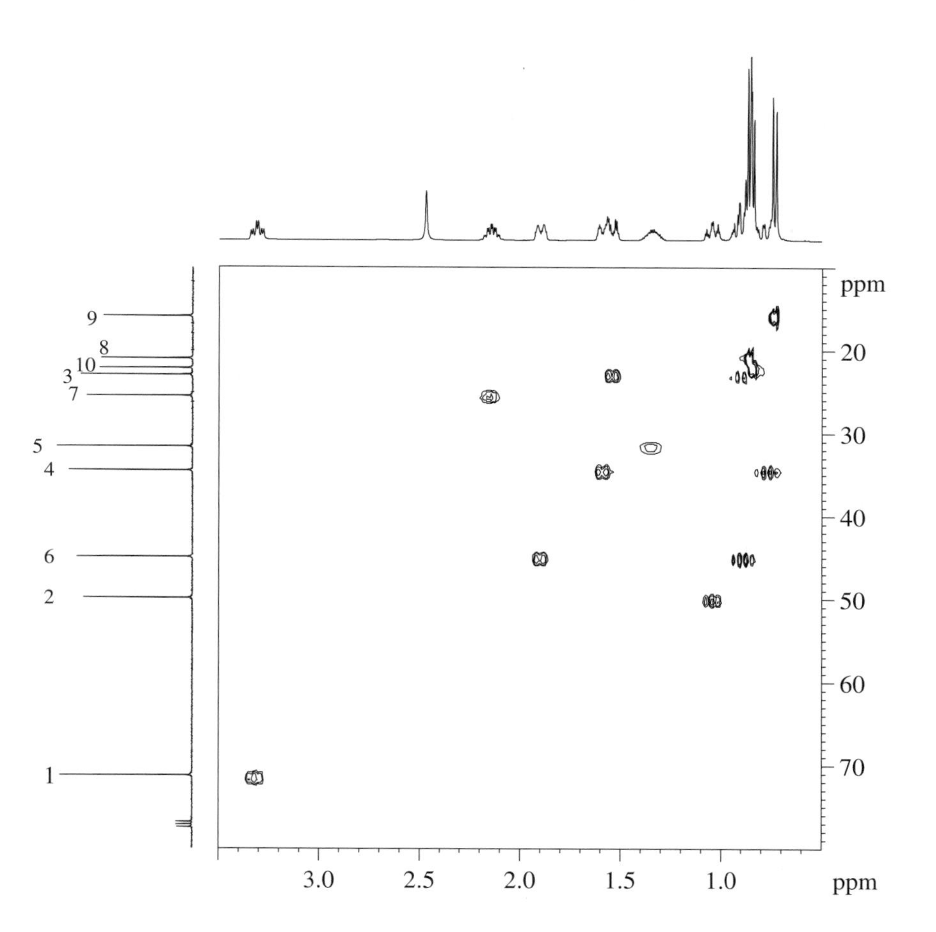

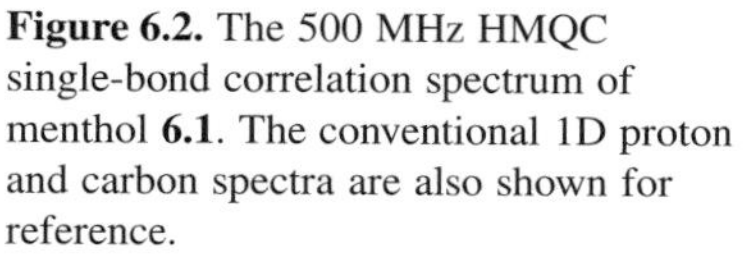

Figure 6.2. The 500 MHz HMQC single-bond correlation spectrum of menthol **6.1**. The conventional 1D proton and carbon spectra are also shown for reference.

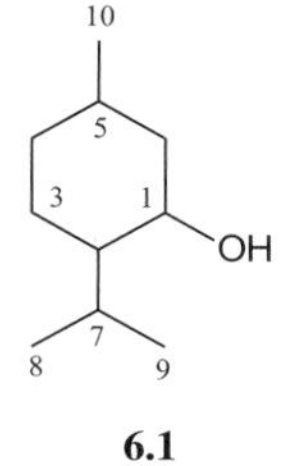

6.1

first is the ability to transfer known proton assignments, determined for example with the methods described in Chapter 5, onto the spectrum of the heteronucleus, so extending characterisation of the molecule. The second is the dispersion of the proton resonances according to the heteronuclear shift, which itself can aid the initial interpretation of the proton spectrum. For example, the region between 0.7 and 1.2 ppm of Fig. 6.2 contains a number of overlapped proton resonances which defy direct analysis even at 500 MHz. A clearer picture of events in this region emerges when these are dispersed along the ^{13}C dimension, revealing the presence of seven distinct groups. This property becomes increasingly valuable as the complexity and overlap within proton spectra increases. The final feature which can prove surprisingly useful in structural assignment is the ability to identify *diastereotopic* geminal pairs. These are not always readily identifiable in COSY spectra owing to the lack of differentiation of geminal and vicinal couplings, and can lead to ambiguity in proton assignment. Only for these geminal pairs will two correlations to a single carbon resonance be observed, as seen for the correlations of C6, C4 and C3 in Fig. 6.2. In contrast to previous homonuclear 2D spectra encountered in this book, heteronuclear shift correlation spectra lack a diagonal and are not symmetrical about $f_1 = f_2$, a simple consequence of there being different nuclides represented in the two dimensions.

The HMQC sequence

Despite the slightly foreboding title, the basic HMQC sequence is rather simple, comprising only four rf pulses (Fig. 6.3), the operation of which is considered here for a simple ^1H–^{13}C spin pair. The sequence starts with proton excitation followed by evolution of proton magnetisation under the influence

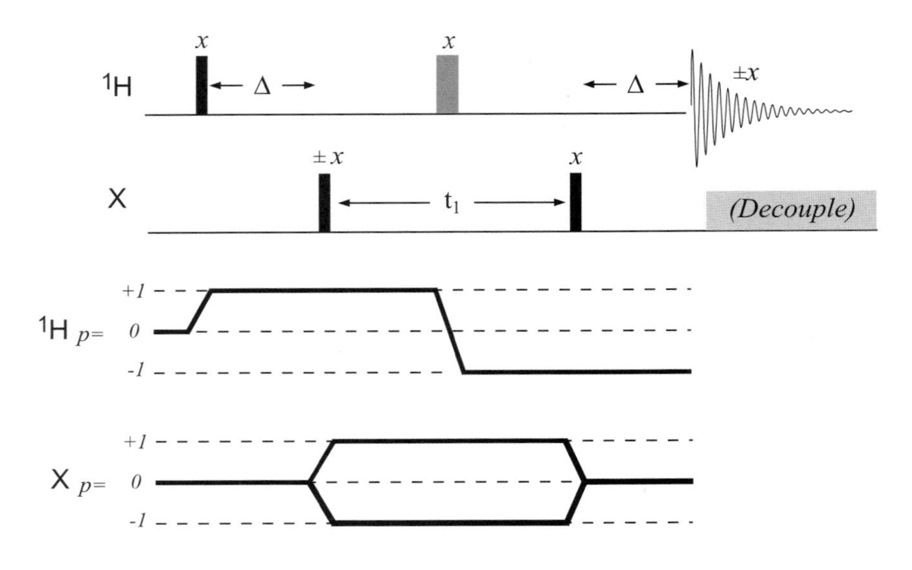

Figure 6.3. The HMQC sequence and associated coherence transfer pathway. The Δ periods are set to $1/2J_{CH}$ to enable to defocusing and subsequent refocusing of the one-bond heteronuclear coupling.

Table 6.2. Typical ranges for one-bond carbon–proton coupling constants

Proton environment	Typical $^1J_{CH}$ range (Hz)
Aliphatic, CH_n–	125–135
Aliphatic, CH_nX (X = N, O, S)	135–155
Alkene	155–170
Alkyne	240–250
Aromatic	155–165

of the one-bond carbon–proton coupling. During a period Δ, antiphase proton magnetisation develops with respect to $^1J_{CH}$ (Table 6.2), Δ is therefore typically set to around 3.3 ms. As in the case of INEPT or COSY, this antiphase magnetisation may be transferred to the coupled partner by the action of a subsequent rf pulse and the role of the first carbon pulse in HMQC is to generate proton–carbon multiple-quantum coherence (hence the title of the experiment). Recall that multiple-quantum coherence was described in Section 5.4 as a pooling of the transverse magnetisation of coupled spins, in this case a proton and its directly bound carbon, that evolves coherently but which cannot be directly observed. If one were to start data collection of the proton signal directly after this carbon pulse there would be nothing to detect, provided the Δ delay was set precisely to $1/2\,J_{CH}$ [1].

This coherence is, in fact, a combination of both heteronuclear double- and zero-quantum coherence, as represented on the coherence transfer pathway of Fig. 6.3. Thus, for example, $^1H_p = +1$ and $^{13}C_p = +1$ corresponds to *double*-quantum coherence ($\Sigma p = 2$), whilst $^1H_p = +1$ and $^{13}C_p = -1$ is *zero*-quantum coherence ($\Sigma p = 0$). The salient point at this stage is how such coherences evolve during the subsequent t_1 period. Since they contain terms for both transverse proton and carbon magnetisation, they will evolve under the influence of *both proton and carbon chemical shifts* (although a feature of multiple-quantum coherences is that they will *not* evolve under the 'active' coupling, so J_{CH} need not be considered at this point). What is ultimately

[1] Notice that this is exactly the procedure described in Section 3.5.1 for the calibration of pulses on the indirectly observed spin and, allowing for the proton 180° pulse, is also closely related to the start of the DEPT sequence (Section 4.4.3) which similarly relies on the generation of multiple-quantum coherence.

required, however, is frequency labelling according to *carbon* shifts only, since this is what one wishes to characterise in the indirectly detected f_1 dimension. To remove the effect of proton shifts during t_1, a spin-echo is incorporated by placing a proton 180° pulse at the midpoint of t_1, so by the end of the evolution period these shifts have refocused and thus have no influence in f_1. Evolution of the carbon shifts is unaffected by the proton pulse, so these remain to produce the desired frequency labelling. The final carbon pulse then reconverts the multiple-quantum coherence back to observable single-quantum proton magnetisation which is once again antiphase with respect to $^1J_{CH}$. When collecting the proton FID it is generally desirable to apply broadband decoupling of the *carbon* spins to remove the J_{CH} doublet fine structure, thus doubling the signal-to-noise ratio. To avoid cancellation of the antiphase proton satellites, a second Δ period is inserted to refocus the proton–carbon coupling, after which the proton magnetisation is detected. Conventional quadrature detection in f_1 is implemented by incrementing the phase of the carbon pulse prior to t_1, according to either the States or TPPI procedures to yield a phase-sensitive display. The result is a two-dimensional spectrum with ^{1}H shifts represented in f_2, ^{13}C shifts in f_1 and crosspeaks indicating one-bond connectivities (Fig. 6.4a and b). One subtle point to be aware of is that although this sequence (and HSQC below) nominally detects single-bond correlations, in exceptional circumstances these may be missing and longer range correlations may appear. This occurs when the actual coupling constant involved is far from the value assumed when calculating Δ, a situation most likely to occur for alkynes where the one-bond and two-bond (**H−C≡C**) couplings are unusually large. A large two-bond coupling (>50 Hz) can be sufficient to produce a crosspeak and care should be taken not to confuse this with a single-bond correlation if dealing with these systems.

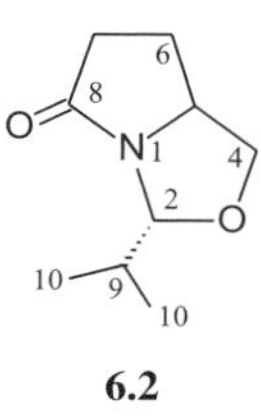

6.2

Figure 6.4. The 400 MHz HMQC spectra of **6.2** recorded (a) with and (b) without carbon decoupling during data collection. In the absence of decoupling, each crosspeak appears with doublet structure along f_2 arising from $^1J_{CH}$. These doublets are merely the usual ^{13}C satellites observed in the 1D proton spectrum. 1K t_2 data points were collected for 256 t_1 increments of 2 transients each. Data were processed with $\pi/2$ shifted sine-bells in both dimensions and presented in phase-sensitive mode. Zero-filling once in t_1 resulted in digital resolutions of 4 and 40 Hz/pt in f_2 and f_1 respectively.

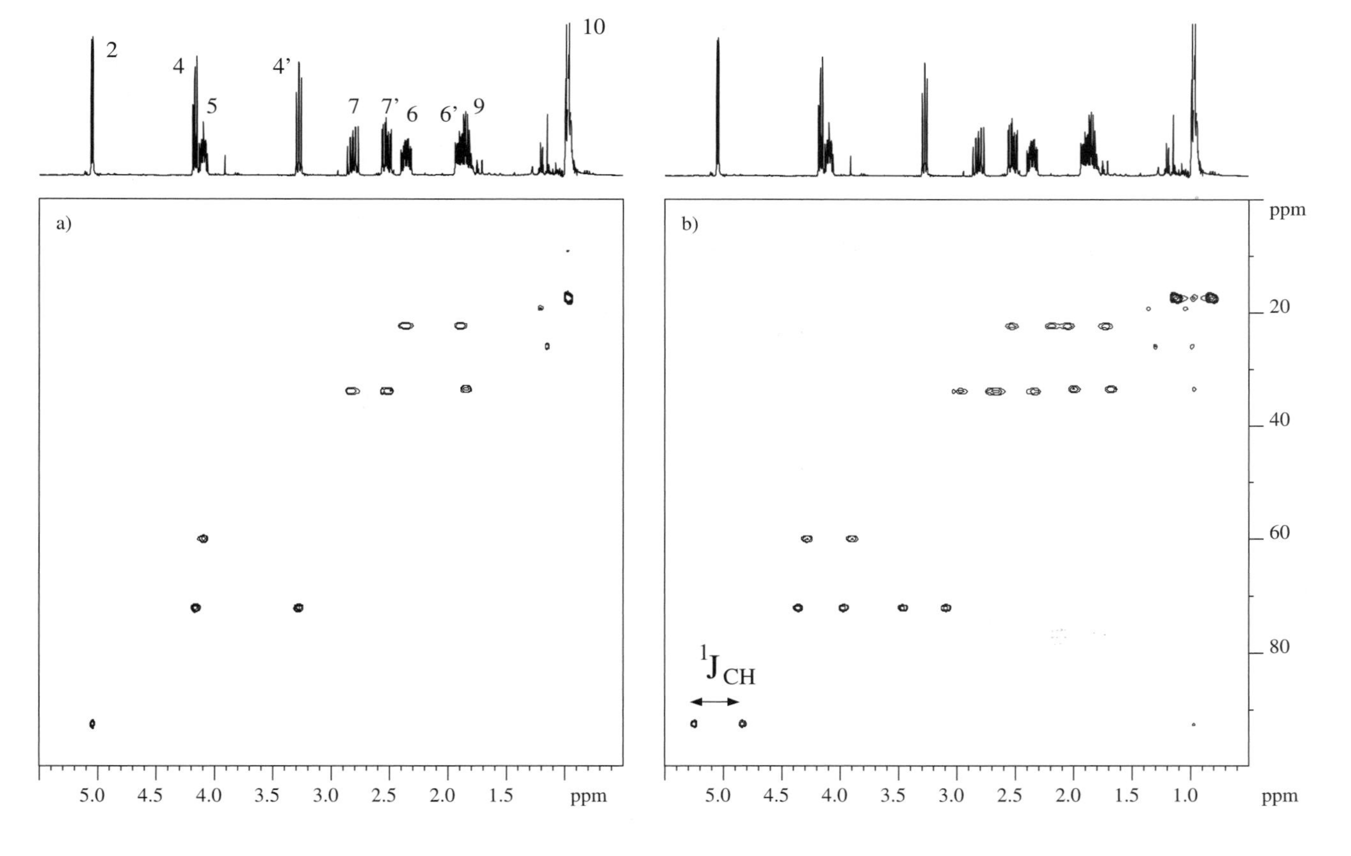

The influence of homonuclear proton couplings

It was noted above that proton magnetisation will also evolve according to its chemical shift during Δ after initial excitation. However, this is exactly refocused during the second Δ period because of the presence of the proton spin-echo, so does not give rise to phase errors in the proton dimension. Of greater concern is the evolution of proton magnetisation in the two Δ periods and of heteronuclear multiple-quantum coherence during t_1, under the influence of homonuclear proton–proton couplings. Since these homonuclear couplings are not refocused by a spin-echo they will evolve in both Δ periods and potentially contribute to unwanted phase errors in the proton dimension. However, Δ is set according to $^1J_{CH}$ which is typically at least an order of magnitude greater than J_{HH}, so that in practice the degree of evolution due to proton–proton coupling is rather small. In other words, Δ is too short for significant evolution to occur and the small phase errors that may arise are rarely troublesome (as described below, this is not the case when seeking correlations through long-range heteronuclear couplings that are comparable in size to homonuclear proton couplings). In contrast, multiple quantum coherence evolves during t_1 under *passive* J_{HH} couplings without being refocused and thus the final carbon resonances are spread by proton–proton couplings along f_1! This may seem a little odd, but it is a consequence of the fact that during t_1 both proton and carbon coherences evolve; we simply choose to remove proton chemical shifts with the spin-echo. In fact, because of the rather low digital resolution used in the carbon-13 dimension, as described below, these proton couplings are rarely resolved, but do contribute to undesirable *broadening* of the resonance along f_1. These homonuclear couplings *do not* appear in the f_1 dimension of HSQC and this feature is the principal difference between the two spectra (see Section 6.3.2 and Fig. 6.7 below).

Interference from ^1H–^{12}C/^1H–^{14}N resonances

The HMQC sequence aims to detect only those protons that are bond to a spin-½ heteronucleus, or in other words only the satellites of the conventional proton spectrum. In the case of ^{13}C, this means that only 1 in every 100 proton spins contribute to the 2D spectrum (the other 99 being attached to NMR inactive ^{12}C) whilst for ^{15}N with a natural abundance of a mere 0.37%, only 1 in 300 contribute. When the HMQC FID is recorded, all *protons* will induce a signal in the receiver *on each scan* and the unwanted resonances, which clearly represent the vast majority, must be removed with a suitable phase cycle if the correlation peaks are to be revealed (the notable exception to this is when pulsed field gradients are employed for signal selection, see Section 6.3.3 below). By inverting the first ^{13}C pulse on alternate scans, the phase of the ^{13}C satellites are themselves inverted whereas the ^{12}C-bound protons remain unaffected (Fig. 6.5). Simultaneous inversion of the receiver will lead to cancellation of the unwanted resonances with corresponding addition of the desired satellites. This two step procedure is the fundamental phase-cycle of the HMQC experiment, as indicated in Fig. 6.3 above.

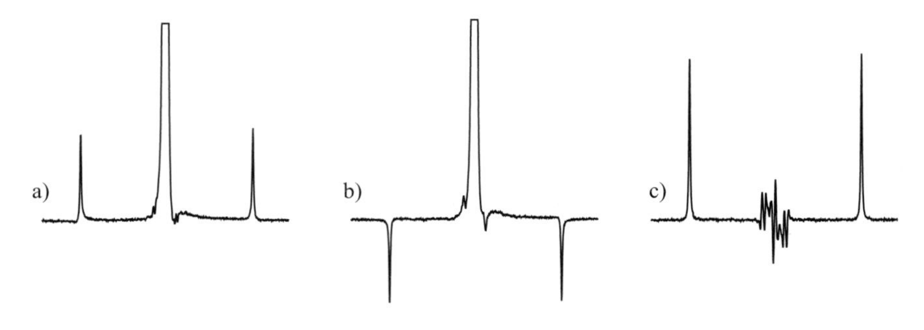

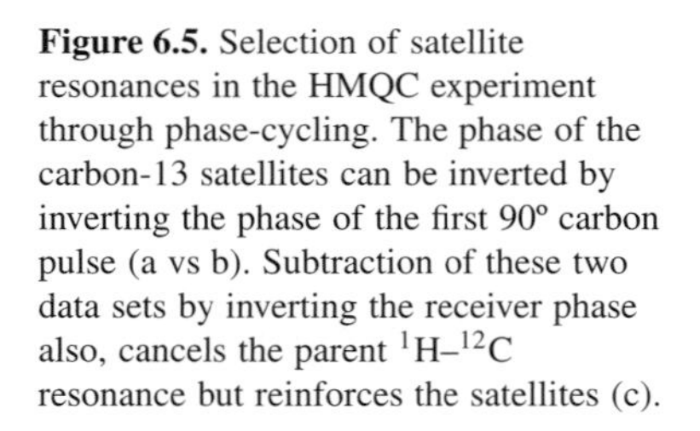

Figure 6.5. Selection of satellite resonances in the HMQC experiment through phase-cycling. The phase of the carbon-13 satellites can be inverted by inverting the phase of the first 90° carbon pulse (a vs b). Subtraction of these two data sets by inverting the receiver phase also, cancels the parent ^1H–^{12}C resonance but reinforces the satellites (c).

The problem with this scheme is that clean suppression of the unwanted resonances is unlikely to be achieved by phase-cycling alone, with residual signals contributing to undesirable bands of t_1-noise in the resulting spectra which may mask the genuine correlations. Extensive time averaging can help to progressively reduce the intensity of these artefacts, but defeats the time-saving advantages otherwise made possible by this higher-sensitivity approach, and the use of a 'double-difference' phase-cycle [4] has been shown to improve the suppression. The detection of all proton resonances on each scan also imposes dynamic range limitations, in that a substantial part of the digitiser collects data that is ultimately discarded. A practically useful approach requires attenuation of the parent proton resonance before data collection begins, and Section 6.3.3 describes two methods that are widely employed to meet this goal.

6.3.2. Heteronuclear single-quantum correlation (HSQC)

The heteronuclear single-quantum correlation experiment (HSQC) [5] also follows the general scheme of Fig. 6.1d but differs from HMQC in that only transverse (single-quantum) magnetisation of the heteronuclear spin evolves during the t_1 period ($\Sigma p = \pm 1$), rather than ^{1}H–X multiple-quantum coherence (Fig. 6.6). The transverse heteronuclear magnetisation is generated by polarisation transfer from the attached protons via the INEPT sequence, exactly as described for the 1D experiment in Section 4.4.2 (compare the first part of Fig. 6.6 with that of Fig. 4.21a). The X-nucleus magnetisation evolves during t_1 with the proton 180° pulse at its midpoint refocusing ^{1}H–X coupling evolution, thus decoupling the ^{1}H–X interaction so only heteronuclear chemical shifts remain in f_1. Following t_1, the heteronuclear magnetisation is transferred back onto the proton by an INEPT step in reverse to produce, once again, in-phase proton magnetisation for detection in the presence of X-spin decoupling. The basic two-step phase cycle is analogous to that of HMQC requiring inversion of the first 90° X pulse with associated inversion of the receiver.

Whilst this ultimately produces similar correlations to those of the HMQC experiment, the crosspeaks do not contain homonuclear ^{1}H–^{1}H couplings along f_1. This is because only X-nucleus magnetisation evolves during t_1 which is not, therefore, influenced by homonuclear proton couplings. This results in improved resolution in this dimension, the principal advantage of HSQC over HMQC for small molecules [6] (Fig. 6.7). The potentially greater f_1 resolution of HSQC makes this approach superior when the heteronuclear spectrum is poorly dispersed, although for many routine small molecule assignments ^{13}C dispersion is not limiting and either approach suffices. The most notable disadvantage of HSQC over HMQC is the greater number of pulses it utilises, especially 180° pulses on the heteronucleus, promoting intensity losses from rf inhomogeneity, pulse miscalibration or off-resonance excitation. This has

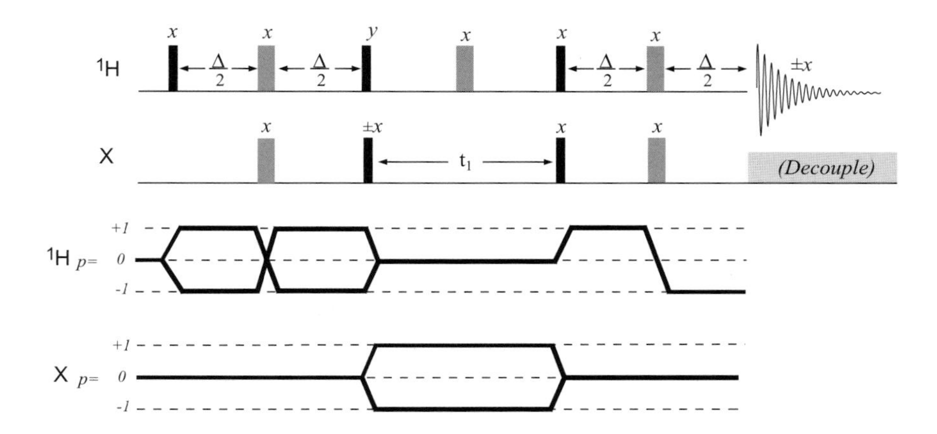

Figure 6.6. The HSQC experiment and associated coherence transfer pathway. The experiment uses the INEPT sequence to generate transverse X magnetisation which evolves and is then transferred back to the proton by an INEPT step in reverse. Notice that, in contrast to HMQC, only single-quantum X coherence evolves during t_1 (see text).

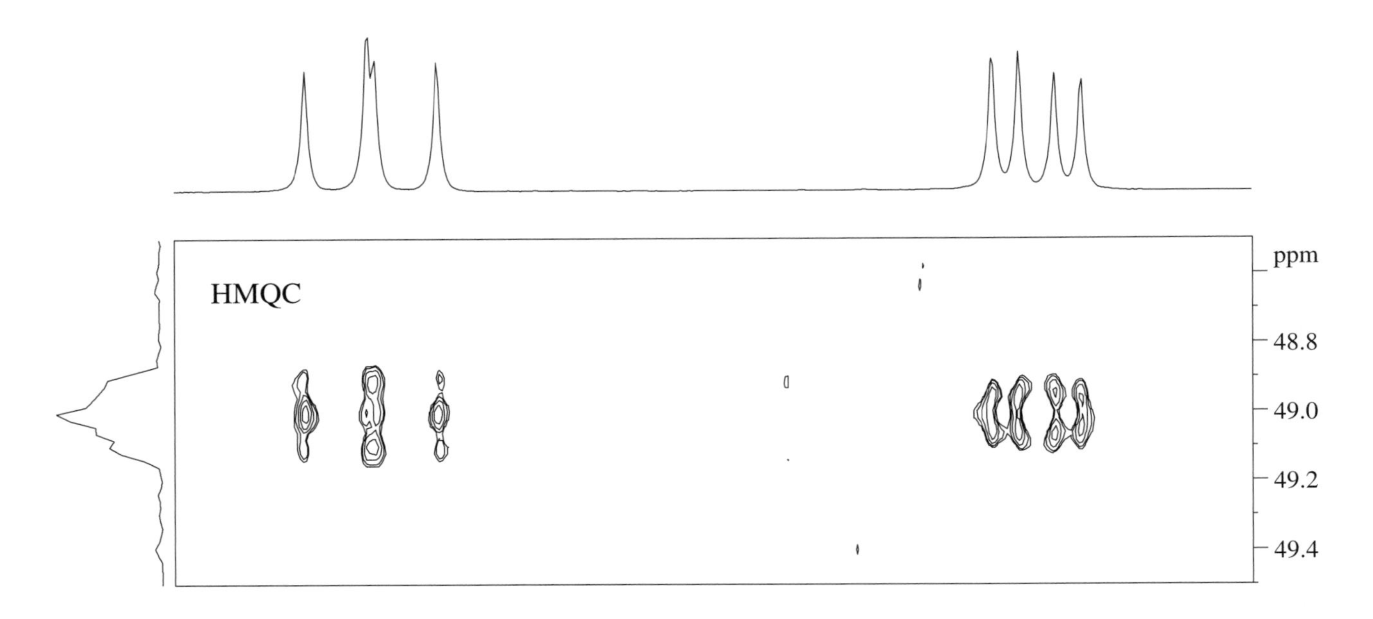

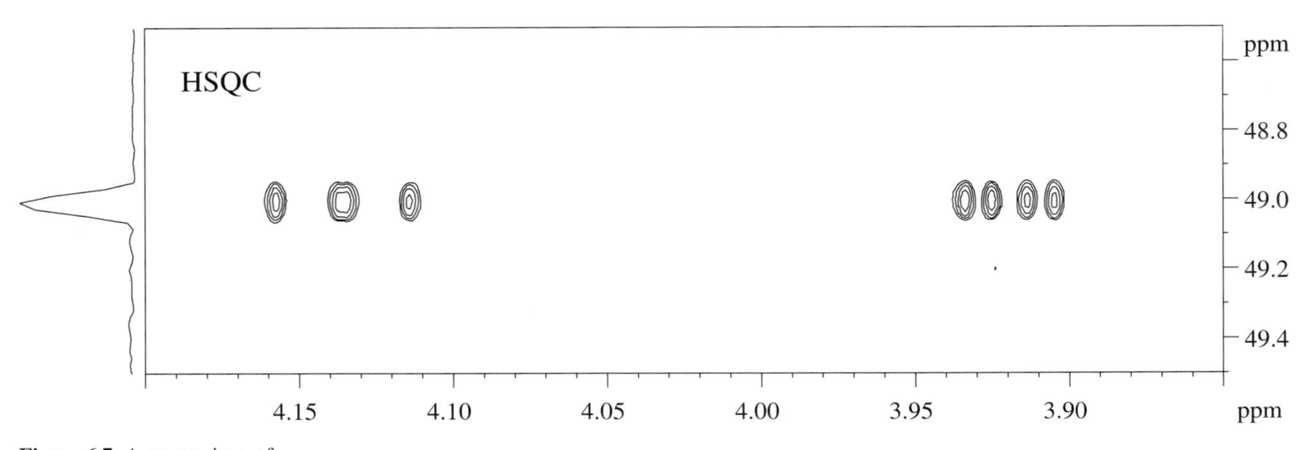

Figure 6.7. A comparison of experimental crosspeaks taken from HMQC and HSQC spectra acquired under identical conditions of high f_1 resolution (2.5 Hz/pt). The upper 1D trace is taken from the conventional 1D proton spectrum, and the vertical traces are f_1 projections from the 2D spectra. The additional broadening in the HMQC spectrum arises from unresolved homonuclear proton couplings in f_1.

greater significance for ^{13}C than ^{15}N owing to the greater frequency spread. Such losses may be minimised by careful probe tuning and, if necessary, the use of composite 180° pulses (Chapter 9). As with HMQC, suppression of the dominant parent resonance is required if the experiment is to be of general utility, as described below.

6.3.3. Practical implementations

To make the HMQC/HSQC sequences generally applicable it is desirable, if not essential, to suppress the resonances of those protons that cannot give rise to correlations in a more effective manner than can be achieved with phase-cycling alone. There are essentially two schemes for achieving this in widespread use. The more recent and certainly most effective approach is through the incorporation of pulse field gradients into the sequences. The older yet still very effective scheme aims to selectively saturate the parent resonances so that they have negligible intensity when detected. The effectiveness of the principal methods for suppressing the parent signal is compared in the 1D ^{1}H–^{13}C HMQC spectra of Fig. 6.8, and discussed further below.

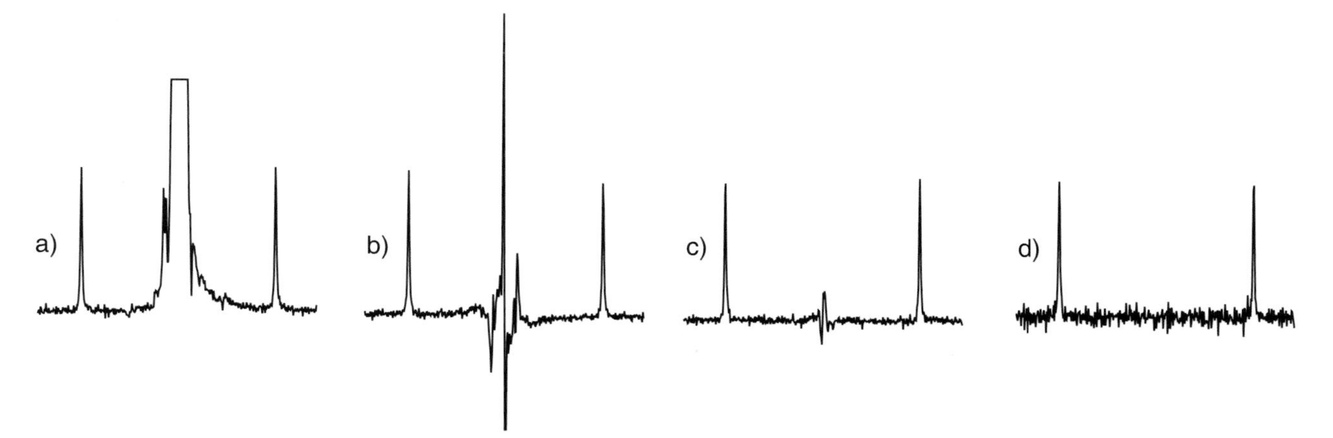

Figure 6.8. A comparison of signal suppression methods used in proton-detected heteronuclear correlation experiments (see descriptions in text). Spectrum (a) is taken from a conventional 1D proton spectrum without suppression of the parent resonance and displays the required ^{13}C satellites. Other spectra are recorded with (b) phase-cycling, (c) optimised BIRD presaturation, and (d) pulsed field gradients to remove the parent line. All spectra were recorded under otherwise identical acquisition conditions and result from two transients. Complete suppression can be achieved with gradient selection, but at some cost in sensitivity in this case (see text).

Gradient-selection [7]

Pulsed field gradients undoubtedly represent the ultimate approach to signal suppression (Fig. 6.8d). Attenuation ratios in excess of 1000:1 are readily achieved in a single scan, meaning the suppression of the parent ^{1}H–^{12}C or ^{1}H–^{14}N signals is complete and conveniently implemented. Phase-cycling is also not essential because signal selection is achieved solely by gradient refocusing, and in situations where sample quantities are not limiting these experiments may be performed within a matter of minutes.

The gradient-selected sequence operates by employing a suitable combination of gradients that refocus only those responses that have followed the desired transfer pathway. The scheme of Fig. 6.9a is suitable for the collection of absolute-value data as this selects only one ^{13}C transfer pathway during t_1 (preferably the echo or N-type), and therefore provides f_1 quadrature detection directly[2]. To understand the signal selection process, consider the action of each gradient in turn, paying due attention to the coherence orders, p, (as represented in the coherence transfer pathway) and magnetogyric ratios of the participating spins. Recall that the coherence transfer pathway represents only the pathway we wish to preserve, others are not shown since they will not be selected. Assuming the gradients have the same profile and are of the same duration but differ only in their strengths, G_n, it is straightforward to summarise the phase induced by the gradients using the shorthand notation of Section 5.5. Thus, the first gradient of Fig. 6.9a acts when both proton and carbon have coherence order $p = +1$ (heteronuclear double-quantum coherence), so the effect of the gradient is written $G_1(\gamma_H + \gamma_C)$. For the second gradient this becomes $G_2(-\gamma_H + \gamma_C)$ and for the third $G_3(-\gamma_H)$. To preserve this pathway, the overall phase induced by the gradients must be zero:

$$G_1(\gamma_H + \gamma_C) + G_2(-\gamma_H + \gamma_C) + G_3(-\gamma_H) = 0 \tag{6.2}$$

Note that only the *ratios* of the γ values become important in heteronuclear experiments and since $\gamma_H/\gamma_C = 4$ the above expression is simplified to:

$$G_1(5) + G_2(-3) + G_3(-4) = 0 \tag{6.3}$$

There are a number of gradient ratios that will satisfy this expression and lead to the desired signal selection, for example 2:2:1, 5:3:4 or 3:5:0.

[2] This retention of only one of the two possible transfer pathways with gradient selection is the reason for the reduced signal-to-noise ratio in Fig. 6.8d (which was acquired with sequence 6.9a with $t_1 = 0$) when compared to the other spectra. The precise details of how gradients are incorporated into pulse sequences will dictate whether such signal losses occur [7]. From a users point of view it is sufficient to be aware that such losses can arise with gradient selection and, when sensitivity is critical, to choose those techniques designed to avoid these.

Figure 6.9. Sequences for the gradient-selected HMQC experiment. Sequence (a) is suitable for the collection of absolute-value data. Sequence (b) provides phase-sensitive data via the echo-antiecho procedure. The N- and P-type pathways are selected with the last gradient whilst the first two gradients are placed within spin-echoes to refocus shift evolution.

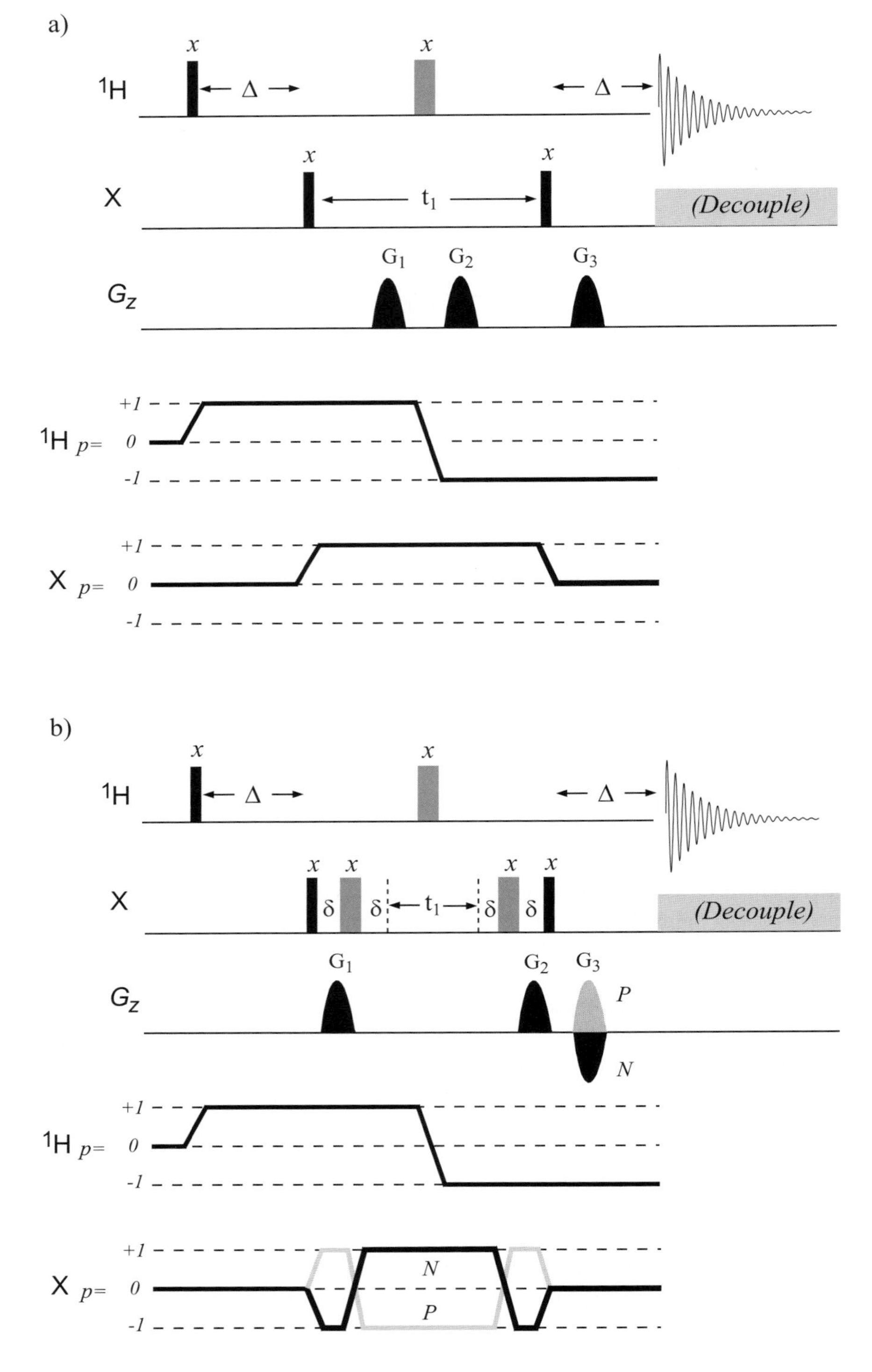

The optimum choice, in the presence of various experimental imperfections, has been considered in some detail [8] and the authors recommend the use of gradients in the ratio 3:5:0 (i.e. using only two gradient pulses). With the appropriate gradient ratio, it remains to ensure the gradients are of sufficient strength to achieve the desired suppression. This absolute-value approach demands a magnitude calculation so is most suited to the fully automated production of routine spectra when resolution and sensitivity are not critical.

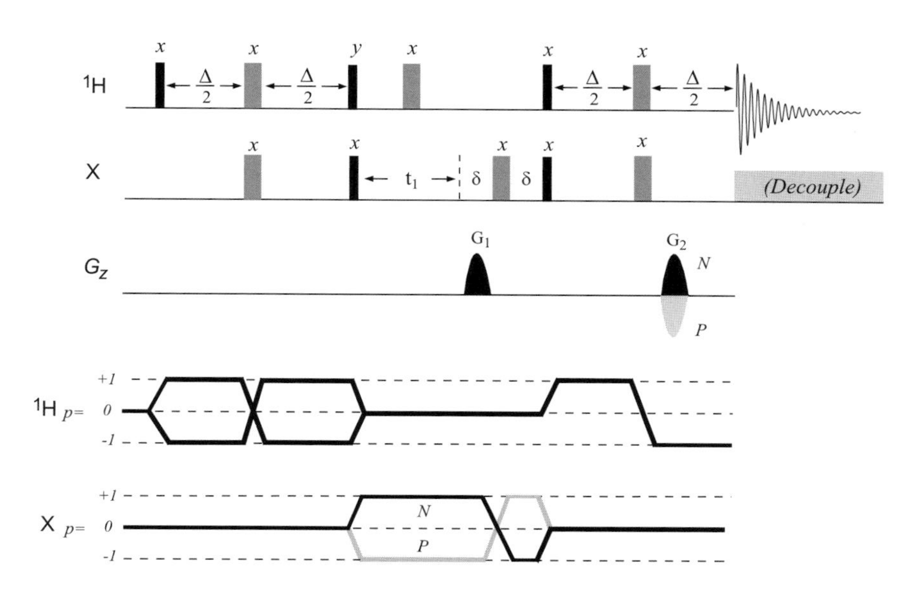

Figure 6.10. A gradient-selected, phase-sensitive HSQC sequence using the echo-antiecho approach. The N- and P-type pathways are selected by the last gradient.

To achieve a phase-sensitive presentation, it is necessary to retain both N and P ^{13}C pathways during t_1. This, however, is fundamentally impossible in a single scan if gradients are applied during the t_1 period since no gradient combination is able to refocus *simultaneously* signals from both the +1 and −1 pathways. For example, the net effect of the 2:2:1 gradient combination on the ^{13}C P-type pathway (coherence order −1 during t_1) will be:

$$G_1(\gamma_H - \gamma_C) + G_2(-\gamma_H - \gamma_C) + G_3(-\gamma_H) =$$
$$(2 \times 3) + (2 \times -5) + (1 \times -4) = -8 \qquad (6.4)$$

and the signal will remain dephased. There are two basic approaches to overcoming this limitation; either to avoid the use of gradients during t_1 altogether [9,10] so allowing use of the conventional States or TPPI methods of quadrature detection, or to collect the N- and P-type signals on alternate scans and combine them via the echo–antiecho method of processing [11–13]. Fig. 6.9b shows the scheme based on the echo–antiecho method. To obtain pure-phase spectra it is necessary to refocus carbon-13 chemical shift evolution that occurs during the gradient pulses, so additional spin-echoes bracket t_1, otherwise the sequence resembles that of Fig. 6.9a. The two different pathways are collected by inverting the sign of the last gradient so, for example, the N-type is refocused with 2:2:−1 and the P-type with 2:2:1 (note this selection is opposite to that of Fig. 6.9a because of the use of additional 180° ^{13}C pulses).

A similar logic to that above applies to gradient selection in the HSQC experiment, for which a variety of different approaches are also possible [7]. A suitable sequence employing the echo–antiecho approach is illustrated in Fig. 6.10, and requires only two gradients in proportion to the magnetogyric ratios of the X and ^{1}H spins since each acts on single-quantum X and ^{1}H magnetisation. Thus, for a ^{1}H–^{13}C correlation experiment, ratios of 4:1 and 4:−1 will select the N- and P-type coherences on alternate scans. Once again, the first gradient is applied within a spin-echo to refocus X-spin chemical shift evolution during the gradient pulse, whereas the second can be applied during the usual INEPT refocusing period.

The ability to completely suppress the parent ^{1}H–^{12}C or ^{1}H–^{14}N resonances produces spectra that are largely devoid of the bands of t_1 noise that may otherwise plague the experiment. This is illustrated in Fig. 6.11 which shows a section of the gradient-selected ^{1}H–^{15}N HSQC spectrum of the carbopeptoid **6.3** recorded with ^{15}N decoupling, plotted conventionally and with the contour levels reduced to show the baseline thermal noise.

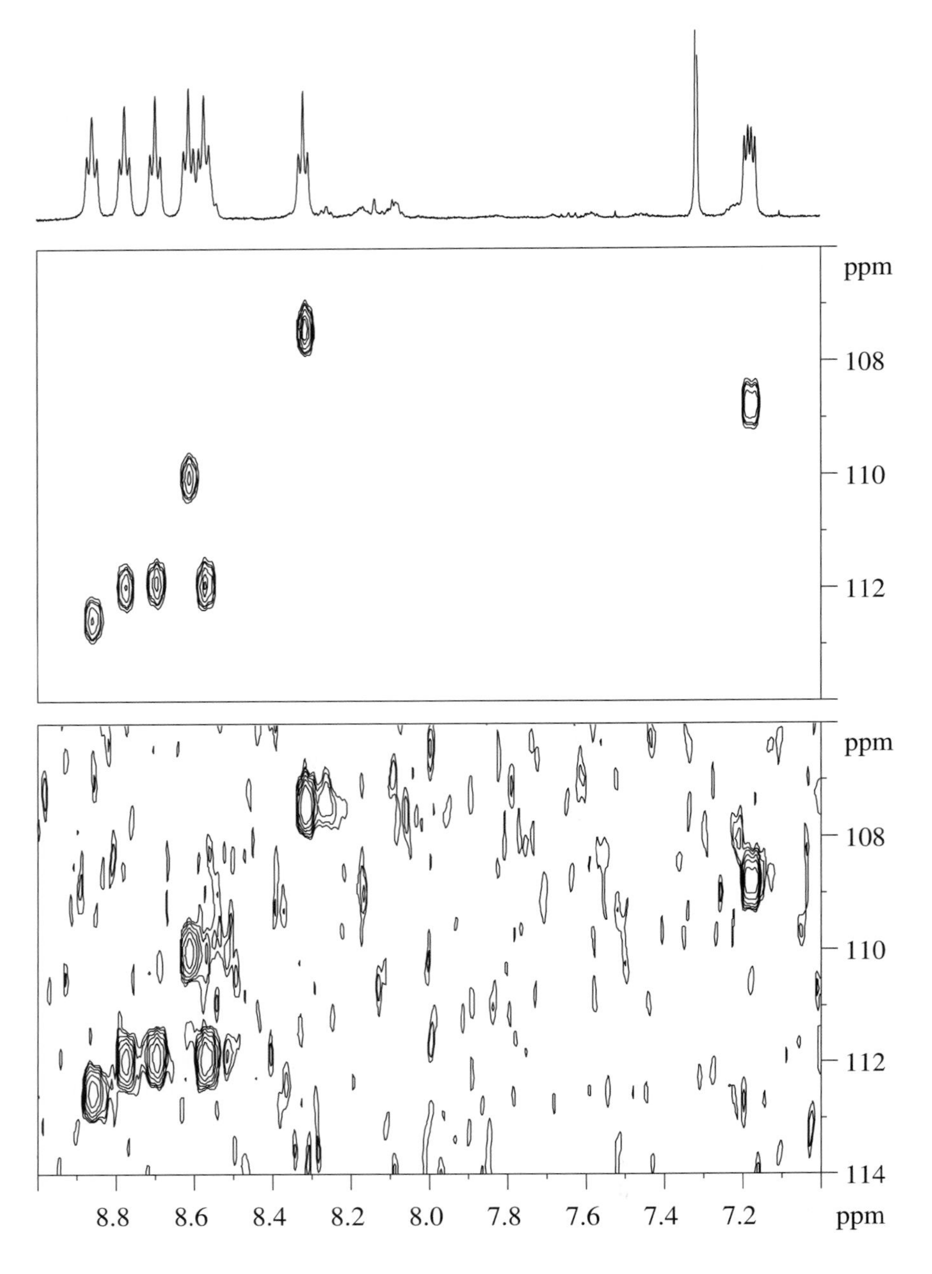

Figure 6.11. A gradient-selected HSQC spectrum of the carbopeptoid **6.3** at natural ^{15}N abundance plotted at high and at low contour levels to show the thermal noise floor. No t_1-noise artefacts remain from the parent ^1H–^{14}N resonances (^{15}N is referenced to external liquid ammonia).

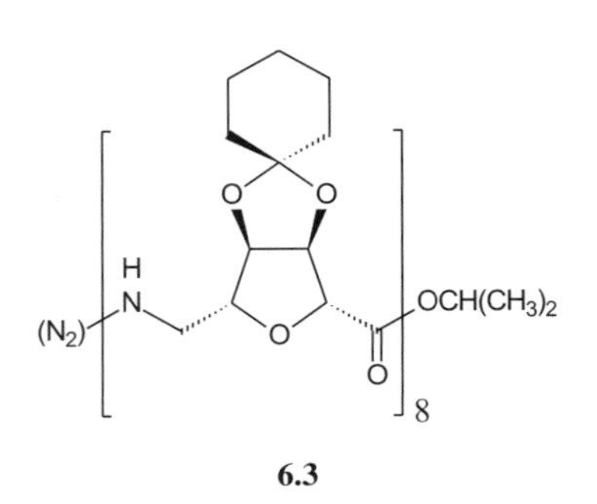

6.3

BIRD-HMQC

Not every NMR spectrometer or probehead is equipped with pulsed field gradient capabilities, yet it may still be possible to execute the basic heteronuclear shift correlation experiments described above. In such cases it is desirable to employ an alternative scheme to remove the interfering parent resonances and produce spectra devoid of objectionable artefacts. In the BIRD (bilinear rotation decoupling) variant (Fig. 6.12), unwanted ^1H–^{12}C resonances are suppressed prior to the HMQC sequence by an ingenious presaturation scheme. This commences with the inversion of *only the carbon-12-bound protons* by the so-called 'BIRD pulse' (the action of which is described below), leaving the carbon-13-bound protons unaffected. Following the inversion, a recovery period, τ, allows the magnetisation vectors to relax back toward the equilibrium $+z$ axis, until they pass through the x–y plane (Fig. 6.13). At this point, the HMQC sequence itself begins and because there exists no longitudinal ^1H–^{12}C magnetisation, no transverse component is ever generated for these

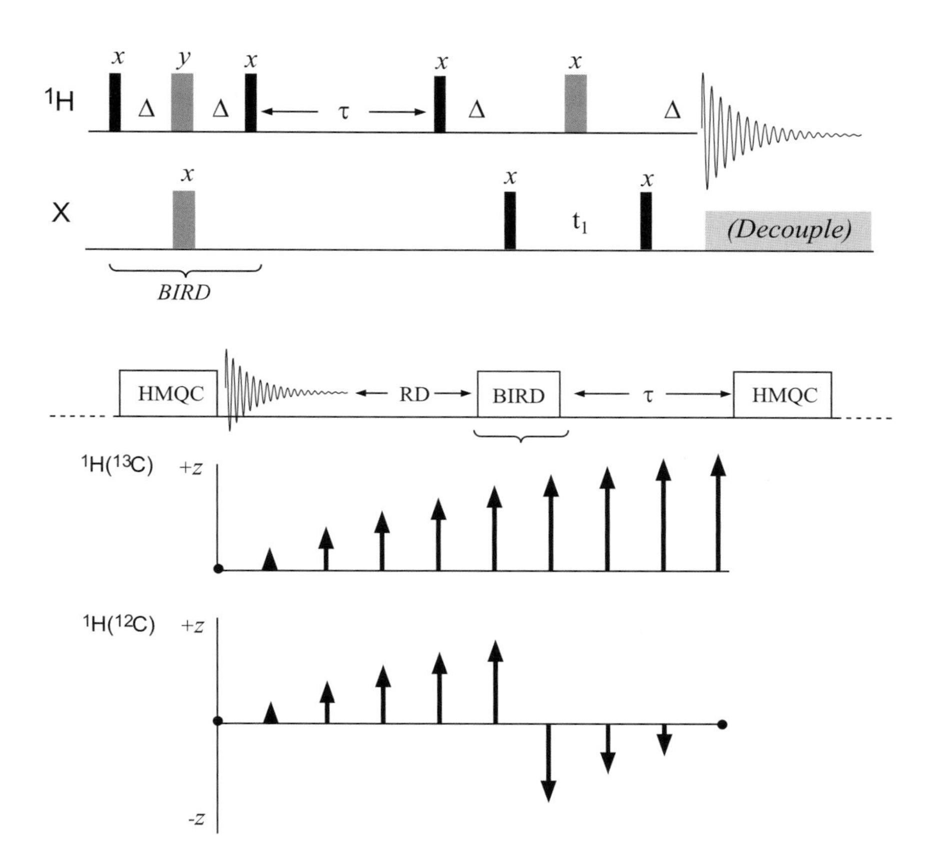

Figure 6.12. The BIRD variant of the HMQC experiment. The conventional HMQC sequence is employed, but is preceded by the BIRD inversion element and an inversion recovery delay, τ. This procedure ultimately leads to saturation of the unwanted parent resonances.

Figure 6.13. Elimination of parent ^1H–^{12}C proton resonances through the BIRD-inversion recovery sequence. At the start of data collection no longitudinal proton magnetisation exists but this reappears during the acquisition period and subsequent recovery delay (RD) through spin relaxation. The BIRD element selectively inverts only those protons attached to carbon-12, which then continue to relax during the inversion recovery delay, τ. With an appropriate choice of τ, the ^1H–^{12}C magnetisation has no longitudinal component when the HMQC sequence starts, so does not contribute to the detected FID.

spins and hence the desired suppression of the resonances is achieved. During detection and the next inversion-recovery period, the ^1H–^{13}C magnetisation simply relaxes back towards its equilibrium value in readiness for the sequence to begin once more since it is not inverted by the BIRD pulse.

The BIRD pulse [14] is in fact a cluster of pulses (Fig. 6.12) used as a tool in NMR to differentiate spins that possess a heteronuclear coupling from those that do not. The effect of the pulse can vary depending on the phases of the pulses within the cluster, so we concentrate here on the selective inversion described above. For illustrative purposes, proton pulse phases of *x, y, x* will be considered as this provides a clearer picture with the vector model, although equivalent results are achieved with phases *x, x, −x*, as in the original publication. The scheme (Fig. 6.14) begins with a proton excitation pulse followed by a spin-echo. Since carbon-12 bound protons have no one-bond

Figure 6.14. Selective inversion of carbon-12 bound protons with the BIRD element. Proton chemical shift evolution is refocused in the sequence but heteronuclear one-bond couplings evolve throughout.

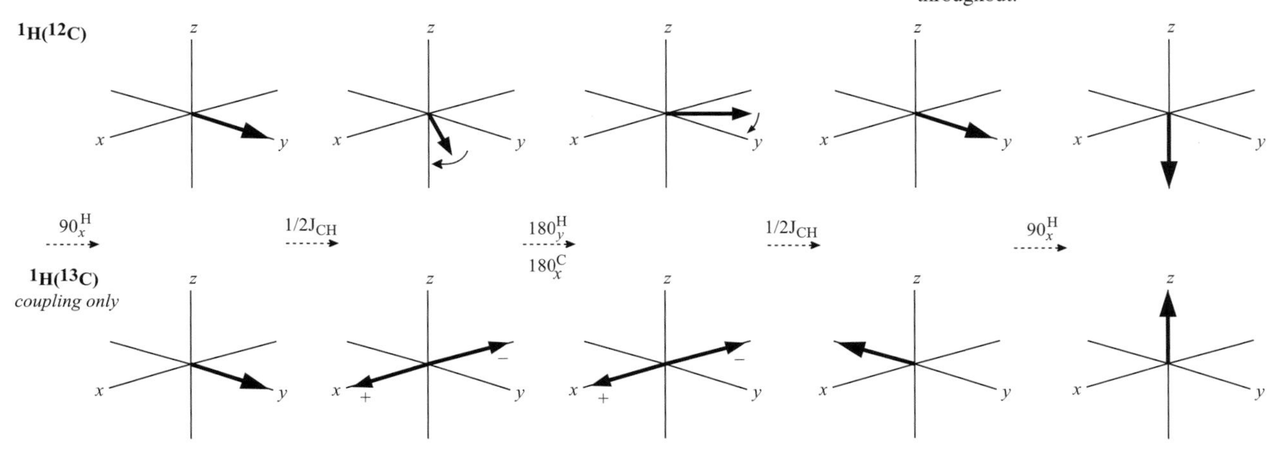

heteronuclear coupling, only their chemical shifts evolve during Δ. These are subsequently refocused by the 180°(^{1}H), so that at the end of the second Δ period, the second 90°(^{1}H) places the vector along the $-z$ axis and produces the desired inversion of the ^{1}H–^{12}C resonances. For those spins that posses a heteronuclear coupling, chemical shifts will refocus as above so we need only consider the effect of the coupling itself. If Δ is set to $1/2J_{CH}$, the two proton vectors will lie along $\pm x$ immediately prior to the 180°(^{1}H, ^{13}C) pulses, after which the coupling will *continue* to evolve. Thus, after a total period 2Δ, the doublet vectors lie along $-y$, and are then returned to $+z$ by the action of the final proton pulse, and for these spins it is as if the BIRD had never appeared.

The success of the BIRD suppression scheme depends crucially on the correct setting of recovery delays and repetition rates, according to the proton T_1 values of the molecule. In practice, one is unlikely to know precisely what these are, and in any case there will exist a spread of these within the molecule, so a best-guess compromise must be made. It turns out [3] that optimum sensitivity is achieved by setting timings according to the *shortest* T_1 value in the molecule, and once again setting the recycle time of the experiment, that is the time between the start of data acquisition of one experiment and the beginning of the HMQC sequence of the next (t_2 + RD + τ of Fig. 6.13), to around 1.3 T_1. The inversion-recovery period, τ, must be set to approximately 0.5 T_1 for efficient suppression, and the proton acquisition time, t_2, set according to the desired digital resolution. The relaxation delay, RD, is then chosen to make up the desired recycle time (t_2 + RD $\approx$ 0.8 T_1). For example, assuming the smallest anticipated T_1 in a molecule to be 600 ms, and that t_2 is set to 200 ms, corresponding to an f_2 digital resolution of 5 Hz/pt, the above guidelines suggests, one selects τ = 300 ms and RD = 280 ms [3]. Since one is generally forced to make a best guess at the T_1 values, the experimental settings are fine tuned to give optimum suppression of the unwanted resonances. This is most conveniently achieved by running the experiment in an interactive set-up mode that allows real-time adjustment of parameters, and altering τ to produce the *minimum* FID.

The BIRD scheme is remarkably efficient at suppressing the troublesome ^{1}H–^{12}C resonances and associated t_1-noise and has been widely employed in the study of small molecules. However, this method is not suitable for the study of very large molecules because during the τ period the *negative* NOE (Chapter 8) generated from the inverted protons causes a *reduction* in the signal intensity of the observed protons and therefore compromises sensitivity. The traditional approach for large molecules in the absence of field gradients [15] has been to apply a strong spin-lock period during the first INEPT sequence of the HSQC experiment. This destroys all magnetisation not aligned with the spin-lock axis and suppresses the unwanted parent signals.

Practical set-up

The high crosspeak dispersion typically associated with heteronuclear shift correlation experiments alongside the lack of any requirement for well defined crosspeak fine structure means HMQC or HSQC experiments can be recorded with rather low digital resolution for routine applications, enhancing their time efficiency. An acquired digital resolution of 5 Hz/pt in the proton f_2 dimension and only 50 Hz/pt in the heteronucleus f_1 dimension are generally sufficient to resolve correlations. Improved f_1 resolution can be achieved by linear

[3] Such rapid pulsing may seem excessively fast for the resonances one wishes to observe, especially for those protons with longer relaxation times. However, one should recall that the proton T_1s generally measured relate to carbon-12 bound protons, whereas in HMQC one is interested in carbon-13 bound protons. This nucleus acts as an additional dipolar relaxation source for the directly attached proton leading to shortening of the proton T_1 value, consistent with the rapid repetition.

prediction of the FIDs when sensitivity allows, and/or the digital resolution enhanced by zero-filling (usually at least once). Alternatively, the use of linear prediction can lead to significant time savings when high sample quantities are available, by reducing the number of t_1 increments that must be recorded and computationally regenerating those missing t_1 data points to provide adequate f_1 resolution [16]. It must be remembered, however, that very high resolution in f_1 is not desirable for the HMQC experiment since one should avoid resolving splittings arising from the homonuclear proton couplings and so degrading the signal-to-noise ratio. Beyond this, processing requires only simple apodisation in both dimensions and either shifted squared sine-bells or Gaussian windows function well for phase-sensitive data sets.

Proton–carbon correlation experiments based on the methods described above can be surprisingly fast to acquire for routine organic samples, especially when pulsed field gradients and probes optimised for proton detection are employed. Repetition rates are dictated by the shorter relaxation times of the ^{13}C bound protons and can therefore be faster than homonuclear COSY spectra, for example, where it is the ^{12}C bound protons that are monitored. They are usually quicker to obtain and often more informative than the 1D carbon spectrum, often accessible in a matter of minutes, and are now routine experiments in the organic laboratory. Accompanying these methods with spectrum editing techniques as described below further extends their utility as structural tools, and may increasingly relegate direct X-observe 1D experiments such as APT and DEPT to the periphery of common structure elucidation methods.

X–Y correlations

It should be remembered that the 2D methods presented here are suitable for the correlation of a variety of different nuclides [17], and do not necessarily require proton detection, so long as the observed nucleus exists at high abundance. As an illustration, Fig. 6.15 shows a phosphorus detected ^{31}P–^{13}C correlation of **6.4**. The basic HMQC sequence was used with additional broadband proton decoupling applied throughout but without carbon-13 decoupling

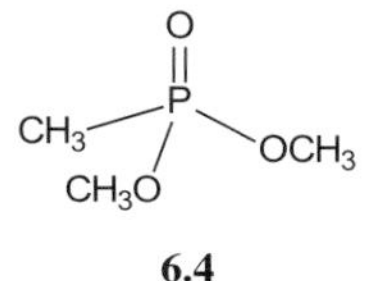

6.4

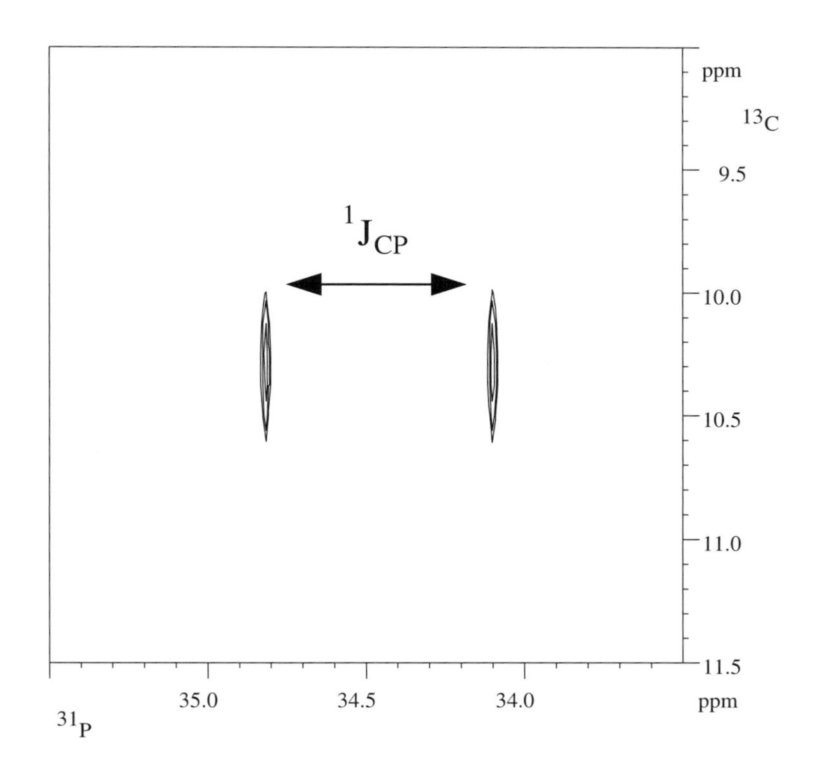

Figure 6.15. Phosphorus detected ^{31}P–^{13}C HMQC spectrum of **6.4**. Proton decoupling was applied throughout but no carbon-13 decoupling was used during detection of the phosphorus FID. These data were acquired using a home-built third channel and a probe modified for triple-channel operation.

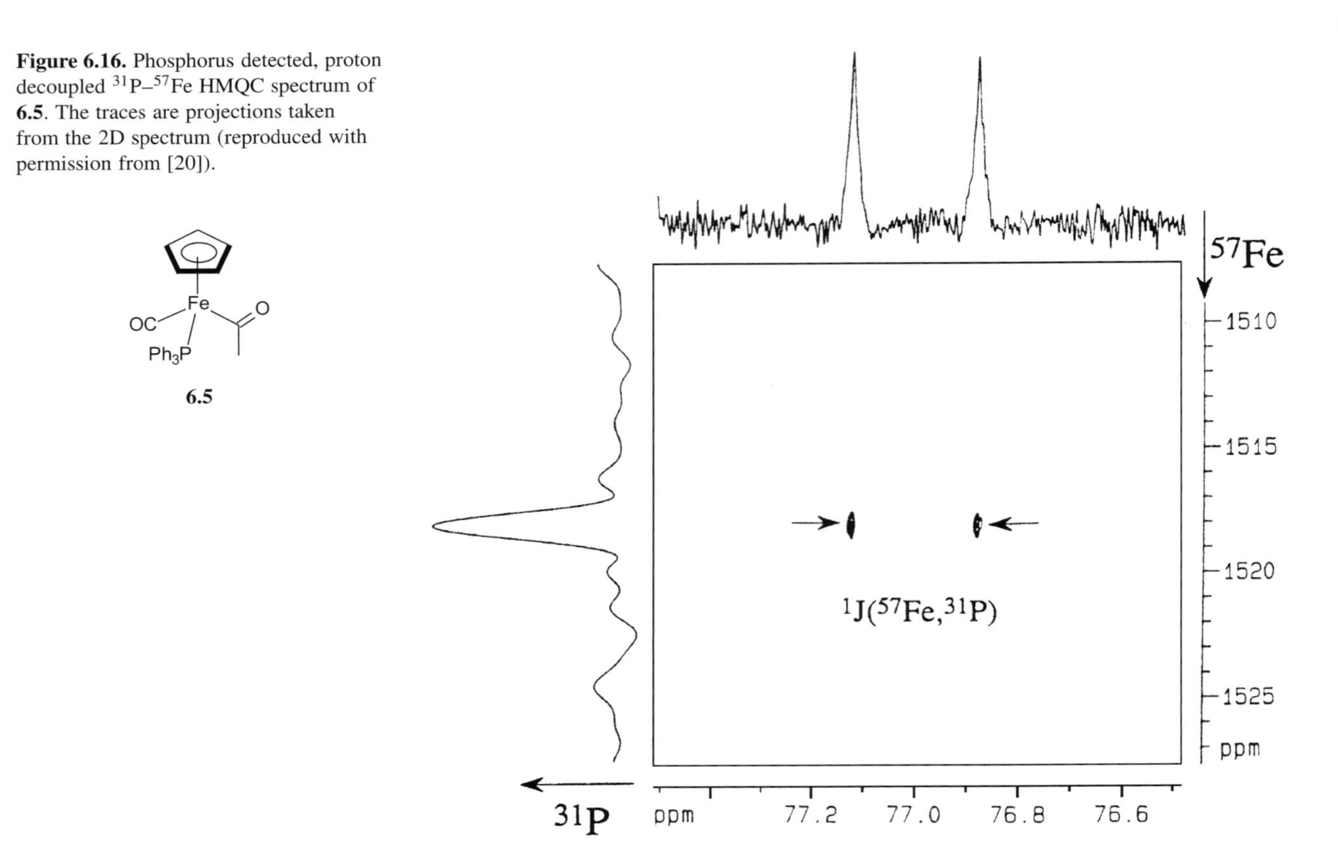

Figure 6.16. Phosphorus detected, proton decoupled ^{31}P–^{57}Fe HMQC spectrum of **6.5**. The traces are projections taken from the 2D spectrum (reproduced with permission from [20]).

during the detection period. The correlation thus appears as a $^{1}J_{CP}$ doublet. Such experiments are often referred to as *heteronuclear X–Y correlations* to indicate the participation of two nuclides other than proton, which itself is usually decoupled. These require the spectrometer to posses three rf channels and a probe that can tune to the nuclei involved so are not accessible on a standard two-channel instrument. However, with the increasing availability of three-channel spectrometers in chemical laboratories, such methods can be expected to provide additional routes to structure determinations and the study of chemical processes. Inverse correlation methods can also provide access to chemical shift data of nuclei that are otherwise too insensitive to be observed directly. This has particular significance for the indirect observation of metals, many of which have very low intrinsic sensitivity but can nevertheless be detected through their spin-coupled neighbours on adjacent ligands, most often proton or phosphorus nuclei. The indirect observation of ^{57}Fe, a notoriously difficult nucleus to observe directly, is one such example [18–20] (Fig. 6.16).

6.3.4. Hybrid experiments

As was alluded to earlier, shift correlation spectra may be further enhanced by combining them with other pulse techniques to produce spectra with altered characteristics or increased information content. This section is necessarily brief and is intended to provide an oversight of the types of techniques that have been developed and which are likely to play an increasingly prominent role in organic structure elucidation. The first two sections below represent two of the more useful methods for structure elucidation and also illustrate the manner in which different techniques may be concatenated to produce 'new' experiments. The third illustrates how the heteronuclear spin may be used in the editing or simplification of 1D proton spectra.

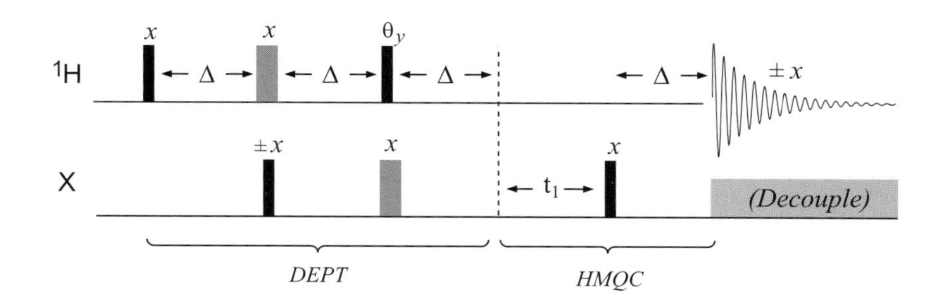

Figure 6.17. The DEPT-HMQC sequence for multiplicity editing within the 2D correlation experiment. The conventional DEPT sequence precedes HMQC and provides editing through the proton θ tip angle. Setting θ = 180° inverts XH_2 responses relative to those of XH and XH_3.

2D multiplicity editing

Short of the ability to identify diastereotopic XH_2 groups, the heteronuclear shift correlation experiments so far encountered provide no direct evidence for the multiplicities of the XH_n groups giving rise to each correlation, analogous to the way in which broadband decoupled heteronuclear spectra provide no multiplicity information. Various methods are now available to provide such editing *within* the 2D heteronuclear correlation experiment itself, so killing two birds with one stone, if you like. These provide more information in routine analysis and require less time than conventional 1D editing methods. Furthermore, such experiments provide access to heteronuclear multiplicity information when small sample quantities preclude the direct observation of X-spin edited spectra, so further extending the range of sample quantities one can consider accessible to structural studies.

Incorporating the DEPT sequence of Chapter 4 within HMQC [21] (Fig. 6.17) offers one method for directly editing the correlation spectrum. The role of the DEPT sequence prior to the HMQC is to generate the heteronuclear multiple-quantum coherence that evolves during t_1. A full analysis [21,22] shows that different multiple-quantum terms are used in this sequence than in 1D DEPT, the net result being that the crosspeak intensities have a different dependence on the $\theta°$ proton editing pulse. The most useful implementation uses θ = 180° to produce spectra with XH_2 correlations inverted relative to those of XH and XH_3 groups, equivalent to the conventional DEPT-135 1D sequence (a phase-sensitive presentation must be used!). Editing may be readily introduced to the HSQC sequence [10,23,24] by the simple addition of a spin-echo (Δ − 180°(^{1}H,X) − Δ) after t_1. During this, only the heteronuclear coupling evolves, exactly as in the pulsed 1D J-modulated spin-echo sequence described in Section 4.3.1. Hence, setting Δ = 1/2J also provides a 2D correlation spectrum in which XH_2 responses are inverted relative to those of XH and XH_3. In one example of this approach, concatenating this echo with the existing t_1 period produces the modified HSQC scheme of Fig. 6.18 [7,25]. The edited HSQC spectrum of the substituted disaccharide **6.6** is shown in Fig. 6.19 and clearly differentiates CH from CH_2 correlations.

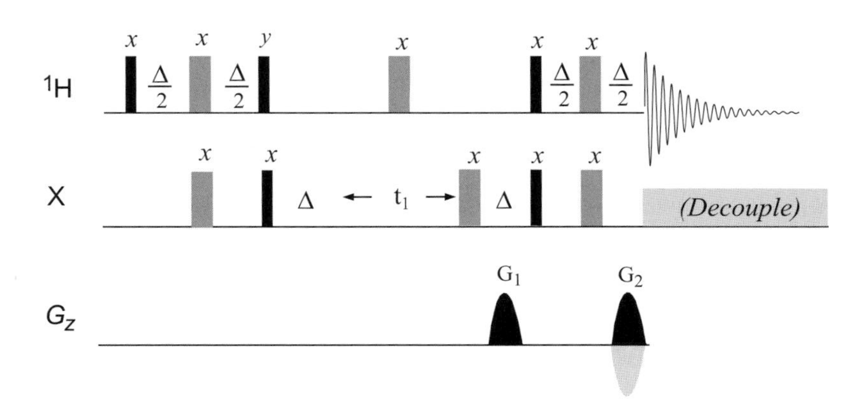

Figure 6.18. The gradient-selected, spin-echo HSQC sequence for multiplicity editing within the 2D correlation experiment. Setting Δ = 1/2J inverts XH_2 responses relative to those of XH and XH_3.

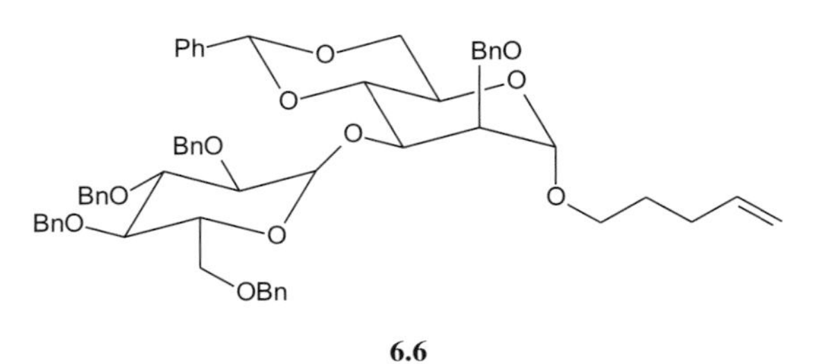

6.6

A potential problem with these editing approaches is the cancellation of overlapping correlations with opposite phases in crowded regions of 2D spectra. Alternative sequences based on editing through multiple-quantum filtration produce spectra containing correlations from *only* XH *or* XH_2 *or* XH_3 groups so the problem of signal cancellation is removed. These methods are too numerous to discuss here, but include REPAY [26], HYSEL [27,28], HmQC [22], MAXY [29,30] and others [10,31]. Even more exotic modifications include the addition of spin-lock transfer [32,33] in a manner similar to that described below. When to turn to such levels of sophistication will be dictated by the nature and complexity of the problem in hand, but the efficiency of such editing schemes with the advent of pulsed field gradients is likely to see these becoming more widely used for the investigation of complex structures.

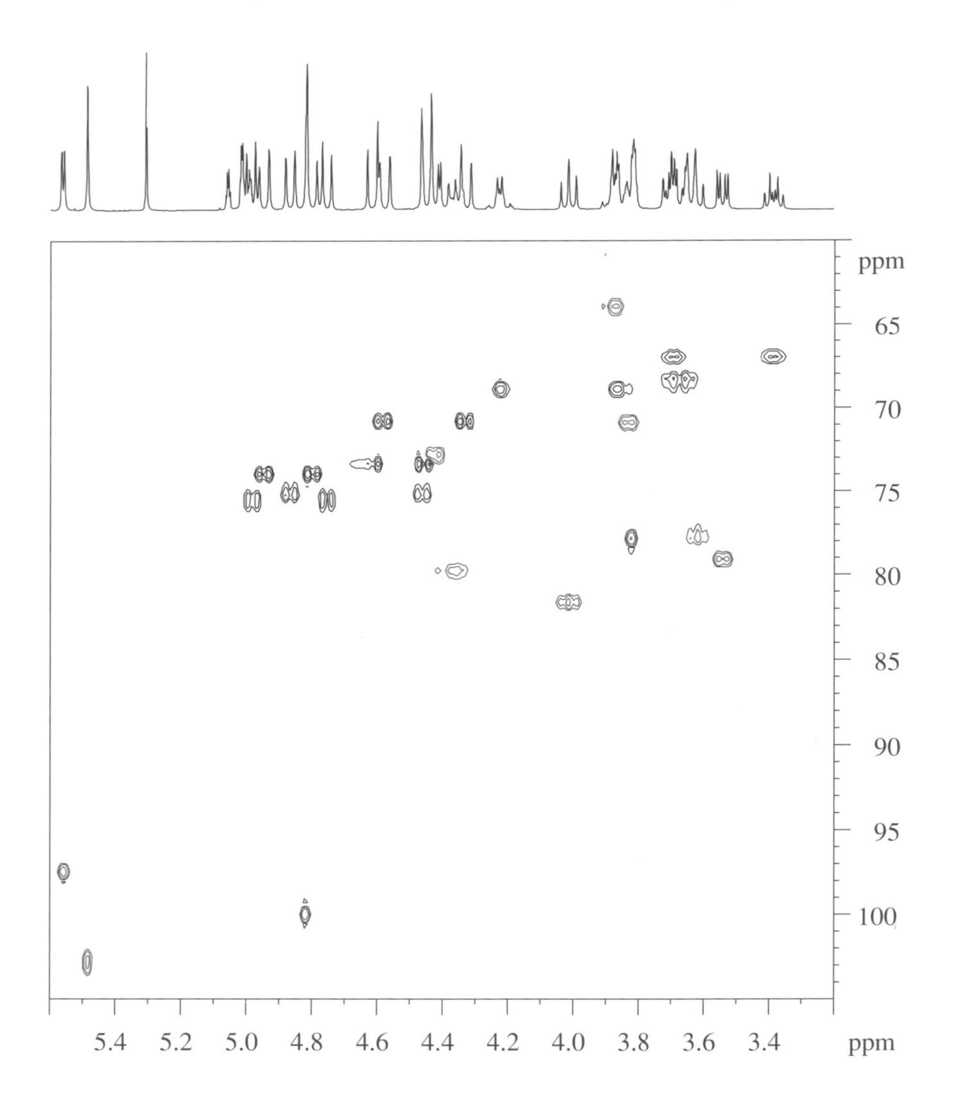

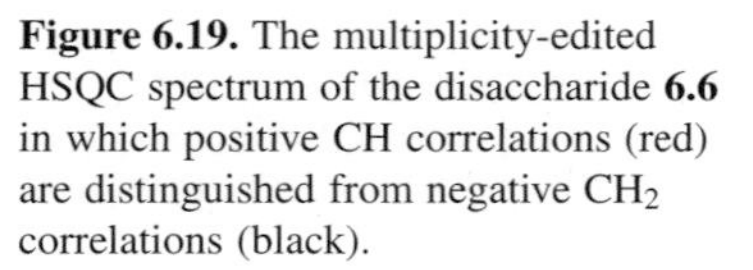
Figure 6.19. The multiplicity-edited HSQC spectrum of the disaccharide **6.6** in which positive CH correlations (red) are distinguished from negative CH_2 correlations (black).

Utilising X-spin shift dispersion

When analysing molecules that display very crowded proton spectra, the 2D homonuclear shift correlation experiments discussed in Chapter 5 may still provide spectra which are too overlapped to allow a complete interpretation. In such instances, the potentially greater dispersion of the parent heteroatom chemical shifts can be used as a means of further separating proton–proton correlations. Furthermore, in cases of exactly overlapping *proton* resonances, the one-bond heteronuclear correlation experiments do not provide unambiguous identification of the parent heteroatom, and one approach to overcoming such problems lies in the transfer of the heteronuclear correlation information onto neighbouring protons. Thus, adding a TOCSY spin-lock mixing period after the HMQC/HSQC sequence and immediately prior to data collection [34,35] transfers magnetisation that has returned to the proton from which it originated onto neighbouring J-coupled protons (Fig. 6.20). Extended mixing periods can again be used to *relay* magnetisation along the proton network, potentially providing a complete proton subspectrum of the molecular fragment to which the heteroatom belongs. The result may be viewed as a X-spin-edited 2D TOCSY experiment, the proton correlations now appearing along rows taken at the hetero-spin f_1 chemical shifts. With sufficient shift dispersion in the X-spin dimension, overlap present in the 2D TOCSY is removed. The HSQC-TOCSY spectrum of the disaccharide **6.7** (Fig. 6.21b) contains numerous additional correlations over the HSQC spectrum (Fig. 6.21a), and can be used to map the proton coupling pathways. This is most clearly seen at the C3 shift of ring B, from which the H3 proton produces TOCSY correlations to all other protons in the ring, with the exception of H1 (which in this case had zero coupling with H2, so causing a breakdown in transfer).

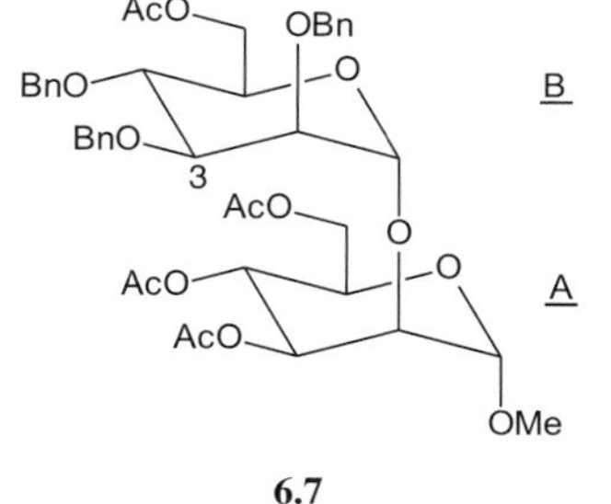

6.7

The enormous simplification of crowded spectra in this manner makes this a very powerful technique. However, a point to bear in mind if considering applying these methods is that crosspeaks in the ^{13}C-edited experiment originate only from ^{13}C satellites because of the initial HMQC/HSQC step, and the experiment is therefore of significantly lower sensitivity than homonuclear 2D TOCSY, in which all protons may participate. The experiment will generally find use after initial stages of investigation of complex spectra using the techniques already presented where ambiguities remain. The use of long-range heteronuclear correlations described shortly should also be considered in such cases.

Further modifications (the addition of a spin-echo) allow the direct and relayed peaks to be differentiated through inversion of the direct correlations [36] and gradient-selected versions of these sequences without [37,38] and with [10,39] such editing have also been proposed. Although the addition of TOCSY

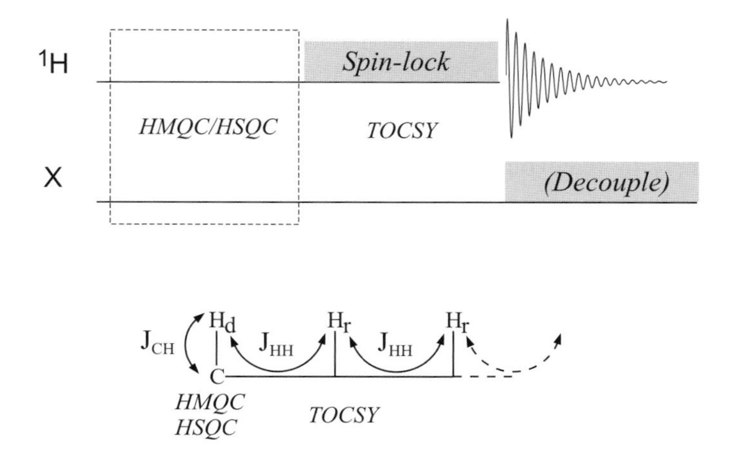

Figure 6.20. The schematic HMQC/HSQC-TOCSY sequence and the coupling pathway it maps. Direct correlations are produced for the proton bound to the spin-$^1/_2$ heteroatom (H_d) as in the basic shift correlation sequence, and further relayed correlations are produced for those protons receiving magnetisation through the TOCSY transfer (H_r).

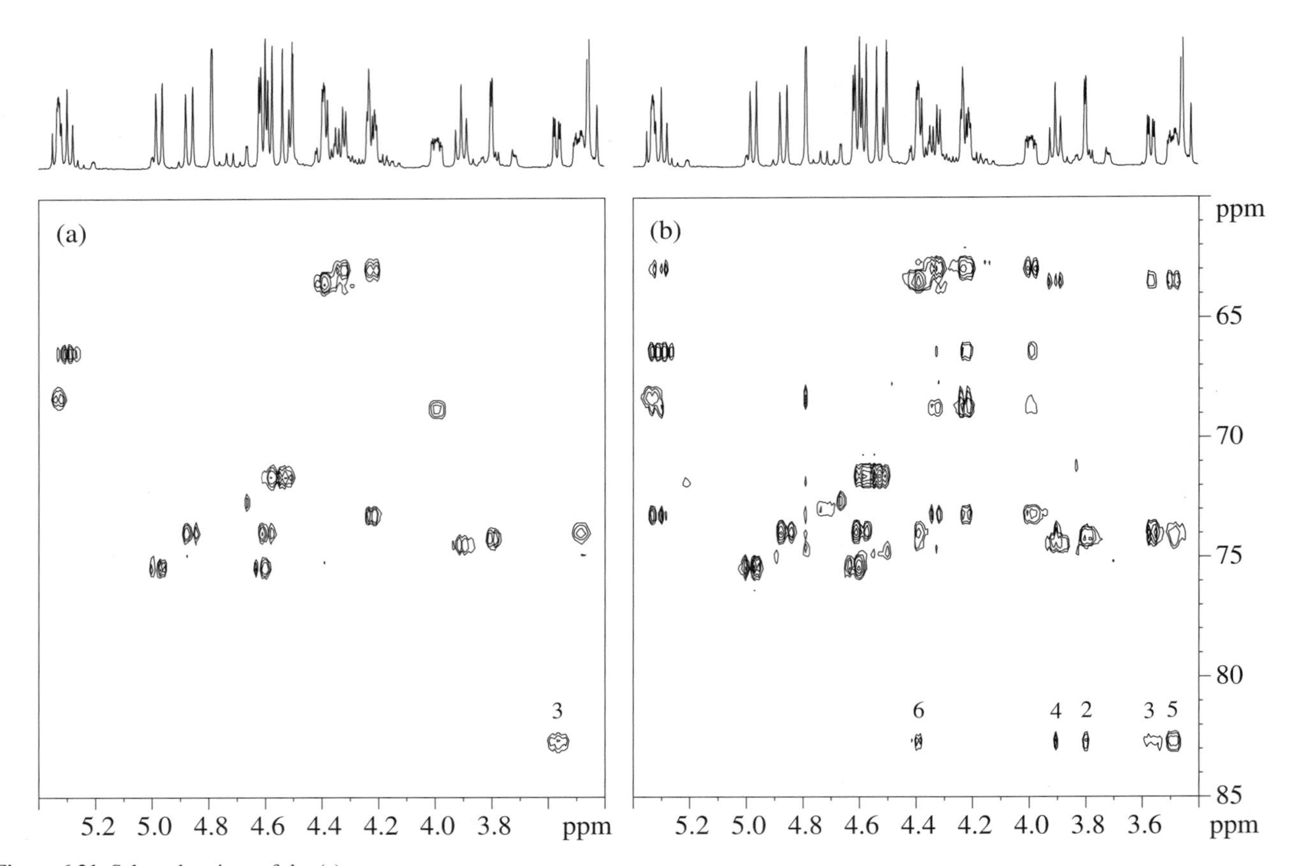

Figure 6.21. Selected regions of the (a) HSQC and (b) HSQC-TOCSY spectra of the disaccharide **6.7**. The proton correlations originating from H3 of ring B are labelled and provide an almost complete map of protons within this ring (only the correlation to H1 is missing).

transfer is probably the most useful extension, any homonuclear mixing scheme can, in principle, be added, including a COSY, NOESY or ROESY step.

Editing and filtering 1D proton spectra

The idea of using the heteroatom to edit the 2D spectrum as described above is equally applicable to the editing of 1D proton spectra. Generally speaking, there are two reasons as to why one may consider doing this. Firstly, one may wish to *edit* the proton spectrum according to the heteroatom multiplicities to produce subspectra that are the proton analogues of, for example, APT or DEPT-edited carbon spectra. These analogues provide a faster and more sensitive route to identifying carbon multiplicities within a molecule, and allow the analysis of smaller sample quantities by virtue of proton detection. Secondly, one may wish to selectively observe only the protons attached to a heteroatom isotope label. In this way, one uses the label to *filter* the proton spectrum, transferring the selectivity associated with the hetero-label onto the more sensitive proton. Whilst the natural abundance proton satellites of unlabelled positions will also pass the filter, they will be present at significantly lower intensity.

The HMQC or HSQC sequences may be transformed into their 1D equivalents simply by removing the incremental t_1 time period (Fig. 6.22), so that

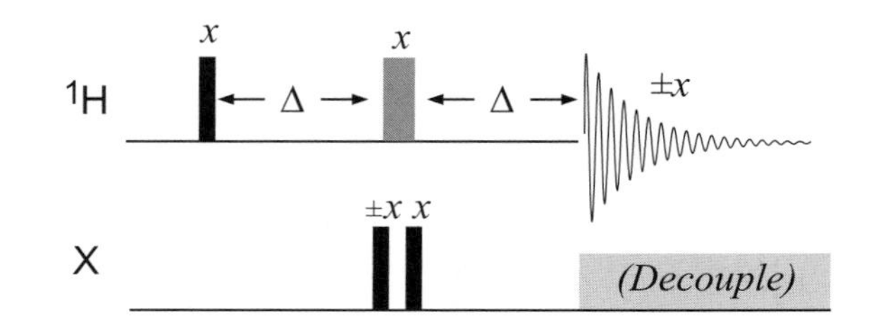

Figure 6.22. The 1D HMQC sequence. This can be used for editing spectra by selecting only those protons bound to NMR-active heteroatoms. Improved suppression of unwanted resonances can be achieved through the incorporation of pulsed field gradients, as for the 2D experiment.

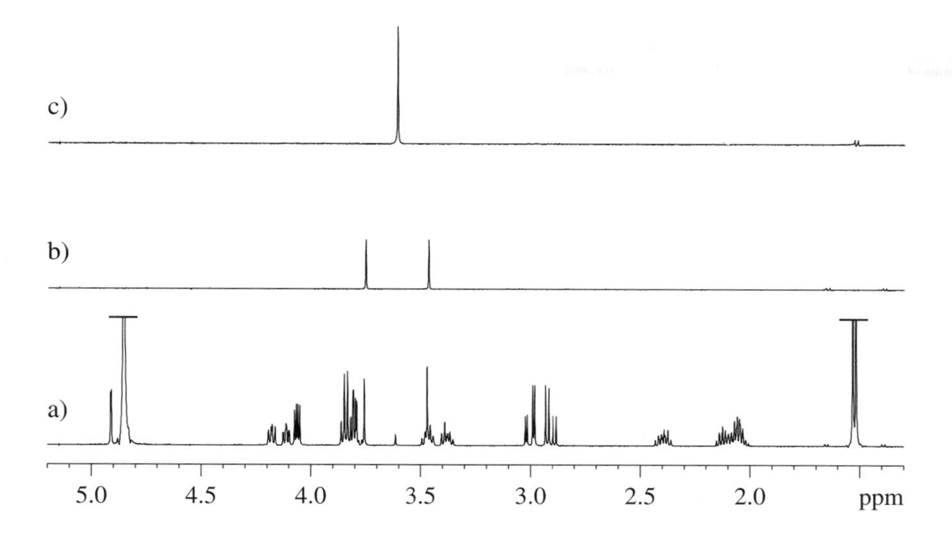

Figure 6.23. The selective observation of protons bound to a carbon-13 label (2–^{13}C–glycine) with a gradient selected 1D HMQC sequence. (a) The 1D proton spectrum and the filtered spectrum recorded (b) without and (c) with carbon-13 decoupling during acquisition.

the experiment becomes just a heteronuclear filter. Only magnetisation that has passed via the X-spin will be observed in the final spectrum and again the suppression of all unwanted signals is greatly improved by the use of pulsed field gradients. The selective observation of ^{13}C-labelled glycine in an aqueous mixture is illustrated in Fig. 6.23.

The classical approach to generating multiplicity-edited proton spectra was to use the INEPT or DEPT sequences in reverse to transfer initial carbon magnetisation onto the proton for detection. These suffered from low sensitivity and poor suppression of ^{1}H–^{12}C resonances, so were never popular. Greater sensitivity may be achieved by using the inverse $^{1}H \rightarrow {}^{13}C \rightarrow {}^{1}H$ approach combined with multiple-quantum filtering [21,27] as for the 2D editing methods discussed above, and methods employing editing via gradient selection now provide far cleaner results [22,28,29,40]. As an example, the selection of a subspectrum containing only methylene proton resonances is shown in Fig. 6.24. Relative to the standard 1D proton spectrum there is a price to pay in sensitivity, because one is again forced to observe the carbon-13 satellites of the conventional proton spectrum. Nevertheless, this simplification may prove useful in the analysis of highly crowded spectra and may provide valuable multiplicity data when small sample quantities preclude the use of carbon-detected editing methods.

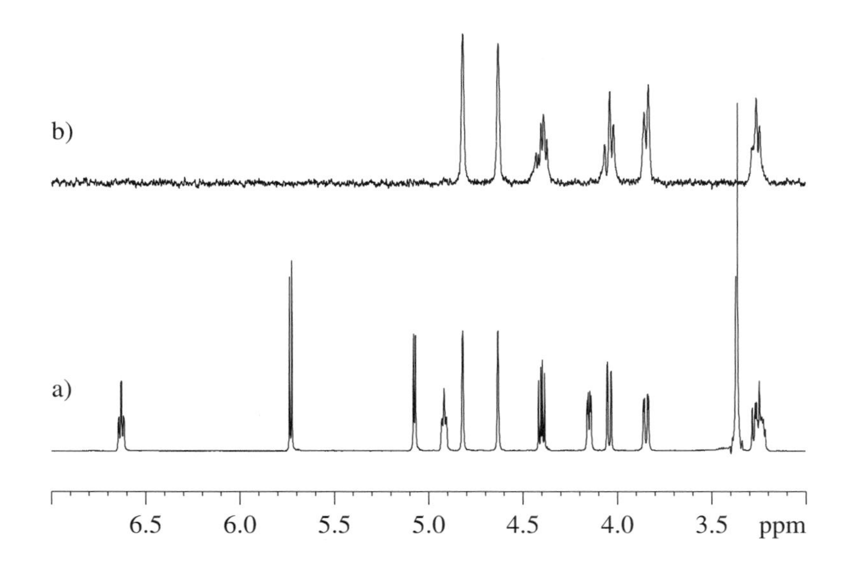

Figure 6.24. Editing of a proton spectrum according to carbon multiplicities. In (b) only those resonances arising from methylene groups have been selected from the conventional spectrum (a). Clean suppression of all other resonances is achieved with pulsed field gradients, although some phase errors remain on the selected signals.

6.4. HETERONUCLEAR MULTIPLE-BOND CORRELATION SPECTROSCOPY

The ^{1}H–X heteronuclear correlation methods presented so far all depend upon the presence of a proton bound to the heteroatom so are therefore unable to provide assignments for non-protonated centres and do not produce unambiguous carbon assignments when proton resonances exactly overlap. An alternative approach in such cases is to establish correlations between carbons and neighbouring protons over more than one bond, so called *long-range* or *multiple-bond* correlations, most commonly through the proton-detected heteronuclear multiple-bond correlation (HMBC) experiment. In the vast majority of cases, this will involve proton–carbon connectivities through couplings over two or three bonds ($^{n}J_{CH}$, n = 2, 3), since those over greater distances are often vanishingly small (see below). The ability to identify ^{1}H–^{13}C correlations across carbon–carbon or carbon–heteroatom linkages presents a wealth of information on the molecular skeleton, providing one of the most powerful approaches to defining an organic structure, perhaps second only to the more difficult to realise direct ^{13}C–^{13}C correlations provided by the INADEQUATE-based methods of Chapter 5. The HMBC spectrum itself closely resembles that of HMQC in appearance, with the long-range correlations of each proton represented in the column taken at its chemical shift. The abundance of information in such spectra is illustrated in the HMBC spectrum of **6.2**, (Fig. 6.25) which should be compared with the HMQC in Fig. 6.4. The complete set of correlations observed in this spectrum are summarised in Table 6.3, but the salient features of the experiment may be appreciated by considering the correlations of proton H2 alone. Firstly notice the breakthrough of the one-bond correlation appearing as the arrowed doublet at 93 ppm. This is exactly equivalent to the correlation seen in Fig. 6.4b, and here serves as a reference point, although these can, at times, be an unwelcome complicating factor. The possibility of such peaks appearing should always be borne in mind when interpreting these spectra. Second, notice that the correlations to C4, C5 and C8 arise from couplings across heteroatoms (N and O), a feature which sometimes seems to surprise those new to the experiment. This type of data can be particularly valuable when proton–proton couplings are absent. Finally, notice that H2 shows a correlation to the carbonyl carbon C8, this centre being unobservable in the one bond correlation experiments. The ability to observe non-protonated centres in this way not only allows their chemical shifts to be determined but also provides valuable connectivity data. Nowadays, the combination of COSY,

Table 6.3. A summary of the long-range correlations observed in the HMBC spectrum (Δ = 60 ms) of **6.2**

Proton	Correlated carbon
2	4, 5, 8, 9, 10
4	2, 5(w), 6(w)
4′	2(w), 5, 6
5	4(w), 7(w), 8
6	4, 5, 7, 8
6′	4, 5, 7, 8
7	5, 6, 8
7′	5, 6, 8
9	2, 10
10	2, 9, 10′
10′	2, 9, 10

Weaker correlations, corresponding to smaller coupling constants, are identified with (w). Not all these are observed in Fig. 6.25 at the contour levels shown.

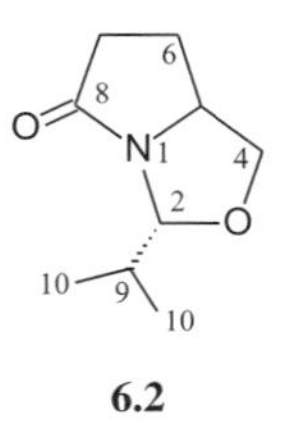

6.2

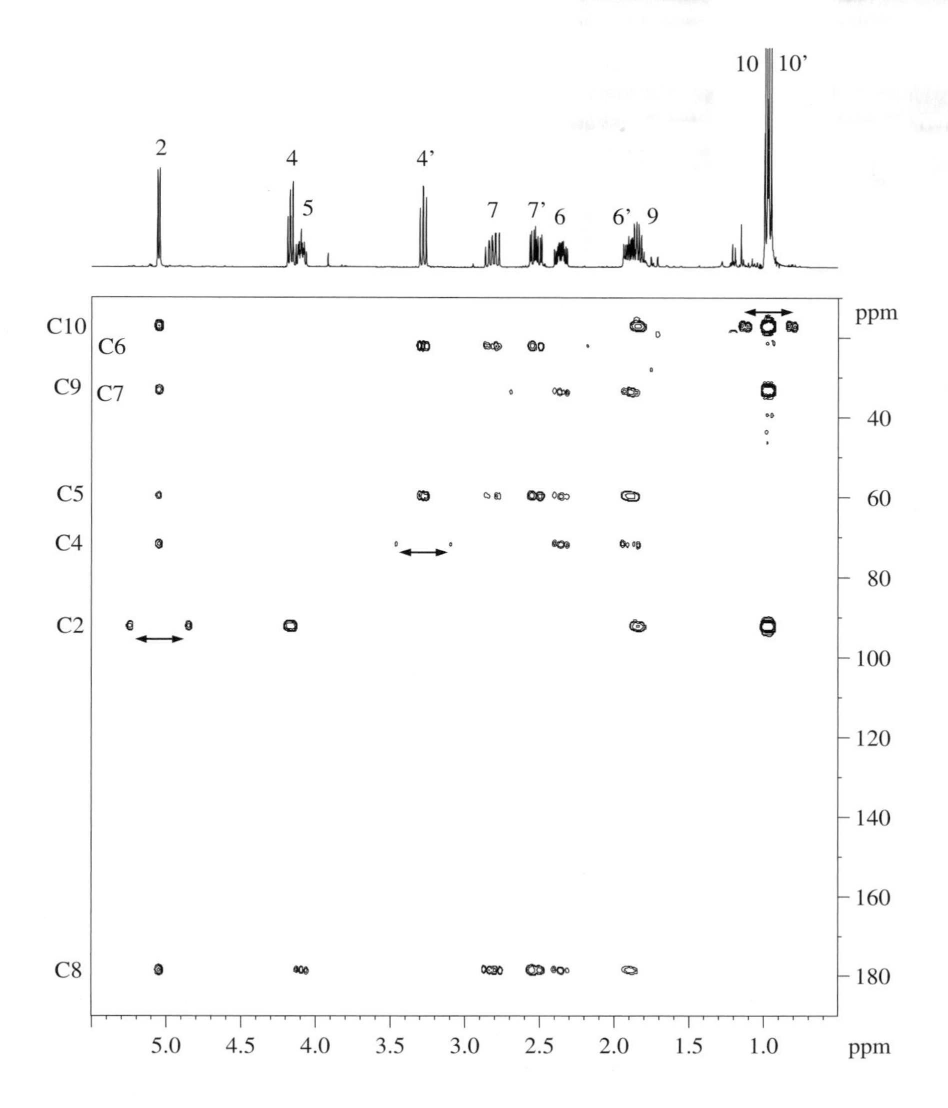

Figure 6.25. The HMBC long-range correlation spectrum of **6.2** recorded with $\Delta = 60$ ms and with gradient selection. The sequence used the low-pass J-filter (Section 6.4.1) to attenuate breakthrough from one-bond correlations (which appear with $^{1}J_{CH}$ doublet structure along f_2 (arrowed)). 1K t_2 data points were collected for 256 t_1 increments of 8 transients each and the data processed with unshifted sine-bells in both dimensions, followed by magnitude calculation. After zero-filling once in t_1 the digital resolution was 4 and 80 Hz/pt in f_2 and f_1 respectively.

HMQC/HSQC and then HMBC typically represent the primary techniques to turn to when addressing problems of molecular connectivity in small organic molecules.

6.4.1. The HMBC sequence

The HMBC experiment [41,42] establishes multiple-bond correlations by again taking advantage of the greater sensitivity associated with proton detection, and in essence is the HMQC sequence 'tuned' to detect correlations via small couplings (Fig. 6.26a). Owing to the close similarity of the two, only differences pertinent to HMBC will be presented here, with the application of the experiment being considered in the following section.

The tuning of the experiment is achieved by setting the Δ preparation period to a sufficiently long time to allow the small long-range proton–carbon couplings to evolve to produce the antiphase displacement of vectors required for the subsequent generation of heteronuclear multiple-quantum coherence. Since long-range ^{1}H–^{13}C couplings are at least an order of magnitude smaller than one-bond couplings (often <5 Hz), Δ should, in principle, be at least 100 ms ($1/2^{n}J_{CH}$), although shorter delays are often used routinely to avoid relaxation losses. During this long Δ period, homonuclear ^{1}H–^{1}H couplings, which are of similar magnitude to the long-range heteronuclear couplings, also evolve and

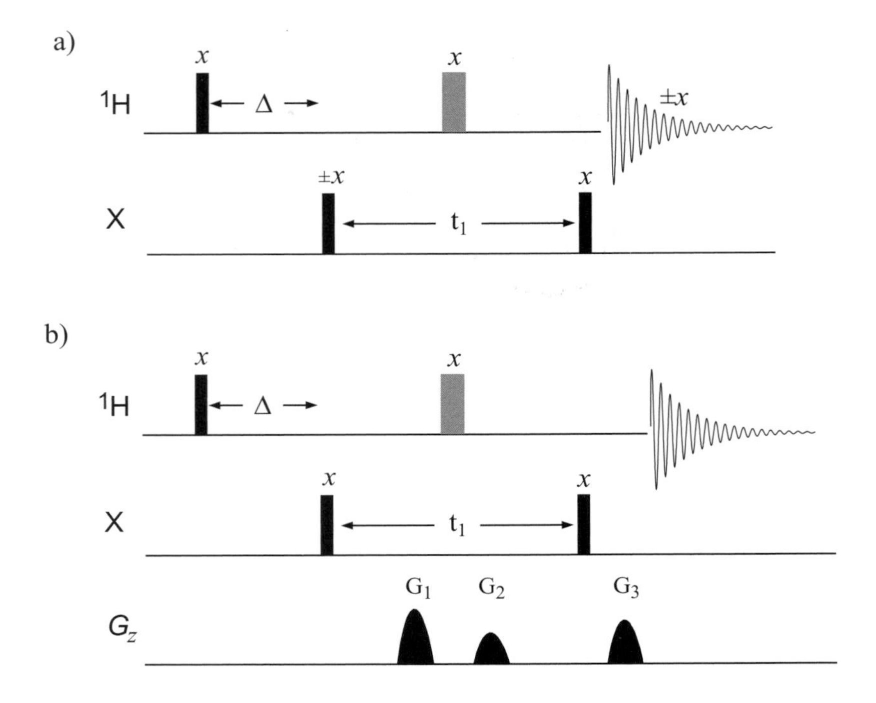

Figure 6.26. The HMBC sequence (a) without and (b) with incorporation of pulsed field gradients. The sequence is closely relate to the HMQC experiment and follows a similar coherence transfer pathway.

introduce phase distortions to the observed crosspeaks (these distortions are small enough to be ignored in HMQC because of the much smaller Δ periods used). Absolute-value presentations are therefore widely used for HMBC spectra to mask these phase errors and it is therefore commonplace to acquire HMBC data sets that are phase-modulated (N-type selection) as a function of t_1. For sensitivity reasons, the refocusing Δ period of HMQC is omitted in HMBC so that long-range heteronuclear couplings are antiphase at the start of t_2, precluding the application of ^{13}C-decoupling.

Although the Δ period is chosen according to long-range couplings, it may also happen to be a multiple of the appropriate setting for one-bond correlations, additionally causing these to appear. Since the FID is acquired without ^{13}C decoupling, these possess distinctive doublet structure in f_2 (as seen in Fig. 6.4b and Fig. 6.25) which aids their identification. These may be considered useful additions or unwanted interferences, depending on your point of view; while they simultaneously provide one-bond correlation data, they may obscure or become confused with long-range correlations. Their suppression is at least partially afforded by the addition of a one-step low-pass J-filter (so-called because it retains or passes only those peaks arising from couplings that are *smaller* than a chosen cut-off value, here the one-bond coupling constant) [41,43] (Fig. 6.27). Here, the Δ_1 period is set according to $^1J_{CH}$, whilst Δ_2 is set according to the long-range coupling, as above. Alternation of

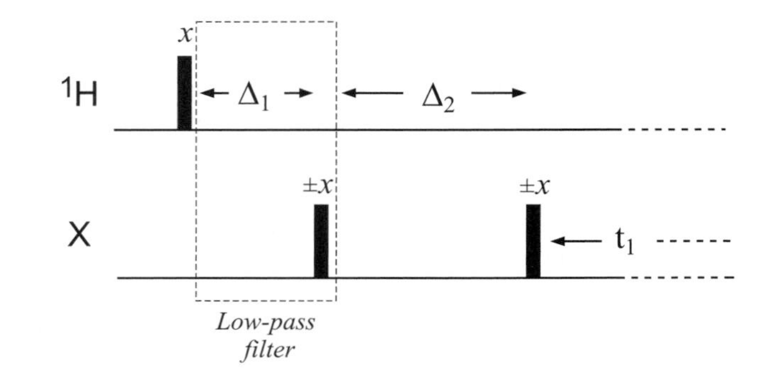

Figure 6.27. The low-pass filter for removing spurious one-bond correlation peaks from HMBC spectra. This comprises an additional Δ_1 period and 90° X pulse at the beginning of the HMBC sequence. Δ_1 is set to $1/^1J_{CH}$ and Δ_2 to $1/^nJ_{CH}$. Inverting the phase of the first X pulse on alternate scans without inverting the receiver phase cancels unwanted one-bond contributions.

the phase of the first carbon pulse *without* changing that of the receiver causes the signals originating from the one-bond correlations to cancel on alternate scans. In practice, the suppression is often not complete particularly when a wide range of $^{1}J_{CH}$ values are present, and a purge scheme in addition to the low-pass J-filter has been shown to provide improved suppression [44]. In routine studies one may omit all filtering schemes and accept the presence of breakthrough from one-bond couplings.

Without any doubt, the greatest problem associated with the HMBC sequence lies in the suppression of the parent ^{1}H–^{12}C signals which may otherwise mask the long-range satellites. Unlike HMQC, the BIRD sequence is not well suited to the removal of these resonances since this is also likely to lead to attenuation of the desired signals. Traditionally, the sequence has relied on phase cycling alone to cancel the intense parent resonance, requiring a very stable spectrometer to be effective. Even so, bands of residual t_1-noise routinely plagued the original HMBC experiment and limited its use as a routine tool in organic chemistry. The introduction of pulsed field gradients [10,45] has revolutionised the applicability of HMBC, since complete suppression of the parent line is readily achieved, along with a dramatic reduction in the associated t_1-noise. The benefits of the gradient selected approach are clearly demonstrated in the spectra of Fig. 6.28. Spectrum 6.28a was recorded with gradient selection and 6.28b with conventional phase cycling under otherwise identical conditions (other than a 256-fold increase in receiver amplification in (a) over that in (b)). Four transients were collected per increment, corresponding to the minimum phase cycle (two steps for signal selection and two for suppression of axial peaks). Long-range correlations are clearly identified in (a) but can barely be observed above the t_1-noise bands in (b). Improved results for the phase-cycled version can be achieved by collecting far more transients at the expense of instrument time.

Long-range correlations based on the HSQC sequence are generally less effective. Significant evolution of ^{1}H–^{1}H couplings during the Δ period leads to unwanted COSY-type transfers among protons by the second 90° proton pulse of the INEPT sequence, a problem not found with HMBC.

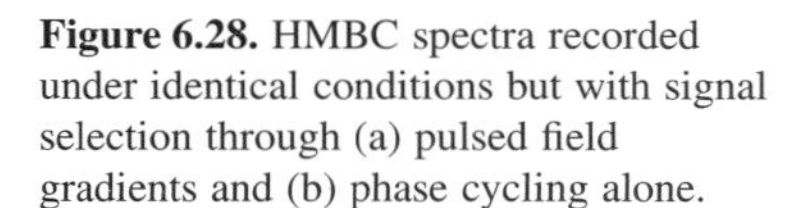

Figure 6.28. HMBC spectra recorded under identical conditions but with signal selection through (a) pulsed field gradients and (b) phase cycling alone.

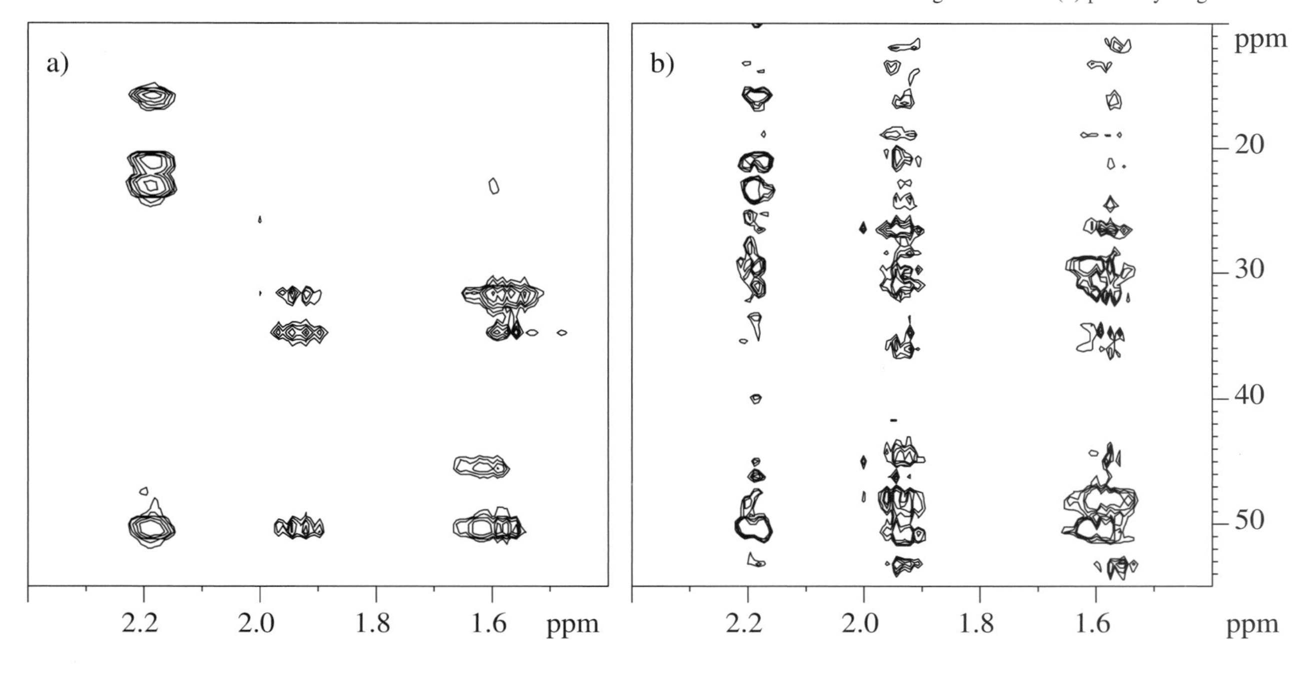

6.4.2. Applying HMBC

The presence of long-range correlations in HMBC spectra is influenced by many factors, both experimental and structural, and an awareness of these points is important for the optimum application and interpretation of the experiment, so are addressed here before progressing to some illustrative applications. Crosspeak intensities depend upon, among other things, both the magnitude of the long-range coupling and on the value selected for Δ, which should optimally be set to $1/2^nJ_{CH}$ (crosspeak intensity $\propto \sin \pi^nJ_{CH}\,\Delta$). Long-range proton–carbon couplings over two or three bonds rarely exceed 25 Hz, and in the absence of unsaturation are more often less than 5 Hz [46,47] (Table 6.4) indicating Δ should be 100 ms or more. Such long delays can lead to relaxation losses prior to detection, especially for larger molecules, so in practice a compromise is met with Δ being set to around 60–80 ms for routine applications. Since small organic molecules tend to have slower relaxation rates, longer delays can be used successfully in the search for more connectivities through smaller couplings, with Δ taking values of up to 200 ms. When using longer delays it also becomes possible to detect peaks that arise from 4-bond correlations which are most likely to occur when the coupling pathway contains unsaturation or when it has the planar zig-zag (*w*-coupling) configuration, as commonly observed in long-range proton–proton couplings also. One should also recognise that three-bond couplings can be, and often are, greater in magnitude than two-bond couplings, displaying a Karplus-type relationship with the dihedral angle [48]. Indeed, one of the limitations when using HMBC data is the lack of differentiation between two- and three-bond connectivities. Furthermore, the lack of a correlation cannot on its own be taken as evidence for the nuclei in question being distant in the structure, since a variety of factors can contribute to $^nJ_{CH}$ being close to zero (for more details see discussions in reference [46]). The most intense correlations are typically observed for methyl groups since magnetisation is detected on three protons simultaneously and because they display simpler coupling structure, if any, whereas the weakest correlations are generally associated with poorly resolved, complex proton multiplets.

Despite the lack of discrimination, protons can potentially correlate to a great many carbon neighbours within two or three bonds, providing a mass of structural data on an unknown molecule. Connectivities may also be traced across heteroatom linkages (as seen in Fig. 6.25) where proton–proton couplings are usually negligibly small and in this the experiment is extremely effective at piecing together otherwise uncorrelated molecular fragments. Thus, the HMBC experiment was able to define the regiochemistry of a dipeptide antibiotic, Tü 1718B, isolated from *Streptomyces* cultures [49]. The two possible structures **6.8a** and **6.8b** were proposed from biosynthetic arguments, but neither proton correlation or carbon chemical shift data could unambiguously identify which was correct. Confirmation was gained

Table 6.4. Typical values of long-range proton–carbon coupling constants

Coupling pathway	$^2J_{CH}$	Coupling pathway	$^3J_{CH}$	Coupling pathway	$^4J_{CH}$
H–C–**C**	(±) ≤5	**H**–C–C–**C**	≤5	**H**–C=C–C=**C**	(±)≤1
H–C=**C**	≤10	**H**–C=C–**C**	≤15[a]	**H**–C–C–C–**C**[b]	≤1
H–C≡**C**	40–60	**H**–C≡C–**C**	≤5		
H–C(=O)–**C**	20–25				

[a] trans > cis, cis usually <10 Hz.
[b] *w*-configuration favoured.
Heteroatoms such as O, N etc may also be included in the coupling pathways illustrated in place of C.

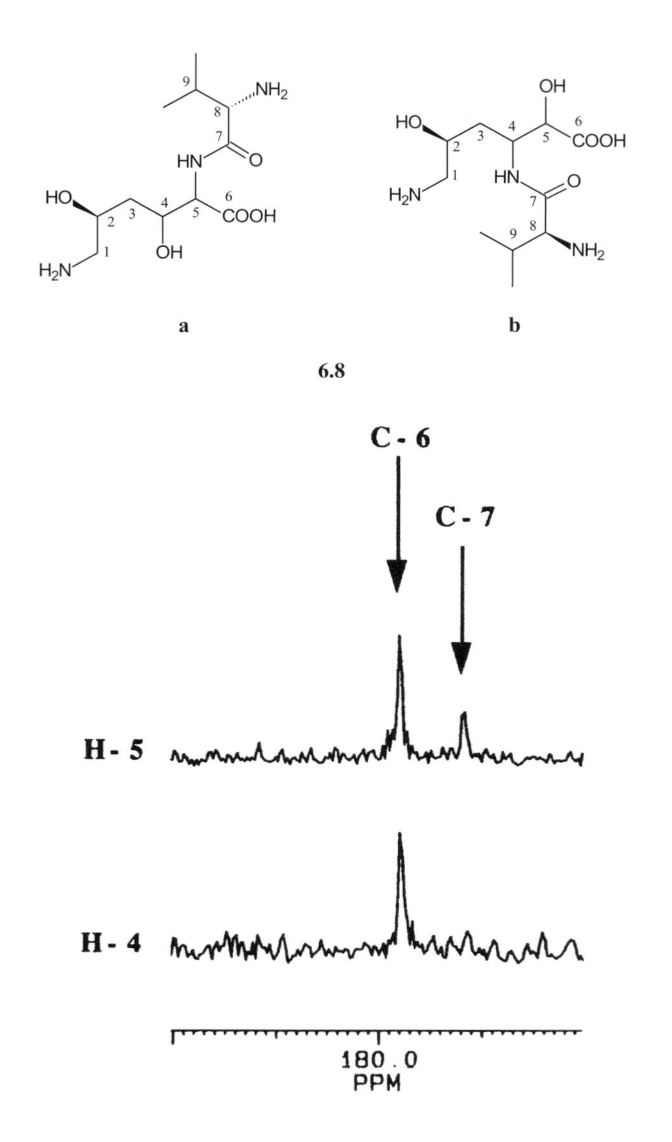

Figure 6.29. Columns taken from the HMBC spectrum of the antibiotic Tü 1718B at the shifts of protons H4 and H5. Only the key correlations to the carbonyl resonances are shown (reproduced with permission from [49]).

from an analysis of the long-range correlations of protons H4 and H5, and in particular those to the carbonyl groups (Fig. 6.29). H5 was observed to correlate to both whereas H4 correlated only to C6, these data being consistent with **6.8a** only. Connectivities across the oxygen linking neighbouring sugar residues in oligosaccharides likewise provides a useful means of identifying neighbouring residues in these compounds. Naturally, the sequence is not limited to proton–carbon connectivities, but can be tailored to any spin-$^1/_2$ pair. Thus, long-range proton–silicon and proton–carbon correlations established through HMBC were used to confirm the structure of an unexpected product from the rearrangement of epoxydisilanes [50] as the silanol **6.9a** rather than **6.9b**. Additional NOE studies identified the stereochemistries across the alkene.

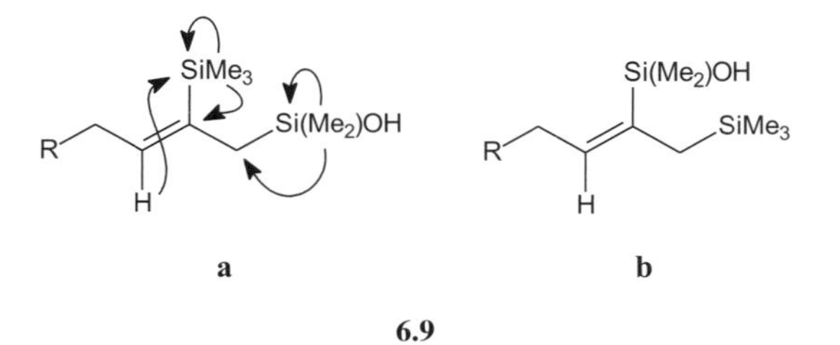

A potentially limiting problem with HMBC is the large spectral width that must be indirectly digitised in f_1. This tends to be a greater problem than with one-bond correlation experiments because of the need to include the carbonyl region in many instances, meaning the data are usually collected with rather low f_1 resolution. In regions of the ^{13}C spectrum where many resonances fall close together, such low resolution may prove insufficient and result in crosspeak overlap. This can be the case with peptide assignments for example, where amide carbonyl resonances all fall within a 6 ppm window. One solution to this is to select only the part of the f_1 dimension that contains the carbonyl resonances, so only these appear in the final spectrum and hence only a small f_1 window need be digitised. This is achieved by selectively exciting only the ^{13}C resonances of interest with a shaped carbon pulse (Chapter 9) in place of the first non-selective carbon pulse of the conventional sequence [51]. Fig. 6.30 compares the carbonyl region of **6.10** from the conventional HMBC with that from the semi-selective HMBC experiment. The goal was to obtain sequence specific assignments for each carbohydrate amino acid residue by identifying long-range ^1H–^{13}C correlations to the carbonyl carbons, thus providing a link between adjacent residues [52]. For Fig. 6.30b the CO region was selectively excited with a 2 ms Gaussian pulse, providing a dramatic increase in f_1 resolution sufficient to identify the required $H2^i$ to CO^i and

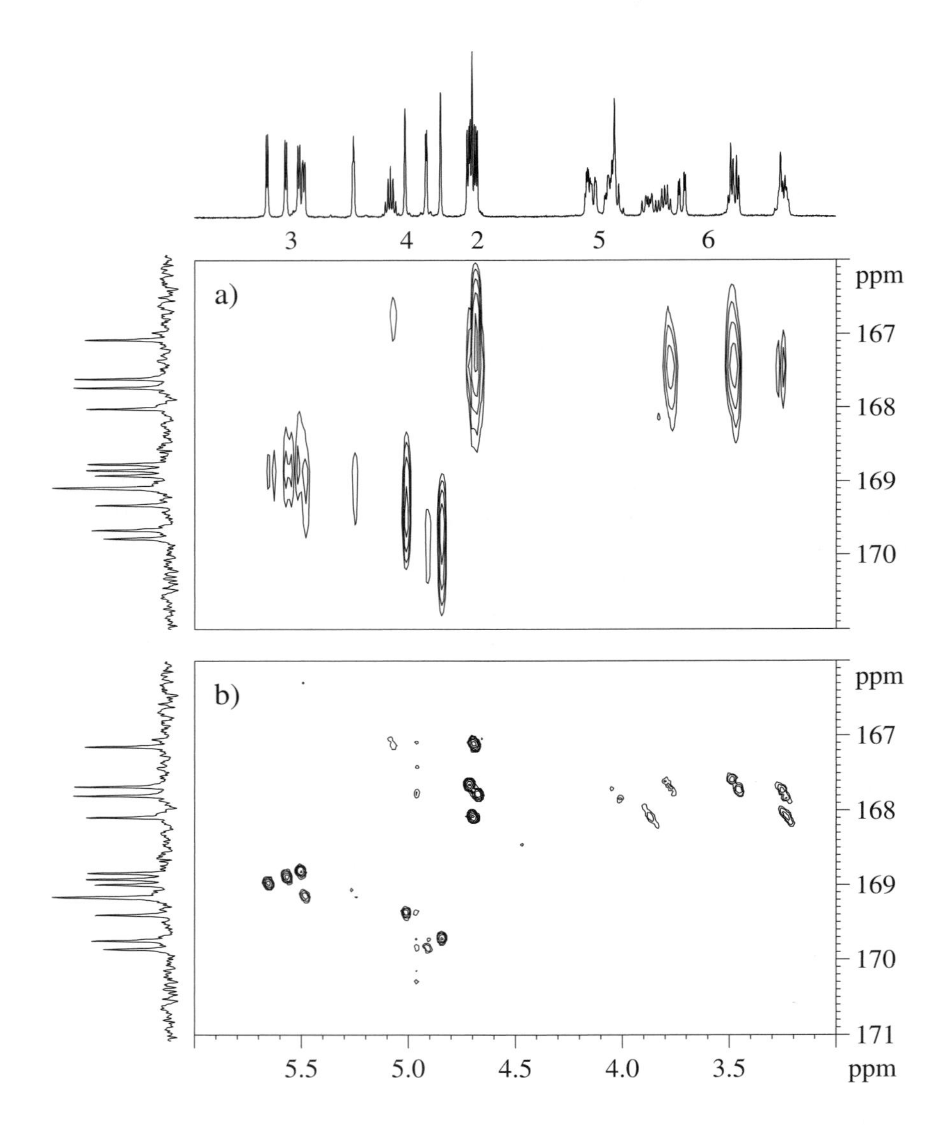

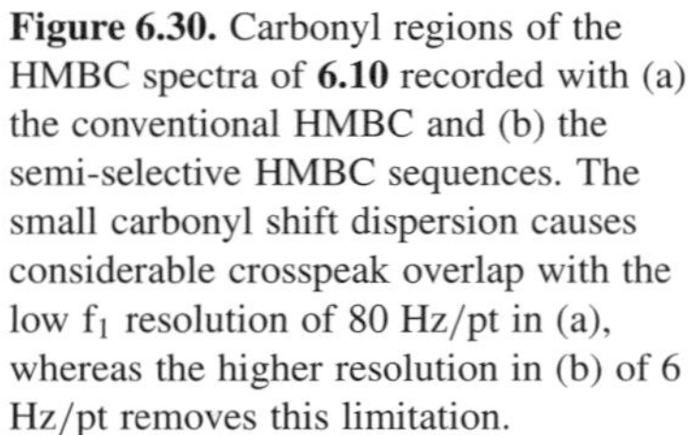

Figure 6.30. Carbonyl regions of the HMBC spectra of **6.10** recorded with (a) the conventional HMBC and (b) the semi-selective HMBC sequences. The small carbonyl shift dispersion causes considerable crosspeak overlap with the low f_1 resolution of 80 Hz/pt in (a), whereas the higher resolution in (b) of 6 Hz/pt removes this limitation.

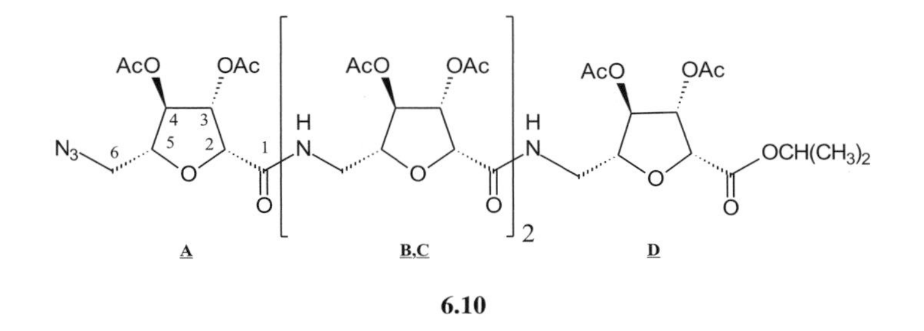

6.10

CO^i to $H6^{i+1}$ long-range correlations and so establish connectivities across the amide linkage.

Practical set-up

Owing to the small size of long-range relative to one-bond couplings, and because of the need for long and often non-optimal Δ delay periods, the sensitivity of the HMBC experiment is somewhat less than that of its HMQC or HSQC cousins. In the absence of pulsed-field gradients, it is necessary to acquire many scans for each increment in an attempt to suppress the intense $^1H–^{12}C$ signal and reveal the correlations of interest, making the experiment time-consuming even when large sample quantities are available (spectral quality being highly dependent on spectrometer stability). Gradient-selected versions are considerably quicker to acquire, being dictated by sensitivity and resolution arguments alone since the suppression of the parent signal is no longer an issue and better use is made of the receiver dynamic-range. As a rule of thumb when using gradient selection, the HMBC experiment will typically take around four times as long to acquire as the corresponding HMQC or HSQC experiment to provide acceptable data. From what was once typically an overnight phase-cycled experiment, high-quality data can now be obtained within a few hours or even tens of minutes; the impact of pulsed field gradients on the HMBC experiment has been profound.

The setting of Δ has been described above and the choice of digital resolution for the two dimensions follows similar arguments as for the one-bond correlation experiments, that is, around 50 Hz/pt in f_1 and around 5 Hz/pt in f_2 should provide acceptable results in most cases. In HMBC the acquired t_2 FID begins as antiphase magnetisation with respect to $^nJ_{CH}$ since the refocusing period is omitted, and the signal builds as refocusing occurs during t_2 itself. Optimal window functions providing close to matched filtering are therefore the unshifted sine-bell or squared sine-bell. In contrast, the t_1 interferograms decay from their maximum values and require only sufficient apodisation to avoid truncation errors that may appear under conditions of high signal-to-noise, so a simple exponential decay function suffices. Improved *digital* resolution of the f_1 dimension can be achieved through zero-filling (at least once is recommended) or the resolution increased by the use of forward linear prediction. Finally, the repetition time of the experiment is dictated by the T_1s of protons *directly* bound to carbon-12 centres. Repetition times are therefore estimated from relaxation times as they would be for *homonuclear* correlation experiments, and are longer than those optimal for HMQC or HSQC.

6.5. TRADITIONAL X-DETECTED CORRELATION SPECTROSCOPY

Traditional methods of heteronuclear shift correlation follow the scheme of Fig. 6.1b above, relying on the direct observation of the X-spin and with the source nucleus (typically protons) detected indirectly. As discussed at the beginning of this chapter, this general approach is less sensitive than that involving the direct observation of the proton and this has led to the general decline in popularity of the X-spin observed techniques. However, one should not shun their existence completely as they have one major advantage over the proton detected counterparts and that is the ability to record the X-spin with high digital resolution. With proton detected methods, the often large X-spin chemical shift range is measured in the indirect dimension and recording this with high resolution entails collecting many t_1 incremented data sets, making the experiments time consuming. With X-spin detection, high resolution may be achieved simply by collecting more data points per FID, making a minimal difference to the total duration of the experiment. When heteronuclear correlations are required for a crowded carbon spectrum, complex aromatic systems for example, the ^{13}C detected approach may thus prove superior. A more mundane reason for turning to these methods may be from purely practical considerations in as much as your spectrometer may be unable to operate in the 'inverse' mode, that is, it cannot be to configured to observe protons and still be able to provide pulsing and decoupling of the heteronucleus. This will be a constraint only for older instruments rather than those of modern architecture. The sensitivity of these methods can also be increased by the use of microcell or microtube techniques [53] in place of standard NMR tubes. The following experiments are described in the context of ^1H–^{13}C, but again can be tailored to other spin pairs.

6.5.1. Single-bond correlations

Methods for establishing heteronuclear connectives are based on the general idea of polarisation transfer from the proton to carbon and as such can be understood with reference to the previously encountered 1D INEPT experiment of Section 4.4.2. Recall that the refocused INEPT sequence allows the transfer of proton populations (polarisation) onto the attached carbon by the application of simultaneous proton and carbon pulses after the heteronuclear coupling has evolved for a period Δ_1. Following the transfer, the carbon magnetisation so created is antiphase with respect to $^1J_{CH}$, so is allowed to evolve under the influence of the heteronuclear coupling for a period Δ_2 until the carbon vectors have realigned. At this point, proton decoupling is applied and the carbon FID recorded (Fig. 6.31a; here the 180° pulses at the midpoint of

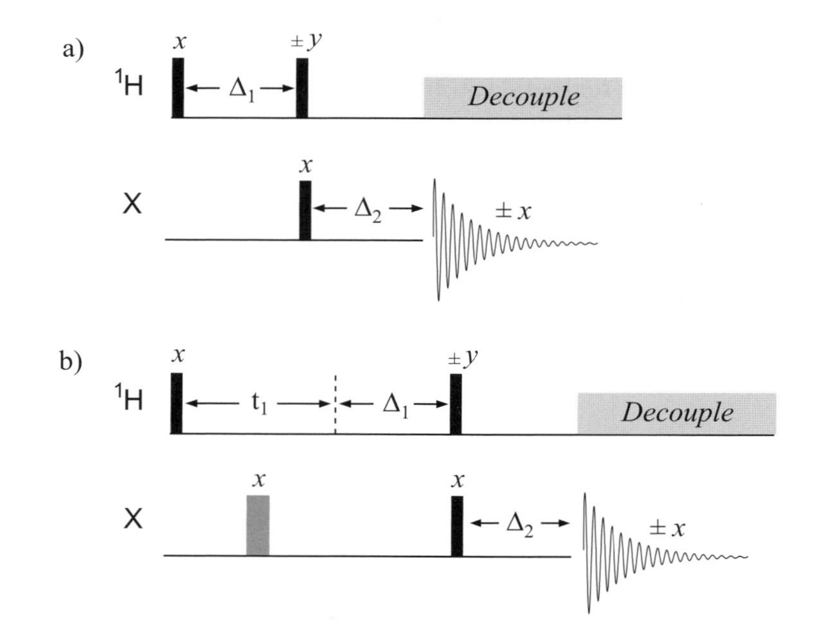

Figure 6.31. (a) The 1D refocused INEPT experiment, shown in simplified form with the refocusing pulses at the midpoints of Δ_1 and Δ_2 removed for clarity. The 2D shift correlation experiment (HETCOR) in (b) is derived from INEPT by the addition of the t_1 evolution period to encode proton chemical shifts prior to polarisation transfer. The 180° carbon pulse at the midpoint of t_1 refocuses heteronuclear coupling evolution and thus provides carbon decoupling in f_1.

Δ_1 and Δ_2 which serve to remove chemical shift evolution are omitted for simplicity). To produce the two-dimensional shift correlation experiment [54], the variable t_1 evolution period is added immediately after the initial proton excitation and prior to the polarisation transfer step so that the detected ^{13}C signal becomes modulated by the proton chemical shift as a function of t_1 (Fig. 6.31b). During this, ^1H–^1H and ^1H–^{13}C coupling will also evolve, leading to the appearance of both homonuclear and heteronuclear splittings in f_1, thereby reducing signal intensities. The removal of ^1H–^{13}C coupling can be achieved by refocusing $^1J_{CH}$ with the insertion of a 180° carbon pulse at the midpoint of t_1 (preferably applied as a composite pulse) to reduce resonance offset effects (Fig. 6.31b). Because there is no net CH coupling evolution in t_1, the antiphase magnetisation required for polarisation transfer only develops during the subsequent Δ_1 period, which is therefore optimised for $\Delta_1 = 1/2^1J_{CH}$ precisely as for INEPT (typically 3.5 ms for ^1H–^{13}C correlations). The resulting spectrum displays crosspeaks correlating carbon chemical shifts in f_2 and proton shifts in f_1 which are further spread by homonuclear proton couplings in f_1. Fig. 6.32 displays a part of the carbon–proton shift correlation spectrum of the palladium complex **6.11**. Despite the extensive crowding in the aromatic region, the carbon shifts are sufficiently dispersed to resolve all correlations (note some resonances are broadened by restricted dynamic processes within the molecule and some are split by coupling to phosphorus).

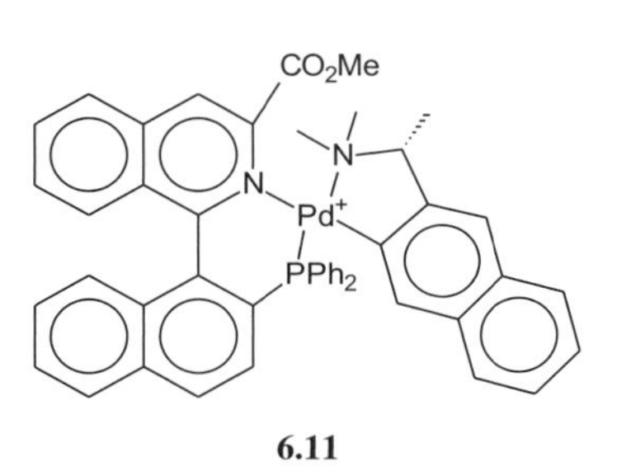

6.11

The scheme of Fig. 6.31b has been widely used to produce absolute-value shift correlation spectra, and is often referred to as HETCOR or hetero-COSY. Conversion to the preferred phase-sensitive equivalent (of which various forms have been investigated [55]) requires the reintroduction of the simultaneous 180°(^1H,^{13}C) pulses into the midpoints of both Δ_1 and Δ_2 to remove chemical shift evolution during these periods, exactly as in the full refocused INEPT. In addition, the incorporation of the States or TPPI phase cycling of the 90° proton pulse of the polarisation transfer step is required. Suppression of axial peaks is through the phase alternation of the final proton pulse together with the receiver

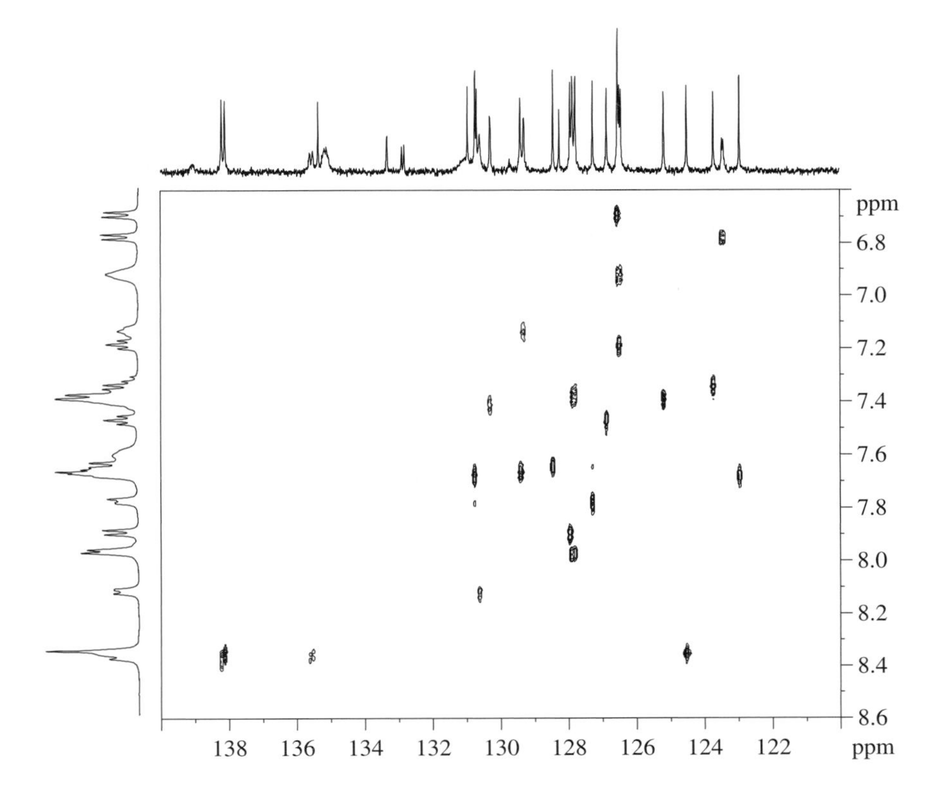

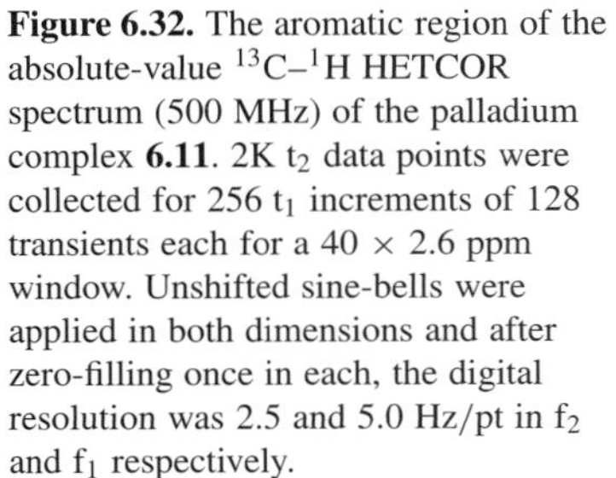
Figure 6.32. The aromatic region of the absolute-value ^{13}C–^1H HETCOR spectrum (500 MHz) of the palladium complex **6.11**. 2K t_2 data points were collected for 256 t_1 increments of 128 transients each for a 40 × 2.6 ppm window. Unshifted sine-bells were applied in both dimensions and after zero-filling once in each, the digital resolution was 2.5 and 5.0 Hz/pt in f_2 and f_1 respectively.

(which is equivalent to that used in INEPT to suppress contributions from the natural X-spin magnetisation). As for the INEPT sequence, editing of the phase sensitive spectrum is possible by judicious choice of the Δ_2 refocusing delay, for example setting $\Delta_2 = 1/1.5J_{CH}$ produces a spectrum in which responses from CH_2 groups are inverted. For routine application, selecting $\Delta_2 = 1/3J_{CH}$ (typically 2.3 ms for ^1H–^{13}C) provides positive intensities for all protonated carbons.

Since polarisation transfer is employed, the experiment repetition rate is dictated by the recovery of the faster relaxing proton spins, and repetition times should be around 1.3 times the proton T_1s. Resolution in the proton dimension can be quite low for routine applications since one does not usually wish to resolve the proton fine-structure and because the homonuclear coupling is in-phase there is no fear of signal cancellation; f_1 digital resolution may therefore be as low as 10 Hz/pt or so, requiring rather few t_1 increments. The number of scans per increment should be set such that the carbon resonances of interest can just be observed in the spectrum of the first recorded FID (which is equivalent to the 1D INEPT experiment).

Homonuclear decoupling in f_1

By introducing a slight modification to the sequences presented above, it is possible to achieve (almost complete) removal of homonuclear proton couplings from f_1 also, so removing the multiplet spread and increasing both sensitivity and f_1 resolution. To do this, a similar approach to that for the removal of the *heteronuclear* coupling in f_1 described above is employed, that is, for each coupled spin pair one must invert, at the midpoint of t_1, *only one* of the two spins so that the coupling evolution is refocused at the end of t_1. For the heteronuclear case this is trivial, since one can apply a 180° carbon pulse confident in the knowledge that its proton partner will be unaffected. However, applying a single proton 180° pulse will fail since both protons experience this pulse and the coupling will continue to evolve, as for a homonuclear spin-echo. Worse still, proton chemical shifts will be refocused during t_1, destroying the proton shift dimension altogether (although undesirable in this application such an approach can be useful if one wishes to analyse only coupling patterns in f_1; see Chapter 7 on J-resolved spectroscopy). What is required then, is a means of inverting only the *remote* J-coupled neighbours of a *directly* ^{13}C-bound proton (which is ultimately involved in the polarisation transfer step). The solution to this seemingly impossible task lies in the application of the BIRD sequence [14] already introduced in Section 6.3.3 as a means of selectively inverting ^{12}C-bound protons whilst leaving ^{13}C-bound protons unaffected [56,57]. The differentiation of protons in the context of f_1 decoupling relies on the very different magnitudes of the one-bond and long-range couplings ($^1J_{CH} \gg {}^{2/3}J_{CH}$) such that the long-range couplings can be considered negligible. Hence, placing the BIRD cluster at the midpoint of t_1 refocuses proton–proton coupling whilst the carbon 180° pulse of BIRD also serves to refocus proton–carbon coupling, as above (hence BIRD; bilinear rotational *decoupling*). The f_1 resonances then appear as singlets with the exception of those from non-equivalent geminal protons which retain their mutual coupling (as neither proton experiences the inversion since both are bound to ^{13}C) [58].

6.5.2. Multiple-bond correlations and small couplings

Numerous X-detected sequences have been presented over the years for establishing long-range ^1H–^{13}C correlations [59] prior to the widespread adoption of the proton-detected counterparts. This section presents the most widely used sequence (COLOC) and briefly mentions more recent sequences that have superior performance in most instances.

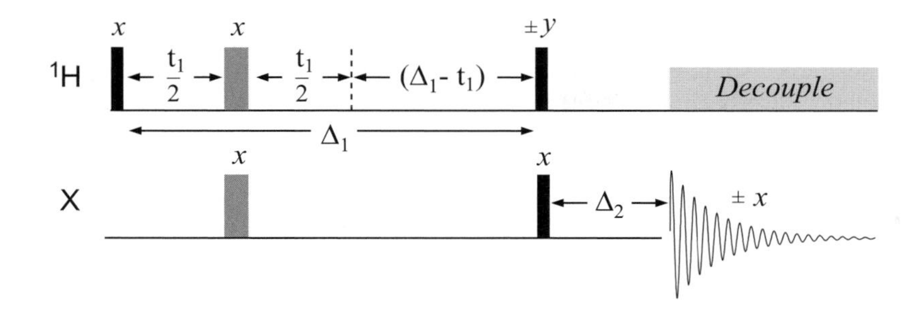

Figure 6.33. The COLOC sequence for establishing long-range correlations through X-spin observation.

A direct way to obtain long-range correlations is to optimise the J-delay periods found in the one-bond HETCOR sequence above for the much smaller long-range couplings. The potential drawback to this approach is the total length of the resulting sequence because of the small couplings and the associated signal losses due to spin-relaxation. An early modification, known as COLOC (COrrelation through LOng-range Coupling), places the t_1 period *within* the Δ_1 delay and thus reduces the overall length of the sequence (Fig. 6.33) [60,61]. The t_1 period is now defined by a pair of 180°(H,C) pulses that move through Δ_1 as t_1 is incremented, such experiments being generally referred to as 'constant-time' experiments. The detected signal is modulated by proton shifts because, although these are refocused during t_1, they continue to evolve in the *remainder* of the Δ_1 period which itself is dependent on t_1, so allowing characterisation of the proton chemical shifts. In contrast, homonuclear proton couplings evolve for the whole of the fixed Δ_1 period regardless of the position of the moving 180° proton pulse meaning these couplings cause no modulation as a function of t_1 and the f_1 dimension conveniently displays only proton chemical shifts. Transfers due to one-bond couplings may also appear in the COLOC spectrum if the Δ_1 and Δ_2 periods happen to be multiples of the appropriate $^1J_{CH}$ values, and, if desired, the inclusion of a low-pass J-filter, as described for HMBC above, may help attenuate these [62].

A greater problem with this sequence is the influence of one-bond couplings during the Δ_2 refocusing period, which can lead to the disappearance of a long-range crosspeak even when the Δ_1 and Δ_2 delays are optimised for the long-range couplings [63]. Following the polarisation transfer step, the vectors associated with the *one-bond* couplings of the carbon nuclei are in-phase (they are *passive* couplings since they have not been responsible for the polarisation transfer process) and during the subsequent Δ_2 period, they will pass in- and out-of-phase a number of times under the influence of $^1J_{CH}$. If these happen, coincidentally, to be have an antiphase disposition when the acquisition starts and proton decoupling is applied, the vectors will cancel and the crosspeak will be lost. In any case, the detected long-range correlation intensity will be modulated by the one-bond H–C coupling of the carbon. To remove these effects, one must ensure these vectors are returned to their original in-phase disposition at the end of Δ_2, but at the same time allow the initially antiphase vectors due to *long-range* couplings to continue to evolve and become in-phase. Once again, this can be achieved by positioning a BIRD sequence at the midpoint of Δ_2, which inverts the carbon and its long-range coupled proton but not the directly coupled proton [64,65]. The use of the BIRD cluster in this manner is recommended for all X-detected long-range correlation methods.

In general, optimisation of the Δ_1 and Δ_2 delays for the COLOC experiment can be problematic. In addition to the one-bond modulation effects already described, the transfer efficiency is also influenced by homonuclear proton couplings evolving during Δ_1. The fixed Δ_1 period also limits $t_{1(max)}$ ($t_{1(max)}/2$

$\leq \Delta_1$) which may dictate fewer t_1 increments be collected than would otherwise be desirable, so reducing the digital resolution of the f_1 dimension. These problems have been addressed in some detail and 1D sequences presented for parameter optimisation [66]. For routine use, it has been suggested [59] that optimising delays for $^nJ_{CH} = 10$ Hz should provide most responses of interest, particularly when using the BIRD sequence within the Δ_2 refocusing delay. This corresponds to delays of 50 ms ($1/2^nJ_{CH}$) and 33 ms ($1/3^nJ_{CH}$) for Δ_1 and Δ_2 respectively.

Various other sequences have been proposed over the years which generally show improved performance over COLOC. If heterospin detected methods are best suited to the study of your molecules, two sequences to note are XCORFE [67] and FLOCK [68], details of which can be found in the original literature.

REFERENCES

[1] L. Müller, *J. Am. Chem. Soc.*, 1979, **101**, 4481–4484.
[2] A. Bax, R.H. Griffey and B.L. Hawkins, *J. Magn. Reson.*, 1983, **55**, 301–315.
[3] A. Bax and S. Subramanian, *J. Magn. Reson.*, 1986, **67**, 565–569.
[4] J. Cavanagh and J. Keeler, *J. Magn. Reson.*, 1988, **77**, 356–362.
[5] G. Bodenhausen and D.J. Ruben, *Chem. Phys. Lett.*, 1980, **69**, 185–188.
[6] W.F. Reynolds, S. McLean, L.-L. Tay, M. Yu, R.G. Enriquez, D.M. Estwick and K.O. Pascoe, *Magn. Reson. Chem.*, 1997, **35**, 455–462.
[7] T. Parella, *Magn. Reson. Chem.*, 1998, **36**, 467–495.
[8] J. Ruiz-Cabello, G.W. Vuister, C.T.W. Moonen, P. van Gelderen, J.S. Cohen and P.C.M. van Zijl, *J. Magn. Reson.*, 1992, **100**, 282–302.
[9] G.W. Vuister, J. Ruiz-Cabello and P.C.M. van Zijl, *J. Magn. Reson.*, 1992, **100**, 215–220.
[10] W. Willker, D. Leibfritz, R. Kerrsebaum and W. Bermel, *Magn. Reson. Chem.*, 1993, **31**, 287–292.
[11] J.R. Tolman, J. Chung and J.H. Prestegard, *J. Magn. Reson.*, 1992, **98**, 462–467.
[12] A.L. Davis, J. Keeler, E.D. Laue and D. Moskau, *J. Magn. Reson.*, 1992, **98**, 207–216.
[13] J. Boyd, N. Soffe, B. John, D. Plant and R. Hurd, *J. Magn. Reson.*, 1992, **98**, 660–664.
[14] J.R. Garbow, D.P. Weitekamp and A. Pines, *Chem. Phys. Lett.*, 1982, **93**, 504–508.
[15] G. Otting and K. Wüthrich, *J. Magn. Reson.*, 1988, **76**, 569–574.
[16] W.F. Reynolds, M. Yu, R.G. Enriquez and I. Leon, *Magn. Reson. Chem.*, 1997, **35**, 505–519.
[17] S. Berger, T. Facke and R. Wagner, *Magn. Reson. Chem.*, 1996, **34**, 4–13.
[18] R. Benn, H. Brenneke, A. Frings, H. Lehmkuhl, G. Mehler, A. Rufunska and T. Wildt, *J. Am. Chem. Soc.*, 1988, **110**, 5661–5668.
[19] E.J.M. Meier, W. Kozminski, A. Linden, P. Lustenberger and W. von Phillipsborn, *Organomet.*, 1996, **15**, 2469–2477.
[20] D. Nanz, A. Bell, W. Kozminski, E.J.M. Meier, V. Tedesco and W. von Phillipsborn, *Bruker Report*, 1996, **143**, 29–31.
[21] H. Kessler, P. Schmeider and M. Kurz, *J. Magn. Reson.*, 1989, **85**, 400–405.
[22] T. Parella, F. Sánchez-Ferrando and A. Virgili, *J. Magn. Reson. (A)*, 1995, **117**, 78–83.
[23] S.-I. Tate, Y. Masui and F. Inagaki, *J. Magn. Reson.*, 1991, **94**, 625–630.
[24] T. Parella, F. Sánchez-Ferrando and A. Virgili, *J. Magn. Reson.*, 1997, **126**, 274–277.
[25] T. Parella, J. Belloc, F. Sánchez-Ferrando and A. Virgili, *Magn. Reson. Chem.*, 1998, **36**, 715–719.
[26] T. Domke and D. Leibfritz, *J. Magn. Reson.*, 1990, **88**, 401–405.
[27] J. Stelten and D. Leibfritz, *J. Magn. Reson.*, 1992, **99**, 170–177.
[28] J. Stelten and D. Leibfritz, *Magn. Reson. Chem.*, 1996, **34**, 951–954.
[29] M. Liu, R.D. Farrant, J.K. Nicholson and J.C. Lindon, *J. Magn. Reson. (B)*, 1995, **106**, 270–278.
[30] M. Liu, R.D. Farrant, J.K. Nicholson and J.C. Lindon, *J. Magn. Reson. (A)*, 1995, **112**, 208–219.
[31] J.M. Schmidt and H. Rüterjans, *J. Am. Chem. Soc.*, 1990, **112**, 1279–1280.
[32] H. Kessler, P. Schmeider and H. Oshkinat, *J. Am. Chem. Soc.*, 1990, **112**, 8599–8600.
[33] U. Wollborn and D. Leibfritz, *Magn. Reson. Chem.*, 1991, **29**, 238–243.
[34] L. Lerner and A. Bax, *J. Magn. Reson.*, 1986, **69**, 375–380.
[35] D.G. Davis, *J. Magn. Reson.*, 1989, **84**, 417–424.
[36] T. Domke, *J. Magn. Reson.*, 1991, **95**, 174–177.
[37] B.K. John, D. Plant, S.L. Heald and R.E. Hurd, *J. Magn. Reson.*, 1991, **94**, 664–669.
[38] G. Mackin and A.J. Shaka, *J. Magn. Reson. (A)*, 1996, **118**, 247–255.
[39] R.C. Crouch, A.O. Davis and G.E. Martin, *Magn. Reson. Chem.*, 1995, **33**, 889–892.

[40] T. Parella, F. Sánchez-Ferrando and A. Virgili, *J. Magn. Reson. (B)*, 1995, **109**, 88–92.
[41] A. Bax, M.F. Summers, *J. Am. Chem. Soc.*, 1986, **108**, 2093–2094.
[42] M.F. Summers, L.G. Marzilli and A. Bax, *J. Am. Chem. Soc.*, 1986, **108**, 4285–4294.
[43] H. Kogler, O.W. Sørensen, G. Bodenhausen and R.R. Ernst, *J. Magn. Reson.*, 1983, **55**, 157–163.
[44] T. Parella, F. Sánchez-Ferrando and A. Virgili, *J. Magn. Reson. (A)*, 1995, **112**, 241–245.
[45] P.L. Rinaldi and P.A. Keifer, *J. Magn. Reson. (A)*, 1994, **108**, 259–262.
[46] P.E. Hansen, *Prog. Nucl. Magn. Reson. Spectrosc.*, 1981, **14**, 179–296.
[47] J.L. Marshall, Carbon–Carbon and Carbon–Proton NMR Couplings: Applications to Organic Stereochemistry and Conformational Analysis, VCH Publishers, Florida, 1983.
[48] V.F. Bystrov, *Prog. Nucl. Magn. Reson. Spectrosc.*, 1976, **10**, 41–81.
[49] J.E. Baldwin, T.D.W. Claridge, K.C. Goh, J.W. Keeping and C.J. Schofield, *Tetrahedron Lett.*, 1993, **34**, 5645–5648.
[50] D.M. Hodgson and P.J. Comina, *Tetrahedron Lett.*, 1996, **37**, 5613–5614.
[51] H. Kessler, P. Schmeider, M. Köck and M. Kurz, *J. Magn. Reson.*, 1990, **88**, 615–618.
[52] M.D. Smith, T.D.W. Claridge, G.E. Tranter, M.S.P. Sansom, G.W.J. Fleet, *J. C. S. Chem. Commun.*, 1998, 2041–2042.
[53] R.W. Dykstra, *J. Magn. Reson. (A)*, 1995, **112**, 255–257.
[54] A. Bax and G.A. Morris, *J. Magn. Reson.*, 1981, **42**, 501–505.
[55] K.A. Carpenter and W.F. Reynolds, *Magn. Reson. Chem.*, 1992, **30**, 287–294.
[56] A. Bax, *J. Magn. Reson.*, 1983, **53**, 517–520.
[57] J.A. Wilde and P.H. Bolton, *J. Magn. Reson.*, 1984, **59**, 343–346.
[58] V. Rutar, *J. Magn. Reson.*, 1984, **58**, 306–310.
[59] G.E. Martin and A.S. Zektzer, *Magn. Reson. Chem.*, 1988, **26**, 631–652.
[60] H. Kessler, C. Griesinger, J. Zarbock and H.R. Loosli, *J. Magn. Reson.*, 1984, **57**, 331–336.
[61] H. Kessler, C. Griesinger and J. Lautz, *Angew. Chem. Int. Ed. Engl.*, 1984, **23**, 444–445.
[62] M. Salazar, A.S. Zektzer and G.E. Martin, *Magn. Reson. Chem.*, 1988, **26**, 28–32.
[63] M.J. Quast, A.S. Zektzer, G.E. Martin and R.N. Castle, *J. Magn. Reson.*, 1987, **71**, 554–560.
[64] C. Bauer, R. Freeman and S. Wimperis, *J. Magn. Reson.*, 1984, **58**, 526–532.
[65] V.V. Krishnamurthy and J.E. Casida, *Magn. Reson. Chem.*, 1987, **25**, 837–842.
[66] M. Perpick-Dumont, R.G. Enriquez, S. McLean, F.V. Puzzuoli and W.F. Reynolds, *J. Magn. Reson.*, 1987, **75**, 414–426.
[67] W.F. Reynolds, D.W. Hughes, M. Perpick-Dumont and R.G. Enriquez, *J. Magn. Reson.*, 1985, **63**, 413–417.
[68] W.F. Reynolds, S. McLean, M. Perpick-Dumont and R.G. Enriquez, *Magn. Reson. Chem.*, 1989, **27**, 162–169.

Chapter 7

Separating shifts and couplings: J-resolved spectroscopy

7.1. INTRODUCTION

Unlike the techniques encountered in the previous two chapters that exploit scalar couplings to *correlate* the chemical shifts of interacting spins, 'J-resolved' experiments aim to separate (or resolve) chemical shifts from scalar couplings, so allowing the chemist to examine one parameter without complications arising from the other. For example, the analysis of crowded proton spectra is often complicated by the overlap of neighbouring multiplets, making the extraction of coupling constants or the accurate measurement of chemical shifts difficult or even impossible. This overlap clearly results from the similarities in chemical shifts of the corresponding protons, so if the individual multiplets could in some way be displayed *independently* of their shifts the overlap would be removed and, in principle, the couplings patterns analysed. In practice J-resolved methods are subject to a number of technical difficulties that may limit their effectiveness in this respect, and as such the methods have found far less use in routine structural work than the shift correlation experiments. That said, one of the major sources of complication when using J-resolved methods (strong-coupling between spins) can be eliminated in many instances, or perhaps reduced to an acceptable level, by the use of higher magnetic field strengths. As these become more routinely available to the research chemist the J-resolved methods are perhaps more likely to re-establish themselves as useful tools in the chemists armoury, rather than fall further into relative obscurity. The methods are all based on two-dimensional spectroscopy in which the shift and coupling parameters are resolved by presenting chemical shifts in f_2 and *only spin couplings* in f_1, which may be heteronuclear or homonuclear couplings depending on the details of the experiment. The principal techniques of this chapter are summarised in Table 7.1. These techniques were the most

Table 7.1. The principal applications of the main techniques described in this chapter

Technique	Principal applications
Heteronuclear J-resolved	Separation of heteronuclear couplings (usually ^{1}H–X) from chemical shifts. Used to determine the multiplicity of the heteroatom or to provide direct measurement of heteronuclear coupling constants.
Homonuclear J-resolved	Separation of homonuclear couplings (usually ^{1}H–^{1}H) from chemical shifts. Used to provide direct measurement of homonuclear coupling constants or to display resonance chemical shifts without homonuclear coupling fine-structure (e.g. 'proton–decoupled' proton spectra).
'Indirect' homonuclear J-resolved	Separation of proton homonuclear couplings according to chemical shift of attached carbon centre. Used to provide direct measurement of homonuclear coupling constants.

widely studied methods in the early development of two-dimensional NMR spectroscopy and may be understood with reference to the vector model, being based on simple spin-echoes. As such, I would recommend familiarity with spin-echoes before proceeding; see Section 2.2.

7.2. HETERONUCLEAR J-RESOLVED SPECTROSCOPY

In the heteronuclear version of the experiment, the chemical shift of the X-spin is presented in f_2 whilst couplings to a heteronucleus, typically protons, are presented in f_1. The f_1 dimension therefore enables an analysis of resonance multiplicity as well as measurement of the heteronuclear coupling constants (J_{XH}) themselves, as described further below.

To reduce the f_1 information content of the 2D spectrum to only couplings, it is necessary to make the detected FIDs insensitive to chemical shift evolution during the t_1 period, which is readily achieved by the use of a spin-echo during t_1 (Fig. 7.1a) [1,2]. Thus, following initial X-spin excitation, simultaneous 180° proton and carbon pulses are applied at the midpoint of t_1, such that X-spin chemical shifts will refocus but the heteronuclear coupling will continue to evolve. There is no 'mixing' step in J-resolved experiments because magnetisation or coherence transfer to other spins is not employed. The sequence is the exact 2D analogue of the J-modulated 1D editing sequence presented in Section 4.2, with the fixed coupling evolution period Δ here being replaced with variable period t_1. Since proton decoupling will invariably be applied during the detection period, heteronuclear X–H couplings do not appear in f_2. The detected signals therefore experience the desired pure amplitude modulation according to the evolution of the heteronuclear couplings in t_1 (Fig. 7.2). In addition, the spin-echo refocuses field inhomogeneity effects since these are merely another source of chemical shift differences within the sample, meaning

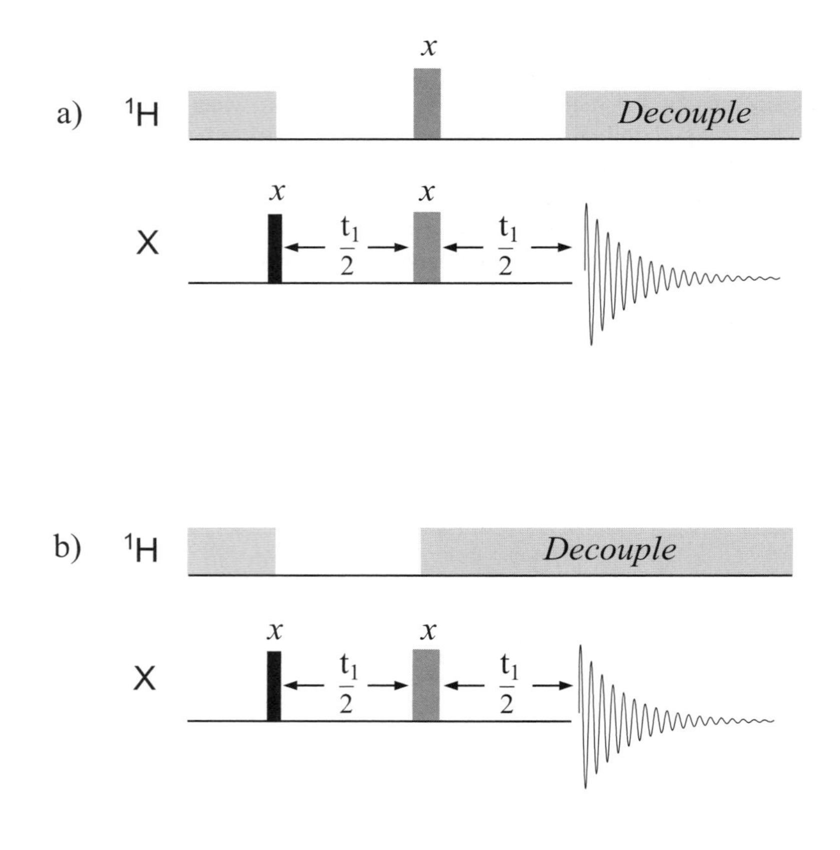

Figure 7.1. The (a) spin-flip and (b) gated decoupling techniques for recording heteronuclear J-resolved spectra. In (b) coupling evolution occurs for half of the t_1 period only, so splittings observed in f_1 appear with half their true $^1J_{XH}$ values.

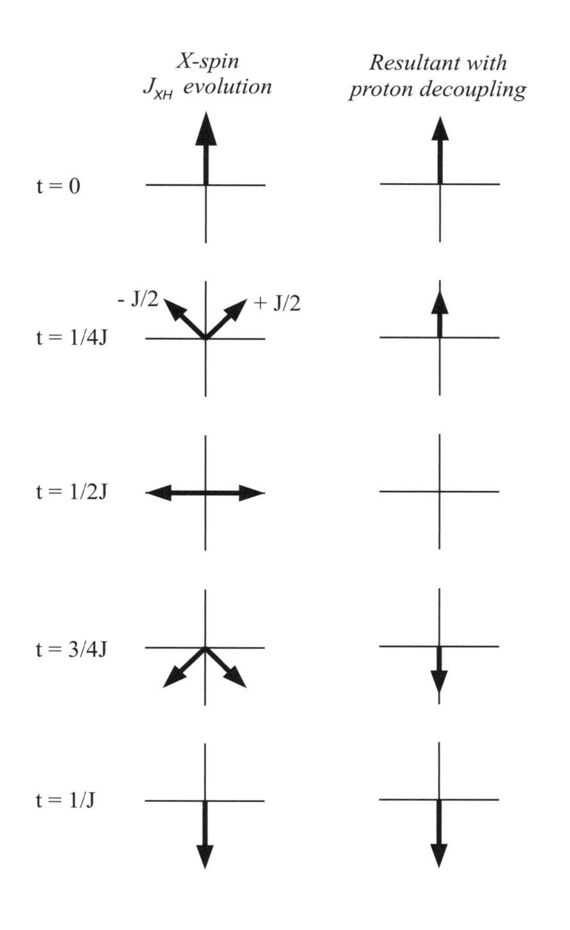

Figure 7.2. Amplitude modulation of the X-spin signal by heteronuclear coupling evolution during t_1, illustrated for a coupled X–H pair. The doublet vectors evolve throughout t_1 but are frozen during t_2 by the application of proton decoupling so the resultant X-spin signal displays pure amplitude modulation according to the heteronuclear coupling constant.

the resonances in f_1 possess *natural* line widths, that is, they are governed by the T_2 rather than T_2^* of the heterospin [3].

The scheme of Fig. 7.1a is referred to as the 'spin-flip' or 'proton-flip' method because of the use of the 180° proton pulse. A widely used alternative is the 'gated decoupler' method [4] (Fig. 7.1b) in which this pulse is omitted and instead proton decoupling is applied for one-half of the evolution period (again analogous to the 1D spectrum editing method). Chemical shifts are again refocused by the X-pulse, but the heteronuclear couplings now evolve for only half the total evolution time, resulting in the f_1 splittings being reduced by a factor of two (be sure which version you are using if you intend to measure J_{XH}!). Note that if neither the proton decoupling or proton 180° pulse were applied in t_1, the effect of the lone X(180°) would be to refocus the heteronuclear coupling in addition to the chemical shifts, so preventing the operation of the experiment altogether. The attraction of the gated decoupler method lies in the simplicity of implementation since no ^{1}H(180°) calibrations are required (a point of greater significance when these methods were initially developed), because the results are better behaved in the presence of strong coupling (see below), and because no artefacts are introduced by inaccuracies in the proton pulse.

Fig. 7.3 shows the ^{1}H–^{13}C J-resolved spectrum of menthol **7.1**. The projection onto the f_2 axis would produce the usual proton decoupled 1D carbon spectrum in which resonances appear as singlets, whilst in f_1 the carbon multiplicities from $^1J_{CH}$ are clearly delineated. Such a spectrum represents a straightforward way of determining multiplicities, but this approach is rarely used nowadays since the 1D spectrum editing methods of Chapter 4 offer a more rapid alternative and the edited heteronuclear shift correlations of Chapter 6 are also more informative. The J-resolved spectrum may prove useful when

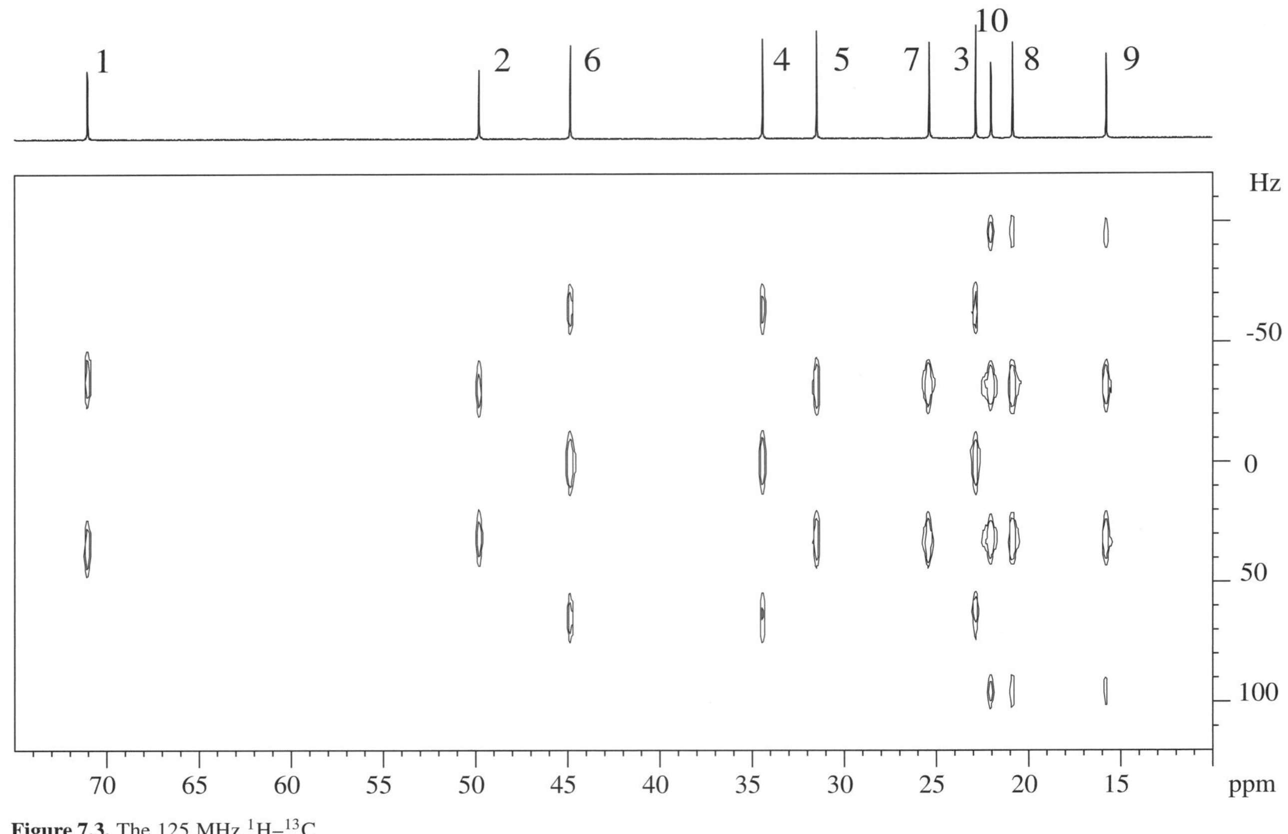

Figure 7.3. The 125 MHz ^{1}H–^{13}C heteronuclear J-resolved spectrum of menthol **7.1**, above which is the conventional 1D carbon spectrum. The gated decoupler method of Fig. 7.1b was used and the splittings in f_1 are therefore half their true $^{1}J_{CH}$ values.

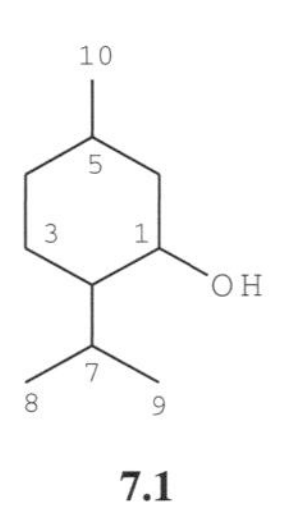

7.1

unusually large $^{1}J_{CH}$ values exist and produce ambiguous results from the spectrum editing methods. Recall that these rely on delays being set according to the estimated J values, and may fail when the timings used are far from optimum, whereas J-resolved methods have no such requirement. A more attractive use for the J-resolved experiment nowadays lies in the measurement of heteronuclear coupling constants themselves, and in particular the measurement of long-range proton–carbon couplings, since these can be of considerable use in conformational or configurational studies. Whilst in principle it is possible to record the f_1 dimension with sufficient digital resolution to resolve the small long-range couplings in the presence of the far greater one-bond couplings (which are typically at least an order of magnitude larger), a more efficient approach is to eliminate $^{1}J_{CH}$ from f_1 and retain only $^{2/3}J_{CH}$, so reducing the f_1 spectral width and allowing finer digitisation for more accurate measurements. A number of approaches toward this are introduced in the following section.

Refinements to the basic sequences of Fig. 7.1 include the use of polarisation transfer sequences prior to the spin-echo to prepare transverse X-spin magnetisation. INEPT, DEPT [5], as well as the more recent PENDANT [6] have been proposed for this purpose. In addition to sensitivity gains, the polarisation transfer sequences allow repetition rates to be dictated by the T_1s of the faster relaxing protons. They may also be favoured when studying nuclei that experience a loss of intensity from the ^{1}H–X nuclear Overhauser effect generated by proton decoupling, that is, those X-spins that have negative magnetogyric ratios, for example ^{15}N or ^{29}Si (see Chapter 8). Avoiding the use of proton decoupling during the relaxation delay minimises NOE build-up and hence signal losses, whilst the addition of the polarisation transfer step enhances sensitivity.

7.2.1. Measuring long-range proton–carbon coupling constants

The use of long-range ^1H–^{13}C coupling constants in the definition of molecular configuration or conformation [7–9] is an increasingly active area, with numerous methods for measuring these developed in recent years. In principle, J-resolved methods are well suited to such measurements, but tend to suffer from the presence of the far greater $^1J_{CH}$ couplings, as mentioned above. The common aim of the sequences described below is to eliminate these less informative one-bond ^1H–^{13}C couplings from f_1 and so reduce the corresponding spectral width. Since $^1J_{CH}$ values are greater than ca. 125 Hz, whilst long-range couplings are typically less than 10 Hz, substantial reductions in the f_1 spectral width can be made which allows higher digital resolution and provides accurate characterisation of the remaining $^nJ_{CH}$ values without the need to collect prohibitively large or time-demanding data sets. The descriptions are deliberately brief but serve to illustrate the most useful approaches.

Semi-selective

By placing a BIRD cluster (Section 6.3) with proton pulse phases x, x, $-x$ at the midpoint of t_1 in place of the simultaneous 180° pulses of the spin-flip method (Fig. 7.4a), it is possible to invert only those protons that share long-range couplings, while leaving those directly bound to a ^{13}C spin unaffected. This results in the selective refocusing of the one-bond couplings by the $^{13}C(180°)$ pulse, and thus their removal from f_1, whereas the long-range couplings remain [10,11]. The f_1 traces display *all* $^nJ_{CH}$ values associated with each carbon, and may thus posses quite complex fine structure since

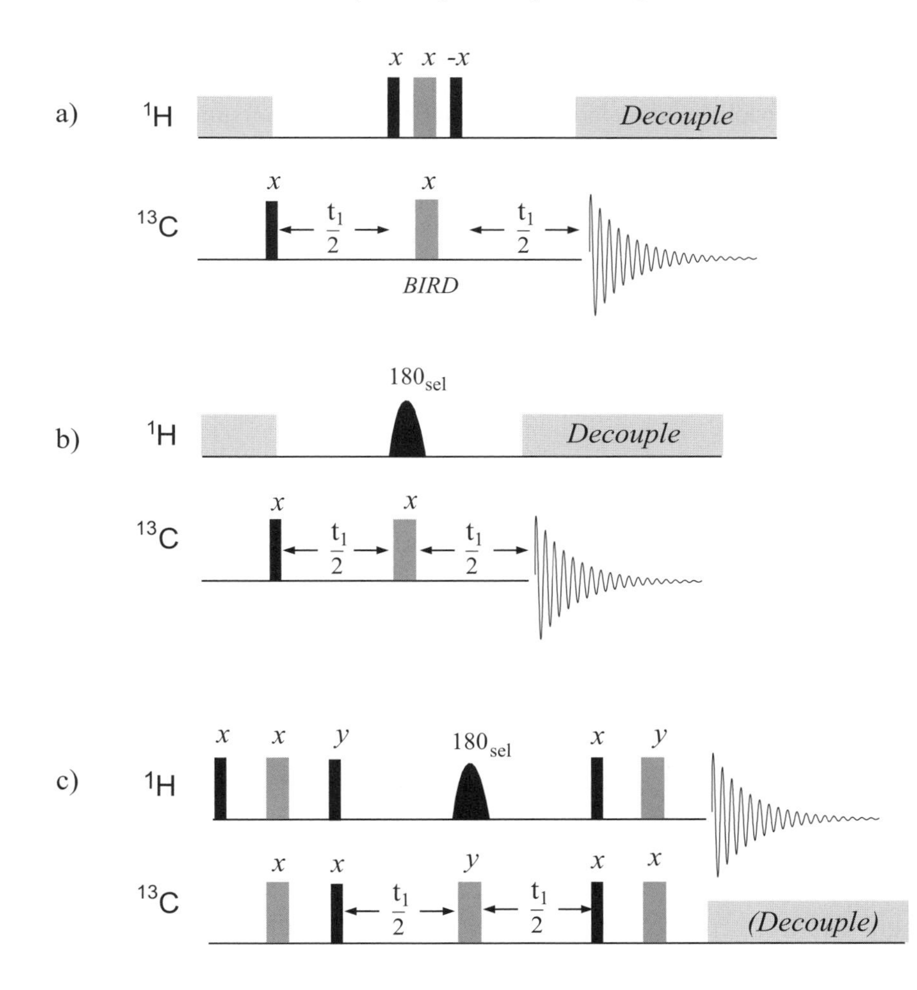

Figure 7.4. Heteronuclear J-resolved sequences for the measurement of long-range proton–carbon coupling constants; (a) semi-selective, (b) selective and (c) selective with proton observation (see text).

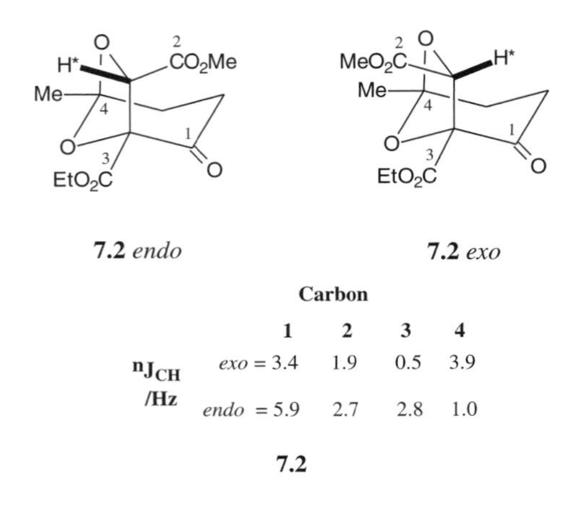

		Carbon			
		1	2	3	4
$^nJ_{CH}$ /Hz	*exo* =	3.4	1.9	0.5	3.9
	endo =	5.9	2.7	2.8	1.0

7.2

each carbon is likely be coupled to many protons. Such complexity may itself preclude the measurement of the coupling constants, and no information on which spin-pair gives rise to a specific coupling is available.

Selective

To simplify the fine-structure appearing in f_1 it is possible to selectively invert only one proton resonance during t_1, taking care to avoid inverting its one-bond ^{13}C satellites. This requires the application of a selective $^1H(180°)$ pulse in place of the usual 'hard' pulse in the spin-flip method (Fig. 7.4b). In this case, all couplings will be refocused and therefore removed from f_1 with the exception of those to the selectively inverted proton(s) [12]. The f_1 dimension therefore displays only simple multiplets (doublets, triplets or quartets depending on the selected proton group) and the coupling pairs giving rise to each are readily identified. This was one approach used for measuring long-range couplings to the bridgehead proton in a series of structures related to **7.2** in an attempt to define the unknown bridgehead stereochemistry and thus differentiate *endo* and *exo* products. A lack of NOEs to the bridgehead protons did not provide an unambiguous definition, so long-range couplings in the unknown products were compared with those in structures of known configuration. Fig. 7.5 shows a section of one such spectrum in which these couplings were measured from the f_1 doublet splittings. This requires the proton(s) of interest to be sufficiently resolved, and numerous experiments will be

Figure 7.5. A region of the selective heteronuclear J-resolved spectrum of **7.2**. Long-range heteronuclear couplings to the selected proton appear as splittings in f_1, the values of which are shown. The vertical trace is taken through the resonance at 199 ppm. 4K t_2 data points were collected for 32 t_1-increments over spectral widths of 100 ppm and 14 Hz respectively. A 100 ms Gaussian 180° pulse was used to select the proton resonance. The final f_1 resolution after zero-filling was 0.1 Hz/pt.

1 2 3 4

J= 5.9 2.7 2.8 1.0

Hz: -2, 0, 2, 4

200 190 180 170 160 150 140 130 120 110 ppm

required if many ${}^{n}J_{CH}$ values are to be determined. This has, nevertheless, been employed extensively in the measurement of long-range coupling constants in monosaccharide units [13], and other variants have incorporated the INEPT [14] and DEPT [15] sequences for sensitivity enhancement.

Selective with proton detection

Long-range couplings can also be observed through the observation of the participating proton, leading to significant sensitivity gains. An elegant way of performing the proton-detected experiment is based on a simple modification of the HSQC sequence presented in Section 6.3, and may be performed as the phase-sensitive experiment with or without pulsed field gradients [6] (Fig. 7.4c). The non-selective ${}^{1}H(180°)$ inversion pulse within t_1 (which serves to refocus heteronuclear couplings in HSQC) is here made selective, and a simultaneous non-selective ${}^{13}C(180°)$ pulse added, producing a spin-echo during t_1 analogous in design and operation to the selective experiment above. The polarisation transfer step following t_1 transfers magnetisation back onto protons for detection through their one-bond proton–carbon coupling. The resulting spectrum thus displays the normal ${}^{1}H$ spectrum in f_2 and ${}^{n}J_{CH}$ doublets along f_1. These couplings are from the selectively inverted proton to the carbon attached to the proton at the f_2 chemical shift (Fig. 7.6), so measurements are restricted to long-range couplings to protonated carbons only. Once again, the spin pair giving rise to the coupling can be identified, assuming the target proton is sufficiently well resolved to be inverted selectively.

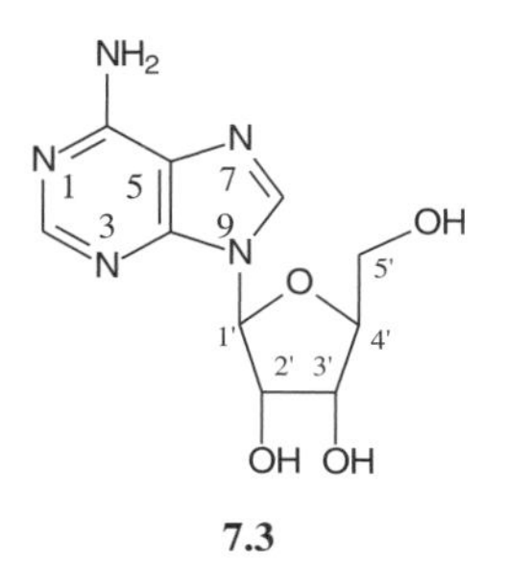

7.3

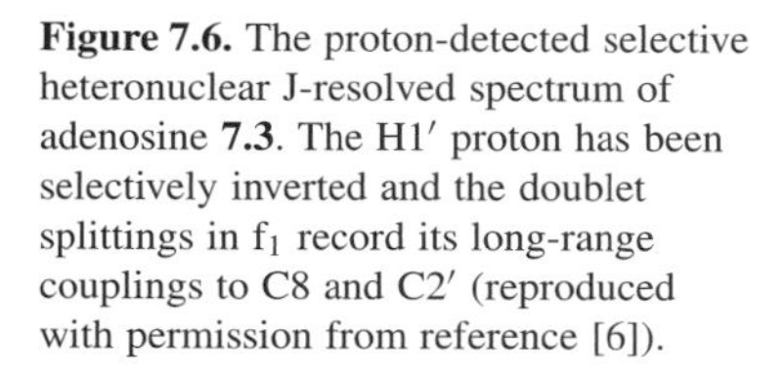

Figure 7.6. The proton-detected selective heteronuclear J-resolved spectrum of adenosine **7.3**. The H1′ proton has been selectively inverted and the doublet splittings in f_1 record its long-range couplings to C8 and C2′ (reproduced with permission from reference [6]).

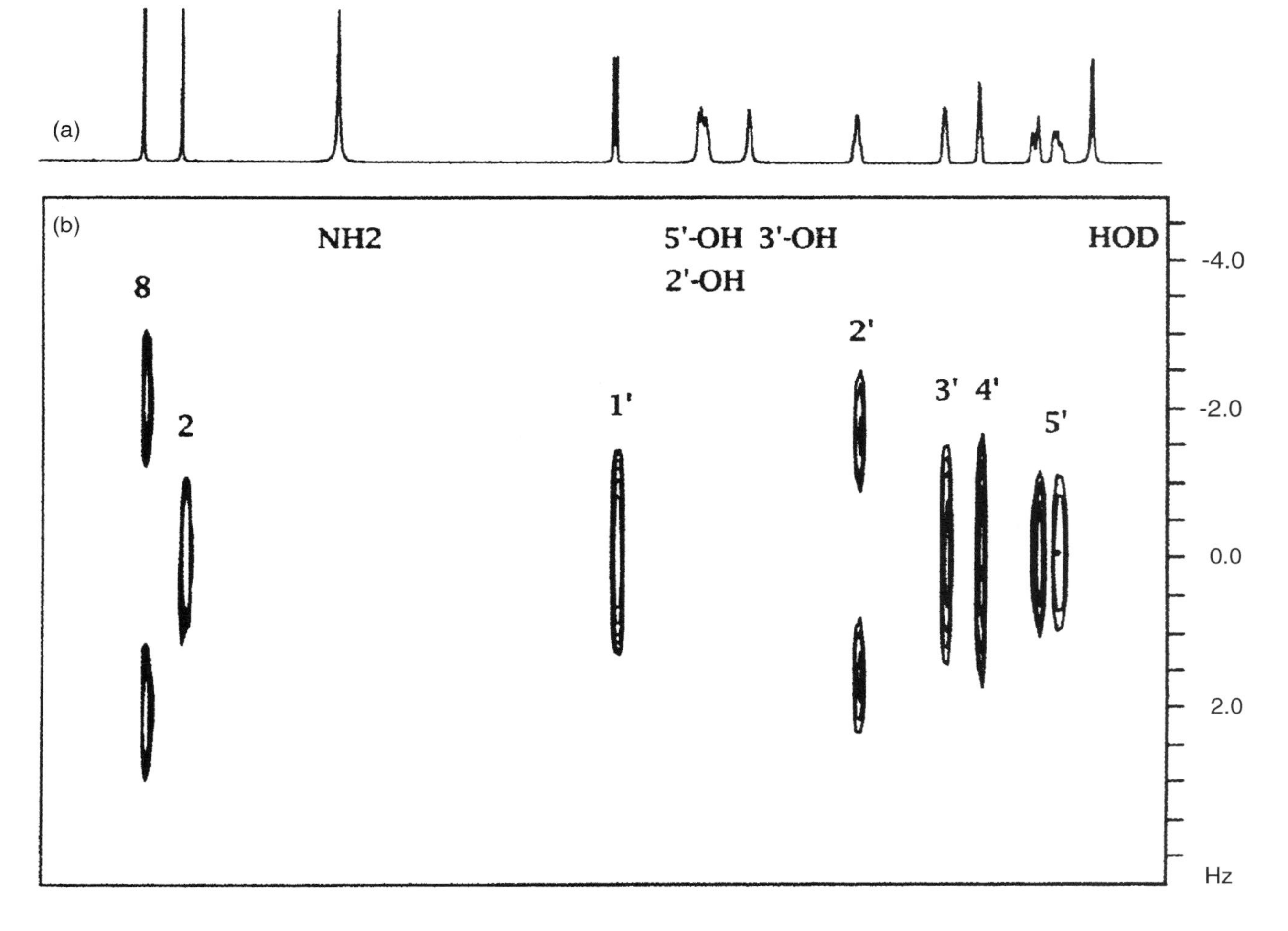

7.2.2. Practical considerations

The requirements for sign discrimination in f_1 for the J-resolved experiments of Fig. 7.1 are readily met in a *first-order system* because the multiplets are symmetrical about their midpoints and thus folding about $f_1 = 0$ causes no confusion. Phase cycling for f_1 quadrature detection is, therefore, unnecessary. Since, in the presence of broadband decoupling, the detected signals experience pure amplitude modulation, pure phase spectra may also be obtained. However, when the system is no longer first order, that is, in the presence of strong coupling *between protons* (more precisely, between those protons giving rise to ^{13}C satellites and others), the H–X multiplets lose their symmetry and the detected signals experience some phase-modulation which in turn introduces dispersion-mode contributions to the spectrum. In this case, the 2D spectrum may be presented in the absolute-value mode to mask the phase distortions or alternatively, following the f_2 transform, columns may be extracted from the 2D data set at the chemical shift of a resonance of interest and processed separately as 1D traces. Despite the overall dispersive contribution to the 2D lineshape, a single column displays absorption-mode characteristics. Further complications arise in the presence of strong-coupling for the spin-flip method [1].

The quality of J-resolved spectra can be seriously compromised by deficiencies in the accuracy of the pulses, (especially the 180° pulses), arising from pulse miscalibration, rf inhomogeneity, and off-resonance effects. The result is the appearance of a variety of additional weak resonances at esoteric positions within the spectrum, referred to as 'ghosts' and 'phantoms'. The ghosts arise from imperfections in the 180° refocusing pulse, such that some transverse magnetisation fails to experience the effect of the pulse and thus fails to be refocused. Phantoms arise from a combined deficiencies in the 90° and 180° pulses causing residual longitudinal magnetisation existing after the 90° pulse to become transverse following the imperfect 180° pulse. These spurious responses may be eliminated with the EXORCYCLE phase cycle [16], which is widely employed in sequences that utilise spin-echoes. This involves stepping the phase of the X-spin refocusing pulse through x, y, $-x$, $-y$ whilst the receiver inverts, that is, steps x, $-x$, x, $-x$. Stepping the refocusing pulse by 90° causes the echo to shift in phase by 180°, as explained in Section 2.2, hence receiver inversion follows the echo, whereas the unwanted responses ultimately cancel. Deficiencies in the proton 180° inversion pulse [17] can be avoided by the use of the gated decoupler method, but where this is not possible, as in the case of the homonuclear J-resolved experiments that follow, use of a composite 180° pulse such as $90_x 240_y 90_x$ is recommended [18].

Practical set-up

Digitisation of the data, in particular the number of t_1 increments that are needed, will be dictated by the information required of the spectrum. If one simply wishes to determine multiplicities, low digital resolution will suffice, allowing few increments and rapid data collection. For carbon-13 the widest multiplets arise from the quartets of methyl groups, so the f_1 spectral width need be about 3.2 times $^1J_{XH}$ (although it would be possible to reduce the spectral width further and deliberately fold in the outer lines of the quartets). This requirement can beneficially be reduced by a factor of two for the gated decoupler method to around 200 Hz, assuming a coupling constant of 125 Hz.

To merely characterise the multiplet structure a digital resolution in f_1 of around 20 Hz/pt should suffice, requiring as little as 20 increments. If one wished to *measure* the value of $^1J_{CH}$ from the multiplet structure, a digital resolution of somewhat less than 5 Hz/pt would be more appropriate. Some 200 increments would be required for 2 Hz/pt leading to a significantly longer experiment, particularly if many scans are required per increment for reasons of sensitivity. If even finer digitisation were required for the measurement of *long-range* couplings, this approach becomes impossible. In contrast, when using one of the selective methods for $^{n}J_{CH}$, one may only need digitise a 10 Hz window, requiring only 40 increments for 0.5 Hz/pt. Following the f_2 transform, columns may be extracted from the 2D data set and treated as 1D FIDs, including the use of zero filling, to further enhance measurements.

7.3. HOMONUCLEAR J-RESOLVED SPECTROSCOPY

The homonuclear version of the J-resolved experiment [19] is most frequently applied in proton spectroscopy, although again is suitable for any abundant nuclide. In principle, the separation of δ and J should reveal proton multiplets in f_1 free from overlap and thus available for analysis, and singlets in f_2 at the corresponding chemical shifts, such that the f_2 projection represents the 'broadband proton–decoupled proton spectrum'. The possibility of generating such a spectrum has obvious appeal, allowing the accurate measurement of chemical shifts in even the most heavily crowded proton spectra. However, a number of technical difficulties must be overcome if one is to achieve this goal which, alas, is not readily accomplished. A second possible application is in the measurement of homonuclear coupling constants themselves, which is possible within the limitations of certain caveats detailed shortly.

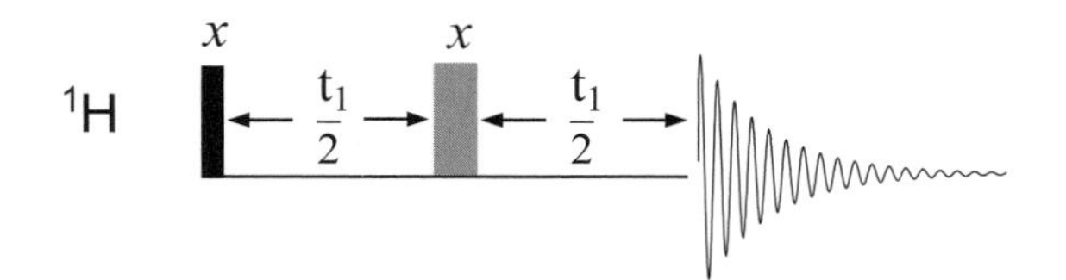

Figure 7.7. The homonuclear J-resolved experiment. The 180° pulse at the midpoint of t_1 refocuses proton shifts but not homonuclear couplings, so only these appear in f_1.

The homonuclear sequence (Fig. 7.7) closely resembles the heteronuclear methods (although restricted to the spin-flip version only), and utilises the EXORCYCLE scheme. The appearance of the homonuclear spectrum is funda-

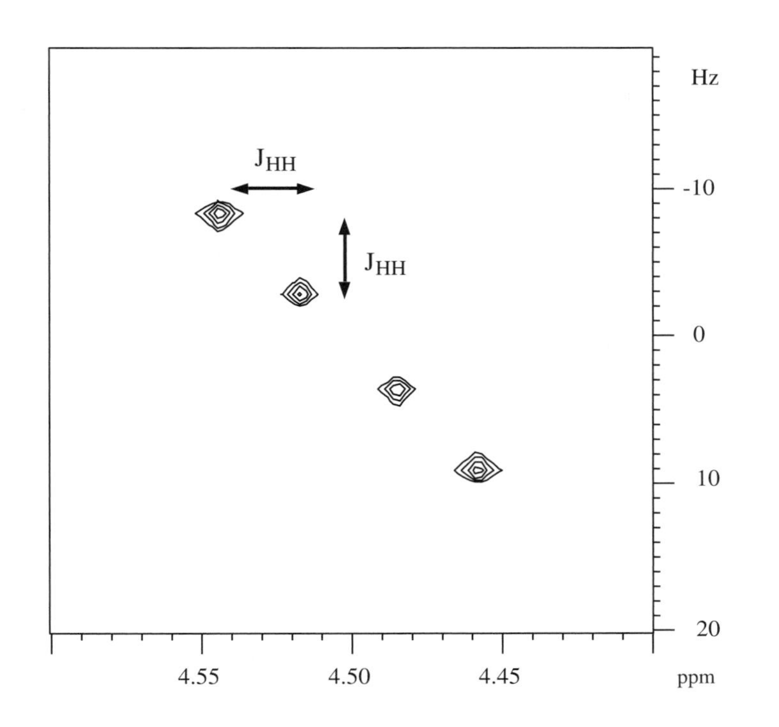

Figure 7.8. A multiplet from the homonuclear J-resolved experiment showing the characteristic tilting brought about by the presence of proton couplings in both f_1 and f_2.

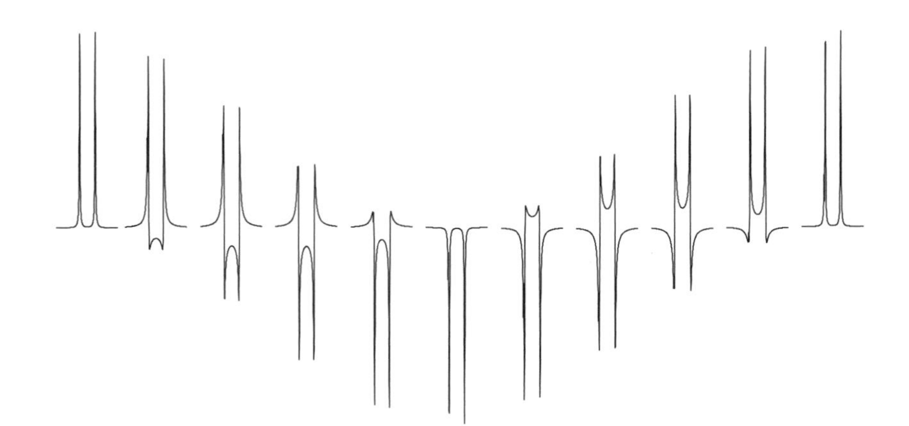

Figure 7.9. Phase modulation of a proton doublet as a function of time. The signals retain full intensity at all times (i.e. they do not experience amplitude modulation) but each line experiences a net 360° phase change over the sequence. The results are shown for a 10 Hz doublet with evolution times from 0 to 200 ms in 20 ms increments.

mentally different from its heteronuclear equivalent in that both chemical shifts *and couplings* appear in f_2 (because one cannot simultaneously *broadband* decouple and observe the proton spectrum). Thus, rather than lying parallel to the f_1 axis, the proton multiplets appear along a slope of -1 (in units of Hz), or, in other words, sit at 45° to either axis (assuming identical plot scaling for both dimensions, Fig. 7.8). Columns parallel to f_1 do not, therefore, display the expected proton multiplets, and the f_2 projection displays both chemical shifts and scalar couplings. To overcome these shortcomings, post-processing techniques are routinely applied, as described in the following section. Furthermore, unlike the heteronuclear case, the detected signals unavoidably experience *phase modulation* as a function of t_1 (Fig. 7.9) resulting in phase-twist lineshapes. The spectra are therefore usually presented as absolute-value data following strong resolution enhancement and a magnitude calculation. Further complications arise in the presence of strong-coupling between protons in the form of additional responses (see below), so the J-resolved experiment is most suitable for overlapped spectra that are still first order.

7.3.1. Tilting, projections and symmetrisation

To reach the ultimate goal of retaining only chemical shifts in f_2, it is possible to eliminate the couplings from this dimension by 'tilting' (or 'shearing') the multiplets through an angle of 45° about their midpoints [20], as illustrated schematically in Fig. 7.10. Software routines for this process are common to NMR processing packages nowadays. The resulting spectrum then has an appearance similar to the heteronuclear analogue, with columns parallel to f_1 reproducing the multiplet structures (providing the magnitude

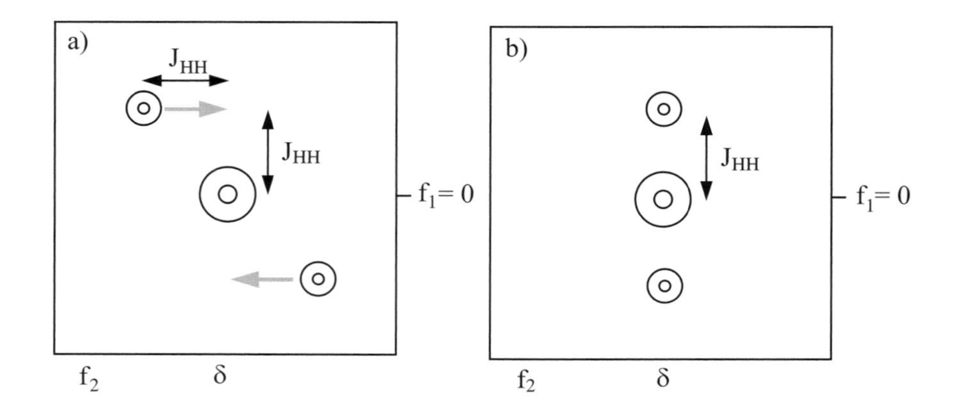

Figure 7.10. A schematic illustration of the tilting procedure for eliminating homonuclear couplings from the f_2 dimension. (a) The original multiplet structure and (b) that following the tilt procedure.

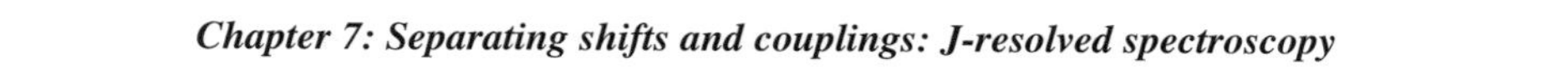

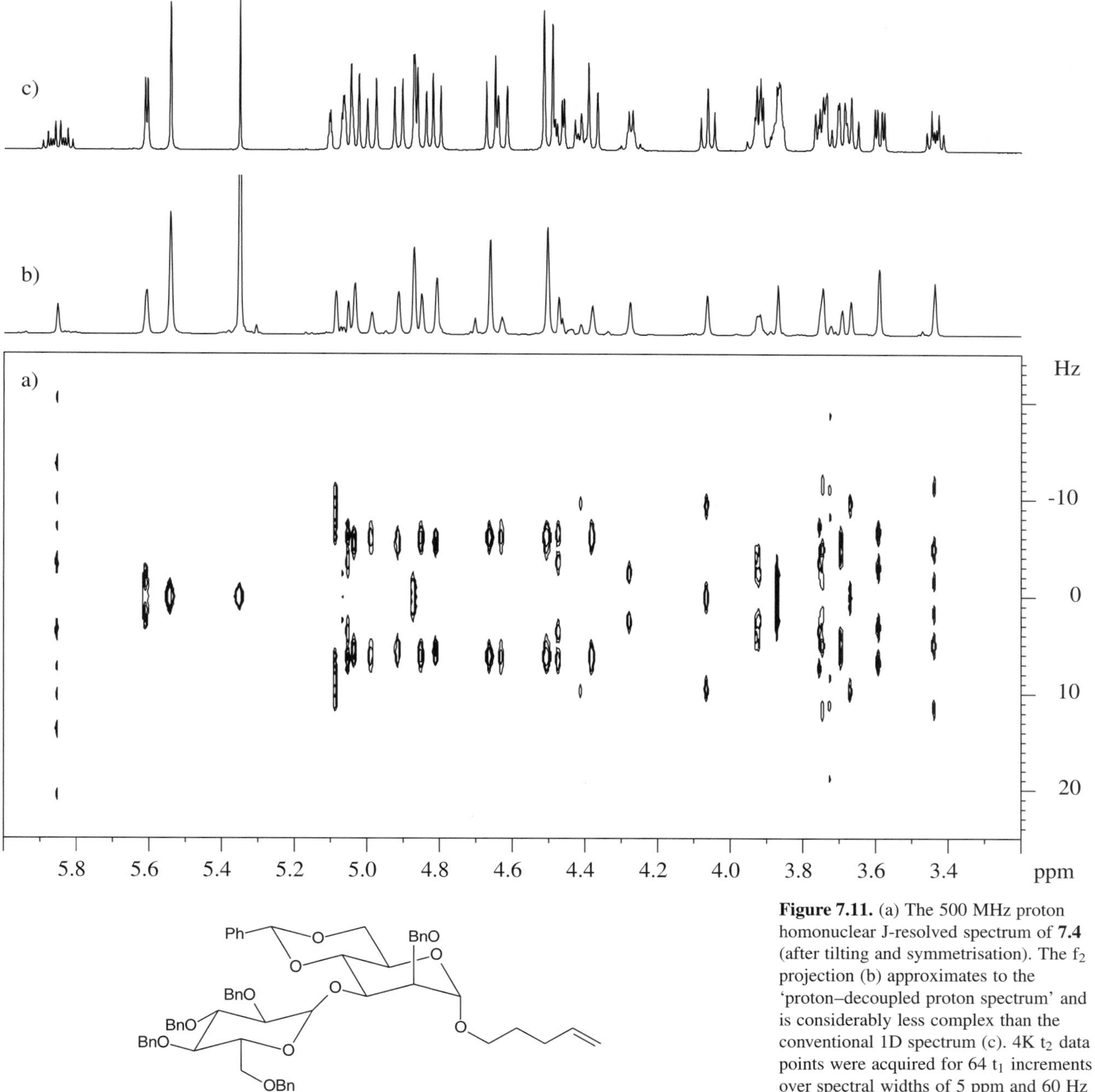

Figure 7.11. (a) The 500 MHz proton homonuclear J-resolved spectrum of **7.4** (after tilting and symmetrisation). The f_2 projection (b) approximates to the 'proton–decoupled proton spectrum' and is considerably less complex than the conventional 1D spectrum (c). 4K t_2 data points were acquired for 64 t_1 increments over spectral widths of 5 ppm and 60 Hz respectively. The final f_1 resolution after zero-filling was 0.5 Hz/pt. Data were processed with unshifted sine-bell windows in both dimensions and are presented in magnitude mode.

calculation has been performed), and the projection onto the f_2 axis producing the broadband decoupled proton spectrum (Fig. 7.11). Fig. 7.12 compares traces taken from this J-resolved spectrum with the equivalent multiplets from the 1D proton spectrum and illustrates the fine-resolution of multiplet structure that can be obtained in the f_1 dimension.

Further improvements in the form of t_1-noise reduction may also be achieved with additional post-processing. Prior to the tilt procedure, bands of t_1-noise will lie parallel to the f_1 axis, as for all 2D experiments, whilst following the tilt they will sit at 45° to it. In first- and higher-order systems, the multiplets will themselves be symmetrical about the line $f_1 = 0$ Hz after tilting [21]. If

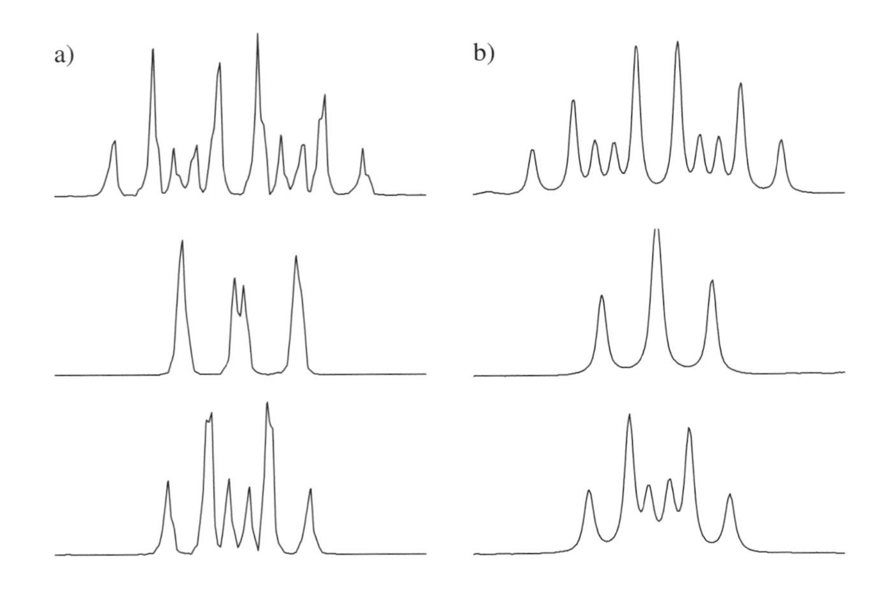

Figure 7.12. (a) Selected f_1 traces taken from the J-resolved spectrum of Fig. 7.11a and (b) the equivalent multiplets from the 1D proton spectrum.

the whole data set were symmetrised about this line, that is, the lower intensity point for symmetrically related data points replacing the higher, the multiplets would be retained whereas the contributions from the sloping t_1-noise would be diminished [22]. The procedure is similar to that introduced in Section 5.3.3 for t_1-noise reduction in absolute-value COSY spectra, although a different symmetrisation procedure is of course required, and again one should be aware of the possibility of introducing artefacts into the spectrum as a result of the symmetrisation routine itself. Again, these routines are common to modern processing packages.

7.3.2. Applications

Perhaps the most obvious application of the homonuclear J-resolved experiment is the separation of overlapping multiplets so that the fine structure within each may be analysed. There is however a strict requirement for the spin-systems to be first-order for the separation of shifts and couplings to be successful [21] because strong-coupling causes unwanted additional responses to appear midway between the shifts of the strongly coupled protons. To reduce the degree of strong-coupling within spectra, the use of the highest available field strength is recommended for J-resolved spectroscopy wherever possible. Fig. 7.13 shows simulated spectra for a three-spin system at field strengths of 200 and 600 MHz. At the lower field (Fig. 7.13a), the coupled protons at 3.7 and 3.8 ppm (J = 12 Hz) experience strong-coupling, as evidenced in the 'roofing' of their 1D resonances, which gives rise to the additional responses between them in the J-resolved spectrum. No such artefacts are associated with the resonance at 4.5 ppm, which experiences only first-order coupling to its partners. In contrast, the higher-field spectrum (Fig. 7.13b) shows no extra responses since all couplings are now (approximately) first-order.

The homonuclear J-resolved spectrum can also assist in the measurement of proton *heteronuclear* couplings, notably when the coupling exists to a high abundance heteronuclide such as ^{31}P or ^{19}F. Since only the proton of a coupled heteronuclear pair will experience the 180° pulse, the heteronuclear coupling is refocused in t_1 and therefore absent in f_1. This coupling will nonetheless be operative during detection of the proton FID and will thus appear in f_2. Since this heteronuclear splitting sits *parallel* to the f_2 axis rather than at 45° to it, it is *not* removed from this dimension by the tilting process and may thus be examined without interference from homonuclear proton couplings along f_2.

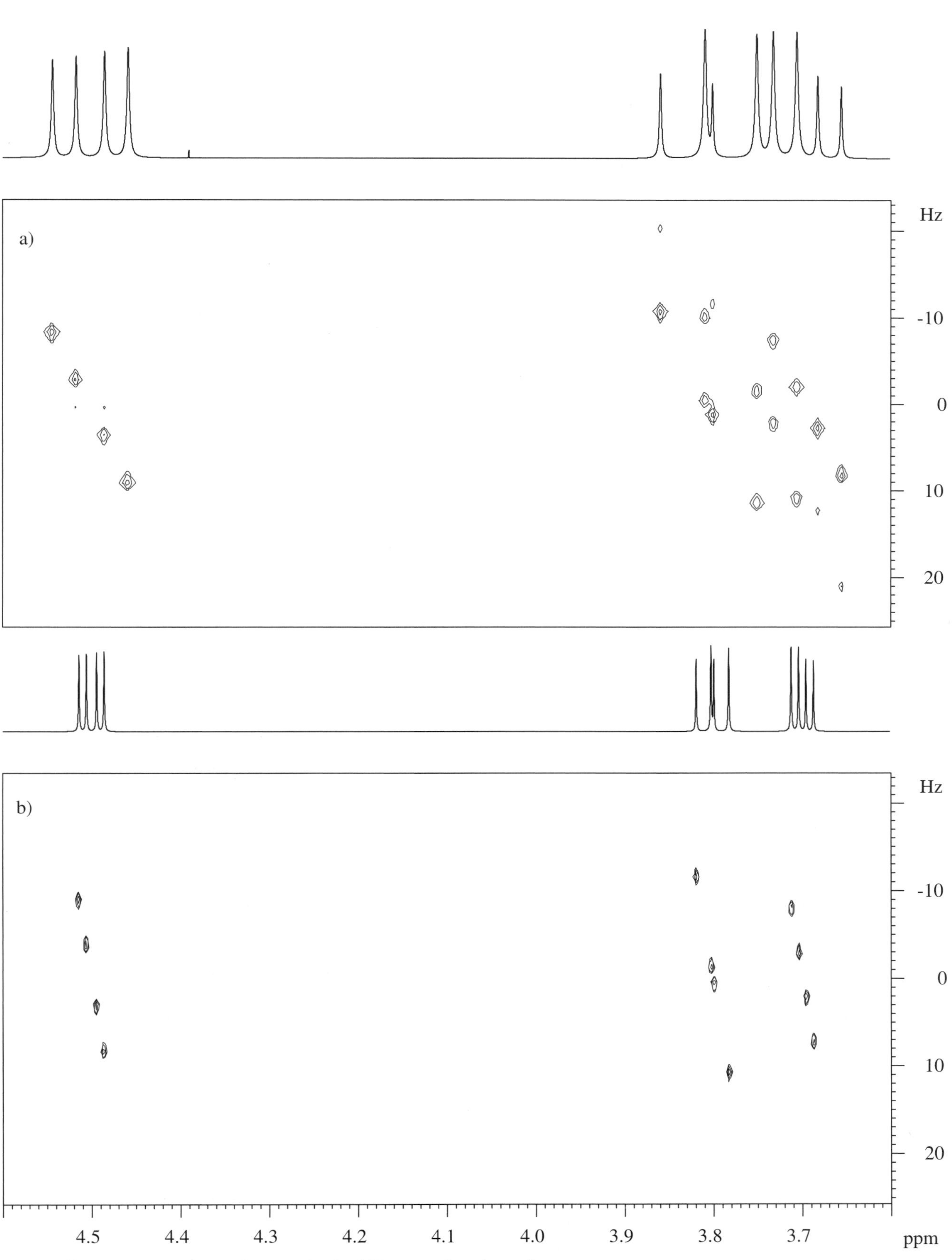

Figure 7.13. Simulated 2D homonuclear J-resolved and 1D spectra for a three-spin proton system at (a) 200 and (b) 600 MHz. The additional artefacts in (a) arise from strong coupling between the two low-frequency protons (no tilting has been applied).

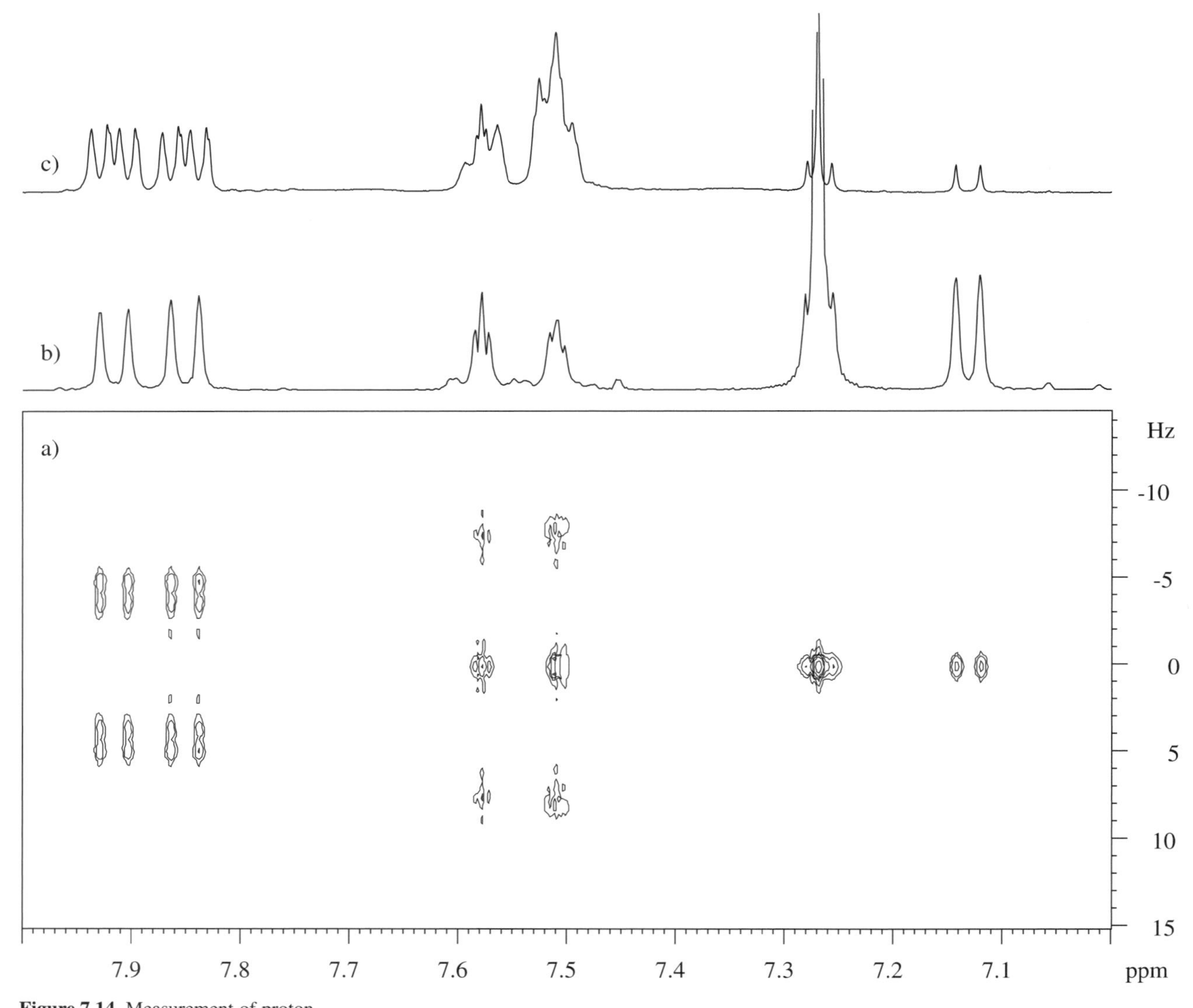

Figure 7.14. Measurement of proton heteronuclear couplings from the tilted homonuclear J-resolved spectrum (a). Spectrum (b) is the f_2 projection of (a) and displays only proton shifts and ^1H–^{31}P coupling. Spectrum (c) is the conventional 1D spectrum displaying shifts and both ^1H–^1H and ^1H–^{31}P couplings.

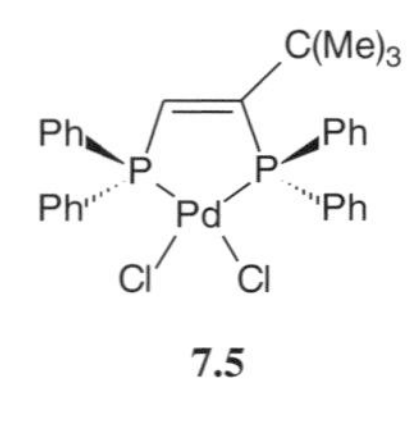

7.5

Fig. 7.14 demonstrates this approach for the palladium phosphine complex **7.5**. The f_2 projection of the tilted spectrum contains splittings for the ^1H–^{31}P couplings only at each proton shift, as is most clearly seen for the phenyl *ortho*-protons at 7.85 and 7.92 ppm. These data can be used to complement the 1D X-spin decoupled proton spectrum in which the proton–proton couplings are observed without interference from H–X spin couplings (see Fig. 4.12 of the same complex for comparison, Section 4.2). If desired, the heteronuclear couplings could also be eliminated from the J-resolved spectrum by the application of broadband X-spin decoupling during the acquisition time. For the measurement of the heteronuclear couplings, rows taken parallel to F_2 through the multiplet components provide better resolution than the projection itself.

The possibility of recording 'broadband homonuclear decoupled spectra' has also be touched on above and illustrated in Fig. 7.11. The simple J-resolved experiment tends to be of limited success in this respect owing to interferences from strong-coupling, the poor lineshapes and limited resolution obtained and because the signal intensities in the projected spectrum bear little relation to signal intensities in the conventional 1D spectrum. Greater success is again expected on higher-field instruments but in any case caution in interpretation of such data is required. More sophisticated data processing [23,24] based on pattern recognition procedures appear to offer some hope of achieving this goal reliably, but the algorithms required have yet to come into widespread availability.

7.3.3. Practical considerations

Many of the practical considerations given above for heteronuclear J-spectroscopy are equally applicable to the homonuclear case, and the selection of digital resolution in f_1 follows similar lines of thinking as before. Most proton multiplets will rarely exceed a width of 50 Hz (although those of other nuclides with many homonuclear couplings may do) so at least for proton spectroscopy, a f_1 spectral width equal to this will suffice. Using 64 or 128 t_1 increments will therefore provide an f_1 resolution of 1.6 and 0.8 Hz/pt respectively. Using many more increments than these causes t_1 to be rather long, meaning relaxation losses become significant. Because of the use of magnitude calculation, strong resolution enhancement functions such as unshifted sine-bells need to be applied in both dimensions.

7.4. 'INDIRECT' HOMONUCLEAR J-RESOLVED SPECTROSCOPY

An alternative approach to resolving proton multiplets is to disperse them according to the chemical shift of the carbon nucleus to which the protons are attached, rather than those of the proton themselves [25]. The advantage of this approach lies in the typically greater dispersion of the carbon chemical shifts, although one must tolerate the reduced sensitivity of carbon observation.

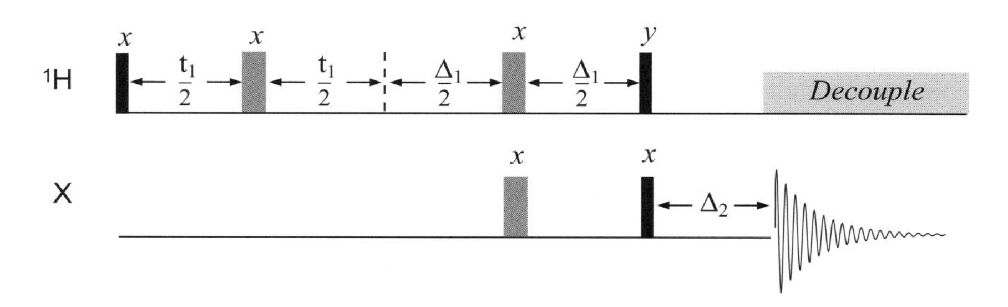

Figure 7.15. The indirect homonuclear J-resolved sequence. Proton couplings are effective during t_1, after which polarisation transfer leads to X-spin observation.

The sequence that achieves this (Fig. 7.15) is a simple variant on the INEPT-based heteronuclear shift correlation sequence of Fig. 6.31 (HETCOR), so the loss in sensitivity is compensated somewhat by the use of a polarisation transfer step. In fact the only difference between the two lies in the net evolution of *only shifts* or *only couplings* for the whole of t_1. The addition of a proton 180° pulse at the midpoint of t_1 here serves to refocus proton chemical shifts and heteronuclear coupling constants (so the X-spin 180° pulse of HETCOR becomes redundant) but leaves the proton homonuclear couplings free to evolve. The resulting spectrum therefore contains only proton multiplets in f_1 dispersed by the corresponding X-spin shifts in f_2 (Fig. 7.16) [1].

---

[1] Notice the similarity here with the approach used in sequence 7.4c in which the basic HSQC shift correlation sequence was adapted for the measurement of long-range heteronuclear couplings. In both cases the chemical shift evolution period of the shift correlation sequence was converted into one in which only couplings were active (through the use of spin-echoes) so producing the J-resolved experiment. Often, subtle changes involving the addition or removal of one or two strategically placed pulses are all that is required to alter the characteristics of the resulting spectrum, producing a 'new' pulse sequence (which may or may not be useful to the chemist). This, in part, contributes to the plethora of NMR pulse sequences found throughout the chemical literature.

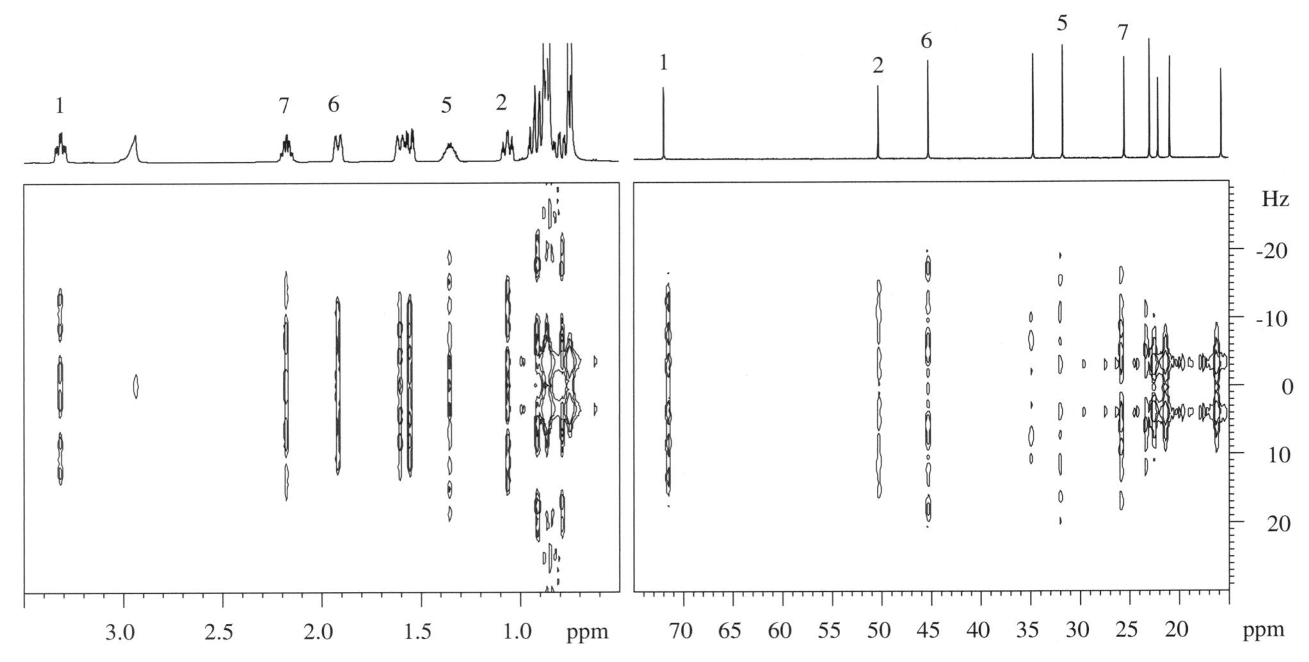

Figure 7.16. Fig. 16.The direct homonuclear J-resolved spectrum and the indirect homonuclear J-resolved spectrum of **7.1**. Both experiments present homonuclear couplings in f_1 dispersed by either the proton or the carbon chemical shift respectively.

Once again, strong coupling, this time between proton satellites, causes problems. Furthermore, because one must consider the *satellites* of each proton rather than the parent resonance itself (since only the satellites can contribute to the polarisation transfer), the multiplet patterns observed for high-order systems may differ from those of the parent resonance in the 1D spectrum. Other complications occur in the case of non-equivalent geminal protons since each may possess different multiplet patterns, yet both will be observed at the same carbon chemical shift, leading to a potentially complex multiplet overlap (this is apparent for the H6 protons in the indirect spectrum of Fig. 7.16) The most likely use of this variant is in the separation of the multiplet patterns of protons having coincidentally degenerate chemical shifts which are not, therefore, separable in the conventional homonuclear version.

REFERENCES

[1] G. Bodenhausen, G.A. Morris, R. Freeman and D.L. Turner, *J. Magn. Reson.*, 1977, **28**, 17–28.
[2] L. Müller, A. Kumar and R.R. Ernst, *J. Magn. Reson.*, 1977, **25**, 383–390.
[3] G. Bodenhausen, R. Freeman, R. Niedermeyer and D.L. Turner, *J. Magn. Reson.*, 1976, **24**, 291–294.
[4] R. Freeman, G.A. Morris and D.L. Turner, *J. Magn. Reson.*, 1977, **26**, 373–378.
[5] V. Rutar and T.C. Wong, *J. Magn. Reson.*, 1983, **53**, 495–499.
[6] M.L. Liu, R.D. Farrant, J.M. Gillam, J.K. Nicholson and J.C. Lindon, *J. Magn. Reson. (B)*, 1995, **109**, 275–283.
[7] V.F. Bystrov, *Prog. Nucl. Magn. Reson. Spectrosc.*, 1976, **10**, 41–81.
[8] P.E. Hansen, *Prog. Nucl. Magn. Reson. Spectrosc.*, 1981, **14**, 179–296.
[9] J.L. Marshall, Carbon–Carbon and Carbon–Proton NMR Couplings: Applications to Organic Stereochemistry and Conformational Analysis, VCH Publishers, Florida, 1983.
[10] A. Bax, *J. Magn. Reson.*, 1983, **52**, 330–334.
[11] V. Rutar, *J. Magn. Reson.*, 1984, **56**, 87–100.
[12] A. Bax and R. Freeman, *J. Am. Chem. Soc.*, 1982, **104**, 1099–1100.
[13] C. Morat, F.R. Taravel and M.R. Vignon, *Magn. Reson. Chem.*, 1988, **26**, 264–270.
[14] T. Jippo, O. Kamo and K. Nagayama, *J. Magn. Reson.*, 1986, **66**, 344–348.
[15] D. Uhrín, T. Liptaj, M. Hricovíni and P. Capek, *J. Magn. Reson.*, 1989, **85**, 137–140.
[16] G. Bodenhausen, R. Freeman and D.L. Turner, *J. Magn. Reson.*, 1977, **27**, 511–514.
[17] G. Bodenhausen and D.L. Turner, *J. Magn. Reson.*, 1980, **41**, 200–206.

[18] R. Freeman and J. Keeler, *J. Magn. Reson.*, 1981, **43**, 484–487.
[19] W.P. Aue, J. Karhan and R.R. Ernst, *J. Chem. Phys.*, 1976, **64**, 4226–4227.
[20] K. Nagayama, P. Bachmann, K. Wüthrich and R.R. Ernst, *J. Magn. Reson.*, 1978, **31**, 133–148.
[21] G. Bodenhausen, R. Freeman, G.A. Morris and D.L. Turner, *J. Magn. Reson.*, 1978, **31**, 75–95.
[22] J.D. Mersh and J.K.M. Sanders, *J. Magn. Reson.*, 1982, **50**, 171–174.
[23] M. Woodley and R. Freeman, *J. Magn. Reson. (A)*, 1994, **109**, 103–112.
[24] M. Woodley and R. Freeman, *J. Magn. Reson. (A)*, 1994, **111**, 225–228.
[25] G.A. Morris, *J. Magn. Reson.*, 1981, **44**, 277–284.

Chapter 8

Correlations through space: The nuclear Overhauser effect

8.1. INTRODUCTION

The previous three chapters in this book have all been concerned with *scalar couplings* between nuclei, that is, the indirect couplings that are transmitted through intermediate electron spins such as those in intervening chemical bonds. We have seen that by application of the appropriate techniques the chemist is able to exploit this coupling information and piece together molecular fragments and, ultimately, gross molecular structures. In this chapter we shall be concerned with a fundamentally different form of interaction between nuclear spins, that of the direct, through-space magnetic interactions (*dipolar couplings*) that give rise to the nuclear Overhauser effect (NOE). This brings about changes in resonance intensities and is, as we shall discover, intimately related to nuclear spin relaxation. The NOE is typically employed during the later stages of a structural investigation when the gross structure of the molecule has been (largely) defined through the application of the various techniques described in the four preceding chapters. The importance of the NOE in modern structure elucidation can hardly be overstated. It is unique in its ability to provide the chemist with information on three-dimensional molecular geometry. Such information can be obtained because the NOE depends upon, amongst other factors, internuclear separations such that only those spins that are 'close' in space are able to demonstrate this effect.

The NOE also finds widespread use as a means of sensitivity enhancement of low-γ spin-$^1/_2$ nuclei, so widespread in fact that it is often taken for granted and its contribution to experiments often overlooked. As described in Chapter 4 and below, the use of broadband proton decoupling during the recording of carbon spectra contributes as much as a three-fold increase in resonance intensity by virtue of the proton to carbon NOE. However, the principal aim of this chapter is to develop an appreciation of the NOE as a tool in structural analysis where it has a unique role to play. The interpretation of NOE measurements does, however, require more care than for those methods that exploit scalar couplings, and is generally more susceptible to erroneous conclusions being drawn. As part of this, it is also important to be conscious of the nature of the NOE measurement being taken, and in particular whether it is a *steady-state* or a *transient* protocol that is used. The first of these is exemplified by the widely employed 'NOE difference' experiment whereas the second is best known as the two-dimensional NOESY technique. The newer gradient-selected one-dimensional NOE experiments, which are making a significant impact on small molecule structural studies, also observe transient

Table 8.1. The principal applications of the main techniques described in this chapter

Technique	Principal applications
NOE difference	Establishing NOEs and hence spatial proximity between protons. Suitable only for 'small' molecules (M_r < 1000), for which NOEs are positive. Observes *steady-state* or *equilibrium* NOEs generated from the *saturation* of a target.
NOESY (2D or 1D)	Establishing NOEs and hence spatial proximity between protons. Suitable for 'small' (M_r < 1000) and 'large' molecules (M_r > 2000) for which NOEs are positive and negative respectively, but may fail for mid-sized molecules (zero NOE). Observes *transient* NOEs generated from the *inversion* of a target. Estimates of internuclear separations can be obtained in favourable cases.
ROESY (2D or 1D)	Establishing NOEs and hence spatial proximity between protons. Suitable for any molecule but often essential for mid-sized molecules; NOEs are positive for all molecular sizes. Observes *transient* NOEs in the rotating-frame, but is prone to interference from other mechanisms so requires cautious interpretation. Estimates of internuclear separations can be obtained in favourable cases.
HOESY	Establishing heteronuclear NOEs and hence spatial proximity between different nuclides, e.g. ^{1}H–^{13}C. Can provide useful stereochemical information when homonuclear NOEs insufficient or inappropriate, but suffers from low sensitivity.
EXSY (2D)	Qualitative mapping of exchange pathways in dynamic systems when exchange rates are slow on the NMR chemical shift timescale, meaning separate resonances are observed for each exchanging species. Quantitative data on exchange kinetics can be obtained in favourable cases.

The molecular masses mentioned provide only approximate ranges over which the experiments are applicable (see main text).

NOEs since they are one-dimensional analogues of NOESY. The nature of the measurement has fundamental implications for how the data should be *interpreted* and indeed *reported*. As such these two fundamentally different approaches to NOE measurements will be treated separately for much of the chapter, although they both share in the same underlying theory. The principal techniques described in this chapter are summarised in Table 8.1.

Which of these two approaches is adopted in the laboratory may be dictated by the motional properties of the molecule(s) under study, and more specifically the rates at which the molecules tumble in solution. Pre-empting what is to follow, it will be shown that the steady-state experiments are only appropriate for molecules that tumble 'rapidly' in solution (we shall also see what defines 'rapidly' in this context). Such measurements have traditionally been the home territory of small organic molecules in relatively non-viscous solutions. In contrast, very much larger molecules that tumble 'slowly' in solution (or smaller molecules in very viscous solutions) can only be meaningfully studied with the transient NOE techniques, which may also suitable for small molecule studies. Between these two extremes of molecular tumbling rates the conventional NOE can become weak and even vanishingly small, a condition most likely to occur for those molecules with masses of around 1000–2000 daltons. It is here that rotating-frame NOE measurements play a vital role, and these shall also be presented below.

The chapter is presented in two parts, the first covering the essential theory that underlies the NOE and the second addressing the practicalities of how one measures NOE enhancements, the experimental steps required to optimise such measurements and how to correctly interpret the data. In keeping with the style of this book, mathematical equations are kept to a minimum and are introduced only when they serve to illustrate a point of fundamental importance. Likewise, the equations are generally presented rather than being derived, and the interested reader is encouraged to read dedicated texts on these topics for further elaboration [1].

Part I: Theoretical aspects

8.2. DEFINITION OF THE NOE

The NOE may be defined as the change in intensity of one resonance when the spin transitions of another are somehow perturbed from their equilibrium populations. The perturbation of interest usually corresponds to either *saturating* a resonance, that is, equalising the spin population differences across the corresponding transitions, or to *inverting* it, in other words, inverting the population differences across the transitions. The magnitude is expressed as a relative intensity change between the equilibrium intensity, I_0, and that in the presence of the NOE, I, such that

$$\eta_I\{S\} = \frac{I - I_0}{I_0} \times 100\,(\%) \tag{8.1}$$

where $\eta_I\{S\}$ indicates the NOE observed for spin I when spin S is perturbed, which shall also be referred to as the NOE *from* spin S *to* spin I. The use of the symbols I and S stems from the original publications on the phenomenon (which, in fact, observed the NOE from electron spins to nuclear spins in a metal) and have become the recognised nomenclature when describing the NOE. However, both S and I have been used to define the *perturbed* spin over the years and even across modern texts both definitions are encountered so one should always be clear as to the terminology in use. Herein, S will always refer to the perturbed (or Source) spin and I to the enhanced (or Interesting) spin. The intensity changes brought about by the NOE can be both positive (an increase) or negative (a decrease) as dictated by the motional properties of the molecules and by the signs of the magnetogyric ratios of the participating spins. Throughout, the term 'enhancement' will be used to refer to the intensity changes.

8.3. STEADY-STATE NOES

In laying down the background to how the NOE arises and what factors dictate its sign and magnitude, we shall restrict our discussion to steady-state NOEs in which the perturbation is brought about by *saturating* S spin transitions by the selective application of weak rf irradiation to the S resonance (Fig. 8.1). It is this form of the NOE that is observed with the popular NOE difference method, which has had such an enormous impact on structural organic chemistry. Further discussions relating to other forms of NOE measurement then follow logically from this background material. We begin by considering the simple case of a homonuclear two-spin system then progress to consider more realistic multispin systems.

8.3.1. NOEs in a two-spin system

Origin of the NOE

Consider a system comprising only two homonuclear spin-$^1/_2$ nuclei, I and S, that exist in a rigid molecule which tumbles isotropically in solution, that is, it has no preferred axis about which it rotates. The two nuclei do not

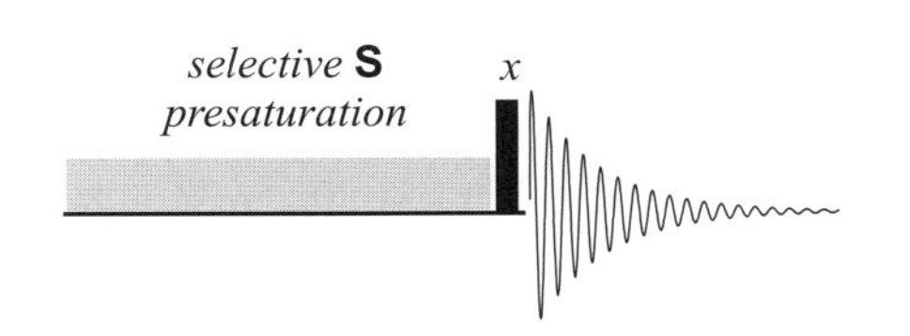

Figure 8.1. The general experimental scheme for observing steady-state NOE enhancements.

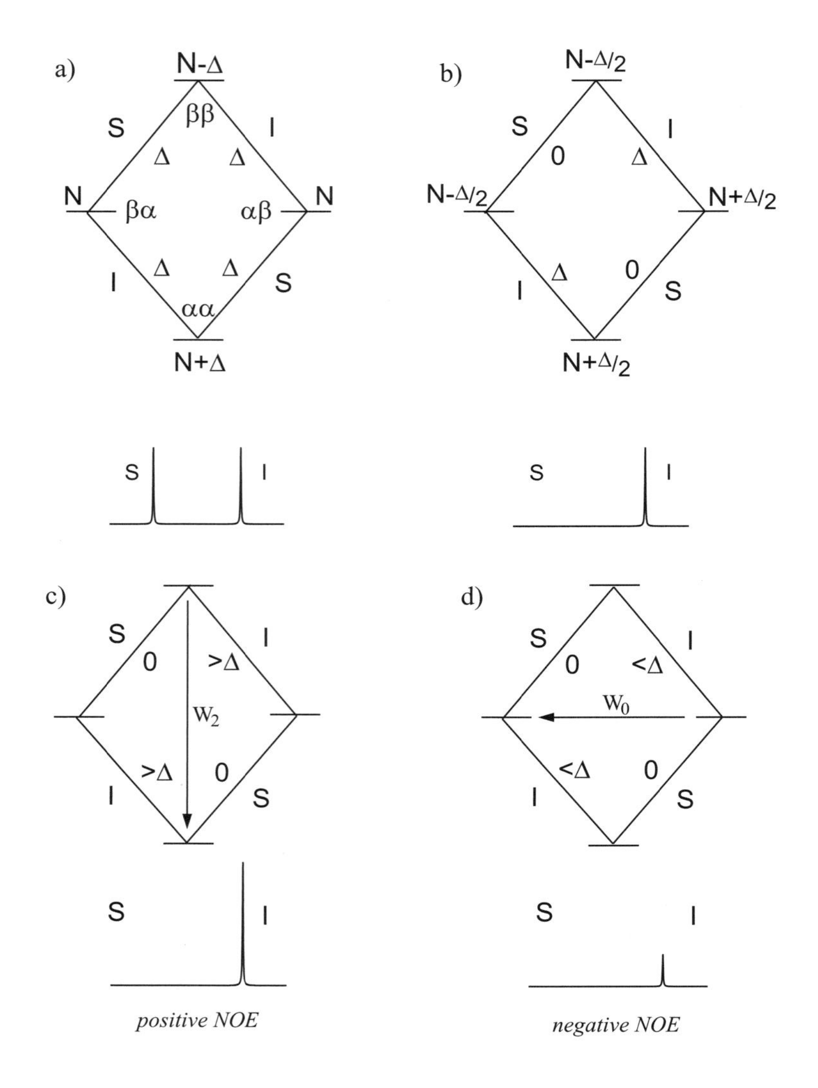

Figure 8.2. Schematic energy level diagrams and population differences for two spins, S and I, which share a dipolar coupling; (a) at equilibrium, (b) after instantaneous saturation of the S-spins, (c) after relaxation via W_2 processes and (d) after relaxation via W_0 processes. Below each are the corresponding schematic spectra.

share a scalar coupling ($J_{IS} = 0$) but are sufficiently close to share a dipolar coupling. This is the direct, through space magnetic interaction between the two spins such that one spin is able to sense the presence of its dipolar coupled partner. This coupling may be viewed as being analogous to the interaction one witnesses when two bar magnets are brought close together, and follows the idea introduced in Chapter 2 that nuclear magnetic dipoles can be viewed as microscopic bar magnets (see below).

The energy level diagram for an ensemble of 4N molecules is shown in Fig. 8.2. Since we are considering a homonuclear system the energies of the I and S transitions will be essentially identical (chemical shift *differences* are negligible relative to Larmor frequencies) and we can therefore assume that the populations of the αβ and βα states are equal at equilibrium. According to the Boltzmann distribution, there will then exist an excess of nuclei in the lower energy αα orientation, and a deficit in the higher energy ββ state. We shall ultimately be interested in the population *differences* across transitions as it is these that dictate the intensity of the NMR resonances, so we shall simply symbolise the population excess as Δ and the deficit as -Δ relative to those of the αβ and βα states. Note that dipolar couplings do not produce observable splittings in solution spectra (see below), so the two transitions associated with each spin are of identical energy. The spectrum in the absence of perturbation therefore contains two singlet resonances of equal intensity (Fig. 8.2a).

Now suppose we instantaneously saturate the S resonance forcing the population differences across the S transitions to zero. The new spin populations are indicated in Fig. 8.2b. Clearly the system has been forced away from the equilibrium population differences so will attempt to regain this by altering its spin populations. The changes of spin states required to achieve this are brought about by longitudinal spin relaxation processes, so we need consider which relaxation pathways are now available to the spins. Ignoring for the moment the mechanism by which these changes may occur, we see that six possible pathways can be identified for a two-spin system (Fig. 8.3). Four of these correspond to the single-quantum transitions, involving the flip of a single spin, for example αα–βα. The W labels represent the 'transition probabilities' for each, or in other the words, the rates at which the corresponding spin flips occur, and the subscripts the magnetic quantum number of the transition. The two other transitions, αβ–βα and αα–ββ, involve the simultaneous flipping of both S and I spins. Although these transitions do occur, they cannot be directly *observed* in an NMR experiment, unlike single spin flips, because the overall change in magnetic quantum, ΔM, does not equal one. They are said to be 'forbidden' by quantum mechanical selection rules. The αβ–βα W_0 process is referred to as the zero-quantum transition ($\Delta M = 0$) whilst the αα–ββ W_2 process is the double-quantum transition ($\Delta M = 2$) (notice the same terminology was used when describing the transitions of scalar-coupled spins in Chapter 5). These are both able to act as relaxation pathways and, in fact, it is only these two that are responsible for the NOE itself. Collectively, they are referred to as *cross-relaxation* pathways, a term suggestive of the simultaneous participation of both spins.

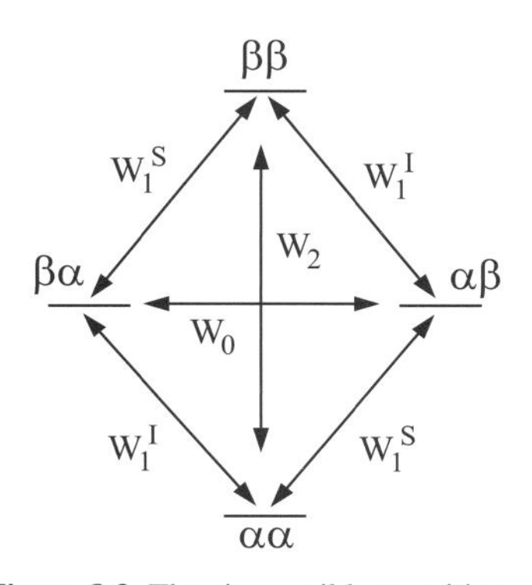

Figure 8.3. The six possible transitions in a two-spin system.

Returning to the diagram of Fig. 8.2 we are now in a position to consider how the various relaxation pathways may be used to re-establish the equilibrium condition, noting that throughout the W_1^S transitions remain saturated by continuous rf energy. The population differences across the I-spin transitions are still Δ, as they were at equilibrium, so the W_1^I processes will play no part in re-establishing equilibrium and thus have no role to play in producing the NOE. The W_2 process will act to remove spins from the ββ state and transfer them to the αα state in an attempt to recover the population differences across the S transitions. In doing so this will *increase* the population *difference* across the two I transitions (Fig. 8.2c). Thus, relaxation via the W_2 process will result in a net increase in the I spin resonance intensities in the spectrum; this is then a *positive NOE*. Likewise, the W_0 process will act to transfer spins from the βα to the αβ state, again in an attempt to recover the population differences across the S transitions. In this case the result will be a *decrease* in the population *difference* across the two I transitions (Fig. 8.2d) so that relaxation via the W_0 process will result in a net reduction in the I spin resonance intensities in the spectrum; this is then a *negative NOE*.

From these qualitative considerations, we can already say a fair amount about how we might expect the NOE to appear. Clearly, the W_2 and W_0 cross-relaxation processes compete with one another, with the dominant pathway dictating the sign of the observed NOE. In addition, the W_1^I pathways will act to re-establish the equilibrium population differences for the I transitions as soon as the NOE begins to develop, so will tend to act against the build-up of the NOE. Thus, if the relaxation mechanisms for the W_1^I pathways happen to be rather more efficient than those of the W_2 and W_0 pathways, then a measurable NOE may never develop; it is, in effect, bypassed altogether. This can have a significant bearing on the experimental measurement of NOEs, as we shall see in due course. The NOE therefore results from a balance between a number of competing relaxation pathways. Saturating the S transitions for a period of time that is long relative to the relaxation times allows a new *steady-state* of populations to arise as a result of this competition, and it is these one eventually measures. A full consideration of the various rate processes involved in the population changes leads to the so-called Solomon equation, which for the steady-state NOE can be used to derive the expression:

$$\eta_I\{S\} = \frac{\gamma_S}{\gamma_I}\left[\frac{W_2 - W_0}{W_0 + 2W_1^I + W_2}\right] \equiv \frac{\gamma_S}{\gamma_I}\left[\frac{\sigma_{IS}}{\rho_{IS}}\right] \tag{8.2}$$

The term σ_{IS} represents the *cross-relaxation rate* for the two spins, whilst ρ_{IS} is the total *dipolar longitudinal relaxation rate* of spin I. The magnetogyric ratios (γ_S and γ_I) are included to take account of the different equilibrium populations that would exist for spins with differing γs; for a homonuclear spin system as considered thus far, these values would obviously be equal and may be ignored. This fundamental expression contains within it the qualitative arguments arrived at above; $W_2 - W_0$ dictates the sign of the NOE whereas W_1^I processes make no contribution to this but serve to reduce its magnitude. To appreciate the size and sign of the NOE, how this relates to molecular motion and, indeed, how this can be related in any way to internuclear distances, it is necessary to define what factors influence the participating rate constants, and for this one needs to consider the spin relaxation processes involved.

Spin relaxation and dipolar coupling

The NOE arises as a result of the redistribution of spin populations and hence flips between spin states. Such redistributions occur as a result of longitudinal spin relaxation (Chapter 2) which do not occur spontaneously but require a suitable stimulus to induce the transitions. This stimulus is a magnetic field fluctuating at the frequency of the corresponding transition (both here and below 'frequency' corresponds to the energy of the transition rather that the rate at which the spin flips occur). Here exists an analogy with the pulse excitation of a spin system initially at equilibrium. The time-dependent magnetic component of the electromagnetic rf radiation interacts with nuclear magnetic moments and is thus able to tip the bulk magnetisation vector into the transverse plane. Only if the rf has a magnetic component oscillating at the Larmor frequency of the spins does excitation occur; this is why one is able to apply excitation pulses to protons whilst leaving, say, carbon spins unaffected.

The magnetic field of relevance to the NOE is the local field experienced by a spin as a result of dipolar interactions with neighbouring magnetic nuclei. These interactions may be visualised using the microscopic bar magnet analogy for spin-$^1/_2$ nuclei in which they are considered to posses a magnetic North and a South pole (Fig. 8.4). Depending on the relative orientation of the two nuclei to one another, the field generated by a neighbouring spin will either reinforce or counteract the applied magnetic field, with the time-dependent fluctuation produced by the rotational motion of the molecule in which these nuclei sit

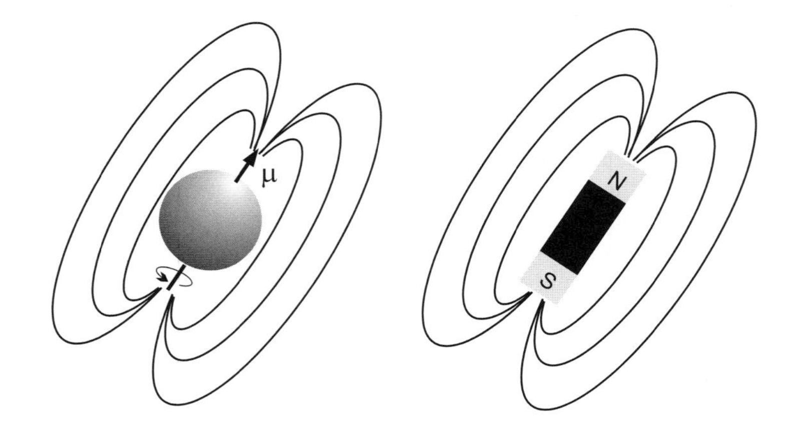

Figure 8.4. The bar-magnet analogy for a spin-$^1/_2$ nucleus in which the magnetic dipole is viewed as possessing a magnetic North and South pole.

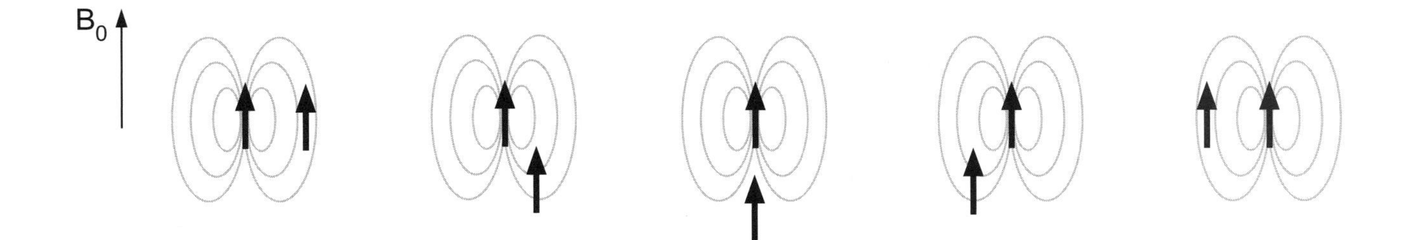

Figure 8.5. The direct, through-space interaction between two near spin-$^1/_2$ nuclei (the dipolar interaction). This fluctuates as the molecule tumbles in solution and can provide a time-dependent field capable of inducing spin transitions.

(Fig. 8.5). It is precisely this that provides the mechanism for relaxation, known as *longitudinal dipole–dipole relaxation*, and this requirement for an interaction with a neighbouring spin intuitively fits with the arguments presented above for the NOE arising from *mutual* spin flips (the W_2 and W_0 processes). The magnitude of dipolar coupling between spins is acutely sensitive to the internuclear separation r, being proportional to r^{-3}, and it is in this that the NOE itself ultimately has a distance dependence.

One should note here that despite providing an important relaxation mechanism, *dipolar couplings do not usually produce observable splittings in solution state NMR spectra*. This is because, although the couplings have a finite value at any instant in time, they are averaged precisely to zero on the NMR timescale by the rapid isotropic tumbling of a molecule.

To induce the spin transitions we have been considering, the molecule must tumble at the appropriate frequency to provide a suitable fluctuating field. The rate at which a molecule tumbles or rotates in solution is typically defined by its rotational *correlation time,* τ_c. This is usually taken to define the average time required for the molecule to rotate through an angle of 1 radian about any axis, meaning rapidly tumbling molecules posses small correlation times whilst slowly tumbling molecules have large correlation times. A *very rough* estimate of this time for a molecule of mass M_r may be obtained from the relationship:

$$\tau_c \approx M_r \times 10^{-12}\ \mathrm{s} \qquad (8.3)$$

The power available within a molecular system to induce transitions by virtue of its molecular tumbling is referred to as the *spectral density* $J(\omega)$ (Section 2.5) and this provides a measure of how the relaxation rates W_0, W_1 and W_2 vary as a function of tumbling rates. This is illustrated schematically in Fig. 8.6 for three different correlation times. An alternative description of the spectral density is that it represents the probability of finding a fluctuating magnetic component at any given frequency as a result of the motion and as such the area under each of the curves of Fig. 8.6 must then be equal. Thus, for a molecule with a short τ_c (rapid tumbling) there exists an almost

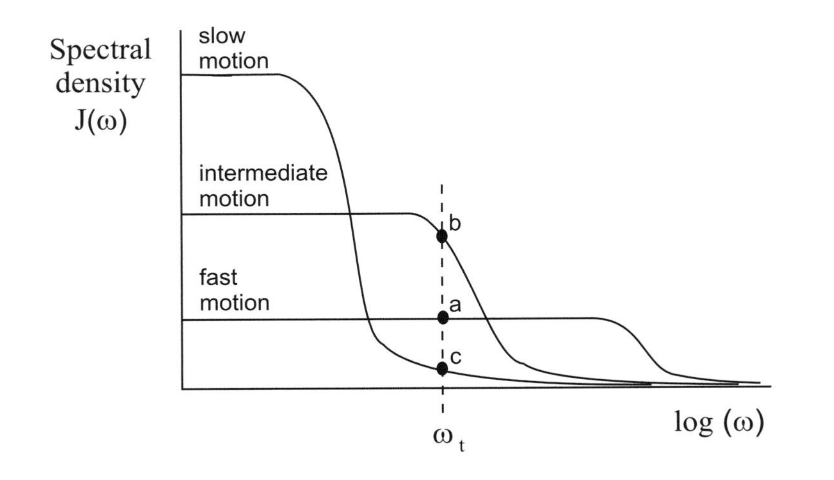

Figure 8.6. Schematic spectral densities for molecules tumbling in three motional regimes as a function of frequency, ω, where ω_t represent the frequency of the spin transition.

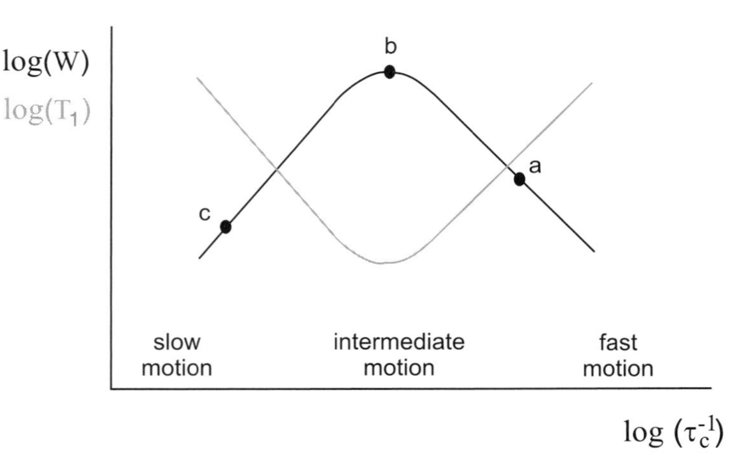

Figure 8.7. The schematic variation in relaxation rates, and hence relaxation times (shown greyed), as a function of molecular tumbling rates.

equal but comparatively small chance of finding components at both high and low frequencies, up to about $1/\tau_c$ at which point the probability falls away rapidly. Conversely, there is only a very small probability that molecules which tumble slowly (*on average*) will generate rapidly oscillating fields, so the corresponding spectral density is concentrated into a smaller frequency window. These curves therefore predict how the relaxation rates will vary with correlation time. For a transition of frequency ω_t the spectral density will be rather low if the molecular tumbling rate is far greater than this (point a) and thus the relaxation rates will also be low. As the rate of tumbling slows and approaches ω_t, the spectral density and hence relaxation rates increases (point b), only to decrease once more as the tumbling rate falls below ω_t. Thus, the dependence of relaxation rates on τ_c may be represented as in Fig. 8.7, with the fastest relaxation occurring when the correlation rate $1/\tau_c$ matches the frequency of the transition. The point of fastest W_1 relaxation also corresponds to the T_1 minimum and vice versa, so T_1 values themselves are also dependent on the rates of molecular motion.

Using these arguments it is possible to predict when the W_0 or W_2 processes will be dominant. For the zero-quantum W_0 processes, the energy differences involved are rather small, being the differences between the I and S frequencies so $\omega_t = |(\omega_I - \omega_S)|$. These transitions will therefore be strongly favoured for a molecule that tumbles slowly in solution. The double-quantum W_2 transitions correspond to the sum of the I and S frequencies, so $\omega_t = (\omega_I + \omega_S)$ and these will be stimulated by rapidly tumbling molecules. As an example of the transition frequencies involved, consider a homonuclear proton system at an observation frequency of 400 MHz. The single-quantum W_{1I} (and W_{1S}) transition frequencies will correspond to approximately 400 MHz. The W_0 transition frequencies are given by the frequency differences of I and S, that is, their chemical shift differences, which will be in the Hz or kHz region. The W_2 transition frequencies are the sums of ω_I and ω_S which will equate to around 800 MHz. Clearly the frequency spread of molecular motions to which the NOE is sensitive is extremely large.

Qualitatively then, one can predict from the previous arguments that *molecules which tumble rapidly in solution* are likely to favour the higher energy W_2 process and hence *exhibit positive NOEs* whilst *those that tumble slowly* will favour the W_0 process and thus *display negative NOEs*; indeed this is what is observed in practice.

Quantitative expressions for the relaxation rates in a dipolar coupled two-spin system have been derived, thus:

$$W_{1I} \propto \gamma_I^2 \gamma_S^2 \left[\frac{3\tau_c}{r^6(1 + \omega_I^2 \tau_c^2)} \right] \tag{8.4a}$$

$$W_0 \propto \gamma_I^2\gamma_S^2\left[\frac{2\tau_c}{r^6\left(1+(\omega_I-\omega_S)^2\tau_c^2\right)}\right] \qquad (8.4b)$$

$$W_2 \propto \gamma_I^2\gamma_S^2\left[\frac{12\tau_c}{r^6\left(1+(\omega_I+\omega_S)^2\tau_c^2\right)}\right] \qquad (8.4c)$$

where the constant of proportionality is the same for each.

Notice that, in addition to the dependence on correlation times, these expressions contain a term for the internuclear separation, r, between spins S and I. Here, at last, one starts to see the origins of the famed 'r^{-6}' distance dependence widely, and sometimes dangerously, associated with NOE interpretations [2]. This distance term is manifested in the degree of dipolar coupling between the two spins. An important point to note at this stage is that this distance dependence actually lies in the relaxation *rates* and, as we shall see, this can have enormous implications for the way in which NOE data are interpreted. The inverse-sixth relationship also means the NOE falls away very rapidly with distance, so in practice significant NOEs will only develop between protons that are within ca. 0.5 nm of each other (naturally, this will be influenced by how sensitive and stable the spectrometer is and hence how small an enhancement one is able to 'see'). Note also the dependence of equations 8.4 upon the square of the magnetogyric ratios of the two spins, so very different rates may occur in heteronuclear systems, depending on the participating spins.

When a molecule tumbles so rapidly in solution such that $\omega\tau_c \ll 1$, all terms in these expressions containing ω become negligible and the rates simplify to:

$$W_{1I} \propto \gamma_I^2\gamma_S^2\frac{3\tau_c}{r^6} \qquad (8.5a)$$

$$W_0 \propto \gamma_I^2\gamma_S^2\frac{2\tau_c}{r^6} \qquad (8.5b)$$

$$W_2 \propto \gamma_I^2\gamma_S^2\frac{12\tau_c}{r^6} \qquad (8.5c)$$

This condition is referred to as the *extreme narrowing limit* since all broadening effects attributable to dipolar interactions are fully averaged to zero under these conditions. This regime typically applies only to small molecules in low viscosity solvents, and the point at which this condition breaks down depends on the correlation time of the molecule as well as the field strength of the spectrometer (through ω).

NOEs and molecular motion

Having taken the trouble to see how the relaxation rates in a two-spin system depend upon molecular motion, we are now in a position to predict the behaviour of the NOE itself as a function of this motion and of internuclear separation. Taking the rate constant equations 8.4 and substituting these into that for the NOE (equation 8.2) produces the curve presented in Fig. 8.8 for the theoretical variation of the *homonuclear* NOE as a function of molecular tumbling rates as defined by $\omega_0\tau_c$, (where ω_0 is the spectrometer observation frequency, approximately equal to ω_I and ω_S). Note this is for a two-spin system which relaxes solely by the dipole–dipole mechanism and as such represents the theoretically maximum possible NOE. The curve has three distinct regions to it, which we shall loosely refer to as the fast, intermediate and slow motion regimes. For those molecules that tumble rapidly in solution (short τ_c, those in the extreme narrowing limit) the NOE has a maximum possible value of +0.5 or 50%. Smaller organic molecules in low viscosity solvents typically fall within this fast motion regime which is traditionally the home ground of steady-state NOE measurements. At the other extreme,

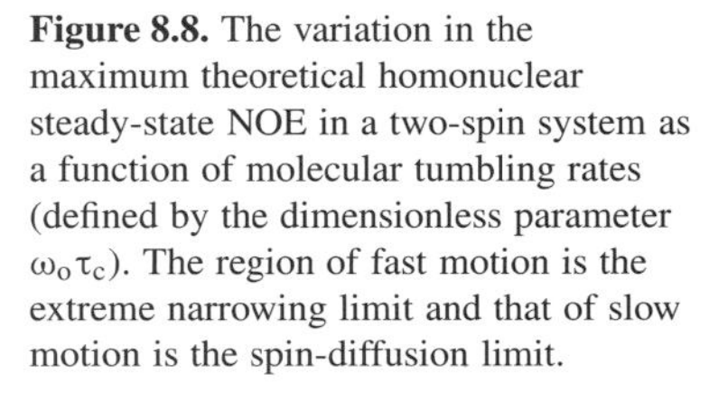

Figure 8.8. The variation in the maximum theoretical homonuclear steady-state NOE in a two-spin system as a function of molecular tumbling rates (defined by the dimensionless parameter $\omega_0\tau_c$). The region of fast motion is the extreme narrowing limit and that of slow motion is the spin-diffusion limit.

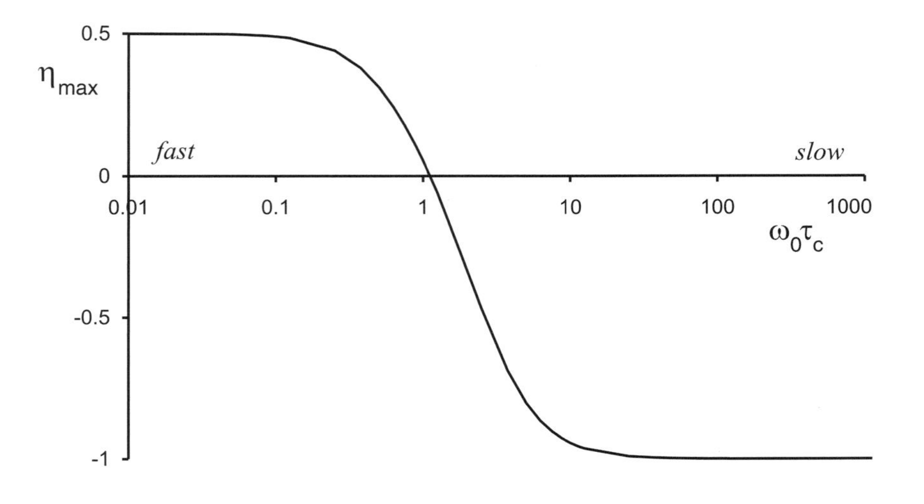

molecules that tumble very slowly in solution experience negative NOEs, as shown above. The maximum enhancement in this motion regime is obtained when W_2 and W_{1I} are both zero, which from equation 8.2 can be seen to be -1 or -100%. For NOE measurements in this region it becomes essential to use transient experiments since those based on steady-state measurements become uninformative, as explained below. This is the region inhabited by (biological) macromolecules and it is studies of these systems that have traditionally made widespread use of transient NOE measurements, principally through the 2D NOESY experiment. Between these two extremes is the intermediate region in which the NOE changes sign and even becomes zero when W_2=W_0. Within this region the magnitude and sign of the NOE is highly sensitive to the rate of molecular motions and can be rather weak, possibly too weak to be observed, clearly a major hindrance to structural studies. The point at which this region is entered will be dependent on a number of factors; the size and shape of the molecule, solution conditions (viscosity, temperature, possibly pH etc) and spectrometer field strength. As a rule of thumb, molecules with a mass of 1000–2000 daltons are likely to fall within this intermediate regime. The increasing interest in larger synthetic molecules in many areas, for example supramolecular chemistry, is likely to mean more molecules routinely handled by the research chemist will fall into this potentially troublesome region. The actual zero cross over point occurs when:

$$\omega\tau_c = \sqrt{5/4} = 1.12 \tag{8.6}$$

or in other words, when the molecular tumbling rate approximately matches the spectrometer observation frequency. This is therefore field dependent, as illustrated in Fig. 8.9 which now shows the variation of the NOE as a function τ_c itself for three different field strengths. The use of a higher field strength increases the likelihood of a relatively 'small' molecule falling within the intermediate regime or perhaps a mid-sized molecule passing from this into the slow motion regime. In some cases therefore, the use of a higher field instrument, often regarded as a panacea for all chemists' woes, may even prove detrimental to the NOE experiment as the 'zero NOE' condition is approached.

The field dependence of the intermediate regime suggests one solution to the problem of 'zero' NOEs; trying a different field strength. This is usually an impossible or impractical answer, so an alternative approach is to alter the solution conditions and hence the rate of molecular tumbling, one of the simplest approaches being to vary the sample temperature. A technically different experimental approach to the problem is to measure NOEs in the *rotating frame* instead. This is described further below, suffice it to say here

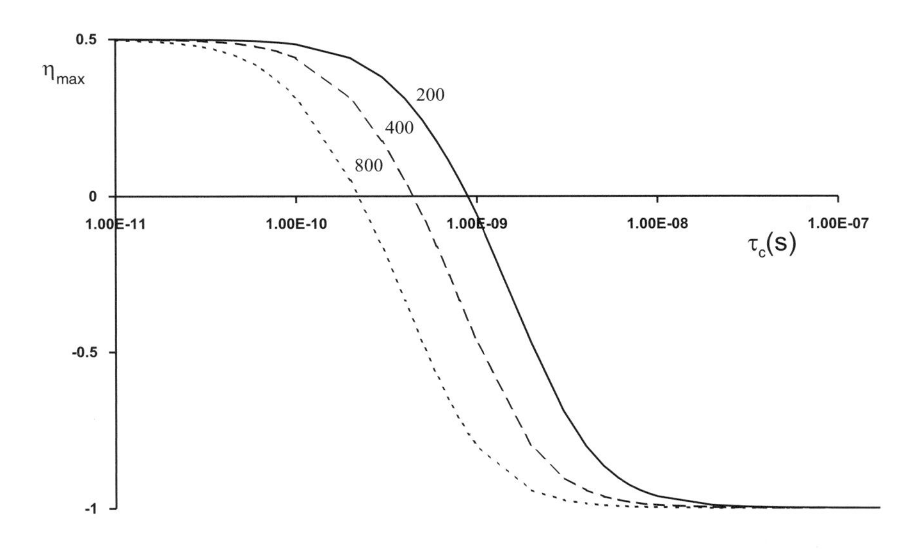

Figure 8.9. The variation in the maximum theoretical homonuclear steady-state NOE in a two-spin system as a function of molecular tumbling rates shown for three spectrometer observation frequencies (MHz).

that these NOEs remain positive for all molecular tumbling rates one is likely to encounter so avoids the 'zero cross-over' problem altogether.

NOEs and internuclear separation

It is now possible to consider how the separation between two spins influences the steady-state NOE enhancement. Assuming the molecule exists in the extreme narrowing regime, then substituting the simplified rate equations 8.5 into equation 8.2 we obtain:

$$\eta_I\{S\} = \left[\frac{\left(\frac{12\tau_c}{r^{-6}}\right) - \left(\frac{2\tau_c}{r^{-6}}\right)}{\left(\frac{2\tau_c}{r^{-6}}\right) + 2\left(\frac{3\tau_c}{r^{-6}}\right) + \left(\frac{12\tau_c}{r^{-6}}\right)}\right] = \left[\frac{12-2}{2+6+12}\right] = \frac{1}{2} \quad (8.7)$$

As in Fig. 8.8 for the maximum NOE above, the enhancement is predicted to be 50%. However, it is also predicted to be *independent* of the internuclear distance. Thus, at least for the hypothetical isolated two-spin system considered here, the magnitude of the steady-state enhancement provides no distance information whatsoever. The important point here is that it is too bold a statement to say that differing NOE enhancements within a molecule scale directly with r^{-6}. In realistic chemical systems various 'other' factors must also be taken into account before any distance dependence is reintroduced; this is further pursued below. Although the *magnitude* of the steady-state enhancement is predicted to be independent of distance in this system, the *rate* at which this is reached is not because of the dependence of relaxation rates on distance (as expressed in equation 8.4) meaning NOEs between closer spins develop more rapidly; this is basis of the transient NOE measurements described in Section 8.4. This also implies that longer-range NOEs will only have significant intensities when long presaturation periods are employed, a point of considerable practical importance.

Heteronuclear NOEs

The equivalent of equation 8.7 for a heteronuclear pair experiencing extreme narrowing is the more general expression:

$$\eta_I\{S\} = \frac{\gamma_S}{2\gamma_I} \quad (8.8)$$

For the common situation of carbon-13 observation in the presence of proton saturation (broadband decoupling), $\gamma_H/\gamma_C \approx 4$ and NOE enhancements can be as much as 200%, equating to a three fold intensity increase. Since the

Table 8.2. Theoretical maximum steady-state heteronuclear NOE enhancements in the presence of proton saturation

X	^{6}Li	^{13}C	^{15}N	^{19}F	^{29}Si	^{31}P	^{57}Fe	^{103}Rh	^{109}Ag	^{119}Sn	^{183}W	^{195}Pt	^{207}Pb
$\eta_X\{^1H\}\%$	339	199	−494	53	−252	124	1548	−1589	−1075	−141	1202	233	239

These numbers assume relaxation exclusively via dipole–dipole interactions, although for the metals in particular chemical shift anisotropy may also be a significant mechanism. ^{6}Li is something of an anomaly in that it is quadrupolar yet can still demonstrate NOEs. It has the smallest quadrupole moment of all such nuclei, however, so the dipole mechanism still makes a significant contribution to relaxation.

relaxation of carbon nuclei is largely dominated by proton dipolar interactions, this maximum is almost met in practice. This is clearly a valuable route to sensitivity enhancement and, at least for the case of ^{13}C, compares favourably for routine acquisitions with the factor of four attainable with the ^{1}H to ^{13}C polarisation transfer sequences of Chapter 4. The maximum NOE enhancements for a variety of nuclei in the presence of proton saturation, denoted X{^{1}H}, are summarised in Table 8.2.

In heteronuclear systems the observed NOE also depends on the signs of the magnetogyric ratios of the cross-relaxing spins, so NOEs from protons to nuclei with negative γs will display negative NOEs *even if the molecule is within the extreme narrowing regime*. The most common examples of this are for ^{15}N and ^{29}Si for which a *reduction* of signal intensity occurs on proton saturation, so much so that the observed resonance can itself become negative. If less than the full (negative) NOE is generated, the resonance may disappear altogether as the NOE cancels the natural signal and because of this it is usual to record the spectra of negative γ species in the absence of the NOE, either by use of the inverse-gated decoupling scheme (Section 4.2.3), by polarisation transfer methods (Section 4.4) or, less commonly, by the addition of a paramagnetic relaxation reagent to quench the NOE. In some instances the heteronuclear NOE may be used more specifically for structural assignment also; see Section 8.9.

Beyond the extreme narrowing condition, the absolute magnitudes of heteronuclear X{^{1}H} NOEs decrease and, in the case of ^{13}C, ^{15}N and ^{29}Si, closely approach zero, although, with the exception of ^{19}F, do not change sign. For very large molecules there is then little sensitivity gain, or loss, arising from the heteronuclear NOE.

8.3.2. NOEs in a multispin system

The previous section considered the NOE for the hypothetical case of a two-spin system in which the spins relax exclusively via mutual dipole–dipole relaxation. In progressing to consider more realistic multispin systems two key issues will be addressed; how the presence of other spins affects the magnitudes of steady-state NOEs and how these reintroduce distance dependence to the NOE. These considerations lead to the conclusion that steady-state NOE measurements must be used in a comparative way to provide structural data, and that they do not generally provide estimates of internuclear distances per se.

Additional relaxation pathways

The NOE arises as a result of dipolar cross-relaxation between two nuclei and hence only the dipole–dipole relaxation mechanism is able to generate the NOE. All other competing mechanisms (with the subtle exception of scalar relaxation in a strongly coupled system, which is rarely of significance) serve to dilute the overall influence of the W_2 and W_0 pathways by stimulating the W_1 relaxation pathway only, and hence reduce the magnitude of the NOE. These various other contributions to spin relaxation may be grouped together

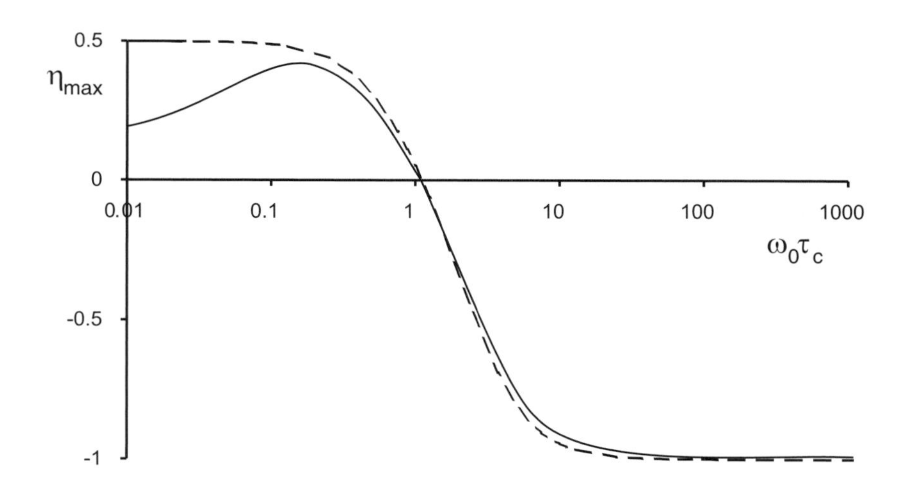

Figure 8.10. Schematic illustration of the maximum homonuclear steady-state NOE in the presence (solid line) and absence (dotted line) of external relaxation sources that compete with cross-relaxation.

and represented by the term ρ_I^*, the external relaxation rate for spin I. Adding this to equation 8.2 gives:

$$\eta_I\{S\} = \frac{\gamma_S}{\gamma_I}\left[\frac{\sigma_{IS}}{\rho_{IS} + \rho_I^*}\right] \tag{8.9}$$

This illustrates the diluting effect of ρ_I^* on the magnitude of the NOE, which is sometimes referred to as the 'leakage' term. The effect of this turns out to be of somewhat greater significance for molecules in the extreme narrowing limit than those in the negative NOE regime. For small, rapidly tumbling molecules the absolute magnitude of σ_{IS} is small when compared with that of slowly tumbling molecules, so the relative contribution to equation 8.9 of ρ_I^* has greater significance. The effect of this is schematically illustrated in Fig. 8.10 which shows that leakage becomes more relevant as molecular tumbling rates increase, meaning very small NOEs may be observed for small molecules. In contrast, leakage effects are less problematic for molecules in the negative NOE regime.

To maximise the size of the NOE for molecules in the extreme narrowing condition it is necessary to minimise ρ_I^*. The most significant contribution to this in routinely prepared solutions is from the paramagnetic oxygen dissolved in solvents. The unpaired spin-$^1/_2$ electron has a magnetic moment that is over 600 times that of the proton, so is able to provide an intense magnetic interaction capable of causing efficient relaxation. Degassing of solutions prior to NOE studies may become necessary when seeking longer-range interactions in small molecules but is otherwise unnecessary for the majority of routine studies (see Section 8.10). Similarly, other paramagnetic impurities, for example some metals, will quench the NOE and must be avoided. Samples to which relaxation agents have been deliberately added to promote relaxation are therefore not suitable candidates for NOE studies. *Intermolecular* dipolar interactions, arising from transient interactions with solute or solvent molecules, are another potential interference. Solvent molecules are deuterated in most cases so the relaxation arising from these nuclei is rather inefficient when compared to protons (due to the γ^2 dependence of longitudinal dipolar relaxation rates) and interactions with other (protonated) solute molecules are generally only likely to be a problem when very concentrated solutions are used, so these are thus best avoided for NOE studies. Other, non-dipole relaxation mechanisms, such as those discussed in Chapter 2, also act to bypass the NOE. Of most significance is the quadrupolar relaxation associated with nuclei with spin > $^1/_2$. This is usually the dominant mechanism for such nuclei so NOEs onto quadrupolar nuclei are very rarely observed (the exception being ^{6}Li as mentioned in Table 8.2).

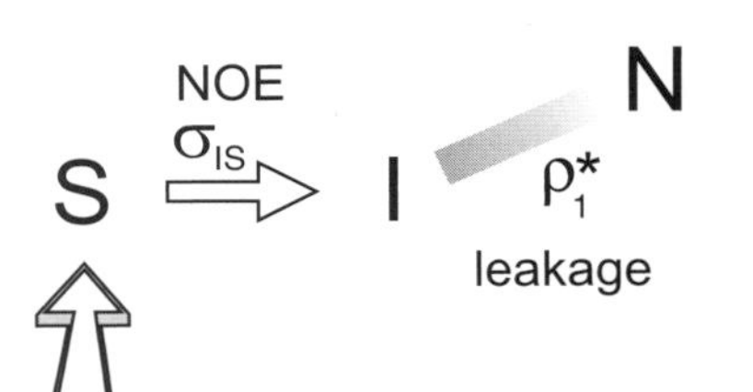

Figure 8.11. A three spin system in which the I-spin neighbour, N, acts as an external relaxation source for spin-I through their mutual dipolar interaction.

Internuclear separations (again)

Perhaps the most obvious 'other' contribution to the relaxation of spin I in a multispin system arises from neighbouring nuclei other than spin S within the same molecule. Dipolar interactions with these spins act to bring about longitudinal relaxation of spin I *independent* of cross-relaxation between I and S, so diluting the I–S NOE. To illustrate the influence of neighbouring spins, consider a hypothetical homonuclear three spin system I, S and N (N = neighbour) in which relaxation arises solely from dipolar interactions ignoring, for convenience, all the other possible contributions to ρ_I^* described above (Fig. 8.11). Assuming, for simplicity, an NOE exists only between I and S, the steady-state NOE may be written:

$$\eta_I\{S\} = \eta_{max}\left[\frac{r_{IS}^{-6}}{r_{IS}^{-6} + r_{IN}^{-6}}\right] \tag{8.10}$$

where η_{max} represents the maximum NOE possible in a homonuclear two-spin system, as previously. Notice that if spin N were not present, the NOE would simply be η_{max} and show no distance dependence, exactly as predicted above for a two-spin system. The effect of introducing an additional spin is to reintroduce distance dependence, yet despite this, the magnitude of the steady-state NOE does not scale simply as r_{IS}^{-6}. In fact, it can be seen that the magnitude of the NOE will be dictated by a balance between the r_{IS} and r_{IN} distances. This is true for all steady-state NOE measurements; the result will always represent a balance between the I–S internuclear separation and all other I–N separations (and in a realistic chemical system there may well be a large number of other neighbouring nuclei). To put it another way, the steady-state NOE arises from a competition between the I–S cross-relaxation and all other relaxation sources of spin I. Equation 8.10 also shows that a reduction in the I–S internuclear distance now does indeed contribute to an increased NOE between I and S since the total contribution to the numerator will be relatively more than to the denominator. The arrival of a neighbouring spin has therefore reintroduced the idea that a smaller internuclear distance can be correlated to some degree with larger NOE enhancements.

This statement must still be treated with some caution, however, as illustrated in Fig. 8.12 in which $\eta_B\{A\}$ is considered for a three-spin system, A, B and C, where the B–C distance is varied (and ignoring any direct AC interaction). When C is distant from B it has little influence on its relaxation, allowing the AB cross-relaxation to dominate, producing close to the maximum NOE. When both A and C are equidistant from B they play an equal role in relaxing B and the NOE is thus half the maximum possible value. As C becomes very much closer to B than is A, it now dominates B-spin relaxation and A–B

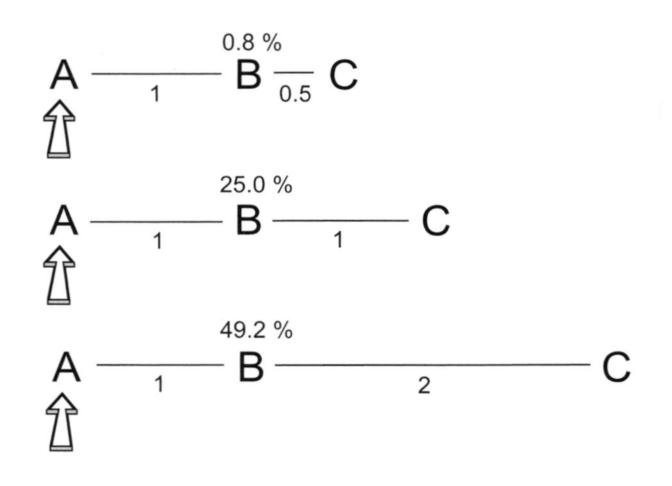

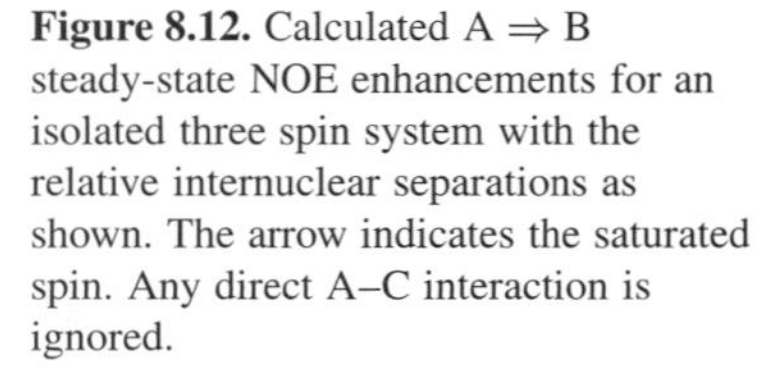

Figure 8.12. Calculated A ⇒ B steady-state NOE enhancements for an isolated three spin system with the relative internuclear separations as shown. The arrow indicates the saturated spin. Any direct A–C interaction is ignored.

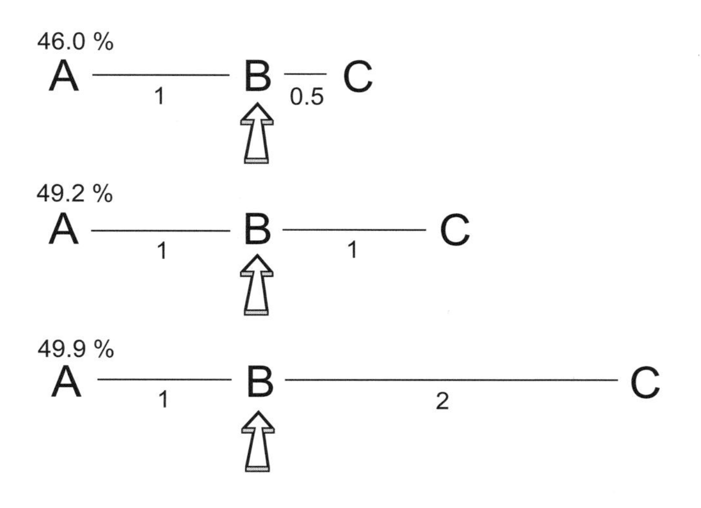

Figure 8.13. Calculated B ⇒ A steady-state NOE enhancements for an isolated three spin system with the relative internuclear separations as shown. The arrow indicates the saturated spin. Any direct A–C interaction is ignored.

cross-relaxation loses out in the competition resulting in a small NOE. Thus, in general, *despite two spins being 'close' to one another*, in that they share a strong dipolar coupling, *they still may not exhibit a large NOE* if the enhanced spin has other near neighbours.

A further important feature emerges if we consider the results from saturating B and studying the effect on A, $\eta_A\{B\}$ (Fig. 8.13) and compare these with those of Fig. 8.12. With C distant from B, the relaxation of A is essentially completely dominated by B so it experiences almost the maximum enhancement. As C approaches B the enhancement on A is reduced only a little since its relaxation is still dominated by the much closer spin B. Thus, in general, *steady-state NOEs between spins are not symmetrical,* that is, $\eta_A\{B\} \neq \eta_B\{A\}$, because the neighbours surrounding A are unlikely to match those surrounding B in both number and proximity.

Indirect effects and spin diffusion

The examples discussed above have been restricted to discussing the direct NOE effects between A and B whilst, for convenience, ignoring effects that may be observed at C itself. If spin C is considered, it is apparent from Fig. 8.14 that it experiences a net decrease in signal intensity when A is saturated. This arises from a relay mechanism in which the population changes on B, brought about by the initial A–B NOE, subsequently alters the population of spin C when this also shares a dipolar coupling and hence also cross-relaxes with B. The *negative* NOE seen at C is a result of the *increase* in the B-spin population differences generated by the A–B NOE. Recall that in the extreme narrowing limit, saturating a resonance (i.e. *decreasing* the population difference across the corresponding transition), causes a *positive* NOE so by the same logic an *increase* in population differences for B (the A–B NOE) will in turn generate a *negative* NOE on its dipolar coupled neighbours. This indirect effect, often referred to as the '*three-spin effect*' should not be confused with direct negative NOEs observed for slowly tumbling molecules.

The magnitudes of negative three-spin enhancements are usually rather small since they rely on the build-up of a sizeable NOE on a neighbouring spin suitable for relaying. Similarly, they also tend to be slow to develop and show

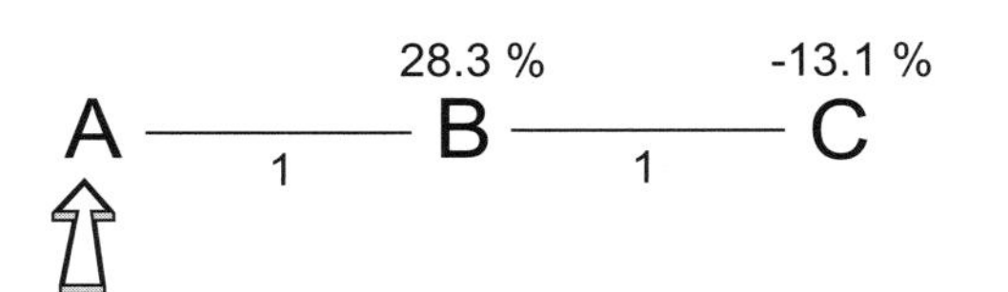

Figure 8.14. The three-spin steady-state effect. The negative enhancement at C arises from an indirect effect via spin B when spin A is saturated. The altered C-spin populations also contribute to the enhanced NOE at B (cf. Fig. 8.12).

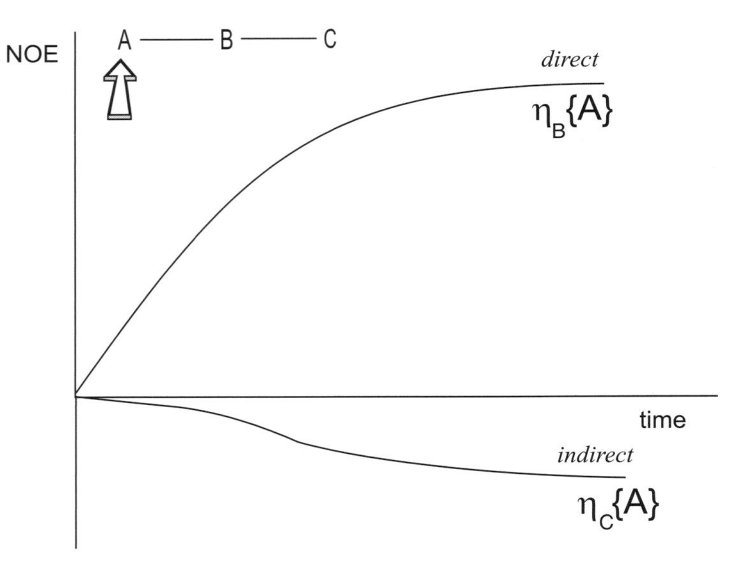

Figure 8.15. Schematic illustration of the build-up of direct and indirect NOEs in a rapidly tumbling three-spin system.

a characteristic lag period before appearing (Fig. 8.15), so tend to be observed only when longer presaturation periods are employed. The illustrations used above produce unrealistically high values for all enhancements since they assume pure dipolar relaxation throughout. Experimental NOEs are more often somewhat less then 20% and three-spin effects are rarely more than a few percent at most. They appear most commonly when B and C are a diastereotopic geminal pair, since the B–C distance is then constrained to be rather short so facilitating the relay. They are also favoured when the three spins have an approximately linear relationship and their appearance can be diagnostically useful when observed. The reason for this geometry being particularly favourable arises from the balance between the negative indirect three-spin effect on C and the positive direct effect between A and C (Fig. 8.16). At small A–B–C angles the direct effect dominates that relayed via B, whilst when linear the opposite applies. In between these extremes the two effects cancel so that *even though A and C may be close in space*, in that they share a strong dipolar coupling, *an NOE between them may be rather small*. This is another important point to be aware of (particularly when longer saturation times are employed) since in any realistic system there may well be a number of competing indirect pathways present. The curves of Fig. 8.16 demonstrate that the zero cross over point varies with internuclear separations and that the slopes of the curves are large at this point, so although the NOE

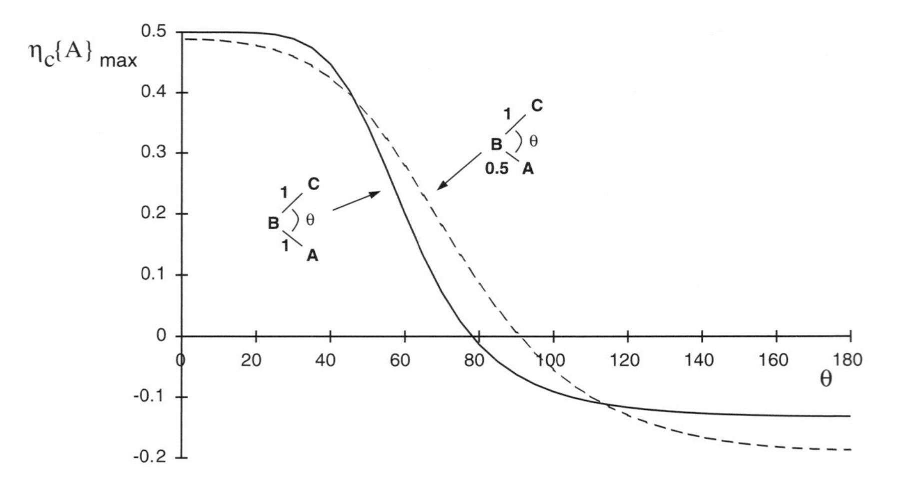

Figure 8.16. Calculated steady-state NOEs for spin C on saturation of spin A, as a function of the A–B–C angle in a rapidly tumbling isolated three-spin system. The resulting NOE is a balance between direct and indirect A–C effects and the two curves illustrate the dependence on relative internuclear separations.

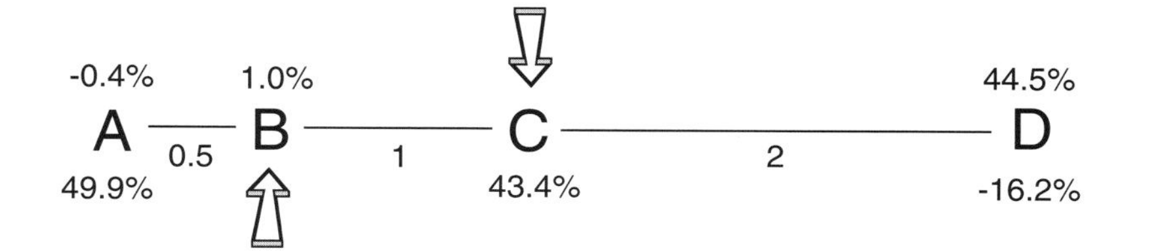

Figure 8.17. Calculated steady-state NOE enhancements in an isolated four spin system with the relative internuclear separations as shown. These enhancements were derived from the equations presented in table 3.1 of [1].

is unlikely to cancel to precisely zero, it may well be less than expected from simple geometrical considerations. Although in principle, 4-spin relay effects can be predicted to give rise to positive NOEs, they are very rarely observed in practice simply because they are so very weak.

The influences of the various factors described above are further illustrated by reference to the four-spin system of Fig. 8.17 which lays testament to the need to consider all neighbouring spin interactions to correctly interpret steady-state NOE data. Clearly the NOE enhancements between B and C differ dramatically despite these effects arising over identical internuclear separations. Furthermore, the enhancements B $\Rightarrow$ A, B $\Rightarrow$ C and C $\Rightarrow$ D are all rather similar despite there being a factor of four difference in distance between the largest and smallest separations. The essence of correctly applying steady-state NOE data is to collect a number of measurements and check for self-consistency within the proposed geometrical arrangement, rather than relying on a single NOE enhancement to provide an answer. If only a single irradiation were performed from C in Fig. 8.17 one might be forced into the erroneous conclusion that D were its closest partner rather than B. A more cautious approach, also measuring the NOEs from both B and D should lead one to question this conclusion. Again, such spectacular differences are unlikely to occur in reality largely because of the effects of the other numerous ρ_I^* relaxation sources we have chosen to ignore, but it should be clear that careful consideration is essential. Generally speaking, only in rather specific cases will a single NOE measurement lead to a definitive structural answer whereas a comparative study of many NOEs within the molecule is more likely to provide a conclusive and correct result.

When a molecule tumbles in solution so slowly that it exhibits negative NOEs, the consequences of indirect effects are more dramatic and far more problematic, so much so that steady-state NOEs become largely useless. The problem is essentially two-fold; firstly these have the same sign as direct effects so cannot be readily distinguished and secondly they grow rapidly and may attain very high intensities. The first of these is a consequence of the fact that a negative NOE (a population decrease) arises from saturation (a forced population decrease) so likewise the indirect effect on the third spin will also be negative, and so on. In the extreme case of extended saturation times, the NOE initially generated between two spins can spread throughout the whole molecule until all nuclei experience the same NOE enhancement (Fig. 8.18). This spreading of information throughout the molecule is often referred to as *spin-diffusion* for fairly obvious reasons, and may be likened to heat diffusing through a conductive solid. The limit of slow-tumbling is also referred to as the *spin-diffusion limit*. Because of this, steady-state NOEs in the negative NOE regime fail to provide reliable distance or proximity information. Instead, it becomes necessary to consider the *rate* at which NOEs grow between spins to glean distance information, dictating the use of kinetic measurements in the form of transient NOE experiments (Section 8.4).

Saturation transfer

Complications arising from the transfer of saturation from one resonance to another by means of chemical exchange (Fig. 8.19) also differ in the

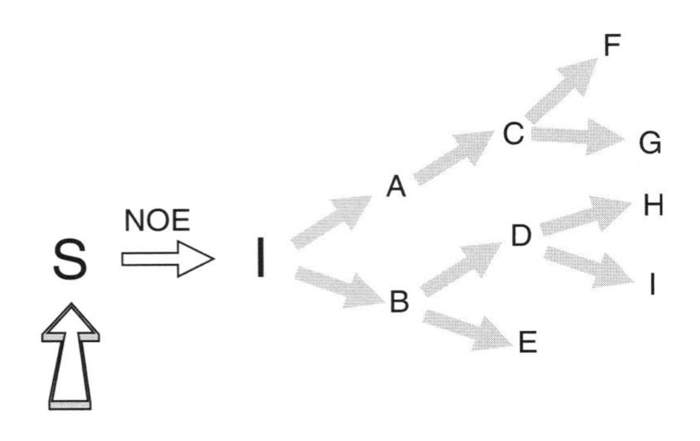

Figure 8.18. Schematic illustration of the spin-diffusion process in which the original S–I NOE is efficiently relayed onto neighbouring nuclei and propagated throughout the molecule.

two extreme motional regimes. Exchange peaks always display the same sign behaviour as the originally saturated resonance so have opposite sign to positive NOEs but the same as negative NOEs. The identification of responses arising from chemical exchange rather than the NOE is a further problem associated with the negative NOE regime; saturation transfer is considered further in the practical sections that follow.

8.3.3. Summary

The previous sections have covered the most important principles at the heart of the NOE. Despite having avoided much of the underlying mathematics, instead relying on pictorial models where possible, these sections are, necessarily, a little more technical than the other chapters in the book. The goal has been to present a reasonably thorough introduction to the NOE that is still compatible with the level of this text, rather than simply presenting the reader with statements of fact with little justification or support. This section reviews the key points that have been presented above, allowing the reader a more immediate reminder as to the most significant features of the NOE.

- The NOE is the change in intensity of the resonance of a nuclear spin, I ('interesting'), when the population differences across the transitions of a near neighbour, S ('source'), are perturbed from their equilibrium values, usually by *saturation* or by *population inversion*. It arises as the perturbed spin system alters its spin populations in an attempt to regain the equilibrium condition.
- *Steady-state* NOEs are those measured after a period of continuous S-spin saturation during which a new 'steady-state' equilibrium condition has developed for the I-spin populations. The NOE enhancement is usually denoted $\eta_I\{S\}$ and quoted as a percentage.

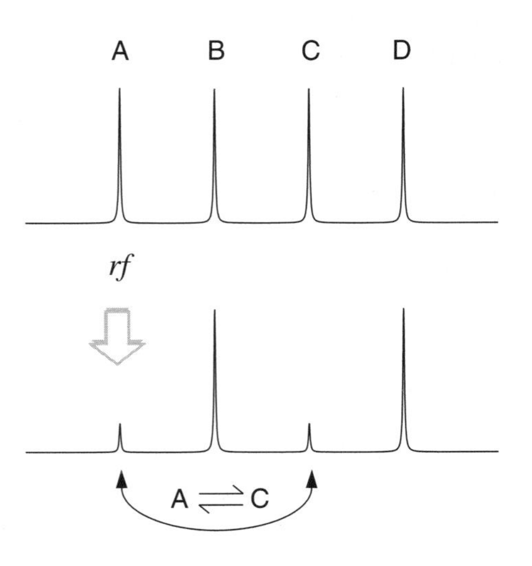

Figure 8.19. Schematic illustration of the saturation transfer process. Saturation of resonance A will lead to the simultaneous saturation of resonance C if nuclei A and C experience mutual chemical exchange during the saturation period.

- Steady-state NOEs can provide information on relative internuclear distances only, not on absolute measurements of internuclear separation.
- The NOE only arises between nuclei that share a mutual *dipolar coupling* (a direct, magnetic through space interaction) and thus relax each other via the dipole–dipole relaxation mechanism. Only this mechanism contributes to the nuclear spin population changes that produce the NOE, which is intimately related to longitudinal spin relaxation. The dependence on internuclear separation that makes the NOE so useful has its origins in the strength of dipolar coupling between two spins, this being inversely proportional to their separation, r_{IS}, (as r_{IS}^{-3}).
- Longitudinal spin relaxation requires a stimulus in the form of a magnetic field fluctuating at a frequency equal to the frequency (energy) of the transition. In the case of dipolar relaxation, this arises from the time-dependent field a spin experiences from its dipolar coupled neighbour as the molecule rotates or tumbles in solution. The NOE in turn is dependent upon the rates of molecular tumbling.
- The magnitude and sign of the NOE results from competition between various relaxation pathways. If the W_2 ($\alpha\alpha \leftrightarrow \beta\beta$) pathway dominates, positive NOEs are observed whereas a dominant W_0 ($\alpha\beta \leftrightarrow \beta\alpha$) pathway leads to the observation of negative NOEs. The W_2 and W_0 relaxation pathways (which involve the mutual flipping of the two spins) are collectively referred to as *cross-relaxation pathways* and it is only these that contribute to the generation of the NOE. All other contributions to relaxation, in the form of W_1 processes ($\alpha\alpha \leftrightarrow \beta\alpha$ and $\alpha\beta \leftrightarrow \beta\beta$, involving the flip of only a single spin), serve to reduce the overall magnitude of the effect, and are thus referred to as *leakage* contributions.
- Rapid molecular tumbling, corresponding to a short correlation time, τ_c, favours the higher energy W_2 process, so small molecules in low viscosity solvents display positive homonuclear NOEs (the *extreme narrowing limit*). In contrast, slow molecular tumbling (long correlation times) favours the lower-energy W_0 process meaning large molecules or smaller molecules in high viscosity solvents, display negative homonuclear NOEs (the *spin-diffusion limit*).
- The maximum possible positive NOE is $\gamma_S/2\gamma_I$, in other words 50% in a *homonuclear* system. Enhancements in *heteronuclear* systems can be far larger, for example 199% from ^{1}H to ^{13}C, and can serve as a useful source of sensitivity enhancement in the observation of the low-γ I-spin. They also become negative if one of the γs is negative, for example −494% from ^{1}H to ^{15}N, and may in some cases lead to severe signal reduction or even disappearance.
- The maximum possible negative NOE in a homonuclear system is −100%.
- Between the extreme narrowing and spin diffusion limits, lies the difficult region in which NOEs can become zero (when $\omega_0\tau_c \approx 1$) or at least rather weak, often demanding a change in experimental conditions or the use of rotating-frame NOE measurements.
- In an isolated homonuclear two-spin system in the extreme narrowing limit relaxing exclusively via the dipole–dipole mechanism, the steady-state NOE is predicted to be +50% and *independent of internuclear separation*, r_{IS}. However, the initial *rate* at which the NOE grows is proportional to r_{IS}^{-6}.
- In a more realistic multispin system, neighbouring nuclei, N, that are close to I can also contribute to its W_1 relaxation pathway. The magnitude of the NOE then becomes dependent on the I–S internuclear separation (inversely as r_{IS}^{-6}), but also has a dependence on the distance(s) between I and its near neighbour(s) (inversely as r_{IN}^{-6}), amongst other factors.
- The direct consequence of this is that to correctly interpret steady-state NOE data, it becomes essential to consider not only the I–S internuclear

separation, but also the proximity of all other nuclei (relaxation sources) surrounding I.

- This also means that steady-state NOEs are rarely symmetrical. That is, the NOE observed at spin A on saturating spin B, $\eta_A\{B\}$, is unlikely to equal the reverse measurement from saturating A and observing B, $\eta_B\{A\}$. This is a consequence of the fact that the neighbours surrounding A are unlikely to match those surrounding B in number and in distance.
- The relaxation of a spin with a very near neighbour will be dominated by this neighbour and as a consequence NOEs onto this spin from a more distant source spin will tend to be small. Conversely, nuclei that have no nearby neighbours will experience relaxation only from distant neighbours and as a result NOEs from these neighbours will be large despite the relatively large internuclear distance involved. Thus, it is to be expected that NOEs over similar distances onto a methylene and onto a methine proton will generally be somewhat smaller for the methylene proton since this will always have at least one near neighbour, its geminal partner.
- Longer-range NOEs build up only slowly, due to the r^{-6} rate dependence, so require long saturation periods before becoming appreciable.
- NOE enhancements may also be *relayed* on to neighbouring spins. For example, an NOE from A to B may be further passed onto a spin C that also cross-relaxes with B, such events being referred to as *indirect effects*. The properties of indirect enhancements differ markedly in the positive and negative NOE regimes.
- When direct (A–B) NOEs are positive, indirect effects at C are weak and negative, are favoured when the three spins have an approximately linear relationship (so may provide useful geometrical information), develop only slowly with a characteristic lag time and are thus also favoured by long presaturation periods. These are referred to as *three-spin effects*. Although further relays are theoretically possible, they are generally too weak to be observed.
- It is possible that positive direct effects and negative indirect effects can cancel or act to reduce the magnitude of the NOE. Thus, despite two spins being close, NOEs between them may be rather small or even negligible (particularly when longer saturation periods are employed).
- When direct (A–B) NOEs are negative, indirect effects are also negative so cannot be distinguished, they spread rapidly and have high intensities. This process is referred to as *spin diffusion* and is fatal for the steady-state NOE since it causes a loss of specificity and hence provides no information on molecular geometry. In this regime, it usually becomes necessary to use kinetic (transient) methods based on the measurement of NOE growth rates.
- Taking into account all the subtleties associated with the steady-state NOE presented above, it should be clear that it is unwise to place too much significance on the absolute magnitudes of steady-state NOE enhancements. In reality, differences of a few percent mean little when taken on their own, and it is generally necessary to consider a collection of enhancements when undertaking structural or conformational analysis to be certain of an unambiguous conclusion. A qualitative interpretation of many measurements is the most appropriate approach to interpreting steady-state NOE data.

8.3.4. Applications

To illustrate the issues described above and how NOE measurements can be used to provide unique data in structural analysis, some specific examples are now presented. As stated in the introduction, the NOE is most often employed during the later stages of structural investigations when the gross structure of the molecule has, at least to a large extent, been defined. For rather small molecules, this may be possible from knowledge of the chemistry used and from one-dimensional spectra, whilst larger or structurally complex molecules may demand the application of various correlation techniques before NOE experiments are considered. In either case, confidence in the accuracy of ones proton assignments is of paramount importance since errors in these are as likely (if not more likely) to lead to erroneous stereochemical conclusions being drawn than they are to being unmasked by the NOE studies themselves.

The examples presented here have all made use of the steady-state NOE difference method and are therefore restricted to 'small' molecules in which key resonances are sufficiently resolved for selective irradiation. This is often the case for many of the molecules routinely handled in the research laboratory when a judicious choice of solvent for optimum signal dispersion has been made. These also provide some indication of the magnitudes of NOE enhancements typically encountered in routine laboratory studies, which often fall short of the theoretical numbers discussed in the previous sections. Additional examples making use of other NOE techniques are presented in later sections.

E vs Z geometry

The differentiation of E and Z alkene isomers is often possible by direct measurement of vicinal proton–proton couplings across the unsaturation, whereby *cis* and *trans* couplings are usually sufficiently different to allow a distinction to be made (typically J_{cis} 7–11 Hz, J_{trans} 12–18 Hz). When only a single alkene proton exists this method can no longer be used and the NOE then offers an alternative approach provided a protonated group exists across the double bond. One such example is in the differentiation of the *E* and *Z* silanols **8.1a** and **8.1b** [3]. The *Z*-isomer was readily identified from the NOEs between the alkene proton and the CH_2Si group and further confirmation was provided by the observation of contrasting NOEs for the other (*E*-) isomer. Asymmetry in the magnitudes of NOE enhancements is also clearly apparent in this example, and should come as no surprise following the preceding discussions.

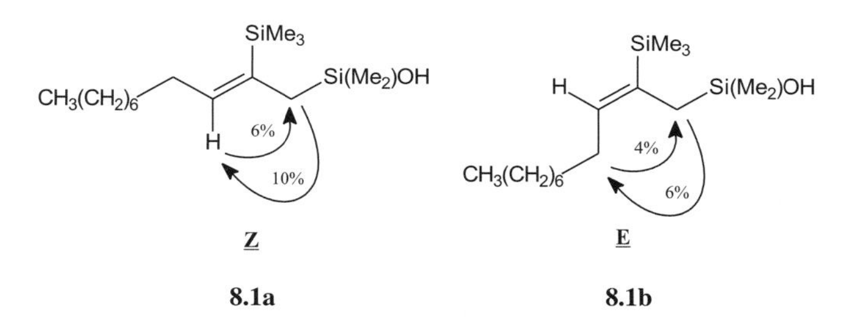

8.1a **8.1b**

Aromatic substitution position

Determining the position of substitution within a molecule can also be problematic when no direct proton–proton couplings exist to link the new moiety, which is often the case when the substitution is made on a heteroatom. Particularly when dealing with aromatic systems, the NOE can often provide unambiguous solutions. These systems are particularly favourable because they are restricted to being planar and hence it is usually safe to assume that nearest neighbours will be those on adjacent positions in the ring. The differentiation of π- and τ-substituted histidines [4] **8.2a** and **8.2b** provides a simple illustration of this. In the π-substituted systems irradiation of the H5 proton enhances only one of the CH_2 groups bound to the imidazole ring whereas in the τ-substituted isomers both were enhanced. Prior to the use of the NOE, differentiation was possible only through chemical degradation or through empirical rules based on differences in the small ($<$1.5 Hz) H2–H5 coupling constant, which could not always be resolved. An alternative approach to consider nowadays in such

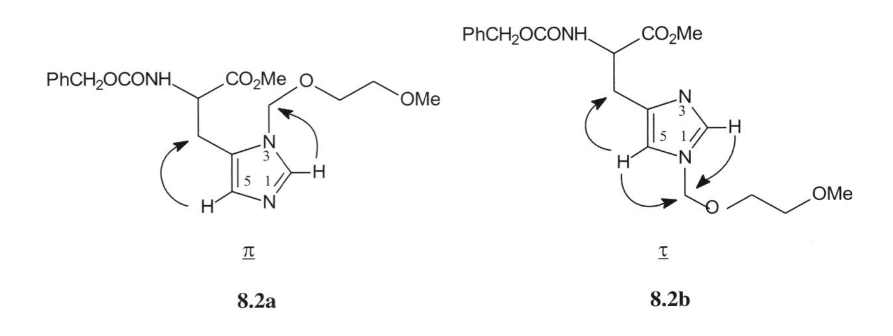

cases would be to establish connectivity by identifying long-range proton–carbon correlations across the heteroatom via the HMBC experiment (Section 6.4).

Substituent configuration

Another widely encountered question, and arguably the most common problem the NOE is used to address in synthetic chemistry, is the relative configuration of substituents on ring systems. This is most often applied to 5- and 6-membered rings, not only because of their ubiquity but also because larger rings tend to have far greater flexibility, making it more difficult to draw unambiguous conclusions. For six-membered rings in particular, direct analysis of proton couplings constants within the ring can in itself be informative because of the generally distinct differences in *axial–axial vs axial–equatorial/equatorial–equatorial* coupling constants in chair conformations (ax–ax ≈ 10–12 Hz, ax–eq/eq–eq ≈ 2–5 Hz; see Section 5.6.2 and Fig. 5.51 for example). However, in the determination of the orientation of the methyl group in **8.3** these would not have been informative, because of the similarity between *ax–eq* and *eq–eq* coupling constants. A collection of NOE enhancements was able to identify the methyl as occupying the equatorial position, in particular the 1,3-diaxial methine proton NOEs.

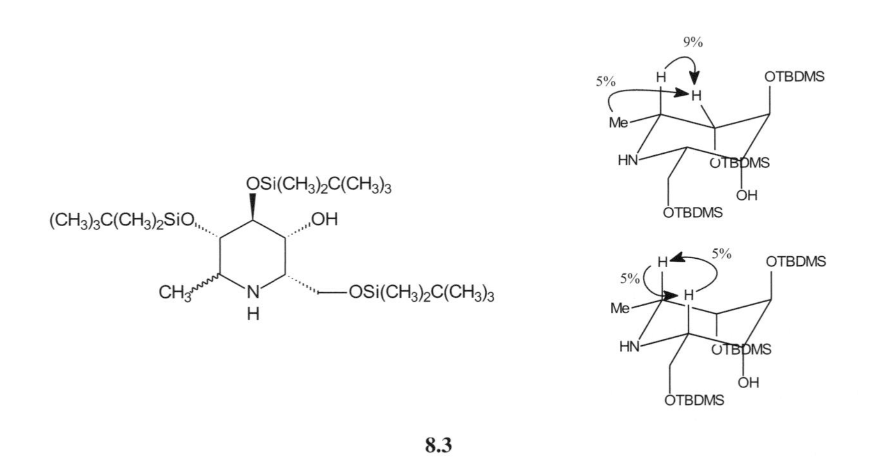

8.3

Vicinal couplings in five-membered rings generally offer greater ambiguity in defining relative configurations and here careful NOE measurements are also required since *cis*- and *trans* NOEs between adjacent protons are often of similar magnitude. In these cases it is wise to collect as many NOE enhancements as is possible for the sample and to ensure self-consistency over all of these within the proposed stereochemistry. These points are exemplified by the configurational assignment of the synthesised epoxyprolines [5] **8.4a** and **8.4b**. The negligible J coupling between H2 and H3 in **8.4a** tentatively suggested these protons to be *trans* and the slightly greater 2.5 Hz coupling

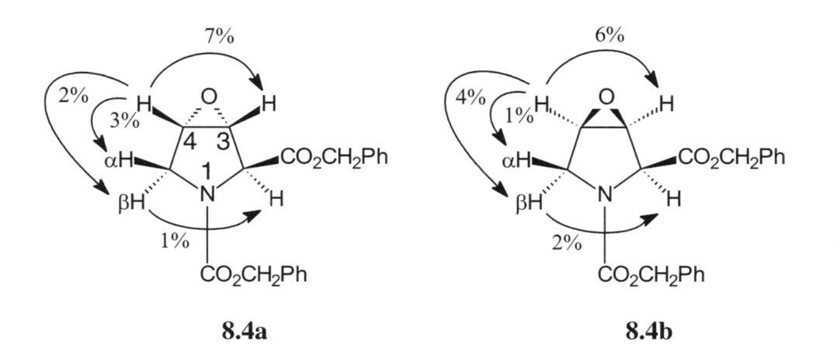

in **8.4b** suggested these share a *cis* relationship. Conclusions drawn from interpretation of the NOE data were consistent with this proposal, but were founded on the comparison of NOEs of rather similar magnitude between the H5 and H4 protons within each. The availability of both isomers allowed a comparison of their H4–H5 NOE patterns, based on identification of H5β from its NOE with H2, and the differences observed between these isomers provided further support for the assignments. Confirmation of these assignments were provided through additional synthetic structural correlations.

In the bicyclic lactam **8.5** it was possible to determine the relative configuration at three stereo-centres based on the known stereochemistry at only one (C_2) through the sequential interpretation of the NOE data. The lack of NOEs between protons on adjacent carbons C5, C6 and C7 suggests the neighbouring protons share *trans* relationships, although the absence of an NOE alone cannot always be considered definitive evidence, as has been stressed in previous discussions. More importantly, this stereochemistry is confirmed by the observation of the additional H6–H4, H6–H2 and H7–H9 NOEs. Again notice the asymmetry in the NOEs between H2–H4, H4–H6 and H4*–H5, with the NOE *onto* the methylene proton being always *less* than the NOE from this onto the methine proton, owing to the close proximity of the geminal neighbours.

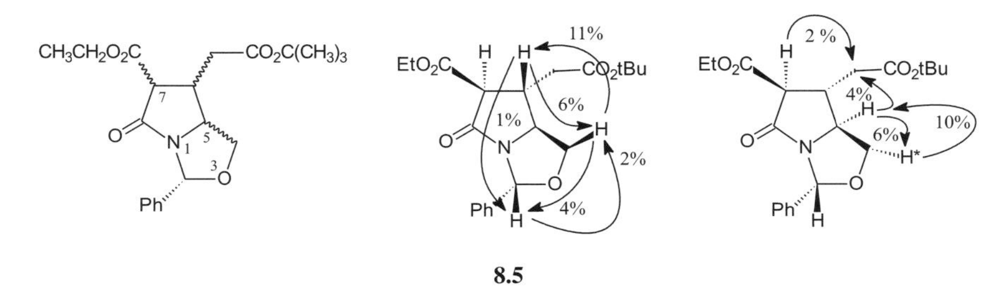

8.5

Resonance assignment

The prerequisite for most NOE studies of configuration or conformation is the assignment of the proton spectrum of the molecule, meaning each resonance can be associated with a unique proton within the gross structure. Such assignments are typically derived from 1D spectra and the various correlation methods described in previous chapters, although in some cases the NOE can itself be used for resonance assignment. It is most likely to be of use for assigning resonances of isolated groups that share no proton scalar couplings, although is by no means limited to this. An example is the assignment of the two methyl resonances in the hepato-protective agent andrographolide **8.6**, which are distant from one another in the molecule and show quite characteristic NOE patterns. The use of long-range heteronuclear correlation experiments should also be considered when addressing such problems, and indeed may be needed in the determination of the gross structure in the first place.

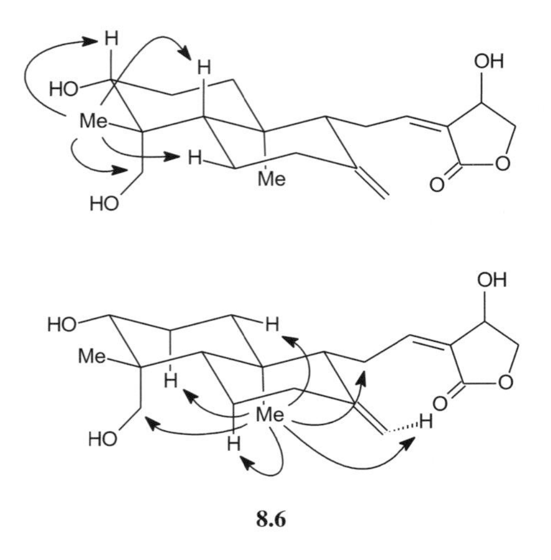

8.6

Endo vs exo adducts

The need to distinguish between *exo* and *endo* adducts in fused ring systems is another commonly encountered challenge that can often be addressed by the NOE, and the determination of the stereochemistry at C5 of the lactam **8.5** above can be viewed as one such example. In such cases, the NOE patterns observed between the proton(s) at the junction and those on the adjacent rings can often provide an unambiguous stereochemical assignment. This was the case for the identification of **8.7**, as the *endo* cycloadduct [6]. The material was synthesised as part of a model study toward the biomimetic synthesis of the manzamine alkaloids, a family of β-carboline alkaloids derived from marine sponges which possess potent antileukaemic and cytotoxic properties. In this, it was important to confirm whether the product stereochemistry was consistent with the proposed biosynthetic hypothesis being investigated, as was shown to be the case. Here the bridgehead proton was in fact too heavily overlapped with an adjacent resonance to be selectively saturated, although NOEs *onto* this were quite distinct. When bridgehead protons cannot be used at all, or if non-existent, the assignment must then rely on the observation of NOEs between ring protons on either side of the junction.

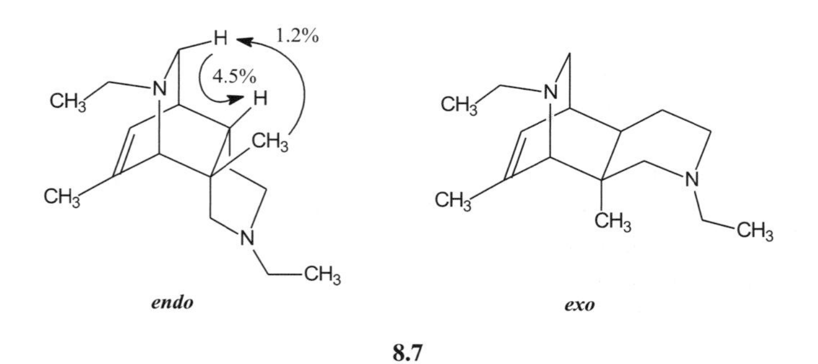

8.7

Conformational preference

The definition of a favoured conformation in small, flexible molecules by use of the NOE represents a far greater challenge, notably because of the rapid interchange between many possible conformations and because the NOE itself will represent only a weighted average of internuclear separations present within the conformers. The detailed investigations required to extract meaningful data in these cases is therefore rarely undertaken. However, when restricted conformational processes lead to one conformation being strongly favoured, sufficient NOE data may be available which allow this to be defined. Obvious examples might be the differentiation of chair and boat conformers of cyclohexanes or of slowly interconverting rotamers. Specific conformations may also be favoured in the presence of steric hindrances or strong hydrogen-bonding interactions, for example. An example of a favoured conformation defined by NOE measurements is structure **8.8** (also produced as part of the biomimetic synthesis of the manzamine alkaloids mentioned above [7]) in which the two heterocyclic rings are approximately orthogonal to one another. The limited and specific NOEs observed at the interface of the two rings are not consistent with free rotation about the single bond linking the rings and the structure appears essentially locked, presumably by the presence of the C8 methyl group. This orthogonal relationship is further supported by the almost negligible coupling between the H7 and H3 protons, indicating that the dihedral angle between them is close to 90°. The combined use of NOEs and coupling constants often represents the optimum approach to questions of conformation.

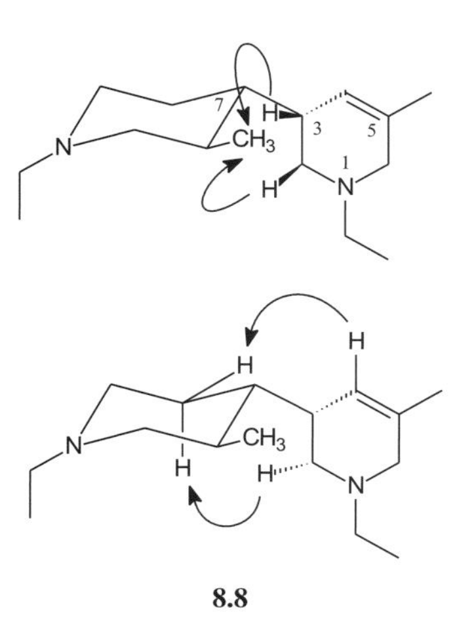

8.8

8.4. TRANSIENT NOES

It has been repeatedly stressed in the preceding sections that the steady-state NOEs measured between two nuclei cannot readily be translated into internuclear separations because they result from a balance between the influences of all neighbouring spins. At best, they provide information on *relative* internuclear distances only. It has also been stressed, however, that the *rate* at which the NOE grows towards this steady-state can be directly related to these distances under appropriate conditions. It has also been shown that for molecules which exhibit negative enhancements, steady-state measurements may fail to provide any reliable information of spatial proximity and here one is forced to consider the kinetics of the NOE. A logical approach to such measurements would be to follow that taken for the measurement of steady-state effects. Saturation of the target resonance for periods that are far less than those needed to reach the steady-state would allow *some* NOE to appear, which is then sampled. Repeating the experiment with progressively incremented saturation periods allows the build-up to be mapped. Owing to the use of shortened saturation periods, the enhancements observed with this method are termed *truncated driven NOEs* or *TOEs*.

Although once popular, this experimental approach is rather less used nowadays and as such shall be considered no further. The more common approach to obtaining kinetic data is to instantaneously perturb a spin system not by saturation but by inverting the target resonance(s) (that is, inverting the population differences across the corresponding transitions) and then allowing the NOE to develop in the absence of further external interference. The new populations are then sampled with a 90° pulse as usual (Fig. 8.20). In this case the NOE is seen initially to build for some time but ultimately fades away as spin relaxation restores the equilibrium condition; these enhancements are thus termed *transient NOEs*.

The measurement of transient NOEs gained widespread popularity, initially in the biochemical community, in the form of the 2D NOESY experiment, which remains an extremely important structural tool in this area and increasingly in the analysis of smaller molecules. More recently, the 1D transient

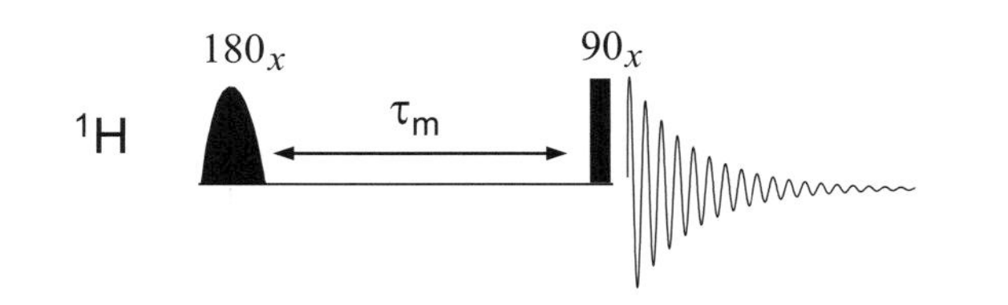

Figure 8.20. A general scheme for observing transient NOEs. Following inversion of a target (source) resonance, the NOE develops during the mixing time, τ_m, after which the system is sampled.

NOE experiment (i.e. 1D NOESY) has been repackaged and remarketed within the chemical community as a gradient-selected sequence capable of providing quite spectacular, high-quality spectra. In this form and as the 2D sequence, transient measurements are destined to be increasingly popular in routine chemical applications. In practice transient experiments, whether 1- or 2-D, are more routinely used qualitatively as 'single-shot' techniques, providing an overview of enhancements within a molecule rather then being employed to map the growth of the NOE. No matter how these are employed, it is necessary to understand something of the kinetics of the NOE to correctly execute and interpret these methods.

8.4.1. NOE kinetics

Following the (assumed) instantaneous inversion of the S-spin resonance, the *initial* growth rate of the NOE at I depends linearly on the cross-relaxation rate between these two spins *even in multispin systems*, such that:

$$\frac{dI_z}{dt} = 2\sigma_{IS}S_z^0 \tag{8.11}$$

(the factor of two here arises simply from the use of inversion of populations in this case rather than saturation, as considered previously). Using equations 8.4 and the approximation $\omega_I = \omega_S$ for a homonuclear system, the cross-relaxation rate $(W_2 - W_0)$ is given by:

$$\sigma_{IS} \propto \gamma^4 \left\{ \frac{6}{1 + 4\omega_0^2\tau_c^2} - 1 \right\} \frac{\tau_c}{r_{IS}^6} \tag{8.12}$$

Unlike the steady-state enhancements, the transient enhancements are influenced by only a single internuclear separation as r_{IS}^{-6}, whilst the so-called *initial rate approximation* is valid. In this the two cross-relaxing spins initially behave as if they were an isolated spin pair and the growth of the NOE has a linear dependence on mixing time. As longer mixing periods are used, the relaxation of spin I begins to compete with cross-relaxation between I and S, so the build-up curve deviates from linearity and the NOE eventually decays to zero (Fig. 8.21). Thus, for the initial rate approximation to be valid, mixing times significantly shorter than the T_1 relaxation time of spin I must be used. Only under these conditions is meaningful distance measurement possible.

If, on the other hand, the goal is to qualitatively identify through-space correlations, as is more often the case in routine work, mixing periods comparable to T_1 provide maximum enhancements. Since transient NOEs develop in the absence of an external rf field they tend to be weaker than steady-state

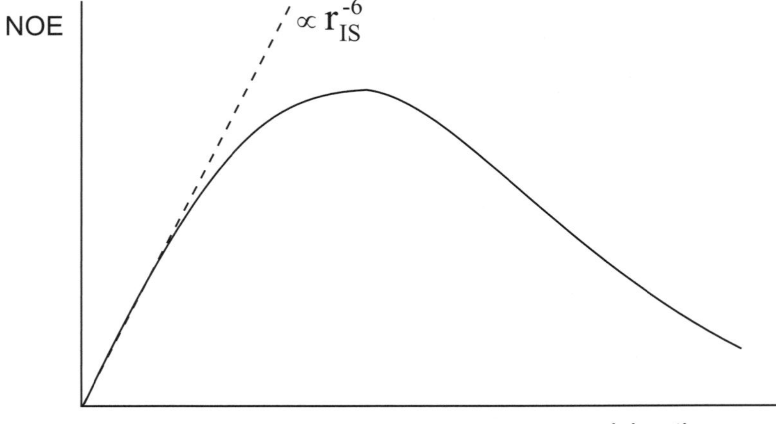

Figure 8.21. Schematic illustration of the development of the NOE between spins I and S as a function of mixing time.

effects. The maximum theoretical homonuclear enhancement is also reduced to 38% (from 50%) for positive NOEs, so careful choice of timing is crucial to the success of transient experiments. In all other respects, the dependence of transient NOEs on correlation times matches that of steady-state effects, with the maximum homonuclear enhancements for the negative NOE regime again being −100% owing to the domination of efficient cross-relaxation. When dealing with molecules which exist within this regime, it is also necessary to be aware of and if possible avoid potential complications arising from the dreaded spin diffusion (indeed, this is the primary reason why transient methods are used) which again demands the use of conservatively short mixing periods.

8.4.2. Measuring internuclear separations

Assuming the initial rate approximation to be valid (the NOE growth linear), the magnitude of an enhancement between two spins, A and B, after a period τ will be proportional to the cross-relaxation rate, which in turn depends on r_{AB}^{-6}:

$$\eta_A\{B\} = k\sigma_{AB}\tau = k' r_{AB}^{-6}\tau \qquad (8.13)$$

The constants of proportionality here contain the molecular correlation time, τ_c, in addition to a number of known physical constants, and {B} is now taken to signify inversion rather than saturation of B. In principle, if τ_c were known, this would directly provide a measure of r_{AB}. Whilst it is possible to determine this (such as from relaxation time measurements) this is rarely done in practice, and it is more common to use a known internal distance as a reference and avoid the need for such laborious measurements. If the NOE between reference nuclei X and Y of internuclear separation r_{XY} is also measured then:

$$\frac{\eta_A\{B\}}{\eta_X\{Y\}} = \frac{r_{AB}^{-6}}{r_{XY}^{-6}} \qquad (8.14)$$

A direct comparison of the two NOE intensities thus provides the unknown internuclear distance. This simple relationship has been extensively used to provide measurements of internuclear separations, particularly in biological macromolecules. From a single experiment, distances can be estimated *assuming the initial rate approximation is valid for all interactions*. This relies on all internuclear vectors in question possessing the same correlation time, which may not be the case where internal motion is present. A consideration of these matter lies beyond the scope of this work; further details may be found in reference [1].

The significance of a single 'internuclear distance' must also be considered carefully. In reality, internuclear separations vary over time with conformational averaging so the concept of a single distance is simply a convenient model of events. Furthermore, in cases of conformational exchange, the calculated distance tends to be heavily weighted toward shorter separations since the NOE is very much more intense for these because of the r^{-6} factor. Consider the case of rapid averaging on the NMR timescale between two equally populated conformers such that only a single resonance is observed for each chemically distinct site. In one conformer the separation between two spins is 0.25 nm whilst in the other it is 0.60 nm (Fig. 8.22). The NOE under conditions of fast conformational exchange averages as $\langle r^{-6}\rangle^{-1/6}$ (where $\langle\ldots\rangle$ indicates the mean value) so that the *apparent* separation as would be calculated from observed NOE intensities is 0.28 nm rather than the mean 0.42 nm one may anticipate.

In general then, NOE measurements will tend to underestimate rather then overestimate distances in such cases, which can be problematic for structure calculations. These problems are most severe in the case of small molecules, where extensive conformational averaging is to be expected, and detailed structure calculations based on *quantitative* distance measurements for flexible

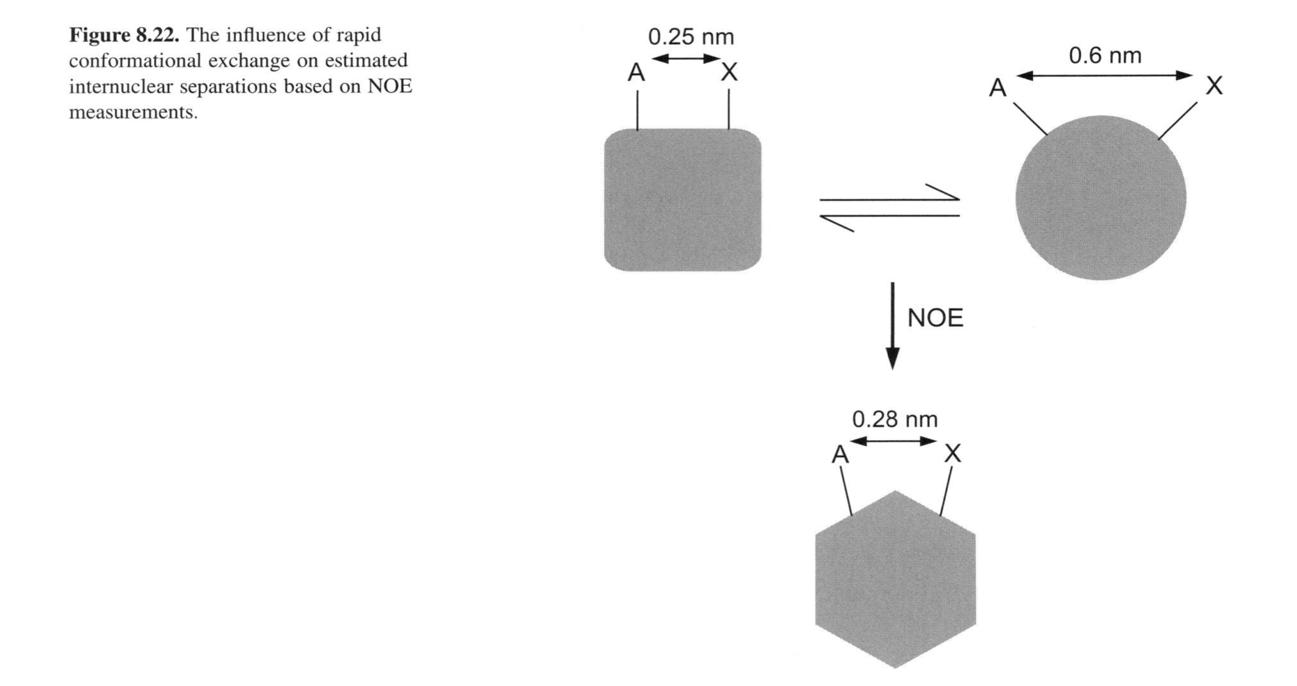

Figure 8.22. The influence of rapid conformational exchange on estimated internuclear separations based on NOE measurements.

small molecules are rather rare. However, this approach has proved enormously successful in the structure calculations of biological macromolecules where, to avoid problems of averaging and errors in intensity measures, semi-quantitative measurements are usually employed (Section 8.7).

8.5. ROTATING-FRAME NOES

The greatest problem associated with the methods described so far is clearly the 'zero-crossing' region around $\omega_0\tau_c \approx 1$ where the conventional (laboratory-frame) NOE observed via steady-state or transient techniques becomes vanishingly small. This typically occurs for mid-sized molecules with masses of around 1000–2000 daltons, depending on solution conditions and spectrometer frequency. With the increasing interest in larger molecules in many areas of organic chemistry research coupled with the wider availability of higher-field instruments, this is likely to be a region visited ever more frequently by the research chemists' molecules. Other than altering solution conditions in an attempt to escape from this, the measurement of NOEs in the *rotating-frame* provides an alternative solution, albeit an experimentally challenging one. These effects are the rotating-frame analogues of the transient NOEs described above and many of the discussions in Section 8.4 relating to their application are relevant here also. In this case, however, the cross-relaxation rate between homonuclear spins is given by:

$$\sigma_{IS} \propto \gamma^4 \left\{ \frac{3}{1+\omega_0^2\tau_c^2} + 2 \right\} \frac{\tau_c}{r_{IS}^6} \tag{8.15}$$

Unlike the corresponding equation for transient NOEs (8.12 above) this expression remains positive for all values of τ_c and the undeniable benefit of rotating-frame NOEs (ROEs) is, quite simply, that they remain positive for all realistic molecular tumbling rates. For small molecules, the magnitude of the ROE matches that of the transient NOE, whilst for larger molecules it reaches a maximum for homonuclear spins of 68%, but under no circumstances does it become zero (Fig. 8.23). Similarly, the NOE and ROE growth rates are identical

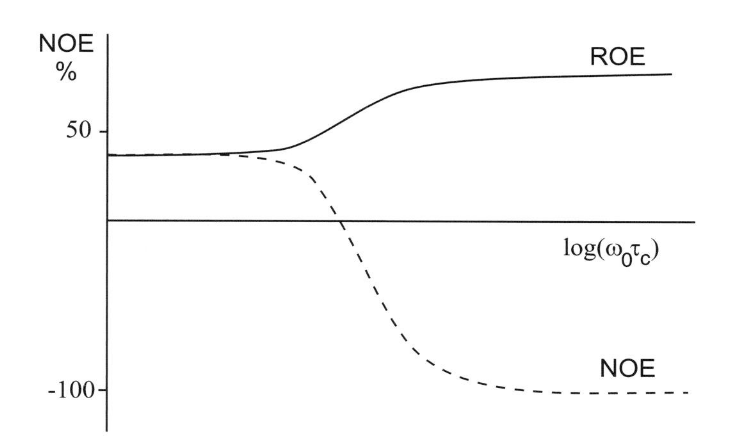

Figure 8.23. Schematic illustration of the dependence of the ROE and transient NOE for an isolated homonuclear two-spin system as a function of molecular tumbling rates.

for small molecules but differ for very large ones. For a small molecule which has $\omega_0\tau_c \ll 1$ both equations 8.12 and 8.15 simplify to:

$$\sigma_{IS} \propto \frac{5\gamma^4\tau_c}{r_{IS}^6} \tag{8.16}$$

In contrast, for very large molecules which have $\omega_0\tau_c \gg 1$, the left hand expression within the brackets of both equations 8.12 and 8.15 becomes negligible, giving:

$$\sigma_{IS}^{NOE} \propto \frac{-\gamma^4\tau_c}{r_{IS}^6} \quad \text{and} \quad \sigma_{IS}^{ROE} \propto \frac{2\gamma^4\tau_c}{r_{IS}^6} \tag{8.17}$$

and hence:

$$\sigma_{IS}^{ROE} = -2\sigma_{IS}^{NOE} \tag{8.18}$$

For very large molecules, the ROE therefore grows twice as fast as the NOE, and has opposite sign [8].

The measurement of ROEs requires a somewhat different experimental approach (Section 8.8). In essence, ROEs develop whilst magnetisation is held static in the *transverse* plane, rather than along the *longitudinal* axis (hence they are sometimes also referred to as transverse NOEs). To generate the required population disturbance of the source spins, the target resonance is subjected to a *selective* 180° pulse prior to the non-selective 90° pulse, such that it experiences a net 270° flip and is thus inverted relative to all others. Transverse magnetisation is then 'frozen' in the rotating-frame by the application of a continuous, low-power *spin-lock* pulse. This is analogous to the spin-lock described in Section 5.7 for the TOCSY experiment and serves the same purpose, that is, to prevent evolution (in the rotating frame) of chemical shifts. The simplest scheme for a 1D sequence is therefore that of Fig. 8.24. The experiment is more frequently performed as the 2D experiment where it is usually termed ROESY (rotating-frame NOE spectroscopy).

The situation during the spin-lock may be viewed as the transverse equivalent of events during the transient NOE mixing time (Fig. 8.25). The action of the spin-lock is to maintain the opposing disposition of magnetisation vectors which would otherwise be lost through differential chemical shift evolution,

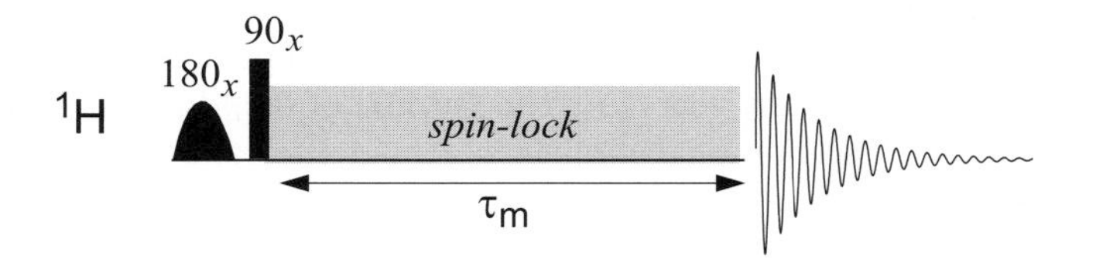

Figure 8.24. A general scheme for observing rotating-frame NOEs. The ROE develops during the long spin-lock pulse which constitutes the mixing period, τ_m.

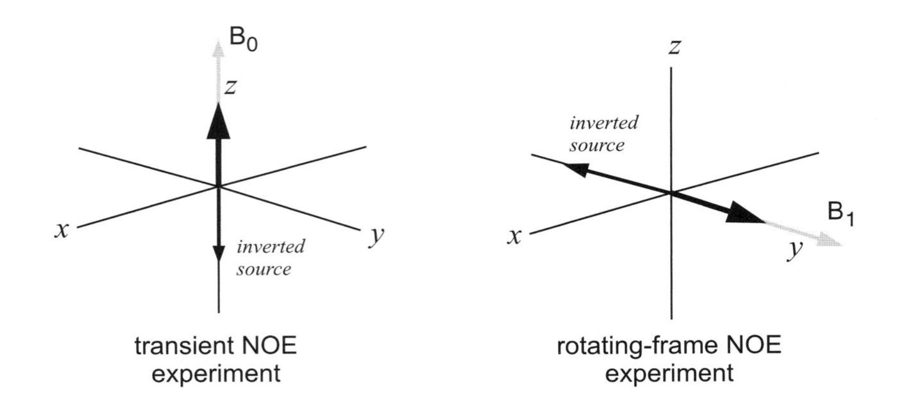

Figure 8.25. The rotating-frame NOE experiment can be viewed as the transverse equivalent of the transient NOE experiment.

and so allows the ROE to develop through cross-relaxation in the transverse plane. Spin relaxation here is characterised by the time constant $T_{1\rho}$ ($\cong T_2$). In utilising the spin-lock one has effectively replaced the static B_0 field of the conventional NOE with the far smaller rf B_1 field and it is this that changes the dynamics of the NOE. Whereas γB_0 typically corresponds to frequencies of hundreds of megahertz, γB_1 is typically only a few kilohertz, meaning $\gamma B_1 \ll \gamma B_0$ and hence ω_1 (the rotating-frame frequencies) $\ll \omega_0$. The consequence of this is that $\omega_1 \tau_c \ll 1$ for all realistic values of τ_c, and *all molecules behave as if they are within the extreme narrowing limit*. Thus, ROEs are positive, any indirect effects have opposite sign to direct effects and tend to be weak, and saturation transfer can be distinguished by sign from ROEs, regardless of molecular size and dynamics.

Set against these obvious benefits are a number of experimental problems, principally TOCSY transfers also occurring during the spin-lock and signal attenuation from off-resonance effects. These issues are further addressed in the practical sections that follow, so it is sufficient to note here that even more care is required when acquiring and using ROEs than is needed for NOEs, so much so that some would regard this only as a specialist's technique.

Part II: Practical aspects

8.6. MEASURING STEADY-STATE NOES: NOE DIFFERENCE

The basic requirement for the observation of steady-state NOEs is a suitable period of presaturation of the target resonance prior to acquisition of the spectrum, as discussed above and indicated in Fig. 8.1. This dictates that steady-state measurements are derived only from one-dimensional spectra, in which spins experiencing the NOE will produce resonances with altered intensities relative to the conventional 1D spectrum. The question then remains as to how best to display and measure these changes. Although, in principle, this may be achieved by direct integration of resonances relative to a control spectrum, in reality this is a non-trivial exercise when enhancements may only be of a few percent. Instead, the universal approach is to use 'difference spectroscopy' and to subtract the control spectrum from the NOE spectrum, yielding a difference spectrum in which, ideally, the only remaining signals are the NOE enhancements and the saturated resonance (Fig. 8.26). This is the so-called NOE difference experiment that has played a pivotal role in structural organic chemistry for many years [9]. The purpose of the 'difference' approach is to make the *observation* of the enhancements *easier* and *more reliable* but imparts no new information to the difference spectrum that was not within the original NOE spectrum.

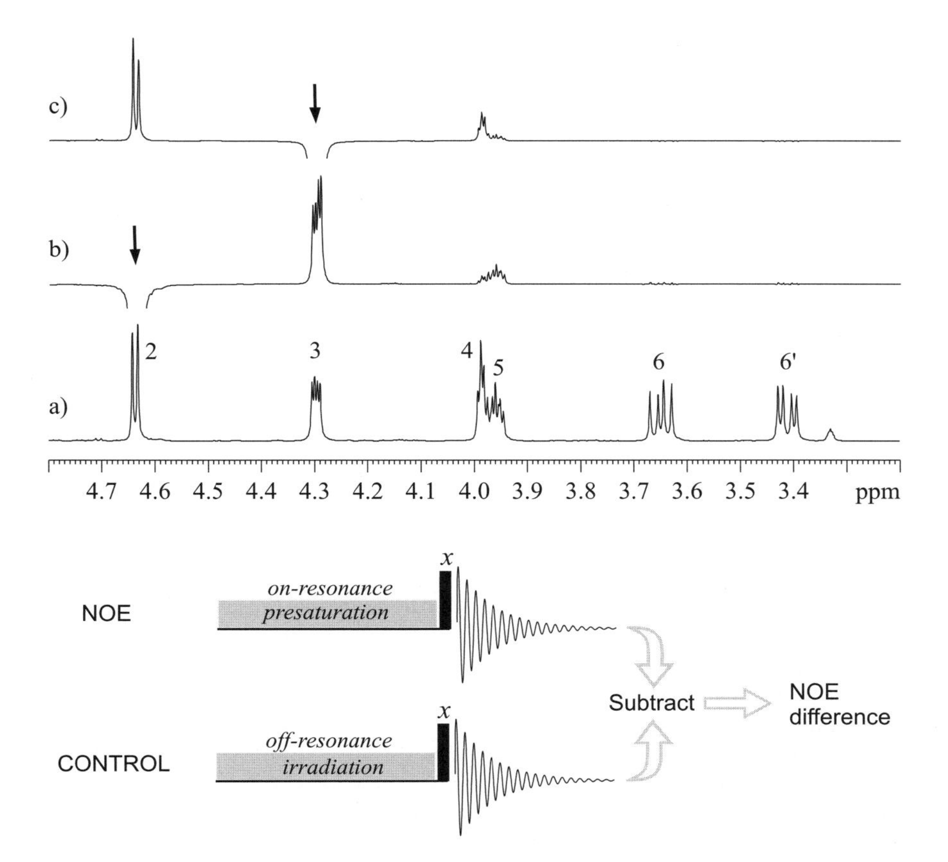

Figure 8.26. The control 1D spectrum (a) and NOE difference spectra (b and c) of **8.9** in MeOD. The difference spectra show only NOE enhancements and the truncated difference signal of the saturated resonance (arrowed). The observed enhancements are consistent with the indicated 2,5-*cis* geometry across the ring oxygen of **8.9**.

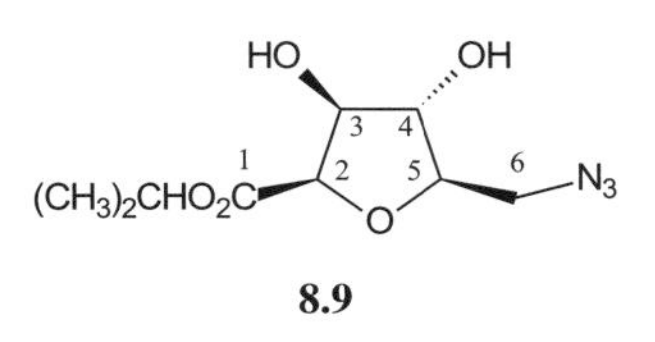

8.9

Figure 8.27. The procedure for generating the NOE difference spectrum.

The NOE difference experiment is illustrated schematically in Fig. 8.27. The NOE spectrum is generated by applying presaturation to the target resonance for a period τ, after which the presaturating rf [1] is gated off and the 1D spectrum acquired with a 90° pulse. The control spectrum is acquired in an *identical* fashion except that presaturation is no longer required. To keep the acquisition conditions for both experiments as similar as possible, which is crucial for a successful difference experiment, the presaturation frequency is moved well away from all resonances (it is placed 'off-resonance') for the control rather than being turned off altogether, typically by placing it at the far edge of the spectrum. Subtracting the resulting spectra (or FIDs followed by FT) yields the difference spectrum. The success of this approach is critically dependent on the two spectra being identical in all respects other than those features introduced by the on-resonance presaturation; if this is not the case spurious difference responses are introduced (see below). In practice, the perfect subtraction of resonances is experimentally very demanding so various procedures have been developed to minimise undesirable variations and hence artefacts.

8.6.1. Optimising difference experiments

Minimising subtraction artefacts

NOE experiments typically require significant spectrometer time because the enhancements being sought are rather small. The NOE difference experiment represents a stringent test of both short- and long-term spectrometer stability, since changes in rf phase or frequency or in magnetic field contribute to

[1] Historically, the presaturating rf as been applied via the 'decoupler channel', traditionally the second rf channel of the spectrometer. However, decoupling is a misleading term in the context of the NOE since spectra are acquired fully J-coupled, hence the term 'presaturating rf' is used throughout

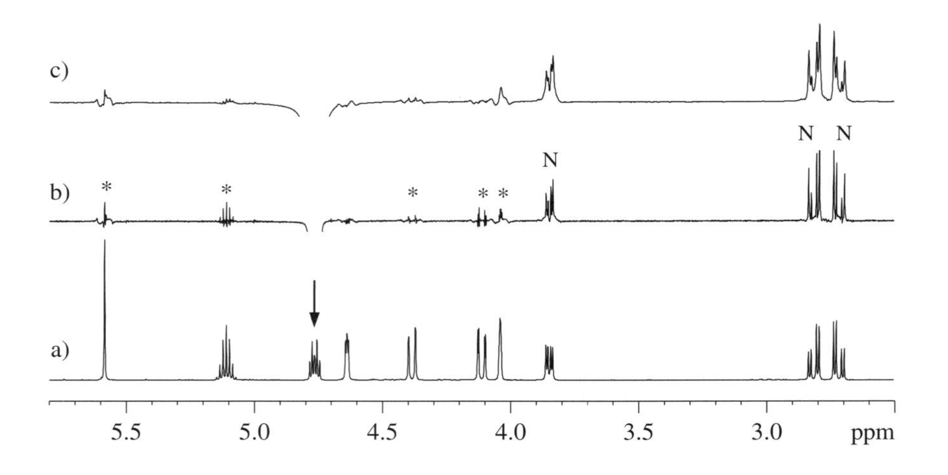

Figure 8.28. NOE difference artefacts shown by an asterisk in the difference spectrum (b), arise when two similar, but not quite identical, signals are subtracted in an attempt to reveal the genuine NOEs (N). The artefacts can be attenuated a little through the use of line-broadening window functions (c, lb = 1 Hz).

difference artefacts (Fig. 8.28). These appear as dispersion-like signals which may mask genuine NOEs, in addition to being unsightly. Improvements in spectrometer design have gone a long way to reducing these instabilities and make the observation of weaker NOEs ever more reliable as artefacts are reduced. Nevertheless, care is still required to achieve optimum results.

The first consideration is a stable sample environment, including sample temperature, a point of particular concern for aqueous solutions and for solutes with temperature dependent shifts. Activity in the vicinity of the magnet has long been considered a contributing factor to poorer difference spectroscopy. How significant this is will depend on the physical location of the instrument, the field strength and on activities in adjacent labs or corridors (including those above and below) but in any case it is clearly wise to minimise disturbances of the field. Acquiring data overnight or at weekends may prove beneficial if such interferences prove to be detrimental.

Short-term random instrumental instabilities are suppressed by acquiring a large number of scans, just as random noise is similarly reduced by signal averaging, and the improved signal-to-noise in the resulting spectra also aids the reliable identification of enhancements. Longer-term instabilities, such as from small field or temperature drifts, are addressed by interleaving acquisitions between the control and the NOE experiments so that over the course of the experiment all spectra experience the same net variation. A typical approach when multiple NOE measurements are required of a sample is to acquire 8 or 16 transients for each experiment and cycle round all irradiation frequencies (usually automatically under computer control). For further signal averaging, the appropriate data sets are co-added to acquire the desired total number of transients which must be identical for each data set collected. For multiple NOE experiments a single control is usually sufficient for all difference calculations. Also included in each acquisition must be 2 or 4 'dummy scans' (in which the recorded data are discarded) to ensure the complete decay of saturation effects from the previous irradiation. Since presaturation is required for each acquisition, the usual relaxation delay between transients is not necessary and is wholly replaced by the presaturation period.

Subtraction artefacts can be further reduced by suitable processing of the data. Difference spectra can be generated by subtracting two FIDs (the NOE and the control) and Fourier transforming the resulting FID, or alternatively by directly subtracting the NOE and the control spectra, in which case they must both be processed with *digitally identical* phase corrections prior to the subtraction. The results of these methods are equivalent except when spectra exhibit a large dynamic range (that is, where very large signals exist in the presence of very small ones) in which case subtracting FIDs may reduce spectrum noise [10]. In either approach, applying mild line-broadening to the spectra, typically of a few Hz, helps reduce difference artefacts (Fig. 8.28c). The rationale behind this is that intensity of the residual signal obtained by subtracting closely overlapping Lorentzian lineshapes is inversely proportional to the resonance linewidths [11].

Optimising presaturation

Selection of the optimum presaturation time τ is highly dependent on the information required of the molecule and on its size and structure, so general guidelines only can be presented here. Since steady-state measurements are only of use in the positive NOE regime, these considerations are limited to small to medium-sized molecules and, in fact, most of what we need to know has already been presented in the previous sections since the choice of τ is dictated by the kinetics of the NOE. Thus, if measurements of genuine quantitative steady-state enhancements are required, then τ must be greater than $5T_1$ of the slowest relaxing spins, which could make signal averaging very time consuming. However, most structural work does not require the full steady-state to be attained so long as a measurable NOE has been able to develop, allowing more transients to be collected in the time available, which itself facilitates the observation of the NOEs. The use of shorter presaturation times then favours the observation of short-range interactions since these are quickest to develop. In contrast, long-range interactions and indirect effects are slow to build and are enhanced with extended presaturation. For the majority of structural studies, information on immediate neighbours is sufficient so relatively short presaturation times of around 4 or 5 seconds (ca. 3 times the longest T_1) tend to be used in routine NOE investigations of small organic molecules ($M_r < 500$). Under such conditions, one should not be surprised to find that longer-range interactions cannot be observed, and if these are expected to be informative, or if indirect effects may provide useful geometrical information, further experiments with extended presaturation ($\gg 5T_1$, possibly tens of seconds) will be required in order to detect these. Clearly it is advantageous to have some prior knowledge of the approximate T_1 values of protons in the molecule, and these may be measured by the quick inversion-recovery method of Section 2.4. This additionally provides supporting data since remote protons that are isolated from dipolar relaxation sources will exhibit unusually long T_1s as well as slow to develop NOEs.

Selective saturation and SPT

To ensure the integrity of NOE data it is essential that only a single resonance is subject to presaturation at any one time. Even a small degree of saturation of a neighbouring resonance caused by 'spillover' of the presaturating rf can be detrimental for structural studies. Such spillover can usually be readily observed in the difference spectrum and should call for great caution when interpreting data. Naturally it is more useful to spot this failing before the experiment is left running for many hours and to take measures to alleviate it. An effective approach to this to directly overlay the NOE and control spectra obtained after the first experiment cycle (e.g. after 8 transients for each) using the spectrometer dual-display mode. Whilst NOE enhancements are unlikely to be apparent at this stage (unless very large), saturation effects close to the target resonance are usually clear and may suggest changes to the experimental set-up are required.

Ideally, each target resonance should be well removed from all others but this criterion cannot always be met. Changes of solvent can be useful here and can sometimes lead to dramatic shifts of resonances which may fortuitously place those of interest in a more exposed position. Beyond this, a variety of experimental procedures can help. The most direct way to reduce the

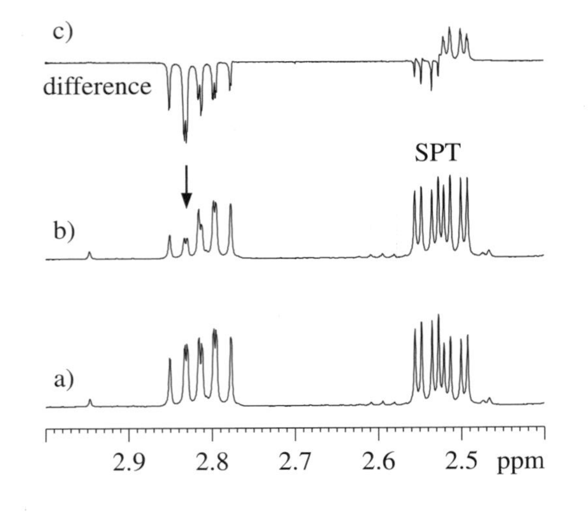

Figure 8.29. SPT artefacts. Unequal saturation of a resonance (b) can cause SPT intensity distortions to appear at its J-coupled neighbours. These appear as antiphase patterns in the difference spectrum b–a (c) which have zero net integral but may mask genuine NOEs.

frequency window over which the rf is effective and hence improve selectivity is to attenuate its power. This also reduces the degree of saturation of the target resonance which in turn reduces the *absolute* magnitude of the NOE enhancement, so a compromise must be sought between the two, usually with the emphasis on selectivity. Additional artefacts may also arise when using low presaturation powers if multiplets are subject to *unequal* perturbation. This arises from so-called *selective population transfer* (SPT) which has already been described in Section 4.4. This is a manifestation of polarisation transfer between J-coupled spins, and is related to the process by which crosspeaks are generated in COSY spectra. The responses appear as *antiphase* multiplets for those spins J-coupled to the saturated spin (Fig. 8.29). The integrated intensity of such antiphase lines is zero if correctly phased, so should not interfere with NOE quantification, although the potentially intense SPT responses could be distracting and may mask genuine NOE responses, so are well worth suppressing.

One approach to achieving even saturation of a multiplet when using weak rf powers to maintain selectivity is to cycle the presaturation frequency between individual lines *within* each multiplet [12]. The process involves irradiating each line for a short period in turn, then repeating the sequence a number of times to achieve the desired total presaturation period. For each irradiation the aim is to saturate only a single line rather than the whole multiplet, so considerably lower rf powers may be used, whilst the cycling ensures approximately equal suppression across the whole multiplet. An important consideration here is the saturation period used for each line; too short and the saturation may be ineffectual (and unwelcome frequency modulation artefacts introduced) whilst too long a period will enable relaxation to become effective leading to uneven saturation. The relaxation behaviour of the spins is again important, and in general periods of 50–300 ms work well for most small molecules; again trial-and-error is the best approach to optimisation (Fig. 8.30). Thus, a five-line multiplet with 200 ms saturation of each line requires 5 cycles to achieve a total presaturation period of 5 s. Such sequences can be readily programmed to operate automatically on modern spectrometers.

An additional approach to suppressing SPT distortions is to collect spectra with an *exact* 90° observation pulse. This effectively spreads saturation evenly throughout all multiplet components and so removes the source of these distortions. The accuracy of the 90° pulse can be improved by the use of a composite pulse (Section 9.1) for which the $270_x 360_{-x} 90_y$ sequence has been suggested [13]. The 90° acquisition pulse also leads to maximum signal and since long presaturation periods are used between transients, this represents a

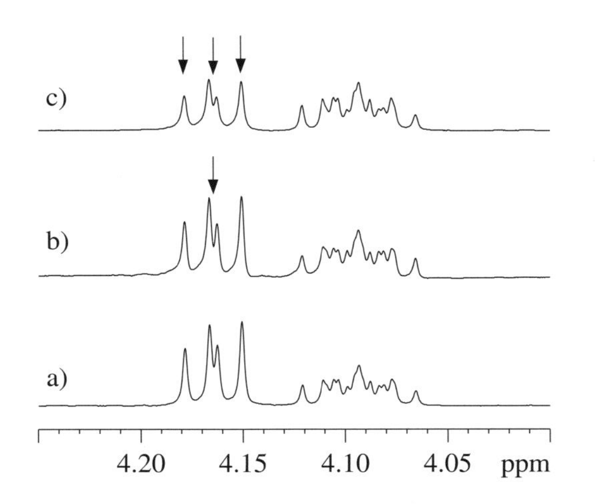

Figure 8.30. Selective multiplet saturation through frequency cycling. (a) Control spectrum, (b) single frequency, low-power presaturation applied to the centre of the multiplet and (c) presaturation using the same power but with frequency cycling over the four lines. The low power ensures the neighbouring multiplet is untouched.

suitable choice for optimum sensitivity and is standard for the NOE difference method. In situations where a multiplet is overlapped by another resonance, the frequency cycling method is unsuitable for uniform multiplet suppression but the use of an accurate 90° pulse means only a single line from the multiplet need be available for saturation for the experiment to work, although the absolute magnitude of the NOE(s) will be small in such cases because of the small degree of population perturbation. In cases of problematic overlap, the alternative is to employ the 2D NOESY experiment described below in which selective irradiation is avoided altogether.

Saturation transfer

An additional mechanism by which saturation of a resonance can occur, other than by direct irradiation, is *saturation transfer* brought about by chemical exchange processes that are slow on the NMR chemical shift timescale. This may be, for example, conformational exchange whereby saturation of a spin transition in one conformer will progressively lead to the saturation of the corresponding spin in another if exchange occurs during the presaturation period (see Fig. 8.19 above). This may be a bonus in some circumstances since it may allow the indirect saturation of an otherwise obscure resonance or it may be studied quite separately as a means of analysing exchange dynamics. In terms of NOE measurements it is more often an additional complication to be aware of since the observed NOEs then result from the saturation of more than one resonance. Similarly, any NOEs that develop may also be transferred to other resonances by the exchange, further complicating matters. Exchange processes are generally more efficient than cross-relaxation, and signals arising from exchange are usually more intense than NOEs and exhibit the same sign behaviour as the originally saturated resonance. Hence they appear with opposite sign to positive NOE enhancements, allowing them to be distinguished, but with the same sign as negative enhancements. This is a further pitfall of working in the negative NOE regime no matter which experimental approach is adopted.

A relatively common feature of NOE experiments performed in organic solvents containing traces of water is transfer of saturation between exchangeable protons and the water (Fig. 8.31). Often these types of resonances can be rather broad and may be unwittingly saturated if they happen to spread under neighbouring signals, which in turn may produce unexpected negative responses elsewhere in difference spectra as a result of exchange. This effect can also be used to identify the resonances of exchangeable protons in a molecule by saturation of the water resonance, as was used in the structure determination of the antibiotic Pulvomycin **8.11** [14] (Fig. 8.32).

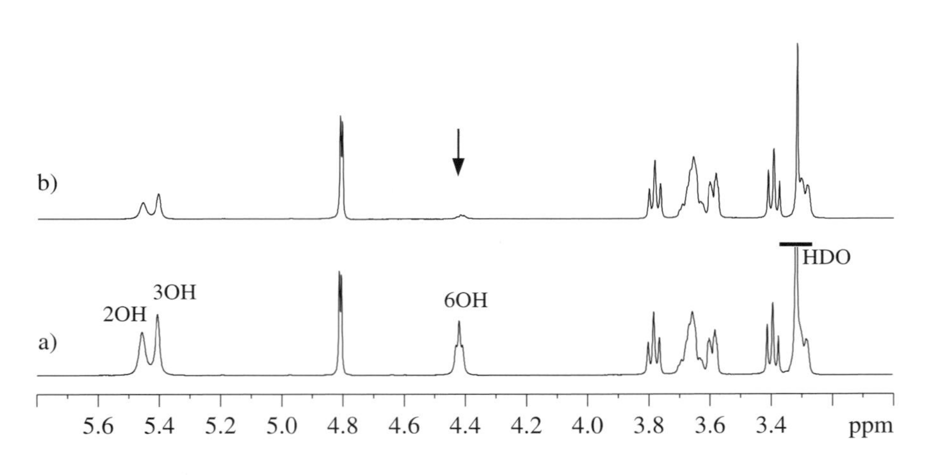

Figure 8.31. An experimental demonstration of saturation transfer. Direct saturation of the 6-OH resonance of α-cyclodextrin **8.10** in DMSO leads to the simultaneous indirect partial saturation of the 2- and 3-OH resonances as well as that of the water (shown truncated in (a)).

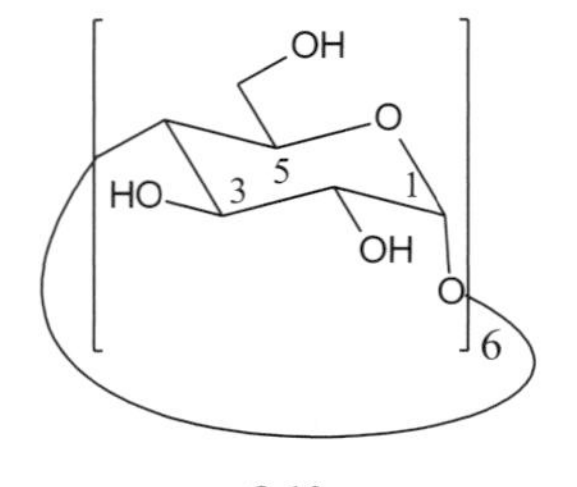

8.10

Quantifying enhancements

The quantification of percentage NOE enhancements can be made most economically by direct analysis of the NOE difference spectrum alone, rather than the perhaps more obvious option of directly comparing integrals between the control and NOE spectra. The saturated peak may be used as an inter-

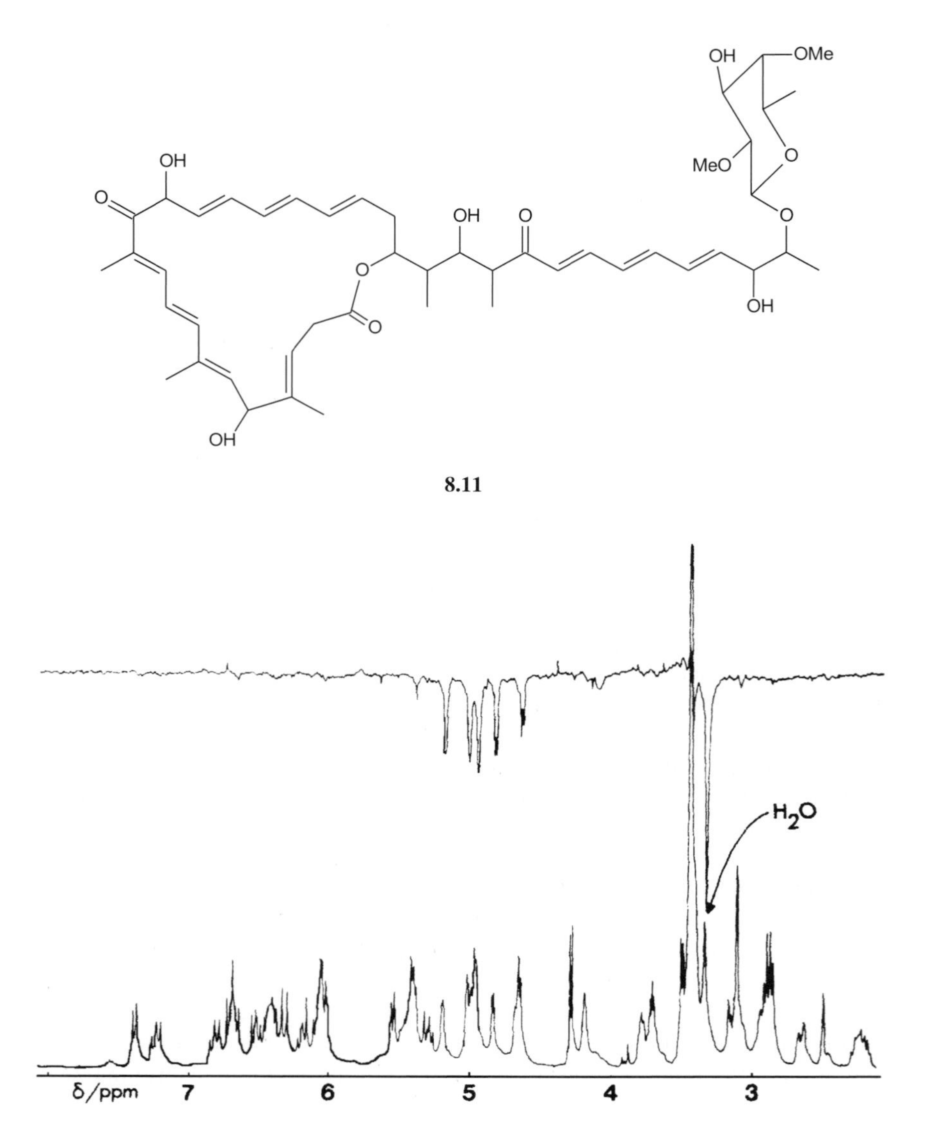

Figure 8.32. The identification of the resonances of exchangeable protons of pulvomycinin **8.11** in a saturation transfer difference experiment. The H_2O peak was selectively saturated, leading to the partial saturation of the five hydroxyl resonances through chemical exchange (reproduced with permission from [14]).

nal reference for these measurements with the peak of a single resonance referenced to −100% or that of a methyl group to −300% and so on (the minus signs arise from the *convention* of plotting enhancements as positive going responses which defines the saturated peak as having negative intensity when NOEs themselves have positive sign). All NOE enhancements are then integrated relative to this reference and the percentages obtained directly. In doing this, one is at the same time compensating for incomplete saturation that may have been brought about by using lower presaturation powers; by scaling the reference peak to assume complete saturation, we likewise scale all enhancements. This procedure is justifiable since absolute NOE enhancements scale in direct proportion to the degree of saturation. The careful use of baseline correction of the difference spectra prior to integration should be considered since baseline errors can make dramatic differences when measuring small enhancements. Even so, very accurate measurements of enhancements are usually of limited use in the final analysis since such small differences cannot be interpreted meaningfully. Quoting results to 1 or perhaps 0.5% is sufficient for most qualitative interpretations where the emphasis should be on interpreting a collection of enhancements rather than relying on a single percentage measurement.

8.7. MEASURING TRANSIENT NOES: NOESY

The most widespread approach for observing transient NOEs is homonuclear two-dimensional NOE spectroscopy (NOESY). This experiment's rise to popularity was driven largely by interests in biological macromolecules, whose NOEs are negative, strong and which grow rapidly, and this area is still very much the home territory of NOESY. It has traditionally been less used for smaller molecules principally because NOEs in the extreme narrowing limit are weaker and tend to grow more slowly, and because the steady-state NOE difference experiment has been a viable, and very effective, alternative. As instrument performance has improved, the NOESY experiment is increasingly finding favour in routine small molecule studies since it may offer benefits over the NOE difference experiment in some circumstances. The most obvious advantage is the non-selective nature of the experiment, alongside the ability, at least in principle, to observe all NOEs within a molecule in a single experiment (these factors parallel the benefits of 2D COSY over 1D selective spin-decoupling). Added to this is the simplicity of setting up NOESY relative to the more time-demanding NOE difference, which has advantages in an automated environment, for example.

Steady-state measurements are well suited to small molecule configurational and conformational studies since the enhancements obtained will be intrinsically larger than those from transient measurements. Energy is continually fed into the spin system in the form of rf saturation and this is able to drive the development of the NOE. Transient enhancements, in contrast, develop solely from the population disturbances brought about by initial resonance inversion so tend to be of lower intensity and of fleeting existence, requiring even more care in their capture. Furthermore, very small molecules ($M_r < 200$) that tumble very rapidly in solution produce extremely weak NOEs because cross-relaxation tends to be inefficient, and the search for transient enhancements may be difficult or even futile in such cases. With larger molecules, more success is to be expected. The recently introduced gradient-selected 1D transient NOE experiments described below are making a significant impact on small molecule NOE studies, in particular because of the exceptional quality spectra they are able to provide. However, their use requires a fundamentally different approach to the analysis of NOE data from that used presently by chemists when interpreting NOE difference spectra because one is measuring *transient* not *steady-stat*e enhancements, as discussed further below.

8.7.1. The 2D NOESY sequence

The NOESY experiment [15] again follows the principles presented in Chapter 5 for the generation of a 2D data set and has a similar appearance to the homonuclear COSY-based correlation spectra, although in this case the crosspeaks of the 2D spectrum indicate NOE interactions between the correlated spins. These peaks arise from the *incoherent* transfer of magnetisation between spins during the mixing time via the NOE, so allowing through-space proximities to be mapped directly. The NOESY spectrum of the naturally occurring terpene andrographolide **8.12** in DMSO is shown in Fig. 8.33 and displays a comprehensive map of close contacts within the molecule, some of which are illustrated on the structure. All NOE crosspeaks have opposite phase to the diagonal, indicating these arise from positive NOE enhancements, as anticipated for a molecule of this size ($M_r = 350$) under ambient conditions. The few crosspeaks sharing the same phase as the diagonal in the 3–6 ppm region are attributable to hydroxyl protons and H_2O and arise from chemical exchange of the protons within these groups (Section 8.7.4).

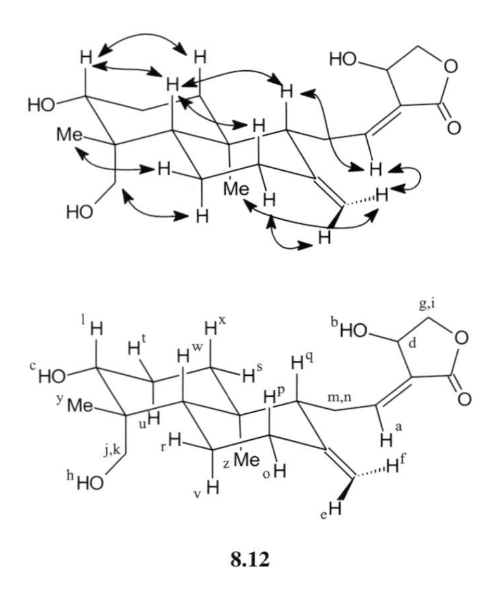

8.12

The NOESY sequence (Fig. 8.34) is closely related to COSY, so here we shall consider only the features specific to the NOE experiment. The most significant of these is the mixing period, τ_m, during which the NOE develops, and to understand the role this plays in a 2D experiment we return to the vector model (Fig. 8.35). Following initial excitation and t_1 evolution, the magnetisation vector exists in the transverse plane. The second 90° pulse places one component of this onto the $-z$ axis and has therefore generated the required population inversion that enables the transient NOEs to develop during the subsequent mixing period (the components that remain in the transverse plane are those detected in the COSY experiment). After a suitable period, τ_m, the new populations are sampled with a 90° pulse and the FID collected; one can readily see a parallel with the 1D 'inversion-mixing-sampling' sequence illustrated in Fig. 8.20. The sequence is repeated to collect sufficient transients for an acceptable signal-to-noise ratio. Whilst, in principle, it should be necessary to leave at least $5\,T_1$ between transients to allow populations to recover, this would lead to excessively long experiments, particularly in the

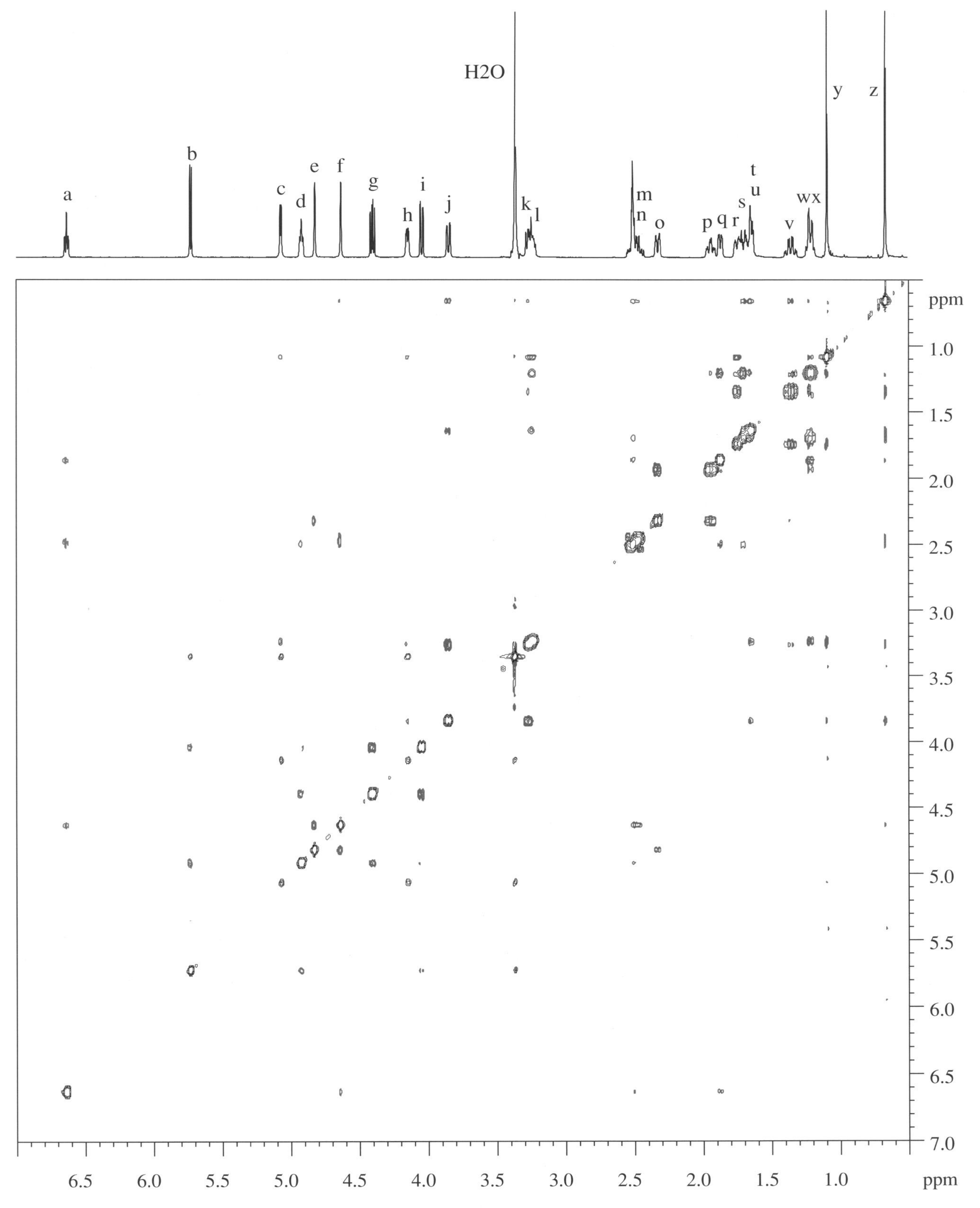

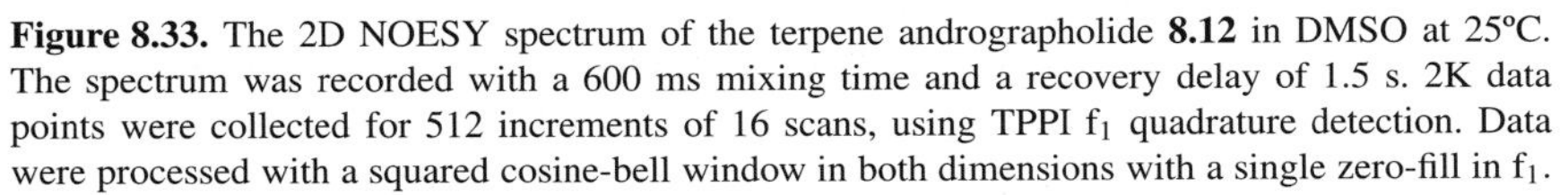

Figure 8.33. The 2D NOESY spectrum of the terpene andrographolide **8.12** in DMSO at 25°C. The spectrum was recorded with a 600 ms mixing time and a recovery delay of 1.5 s. 2K data points were collected for 512 increments of 16 scans, using TPPI f_1 quadrature detection. Data were processed with a squared cosine-bell window in both dimensions with a single zero-fill in f_1.

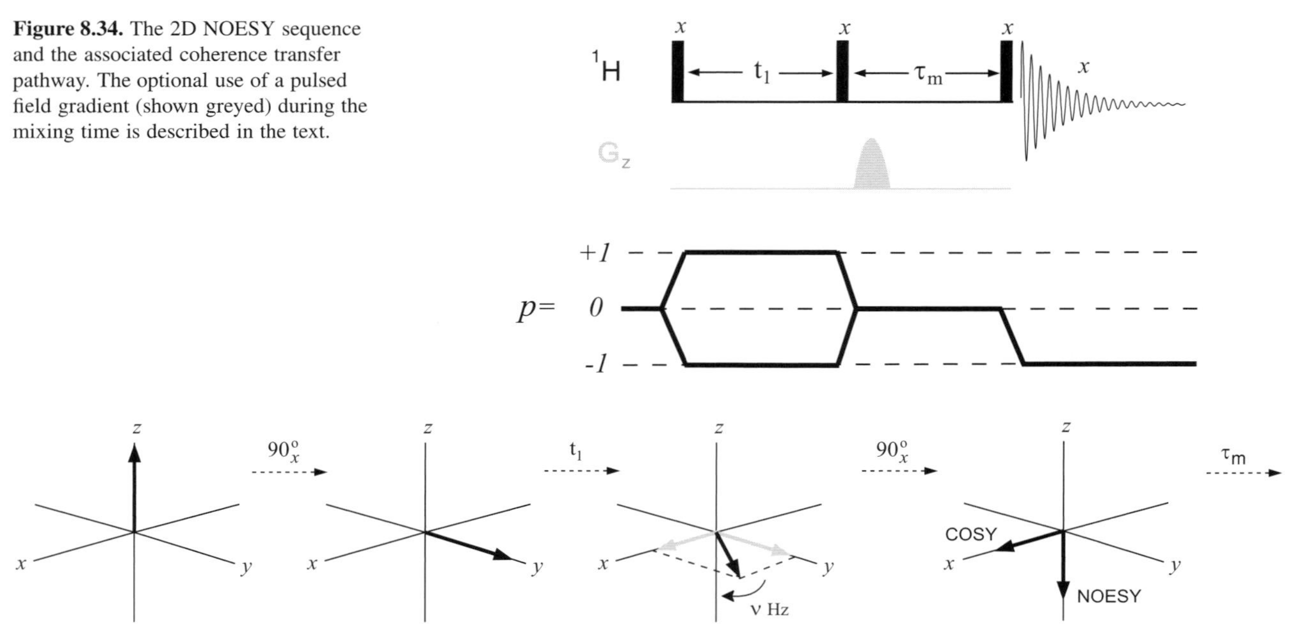

Figure 8.34. The 2D NOESY sequence and the associated coherence transfer pathway. The optional use of a pulsed field gradient (shown greyed) during the mixing time is described in the text.

Figure 8.35. The evolution of magnetisation vectors during NOESY. The inverted magnetisation generates the NOE during the mixing time, whilst the remaining transverse components are removed by phase-cycling or pulsed field gradients.

case of small molecules. In practice, shorter recovery delays of typically 1–3 s for small molecules ($\approx$1–2 T_1 of the slowest relaxing spins) represents an acceptable compromise [16].

The NOESY sequence, aside from details of the timings used, is identical to the DQF–COSY sequence, so it remains to select the signals arising from NOEs and suppress all others by means of a suitable phase-cycle (Table 8.3). Since NOEs are associated only with spin populations, that is *z*-magnetisation of coherence order zero, they are insensitive to rf phase changes and a 4-step cycle is sufficient to cancel signals originating from single- and double-quantum coherences present during τ_m. The cycle is repeated with additional phase inversion of the preparation pulse and receiver to cancel axial peaks and produces a final 8-step cycle. Since chemical exchange between sites is also associated with longitudinal magnetisation exchange between resonances, this experiment will also detect these processes, in an analogous fashion to the saturation transfer described for the 1D experiments above, and for this reason this exact sequence is also termed EXSY (exchange spectroscopy) when used to study dynamic exchange processes (see Section 8.7.3). The use of a phase-sensitive presentation is highly recommended for NOESY since this allows the discrimination of chemical exchange from positive NOEs on the basis of sign, aids the identification of some of the artefacts described below and allows quantitative measurements [17], in addition to the usual benefits of

Table 8.3. The basic NOESY phase-cycle for the three pulses (P_N) and the receiver (P_R), including the suppression of axial peaks

Transient	P_1	P_2	P_3	P_R
1	x	x	x	x
2	x	x	y	y
3	x	x	−x	−x
4	x	x	−y	−y
5	−x	x	x	−x
6	−x	x	y	−y
7	−x	x	−x	x
8	−x	x	−y	y

Table 8.4. Responses observed in the NOESY experiment

Sign of diagonal	Origin of crosspeak	Sign of crosspeak
	Positive NOE	Negative
Positive	Negative NOE	Positive
	Chemical exchange	Positive

By convention the diagonal is phased to be positive and the sign of all other signals are given relative to this.

a higher-resolution display. NOESY crosspeaks are then in-phase, have pure absorption lineshapes and appear symmetrically about the diagonal [18] (since transient enhancements are symmetrical in nature). Their sign with respect to the diagonal is dictated by the molecule's motional behaviour and hence the sign of the NOE itself (Table 8.4).

Selection of the desired coherences may also performed via pulsed field gradients so avoiding the need for phase-cycling [19–21]. This may be achieved simply by inserting a single z-gradient within the mixing time such that all coherences are dephased and only the desired z-magnetisation remains (Fig. 8.34). The time-saving benefits of such an approach are likely to be significant only when large sample quantities are available and rather few scans per increment provides sufficient sensitivity. When working with aqueous solutions, additional solvent suppression schemes may be added to this basic sequence; these are described in Section 9.4.

Complications with NOESY

The principal unwanted signals remaining from the NOESY sequence are those arising from zero-quantum coherences (ZQCs) that existed during the mixing time and which are subsequently transformed into observable signals by the last 90° pulse. Since these also possess coherence order zero they are not removed by the phase-cycling or gradient selection procedures (both of which act as a zero-quantum filter), and thus may contaminate the final spectrum. These coherences arise between J-coupled spins and so give rise to COSY-like peaks which are the zero-quantum analogues of the signals detected in the double-quantum filtered COSY sequence and have a similar antiphase multiplet appearance. If both a ZQC and NOE peak are coincident, that is if J-coupled spins also demonstrate an NOE between them, the NOE peak may appear somewhat distorted by this superposition. For large molecules with broad lines, ZQC peaks, being antiphase, tend to cancel, but in small or mid-size molecules, their active removal may be beneficial.

A number of approaches have been suggested for the suppression of ZQC peaks [15,22,23], but probably the most widely used is to introduce a small, random variation in the mixing time between transients or, more commonly, between one t_1-increment and the next. Whereas during the mixing time ZQCs oscillate with a frequency equal to the chemical shift difference of the two J-coupled spins, $(\nu_A - \nu_X)$, NOEs grow progressively (Fig. 8.36). Thus, random variation of τ_m will cause the ZQCs to average away whilst NOE intensities are little affected (in fact, this randomisation of the ZQC signals merely makes them appear as t_1-noise rather than discrete crosspeaks). The degree of variation required must be at least comparable to the inverse of the smallest shift differences, $|(\nu_A - \nu_X)|^{-1}$, so assuming this to be, say, 50 Hz, a random fluctuation of $\pm$10 ms is required. Even with such an approach, it is not uncommon to observe some residual COSY-like structure within crosspeaks very close to the diagonal. Additional artefact peaks can also arise between J-coupled spins when pulse widths deviate from 90° [24], so careful calibration of these is required.

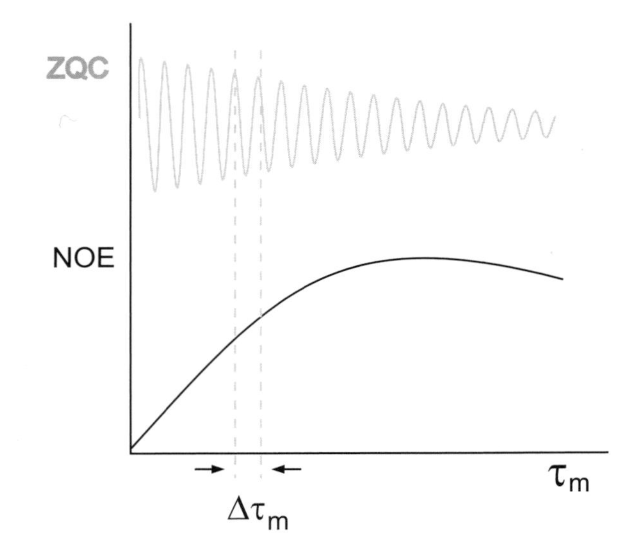

Figure 8.36. The variation of NOE and ZQC intensities during the NOESY mixing time. The ZQC contributions may be suppressed by making small random variations to τ_m, whereas this has negligible effect on the NOE intensities.

Optimum choice of mixing time

The appropriate choice of the mixing time, τ_m, is critical to the success of transient NOE experiments. Incorrect choice of τ_m can cause complete absence of observable signals since too short a value means the enhancements have yet to grow to a detectable level whilst an excessively long τ_m can mean the enhancements have decayed through relaxation. The motional properties of the molecule and hence the longitudinal relaxation times of the nuclei play the most significant role in dictating NOE growth and in turn dictate the selection of τ_m (Fig. 8.37). This choice is also very dependent on the type of information required of the spectrum. If the desire is to establish a qualitative map of NOEs within a molecule, as is often the case in small molecule work, then the optimum mixing time will be where the NOEs have their maximum intensities. If the data were to be used for quantitative or semi-quantitative distance measurements for use in structural calculations, then it is necessary to ensure one is working within the linear growth region of the NOE development, and mixing times well short of the NOE maxima will be required.

Selection of τ_m for qualitative studies can be estimated from knowledge of longitudinal relaxation times, T_1. Often these are not known prior to running the experiment, so a quick measurement may assist here, or failing that an estimate based on previous knowledge. Precise measurements are little benefit since there will inevitably exist a spread of values within the molecule an a compromise value for τ_m is required in any case. For small molecules, a mixing

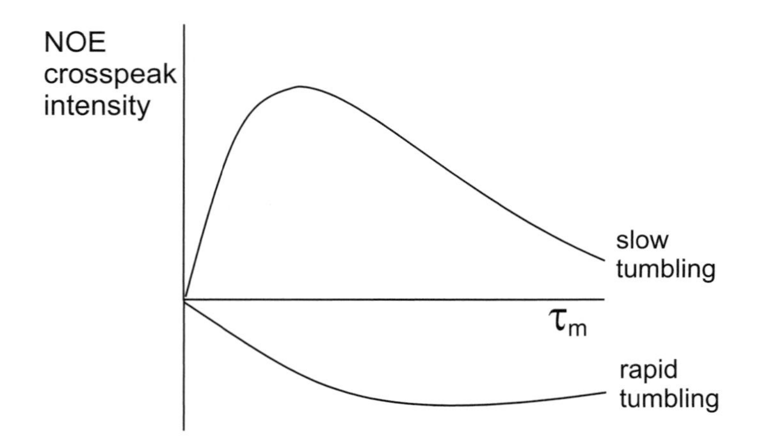

Figure 8.37. Schematic illustration of NOE crosspeak growth for large, slowly tumbling and small, rapidly tumbling molecules. The diagonal peak is assumed to be positive.

time of around T_1 will produce near maximum intensities and for many routine organic molecules with masses of <500 daltons in low-viscosity solvents, T_1 values typically range from around 0.5 to 2 s. As molecular tumbling rates decrease (owing to molecular mass or solution viscosity increases, for example), T_1s likewise decrease and correspondingly shorter mixing times are, in principle, desirable. In such cases one should be aware of approaching the dreaded $\omega_0\tau_c \approx 1$ region where efficacy of the experiment will diminish. For molecules in the negative NOE regime shorter values of around 0.5 T_1 are more appropriate to avoid complications arising from spin diffusion.

Distance measurement

Whilst a comprehensive discussion of this matter lies beyond the scope of this book, the principles involved are very briefly introduced here, the idea being that the reader should at least be able to follow discussions on structure calculations based on NOE measurements found in the chemical literature. The vast majority of work in this area has been applied to biomolecular structures and the protocols developed with macromolecules in mind, which will not translate directly to quantitative measurements in small molecules [25], although the general principles remain the same.

The most used approach to distance measurements stems from equations 8.13 and 8.14 above and relies on a known reference distance, r_{XY}, from which others may be calculated [26], and the assumption of uniform isotropic molecular tumbling. Recall the basic equation was:

$$\frac{\eta_A\{B\}}{\eta_X\{Y\}} = \frac{r_{AB}^{-6}}{r_{XY}^{-6}} \tag{8.19}$$

The ratio of the NOEs is determined by one of two general approaches. The first (Fig. 8.38a) involves determining the NOE growth rate for both the reference and unknown distances, by recording a series of NOESY spectra over a range of τ_m values and monitoring the crosspeak build-up intensities through their volume integrals [27]. The second, simpler approach (Fig. 8.38b) directly compares crosspeak intensities I_{XY} and I_{AB} measured at a single mixing time *that is known to lie within the linear growth regime* for both spin-pairs. The success of these approaches relies, in part, on the insensitivity of the calculated distance on the accuracy of the experimentally measured NOEs, due to the r^{-6} dependence. Thus, a factor of two error in the growth rate corresponds to only ≈10% error in the final distance estimate. Nevertheless, the measurement of a single 'accurate distance' does not allow for internal flexibility so has little meaning in a solution-state structure, and a more general and widely used approach has been to make use of semi-quantitative distance measurements in structure calculations. This involves categorising peak intensities as strong, medium and weak relative to the reference peak, which in turn are taken to

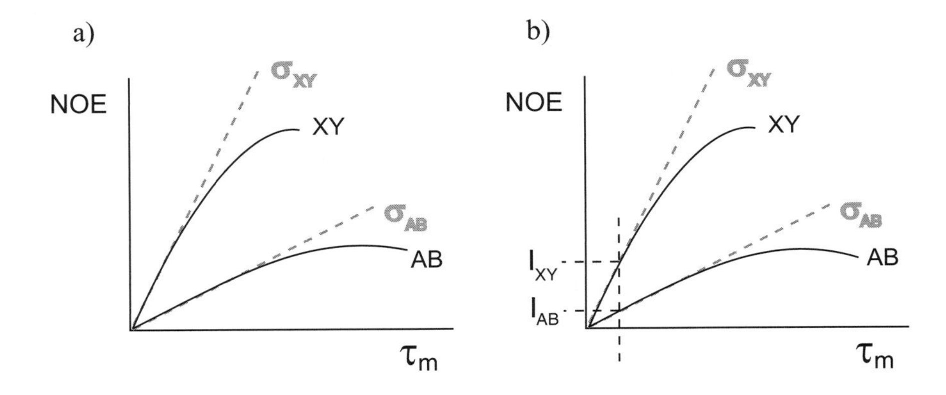

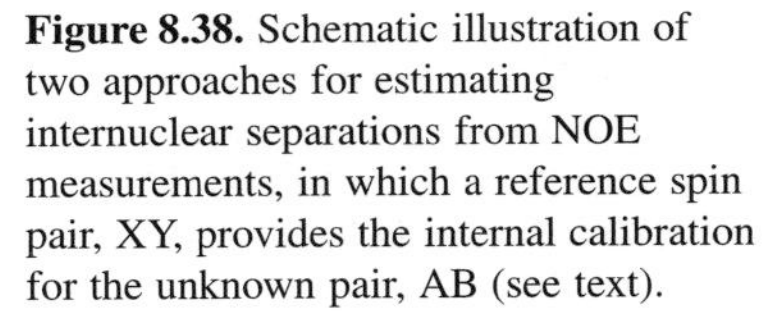
Figure 8.38. Schematic illustration of two approaches for estimating internuclear separations from NOE measurements, in which a reference spin pair, XY, provides the internal calibration for the unknown pair, AB (see text).

indicate upper (and perhaps lower) distance bounds on internuclear separations of, typically, ≤0.25 nm, ≤0.35 and ≤0.5 nm respectively. The internal reference distance measurement for proton NOEs is typically that between diastereotopic geminal protons (0.175 nm) or between ortho aromatic protons (0.28 nm). Although these constraints are lax, the combined application of many of them is able to produce well defined solution conformations, as has been amply demonstrated for numerous protein structures [28,29]. An alternative approach that makes use of a single NOESY spectrum utilises the ratio of crosspeak and diagonal peak intensities in place of growth rates [30]; both methods have been applied to the determination of the solution conformation and configuration of a small molecule and shown to produce similar results [31].

8.7.2. 1D NOESY sequences

To date 1D transient NOE experiments, often referred to as 1D NOESY, have found limited use in routine structural work, finding application principally in situations where quantitative distance measurements are sought. Such work tends to be time consuming because of the need to collect experiments over a range of τ_m values. The simplest 1D scheme (Fig. 8.39a) requires a control experiment recorded with $\tau_m = 0$ s and the use of difference spectroscopy to reveal NOE enhancements, and thus also suffers from the dreaded difference artefacts. For small molecule work, the steady-state NOE difference method is usually more efficient whilst larger, more complex molecules generally demand the 2D approach. However, the recently introduced 1D gradient-selected experiments [32–34] are finding increasing use in qualitative applications, and are considered here. These experiments do not depend upon difference spectroscopy to reveal the NOE enhancements, and cleanly remove those signals not arising from the NOE, so providing very high-quality spectra in short times. These are most useful when a collection of enhancements are used qualitatively to provide structural answers, and present a more rapid alternative to the NOE difference experiment.

1D gradient NOESY

The most robust experiment to date [34] (Fig. 8.39b) is based upon the 'double pulsed field gradient spin-echo' (DPFGSE) sequence as a means of selecting a target resonance from which the NOEs ultimately develop. This

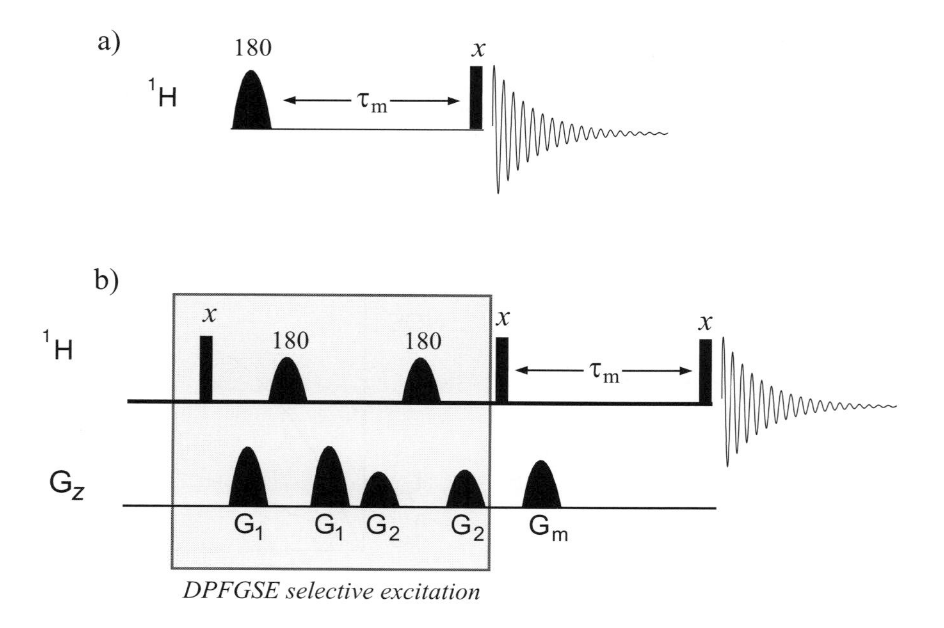

Figure 8.39. 1D NOESY schemes. The basic scheme (a) requires difference spectroscopy to reveal the NOEs whereas the gradient-selected DPFGSE–NOE (b) reveals NOEs without interference from difference artefacts.

approach to selective excitation is described in Section 9.3.3, so here it is sufficient to note that it serves to selectively refocus the target spin magnetisation in the transverse plane, leaving that of all other spins dephased and thus unobservable. The subsequent 90° pulse places the target magnetisation along the $-z$ axis, after which the NOE enhancements grow, with a purge gradient (G_m) applied to remove any residual transverse components that remain. Finally, the magnetisation is sampled in the usual way and the recorded spectrum contains only the inverted target resonance and the NOEs that have developed from it, *without the need for a difference spectrum to be calculated.* High-quality spectra can therefore be collected that are free from difference artefacts; this is the principal benefit of the gradient-selected schemes. The experiments also tend to be quick to perform since extensive signal averaging is no longer required for artefact reduction, and because optimum use may be made of the receiver dynamic range since only the desired signals are ever digitised.

In practice, relaxation processes operating during τ_m will cause the unwanted dephased magnetisation components to recover and produce small observable signals in the final spectrum for all resonances. These can be kept to close to zero intensity by the insertion of typically 1 or 2 non-selective 180° pulses (bracketed by opposing purge gradients) spaced judiciously within the mixing period [34] with any small signals that may remain cancelled by a 4-step difference phase-cycle (EXORCYCLE). Thus, although difference spectroscopy is being used, in the form of the phase-cycle, the signals to be cancelled have close to zero intensity and are readily removed, in contrast to conventional difference spectroscopy where the signals have essentially maximum intensity. For small and mid-sized molecules, experience suggests that one inversion pulse during τ_m is sufficient to reduce the background signals to undetectable levels with mixing times of ca. 500 ms or less, whereas 2 are preferred for a τ_m of up to 1.5 s.

Artefacts can again arise between J-coupled spins, equivalent to the SPT effects described for the steady-state difference experiments and the ZQC peaks described for NOESY, and more involved experimental procedures have been described for their removal [34]. However, in many cases the great benefit of the gradient approach is the ability to observe, with confidence, long-range NOE enhancements that would otherwise have been undetectable, or at least unreliable, in the presence of difference artefacts. The presence of large scalar couplings between such distant spins is, in most cases, unlikely, so the problem of J-coupling interference does not arise and the simpler schemes often suffice. A typical result using the sequence of Fig. 8.39b is shown in Fig. 8.40 for the small bicyclic lactam **8.13**. Each resonance was selected with a 40 ms Gaussian 180° pulse with two non-selective 180° pulses applied during the 800 ms mixing time. Each spectrum was the result of only 12 minutes data collection on a 10 mg sample yet are perfectly acceptable for qualitative stereochemical assignments. The observed signals may be attributed to NOE enhancements or to SPT-type artefacts, with an otherwise featureless baseline where these do not arise.

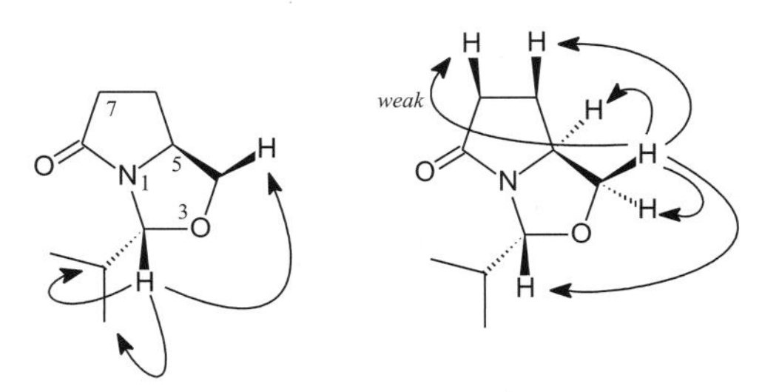

8.13

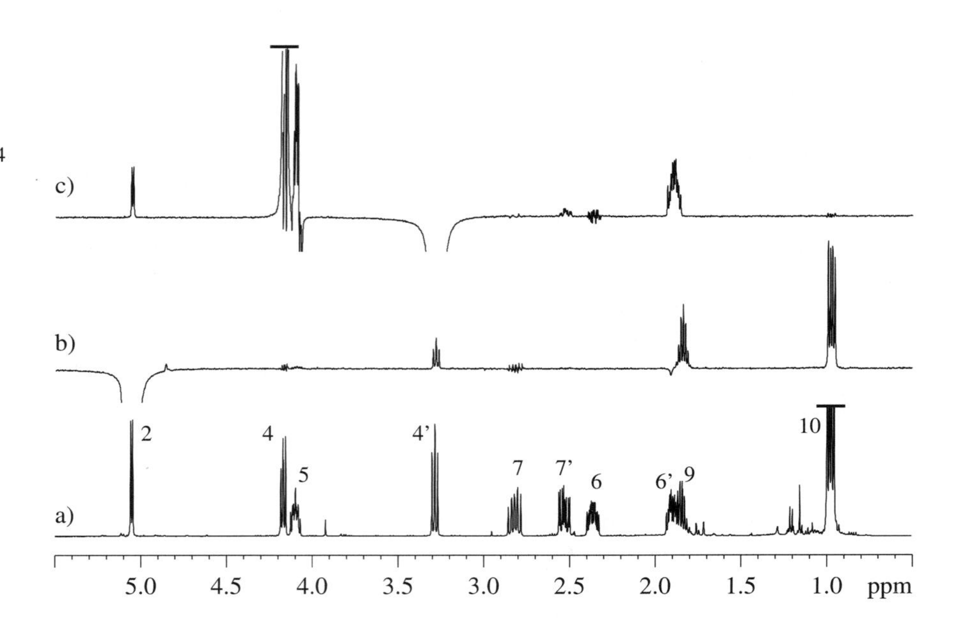

Figure 8.40. Selected 1D gradient NOESY spectra of the bicyclic lactam **8.13** recorded with a mixing time of 800 ms. Spectrum (a) is the conventional 1D spectrum and the strong geminal H4′–H4 NOE in (c) has been truncated.

Interpreting transient NOEs

The clear advantages of the gradient-selected NOE experiment over the conventional steady-state NOE difference means this is becoming a popular tool in small molecule structural studies. However, there are fundamental differences between the data presented by the two experimental protocols, with *steady-state* experiments observing *equilibrium* NOEs and *transient* experiments observing *kinetic* NOEs. As a consequence, 1D NOESY experiments demand a somewhat different approach to data interpretation over that currently adopted for steady-state NOE difference measurements, some of the key considerations include:

- The NOE enhancement will be acutely sensitive to the choice of mixing time, and may vary markedly with changes in this.
- The NOE enhancement will also depend on the degree of target inversion, which may well be less than complete owing to pulse imperfections and other experimental shortcomings.
- Experimentally, the measurement of percentage enhancements is more complex. The target resonance cannot be used as a reference, as it conveniently is in steady-state difference spectra, since it is reduced by relaxation during τ_m. A reference spectrum with $\tau_m \approx 0$ s is required from which enhancements can be quantified and the degree of inversion of *each* target resonance must also be determined. The experimentally measured NOE enhancements must then be scaled accordingly for '100% inversion'.
- The *absolute* percentage enhancements from transient experiments will be smaller than steady-state enhancements, even though they may be clearer to see. The perception of what is considered a 'reliable' or 'measurable' enhancement must therefore be adjusted if percentages are reported.

The experimental complexities associated with the correct measurements of percentage enhancements are likely to mean that these figures will *not* be routinely determined for the qualitative structural studies commonly undertaken (and anyone purporting to have measured these should give details of the approach taken). A more realistic approach to reporting results would seem to be that adopted in macromolecular NOE studies using semi-quantitative classification of enhancements (small, medium and weak). Assuming mixing times are kept short such that enhancements lie within, or at least close to, the linear growth regime, a comparison *within* each trace provides approximate relative distance relationships between the target spin and those exhibiting the enhancements. Piecing together data from a number of experiments to produce a self-consistent argument is again advisable.

8.7.3. Applications

Despite finding its origins largely in the hands (or labs.) of biological spectroscopists, there is nowadays an increasing use of NOESY methods for structural and conformational analysis of the small to mid-sized molecules encountered in the chemical laboratory. The examples presented here demonstrate a variety of systems to which these experiments have been applied and also serve to illustrate some the benefits of using these in preference to the steady-state experiment previously encountered.

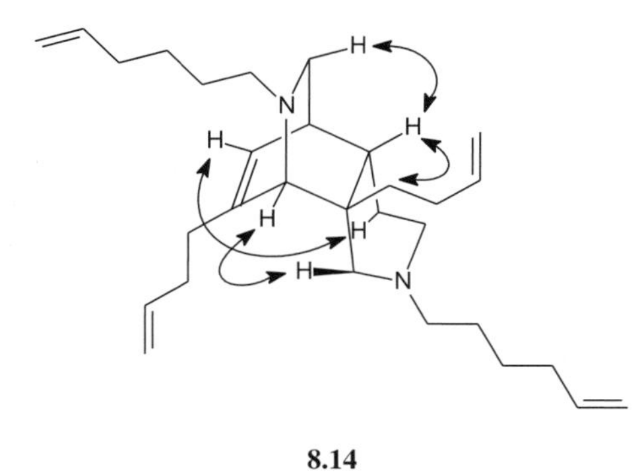

8.14

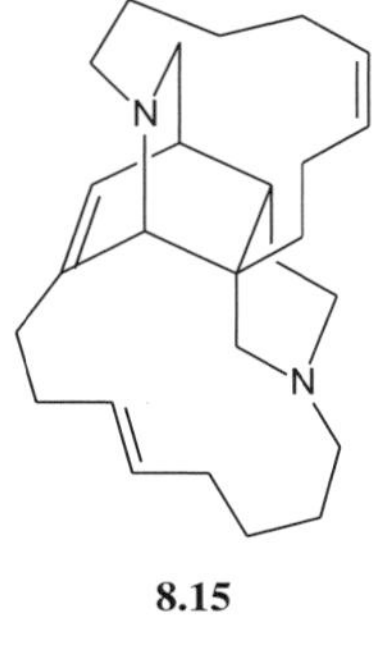

8.15

In the first example, 2D NOESY spectra were used to define the stereochemistry in the synthetic cycloadduct **8.14** [7], a potential biomimetic precursor to the naturally occurring marine-sponge alkaloid Keramaphidine B, **8.15**. This problem is essentially the same as that addressed for **8.7** above using the NOE difference experiment, but in this case the additional unsaturated sidechains caused extensive overlap in the proton spectrum and precluded the use of selective presaturation. Sufficient characteristic NOEs present in a 600 ms NOESY spectrum gave conclusive proof of the *endo* stereochemistry, as shown. Only positive NOEs were observed, consistent with a molecule of mass 436 daltons in chloroform. NOESY spectra have also been successfully applied to the structure elucidation of molecules for considerably greater mass and complexity, as illustrated by the cytotoxic macrolide cinachyrolide A, **8.16** [35], also from a marine sponge. The structure of the molecule was determined through extensive 600 MHz 2D NMR experiments, of which NOESY played

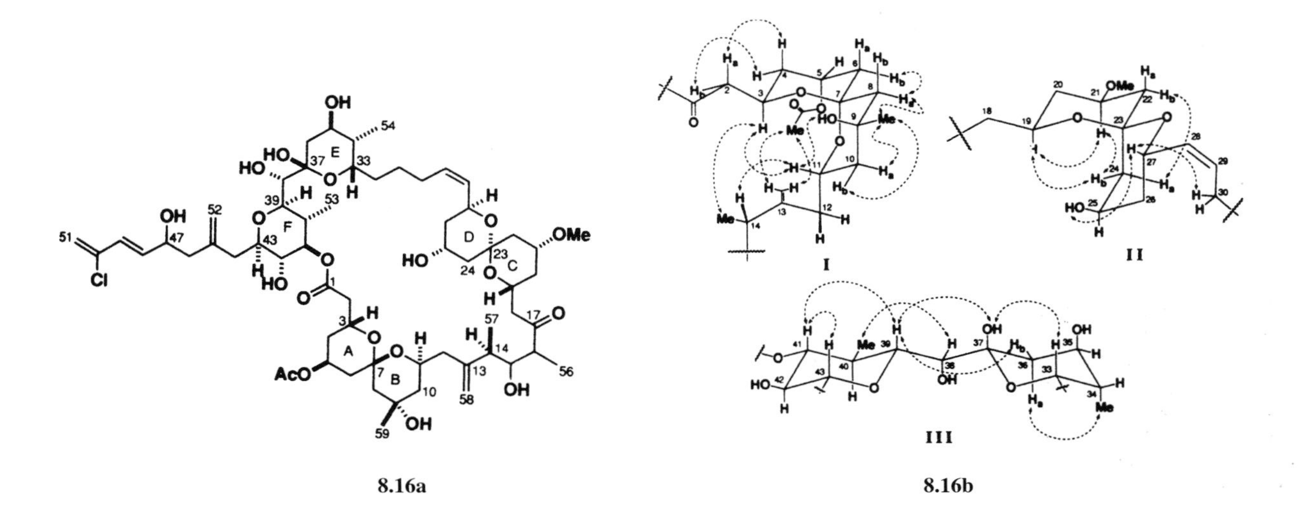

8.16a **8.16b**

a crucial role in defining the relative stereochemistry of the six oxane rings, as shown (reproduced with permission from [35]). Mixing times of up to 700 ms were employed, and the relatively large mass of the molecule, approximately 1200 daltons, may have been close to the cut-off for the successful observation of NOEs. Probably for this reason, supporting evidence also came from 200 ms ROESY spectra.

A common area in which 2D methods have been employed to help provide resonance assignments is in the study of peptides or small proteins, specifically when the same amino-acids occur more than once in the peptide sequence. The so-called *sequential assignment* process is used to define the position of a specific amino-acid residue within a peptide, and relies upon the observation of NOEs between protons in adjacent residues (Fig. 8.41). Typically these will be between an alpha proton and the amide NH of the following residue. The identification of neighbouring amino-acids in this way can be used to string these units together, which may then be mapped onto the usually known peptide sequence. Once all residues have been sequentially identified, this can provide the basis for conformational studies through longer-range NOE contacts [36–38]. As an illustration, the stepwise identification of neighbouring resonances is mapped in a NOESY spectrum for the cyclic decapeptide gramicidin-S **8.17** in Fig. 8.42.

Whilst most studies rely on the observation of *intra*molecular NOEs, *inter*molecular NOEs can also be used to define relationships between molecules

Figure 8.41. The intraresidue NOEs used to identify neighbouring residues in a peptide sequence.

Pro—Val—Orn—Leu—DPhe
DPhe—Leu—Orn—Val—Pro

8.17

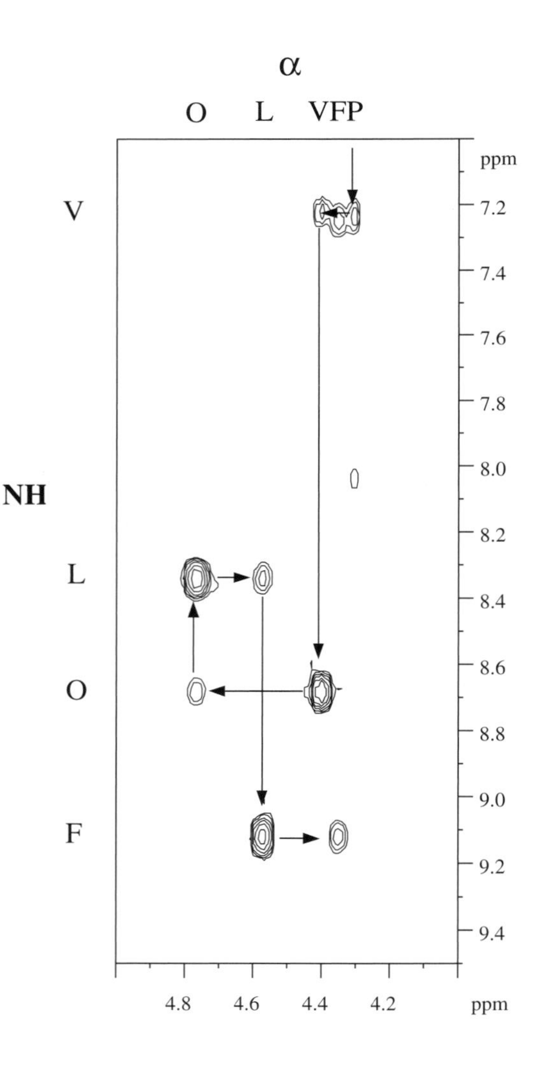

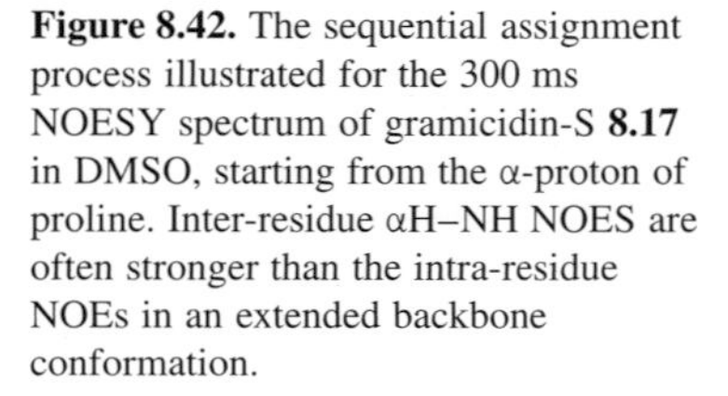

Figure 8.42. The sequential assignment process illustrated for the 300 ms NOESY spectrum of gramicidin-S **8.17** in DMSO, starting from the α-proton of proline. Inter-residue αH–NH NOES are often stronger than the intra-residue NOEs in an extended backbone conformation.

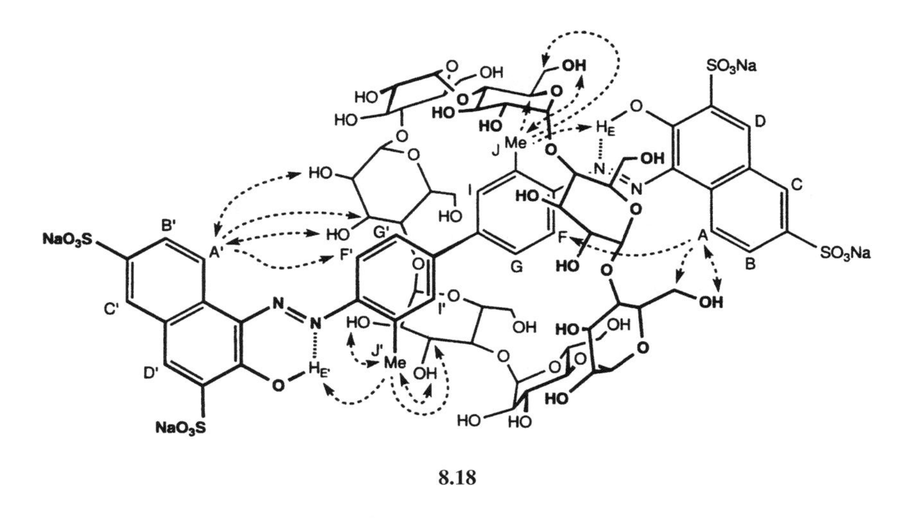

8.18

which share close proximity. This can be of particular relevance in the study of 'host–guest' complexes where one wishes to determine how the guest sits within the host, and here scalar coupling information can play no direct part. One example of this is the characterisation of rotaxanes, extended conjugated systems sitting within protective encapsulating molecular jackets. The azo-dye rotaxane **8.18** uses α-cyclodextrin as the sheath molecule, which is inherently asymmetric with a smaller cavity entrance at the 5–6 rim of the molecule. The 1D gradient NOESY experiment described above was employed to map long-range host–guest NOEs [39] which differentiated the two ends of the complex and presented a picture of how the α-CD sat. NOEs from both HG′ and HI to the 3 and 5 protons on the inner face of the α-CD confirmed that the central portion of the guest was encapsulated within the hydrophobic cavity (structure reproduced with permission from reference 39). All NOEs for these molecules in DMSO were negative due to the limited mobility of these bulky complexes.

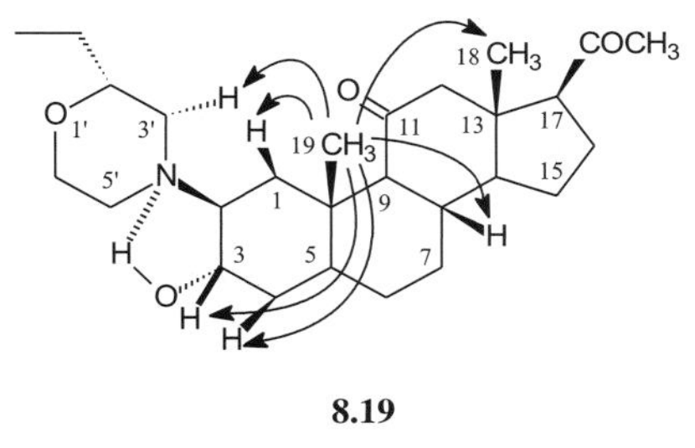

8.19

Both 1D gradient and 2D NOESY experiments have also been used to asses the conformations of anaesthetic steroids [40]. These possess crowded proton spectra requiring the use of 2D experiments and mean 1D methods can only be applied to a limited set of resonances, which nevertheless prove significant. In **8.19** in particular, long-range NOEs in the 1D experiment were observed from methyl 19 specifically to the H3′ equatorial proton of the attached morpholine ring but not to other protons within this moiety. This indicated free rotation of the ring was restricted (by an intermolecular hydrogen bond) and that it occupied a fixed position relative to the steroid skeleton. Knowledge of such conformations prove useful in elucidating the mechanisms by which these molecules cause anaesthesia.

The final example specifically illustrates some advantages of the 1D gradient NOESY experiment over the conventional NOE difference method. One goal of this work was to compare the solution structure of the diphenylallyl palladium complex **8.20** with the crystal structure and from this derive mechanistic insights [41]. The proton spectrum was assigned through combined analysis of phosphorus-decoupled proton, DQF–COSY and ROESY spectra recorded with a 500 ms mixing time. Specific close proximities were also probed through 1D NOE methods where suitably resolved resonances were available, selected examples of which are shown in Fig. 8.43. Some specific points are worthy of comment, the first of which is the clarity in the NOE spectra which are not confused by difference artefacts, thus giving one confidence that the observed enhancements are genuine. Secondly, these spectra were collected in only fourteen minutes each, whilst the equivalent difference experiments required data collection over many hours to adequately suppress resonances in the cluttered aromatic region. Finally notice the enhancements of the broad

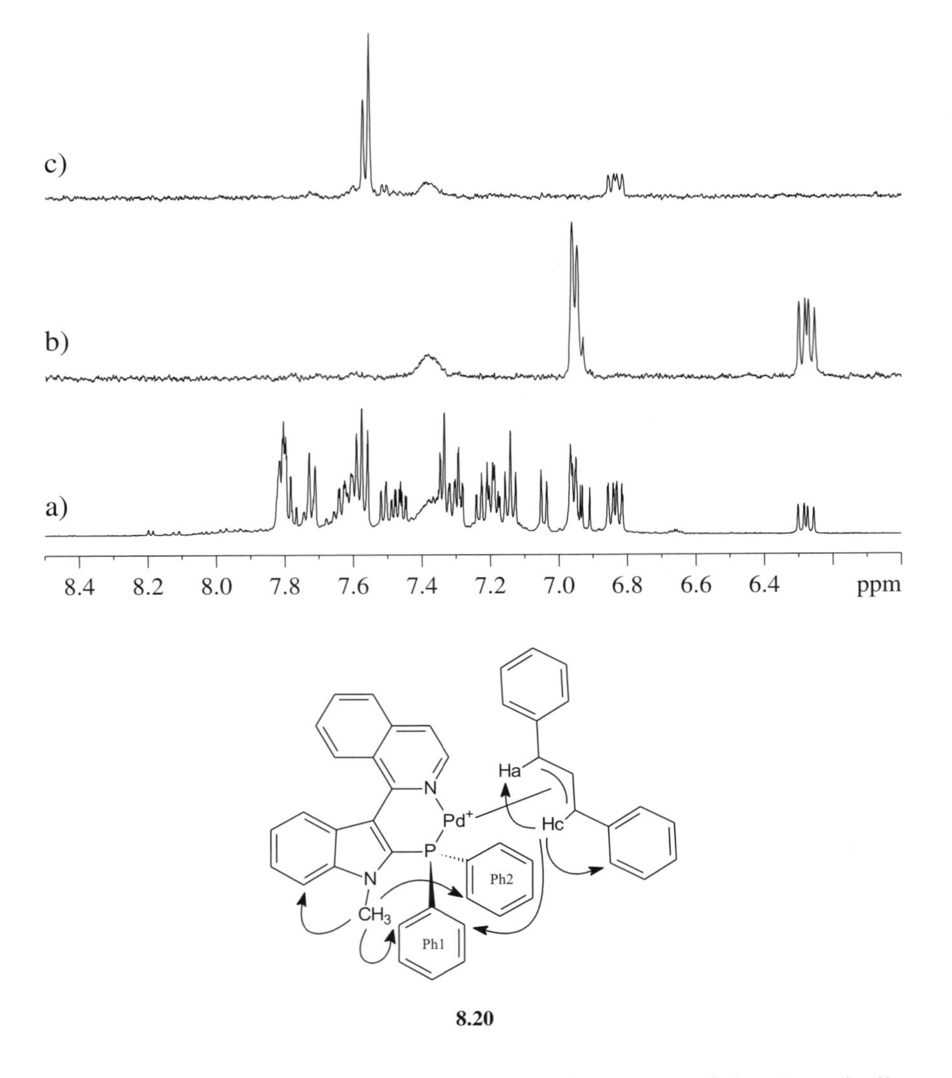

Figure 8.43. Selected 1D gradient NOESY spectra of the palladium complex **8.20** recorded with a mixing time of 800 ms in each case. (a) Parent 1D spectrum, (b) Hc selected (not shown) and (c) N–Me selected (not shown).

resonance at 7.4 ppm, associated with the *ortho* protons of the dynamically restricted phosphine–Ph[1] group. These were distinctive and highly informative but could not be reliably distinguished in either the 1D difference or 2D experiments. The ability to extract data of this sort with confidence makes the 1D gradient NOESY experiment an extremely powerful tool in modern structure elucidation.

8.7.4. Measuring chemical exchange: EXSY

Although the mechanisms of chemical exchange and the NOE are quite unrelated, they share in common the transfer of longitudinal magnetisation and as such can be detected with the same 1D or 2D NMR experiments. In fact, the NOESY and EXSY (Exchange Spectroscopy [42]) pulse sequences are one and the same and the 1D transient NOE sequence can equally well be applied to the measurement of rate constants [43,44]. Furthermore, the presaturation sequence used for the 1D steady-state experiment also serves as a simple method to provide evidence of slow conformational exchange processes, particularly in proton spectra, by virtue of saturation transfer effects, provided the exchange is fast relative to the spin relaxation rates. The practical difference lies in the EXSY dependence on exchange rates rather than spin relaxation rates. Before proceeding, these experiments are now considered in the context of observing exchange processes, and some basic understanding of the exchange phenomenon is assumed here. More extensive discussions of 1- and 2-D methods for studying dynamic processes may be found in reviews [45–47] and in dedicated texts [48,49].

In parallel with studies of the NOE, exchange experiments can be used both qualitatively, to map exchange pathways, and quantitatively to determine rate constants. In systems of multi-site exchange, the 2D EXSY experiment proves particularly powerful in the measurement of these for all pathways. However, for either application the exchange processes studied must be slow on the NMR chemical shift timescale since the exchanging resonances must be resolved in order to observe transfer between them. Furthermore, this exchange must not be *too* slow otherwise relaxation processes occurring during τ_m will remove all memory of the exchange process. Thus, magnetisation transfer experiments are suitable for only a limited range of exchange rates and as a rule of thumb, these must be *at least* comparable to the longitudinal relaxation rates ($k_{ex} \geq 1/T_1$). In practice, this means these experiments are sensitive to exchange processes with $k_{ex} \approx 10^2$–10^{-2} s^{-1}. The study of different nuclei provides a window on different rates within this range due to differences in relaxation times, with slower relaxing spins able to probe slower exchange processes. Fast exchange processes that lead to resonance coalescence can be studied by lineshape analysis (often referred to as 'bandshape analysis') [46,48] which generally requires the use of computer simulation [50]. As an illustration of the 2D method, Fig. 8.44 shows the boron-11 EXSY spectrum of a 1 : 1 mix of BCl_3 and BBr_3. The exchange crosspeaks demonstrate the presence of ligand scrambling processes and indicates the mechanism involves the exchange of a single halogen atom only [51]. A quantitative analysis of these data also provided (pseudo) first-order rate constants for each exchange process (see below).

The correct choice of the mixing time, τ_m, is again important for the

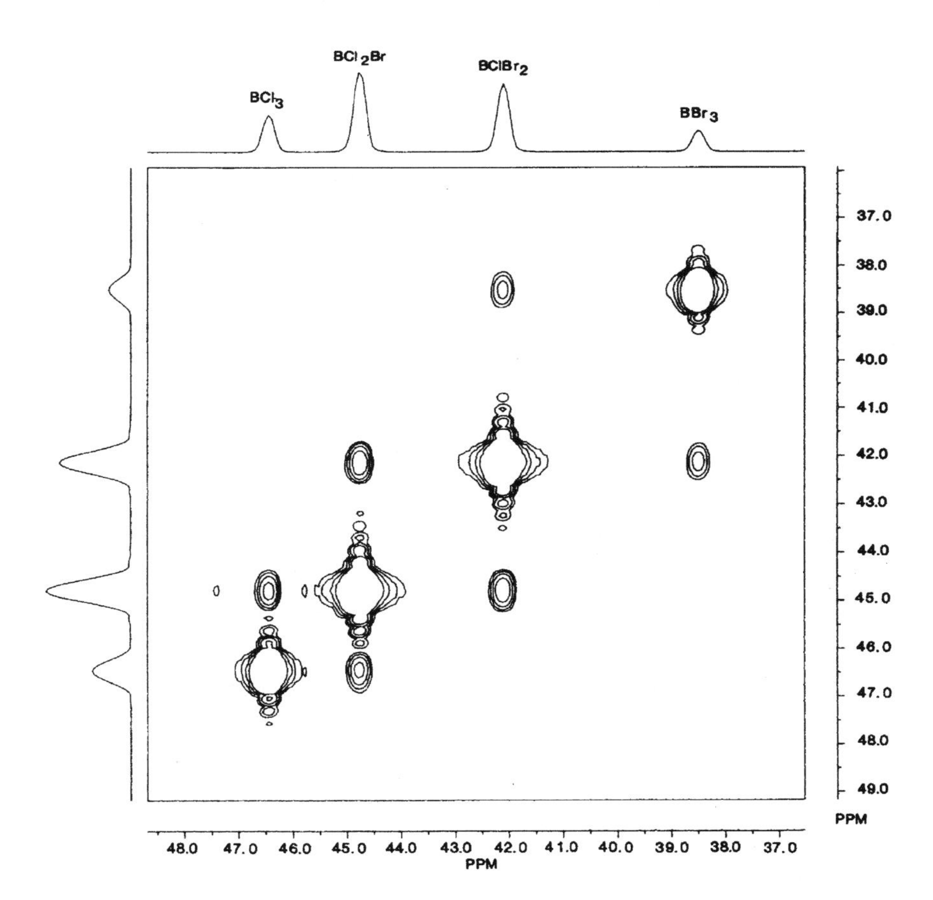

Figure 8.44. The 128 MHz boron-11 2D EXSY spectrum of a 1 : 1 mix of BCl_3 and BBr_3 at 400 K. The spectrum was recorded with a mixing time of 50 ms (reproduced with permission from [51].

successful application of EXSY, and is dependent upon the exchange rate constants. Since these are unlikely to be known with any accuracy (since there would be little point in performing the experiment if they were) and as there may exist a spread of these within the kinetic system being studied, some compromise is again required. In the absence of known rates, an upper limit can be defined (at least for the case of symmetrical two-site exchange) from the knowledge that the rate constant at coalescence, k_c, is given by $\pi\delta\nu/\sqrt{2}$, where $\delta\nu$ is the shift difference (in Hz) in the slow exchange limit, and hence $k_{ex} < k_c$. In qualitative work the aim is to achieve maximum crosspeak intensity and setting $\tau_m \approx 1/k_{ex}$ is recommended. Under such compromise conditions, multi-step transfers may operate when suitable pathways are possible, producing 'indirect' exchange peaks, which may be confused with direct peaks from relatively slow processes. For example, this has been observed for the ligand scrambling reactions of tin halides [52,53], boron halides [51] and of lead(IV) tetracarboxylates [54].

The choice of τ_m for quantitative work will be dictated by the method used to calculate k_{ex}, of which there are essentially two widely used approaches [51]. The simplest involves collecting EXSY spectra over a range of τ_m values and following the initial linear growth of exchange crosspeaks (the initial rate approximation again, by exact analogy with the approach taken for NOESY analysis). The great disadvantage to this is clearly the need to collect many 2D spectra, which may place unreasonable demands on instrument time. The second approach is to determine rate constants from a single EXSY spectrum and for this the appropriate choice of τ_m is absolutely critical; too small and crosspeaks will have weak intensities which are subject to significant error, too long and the intensities become insensitive to the kinetic parameters. Methods for determining the optimum τ_m in such cases have been described [55,56]. Having obtained suitable data, computational analysis (for example with the program D2DNMR [57]) may be employed to calculate exchange rates for multi-site systems. For the simpler case of equally populated 2-site exchange, explicit equations have been presented for the estimation of rate constants from diagonal and crosspeak intensities [58].

One- and two-dimensional magnetisation exchange experiments find greatest use in the study of inorganic and organometallic systems, for which a high degree of fluxionality often exists, and extensive reviews of these areas have been presented [46,47,59] in which a wide variety of example applications may be found. The study of low natural abundance nuclei, as is most often undertaken, has the significant benefit of avoiding potential complications from NOE effects that may also be operative. Dynamic proton studies are most likely to be subject to such interferences.

8.8. MEASURING ROTATING-FRAME NOES: ROESY

The benefits of recording rotating-frame NOEs have been described in earlier sections, and stem from the fact that they are positive for all molecular tumbling rates and hence all molecular sizes, and prove particularly advantageous in the study of mid-sized molecules which exhibit very small conventional NOEs owing to their motional properties ($\omega_0\tau_c \approx 1$). Since the observation of an ROE involves the measurement of a transient enhancement, their application and interpretation largely parallels that for NOESY and need not be repeated here. The principal concern with ROESY experiments (originally termed CAMELSPIN [60] owing to the similarity of the motion of a figure skater during this manoeuvre and the behaviour of magnetisation vectors during the experiment) is the fact that they are susceptible to a number of processes other than cross-relaxation and hence may contain a variety of interfering and potentially confusing responses, in addition to the desired ROEs. An appreciation of these effects, and how to deal with them, is therefore mandatory for anyone wishing to employ these methods and some caution is required in the application of ROE experiments.

8.8.1. The 2D ROESY sequence

The original and simplest ROESY sequence (Fig. 8.45) has the mixing period defined by the duration of a continuous, low-power rf spin-locking pulse during which the ROE develops. Magnetisation not parallel to the spin-lock axis is dephased by rf field inhomogeneity (Fig. 8.46) so the only phase-cycling required is that for axial peak suppression and f_1 quadrature-detection. Gradient-selected variants have also been presented [20]. For small and mid-sized molecules the selection of τ_m follows that presented for NOESY with, typically, $\tau_m \leq 600$ ms, whereas for large molecules in the spin-diffusion limit, shorter values are more appropriate since the ROE growth rate is twice that of the NOE (Section 8.5). The 2D spectrum maps through-space interactions through crosspeaks that have *opposite* phase to the diagonal.

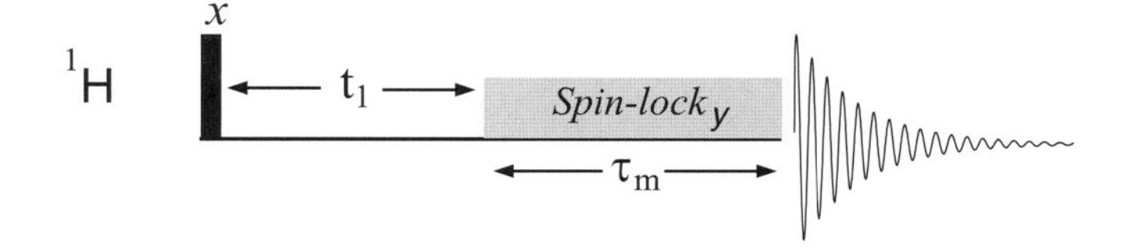

Figure 8.45. The 2D ROESY sequence. The mixing time, τ_m, is defined by the duration of the low-power spin-lock pulse.

Complications with ROESY

As stated previously, crosspeaks may arise in ROESY spectra as a result of processes that occur during the spin-lock other than cross-relaxation between spins. The principal complications that can arise are [61]:

- COSY-type crosspeaks between J-coupled spins
- TOCSY transfers between J-coupled spins, and
- crosspeak attenuation from rf off-resonance effects.

The first of these arises when the long spin-lock pulse acts in an analogous fashion to the last 90° pulse of the COSY experiment so causing coherence transfer between J-coupled spins. The resulting peaks display the usual antiphase COSY peak structure and tend to be weak so are of least concern. A far greater problem arises from TOCSY transfers which arise because the spin-lock period in ROESY is similar to that used in the TOCSY experiment (Section 5.7). This may, therefore, also induce coherent transfers between J-coupled spins when these experience similar rf fields, that is, when the Hartmann-Hahn matching condition is satisfied. Since the ROESY spin-lock is not modulated (i.e. not a composite pulse sequence), this match is restricted to mutually coupled spins with similar chemical shift offsets or to those with equal but opposite

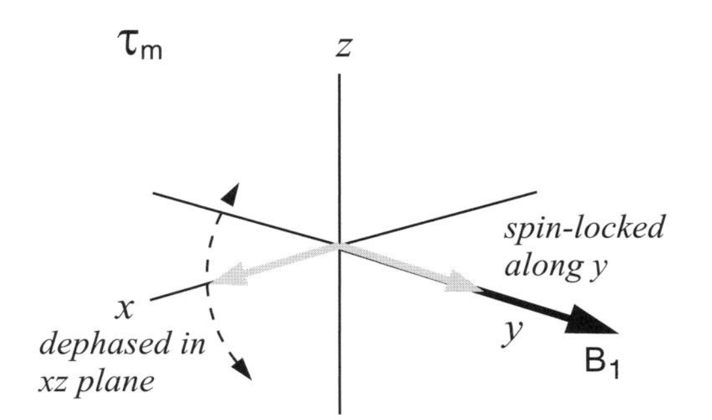

Figure 8.46. The ROESY spin-lock. During this, magnetisation parallel to the B_1 field remains spin-locked whereas orthogonal components are driven about this (here in the *xz* plane) and eventually dephase through B_1 field inhomogeneity.

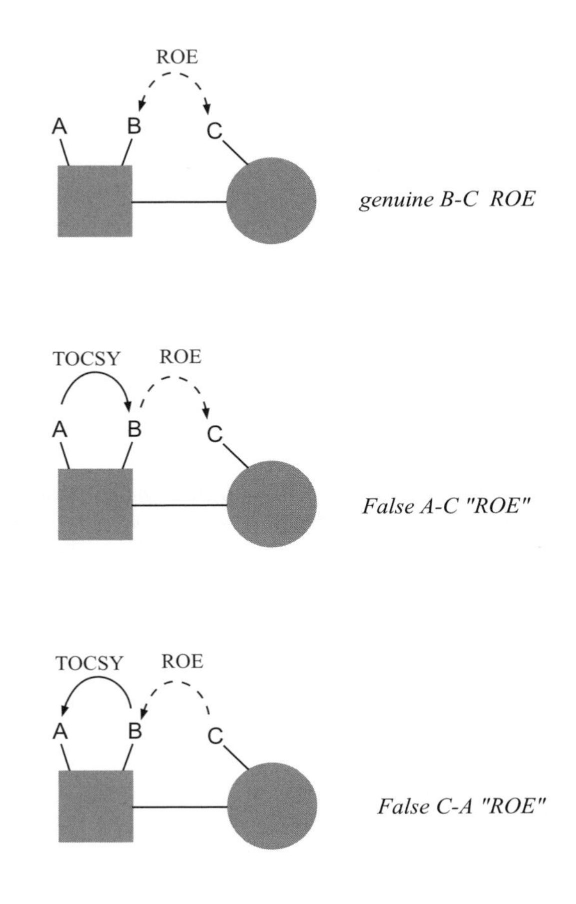

Figure 8.47. The generation of 'false' ROE peaks in ROESY spectra may arise from combined ROE and TOCSY mechanisms.

offsets from the transmitter frequency. These conditions are often met in natural products such as carbohydrates, nuclei acids, peptides, alkaloids and steroids for which resonances cluster about the centre of the spectrum. In addition to unwanted direct TOCSY effects, multistep transfers involving both TOCSY and ROE stages can lead to 'false' ROEs, crosspeaks that, a priori, appear to arise from a ROE between two correlated spins but in fact do not [62]. Thus, for example, a TOCSY transfer from spin A to B followed by ROE transfer from B to C in the system of Fig. 8.47 would produce an apparent crosspeak suggesting a ROE between A and C. Likewise the reverse ROE–TOCSY steps would produce a similar signal, but the conclusion would be wrong in either case! Since TOCSY transfer retains signal phase, its involvement in crosspeak generation is not obvious and can lead to drastically incorrect conclusions. Furthermore, if TOCSY and ROE transfers occur simultaneously between two spins, the ROE peak may be reduced in magnitude or even cancelled owing to opposite peak phases. The various transfer pathways that may occur during ROESY are summarised in Table 8.5.

Considerable attention has been given to ways of avoiding TOCSY transfer during ROESY, the simplest of which is through limiting the Hartmann-Hahn match with judicious positioning of the transmitter frequency to ensure coupled spins either side of this are not symmetrically disposed [63]. Alternatively, recording two spectra with differing transmitter offsets leaves genuine ROE peaks little changed but should significantly alter the intensities of those involving a TOCSY step and so facilitate their identification [62]. Modification of the spin-lock itself also leads to a reduction in TOCSY efficiency (Fig. 8.48). Using low-power rf (Fig. 8.48a) with γB_1 comparable to the maximum resonance frequency offset [61] (typically 2–3 kHz) or a pulsed spin-lock [64] (Fig. 8.48b) offer similar attenuation [65], neither of which is complete in real-

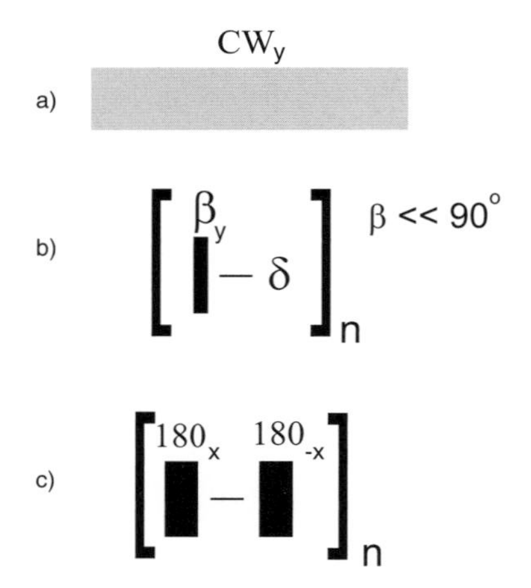

Figure 8.48. Practical mixing schemes for the ROESY experiment. (a) a single, low-power pulse, (b) a pulsed spin-lock comprising a repeated sequence of a small tip angle pulse followed by a short delay, and (c) the Tr–ROESY alternating-phase spin lock.

Table 8.5. Responses observed in the ROESY experiment

Sign of diagonal	Origin of cross-peak	Sign of cross-peak
	Direct ROE	Negative
	Indirect ROE (3-spin effect)	Positive (weak)
	TOCSY	Positive
Positive	TOCSY–ROE	Negative (false ROE)
	ROE–TOCSY	Negative (false ROE)
	Chemical exchange	Positive
	COSY-type	Antiphase/mixed phase

By convention the diagonal is phased to be positive and the sign of all other signals are given relative to this.

ity [2]. More recently, an alternating-phase spin-lock (Fig. 8.48c) has been shown to be effective at reducing TOCSY transfer [66–68], and has been termed Tr–ROESY (transverse ROESY). This approach destroys the Hartman-Hahn match between coupled spins, so suppressing TOCSY transfer, and measures an average of the ROE and NOE since the magnetisation vectors spend time in both the transverse plane and along the longitudinal axis as they follow a swinging 'tic-toc' trajectory (Fig. 8.49). The drawback is a potential reduction in crosspeak intensity relative to conventional ROESY due to this averaging process. In small molecules ($\omega_0\tau_c \ll 1$) there is no theoretical loss, this becomes a factor of 2 for mid-size molecules ($\omega_0\tau_c \approx 1$) and a factor of 4 for very large molecules ($\omega_0\tau_c \gg 1$). Hence there is something of a compromise for molecules in the $\omega_0\tau_c \approx 1$ region for which rotating-frame measurements prove most beneficial. This method generally requires γB_1 to be twice the maximum resonance frequency offset (γB_1 typically 4–6 kHz) and its success is highly dependent on the use of accurately calibrated 180° pulses. The ability of this scheme to suppress interference from the TOCSY mechanism is illustrated in Fig. 8.50 for a tetrameric carbopeptoid **8.21** in which the disappearance of

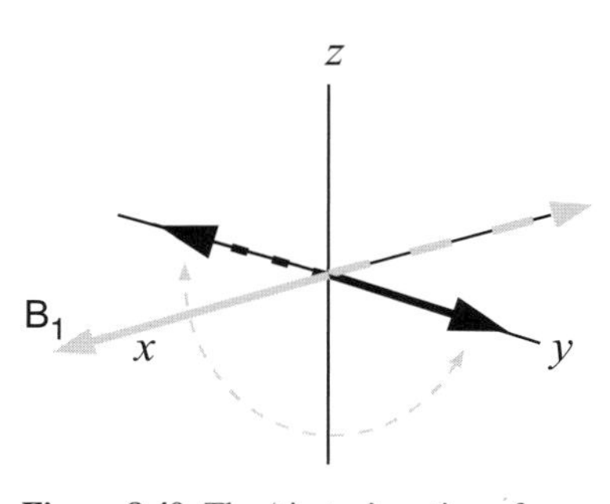

Figure 8.49. The 'tic-toc' motion of spin-locked vectors during the Tr–ROESY mixing sequence.

Figure 8.50. ROESY spectra of a tetrameric carbopeptoid **8.21** recorded with (a) a single 2.6 kHz continuous spin-lock pulse and (b) a 3.7 kHz phase-alternating Tr–ROESY spin-lock at 4.5 ppm. Spectrum (a) is dominated by TOCSY peaks which share the same phase as the diagonal, whereas these have been largely suppressed in (b), so revealing the genuine ROE peaks.

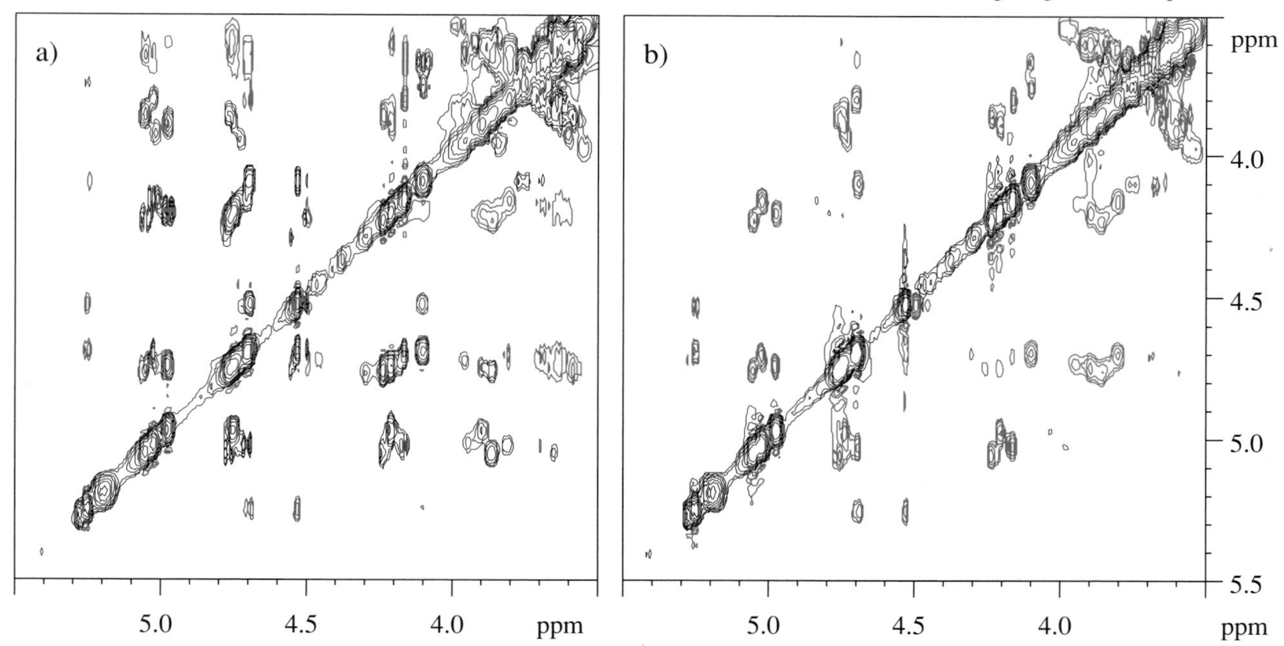

[2] Considerable care should be taken in the rf power used for the spin-lock as damage to transmitters and/or probes may result if high powers are applied for excessively long periods

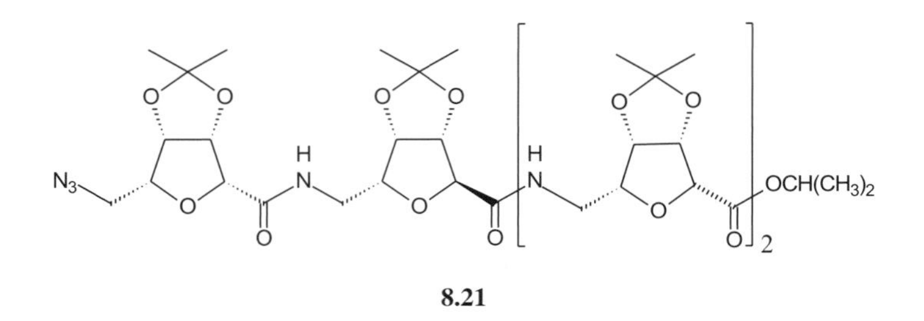

both TOCSY and false-ROE peaks alongside the revelation of genuine ROEs in (b) is clear.

The third complicating factor specific to ROESY is the attenuation of cross-peak intensities as a function of resonance offset from the transmitter frequency [69]. Off-resonance spins experience a spin-lock axis that is tipped out of the x–y plane (Section 3.2.1) resulting in a reduction in observable transverse signal in addition to a reduction in cross-relaxation rates. This is more of a problem for quantitative measurements, although fortunately mid-sized molecules show the weakest dependence of ROE cross-relaxation rates on offset. The so-called compensated ROESY sequence [69] eliminates these frequency-dependent losses should quantitative data be required.

8.8.2. 1D ROESY sequences

In parallel with the 1D NOESY sequences above, the 2D ROESY experiment also has its 1D equivalent (in fact, this was the original ROE experiment [60]) and gradient-selected analogues [70–72] all of which incorporate selective excitation of the target spin. These can be derived from 1D NOESY sequences by incorporation of a suitable spin-lock in place of the 90–τ_m–90 segment of NOESY, and thus require no further elaboration.

8.8.3. Applications

In view of the various complicating factors associated with the ROESY experiment, it is perhaps prudent to avoid using the technique whenever possible, and instead select a steady-state or conventional transient experiment as first choice. Certainly for small molecules within the extreme-narrowing limit there is no theoretical difference between NOEs and ROEs, so there seems little to be gained from undertaking potentially more complex ROE investigations. Nevertheless, situations arise when the use of rotating-frame measurements are unavoidable because conventional experiments yield negligible enhancements ($\omega_0\tau_c \approx 1$), and it is not surprising that most applications of ROESY have involved larger molecules, notably macrocyclic natural products, peptides, oligosaccharides and host–guest complexes, as illustrated by the selected examples below.

In the structure elucidation of the cytotoxic dimeric steroid crellastatin A [73] **8.22**, the basic skeleton of the molecule was derived principally from numerous long-range proton–carbon correlations observed in HMBC spectra. The relative stereochemistry of the bicyclic system at the junction of the two steroid units was derived from analysis of ROESY spectra recorded at 600 MHz with a 400 ms mixing time, some key enhancements being illustrated. The final configuration at C22′ was determined by comparison of the lowest energy molecular mechanics conformers of the two possible stereoisomers with opposite configurations at this position. Only the C22′–R configuration placed the 21′ and 26′ methyl groups in the proximity required for the observation of an ROE between them. The large mass of the molecule, 933 daltons, no doubt precluded the use of NOESY measurements, and the literature contains many

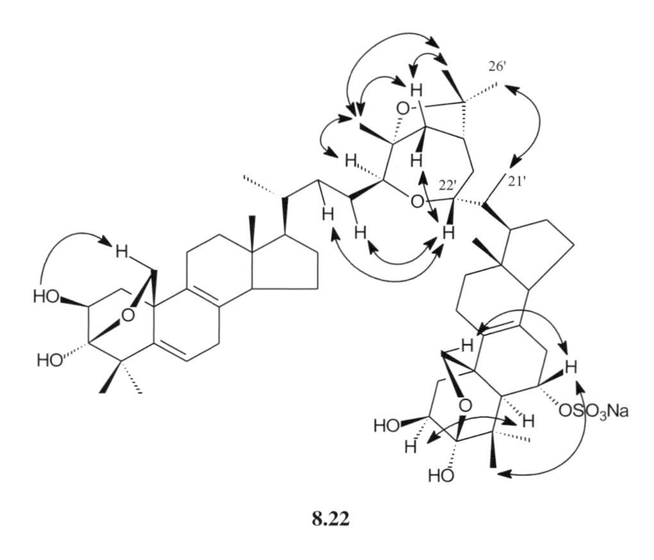

8.22

examples of this sort where rotating-frame measurements have played a crucial role in characterising novel natural products. Conformational studies of peptides have also benefited from the ROESY experiment since it is usually larger peptides that possess defined conformations and hence potentially interesting pharmacological properties. Recent studies of oligomers of β-peptides have identified synthetic hexameric sequences with helical structures in solution that have been characterised by NMR and CD measurements [74]. In the sequence **8.23** distinctive ROEs were observed between the amide proton of residue *i*

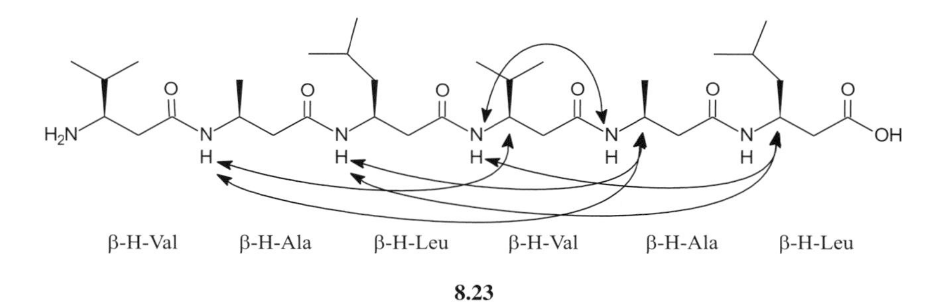

8.23

and the H^{β} protons of residues $i + 2$ and $i + 3$ along the sequence, and were characteristic of the structure adopted by the peptide. These and additional ROEs were used to define 18 distance constraints for use within in molecular dynamics protocols by classifying their intensities as weak, medium and strong (Section 8.7.1), enabling the conformation of the hexapeptide to be calculated. The 14 lowest energy structures derived from this are shown in Fig. 8.51, and clearly illustrate the well defined helical structure adopted in pyridine.

Host–guest inclusion complexes are also frequently restricted to study by rotating-frame measurements because of the requirement for a sufficiently large host to be able to accommodate the guest, and one of the best studied host systems appears to be the cyclodextrins, CD (see structure **8.10** above). One such investigation was into the inclusion of constituents of the natural sex pheromone mix of the olive fruit fly as a means of controlling the release of these volatile substances [75]. Characterisation of the complexes so formed, such as ethyl dodecanoate encapsulated within permethylated α-CD 'cups' **8.24**, was necessarily made from intermolecular ROEs since NOESY enhancements were negligible. Interference from TOCSY transfer were also evident in this work in the form of false ROEs from the guest molecules to the H4 protons of the cyclodextrin. Since these protons sit on the outer face of the CD

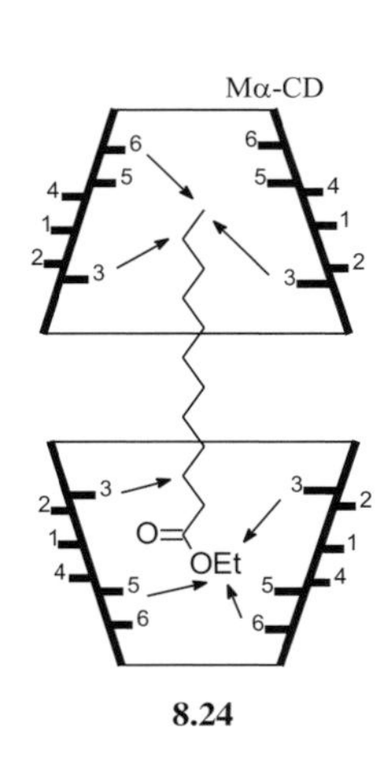

8.24

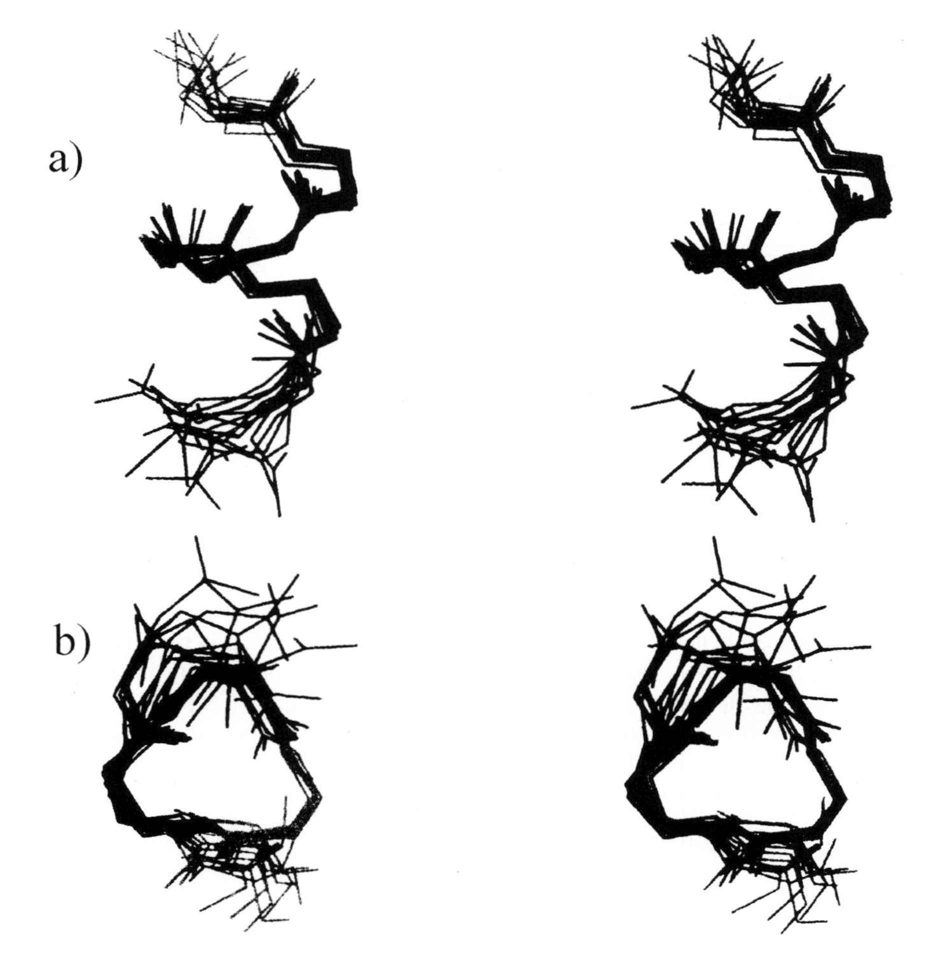

Figure 8.51. Stereo representations of the 14 lowest-energy calculated structures of the hexameric β-peptide **8.23** based on distance restraints derived from ROESY spectra. Side and top views are shown in (a) and (b) respectively, with the side chains of the β-amino acids omitted for clarity (reproduced with permission from [74]).

cup, the ROEs are unlikely to arise from direct effects but instead are produced by transfer of the genuine ROEs between the guest and the inner H3 and H5 protons. Rotating-frame measurements have also been used extensively in the characterisation of relatively bulky organometallic complexes, such as **8.20** in Section 8.7.3 above. A second example is the definition of the absolute configuration of the P-chiral diphosphine, [76] **8.25**. One aspect of this was the identification of the complex as either regio-isomer **8.25a** or **8.25b** prior to further investigation, and this was shown to be **8.25a** through the observation of ROEs from the phenyl rings, as shown. Analysis of the ROESY data further distinguished between two possible diastereomers and so defined the absolute stereochemistry.

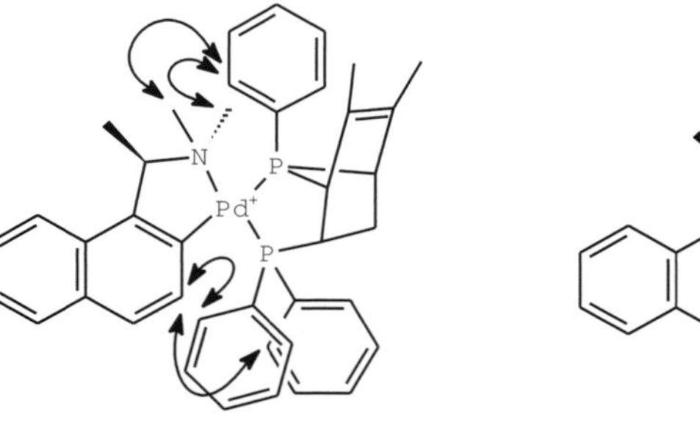

8.25a

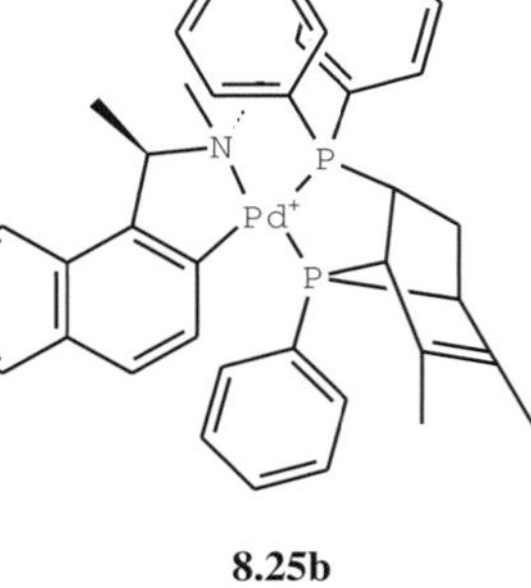

8.25b

8.9. MEASURING HETERONUCLEAR NOES

Beyond its general use as a means of sensitivity enhancement of spin-½ nuclei with low magnetogyric ratios, the specific heteronuclear NOE has occasionally been used as a tool in structural studies [77] and is capable of providing a unique source of structural information in favourable circumstances. Techniques for its observation largely parallel those for homonuclear experiments in the form of 1D steady-state difference and 2D transient experiments, so there is nothing fundamentally new to understand here. The main limitation with these approaches is the low sensitivity associated with the observation of the low-γ spin, meaning heteronuclear NOEs tend to be far less used than their homonuclear proton counterparts. The most widespread applications have involved the ^{1}H–^{13}C NOE, and more recently the ^{1}H–^{6}Li NOE [78], in which the proton is saturated and the heteroatom observed. As the relaxation of a proton bearing carbon is dominated by its dipolar interactions with this proton, only quaternary carbons tend to show useful specific long-range NOEs. Furthermore, the selective irradiation of a proton resonance is often restricted to the parent ^{12}C line leaving the ^{13}C satellites in the spectrum unaffected, which means the ^{13}C centre associated with the target proton usually does *not* show an enhancement under these conditions.

The 1D NOE difference sequence (Fig. 8.52a) is essentially identical to the homonuclear equivalent, except for the addition of broadband proton decoupling during the acquisition. Since this generates non-specific NOE enhancements, a sufficient recovery period must be left between transients to allow this to decay. The presaturation period is now dictated by the potentially long relaxation times of the heteroatom, but beyond this consideration, similar steps are required for optimisation as for the homonuclear experiment. Problems associated with selective presaturation of the target proton resonance may be handled in a similar fashion, although here SPT interferences are suppressed by the use of broadband decoupling. More recently, a 1D gradient-selected transient heteronuclear NOE experiment has been presented that makes use of selective excitation in the generally better dispersed carbon dimension and which uses proton detection for enhanced sensitivity [79].

The 2D sequence [80,81], referred to as HOESY (heteronuclear Overhauser spectroscopy, Fig. 8.52b), avoids the need for selective proton presaturation but naturally suffers from low sensitivity. The sequence parallels that of NOESY,

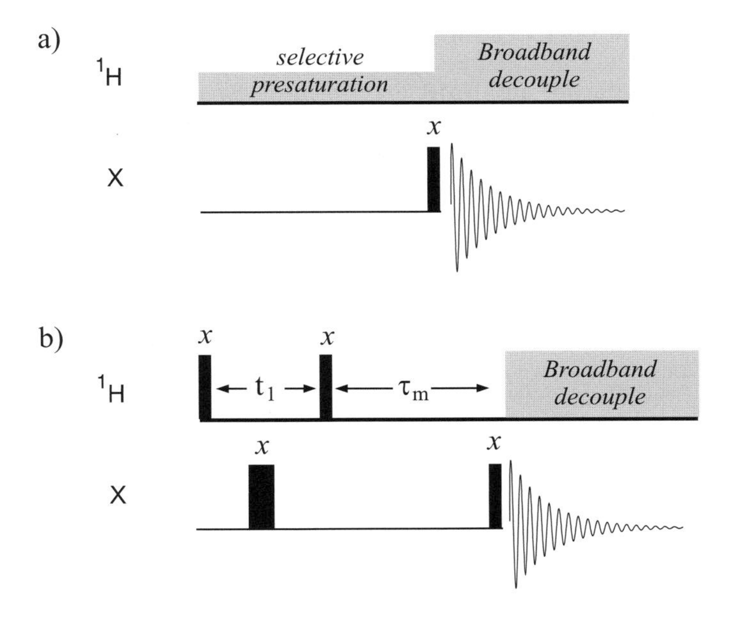

Figure 8.52. Sequences for measuring heteronuclear NOE enhancements. (a) The steady-state experiment and (b) the 2D transient experiment (HOESY).

with the additional 180° pulse during t_1 serving to refocus the $^1J_{CH}$ coupling evolution and hence provide carbon decoupling in f_1. Equations have been presented to allow an estimation of the optimum τ_m for observing structurally informative long-range NOEs to non-protonated nuclei [82]. For the case of the ^{1}H–^{13}C NOE where the carbon relaxation times are considerably longer than those of the proton, maximum crosspeak intensity is achieved when $\tau_m \approx 2T_1$ of the proton. Negative three-spin $^1H \rightarrow ^1H \rightarrow ^{13}C$ effects may also become apparent [83]. A gradient-enhanced, proton-detected version has also been presented and demonstrated for ^{1}H–^{31}P and ^{1}H–^{7}Li spin pairs [84].

Heteronuclear experiments have found use in ^{13}C NMR spectroscopy for the assignment of quaternary carbons and for the determination of stereochemistry in situations where the ^{1}H–^{1}H NOE cannot be applied. Quaternary assignments are more often made via long-range correlation techniques such as HMBC, although the inability to distinguish a priori 2- and 3-bond correlations may cause ambiguity. The short-range nature of the NOE means it can prove useful in distinguishing nearby non-protonated centres (that is, those two bonds away [85]) from more distant ones [86,87]. The use of heteronucelar NOEs in the determination of stereochemistry is illustrated by the example of compound **8.26** [88]. Homonuclear proton NOEs could not be used to distinguish *E* and *Z* isomers of the oxazolones shown, although this was possible from the heteronuclear NOEs generated at the carbonyl carbon following saturation of the methyl protons.

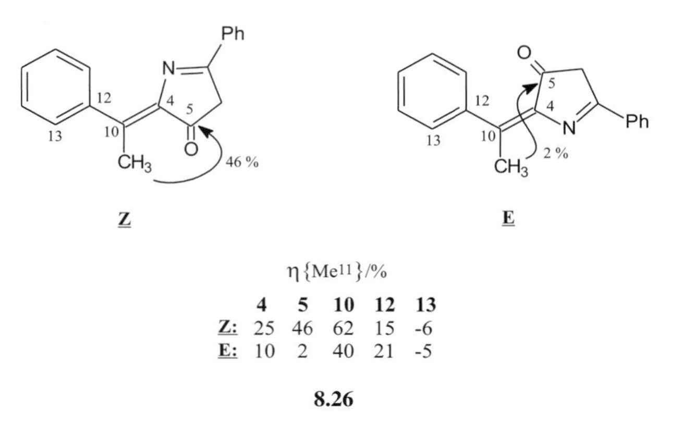

$\eta\{Me^{11}\}/\%$

	4	5	10	12	13
Z:	25	46	62	15	-6
E:	10	2	40	21	-5

8.26

8.10. EXPERIMENTAL CONSIDERATIONS

The prerequisite for any NMR experiment is, of course, a well prepared sample (Section 3.3). In addition to this there are a number of experimental factors that have particular significance regarding NOE measurements, especially for the NOE difference method which is quite intolerant of poor experimental set-up. A stable magnetic field throughout the course of the experiment is essential, which means the lock system that is responsible for the field-frequency regulation must do its job well. Field-drift is especially noticeable and detrimental in the NOE difference method and optimal performance can be achieved by selecting solvents with a strong and sharp lock signal where possible, for example acetone. Solvents with weak deuterium signals such as $CDCl_3$, or broad deuterium resonances such as D_2O, are likely to perform less well in this respect. The choice of solvent and sample temperature can also be used to good effect to control the molecular tumbling rates (correlation times) which can be particularly useful for mid-sized molecules close to the $\omega_0\tau_c \approx 1$ zero-NOE condition. Use of a low-viscosity solvent and/or higher temperatures will increase tumbling rates and move motion toward the extreme-narrowing limit whilst high-viscosity solvents and/or lower temperatures will encourage a move toward the spin-diffusion limit. Lock solvents with temperature dependent shifts, notably D_2O, call for particular attention regarding temperature stability and will almost certainly require active temperature regulation. Drift in the lock resonance frequency arising from sample temperature changes will cause a shift of *all* resonance frequencies in the spectrum as the field follows this drift, disastrous for difference experiments. If no temperature regulation is available or if it performs inadequately then a small quantity of an additional deuterated solvent (5–10% d_6-acetone is a good choice) may be used for locking and usually provides superior results (for the same reason, one should always lock on the CD_3 resonance rather than the OD resonance of d_4-methanol).

It is also worth avoiding very high solute concentrations in NOE samples since transient interactions between protonated molecules promote *intermolecular* dipole–dipole relaxation which competes with the generation of the NOE itself. Another external relaxation source that can quench the NOE is paramagnetic impurities, notably certain metal ions and molecular oxygen. Whilst these are an unwelcome addition to any NMR sample, their presence can be particularly detrimental to NOE studies since paramagnetic relaxation dominates cross-relaxation. Metal ions can be removed by filtration through suitable chelating resins whereas oxygen may be removed with the freeze–pump–thaw method described in Section 3.3. Whether it is worth the effort to remove oxygen traces depends very much on the sample being studied. For very large molecules, efficient cross-relaxation means the NOE builds rapidly to large values and as such oxygen induced paramagnetic relaxation has relatively little significance and degassing little effect. In contrast, external relaxation for smaller molecules that tumble rapidly causes the NOE to be drastically reduced (see Fig. 8.10) and degassing is likely to be of greater benefit. Likewise weak, slowly developing NOEs arising from long-range enhancements or indirect effects will also be enhanced if external relaxation is minimised. In general, the smaller the molecule and the further the internuclear separations being investigated, the greater the gain from solvent degassing. In practice, however, *routine* NOE experiments *are* successful in non-degassed solutions for all but the smallest molecules and, with the above caveats in mind, it is not essential to degas for the majority of organic molecules.

Finally, sample spinning can also be detrimental to NOE experiments (or indeed any NMR experiment when small responses are sought) particularly through the process known as 'Q-modulation', the variation in coupling between the rf coil and the sample caused by a slight wobbling of the sample as it rotates in the probe [3]. With the improved non-spinning lineshapes attainable with modern shim assemblies, sample spinning can no longer be recommended for either 1- or 2-D NOE experiments.

REFERENCES

[1] D. Neuhaus and M.P. Williamson, The Nuclear Overhauser Effect in: Structural and Conformational Analysis, VCH Publishers, New York, 1989.

[2] R.A. Bell and J.K. Saunders, *Can. J. Chem.*, 1970, **48**, 1114–1122.

[3] D.M. Hodgson and P.J. Comina, *Tetrahedron Lett.*, 1996, **37**, 5613–5614.

[4] R. Colombo, F. Colombo, A.E. Derome, J.H. Jones, D.L. Rathbone and D.W. Thomas, *J. C. S. Perkin Trans.* **1**, 1985.

[3] This is sometimes evident as a modulation or jumping of the probe tuning profile and by the appearance of responses in 1D spectra that are symmetric about main resonances at frequencies equal to the spinning speed yet are indifferent to shimming and of random phase, both of which distinguishes then from 'spinning sidebands' caused by field inhomogeneity. Owing to their random properties they tend to be averaged away over many transients but are nonetheless undesirable artefacts

[5] J.K. Robinson, V. Lee, T.D.W. Claridge, J.E. Baldwin and C.J. Schofield, *Tetrahedron*, 1998, **54**, 981–996.
[6] J.E. Baldwin, T.D.W. Claridge, F.A. Heupel and R.C. Whitehead, *Tetrahedron Lett.*, 1994, **35**, 7829–7832.
[7] J.E. Baldwin, L. Bischoff, T.D.W. Claridge, F.A. Heupel, D.R. Spring and R.C. Whitehead, *Tetrahedron*, 1997, **53**, 2271–2290.
[8] B.T. Farmer, S. Macura and L.R. Brown, *J. Magn. Reson.*, 1988, **80**, 1–22.
[9] J.K.M. Sanders and J.D. Mersh, *Prog. Nucl. Magn. Reson. Spectrosc.*, 1982, **15**, 353–400.
[10] J.C. Lindon and A.G. Ferrige, *Prog. Nucl. Magn. Reson. Spectrosc.*, 1980, **14**, 27–66.
[11] D. Neuhaus, G. Wagner, M. Vasák, J.H.R. Kági and K. Wüthrich, *Eur. J. Biochem.*, 1985, **151**, 257–273.
[12] M. Kinns and J.K.M. Sanders, *J. Magn. Reson.*, 1984, **56**, 518–520.
[13] A.J. Shaka, C. Bauer and R. Freeman, *J. Magn. Reson.*, 1984, **60**, 479–485.
[14] R.J. Smith, D.H. Williams, J.C.J. Barna, I.R. McDermott, K.D. Haegele, F. Piriou, J. Wagner and W. Higgins, *J. Am. Chem. Soc.*, 1985, **107**, 2849–2857.
[15] S. Macura, Y. Huang, D. Suter and R.R. Ernst, *J. Magn. Reson.*, 1981, **43**, 259–281.
[16] N.H. Andersen, K.T. Nguyen, C.J. Hartzell and H.L. Eaton, *J. Magn. Reson.*, 1987, **74**, 195–211.
[17] E.T. Olejniczak, J.C. Hoch, C.M. Dobson and F.M. Poulsen, *J. Magn. Reson.*, 1985, **64**, 199–206.
[18] M.P. Williamson and D. Neuhaus, *J. Magn. Reson.*, 1987, **72**, 369–375.
[19] V. Dötsch, G. Wider and K. Wüthrich, *J. Magn. Reson. (A)*, 1994, **109**, 263–264.
[20] T. Parella, F. Sánchez-Ferrando and A. Virgili, *J. Magn. Reson.*, 1997, **125**, 145–148.
[21] R. Wagner and S. Berger, *J. Magn. Reson. (A)*, 1996, **123**, 119–121.
[22] G. Otting, *J. Magn. Reson.*, 1990, **86**, 496–508.
[23] L. Mitschang, J. Keeler, A.L. Davis and H. Oschkinat, *J. Biomol. NMR*, 1992, **2**, 545–556.
[24] G. Bodenhausen, G. Wagner, M. Rance, O.W. Sorensen, K. Wuthrich and R.R. Ernst, *J. Magn. Reson.*, 1984, **59**, 542–550.
[25] N.H. Andersen, H.L. Eaton and X. Lai, *Magn. Reson. Chem.*, 1989, **27**, 515–528.
[26] J.W. Keepers and T.L. James, *J. Magn. Reson.*, 1984, **57**, 404–426.
[27] A. Kumar, G. Wagner, R.R. Ernst and K. Wüthrich, *J. Am. Chem. Soc.*, 1981, **103**, 3654–3658.
[28] J.N.S. Evans, Biomolecular NMR Spectroscopy, Oxford University Press, Oxford, 1995.
[29] J. Cavanagh, W.J. Fairbrother, A.G. Palmer and N.J. Skelton, Protein NMR Spectroscopy: Principles and Practice, Academic Press, San Diego, 1996.
[30] G. Esposito and A. Pastore, *J. Magn. Reson.*, 1988, **76**, 331–336.
[31] M. Reggelin, H. Hoffman, M. Köck and D.F. Mierke, *J. Am. Chem. Soc.*, 1992, **114**, 3272–3277.
[32] J. Stonehouse, P. Adell, J. Keeler and A.J. Shaka, *J. Am. Chem. Soc.*, 1994, **116**, 6037–6038.
[33] K. Stott, J. Stonehouse, J. Keeler, T.L. Hwang and A.J. Shaka, *J. Am. Chem. Soc.*, 1995, **117**, 4199–4200.
[34] K. Stott, J. Keeler, Q.N. Van and A.J. Shaka, *J. Magn. Reson.*, 1997, **125**, 302–324.
[35] N. Fusetani, K. Shinoda and S. Matsunaga, *J. Am. Chem. Soc.*, 1993, **115**, 3977–3981.
[36] K. Wüthrich, NMR of Proteins and Nucleic Acids, Wiley, New York, 1986.
[37] H.J. Dyson and P.E. Wright, *Ann. Rev. Biophys. Biophys. Chem.*, 1991, **20**, 519–538.
[38] M.P. Williamson and J.P. Waltho, *Chem. Soc. Rev.*, 1992, **21**, 227–236.
[39] S. Anderson, T.D.W. Claridge and H.L. Anderson, *Angew. Chem. Int. Ed. Engl.*, 1997, **36**, 1310–1313.
[40] L. Fielding, N. Hamilton and R. McGuire, *Magn. Reson. Chem.*, 1997, **35**, 184–190.
[41] T.D.W. Claridge, J.M. Long, J.M. Brown, D. Hobbs and M.B. Hursthouse, *Tetrahedron*, 1997, **53**, 4035–4050.
[42] J. Jeener, B.H. Meier, P. Bachmann and R.R. Ernst, *J. Chem. Phys.*, 1979, **71**, 4546–4553.
[43] I.D. Campbell, C.M. Dobson, R.G. Ratcliffe and R.J.P. Williams, *J. Magn. Reson.*, 1978, **29**, 397–417.
[44] J.J. Led and H. Gesmar, *J. Magn. Reson.*, 1982, **49**, 444–463.
[45] K.G. Orrell, V. Šik and D. Stephenson, *Prog. Nucl. Magn. Reson. Spectrosc.*, 1990, **22**, 141–208.
[46] K.G. Orrell and V. Šik, *Ann. Rep. NMR. Spectrosc.*, 1993, **27**, 103–171.
[47] C.L. Perrin and T.J. Dwyer, *Chem. Rev.*, 1990, **90**, 935–967.
[48] J. Sandström, Dynamic NMR Spectroscopy, Academic Press, London, 1982.
[49] M. Oki, Applications of Dynamic NMR Spectroscopy to Organic Chemistry, VCH Publishers, Weinheim, 1985.
[50] D.S. Stephenson and G. Binsch, *J. Magn. Reson.*, 1978, **32**, 145–152.
[51] E.F. Derose, J. Castillo, D. Saulys and J. Morrison, *J. Magn. Reson.*, 1991, **93**, 347–354.
[52] R. Ramachandran, C.T.G. Knight, R.J. Kirkpatrick and E. Oldfield, *J. Magn. Reson.*, 1985, **65**, 136–141.

[53] I. Pianet, E. Fouquet, M. Pereyre, M. Gielen, F. Kayser, M. Biesemans and R. Willem, *Magn. Reson. Chem.*, 1994, **32**, 613–617.
[54] J.E.H. Buston, T.D.W. Claridge and M.G. Moloney, *J. C. S. Perkin Trans*, 1995, **2**, 639–641.
[55] C.L. Perrin, *J. Magn. Reson.*, 1989, **82**, 619–621.
[56] V.S. Dimitrov and N.G. Vassilev, *Magn. Reson. Chem.*, 1995, **33**, 739–744.
[57] E.W. Abel, T.P.J. Coston, K.G. Orrell, V. Sik and D. Stephenson, *J. Magn. Reson.*, 1986, **70**, 34–53.
[58] G. Bodenhausen and R.R. Ernst, *J. Am. Chem. Soc.*, 1982, **104**, 1304–1309.
[59] K.G. Orrell and V. Šik, *Ann. Rep. NMR. Spectrosc.*, 1987, **19**, 79–173.
[60] A.A. Bothner-By, R.L. Stephens, J. Lee, C.D. Warren and R.W. Jeanloz, *J. Am. Chem. Soc.*, 1984, **106**, 811–813.
[61] A. Bax and D.G. Davis, *J. Magn. Reson.*, 1985, **63**, 207–213.
[62] D. Neuhaus and J. Keeler, *J. Magn. Reson.*, 1986, **68**, 568–574.
[63] T.M. Chan, D.C. Dalgarno, J.H. Prestegard and C.A. Evans, *J. Magn. Reson.*, 1997, **126**, 183–186.
[64] H. Kessler, C. Griesinger, R. Kerssebaum, K. Wagner and R.R. Ernst, *J. Am. Chem. Soc.*, 1987, **109**, 607–609.
[65] A. Bax, *J. Magn. Reson.*, 1988, **77**, 134–147.
[66] T.L. Hwang and A.J. Shaka, *J. Am. Chem. Soc.*, 1992, **114**, 3157–3159.
[67] T.L. Hwang, M. Kadkhodaei, A. Mohebbi and A.J. Shaka, *Magn. Reson. Chem.*, 1992, **30**, 24–34.
[68] T.L. Hwang and A.J. Shaka, *J. Magn. Reson. (B)*, 1993, **102**, 155–165.
[69] C. Griesinger and R.R. Ernst, *J. Magn. Reson.*, 1987, **75**, 261–271.
[70] P. Adell, T. Parella, F. Sánchez-Ferrando and A. Virgili, *J. Magn. Reson. (B)*, 1995, **108**, 77–80.
[71] C. Dalvit and G. Bovermann, *Magn. Reson. Chem.*, 1995, **33**, 156–159.
[72] M.J. Gradwell, H. Kogelberg and T.A. Frenkiel, *J. Magn. Reson.*, 1997, **124**, 267–270.
[73] M.V. D'Auria, C. Giannini, A. Zampella, L. Minale, C. Debitus and C. Roussakis, *J. Org. Chem.*, 1998, **63**, 7382–7388.
[74] D. Seebach, M. Overhand, F.N.M. Kühnle, B. Martinoni, L. Oberer, U. Hommel and H. Widmer, *Helv. Chim. Acta*, 1996, **79**, 913–941.
[75] A. Botsi, K. Yannalopoulou, B. Perly and E. Hadjoudis, *J. Org. Chem.*, 1995, **60**, 4017–4023.
[76] B.-H. Aw, S. Selvaratnam, P.-H. Leung, N.H. Rees and W. McFarlane, *Tetrahedron Asymmetry*, 1996, **7**, 1753–1762.
[77] K.E. Kövér and G. Batta, *Prog. Nucl. Magn. Reson. Spectrosc.*, 1987, **19**, 223–266.
[78] W. Bauer, in: Lithium Chemistry, ed. A.-M. Sapse and P.V.R. Schleyer, Wiley, New York, 1995.
[79] K. Stott and J. Keeler, *Magn. Reson. Chem.*, 1996, **34**, 554–558.
[80] P.L. Rinaldi, *J. Am. Chem. Soc.*, 1983, **105**, 5167–5168.
[81] C. Yu and G.C. Levy, *J. Am. Chem. Soc.*, 1983, **106**, 6533–6537.
[82] K.E. Kövér and G. Batta, *J. Magn. Reson.*, 1986, **69**, 344–349.
[83] K.E. Kövér and G. Batta, *J. Magn. Reson.*, 1986, **69**, 519–522.
[84] W. Bauer, *Magn. Reson. Chem.*, 1996, **34**, 532–537.
[85] F. Sánchez-Ferrando, *Magn. Reson. Chem.*, 1985, **23**, 185–191.
[86] F.J. Leeper and J. Staunton, *J. C. S. Chem. Commun.*, 1982, 911–912.
[87] M.F. Aldersley, F.M. Dean and B.E. Mann, *J. C. S. Chem. Commun.*, 1983, 107–108.
[88] C. Cativiela and F. Sánchez-Ferrando, *Magn. Reson. Chem.*, 1985, **23**, 1072–1075.

Chapter 9

Experimental methods

This final chapter principally considers a collection of experimental methods that find widespread use in modern high-resolution NMR yet cannot be considered as individual techniques in their own right. Rather they are sequence segments that are used within, or may be added to, the techniques already encountered to enhance their information content or to overcome a range of experimental limitations. In addition to the familiar simple rf pulses, they are the components used to construct modern NMR experiments. Depending on context, they may be considered as essential to the correct execution of the desired experiment or viewed as an optional extra to enhance performance. For example, the broadband decoupling of protons during carbon acquisition routinely makes use of so-called 'composite pulse decoupling' schemes to achieve efficient removal of all proton couplings, and is nowadays considered essential. Pulsed field gradients (PFGs) could equally well have been included in this chapter since in many cases they serve as an alternative means of signal selection to traditional phase-cycling procedures. The fact that PFGs have been described in a previous chapter lays testament to the manner in which these have pervaded modern NMR sequences and may now be considered routine for many multi-dimensional experiments. The chapter concludes with a brief introduction to some relatively new experimental methods (at least new in the context of high-resolution NMR) which are finding increasing popularity.

9.1. COMPOSITE PULSES

The plethora of NMR multipulse sequences used throughout modern chemical research all depend critically on nuclear spins experiencing rf pulses of precise flip angles for their successful execution. Careful pulse width calibration is essential in this context yet despite this a number of factors invariably conspire against the experimentalist and produce pulses that deviate from these ideals, so leading to degraded experimental performance. Beyond the limitations imposed by the capabilities of the spectrometer itself with regard to the accuracy of pulse timing, powers, rf phase shifts and so on (which the operator can do little about), there are two notable contributions to pulse imperfections that can be addressed experimentally, namely rf (B_1) inhomogeneity and off-resonance effects.

Inhomogeneity in the applied rf field means not all nuclei within the sample volume experience the desired pulse flip angle (Fig. 9.1b), notably those at the sample periphery. This is similar in effect to the (localised) poor calibration of pulse widths and references to 'rf (or B_1) inhomogeneity' below could equally read 'pulse width miscalibration'. Modifications that make sequences

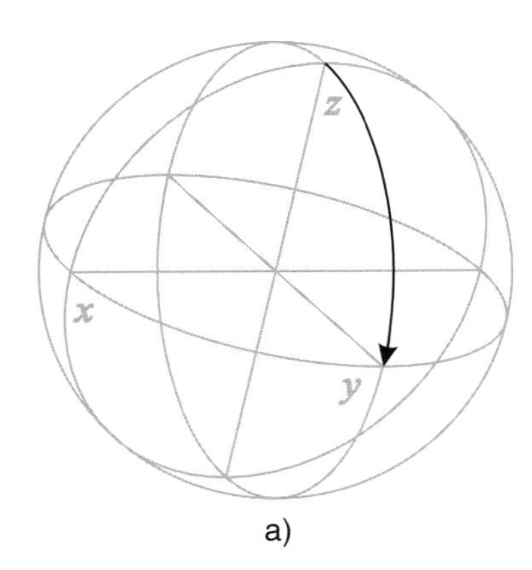

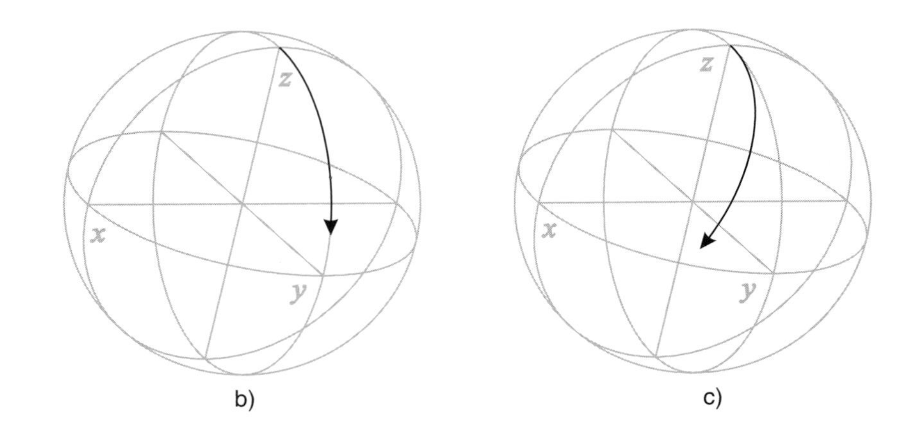

Figure 9.1. Imperfections in pulse excitation. (a) A perfect $90°_x$ pulse applied at equilibrium, (b) the effect of B_1-inhomogeneity and (c) the effect of off-resonance excitation.

more tolerant of rf inhomogeneity therefore also provide the experimentalist with some leeway when setting up pulse experiments. Off-resonance excitation arises when the transmitter frequency does not exactly match the Larmor frequency of the spin, and has already been introduced in Section 3.2. To recap briefly, off-resonance magnetisation vectors experience an effective rf field in the rotating frame that is tilted out of the x–y plane, rather than the applied B_1 field itself (Fig. 9.2). These vectors are driven about this effective field and thus do not following the trajectory of the ideal on-resonance case (Fig. 9.1c). The problem stems from the fact that pulse NMR instruments use monochromatic rf radiation when polychromatic excitation is required.

Combating the effects of B_1 inhomogeneity lies very much with the design of the rf coil and this is in the hands of probe manufacturers who have made considerable progress. Overcoming off-resonance effects directly requires the use of higher power transmitters that are able to excite over wider bandwidths, but there are two principal problems with this. First is the inability of probe circuitry to sustain such high powers without being damaged and, second, is the insidious effect of rf sample heating which may damage precious samples. Improvements in probe circuitry that minimise this heating have largely been paralleled by increases in static field strengths so despite technical developments the problem remains significant. This has greater relevance at higher field strengths and for those nuclei that exhibit a wide chemical shift range, such as ^{13}C or ^{19}F. Consider data acquisition with a simple 90° pulse which has a uniform excitation bandwidth of around $\pm\gamma B_1$ Hz or $2\gamma B_1$ Hz in total, where γB_1 is the rf field strength (Section 3.5). Taking a typical value of 10 μs for the 90° pulse, this corresponds to a γB_1 of 25 kHz. For ^1H excitation at 400 MHz, the total excited bandwidth is thus 125 ppm which is clearly ample, and for ^{13}C is a satisfactory 500 ppm. In contrast, a 180° inversion pulse has a total bandwidth (here taken to mean >90% effective) of only ca. 0.4 γB_1, which in this example corresponds to 25 ppm for proton, again

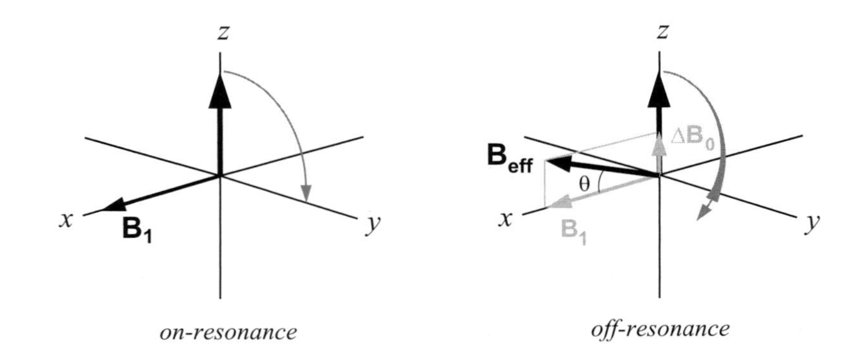

Figure 9.2. Off-resonance excitation causes the bulk vector to rotate about an effective rf field, tipped out of the transverse plane.

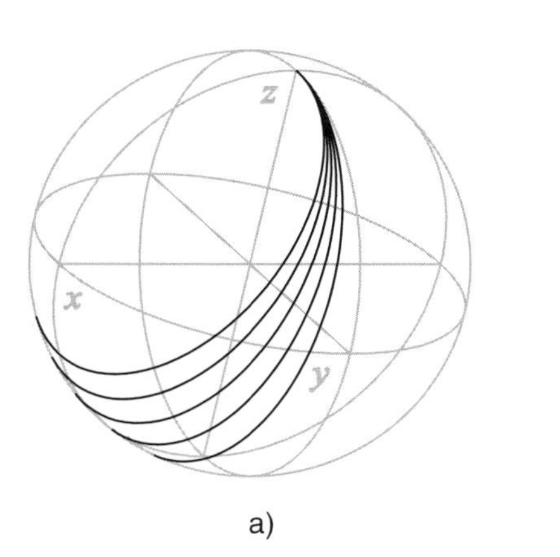

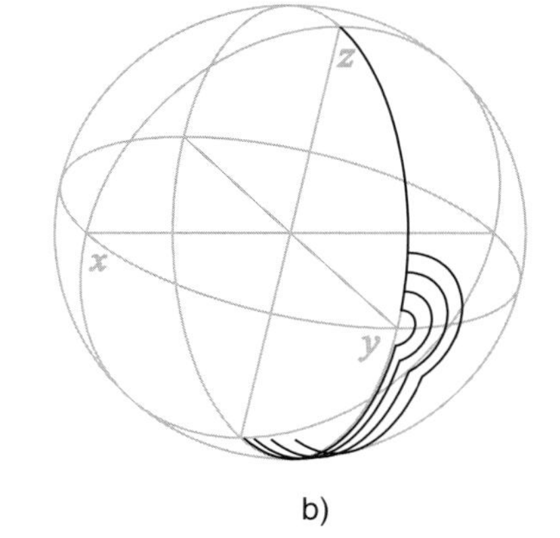

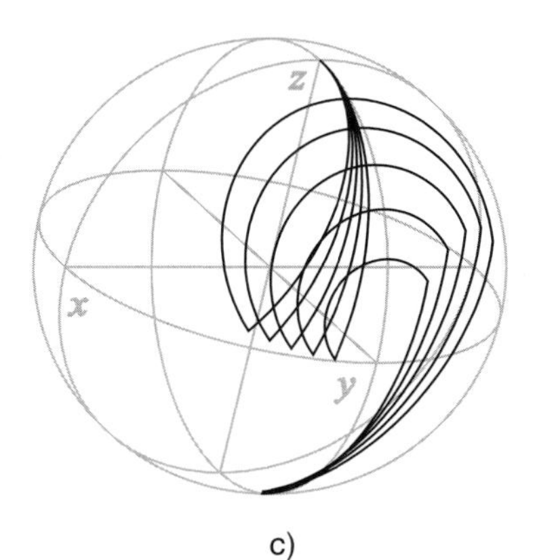

Figure 9.3. The simulated effect of (a) resonance offset (of 0.2–0.6γB_1) on the inversion properties of a single $180°_x$ pulse, and the compensating effect of the $90_x180_y90_x$ composite 180° pulse on (b) B_1-inhomogeneity and (c) resonance offset.

acceptable, but only 100 ppm for ^{13}C. Carbon nuclei that resonate outside this bandwidth will experience reduced inversion efficiency. Off-resonance errors for a 180° inversion pulse are illustrated in Fig. 9.3a in which it is clear that the trajectories of the magnetisation vectors are far from ideal and terminate further from the 'South Pole' as the offset increases, so generating unwanted transverse magnetisation. The combined effect of such pulse imperfections in multipulse sequences is to degrade sensitivity and to potentially introduce spurious artefacts, and so it is advantageous to somehow compensate for these.

An elegant approach to compensating deficiencies arising from B_1 inhomogeneity or resonance offset is the use of a *composite pulse* [1]. This is, in fact, a cluster of pulses of varying duration and phase that are used in place of a single pulse and have a net rotation angle equal to this but have greater overall tolerance to these errors. The first composite pulse [2], and still one of the most widely used, is the three pulse cluster $90_x180_y90_x$, equivalent to a $180°_y$ pulse. The self-compensating abilities of this sequence for spin inversion are illustrated by the trajectories of Fig. 9.3. For both B_1 inhomogeneity (Fig. 9.3b) and for resonance offset (Fig. 9.3c) the improved performance is apparent from the clustering of vectors close to the South Pole. The effective bandwidth of the $90_x180_y90_x$ composite pulse is ca. $2\gamma B_1$, five times greater than the simple 180° pulse, a considerable improvement from such a simple modification. A single 90° excitation pulse, in contrast, is itself effective over ca. $2\gamma B_1$ owing to a degree of 'in-built' self-compensation for off-resonance effects when judged by its ability to generate transverse magnetisation. Here, the increased effective field experienced by off-resonance spins tends to drive the vectors further toward the transverse plane (as described in Section 3.2.1) at the expense of frequency dependent phase errors (Fig. 9.4). For high-power pulses, these errors are approximately a linear function of frequency and are readily removed through phase correction of the spectrum. Compensation for B_1 inhomogeneity may be achieved with the composite 90_x90_y (or better still 90_x110_y [1]) sequence which places the magnetisation vectors closer to the transverse plane than a single pulse (Fig. 9.5). This may prove beneficial in situations where the elimination of z-magnetisation is of utmost importance, such as the NOE difference experiment (see discussions in Section 8.6.1; the $270_x360_{-x}90_y$ composite 90° pulse has been suggested as a better alternative in this case [3]).

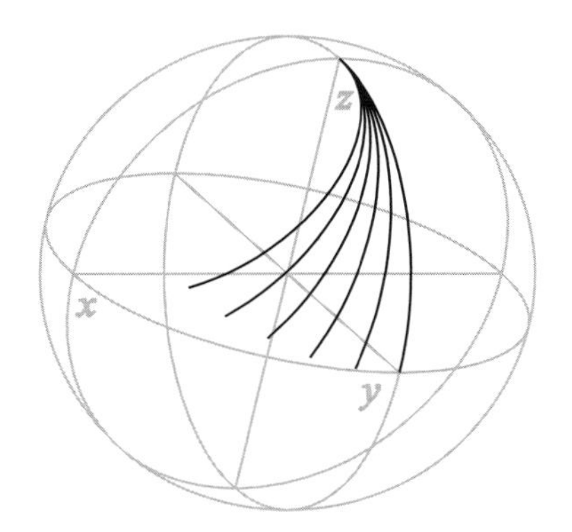

Figure 9.4. The 90° pulse is self-compensating with respect to resonance offset in its ability to generate transverse magnetisation, although there is no compensation for phase errors.

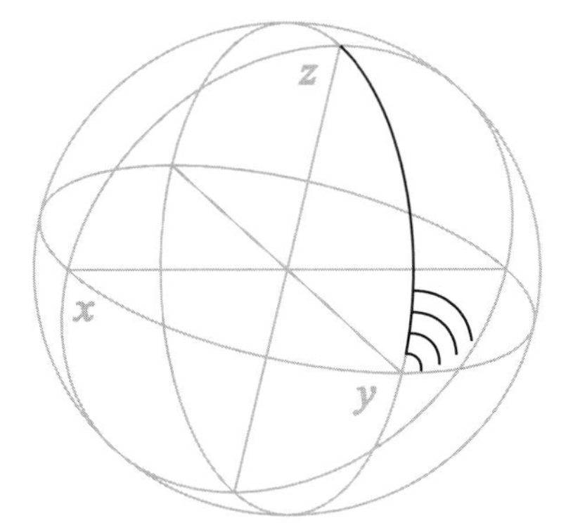

Figure 9.5. The 90_x90_y composite 90° pulse compensates for B_1 inhomogeneity by placing the vector closer to the transverse plane.

A composite pulse may be included within a pulse sequence directly in place of a single pulse. The *relative* phase relationships of pulses within each cluster must be maintained but are otherwise stepped according to the phase-cycling associated with the single pulse they have replaced. The selection of a suitable composite pulse is not always a trivial process, as discussed below.

Table 9.1. Properties of a single 180° pulse and some composite 180° pulses

Composite pulse	Duration (×180°)	Bandwidth [1] (γB_1)	Properties	Ref.
Inversion				
180_x	1	0.4		
$90_y180_x90_y$	2	2.0		[2]
$90_y240_x90_y$	2.3	1.2	More uniform inversion profile than $90_y180_x90_y$	[1]
$360_x270_{-x}90_y360_{-y}270_y90_x$	8	2.0	Compensated for B_1 inhomogeneity (±25%) and resonance offset	[5]
$38_x111_{-x}159_x250_{-x}$	3.1	2.6		[6]
$151_{247}342_{182}180_{320}342_{182}151_{247}$	6.5	2.0	Requires small rf phase shifts[2]	[7,8]
Refocusing				
180_0	1	0.5		
$90_y180_x90_y$	2	0.6	Introduces phase errors to spin-echoes	
$90_y240_x90_y$	2.3	1.0	Introduces phase errors to spin-echoes	[9]
$360_x270_{-x}90_y360_{-y}270_y90_x$	8	1.0	Compensated for B_1 inhomogeneity (±25%) and resonance offset	[5]
$336_x246_{-x}10_y74_{-y}10_y246_{-x}336_x$	7	1.2	Phase distortionless spin-echoes	[10]
$151_{247}342_{182}180_{320}342_{182}151_{247}$	6.5	2.0	Phase distortionless spin-echoes, requires small rf phase shifts [2]	[7,8]

Only a selection of available pulses is presented with the emphasis on compensation for resonance offset effects and on sequences of short total duration. Individual pulses are presented in the form XX_{yy} where XX represents the nominal pulse flip angle and yy its relative phase, either in units of 90° (x, y, $-x$, $-y$) or directly in degrees where appropriate. Sequences are split into those suitable for (A) population inversion (act on M_z) or (B) spin-echo generation (act on M_{xy}).

[1] Bandwidths are given as fractions of the rf field strength and represent the total region over which the pulses have ca. 90% efficacy.

[2] Small rf phase shifts refer to those other than multiples of 90°.

9.1.1. A myriad of pulses

The design of composite pulses has been a major area of research in NMR for many years and a huge variety of sequences have been published and many reviewed [4]. In reality only a rather small subset of these have found widespread use in routine NMR applications, some of which are summarised in Table 9.1. One general limitation is that the sequences which provide the best compensation tend to be the longest and most complex, often many times longer than the simple pulse they have replaced, and as such may not be readily implemented or are unsuitable for use within a pulse sequence. Furthermore, most composite pulses are effective against *either* B_1 inhomogeneity *or* offset effects, but not both simultaneously, so some compromise must be made, usually with the emphasis on offset compensation in high-resolution NMR (in other applications, such as in vivo spectroscopy, B_1 compensation has far greater importance because of the heterogeneous nature of the samples studied). In some cases, dual compensation has been included [5], although again at the expense of longer sequences (Table 9.1). More recently, numerical methods have been applied to the design of new sequences which aim to keep the total duration acceptably small, most of which also make use of small rf phase shifts, that is, those other than 90°. A comparison of the performance of some composite inversion pulses is illustrated in Fig. 9.6, and clearly demonstrates the offset compensation they provide relative to a single 180° pulse.

9.1.2. Inversion vs. refocusing

One complicating factor associated with the implementation of composite pulses arises from the fact that many composite sequences have been designed with a particular initial magnetisation state in mind and may not perform well, or give the expected result, when the pulses act on other states. The two principal applications are the use of composite 180° pulses for population *inversion*, for example in the inversion-recovery experiment, or in the ubiquitous spin-echo for *refocusing*, in which they act on longitudinal and transverse

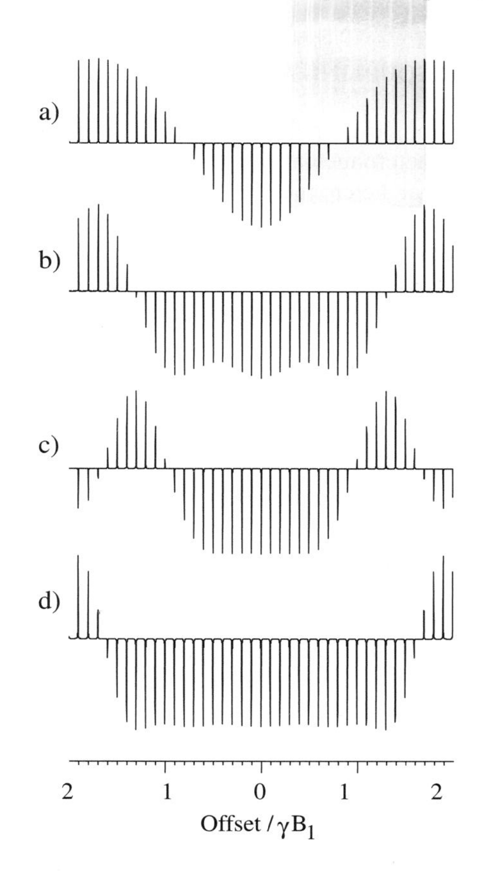

Figure 9.6. Simulated inversion profiles for (a) 180_x, (b) $90_y180_x90_y$, (c) $90_y240_x90_y$ and (d) $38_x111_{-x}159_x20_{-x}$ (see Fig. 9.3 and Table 9.1 also).

magnetisation respectively. Thus, for example, the $90_x180_y90_x$ sequence provides little offset compensation when used as a refocusing pulse (Table 9.1). Although it offers compensation for B_1 inhomogeneity and thus makes the magnitude of the echoes less sensitive to this by returning vectors closer to the x–y plane, it introduces errors in the *phase* of the echoes which may be detrimental to the overall performance of the experiment. The $90_x240_y90_x$ sequence has been proposed as a better refocusing element for offset compensation [9], and more sophisticated sequences have been generated that exhibit low phase-distortion [10,11]. Recent sequences designed by numerical optimisation methods have been designed without a predefined initial state and are therefore equally effective as inversion or phase-distortionless refocusing elements [7,8].

The moral of all this is that considerable care must be taken when composite pulses are introduced into pulse sequences. It is wise to test the compensated pulse sequences versus the original uncompensated sequences on a known sample to see if improvements are achieved in reality. This offers a check on whether the composite elements have been correctly introduced and on whether the spectrometer is capable of correctly executing the desired sequences, many of which demand accurate control of rf amplitudes and phases for extended periods. The experimental performance of the composite pulse itself is best tested with a simple experiment, such as an inversion sequence ($180_{x,-x}$–90_x–FID_x) or spin-echo sequence (90_x–$180_{x,y,-x,-y}$–$\mathrm{FID}_{x,-x}$) with a composite 180° pulse, and by analysing the results for a single resonance. Offset compensation may be investigated by stepping the transmitter frequency for the 180° composite pulse away from the initial on-resonance position (but setting it to be on-resonance for the 90° pulse) and B_1 inhomogeneity may be simulated by reducing pulse flip angles.

9.2. BROADBAND DECOUPLING AND SPIN-LOCKS

An area in which composite pulses have been applied routinely with great success is broadband heteronuclear decoupling. Their use in this area arises from the realisation that heteronuclear decoupling may be achieved by the continuous application of a train of 180° inversion pulses on the decoupled spin. This is illustrated in Fig. 9.7 for a X–H pair in the presence of proton decoupling. Following excitation of X, doublet vectors diverge for a period τ according to J_{XH}. A 180° pulse applied to protons at this time inverts the proton α and β states thus reversing the sense of precession of the X-spin vectors which refocus after a further period τ. If this procedure is repeated during the observation of X at a fast rate relative to the magnitude of J_{XH}, the action of the coupling is suppressed and hence the spectrum appears decoupled.

From discussions in the previous section it should be apparent that the application of a sequence of simple 180° pulses is unlikely to give effective decoupling over a wide bandwidth or in the presence of B_2 inhomogeneity or pulse miscalibration (note the use of the symbol B_2 with reference to a *decoupling* rf field). Further, by reducing the period τ to be infinitely small, the sequence becomes a single, continuous decoupler pulse (so-called *continuous wave* decoupling) and by the same arguments this is also a poor decoupling sequence, particularly off-resonance. One solution to achieving decoupling over wider bandwidths is to employ composite inversion pulses that display superior off-resonance performance, and modern *composite pulse decoupling* (CPD) has evolved from the original $90_x180_y90_x$ and $90_x240_y90_x$ sequences.

Simply inserting these in place of single 180° pulses turns out not to be the full answer, however. Defining the composite inversion cluster $90_x180_y90_x$ as an element R, then we can see that the sequence RR (two sequential composite pulses) should return spin vectors back to the +z axis for the rotation sequence to begin again. However, these elements are themselves not perfect (see Fig. 9.3), leaving small errors which will accumulate if the sequence is simply repeated. The trick is to repeat the process but in reverse, with all rf phases inverted so as to counteract these errors, giving rise to the element $\overline{R}$ ($90_{-x}180_{-y}90_{-x}$) and the so-called 'magic-cycle' $RR\overline{RR}$. This original decoupling sequence is termed MLEV-4. It was subsequently realised that small residual errors from the magic-cycle could be compensated by further nesting of these elements to produce 'super-cycles', such as $RR\overline{RR}\,\overline{R}RR\overline{R}\,\overline{RR}RR\,R\overline{RR}R$, giving, in this case, MLEV-16. These cycles prove more effective over greater bandwidths without the requirement of excessive rf powers.

Numerous such composite pulse decoupling sequences have been developed along these lines over the years (Table 9.2), the most widely used being WALTZ-16 whose basic inversion element is $90_x180_{-x}270_x$ (or more succinctly $1\overline{2}3$, hence the name). This provides very effective decoupling over a band-

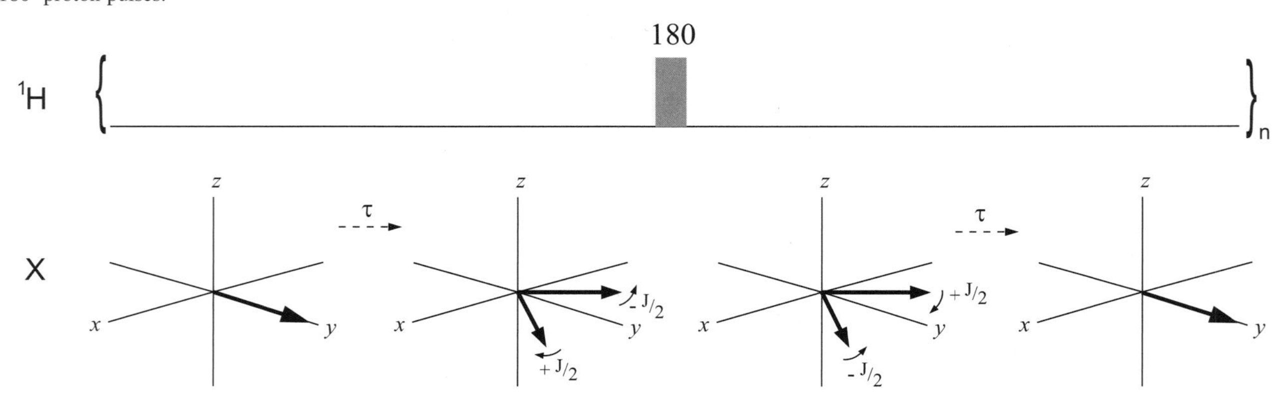

Figure 9.7. Heteronuclear broadband proton decoupling with a sequence of 180° proton pulses.

Table 9.2. Selected composite-pulse sequences for broadband decoupling and spin-locking

Sequence	Bandwidth (γB_2)	Application	Ref.
Decoupling			
Continuous wave	<0.1	Selective decoupling only	
MLEV-16	1.5	^{1}H decoupling	[12]
WALTZ-16	2.0	High-resolution ^{1}H decoupling	[13,14]
DIPSI-2	1.2	Very high-resolution ^{1}H decoupling	[15]
GARP	4.8	X-nucleus decoupling	[16]
SUSAN	6.2	X-nucleus decoupling	[17]
Spin-locking			
MLEV-17	0.6	TOCSY mixing scheme	[18]
DIPSI-2	1.2	TOCSY mixing scheme	[19]
FLOPSY-8	1.9	TOCSY mixing scheme	[20]

The technicalities of broadband decoupling are extensively discussed in [21].

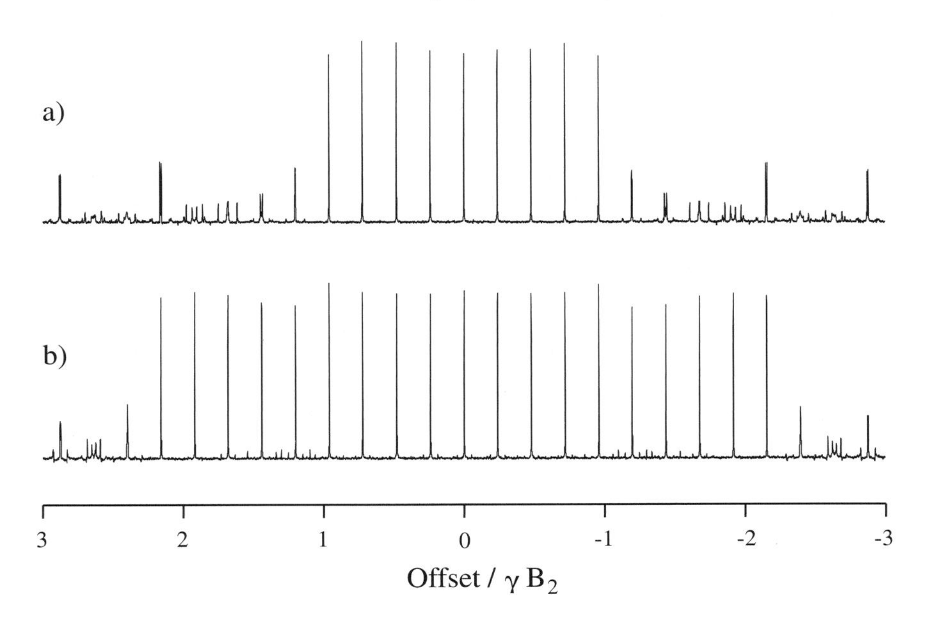

Figure 9.8. Experimental comparison of the decoupling offset profiles of (a) WALTZ-16 and (b) GARP relative to the decoupling rf field strength γB_2. The sample was carbon-13 labelled methanoic acid in D_2O.

width of $2\gamma B_2$ leaving only small residual splittings (<0.1 Hz) and is thus favoured in high-resolution applications such as broadband proton decoupling of heteronuclear spectra. This bandwidth can, however, be limiting when decoupling nuclei with greater chemical shift dispersion, for example ^{13}C or ^{31}P, as is required in proton-detected heteronuclear shift correlation sequences such as HMQC or HSQC. For such cases, other CPD sequences have been developed through computational optimisation, which have less stringent requirements for residual linewidths (<1.5 Hz). The most popular of these has been GARP (Table 9.2) which is effective over $4.8\gamma B_2$ (Fig. 9.8). With the continued increase in static fields, even greater decoupling bandwidths become necessary and the most effective approach so far appears to be the use of adiabatic decoupling methods, introduced briefly below.

9.2.1. Spin-locks

The ability to 'spin-lock' magnetisation along a predefined axis plays an essential role in the TOCSY experiment, and Section 5.7.1 has already introduced the idea that a single continuous pulse or a series of closely spaced

180° pulses can, in principle, be used in this context to repeatedly refocus chemical shift evolution whilst allowing *homonuclear* couplings to evolve. Following from the above discussions, a better approach is clearly to use a series of 180° composite pulses which offer compensation for resonance offset and rf inhomogeneity and many of the sequences originally designed as heteronuclear decoupling sequences (which require repeated spin inversions) have since been applied as spin-lock sequences (which require repeated spin refocusing). The original and most widely used is based on the MLEV-16 sequence, in which each cycle is followed by a 60° pulse to compensate errors that would otherwise accrue during extended mixing, producing the popular MLEV-17 spin-lock (Table 9.2). More sophisticated sequences have since been developed that allow for the influence of homonuclear couplings, a factor not considered during the design of the early heteronuclear decoupling sequences. In particular, the DIPSI-2 sequence has better performance than MLEV or WALTZ in these circumstances, and is thus also widely used in the TOCSY experiment.

9.2.2. Adiabatic pulses

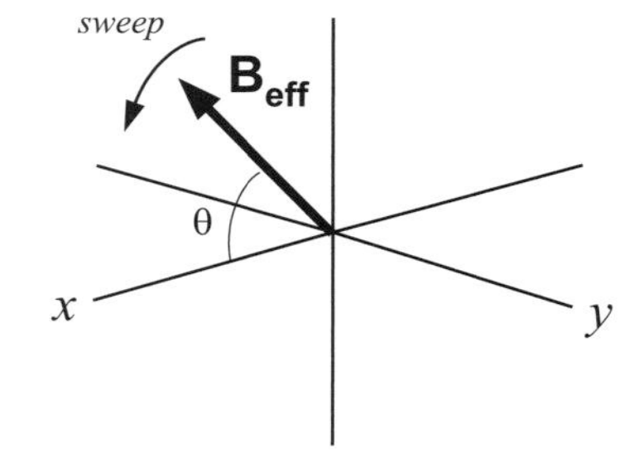

Figure 9.9. The adiabatic inversion pulse. An rf frequency sweep during the pulse causes the effective rf field experienced by the spins to trace an arc from the $+z$-axis to the $-z$-axis, dragging with it the bulk magnetisation vector.

An alternative approach to spin inversion that is becoming a popular route to very efficient broadband decoupling is offered by adiabatic pulses. Rather than pulses applied at a single frequency, these employ a frequency-sweep *during* the pulse that begins far from resonance at a positive offset, passes through the resonance condition and finally terminates far from resonance with a negative offset. During this, the effective rf field, B_{eff}, experienced by the spins begins along the $+z$-axis, traces an arc which passes through the x–y plane at the on-resonance condition and finishes along the $-z$-axis (Fig. 9.9). If the sweep is sufficiently 'slow', magnetisation vectors initially at equilibrium will continually circle about B_{eff} during the pulse and will be 'dragged along' by the effective field thus also terminating at the South Pole and experiencing the desired inversion. More formally, the sweep must be slow enough to satisfy the adiabatic condition:

$$\left|\frac{d\theta}{dt}\right| \ll \gamma B_{eff} \tag{9.1}$$

(where θ represents the angle between B_{eff} and the x-axis, as above) but should be fast with respect to spin relaxation (hence it is referred to as *adiabatic rapid passage*). So long as the adiabatic condition is met, precise inversion is achieved regardless of the magnitude of the applied rf field and hence these pulses are very tolerant of B_1 inhomogeneities. Furthermore, they are effective over very wide bandwidths and are thus well suited to broadband decoupling.

To ensure B_{eff} genuinely begins and ends along the $\pm z$-axis when finite offsets are used, the pulse amplitude may be gently truncated at each end which, in one example, produces a sausage-shaped amplitude profile leading to the name WURST [22]. Placing adiabatic inversion pulses within suitable supercycles has resulted in decoupling schemes that are effective over far greater bandwidths ($>20\ \gamma B_2$) than those attainable with conventional composite pulse sequences [23–26]. These are therefore suitable for the decoupling of heteronuclei on the highest field instruments currently available, and may be expected to become more widely employed.

9.3. SELECTIVE EXCITATION AND SHAPED PULSES

So far discussions in this chapter have dealt solely with the use of so-called *hard* pulses, that is, pulses that (ideally) are equally effective over the whole chemical shift range. We have also seen tools, in the form of composite pulses, that help one approach more closely this ideal. In some instances, however, it can be a distinct advantage if only a selected region of the spectrum is influenced by a so-called *soft* pulse, and a number of examples utilising such *selective excitation* have been presented in Chapters 5, 6 and 7. Principal applications in a chemical context include:

- Reduced dimensionality of nD sequences, for example 1D analogues of 2D experiments [27].
- Selective removal of unwanted resonances, for example solvent suppression [28].
- Extraction of specific pieces of information, for example the measurement of specific long-range ^{1}H–^{13}C coupling constants [29,30].
- 1D sequences which intrinsically require the selection of a single resonance, for example inversion (saturation) transfer experiments in studies of chemical dynamics [31].

Replacing hard pulses with their selective counterparts within a multidimensional sequence leads to a spectrum of lower dimensionality having a number of advantages over its fully fledged cousin. [32] 1D analogues of more conventional 2D experiments allow greater digital resolution and thus a more detailed insight into fine structure; these equate to high-resolution slices through the related 2D experiment at the shift of the selected spin, for example the 1D TOCSY experiment of Section 5.7.3. They may also benefit by being quicker to acquire and process, and demand less storage space. In short, such methods may provide faster and more detailed answers when a specific structural question is being addressed, and discussions on the use of selective pulses in high-resolution NMR have featured in a number of extensive reviews [32–35].

The basic approach to producing a *selective* pulse is to reduce the rf power and increase the duration of the pulse. By reducing B_1, the frequency spread over which the pulse is effective likewise diminishes. At the same time, the speed of rotation of vectors being driven by the rf also decreases meaning longer pulses are required to achieve the desired tip angle. The simplest selective pulse is therefore a long, weak rectangular pulse, by analogy with the usual rectangular hard pulse. Unfortunately, this has a rather undesirable excitation profile in the form of side-lobes that extend far from the principal excitation window (Fig. 9.10a), resulting in rather poor selectivity. The origin of the undesirable side-lobes lies in the sharp edges of the pulse which introduce a sinc-like oscillation to the profile; notice the similarity with the 'sinc wiggles' in spectra caused by the premature truncation of an FID (Section 3.2.3). These may be suppressed by smoothing the edges of the pulse (Fig. 9.10b), just as apodisation of the truncated FID reduces the wiggles, and such *pulse shaping* has given rise to a multitude of selective pulses with various characteristics that find use in modern NMR. A small selection of these are considered

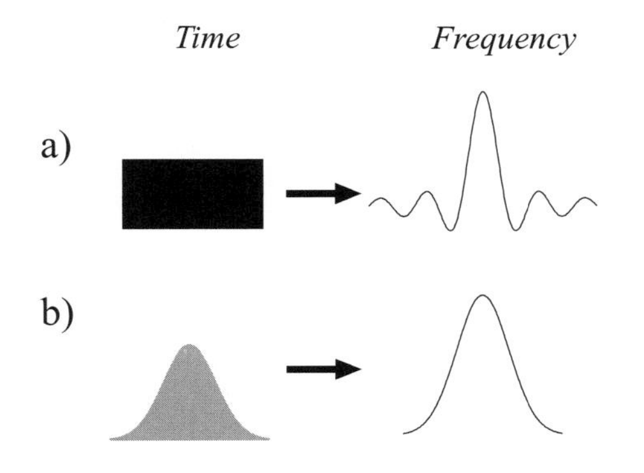

Figure 9.10. Schematic excitation profiles for (a) a low-power rectangular pulse and (b) a smoothly truncated *shaped* pulse.

below, which have been chosen for their general applicability and/or otherwise desirable properties for high-resolution spectroscopy.

9.3.1. Shaped soft pulses

Associated with the use of selective pulses are a number of experimental factors that have a considerable bearing on the selection (and design) of a soft pulse. In short, the key features are its:

- Duration
- Frequency profile, and
- Phase behaviour

Soft pulses are typically 1–100 ms long, three orders of magnitude longer than hard pulses, which may have implications with regard to chemical shift and coupling evolution and to relaxation losses during the pulse. Therefore, it is desirable to have a soft pulse that is as short as possible but still able to deliver the desired selection. This selection is defined by the frequency bandwidth over which the pulse is effective. Ideally, the bandwidth profile should be rectangular, with everything outside the desired window insensitive to the pulse; such a profile is sometimes referred to [36] as the 'top-hat' for obvious reasons. The phase of the excited resonances should also be uniform over the whole excitation window, although in reality this can be difficult to achieve since off-resonance effects are severe with the weak B_1 fields used. Unless corrected, this results in considerable phase distortion away from exact resonance that can be detrimental to multipulse sequences where the precise control of magnetisation is often required. These primary considerations are addressed below for some commonly encountered shaped pulses, whilst their practical implementation is described in Section 9.3.4.

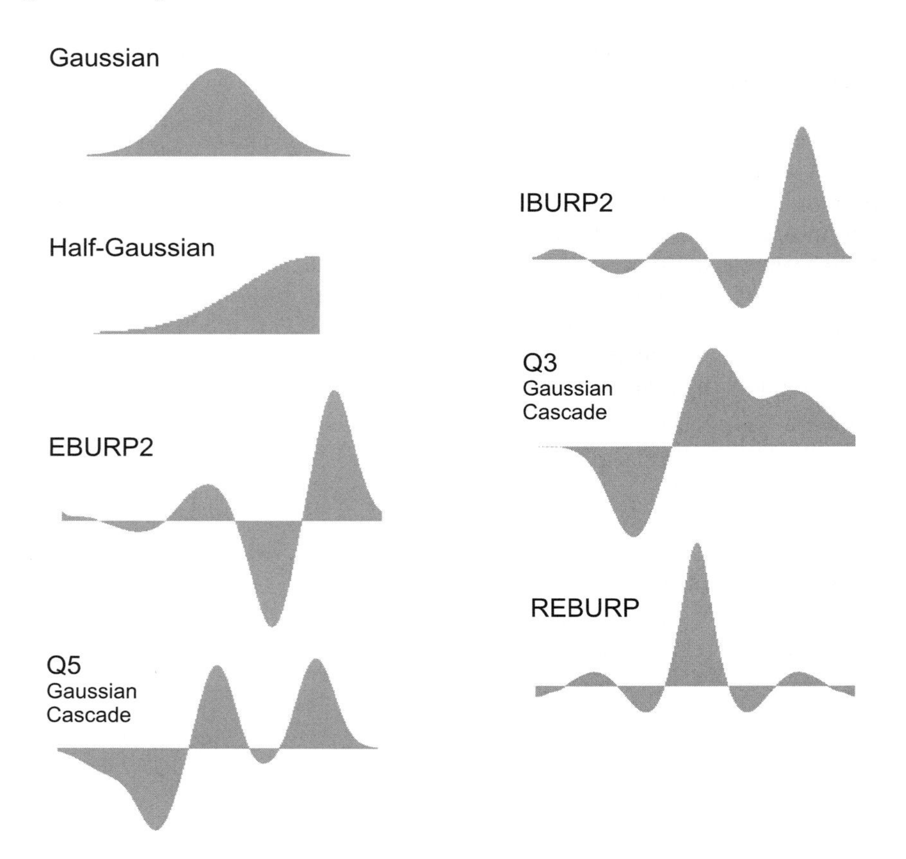

Figure 9.11. Time domain profiles of some common shaped, selective excitation pulses.

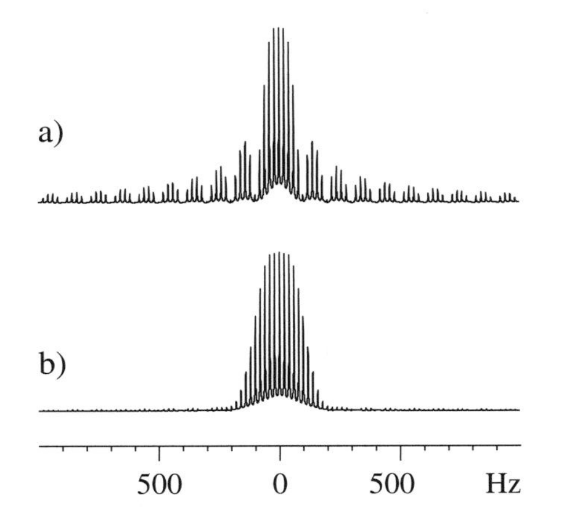

Figure 9.12. Absolute value frequency domain excitation profiles for (a) a rectangular pulse and (b) a Gaussian shaped pulse.

Gaussian pulses

The original and experimentally simplest shaped pulse has the smooth Gaussian envelope [37] (Fig. 9.11). The (absolute-value) frequency excitation profile is also Gaussian, tailing away rapidly with offset and, although not an ideal top-hat profile, is clearly superior to the rectangular pulse (Fig. 9.12). Just outside the principal excitation window, magnetisation vectors also feel the effect of the pulse but tend to be driven back toward the starting position so experience no *net effect*, as can be seen in the trajectories of Fig. 9.13a. This figure also illustrates the problem of phase dispersion across the bandwidth for different offsets. For a pulse of phase x both the desired M_y and unwanted M_x components (absorptive and dispersive responses, respectively) are created by the pulse, producing a significant phase gradient which can be corrected in the simplest of experiments, but in general may be problematic. In addition, inverted responses are also generated (those terminating in the $-y$ hemisphere) and thus the profile retains some undesirable oscillatory behaviour (this is hidden in Fig. 9.12 because an absolute-value display was used, but can be seen in Fig. 9.14 below)

Better alternatives for the excitation of longitudinal magnetisation are the half-Gaussian [38] which, as the name suggests, is simply a Gaussian profile terminated at the midpoint, and the 270° Gaussian [39]. The half-Gaussian pulse does not produce negative side-lobes because vectors never reach the $-y$ hemisphere, although it still generates a considerable dispersive component, M_x, as is apparent from the trajectories of Fig. 9.13b. The dispersive responses may be removed by a phase-alternated hard 'purge' pulse applied orthogonally to and immediately after the half-Gaussian. This cancels any M_x components, retaining only the desired M_y components and so removing the phase gradient.

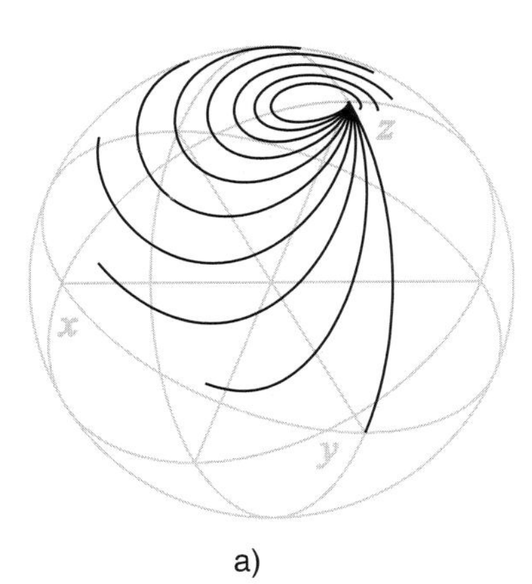

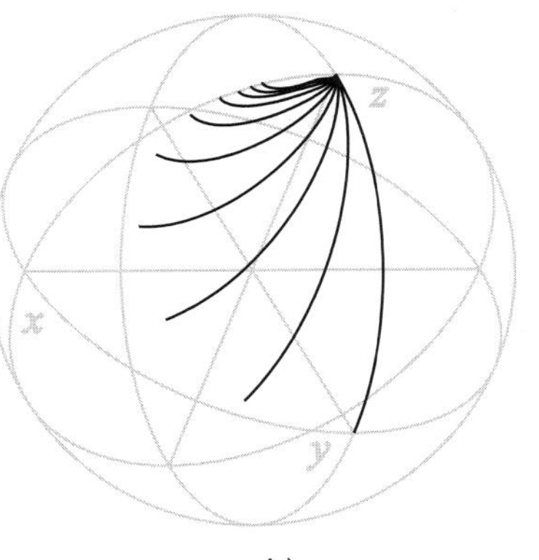

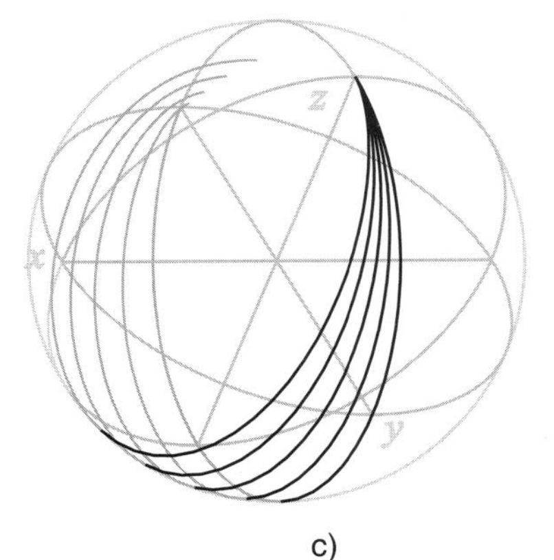

Figure 9.13. Simulated excitation trajectories as a function of resonance offset for (a) the Gaussian, (b) the half-Gaussian and (c) the Gaussian-270.

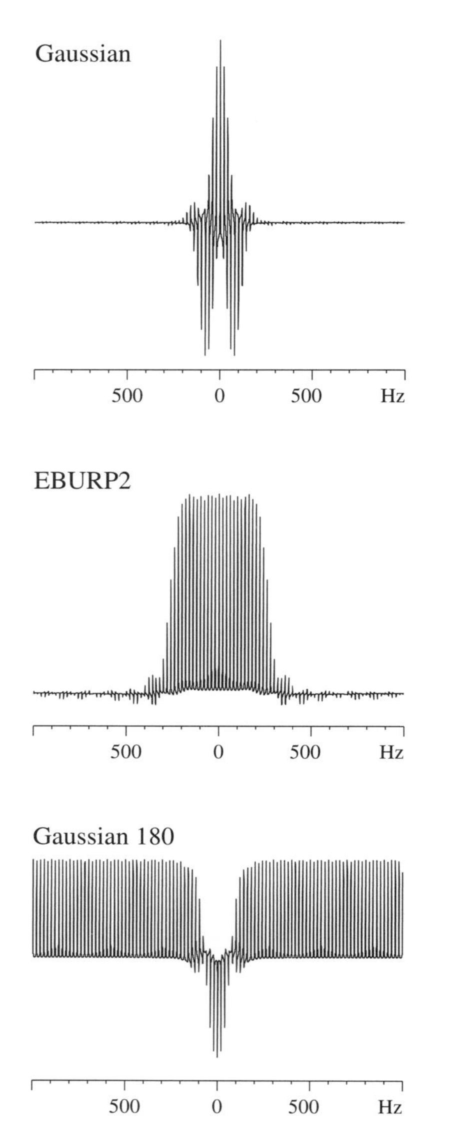

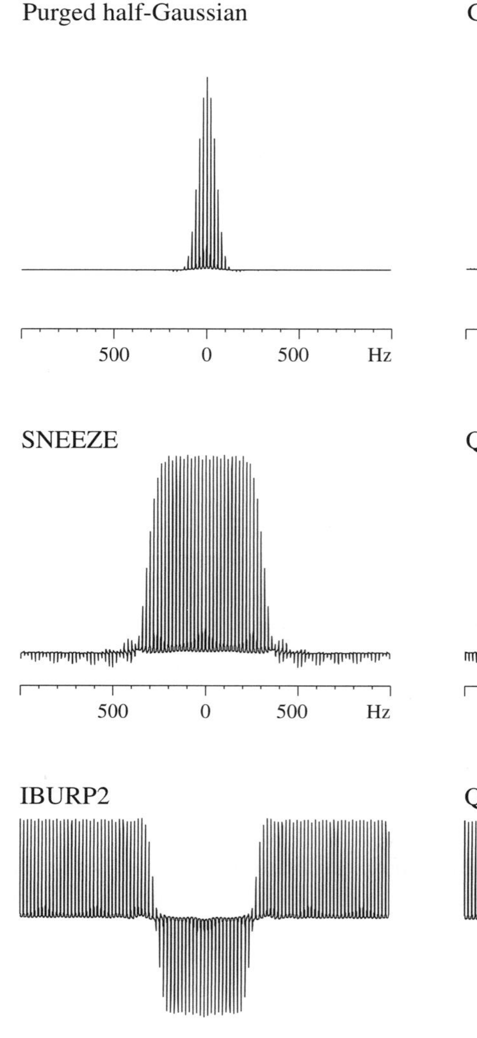

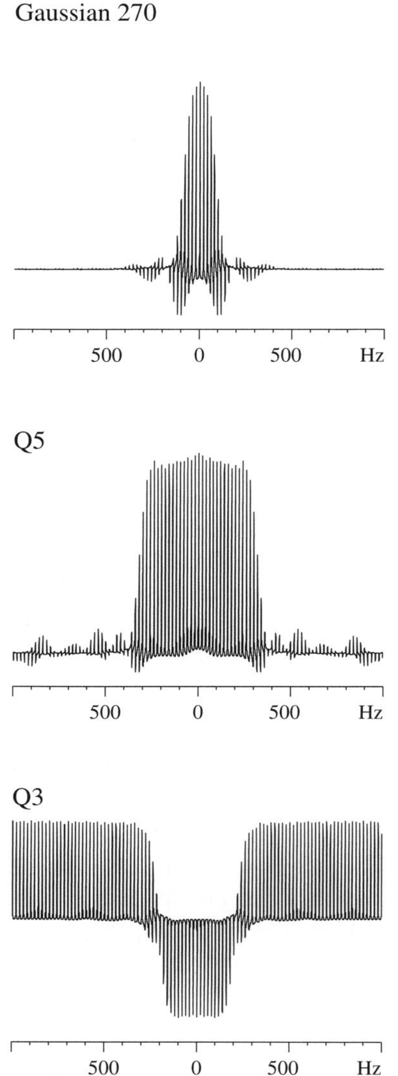

Figure 9.14. Simulated excitation profiles of selected shaped pulses (of 10 ms duration); see Table 9.3. The inversion profiles (lower trace) were simulated with a 180(soft)–90(hard) sequence.

The improved phase properties of this scheme gives rise to the name 'purged half-Gaussian' [40] (Fig. 9.14). The 270° Gaussian pulse equates to a net −90° rotation and is identical to the 90° pulse except for a three-fold amplitude difference. The interesting feature of this pulse is that for spins close to resonance it has a self-refocusing effect on both chemical shifts and coupling constants (Fig. 9.13c) and thus has better phase properties than the 90° Gaussian pulse (Fig. 9.14). As it does not require purging it is simple to implement and is the excitation pulse of choice when high selectivity is not critical.

Pure-phase pulses

More elaborate pulse shapes have been developed over the years which aim to produce a near top-hat profile yet retain uniform phase for all excited resonances within a predefined frequency window. These operate without the need for purging pulses or further modifications, allowing them to be used directly in place of hard pulses. They are typically generated by computerised procedures which result in more exotic pulse envelopes (and acronyms! Fig. 9.11) that drive magnetisation vectors along rather more tortuous trajectories than the simpler Gaussian-shaped cousins. Trajectories are shown in Fig. 9.15

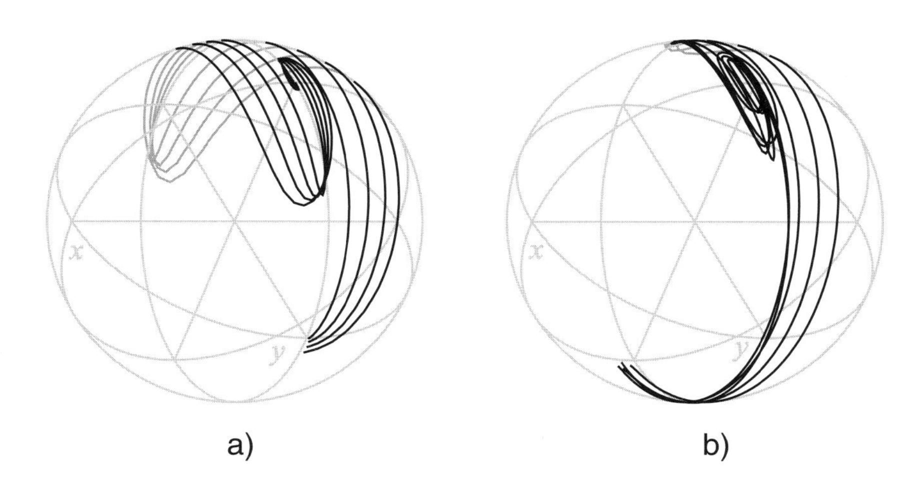

Figure 9.15. Simulated trajectories of nuclear spin vectors as a function of resonance offset during the (a) EBURP-2 90° excitation pulse and (b) the IBURP2 180° inversion pulse.

for two members of the BURP family of pulses (Band-selective, Uniform Response, Pure phase pulses) [41,42] namely the EBURP2 excitation pulse and the IBURP2 inversion pulse (see Section 9.3.4 and Table 9.3 below). Despite the globetrotting journeys undertaken, trajectories for all offsets within the effective window terminate close to the ideal endpoint with only small intensity or phase errors. The corresponding profiles for these pulses are also shown in Fig. 9.14 and illustrate the uniform, pure-phase behaviour (although experimentally these depend *critically* on accurate pulse calibrations). An experimental demonstration of selective excitation with the EBURP2 pulse is shown in Fig. 9.16 for two different excitation bandwidths. Interestingly, a different design approach has led to a similar family of pulses with largely similar properties, the so-called Gaussian cascades [43,44] (Table 9.3 and Fig. 9.11). As the name implies, these are composed of clusters of Gaussian pulses whose net effect is closer to the ideal than any single element, reminiscent of the composite hard pulses of Section 9.1.

One feature of these more elaborate envelopes is that they have often been designed with a particular function in mind and may not perform well at anything else. For example, the EBURP pulses are 90° *excitation pulses* designed to act on longitudinal magnetisation and will not perform as desired on transverse magnetisation. Likewise, selective 180° inversion pulses may not work as refocusing elements in the generation of spin-echoes. Those designed to act on *any* initial magnetisation state are referred to as *universal pulses*. More complex profiles also tend to be longer than their simpler counterparts for a given excitation bandwidth and tip angle, meaning relaxation effects may be problematic, particularly for the case of larger molecules or for very long,

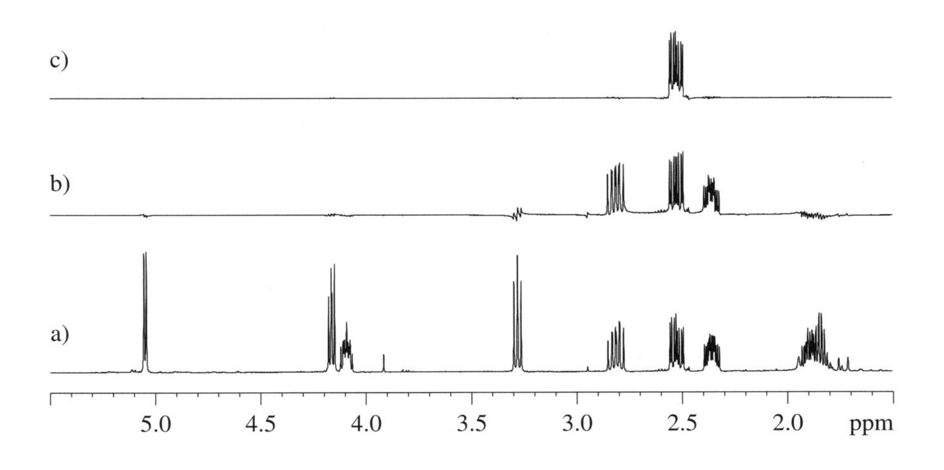

Figure 9.16. Selective excitation with the EBURP2 band-selective pulse, using a pulse duration of (b) 14 ms and (c) 100 ms.

Table 9.3. Properties of some shaped soft pulses

Shaped pulse	Application	Bandwidth factor	Attenuation factor/-dB	Ref.
Rectangular	Universal	1.1	–	–
90° pulses				
Gaussian	Universal	2.1	7.7	[37]
Purged half-Gaussian	Excitation only	0.8	7.7	[38,40]
Gaussian-270	Excitation only	1.3	16.7	[39]
EBURP2	Excitation only, Pure-phase	4.9	24.3	[42]
SNEEZE *	Excitation only, Pure-phase	5.8	26.6	[57]
UBURP	Universal, Pure-phase	4.7	32.2	[42]
G4 Gaussian Cascade	Excitation only, Pure-phase	7.8	25.4	[43]
Q5 Gaussian Cascade	Universal, Pure-phase	6.2	25.3	[44]
180° pulses				
Gaussian-180	Universal	0.7	13.7	[37]
IBURP2	Inversion only, Pure phase	4.5	25.9	[42]
REBURP	Universal, Pure phase	4.6	28.0	[42]
G3 Gaussian Cascade	Inversion only, Pure phase	3.6	23.1	[43]
Q3 Gaussian Cascade	Universal, Pure phase	3.4	22.4	[44]

Universal pulses act equally on any initial magnetisation state whereas excitation and inversion pulses are designed to act on longitudinal magnetisation only. The bandwidth factor is the product of the pulse duration, Δt, and the excitation bandwidth, Δf, which is here defined as the excitation window over which the pulse is at least 70% effective (net pulse amplitude within 3 dB of the maximum; other publications may define this value for higher levels and so quote smaller bandwidth factors). Use this factor to estimate the appropriate pulse duration for the desired bandwidth. The attenuation factor is used for approximate power calibration and represents the amount by which the transmitter output should be increased over that of a soft rectangular pulse of equal duration. The Gaussian based profiles are truncated at the 1% level.
* The SNEEZE pulse produces more uniform excitation of M_z than EBURP2.

highly selective pulses. The influence of relaxation on their performance has been addressed [45,46] and more tolerant profiles suggested, such as SLURP [47].

Implementing shaped pulses

On modern high-resolution instruments the control of pulse amplitudes 'on-the-fly' is a standard (if not, optional) feature. This is performed by a so-called waveform generator which controls the rf envelope prior to power amplification [48]. Typically the envelope is defined in a series of discrete steps as a histogram, with each element having a defined amplitude and phase. To provide a close match to the smooth theoretical pulse envelope, a sufficient number of elements must be defined, which is typically 256 (but is dependent upon the available waveform memories). Many pulse envelopes come predefined on modern instruments, allowing their direct implementation.

9.3.2. DANTE sequences

In the early days of selective excitation, spectrometers were not equipped to generate amplitude modulated rf pulses and the DANTE method (Delays Alternating with Nutation for Tailored Excitation) was devised, [49] requiring only short, hard pulses. Although largely superseded by the amplitude modulated soft pulses, DANTE may still be the method of choice on older instrumentation or on those newer instruments which lack waveform generators.

The basic DANTE sequence is composed of a series of N short hard pulses of tip angle α where $\alpha \ll 90°$, interspersed with fixed delays, τ, for free precession:

$$\text{DANTE: } [\alpha-\tau-]_N \text{ or more correctly } [\tau/2-\alpha-\tau/2]_N$$

The total length of the selective pulse is the product $N\tau t_p$ where t_p is the duration of each hard pulse, and the net on-resonance tip angle is the sum of

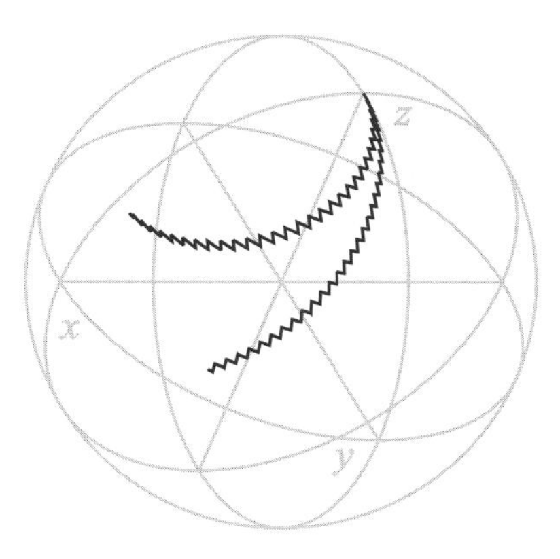

Figure 9.17. Simulated zig-zag excitation trajectories of nuclear spin vectors at different resonance offsets for the DANTE sequence. 30 pulses of 3° each were separated by delays of 0.33 ms, producing an effective 10 ms selective pulse.

the individual pulses. The effect of such a sequence is illustrated in Fig. 9.17. Following each hard pulse, spins are able to precess during the short delay τ according to their offset from resonance before being pulsed once again. On-resonance trajectories show no precession and are thus driven directly toward the $+y$ axis whilst those close to resonance to follow a zig-zag path. If N is sufficiently large (typically $\geq$20) the result closely resembles the smooth path taken under the influence of a soft rectangular pulse of equivalent total duration, and similar selectivity results.

Pulse 'shaping' in the DANTE approach is achieved by keeping the pulse *amplitude* constant but varying the *duration* of each hard pulse throughout the sequence to match the desired envelop. Thus for example, a Gaussian envelope may be emulated by varying pulse durations according to a Gaussian profile. Limitations with this approach arise when very small pulse durations are required (<1 μs) since pulse transmitters are then unable to deliver the necessary precision. In such cases it may be necessary to add fixed attenuation to the transmitter output to allow the use of longer but more accurate hard pulses.

The major difference between soft shaped pulses and DANTE methods is the occurrence of strong sideband excitation windows either side of the principal window with DANTE. These occur at offsets from the transmitter at multiples of the hard-pulse frequency, $1/\tau$. They arise from magnetisation vectors that are far from resonance and which precess full circle during the τ period. Since this behaviour is precisely equivalent to no precession, they are excited as if on-resonance. Further sidebands at $\pm 2/\tau$, $3/\tau$ and so on also occur by virtue of trajectories completing multiple full circles during τ. Such multisite excitation can at times be desirable [50,51] but if only a single excitation window is required, the hard pulse repetition frequency must be adjusted by varying τ to ensure the sideband excitations do not coincide with other resonances.

9.3.3. Excitation sculpting

Recent methods of selective excitation have combined shaped pulses with pulsed field gradients to produce experimentally robust excitation sequences with a number of desirable properties [34]. These sequences are based on either single or double pulsed field gradient spin-echoes (Fig. 9.18), which may be understood with reference to the single echo sequence (Fig. 9.18a). This may be represented G_1–S–G_1 where S represents *any* selective 180° pulse (or pulse train) and the bracketing gradients are identical. For those spins that experience the selective inversion pulse the two gradients act in opposition and thus refocus this selected magnetisation, hence this is known as a gradient-echo. Spins that do not experience this pulse, that is, those outside its effective bandwidth, only feel the cumulative effect of both gradients so remain fully dephased in the

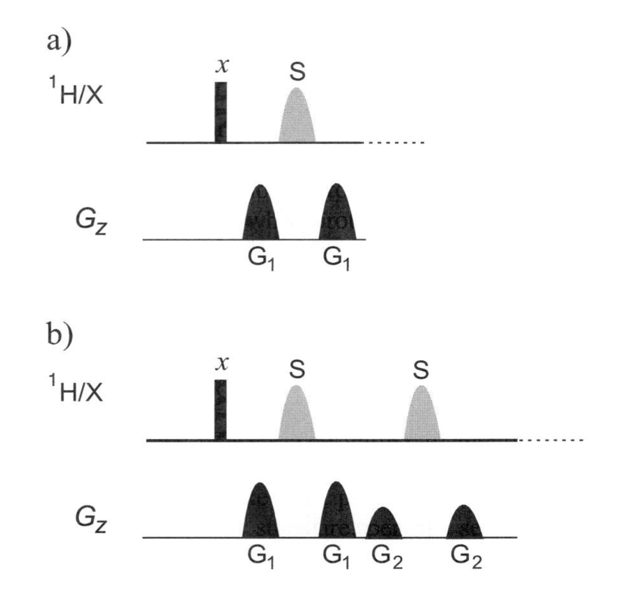

Figure 9.18. Selective excitation sequences based on (a) a single and (b) a double pulsed field gradient spin-echo. The element S represents any selective 180° inversion pulse or pulse sequence.

transverse plane and thus unobservable. This single gradient-echo therefore achieves clean resonance selection according to the profile of pulse S.

The *phase* profile of the selected resonance(s) is also dictated by the phase properties of the selective pulse S, which may not be ideal. Repeating the gradient-echo once again (Fig. 9.18b, with a different gradient strength G_2 to avoid accidental refocusing of the previously dephased unwanted magnetisation) exactly cancels any remaining phase errors and the resulting *pure-phase* excitation profile depends only on the *inversion* properties of the selective pulse. Experimentally this is an enormous benefit because it makes implementation of the selective sequence straightforward and because the field gradients ensure excellent suppression of unwanted resonances. It is also very much easier to select (and design) a pulse with a desirable 'top-hat' inversion profile when its phase behaviour is of no concern. The resulting excitation profile of the double PFG spin-echo (DPFGSE) sequence is dictated by the cumulative effect of the repeated inversion pulses, resulting in a 'chipping away' of magnetisation by the series of gradient-echoes, hence the term *excitation sculpting* [28,52]. An example of the clean, pure-phase selective excitation that can be achieved with this sequence is illustrated in Fig. 9.19. This could represent the starting point for a variety of selective 1D experiments, including TOCSY [53,54] and NOESY [52,55] (see for example the 1D gradient NOESY experiment of Section 8.7.2). The use of a second DPFGSE sequence after one transfer step leads

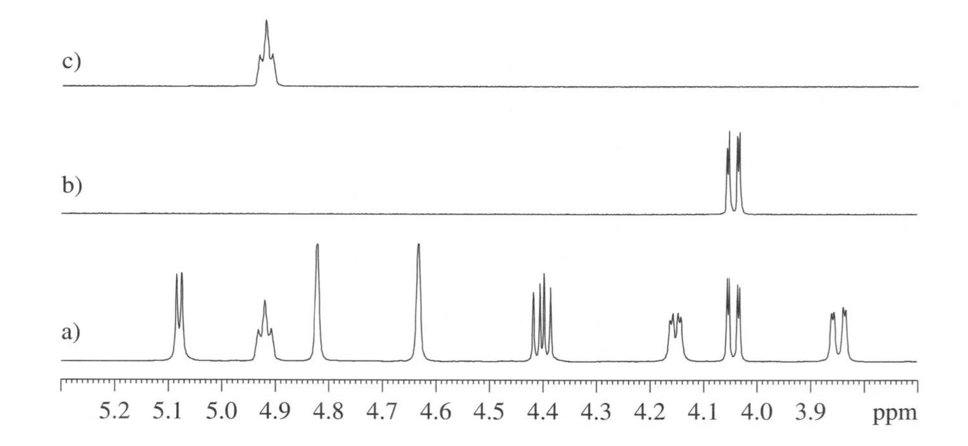

Figure 9.19. Clean selective excitation with the double pulsed field gradient spin-echo sequence using a 40 ms Gaussian 180° pulse and gradients of $0.07:0.07:0.03:0.03$ T m^{-1}.

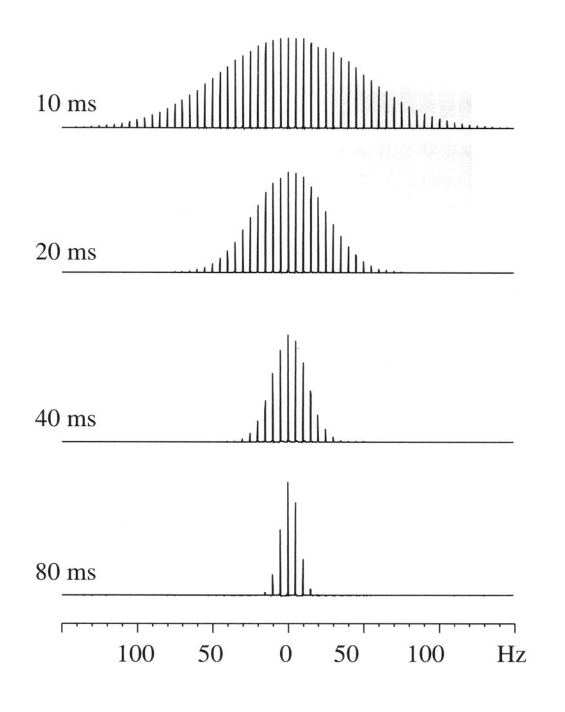

Figure 9.20. Experimental excitation profiles for the double pulsed field gradient spin-echo excitation sequences in which the element S is a 180° Gaussian pulse (truncated at 1%) of the duration shown.

to more elaborate 'doubly-selective' experiments [56] whilst a slight modification produces an extremely effective solvent suppression scheme, described below in Section 9.4.3. In contrast, the shorter single-echo excitation may be more appropriate for the study of very large molecules where T_2 relaxation losses can be significant.

For use in the laboratory, it is convenient to choose a simple, robust inversion pulse as the element S, and the Gaussian pulse is well suited to routine use. Example excitation profiles for this are illustrated in Fig. 9.20 and offer guidance on selection of pulse duration for a desired excitation window. For proton spectroscopy, a Gaussian pulse of around 40 ms proves suitable for many applications.

9.3.4. Practical considerations

This section addresses some of the practical consequences of employing soft pulses which, in general, demand rather more care and attention in their implementation than hard pulses and so typically require operator intervention for their success. The general rule for soft pulse selection is to choose the simplest pulse that will provide the performance characteristics required. More complex pulses shapes, although providing improved profiles, tend to require greatest care in calibration and are thus rather more awkward to use. For example, the BURP pulses are extremely sensitive to pulse width miscalibration which leads to significant distortion of otherwise 'uniform' profiles [42]. From a purely practical perspective, the simple Gaussian pulses are most robust and easiest to employ and represent a suitable initial choice for many applications [32]. If very high selectivity is required, the pure-phase pulses (BURP or Gaussian cascade families) prove more suitable and the choice of a specific or a universal pulse (Table 9.3) becomes significant, according to the application.

The excitation profile of soft pulses is defined by the duration of the pulse, these two factors sharing an inverse proportionality. More precisely, pulse shapes have associated with them a dimensionless *bandwidth factor* which is the product of the pulse duration, Δt, and its effective excitation bandwidth, Δf, for a correctly calibrated pulse. This is fixed for any given pulse envelope, and represents its time efficiency. It is used to estimate the required pulse duration for a desired effective bandwidth; Table 9.3 summarises these factors for some common pulse envelopes. Thus, an excitation bandwidth of 100 Hz requires a 21 ms 90° Gaussian pulse but a 49 ms EBURP2 pulse; clearly the Gaussian pulse is more time efficient.

Having determined the necessary pulse duration, the transmitter *power* must be calibrated so that the pulse delivers the appropriate tip angle. This procedure differs from that for hard pulses where one uses a fixed pulse amplitude but varies its duration. For practical convenience, amplitude calibration is usually based on previously recorded calibrations for a soft rectangular pulse (as described below), from which an estimate of the required power change is calculated. Table 9.3 also summarises the necessary changes in transmitter attenuation for various envelopes of equivalent duration, with the more elaborate pulse shapes invariably requiring increased rf peak amplitudes (decreased attenuation of transmitter output).

As an illustration, suppose one wished to excite a window of 100 Hz with a SNEEZE pulse and had previously determined that a soft 90° rectangular pulse of 10 ms required an attenuation of 68 dB. From the bandwidth factor, one can determine the pulse duration must be 58 ms. From the power ratio equations given in Section 3.5.1, one may calculate that a soft *rectangular* pulse of this duration requires 15.3 dB *greater attenuation* than the 10 ms pulse (20 log 5.8), simply because it is longer. Table 9.3 shows that the SNEEZE envelop requires 26.6 dB *less* attenuation than a rectangular pulse of equal duration. The SNEEZE pulse therefore requires 11.3 dB (26.6–15.3) *less attenuation* than that of the reference 10 ms soft rectangular pulse, and the transmitter amplitude setting becomes 56.7 dB. Fine tuning may then be required for optimum results.

Amplitude calibration

The amplitude calibration of a soft pulse essentially follows the procedures introduced for hard pulses in Section 3.5.1 and the descriptions below assume familiarity with these. For soft pulses on the observe channel, the transmitter frequency should be placed on-resonance for the target spin and the 90° or 180° condition sought directly by variation of transmitter power, staring with very low values (high attenuations) and progressively increasing (note there is a logarithmic not linear dependence on dB). If calibrations must be performed indirectly, for example on the decoupler channel of older instruments, a slight variation on the method of Fig. 3.52 (Chapter 3) for a 2-spin AX pair is used (Fig. 9.21a). In this an additional spin-lock pulse is applied to the observe channel during the soft pulse to decouple the AX interaction [58]. This collapse of the AX doublet means the soft pulse can be applied to the centre of the A-spin doublet. Variation of the transmitter power produces results similar to those of Fig. 3.53 (Chapter 3) when the 90° or 180° condition is satisfied. With modern instrumentation, however, calibrations determined directly on the observe channel are equally valid if this is later used as the indirect (decoupler) channel in subsequent experiments, so bypassing the need for such indirect calibrations.

Phase calibration

Precise control of the *relative* phase between pulses is crucial to the success of many multi-pulse NMR experiments and some correction to the phase of a soft pulse may be required to maintain these relationships when both hard and soft pulses are to be applied to the same nucleus. When soft pulses are used on the observe channel the phase difference (which may arise because of the potentially different rf paths used for high and low power pulses) may be determined by direct inspection of two separate 1D pulse-acquire spectra recorded with high- and low-power pulses but under otherwise identical

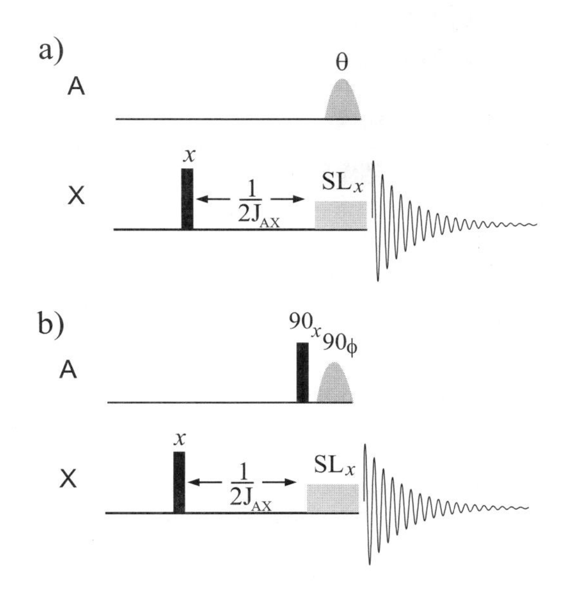

Figure 9.21. Soft pulse calibration sequences for (a) amplitude and (b) hard/soft phases differences for selective pulses applied on the indirect (decoupler) channel. SL is a spin-lock applied to decouple A and X during the soft pulse.

conditions. Using only zero-order (frequency-independent) phase correction of each spectrum, the difference in the resulting phase constants (soft *minus* hard) represents the phase difference between the high- and low-power rf routes. Adding this as a constant offset to the soft pulse phase should yield spectra of identical phase to that of the hard pulse spectrum when processed identically and this correction can be used in all subsequent experiments, *provided the soft pulse power remains unchanged.*

If phase differences must be determined indirectly, the sequence of Fig. 9.21b is suitable in which an additional hard pulse is used on the indirect channel prior to the soft. When two 90° pulses are of the same phase their effect is additive and the sequence behaves as if a net 180° pulse has been applied to the selected spin, resulting in an inversion of the antiphase AX doublet. Similarly, two 180° pulses of similar phase behave as a net 360° pulse and have no effect on the doublet.

9.4. SOLVENT SUPPRESSION

Fortunately the majority of solvents used in organic NMR spectroscopy are readily available in the deuterated form. For proton spectroscopy in particular, this allows the chemist to focus on the solute spectrum, undisturbed by the solvent that is present in vast excess. Unfortunately, most molecules of biochemical or medicinal interest, notably biological macromolecules, must be studied in water and in order to observe all protons of interest within these molecules, including the often vitally important labile protons, protonated water must be employed as the solvent (containing 10% D_2O to maintain the field-frequency lock). Whereas H_2O is 110 M in protons, solute concentrations are more typically in the millimolar region and the 10^4–10^5 concentration difference imposes severe experimental difficulties which demand the attenuation of the solvent resonance. In addition, the recently developed area of LC–NMR often favours the use of protonated solvents for reasons of economy, also making efficient and robust solvent suppression essential. At a more mundane level, when using D_2O or MeOD for routine spectroscopy a considerable residual HDO resonance may remain which may limit the useable receiver amplification and appear aesthetically unappealing.

The principal reason to suppress a large solvent resonance in the presence of far smaller solute resonances is so that the dynamic range of the NMR signals lies within the dynamic range of the receiver and ADC (Section 3.2.6). Further concerns include the baseline distortions, t_1-noise in 2D experiments, radiation damping and potential spurious responses that are associated with very intense signals. Radiation damping leads to severe and undesirable broadening of the water resonance which may mask the solute resonances. It arises because the intense NMR signal produced on excitation of the water generates a strong rf current in the detection coil (which ultimately produces the observed resonance). This current in turn generates its own rf field which drives the water magnetisation back toward the equilibrium position; if you like, it acts as a water-selective pulse. The loss of transverse magnetisation is therefore accelerated, producing a reduced apparent T_2 and hence a broadened resonance [1]. In general, good lineshape is a prerequisite for successful solvent suppression and whilst efforts to optimise shimming pay dividends, the intense water singlet of H_2O is ideally suited to gradient shimming (Section 3.4.4). Improved results have also been reported when microscopic air bubbles have been removed from the sample by ultrasonic mixing.

The goal of solvent suppression is therefore to reduce the magnitude of the solvent resonance before the NMR signal reaches the receiver. This seemingly simple requirement has generated an enormous research area [59, 60], emphasising the fact that this is by no means a trivial exercise. The more widely used approaches can be broadly classified into three areas:

- methods that saturate the water resonance
- methods that produce zero net excitation of the water resonance, and
- methods that destroy the water resonance with pulsed field gradients.

The following sections illustrate examples from these areas that have proved most popular but represent only a small subsection of available methods (see, for example, Chapter 2 of reference [61]). All schemes inevitably involve some loss of signal intensity for those resonances close to that of the solvent, and careful adjustment of solution conditions, the simplest being sample temperature, can prove useful in avoiding signal loses by shifting the water resonance relative to solute signals (Fig. 9.22).

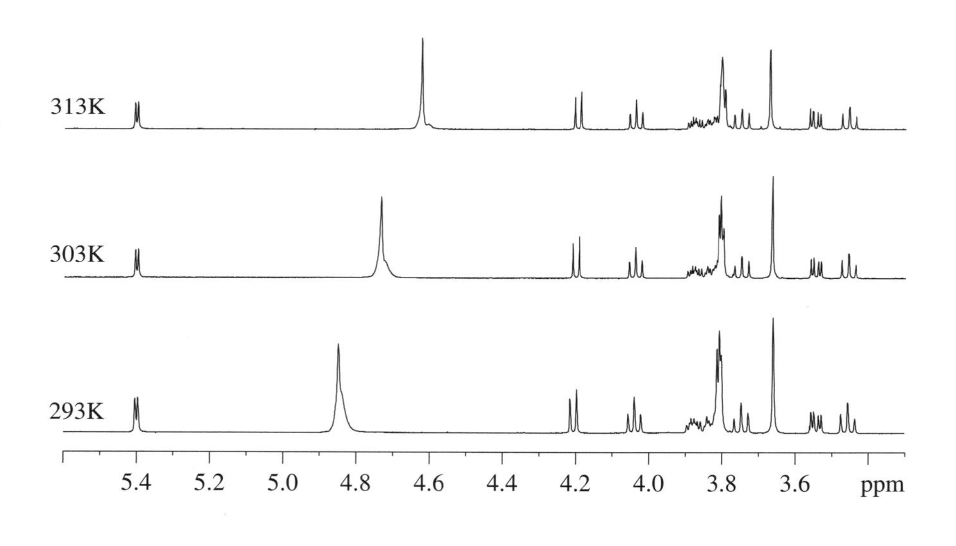

Figure 9.22. The temperature dependence of HDO (here partially suppressed). The shift corresponds to approx. 5 Hz/K at 500 MHz. Spectra are referenced to internal TSP.

[1] A test for radiation damping involves detuning the proton coil and re-acquiring the proton spectrum. This degrades the rf coupling so reducing the back electromotive force (emf) and sharpening the water resonance.

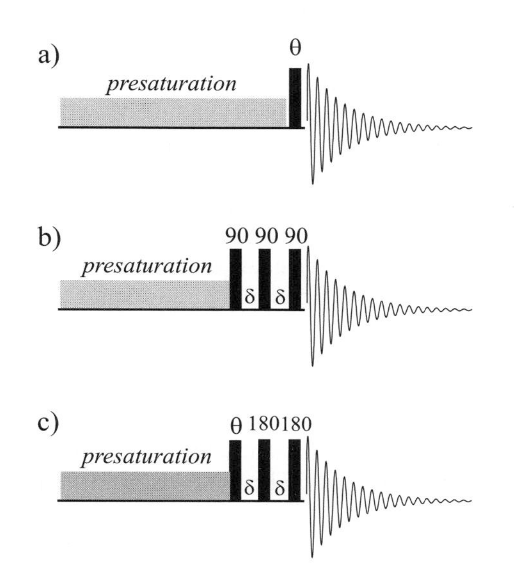

Figure 9.23. Solvent suppression schemes based on presaturation; (a) presaturation alone, (b) 1D NOESY and (c) FLIPSY. Sequence (b) makes use of the conventional NOESY phase cycle whereas FLIPSY uses EXORCYCLE on one (or both) of the 180° pulses (i.e. pulse = x, y, $-x$, $-y$, receiver = x, $-x$, x, $-x$).

9.4.1. Presaturation

The simplest, most robust and most widely used technique is presaturation of the solvent [62]. This is simple to implement, may be readily added to existing experiments and leaves (non-exchangeable) resonances away from the presaturation frequency unperturbed. It involves the application of continuous, weak rf irradiation at the solvent frequency prior to excitation and acquisition (Fig. 9.23a), rendering the solvent spins saturated and therefore unobservable (Fig. 9.24). Invariably resonances close to the solvent frequency also experience some loss in intensity, with weaker irradiation leading to less spillover but reduced saturation of the solvent. Longer presaturation periods improve the suppression at the expense of extended experiments so a compromise is required and typically 1–3 s are used; trial and error usually represents the best approach to optimisation. Wherever possible the same rf channel should be used for both the presaturation and subsequent proton pulsing, with appropriate transmitter power switching.

A feature of simple presaturation in protonated solutions (particularly with older probes which may also lack appropriate shielding of the coil leads [63]) is a residual 'hump' in the 1D spectrum that originates from peripheral regions

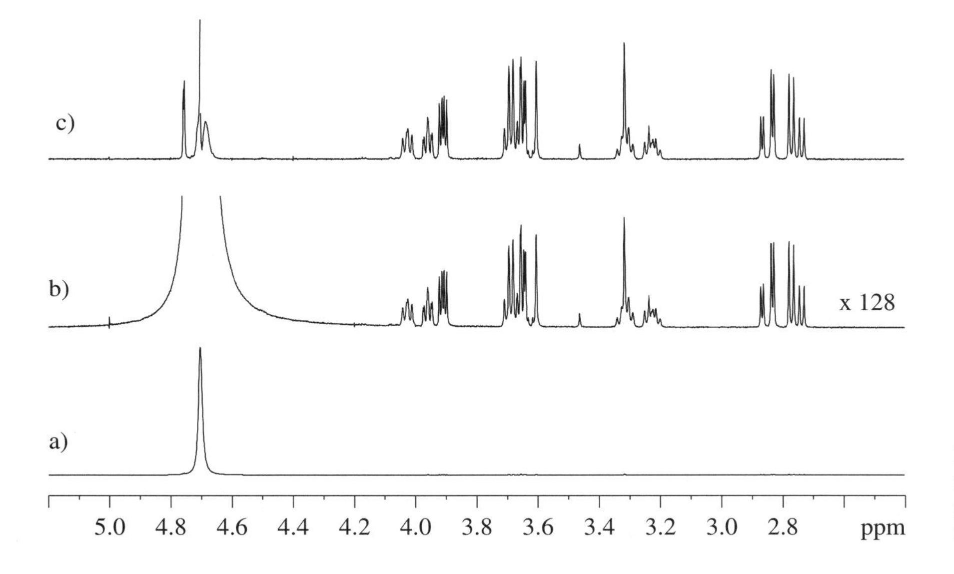

Figure 9.24. Solvent presaturation (2 s) reveals a resonance at 4.75 ppm in (c) previously masked by the HDO resonance in (a) and its expansion (b).

of the sample that suffer B_0 and B_1 inhomogeneity. An effective means of suppressing this for 1D acquisitions is the so-called NOESY-presat sequence which simply employs a *non-selective* 1D NOESY with zero mixing time, or in other words, the first increment of a 2D NOESY experiment with $\tau_m = 0$ (Fig. 9.23b). The usual NOESY phase-cycle suppresses the hump, but the sequence is restricted to acquisition with 90° pulses which is not optimal for signal averaging or spectrum integration. A recent variation, termed FLIPSY [64] (flip-angle adjustable one-dimensional NOESY, Fig. 9.23c), allows the use of an arbitrary excitation pulse tip angle θ and is thus better suited to routine use.

The principal disadvantage of all presaturation schemes is that they also lead to the suppression of exchangeable protons by the process of saturation transfer. In favourable cases solution conditions may be altered (temperature and pH) to slow the exchange sufficiently to reduce signal attenuation, but this approach has limited applicability. The methods presented below largely avoid such losses.

9.4.2. Zero excitation

The second general approach strives to produce no net excitation of the solvent resonance. In other words, the sequence ultimately returns the solvent magnetisation to the $+z$ axis whilst at the same time placing all other magnetisation in the transverse plane prior to acquisition. Section 9.4.3 describes a related approach used in conjunction with pulsed field gradients.

Jump-return

To appreciate the principal behind zero excitation, consider the simplest example, the 'jump-return' sequence [65] 90_x–τ–90_{-x} which subsequently spawned many others. The transmitter frequency is placed on the solvent resonance and all magnetisation is tipped into the transverse plane by the first hard 90° pulse. During the subsequent delay τ (typically a few hundred μs long) all vectors fan out in the transverse plane according to their offsets, except that of the solvent which, being on-resonance, has zero frequency in the rotating frame. Thus, the solvent resonance is tipped back to the $+z$-axis as the second 90° pulse rotates the xy plane into the xz plane. The only remaining transverse magnetisation is the $\pm x$-component of all vectors prior to the second pulse. Hence, this produces a sine-shaped excitation profile (Fig. 9.25a) with maximum amplitude at offsets from the transmitter frequency of $\pm 1/4\tau$ Hz and with resonances either side of the transmitter displaying opposite phase (owing to their $\pm x$ orientations) but without additional phase dispersion. In practice this simple scheme produces a rather narrow null at the transmitter offset and a significant solvent resonance typically remains, so more sophisticated sequences have been investigated.

Binomial sequences

The binomial sequences aim to improve the zero excitation profile and provide schemes that are less sensitive to spectrometer imperfections. The series may be written $1\text{–}\bar{1}$, $1\text{–}\bar{2}\text{–}1$, $1\text{–}\bar{3}\text{–}3\text{–}\bar{1}$ ... and so on, where the numbers indicate the relative pulse widths, each separated by a delay τ, and the overbar indicates phase inversion of the pulse. For *off-resonance spins* the pulse elements are *additive* at the excitation maximum so for example, should one require 90° off-resonance excitation, $1\text{–}\bar{1}$ corresponds to the sequence 45_x–τ–45_{-x}. Of this binomial series, it turns out that the $1\text{–}\bar{3}\text{–}3\text{–}\bar{1}$ sequence [66] has good performance and is most tolerant of pulse imperfections by virtue of its symmetry [67]. The trajectory of spins with frequency offset $1/2\tau$ from the transmitter for a net 90° pulse ($1 \equiv 11.25°$) is shown in Fig. 9.26. During each

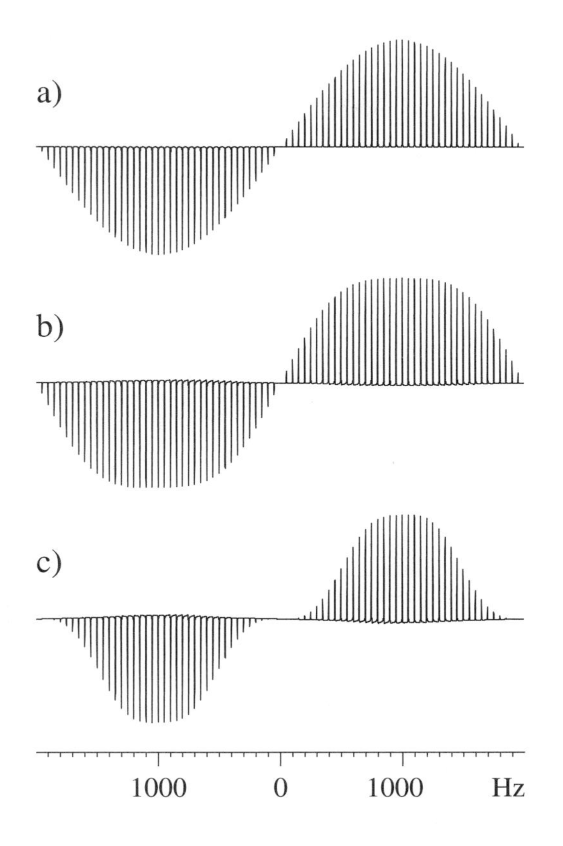

Figure 9.25. Simulated excitation profiles for (a) the jump-return sequence, (b) the 1–$\bar{1}$ and (c) 1–$\bar{3}$–3–$\bar{1}$ binomial sequences. The jump-return sequence used $\tau = 250$ and the binomial sequences used $\tau = 500$ μs.

τ period the spins precess in the rotating frame by half a revolution so the effect of the phase inverted pulses is additive and the magnetisation vector is driven stepwise into the transverse plane. As before, the on-resonance solvent vector shows no precession so is simply tipped back and forth, finally terminating at the North Pole.

Whilst maximum excitation occurs at $\pm 1/2\tau$ Hz from the transmitter offset, further nulls occur at offsets of $\pm n/\tau$ (n = 1, 2, 3, ... corresponding to complete revolutions during each τ) so a judicious choice of τ is required to provide excitation over the desired bandwidth. The excitation profiles of the 1–$\bar{1}$ and 1–$\bar{3}$–$\bar{1}$ sequences are shown in Fig. 9.25b and c. Clearly the excitation is non-uniform, so places limits on quantitative measurements, and once again there exists a phase inversion either side of the solvent. Both provide an effective null at the transmitter offset and suppression ratios in excess of 1000-fold can be achieved.

When implementing this sequence it may be necessary to add attenuation to the transmitter to increase the duration of each pulse so that the shorter elements do not demand very short (<1 μs) pulses (note the similarity with the requirements for the DANTE hard-pulse selective excitation described above). The binomial sequences can be adjusted to provide an arbitrary overall tip angle by suitable adjustment of the tip angles for each element. For example, *inversion* of all off-resonance signals can be achieved by doubling all elements relative to the net 90° condition. Exactly this approach has been exploited in the gradient-echo methods described below.

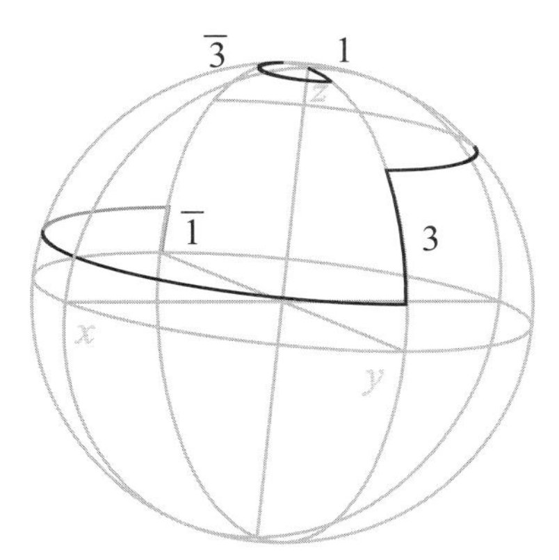

Figure 9.26. The trajectory of a magnetisation vector during the 1–$\bar{3}$–3–$\bar{1}$ binomial sequence. The trace is shown for a spin with an offset from the transmitter of $+1/2\tau$ Hz, corresponding to the excitation maximum.

9.4.3. Pulsed field gradients

The most effective approach to date to solvent suppression is the destruction of the net solvent magnetisation by pulsed field gradients (PFGs), so ensuring nothing of this remains observable immediately prior to acquisition. The

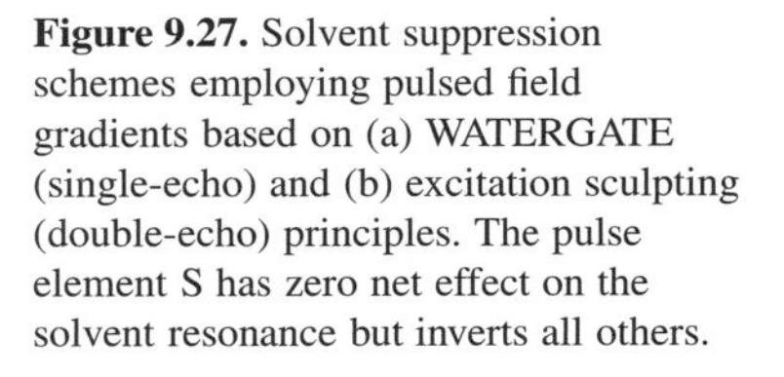
Figure 9.27. Solvent suppression schemes employing pulsed field gradients based on (a) WATERGATE (single-echo) and (b) excitation sculpting (double-echo) principles. The pulse element S has zero net effect on the solvent resonance but inverts all others.

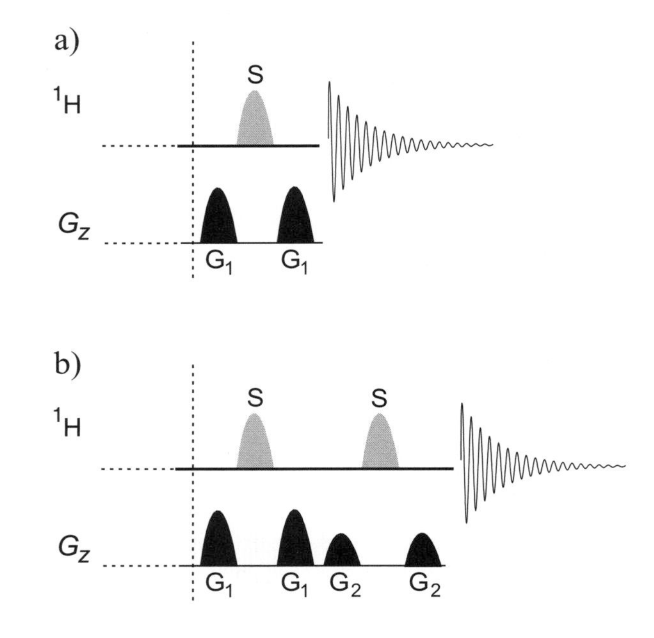

gradient schemes described here act on transverse magnetisation so are readily appended to existing one- and multi-dimensional sequences [68].

One popular approach, termed WATERGATE [69,70] (water suppression by gradient-tailored excitation) makes use of a single PFG spin-echo, G_1–S–G_1, (Fig. 9.27a) in which both gradients are applied in the same sense. The element S is chosen to provide zero net rotation of the solvent resonance but to provide a 180° *inversion* to all others. This results in the solvent magnetisation experiencing a cumulative dephasing by the two gradients leading to its destruction whilst all others are refocused in the spin-echo by the second gradient and are therefore retained.

As apparent from the previous section, a binomial sequence has a suitably tailored profile for the element S, and the series 3α–9α–19α–$\overline{19\alpha}$–$\overline{9\alpha}$–$\overline{3\alpha}$ (Fig. 9.28a, with $26\alpha = 180°$ and a delay τ between pulses, here termed W3 [71]) has a desirable off-resonance inversion profile for this purpose. The WATERGATE excitation profile for this is shown in Fig. 9.29a. Once again characteristic nulls also occur at offsets of $\pm n/\tau$ Hz, but between these the excitation is quite uniform and does not suffer the phase inversion of the unaccompanied 90° binomials. More recently, extended binomial sequences have been shown to provide a narrower notch at the transmit-

a) 1H 3 9 19 $\overline{19}$ $\overline{9}$ $\overline{3}$

b) 1H 90_x 180_x 90_x

"S"

Figure 9.28. Two possible versions of the off-resonance inversion element S based on (a) a binomial-type hard pulse sequence and (b) a combination of soft and hard pulses.

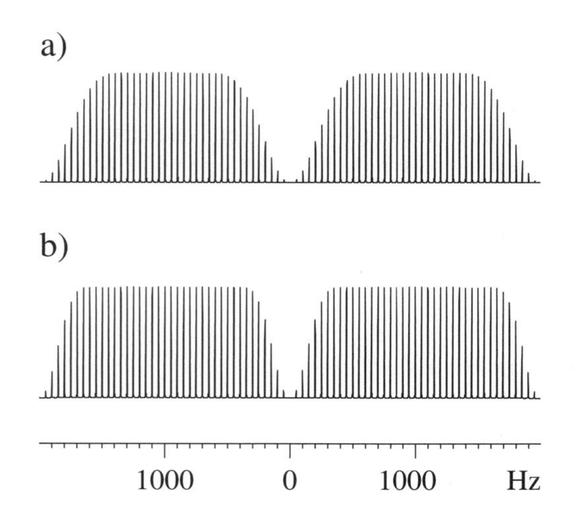

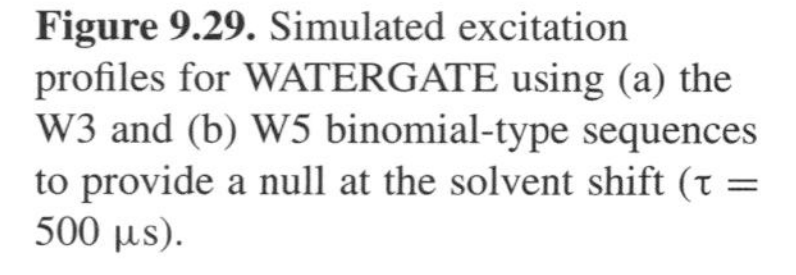
Figure 9.29. Simulated excitation profiles for WATERGATE using (a) the W3 and (b) W5 binomial-type sequences to provide a null at the solvent shift (τ = 500 μs).

ter offset, so reducing the attenuation of solute signals close to the solvent, and have a wider region of uniform excitation [71]. The improved profile of the so-called W5 sequence is shown in Fig. 9.29b. This sequence alas, lacks the elegant shorthand form of its predecessor and is thus represented 7.8–18.5–37.2–70.0–134.2–$\overline{134.2}$–$\overline{70.0}$–$\overline{37.2}$–$\overline{18.5}$–$\overline{7.8}$, where each number represents the pulse tip angle in degrees.

The element S may also be provide by a combination of hard and soft pulses [28,69] (Fig. 9.28b). Here, the soft pulses act only on the water resonance causing it to experience a net 360° rotation whilst all others undergo the desired inversion by virtue of the hard pulse. In this approach, the suppression profile is dictated by the inversion properties of the soft pulse combination, and off-resonance nulls no longer occur (although correction for any phase difference between the hard and soft pulses may be required).

An improved approach to gradient suppression employs the method of 'excitation sculpting' described in Section 9.3.3 and applies a double PFG spin-echo instead of one [28]. This has the advantage of refocusing the evolution of homonuclear couplings and can produce spectra with improved phase properties and less baseline distortion. The elements S are exactly as above, except with the double-echo the suppression notch is wider since it is applied twice. The water suppression spectrum of the 2 mM sucrose in 90% H_2O using this approach is shown in Fig. 9.30a and illustrates the impressive results now routinely available. The residual solvent signal can be further removed (Fig. 9.30b) by subtracting low-frequency components from the FID prior to Fourier transformation [72], which requires the transmitter frequency to be placed on the solvent resonance. The excitation sculpting sequence can also be readily tailored to achieved multi-site suppression [73,74].

One final point regarding PFGs is that their use in certain experiments for coherence selection also leads to solvent suppression without further modification. For example, the DQF-COSY experiment inherently filters out uncoupled spins (i.e. singlets) and when PFGs are used to provide this filtering the singlet water resonance is also removed at source. In fact, the original publication that stimulated widespread use of field gradients in high-resolution NMR impressively demonstrated such suppression [75]. Similarly, heteronuclear correlation

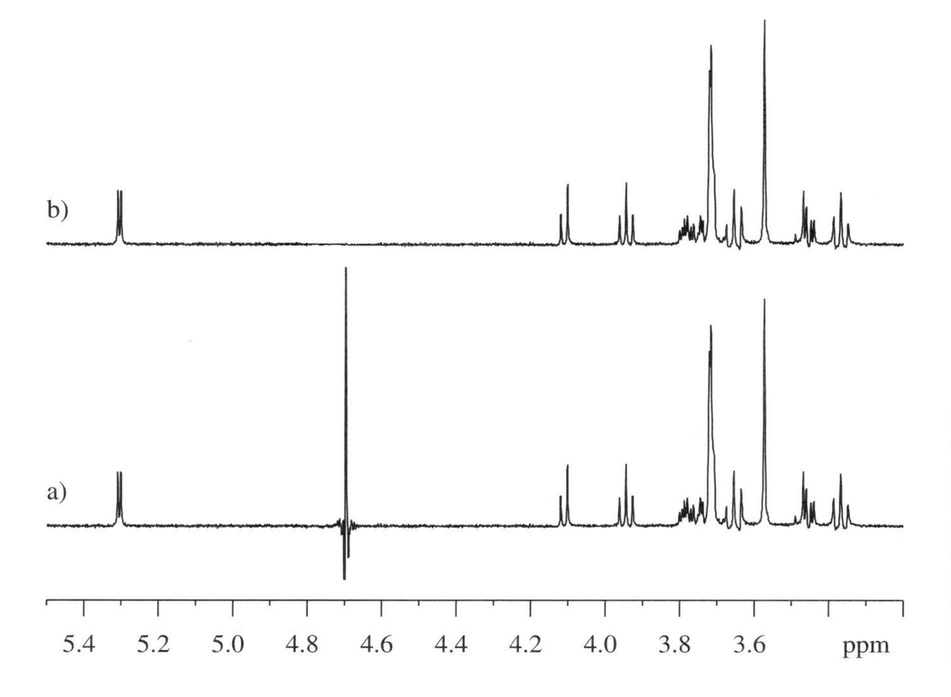

Figure 9.30. Solvent suppression with the excitation sculpting scheme using the approach of Fig. 9.28b with 4.1 ms 90° Gaussian pulses and gradients of 0.1 : 0.1 : 0.03 : 0.03 T m^{-1}. The sample is 2 mM sucrose in 9 : 1 H_2O : D_2O. In (b) the small residual solvent signal has been completely removed through additional processing of the FID (see text).

experiments such as HMQC and HSQC intrinsically select protons bound to, say, ^{13}C or ^{15}N and thus also reject water. Whilst these methods on their own are unlikely to prove sufficient for 90% H_2O solutions (in which case the above gradient-echo methods may be appended), they are often sufficient for 'wet' D_2O samples.

9.5. RECENT METHODS

This final section briefly considers two areas that have existed as NMR methods in their own right for over three decades, yet have only relatively recently come into use in mainstream high-resolution NMR. This transition has been prompted by technical developments in instrumentation, more specifically in probehead design. The first method employs so-called magic-angle spinning of (partially) fluid heterogeneous samples, a technique that has been used extensively over the years for solid-state materials and which demands very fast sample rotation, typically at many kilohertz. The second studies the self-diffusion properties of molecules in solution and relies on field gradient pulses to map their motion.

9.5.1. Heterogeneous samples and MAS

Recent years has seen a burgeoning interest in solid-phase organic synthesis protocols for the production of combinatorial libraries of new molecules [76,77]. A challenging factor in this approach is the analysis of the newly synthesised materials for which, ideally, direct structural analysis of the material would be carried out whilst still attached to the solid support and thus still available for further chemistry. The direct NMR analysis of such materials, even when solvated, is complicated by two principal factors which can severely degrade spectrum resolution:

- restricted motion of the tethered analyte, and
- physical heterogeneity within the sample.

The restricted motion of the tethered analytes relative to that of free molecules in solution may mean that dipolar couplings, D, are not fully averaged to zero. These have a time-averaged angular dependence:

$$D \propto r^{-3}(3\cos^2\theta - 1) \tag{9.2}$$

where r is the internuclear separation and θ the angle between the static field and the internuclear vector. For isotropically tumbling molecules in low-viscosity solutions, the angular term is averaged to zero as the molecule freely rotates so dipolar couplings are not observed. When motion is sufficiently restricted, some residual dipolar coupling may be reintroduced, which will contribute to broadening of the NMR resonances. The use of long linker chains allows greater motion of the terminal analyte in solution and can therefore reduce this effect. A potentially greater problem lies in the sample heterogeneity. Solutions containing solid phase resins, usually in the form of beads, suffer from local field inhomogeneity at the bead-solvent interfaces due to magnetic susceptibility differences and this leads to severe line broadening (Fig. 9.31a). Although some success has been reported with carbon-13 gel-phase NMR [78,79], these deleterious effects mean high-resolution spectra of solvated resins cannot be obtained with conventional solutions probes, a particular problem for proton spectroscopy.

Magnetic susceptibility has a similar angular dependence to that above, and line-broadening from both dipolar couplings and susceptibility discontinuities can be eliminated by rapid rotation of the sample about the so-called 'magic

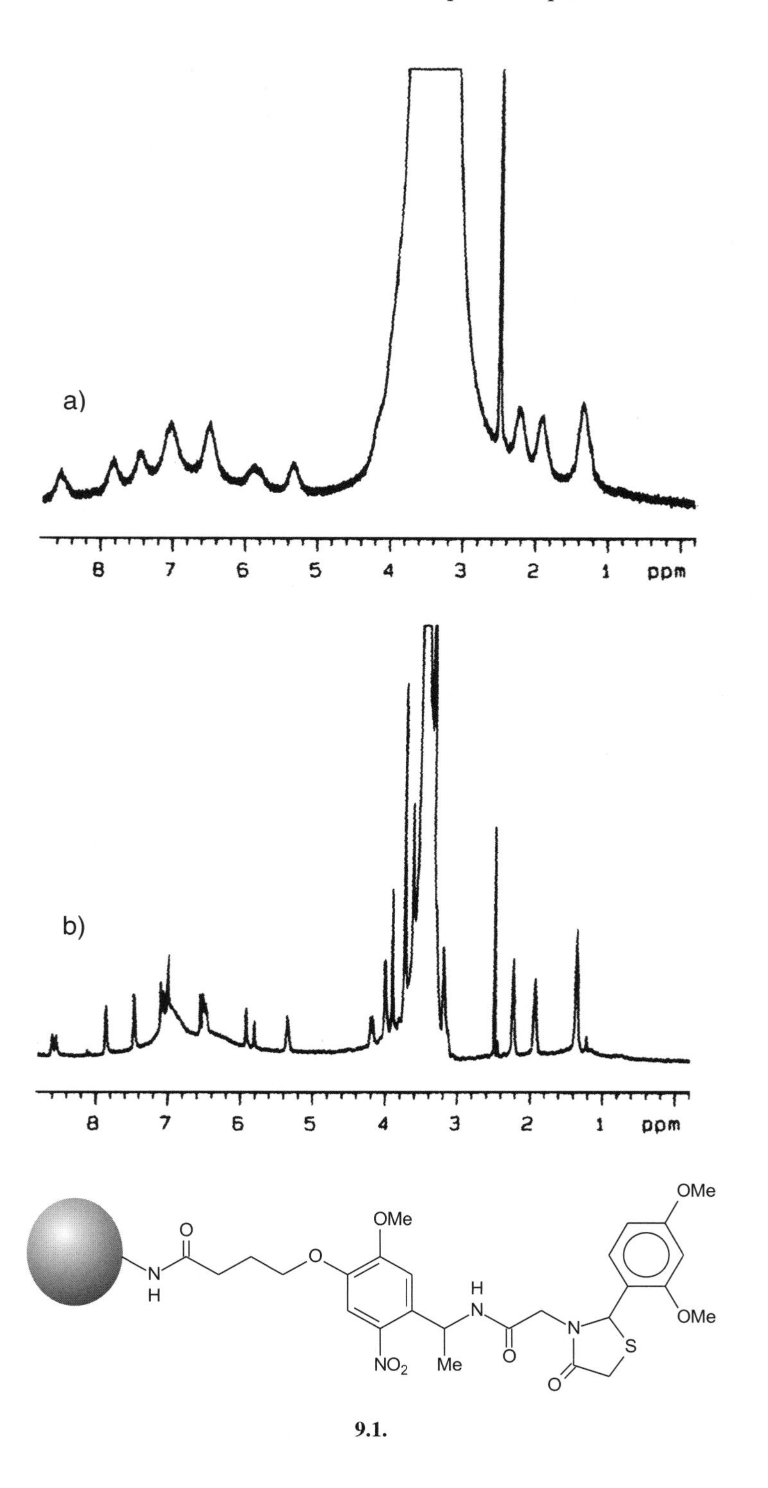

Figure 9.31. (a) The conventional ^{1}H spectrum of the analyte **9.1** bound to TentaGel resin beads solvated in DMSO and (b) the MAS (2 kHz) spectrum of the same sample. One tenth of the mass of the sample in (a) was used in (b) and both were collected in 16 transients. The large truncated signals arise from the resin itself (reproduced with permission from reference [81]).

angle' $\theta = 54.7°$, such that the above angular term becomes zero and the overall molecular motion emulates that in solution. This *magic angle spinning* (MAS) (Fig. 9.32) has long been used in solid-state NMR spectroscopy to average the effects of chemical shift anisotropy [80] and has now become a standard technique for the direct analysis of resin-bound entities. The substantial gain in resolution and hence sensitivity provided by MAS over a conventional solutions probe is illustrated in Fig. 9.31 [81]. The MAS spectrum shows a three-fold signal-to-noise gain yet uses one tenth of the sample relative to the conventional solution spectrum, and has considerably greater information

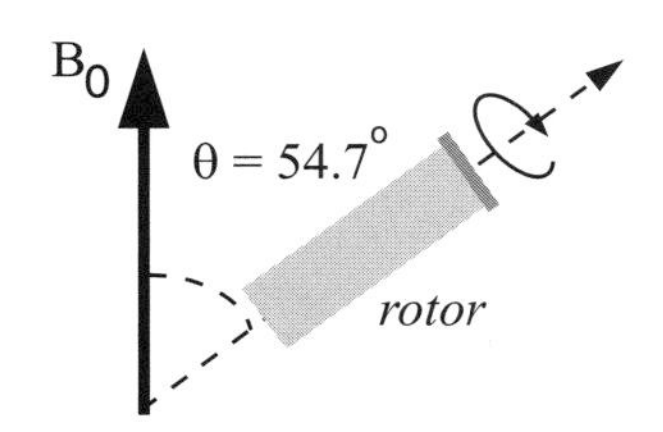

Figure 9.32. Magic angle spinning involves rapid rotation of a sample inclined at the magic angle of 54.7° to the static field.

content. To achieve such results, samples are typically spun at rates of 2–5 kHz, in purpose-built rotors requiring access to the appropriate high-resolution MAS probehead [82] and spin regulation hardware. The full range of high-resolution one- and two-dimensional techniques can now be applied to resin-bound samples under MAS conditions for the complete, non-destructive identification of new synthetic products [83–86]. The influence of various solvent and resin combinations has also been investigated and shown to have considerable bearing on the quality of spectra one obtains [87].

9.5.2. Diffusion-ordered spectroscopy

The study of molecular diffusion in solution offers insights into a range of physical molecular properties including molecular size, shape and aggregation states, and NMR-based measurements have been applied to many areas of chemistry for over three decades [88]. Different mobility rates or *diffusion coefficients* may also be used as the basis for the separation of the spectra of mixtures of compounds in solution, this procedure being referred to as *diffusion-ordered spectroscopy* or DOSY. The application of NMR diffusion measurements to the separation of small-molecule mixtures in this way is a relative newcomer to high-resolution NMR and is a developing area that is sure to find increasing use in the research laboratory, so is briefly introduced in this section. All modern NMR-based diffusion measurements rely on the application of pulsed field gradients to map the physical location of a molecule in solution and have recently been made possible on conventional high-resolution NMR spectrometers through the provision of actively shielded PFG probeheads. Molecular diffusion is then characterised along the direction of the applied field gradient, which is typically along the *z*-axis of conventional gradient probeheads.

The PFG spin-echo

The basic scheme for the characterisation of diffusion is the pulsed field gradient spin-echo [89] (Fig. 9.33a). In the absence of gradient pulses, this will refocus chemical shift evolution such that the detected signal is attenuated only by transverse relaxation during the 2τ period. When pulsed field gradients are employed, *complete* refocusing of the signal will *only* occur when the local field experienced by a spin is identical during the two gradient pulses. Since a field

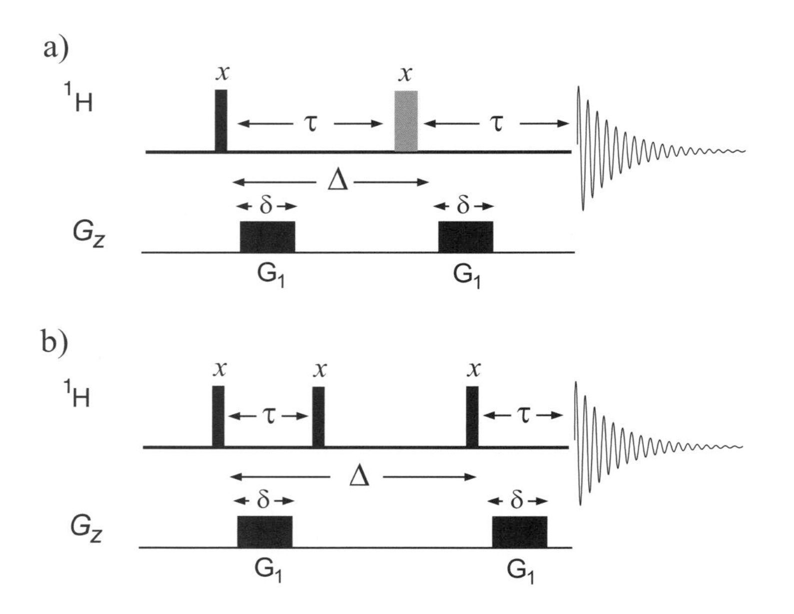

Figure 9.33. Basic sequences for measuring molecular diffusion based on (a) the PFG spin-echo and (b) the PFG stimulated-echo. The diffusion during the period Δ is characterised by a series of measurements with increasing gradient strengths.

gradient is used, the *local* field is spatially dependent meaning this refocusing condition is only met if the spin remains in the same physical location when the two PFGs are applied. If the molecule were to diffuse away from its initial position during the diffusion delay Δ, then the local field experienced during the second PFG would not exactly match that of the first and only partial refocusing of the signal would occur (Fig. 9.34). The detected signal would therefore be attenuated by an amount dictated by how far the molecule moved during the period Δ, and hence by its diffusion coefficient. To characterise diffusion rates, it is possible to progressively alter the delay Δ, the length of the gradient pulses or the strength of the gradient pulses and to monitor the corresponding signal decay. However, changes made to the overall length of the echo sequence will introduce additional complications arising from increasing relaxation losses, so it is universal practice to increase gradient strengths whilst keeping all time periods invariant. Whilst T_2 relaxation losses still occur in this case, they are constant for all experiments and thus do not contribute to the progressive signal attenuation that is monitored (Fig. 9.35).

The observed signal intensity, I, for the basic PFG spin-echo experiment is given by:

$$I = I_0 \exp\left(\frac{-2\tau}{T_2} - (\gamma\delta G)^2 D \left(\Delta - \frac{\delta}{3}\right)\right) \tag{9.3}$$

where I_0 is the signal intensity at zero gradient strength, G is the gradient strength, D is the diffusion coefficient and the delays Δ and δ are as in Fig. 9.33. Plotting $\ln(I/I_0)$ vs. G^2, for example, yields a linear plot whose slope is proportional to D (Fig. 9.36) and since the constants γ, δ and Δ are known, the diffusion coefficient may be calculated.

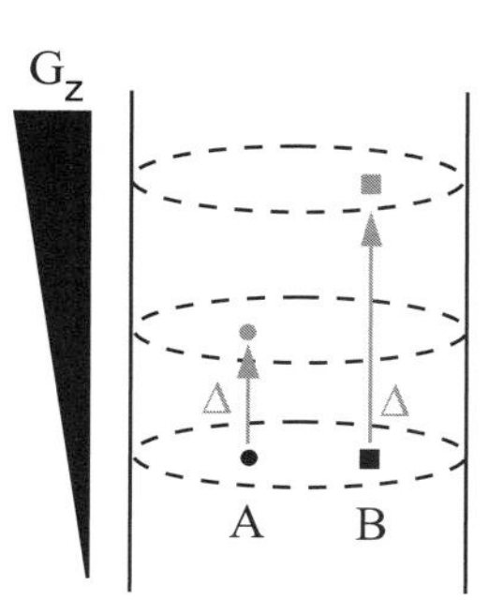

Figure 9.34. A schematic representation of signal attenuation through molecular diffusion. The local field experienced by molecule A during the first gradient pulse (molecules shown in black) does not precisely match that experienced during the second gradient pulse (molecules shown in grey) due to diffusion during the delay Δ. The signal of A does not fully refocus and its response is attenuated. Greater attenuation is observed for the faster moving molecule B due to the greater difference in local fields it experiences during the two gradient pulses.

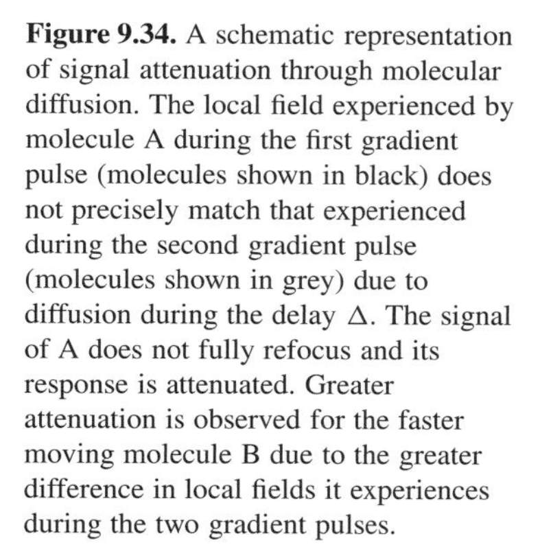

G^2/T^2m^{-2}

The PFG stimulated-echo

The PFG spin-echo sequence above is limited in practice by the aforementioned relaxation losses and is now little used for diffusion measurements. Because magnetisation is transverse during the diffusion period, these losses are dictated by the transverse (T_2) relaxation rates which themselves increase with molecular size. Since larger molecules require longer diffusion periods to move

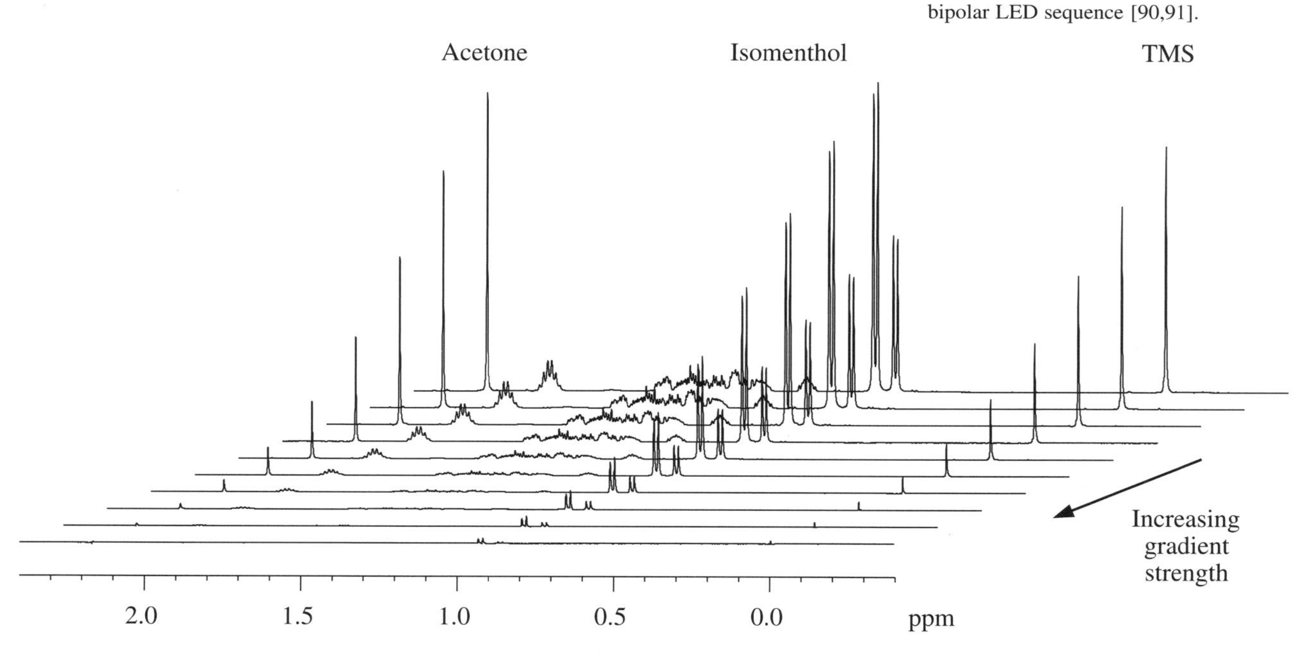

Figure 9.35. Diffusion-weighted proton spectra from a mix of acetone, TMS and isomenthol in $CDCl_3$ at 298K. The diffusion delay Δ was 30 ms and gradient strengths were progressively increased from 0.025 to 0.25 T m^{-1} in a bipolar LED sequence [90,91].

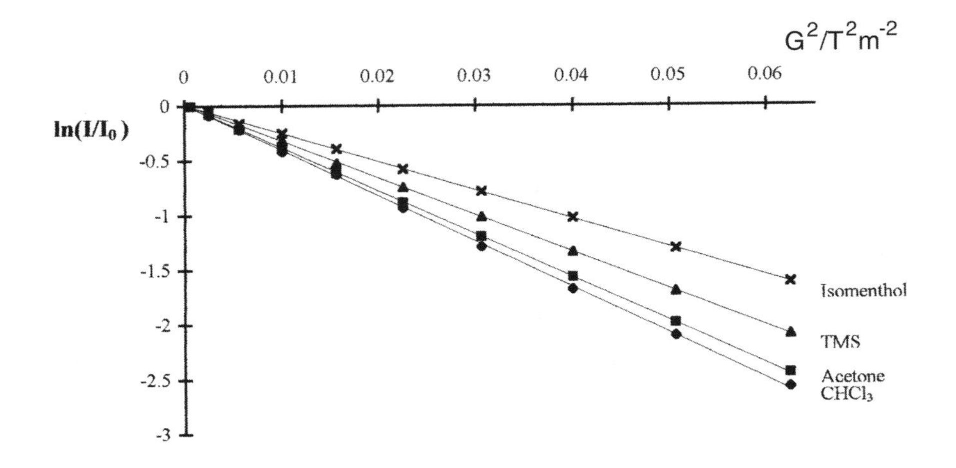

Figure 9.36. Diffusion plot for the decays shown in Fig. 9.35, including the decay data for the residual $CHCl_3$ of the solvent. The slopes indicate relative mobility rates of 1.0 : 1.3 : 1.5 : 1.6 for isomenthol : TMS : acetone : chloroform.

significant distances, the use of long Δ delays (which are typically in the region of 10 ms to 1s) can lead to unacceptable signal-to-noise degradation. In the stimulated-echo sequence [92] (Fig. 9.33b) magnetisation is longitudinal during the diffusion period, by virtue of the second 90° pulse, meaning the sequence is now limited by slower longitudinal (T_1) relaxation rates instead. Following the diffusion period, the magnetisation is returned to the transverse plane by the third 90° pulse for refocusing and detection. All recently introduced diffusion sequences [90,91,93] are derived from this basic stimulated-echo sequence, for which variations on equation Eq. 9.3 must be employed.

2D DOSY

Once diffusion coefficients have been determined from appropriate data-fitting routines, it becomes possible to use these data to generate a diffusion dimension within a pseudo-2D spectrum [93–96]. Since each resonance for a given molecule should be associated with the same diffusion coefficient, these may be dispersed along the diffusion dimension of a DOSY plot (Fig. 9.37). This diffusion-based spectrum editing can also be applied to conventional two-dimensional sequences, yielding pseudo-3D plots [97]. The editing of mixtures of small molecules in this way is experimentally very demanding because of the need to resolve small differences in similar diffusion coefficients, and the technical difficulties associated with this have been discussed in some detail [93]. Nevertheless, this clearly has great potential for the direct analysis of

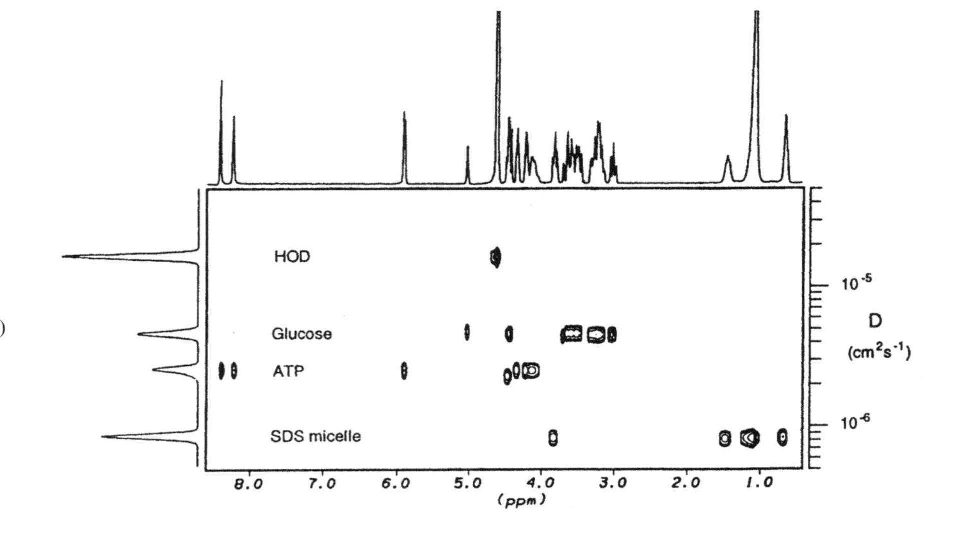

Figure 9.37. The 2D diffusion-ordered spectrum (DOSY) of a mixture of glucose, adenosine 5′-triphosphate (ATP) and sodium dodecyl sulphate (SDS) micelles in D_2O. The horizontal axis displays the conventional proton spectrum which is dispersed along the vertical dimension by individual diffusion coefficients (reproduced with permission from reference [95]).

mixtures from a wide variety of sources and is clearly an area for future development.

REFERENCES

[1] R. Freeman, S.P. Kempsell and M.H. Levitt, *J. Magn. Reson.*, 1980, **38**, 453–479.
[2] M.H. Levitt and R. Freeman, *J. Magn. Reson.*, 1979, **33**, 473–476.
[3] A.J. Shaka, C. Bauer and R. Freeman, *J. Magn. Reson.*, 1984, **60**, 479–485.
[4] M.H. Levitt, *Prog. Nucl. Magn. Reson. Spectrosc.*, 1986, **18**, 61–122.
[5] A.J. Shaka and R. Freeman, *J. Magn. Reson.*, 1983, **55**, 487–493.
[6] M. Keniry and B.C. Sanctuary, *J. Magn. Reson.*, 1992, **97**.
[7] N.S. Bai, M. Ramakrishna and R. Ramachandran, *J. Magn. Reson. (A)*, 1993, **102**, 235–240.
[8] R. Ramachandran, *J. Magn. Reson. (A)*, 1993, **105**, 328–329.
[9] M.H. Levitt and R. Freeman, *J. Magn. Reson.*, 1981, **43**, 65–80.
[10] R. Tycko, H.M. Cho, E. Schneider and A. Pines, *J. Magn. Reson.*, 1985, **61**, 90–101.
[11] A.E. Derome, *J. Magn. Reson.*, 1988, **78**, 113–122.
[12] M.H. Levitt, R. Freeman and T. Frenkiel, *J. Magn. Reson.*, 1982, **47**, 328–330.
[13] A.J. Shaka, J. Keeler and R. Freeman, *J. Magn. Reson.*, 1983, **53**, 313–340.
[14] A.J. Shaka, J. Keeler, T. Frenkiel and R. Freeman, *J. Magn. Reson.*, 1983, **52**, 335–338.
[15] A.J. Shaka, C.J. Lee and A. Pines, *J. Magn. Reson.*, 1988, **77**, 274–293.
[16] A.J. Shaka, P.B. Barker and R. Freeman, *J. Magn. Reson.*, 1985, **64**, 547–552.
[17] N.S. Bai, N. Hari and R. Ramachandran, *J. Magn. Reson. (A)*, 1994, **106**, 241–244.
[18] A. Bax and D.G. Davis, *J. Magn. Reson.*, 1985, **65**, 355–360.
[19] S.P. Rucker and A.J. Shaka, *Mol. Phys.*, 1989, **68**, 509–517.
[20] M. Kadkhodaie, O. Rivas, M. Tan, A. Mohebbi and A.J. Shaka, *J. Magn. Reson.*, 1991, **91**, 437–443.
[21] A.J. Shaka and J. Keeler, *Prog. Nucl. Magn. Reson. Spectrosc.*, 1987, **19**, 47–129.
[22] E. Kupce and R. Freeman, *J. Magn. Reson. (A)*, 1995, **115**, 273–276.
[23] E. Kupce and R. Freeman, *J. Magn. Reson. (A)*, 1995, **117**, 246–256.
[24] E. Kupce and R. Freeman, *J. Magn. Reson. (A)*, 1996, **118**, 299–303.
[25] R. Fu and G. Bodenhausen, *J. Magn. Reson. (A)*, 1995, **117**, 324–325.
[26] R. Fu and G. Bodenhausen, *J. Magn. Reson. (A)*, 1996, **119**, 129–133.
[27] H. Kessler, H. Oschkinat, C. Griesinger and W. Bermel, *J. Magn. Reson.*, 1986, **70**, 106–133.
[28] T.L. Hwang and A.J. Shaka, *J. Magn. Reson. (A)*, 1995, **112**, 275–279.
[29] A. Bax and R. Freeman, *J. Am. Chem. Soc.*, 1982, **104**, 1099–1100.
[30] M.L. Liu, R.D. Farrant, J.M. Gillam, J.K. Nicholson and J.C. Lindon, *J. Magn. Reson. (B)*, 1995, **109**, 275–283.
[31] J.M. Brown, P.A. Chaloner and G.A. Morris, *J.C.S. Perkin Trans.*, 1987, **2**, 1583–1588.
[32] H. Kessler, S. Mronga and G. Gemmecker, *Magn. Reson. Chem.*, 1991, **29**, 527–557.
[33] R. Freeman, *Chem. Rev.*, 1991, **91**, 1397–1412.
[34] S. Berger, *Prog. Nucl. Magn. Reson. Spectrosc.*, 1997, **30**, 137–156.
[35] R. Freeman, *Prog. Nucl. Magn. Reson. Spectrosc.*, 1998, **32**, 59–106.
[36] H. Geen, S. Wimperis and R. Freeman, *J. Magn. Reson.*, 1989, **85**, 620–627.
[37] C. Bauer, R. Freeman, T. Frenkiel, J. Keeler and A.J. Shaka, *J. Magn. Reson.*, 1984, **58**, 442–457.
[38] J. Friedrich, S. Davies and R. Freeman, *J. Magn. Reson.*, 1987, **75**, 390–395.
[39] L. Emsley and G. Bodenhausen, *J. Magn. Reson.*, 1989, **82**, 211–221.
[40] H. Kessler, U. Anders, G. Gemmecker and S. Steuernagel, *J. Magn. Reson.*, 1989, **85**, 1–14.
[41] H. Geen and R. Freeman, *J. Magn. Reson.*, 1990, **87**, 415–421.
[42] H. Geen and R. Freeman, *J. Magn. Reson.*, 1991, **93**, 93–141.
[43] L. Emsley and G. Bodenhausen, *Chem. Phys. Lett.*, 1990, **165**, 469–476.
[44] L. Emsley and G. Bodenhausen, *J. Magn. Reson.*, 1992, **97**, 135–148.
[45] P.J. Hajduk, D.A. Horita and L.E. Lerner, *J. Magn. Reson. (A)*, 1993, **103**, 40–52.
[46] D.A. Horita, P.J. Hajduk and L.E. Lerner, *J. Magn. Reson. (A)*, 1993, **103**, 53–60.
[47] J.-M. Nuzillard and R. Freeman, *J. Magn. Reson. (A)*, 1994, **107**, 113–118.
[48] N.L. Gregory, T.D.W. Claridge and A.E. Derome, *Meas. Sci. Technol.*, 1994, **5**, 1359–1365.
[49] G.A. Morris and R. Freeman, *J. Magn. Reson.*, 1978, **29**, 433–462.
[50] S.L. Patt, *J. Magn. Reson.*, 1992, **96**, 94–102.
[51] E. Kupce and R. Feeman, *J. Magn. Reson. (A)*, 1993, **105**, 234–238.
[52] K. Stott, J. Stonehouse, J. Keeler, T.L. Hwang and A.J. Shaka, *J. Am. Chem. Soc.*, 1995, **117**, 4199–4200.
[53] G.Z. Xu and J.S. Evans, *J. Magn. Reson. (B)*, 1996, **111**, 183–185.
[54] K.E. Kövér, D. Uhrín and V.J. Hruby, *J. Magn. Reson.*, 1998, **130**, 162–168.
[55] K. Stott, J. Keeler, Q.N. Van and A.J. Shaka, *J. Magn. Reson.*, 1997, **125**, 302–324.

[56] M.J. Gradwell, H. Kogelberg and T.A. Frenkiel, *J. Magn. Reson.*, 1997, **124**, 267–270.
[57] J.-M. Nuzillard and R. Freeman, *J. Magn. Reson. (A)*, 1994, **110**, 252–256.
[58] J.-M. Bernassau and J.-M. Nuzillard, *J. Magn. Reson. (B)*, 1994, **103**, 77–81.
[59] P.J. Hore, *Method. Enzymol.*, 1989, **176**, 64–77.
[60] M. Guéron, P. Plateau and M. Decorps, *Prog. Nucl. Magn. Reson. Spectrosc.*, 1991, **23**, 135–209.
[61] W.E. Hull, in Two-dimensional NMR Spectroscopy; Applications for Chemists and Biochemists, eds. W.R. Croasmun and R.M.K. Carlson, VCH Publishers, New York, 1994.
[62] D.I. Hoult, *J. Magn. Reson.*, 1976, **21**, 337–347.
[63] R.W. Dykstra, *J. Magn. Reson.*, 1987, **72**, 162–167.
[64] D. Neuhaus, I.M. Ismail and C.-W. Chung, *J. Magn. Reson. (A)*, 1996, **118**, 256–263.
[65] P. Plateau and M. Guéron, *J. Am. Chem. Soc.*, 1982, **104**, 7310–7311.
[66] P.J. Hore, *J. Magn. Reson.*, 1983, **54**, 539–542.
[67] P.J. Hore, *J. Magn. Reson.*, 1983, **55**, 283–300.
[68] D. Callihan, J. West, S. Kumar, B.I. Schweitzer and T.M. Logan, *J. Magn. Reson. (B)*, 1996, **112**, 82–85.
[69] M. Piotto, V. Saudek and V. Slenár, *J. Biomol. NMR*, 1992, **2**, 661–665.
[70] V. Sklenár, M. Piotto, R. Leppik and V. Saudek, *J. Magn. Reson. (A)*, 1993, **102**, 241–245.
[71] M. Liu, X. Mao, C. Ye, J.K. Nicholson and J.C. Lindon, *J. Magn. Reson.*, 1998, **132**, 125–129.
[72] D. Marion, M. Ikura and A. Bax, *J. Magn. Reson.*, 1989, **84**, 425–430.
[73] T. Parella, P. Adell, F. Sanchez-Ferrando and A. Virgili, *Magn. Reson. Chem.*, 1998, **36**, 245–249.
[74] C. Dalvit, G. Shapiro, J.-M. Böhlen and T. Parella, *Magn. Reson. Chem.*, 1999, **37**, 7–14.
[75] R.E. Hurd, *J. Magn. Reson.*, 1990, **87**, 422–428.
[76] G. Lowe, *Chem. Soc. Rev.*, 1995, **24**, 309–317.
[77] D. Obrecht and J.M. Villalgordo, Solid-supported Combinatorial and Parallel Synthesis of Small-Molecular-Weight Compound Libraries, Pergamon, Oxford, 1998.
[78] E. Giralt, J. Rizo and E. Pedroso, *Tetrahedron*, 1984, **40**, 4141–4152.
[79] G.C. Look, C.P. Holmes, J.P. Chinn and M.A. Gallop, *J. Org. Chem.*, 1994, **59**, 7588–7590.
[80] E.R. Andrew, *Prog. Nucl. Magn. Reson. Spectrosc.*, 1971, **8**, 1–39.
[81] W.L. Fitch, G. Detre, C.P. Holmes, J.N. Shoolery and P.A. Keifer, *J. Org. Chem.*, 1994, **59**, 7955–7956.
[82] P.A. Keifer, L. Baltusis, D.M. Rice, A.A. Tymiak and J.N. Shoolery, *J. Magn. Reson. (A)*, 1996, **119**, 65–75.
[83] R.C. Anderson, M.A. Jarema, M.J. Shapiro, J.P. Stokes and M. Ziliox, *J. Org. Chem.*, 1995, **60**, 2650–2651.
[84] R.C. Anderson, J.P. Stokes and M.J. Shapiro, *Tetrahedron Lett.*, 1995, **36**, 5311–5314.
[85] R. Jelinek, A.P. Valente, K.G. Valentine and S.J. Opella, *J. Magn. Reson.*, 1997, **125**, 185–187.
[86] C. Dhalluin, C. Boutillon, A. Tartar and G. Lippens, *J. Am. Chem. Soc.*, 1997, **119**, 10494–10500.
[87] P.A. Keifer, *J. Org. Chem.*, 1996, **61**, 1558–1559.
[88] P. Stilbs, *Prog. Nucl. Magn. Reson. Spectrosc.*, 1987, **19**, 1–45.
[89] E.O. Stejskal and J.E. Tanner, *J. Chem. Phys.*, 1965, **42**, 288–292.
[90] S.J. Gibbs and C.S. Johnson Jr., *J. Magn. Reson.*, 1991, **93**, 395–402.
[91] D. Wu, A. Chen and C.S. Johnson Jr., *J. Magn. Reson. (A)*, 1995, **115**, 260–264.
[92] J.E. Tanner, *J. Chem. Phys.*, 1970, **52**, 2523–2526.
[93] M.D. Pelta, H. Barjat, G.A. Morris, A.L. Davis and S.J. Hammond, *Magn. Reson. Chem.*, 1998, **36**, 706–714.
[94] K.F. Morris and C.S. Johnson Jr., *J. Am. Chem. Soc.*, 1992, **114**, 3139–3141.
[95] K.F. Morris and C.S. Johnson Jr., *J. Am. Chem. Soc.*, 1993, **115**, 4291–4299.
[96] H. Barjat, G.A. Morris, S. Smart, A.G. Swanson and S.C.R. Williams, *J. Magn. Reson. (B)*, 1995, **108**, 170–172.
[97] H. Barjat, G.A. Morris and A.G. Swanson, *J. Magn. Reson.*, 1998, **131**, 131–138.

Appendix

Glossary of acronyms

Acronym/term	Translation	Section
ACOUSTIC	Alternate compound 180s used to suppress transients in the coil (for quadrupolar nuclei)	4.5
APT	Attached proton test (spectrum editing)	4.3.2
BIRD	Bilinear rotation decoupling	6.3.3
BURP	Band-selective, uniform-response, pure-phase (selective) pulse	9.3.1
CAMELSPIN	Cross-relaxation appropriate for minimolecules emulated by locked spins (aka ROESY)	8.8
COLOC	Correlation through long-range coupling	6.5.2
COSY	Correlation spectroscopy	5.2
CPD	Composite pulse decoupling	9.2
CPMG	Carr–Purcell–Meiboom–Gill T_2-dependent spin-echo sequence	2.4.4
CSA	Chemical shift anisotropy	2.5.3
CYCLOPS	Cyclically-ordered phase-sequence (for suppressing quadrature artefacts)	3.2.5
DANTE	Delays alternating with nutation for tailored (selective) excitation	9.3.2
DEPT	Distortionless enhancement by polarisation transfer (spectrum editing)	4.4.3
DIPSI	Decoupling in the presence of scalar interactions	9.2
DOSY	Diffusion-ordered spectroscopy	9.5.2
DPFGSE	Double pulsed field gradient spin-echo (selective excitation)	9.3.3
DQF-COSY	Double-quantum filtered correlation spectroscopy	5.6.2
EXORCYCLE	Phase cycle to suppress 'ghost' and 'phantom' artefacts in spin-echo sequences	7.2.2
EXSY	Exchange spectroscopy	8.7.4
FID	Free induction decay	2.2.2
FLIPSY	Flip-angle adjustable one-dimensional NOESY (solvent suppression)	9.4.1
FLOPSY	Flip-flop mixing sequence (for total correlation spectroscopy)	9.2
GARP	Globally-optimised, alternating phase, rectangular pulses (for broadband decoupling)	9.2
HEHAHA	Heteronuclear Hartmann–Hahn spectroscopy	5.7
HETCOR	Heteronuclear correlation	6.5.1
HMBC	Heteronuclear multiple bond correlation	6.4
HMQC	Heteronuclear multiple quantum correlation	6.3.1
HOESY	Heteronuclear Overhauser effect spectroscopy	8.9
HOHAHA	Homonuclear Hartmann–Hahn spectroscopy (aka TOCSY)	5.7
HSQC	Heteronuclear single-quantum correlation	6.3.2
INADEQUATE	Incredible natural-abundance double-quantum transfer experiment	5.8
INEPT	Insensitive nuclei enhanced by polarisation transfer	4.4.2
J-MOD	J-modulated spin-echo	4.3.1
LP	Linear prediction	3.2.3
MAS	Magic angle spinning	9.5.1
MLEV	Broadband decoupling cycle from Malcolm Levitt	9.2
MQF	Multiple-quantum filter	5.6.2
NOE	Nuclear Overhauser effect	8.2
NOESY	Nuclear Overhauser effect spectroscopy	8.7
PENDANT	Polarisation enhancement nurtured during attached nucleus testing (spectrum editing)	4.4.4

Acronym/term	Translation	Section
PFG	Pulsed field gradient	5.5
RIDE	Ring-down delay sequence (for quadrupolar nuclei)	4.5
ROESY	Rotating-frame Overhauser effect spectroscopy	8.8
SEFT	Spin-echo Fourier transform (spectrum editing)	4.3.1
SPFGSE	Single pulsed field gradient spin-echo (selective excitation)	9.3.3
SPT	Selective population transfer	4.4.1
SUSAN	Spin decoupling employing ultra-broadband inversion sequences generated via simulated annealing (broadband decoupling sequence)	9.2
TOCSY	Total correlation spectroscopy	5.7
TPPI	Time-proportional phase incrementation	5.3.1
Tr-ROESY	Transverse rotating-frame Overhauser effect spectroscopy	8.8
WALTZ	Wideband, alternating-phase, low-power technique for zero-residual splitting (broadband decoupling sequence)	9.2
WATERGATE	Water suppression through gradient tailored excitation	9.4.3
WATR	Water attenuation by transverse relaxation	2.4.4
WURST	Wideband, uniform rate and smooth truncation (broadband decoupling sequence)	9.2

Index

^{11}B
coupling 43
2D EXSY 327
linewidths 41

^{13}C
satellites in ^{13}C spectra 211
satellites in ^{1}H spectra 116, 224, 228, 244
satellites, selecting 228, 243

^{14}N
coupling 44
linewidths 41

^{15}N
INEPT 136, 137
HSQC 233

^{17}O
backward linear prediction 144
RIDE 144

180° pulse 18

^{195}Pt
field dependence, satellites 39

^{19}F
coupling to ^{11}B 43
quantitative 55

1D NMR, *see* One-dimensional NMR

$^{1}J_{CC}$
measuring 214
typical values 216

$^{1}J_{CH}$
typical values 128, 226

$^{2/3}J_{CH}$
typical values 248

^{29}Si
HMBC 249

2D NMR, *see* Two-dimensional NMR

^{2}H
coupling 42
lock 85–87

3-(trimethylsilyl)propionate
shift reference 78

^{31}P
^{13}C HMQC 237
^{31}P COSY 159
^{57}Fe HMQC 238
decoupling 124

3D NMR, *see* Three-dimensional NMR

^{57}Fe
^{31}P HMQC 238

^{77}Se
CSA relaxation 39

90° pulse 18

A

Absolute-value
display, 2D 161, 164, 173, 268
gradient spectroscopy 183

Absorption mode 19

Acetylisomontanolide
^{13}C INADEQUATE 214

ACOUSTIC 111, 144

Acoustic ringing 143

Acquisition parameters
2D NMR 170–172

Acquisition time 54–56
2D NMR 170–172

Acronyms
glossary 373, 374

Active coupling 192–194

ADC, *see* Analogue-to-digital converter

ADEQUATE 217

Adiabatic
decoupling 347, 348
pulses 348

Aliasing 53, 62
2D NMR 163
folded signals 62, 163
wrapped signals 62, 163

Alkynes, unusual behaviour
DEPT 142
HMQC/HSQC 227

Allylic coupling
in COSY 199

Alumina, activated 76

Amplitude modulation 150, 161, 178

Analogue filters 53, 54

Analogue-to-digital converter 47, 52, 65–70
overload 66
resolution 65

Andrographolide
DQF-COSY 196
NOEs 300
NOESY 315

Anisotropy
in chemical bonds 38

Antiphase
splittings, in COSY 162, 192–197
vectors 21

Apodisation 57, 70

APT, *see* Attached proton test

Artefacts
axial peaks 167
clipped FID 66
ghosts 266
phantoms 266
quad images 64, 168
SPT 131, 309
subtraction 307
t_1-noise 168, 234, 247
truncated FID 56

ASTM
test sample 107, 108

Attached proton test 111, 128

Audio frequencies 51, 61

Axial peaks 167

B

B_0, static magnetic field 13, 49
inhomogeneity 30, 46
optimising 87–94

B_1, rf magnetic field, transmitter 18, 49
calibrating 95–98
inhomogeneity 341–343

B_2, rf magnetic field, decoupler 346, 347
calibrating 98

Bandpass filter 52, 53

Bandwidth
decoupling 347
excitation 344

Bar magnet analogy 14, 38
B_{eff}, effective rf field 49, 342, 348
Benzene
 test sample 107
Bicycle dynamo 129
Bilinear rotation decoupling 235
 HMQC 234–236
 HETCOR 254, 255
 J-resolved spectroscopy 263
BIRD pulse, *see* Bilinear rotation decoupling
Bloch–Siegert shift 98, 118
Boiling points
 deuterated solvents 76
Boltzmann distribution 15, 129, 280
Boronic acids 41
Broadband decoupling 120, 346–348
 of proton spectra 267, 269, 272
Bulk magnetisation vector 15
BURP pulse 350, 352–354
 excitation trajectory 353
Butanol
 ^{13}C INADEQUATE 213

C

Calibrations
 decoupler pulses 97
 pulsed field gradients 99–104
 pulse widths 94–99
 radiofrequency fields 95, 98
 shaped pulses 358, 359
 temperature 104
CAMELSPIN, *see* ROESY
Carbopeptoids
 COSY 156–158
 HMBC 250
 HSQC, ^{15}N 233
 TOCSY 202, 211
 Tr-ROESY 331
Carousel analogy
 rotating frame 17
Carr–Purcell sequence 33
Carr–Purcell–Meiboom–Gill sequence 33, 34
Chemical exchange
 acidic protons 77
 EXSY 326–328
 solvent attenuation 34
Chemical shift
 anisotropy 38, 39
 early observations of 2
 refocusing 22–24
 visualising in vector model 20, 21
Chemical shift anisotropy relaxation 38
Chloroform
 test sample 106
Chromium(III) acetylacetonate 38, 113, 216
Cleaning, NMR tubes 80
Clipping, FID 66
Coherence 174–178
 double-quantum 175, 189, 212
 multiple-quantum 176
 order 175
 phase- 18, 175
 single-quantum 175
 transfer pathways 177, 178
 zero-quantum 175, 317
Coherence transfer 174–178
 echoes, 2D 165
 in COSY 155
 in TOCSY 203
Coherence transfer pathways 177, 178
 in COSY 177
 in DQF-COSY 189
 in HMQC 226
 in HSQC 229
 in INADEQUATE 212
 in NOESY 316
COLOC 254
Composite pulse decoupling 120, 346–348
 schemes for 346, 347
Composite pulses 51, 310, 341–345
 broadband decoupling 346–348
 inversion *versus* refocusing 344, 345
 properties of, selected 344
 spin locks 208, 347, 348
Conformational averaging, of NOE 303, 304
Constant-time experiments 255
Contour plot 153, 166, 174
Cooley–Tukey algorithm
 fast Fourier transform 24
Correlation spectroscopy, COSY 153–160, 187–201
 gradient selected 182, 190
 interpreting 156–159, 192–197
 phases in 162, 189
 small couplings 199, 200
 which method? 188, 189
Correlation time, τ_c 36, 283
COSY, *see* Correlation spectroscopy
COSY-45, *see* COSY-β
COSY-60, *see* COSY-β
COSY-90 148, 153, 188
COSY-β 148, 158, 188, 197–199
 geminal *versus* vicinal couplings 198
Coupling constants
 from DQF-COSY 194–197
 in cyclohexanes 196
CPMG, *see* Carr–Purcell–Meiboom–Gill
Crosspeak 155, 159, 160
 disappearance, in COSY 194
 tilting, in COSY 198
Cross-polarisation 201
Cross-relaxation 281, 288, 306
 rates 282, 289, 302
Cryo-probes 83
CSA, *see* Chemical shift anisotropy relaxation
Cyclodextrins
 NOESY 325
 ROESY 333
Cyclohexanes
 couplings, from COSY 196
CYCLOPS 64
 in 2D NMR 168, 215

D

D_2O exchange 77
DANTE 354, 355
 excitation sidebands 355
dB scale 95
Decimation 69
Decoupler
 calibrating 97–99
 gating 119, 121
 leakage 119
 radiofrequency 47
Decoupling 116–124, 346–348
 adiabatic 347, 348
 bandwidths 120, 347
 basis of 117
 broadband 120, 124, 346, 347
 composite pulse 120, 346, 347
 gated 121
 heteronuclear 120, 124
 homonuclear 117
 selective 117, 121
Degassing
 NOE 289, 337
 techniques for 80
Delayed-COSY 148, 188, 199, 200
DEPT 111, 139–142
 editing with 140
 optimising sensitivity 142
 sequence 139
DEPT-Q 111, 141, 143
Detection 51–56, 59–63
 2D NMR 149
 single channel 59
 two channel, quadrature 60–63
Deuterated solvents
 changing 77
 properties 76
Developments, in NMR 1–4
Diagonal peak 155, 159
Diastereotopic protons
 ^{1}H–^{13}C correlations 225
Difference spectroscopy 118
 NOE 306–313

Diffusion
coefficient 369, 370
losses 32, 186
ordered spectroscopy 368–370
Digital filters 69
Digital resolution 54, 55
Digital signal processor 52, 69
Digitisation 51, 65
2D NMR 170–172
noise 66
Digitiser, *see* Analogue-to-digital converter
Dilute spins
correlating 211–218
Dioxane
shift reference 78
test sample 107
Dipolar interactions
coupling 38, 277, 280, 282–285, 366
electrons 38
Dipole–dipole relaxation 37, 38, 282–285
DIPSI-2 209, 347
Directional coupler 85
DOSY 368–370
Double difference phase-cycle 229
Double pulsed field gradient spin-echo 355–357
1D NOESY 320
1D TOCSY 211
Double-quantum filter
1D 191
2D 189–192
INADEQUATE 212, 215
DPFGSE, *see* Double pulsed field gradient spin-echo
DQF, *see* Double-quantum filter
DQF-COSY 148, 188–197
multiplet structures 192–196
sequence 189
versus COSY 188, 190, 191
Drying, samples 76
DSP, *see* Digital signal processor
Dummy scans 171
Duty cycle 119
Dwell time 52
Dynamic range 65–70, 359

E

EBURP, *see* BURP pulse
Echo-antiecho method 184
Eddy currents 186
Editing
1D proton spectra 242
background suppression 34
by molecular size 34, 368, 370
heteronuclear 2D correlations 239, 240
multiplicity 125–129, 138–143, 239, 240
T_2 filter 34
Electric field gradients 41
Enantiomeric excess
improving accuracy 55, 56, 116
Equilibrium
establishing 26, 27
Equilibrium NOE, *see* Steady-state NOE
Ernst angle 112
Ethanediol
temperature calibration 104
Ethanol, first high-resolution ^{1}H spectrum 1
Ethylbenzene
test sample 107
Ethylene glycol, *see* Ethanediol
Evolution time
2D NMR 149
Excitation bandwidth 50, 342, 357
Excitation profiles 49, 345, 351, 352
binomial pulses 363
inversion 345
shaped pulses 351, 352
Excitation sculpting 355–357
1D NOESY 320
solvent suppression 365
Excitation trajectories
binomial pulses 363
composite pulses 343
DANTE 355
imperfect pulses 342
shaped pulses 351, 353, 355
EXORCYCLE 215, 266, 321, 361
Experiment selection, guidelines for 8–12
Exponential
decay 71, 106
multiplication 70, 71
recovery 26, 27
EXSY 278, 316, 326–328
exchange rates 327, 328
mixing time 327
sequence 316
Extreme narrowing limit 36, 285, 306

F

f_1, 2D NMR 8, 149
f_2, 2D NMR 8, 149
Field-gradient pulses, *see* Pulsed field gradients
Filling factor 224
Filtering
1D proton spectra 242, 243
low-pass, HMBC 246
noise 53, 54
sample solutions 80
FLIPSY 362
FLOCK 256
FLOPSY-8 347
Folding, *see* Aliasing
Fourier transformation
complex 60, 61
introduction, to NMR 2
in 2D NMR 151, 173
mathematical expression 24
real 60, 62
FRED™ 216
Free induction decay 20
Freeze-pump-thaw method 81, 337
Frequency domain 24
Frequency labelling 150–152
Frequency sweep
adiabatic 348

G

Gadolinium
triethylene-tetraamine-hexaacetate 216
GARP 347
Gated decoupling 121, 122
in J-resolved spectroscopy 260
inverse 115, 121, 288
power 121
Gaussian pulse 350–352, 354
1D NOESY 321
cascades 350, 352–354
semi-selective HMBC 250
Gel-phase NMR 366
Geminal couplings
in COSY 158, 198
Gradient echo
gradient calibrations 100–103
selective excitation 355–357
shimming 93
signal selection 179–183
Gradient image profiles 93, 102
Gradient shimming 92–94
Gradient-accelerated spectroscopy, *see* Gradient-selected spectroscopy
Gradient-enhanced spectroscopy, *see* Gradient-selected spectroscopy
Gradient-selected spectroscopy 178–187
advantages and limitations 185
Gramicidin-S
TOCSY 205, 208
NOESY 324
Gyromagnetic ratio, *see* Magnetogyric ratio

H

Hard pulse 49, 348

Hartmann–Hahn match 201, 204
  in ROESY 329

Heating, rf induced 120, 342

HEHAHA 205

Heisenberg Uncertainty principle
  pulse excitation 48
  relaxation 26

HETCOR 221, 252–254, 273
  homonuclear f_1 decoupling 254

Heterogeneous samples 366–368

Heteronuclear multiple-bond correlations
  proton detected 244–251
  X-detected 254–256

Heteronuclear single-bond correlations
  proton detected 224–243
  BIRD selection 234–236
  breakthrough in HMBC 246
  gradient selection 231–234
  HMQC 224–229
  HSQC 229, 230
  hybrid experiments 238–243
  X-detected 251–254

History of NMR 1–4

HMBC 221, 244–251
  applications 248–251
  semi-selective 250
  sequence 245

HMQC 221, 224–229
  edited, multiplicity 239
  for X–Y correlations 237
  sequence 225
  TOCSY 241
  versus HSQC 224, 229

HOESY 278, 335, 336
  mixing time 336
  sequence 335

HOHAHA, *see* TOCSY

Homospoil pulse 185

HSQC 221, 229, 230
  edited, multiplicity 239
  TOCSY 241

Hybrid experiments 238–243

HYSEL 240

I

IBURP, *see* BURP pulse

Imaginary spectrum 24, 162

Imaging
  gradient shimming 93

INADEQUATE 148, 211–218
  1D, measuring J_{CC} 213
  2D 212, 213
  implementing 215
  proton detected 216–218

INEPT 111, 132–139, 217
  editing with 138, 139
  in ^{1}H–X correlations 229, 252
  multiplet intensities in 135
  optimising sensitivity 134
  refocused 133–136
  signal averaging 137

INEPT$^+$ 135

INEPT-INADEQUATE 216–218

Inhomogeneity
  B_0 field 30, 46
  B_1 field 341–343

INSIPID 217

Instrumentation 45–48

Integration 38, 114–116

Interferogram 150

Inverse spectroscopy 222–224

Inversion recovery
  for measuring T_1 27–30
  in BIRD-HMQC 235, 236
  sequence 28

Isochromats 32

Isotropic mixing, *see* Spin-lock

I-spin, in NOE 279

J

J-modulated spin-echo 111, 125–128
  2D analogue, J-resolved 260–262
  editing in HSQC 239

J-resolved spectroscopy 259–274
  heteronuclear 259, 260–267
  homonuclear 267–273
  indirect homonuclear 273, 274

L

Laboratory frame of reference 16

Larmor
  frequency 15
  precession 14

LC–NMR 3, 148, 359

Leu-enkephalin
  1D DQF 191
  DQF-COSY 194

Ligand exchange, EXSY 327

Linear prediction 57–59
  backward 59, 144
  forward 58
  in 2D NMR 58, 59, 173

Line-broadening functions 70, 71

Lineshape
  2D 161–163, 165, 166
  absorption mode 19
  defects 90
  dispersion mode 19
  Gaussian 72
  Lorentzian 31, 72, 116
  tests 106, 107

Linewidth
  half-height 31

Lock
  optimising 86
  parameters 86
  system 85–87

Longitudinal spin relaxation 26–30, 281

Long-range COSY, *see* Delayed-COSY

Long-range couplings
  ^{1}H–^{13}C 123, 244, 254, 263–265
  in COLOC 254
  in COSY 159, 199
  in HMBC 244, 245, 248
  measuring $^nJ_{CH}$, J-resolved 263–265

Low-pass J-filter 246

LP, *see* Linear prediction

M

Magic angle spinning 366–368

Magic-cycle, decoupling 346

Magnetic field, B_0 13
  optimising 87–94

Magnetic flux density 13

Magnetic moment
  molecular 39
  nuclear 13

Magnetic susceptibility
  matched NMR tubes 79
  sample preparation 79, 366

Magnetogyric ratio, γ 13, 129, 222

Magnets, superconducting 2, 45, 46, 88
  progress 46
  shielded 3

Magnitude calculation 165

Manganese(II) chloride 38

Marine alkaloids 205, 323

MAS, *see* Magic angle spinning

Matching
  probehead 83

MAXY 240

Melting points
  deuterated solvents 76

Menthol
  INEPT-INADEQUATE 218
  J-resolved 261

Methanol
  temperature calibration 104

Mixing
  2D NMR 149

Mixing time, τ_m
  in NOESY 318

in ROESY 305, 329
in TOCSY 202, 204, 206
MLEV-16 209, 346, 347
MLEV-17 208, 347
Molecular sieves
drying samples 76
Monochromatic radiation
pulsed excitation with 48, 342
Mosher's acid derivatives
ee determinations 55
Multiple-quantum coherence 174–176
DEPT 139
HMQC 226
simplified picture 176
Multiplets
visualising in vector model 22
Multiplicities from
DEPT 138–143
HMQC/HSQC 239, 240
INEPT 125–129
J-resolved spectroscopy 261, 262

N

Natural abundance
of quadrupolar nuclei, selected 40
of spin-$^1/_2$ nuclei, selected 14
NMR spectrometer, schematic 45
NOE difference 278, 306–313, 335
heteronuclear 335
optimising 307–313
quantifying enhancements 312
SPT artefacts 309
subtraction artefacts 307
NOE, *see* Nuclear Overhauser effect
NOESY 278, 313–323
1D, gradient 320–322
complications with 317
crosspeaks, origins 317
internuclear separations 319
mixing time, optimum 318
presat 362
sequence 316
Noise
filtering 53
digitisation- 66, 68
quantisation- 66, 68
N-type
echoes 165
pathway 177, 183
signals, 2D 164, 165, 177, 178, 183
Nuclear Overhauser effect (*see also* Steady-state NOE; Transient NOE; Rotating frame NOE) 37, 277–337
applications, examples 296–301, 323–326, 332–334
conformational averaging 303
cross-relaxation 81, 282, 288, 306
definition 279
distance dependence 283, 285, 287
experimental aspects 336, 337
field strength dependence 286, 287
heteronuclear 136, 287, 335, 336
homonuclear 279, 285
initial rate approximation 302, 303
internuclear separations 303, 319, 320
introduction of, to NMR 2
kinetics 302, 303
leakage of 289
molecular motion 285–287
negative 281, 288, 293
positive 281
sensitivity enhancement 2, 136
steady-state 277, 279–301, 306–313
rotating frame 286, 304–306, 328–334
transient 277, 301–304, 313–328
Nuclear spin 13
Nutation angle 18
Nyquist condition 52, 53
2D NMR 170
oversampling 68

O

Off-resonance
effects 49, 341, 342
excitation 362
One-dimensional NMR
techniques 111–145, 191, 210, 213, 242, 307, 320, 332
On-resonance
excitation 20, 48, 49
Oversampling 68–70
decimation 69
Overview, of modern NMR methods 8–12

P

Palladium phosphine complexes
$^1H\{^1H\}$ decoupled 272
$^1H\{^{31}P\}$ decoupled 124
HETCOR 253
NOESY 325
ROESY 334
Paramagnetic relaxation 34, 38
agents for 38, 113
NOE, quenching 289, 337
removing O_2 80, 337
Pascal's triangle 193
Passive coupling 192–194
PENDANT 111, 142, 143
PFG spin-echo, diffusion 368
PFG stimulated echo, diffusion 369
Phantom
in gradient calibrations 101
Phase
correction, 1D 73, 74
correction, 2D 174
coherence 18, 175
real and imaginary data 24
Phase cycling 63–65, 179
axial peak suppression 167
CYCLOPS 64, 168
double-difference 229
EXORCYCLE 266
multiple-quantum filtration 190
Phase errors
aliased signals 62
correction 73–75, 174
zero and first order 73
Phase modulation 164, 178
in J-resolved spectroscopy 268
Phase-sensitive
display, 2D 161, 162, 166
gradient spectroscopy 183
Phase-sensitive detector 61
Phase-twist lineshape 165, 166
in J-resolved spectroscopy 268
Polarisation transfer 130–132
in COSY 155, 160
versus NOE, sensitivity 136
Population differences 15, 18, 19
equalising 18
inverting 19
Population inversion 19, 130–132
Pre-acquisition delay
phase errors 74
quadrupolar nuclei 143
Precession
gyroscope 15
nuclear, Larmor 14, 15
Preparation
2D NMR 149
Probeheads 47, 82
acoustic ringing 143
actively shielded 187
broadband 47, 83
cryogenic 83
inverse 83, 224
magic angle spinning 368
micro 82
nano 82
Q-modulation 337
selective 47
sizes 47, 82
tuning and matching 83–85
Processing parameters
2D NMR 172
general scheme, 2D 173
Propagation, magnetisation
with spin-lock 204, 208
Protocol, for structure confirmation 10
Pseudo-diagonal 213
P-type
anti-echoes 165
pathway 177, 183
signals, 2D 164, 177, 178, 183
Pulse excitation 18–20, 48–51
bandwidth 50, 349

tip or flip angle 18
vector model 18–20

Pulse imperfections
B_1 inhomogeneity 341–343
compensating for 341–345
excitation trajectories 342
in HSQC 229
off-resonance effects 49, 341–343

Pulse sequence nomenclature 7

Pulse width
calibration 94–99
definition 49
miscalibration 341

Pulsed field gradients 178–187
implementing 186
introduction of, to NMR 3
in high-resolution NMR 184–186
shimming with 92–94
signal selection 179–182
symbols used for 7

Pulses
gradient 179–183
hard, rf 49, 354
shaped/selective, rf 348–354
soft, rf 49, 348–354
symbols used for 7

Pure-phase pulse, shaped 352–354

Q

Q3, shaped pulse 352

Q5, shaped pulse 352

Q-modulation 337

Quad-images 63–65, 168
compensating for, CYCLOPS 64, 168

Quadrature detection 59–63
aliasing 62, 163
echo-antiecho method, 2D 184, 233
images 63–65
in 2D NMR 161–166
States method, 2D 162
TPPI method, 2D 162

Quadrupolar nuclei
coupling with spin-$^1/_2$ nuclei 42, 43
linewidths 40
natural abundance of, selected 40
observing 143
sensitivity 40

Quadrupolar relaxation 40–43
NOE, quenching 289

Quantitative NMR 114–116
^{19}F 55

R

Radiation damping 85, 360

Radio-frequency pulses (*see also* Pulses)
symbols for 7

Real spectrum 24, 162

REBURP, *see* BURP pulse

Receiver 45
gain 66, 67

Receiver imbalance
compensating for, CYCLOPS 64

Recovery delay, *see* Relaxation delay

Reference compounds 77, 78

Reference deconvolution 169

Reference distances, NOE 320

Referencing spectra
internal 77
external 78
Ξ scale 78

Refocusing
chemical shifts 22–24

Relative sensitivity
of spin-$^1/_2$ nuclei, selected 14
of quadrupolar nuclei, selected 40

Relaxation 25–44
free induction decay 20
longitudinal 26–30
mechanisms 35–44
spin-lattice 26
spin-spin 31
transverse 30–35

Relaxation delay 111–114

Relayed-COSY 200, 201

Relaying of information
in TOCSY 203–205

REPAY 240

Repetition rate
Ernst angle 112
optimum, in 1D NMR 112–114
optimum, in 2D NMR 172
optimum, with 90° pulse 114

Resolution
tests 106, 107

Resolution enhancement 71–73
linear prediction 57–59

RIDE
quadrupolar nuclei 111, 144

Robotic sample changer 48

ROE, *see* Rotating frame NOE

ROESY 278, 328–334
1D 332
compensated 332
complications with 329–332
crosspeaks, origins 331
mixing time 305, 329
sequence 329
spin-locks 329–331
transverse, Tr- 331

Rotating frame NOE 304–306, 328–334
appearance in TOCSY 207
applications, examples 332–334
false 330, 331
mixing time 305
spin-lock 305, 329–332

Rotating frame of reference 16–18

Rotaxanes
NOESY 325

S

Saccharides
HSQC, edited 240
HSQC-TOCSY 242
J-resolved 269

Sample preparation 75–81, 360

Sample spinning 46, 337
2D NMR 169
introduction of, to NMR 2
shimming 88
sidebands 90

Sampling, data 52–56
sequential 62, 162
simultaneous 61, 162

Satellites 39, 116, 211, 224, 228, 244

Saturation 19, 38, 113
transfer 293, 311, 362

Sealing tubes 81

SEFT 125

Selective detection 69

Selective excitation 49, 348–359
DANTE 354, 355
gradient echoes 355–357
shaped pulses 351–354

Selective population inversion 131

Selective population transfer 131, 132
artefacts 131, 309

Sensitivity 129, 130, 222–224
enhancement 70, 71
heteronuclear correlations 222
optimising, in 1D NMR 112–114
oversampling 69
tests 107, 108

Sequential assignment, peptides 323, 324

Shaped pulse 349–354
bandwidth factor 354, 357
calibrating 358, 359
DANTE 355
Gaussian 350–352
properties of, selected 354
pure phase 352–354

Shift correlation
chemical exchange 326–328
dipolar couplings 313–320
heteronuclear couplings 221–256
homonuclear couplings 147–218
overview 10

Shim coils 46, 88

Shimming 46, 87–94
automatic 89
defects 90
gradient 92–94
on the FID 91

Signal averaging 67, 68
NOE difference 307–309

Signal-to-noise ratio
measuring, tests 107, 108
in signal averaging 67, 68

Single-channel detection 59, 62
TPPI 162
SNEEZE, shaped pulse 354, 358
Soft pulse (*see also* Shaped pulse) 49, 349–354
Solvent suppression 359–366
binomial sequences 362, 364
via chemical exchange 34
jump-return 362
presaturation 361
pulsed field gradients 363–366
test sample, sucrose 109, 365
Solvents 75, 336, 359
Spectral density 36, 283
Spectral width 52
Spectrum editing 125–143
with spin echoes 34, 35, 125–129
Spin coupling
visualising in vector model 21–24
Spin quantum number, I 13, 40
Spin rotation relaxation 39, 40
Spin-diffusion 293, 294, 303
Spin-echoes 21–24
experimental observation 31–34
homonuclear 23, 203
heteronuclear 23
in J-resolved spectroscopy 260
in spin-locks 203
J-modulated 125–129
T_2 measurement 31–34
visualising in vector model 21
Spin-lattice relaxation, *see* Longitudinal spin relaxation
Spin-lock 347, 348
in ROESY 305, 329–332
in TOCSY 203, 347
Spinning sidebands 90
Spin-spin relaxation, *see* Transverse spin relaxation
Spin-spin-coupling
early observations of 2
Spontaneous emission 27
S-spin, in NOE 279
Stacked plot 153, 166
States method
2D quad detection 162
axial peaks 167
Steady-state
2D NMR 171
magnetisation 112, 113
Steady-state NOE 279–301
applications, examples 296–301
distance dependence 287, 290
indirect, 3-spin 291
measuring 306–313
multispin system 288–294
origin 279
spin-diffusion 293
summary of, key points 294–296
two-spin system 279–288
versus transient NOE 313
Steroids
NOESY 325
ROESY 332
Stimulated emission 27
Strong coupling
in J-resolved spectroscopy 266, 270
spin-locks, TOCSY 203
Sucrose
test sample 109, 365
Superconducting magnets, *see* Magnets
Supercritical fluids
as NMR solvents 41
Super-cycle, decoupling 346
SUSAN 347
Symmetrisation
in COSY 169
in J-resolved spectroscopy 268
Symmetry
in complexes 41
quadrupolar lineshapes 41

T

t_1, 2D NMR 8, 149
noise 168, 233, 247
T_1, relaxation time constant
definition 26
dependence on tumbling rates 283, 284
measuring 27–30
T_2^*, relaxation time constant
definition 30
t_2, 2D NMR 8, 149
T_2, relaxation time constant
definition 30
measuring 31–34
Temperature calibration
high, ethanediol 104
low, methanol 104
TentaGel resin 367
Tests, spectrometer 105–109
gradient linearity 102, 103
lineshape 106, 107
resolution 106, 107
sensitivity 107, 108
solvent suppression 109
Tetramethylsilane 77
Three-dimensional NMR
DOSY 370
schematic sequence 152
Three-spin effects, NOE 291–293
Tilting
of J-resolved spectra 268
within COSY crosspeaks 198
Time domain 24
extension, linear prediction 57–59
extension, zero-filling 56, 57
Time profile
of shaped pulses 350
Time proportional phase incrementation
2D quad detection 162
axial peaks 167
TMS, *see* Tetramethylsilane
TOCSY 148, 188, 201–211
1D, selective 210
applications 205–208
breakthrough in ROESY 329–331
gradient-selected 209
HMQC/HSQC 241
implementing 208
mixing schemes for 346–348
sequence 202
spin-locks 202, 347
TOE, *see* Truncated driven NOE
Top-hat profile 350, 352
TPPI, *see* Time proportional phase incrementation
Transient NOE 277, 301–304, 313–328
applications, examples 323–326
initial rate approximation 303
internuclear separations 303, 319
interpreting, comments 322
mixing time 301, 314–316, 318, 319
versus steady-state NOE 313
Transition probabilities 281
Transitions
directly connected 197
remotely connected 198
Transmitter
attenuation 95, 96
radiofrequency 47
Transverse magnetisation
observable 18–20
loss of 30, 31
Transverse spin relaxation 30–35
Trim pulses 208
Triple-quantum filter 196
Tr-ROESY 331
Truncated driven NOE 301
Truncation artefacts 56, 57
sinc wiggles 57, 71, 173
suppressing 57, 70, 73
TSP, *see* 3-(trimethylsilyl)propionate
Tubeless-NMR 3
Tubes 78–80
cleaning 80
sapphire 41
sealing 81
susceptibility matched 79
Tumbling rates
in solution 36–43, 282–287
correlation time 36, 283
Tuning
probehead 83–85

Two-dimensional NMR
  introduction, to NMR 3
  introduction to 148–153
  practical aspects 160–174
  schematic sequence 152

U

Ultrasonic mixing 360
Universal pulse, shaped 353

V

Vector model 16–24
Vibrations, floor 47
Vicinal couplings
  in COSY 158, 198
Volume integrals, 2D 319
Volumes
  NMR samples 78, 79

W

WALTZ-16 209, 346, 347
Water
  temperature dependence 360
WATERGATE 364
WATR 34
Waveform generator 354
Window functions 70–73
  2D NMR 165, 166, 172–174
  exponential multiplication 70, 71
  Gaussian multiplication 71
  Lorentz–Gauss transformation 71
  matched filter 71
  resolution enhancement 71–73
  sensitivity enhancement 70, 71
  sine-bell 73, 165
  Traficante 72
  trapezoidal 73
WURST 348

X

XCORFE 256
X–Y heteronuclear correlations 237, 238

Z

Zero-crossing, NOE 286, 304, 305
Zero-filling 56, 57, 173
  2D NMR 171–173
z-filtration 209

α-pinene
  ^{13}C spectra 122
  Proton T_1s 29
β-peptides
  ROESY 333, 334
γ, *see* Magnetogyric ratio
τ_m, *see* Mixing time
τ_{null}
  inversion recovery 29